Heat and Mass Transfer in the System of Thermal Modes of Energy – Technical and Technological Equipment (HMTTSC-2016)

MATEC Web of Conferences Volume 72 (2016)

Tomsk, Russia
19 - 21 April 2016

Editors:

G.V. Kuznetsov
P.A. Strizhak
E.E. Bulba
A.O. Zholanova

ISBN: 978-1-5108-2870-4

Printed from e-media with permission by:

Curran Associates, Inc.
57 Morehouse Lane
Red Hook, NY 12571

Some format issues inherent in the e-media version may also appear in this print version.

Printed by Curran Associates, Inc. (2016)

For additional information, please contact EDP Sciences – Web of Conferences
at the address below.

EDP Sciences – Web of Conferences
17, Avenue du Hoggar
Parc d'Activité de Courtabœuf
BP 112
F-91944 Les Ulis Cedex A
France

Phone: +33 (0) 1 69 18 75 75
Fax: +33 (0) 1 69 28 84 91

contact-edps@webofconferences.org

Additional copies of this publication are available from:

Curran Associates, Inc.
57 Morehouse Lane
Red Hook, NY 12571 USA
Phone: 845-758-0400
Fax: 845-758-2633
Email: curran@proceedings.com
Web: www.proceedings.com

TABLE OF CONTENTS

TEMPERATURE AND CONCENTRATION TRACES OF SPRAY FLOWS DURING MOTION IN A FLAME 1
Antonov Dmitry V., Zhdanova Alena O., Kats Mark D.

METHOD OF COMPREHENSIVE ASSESSMENT OF EFFICIENCY OF DECREASE OF FIRE AND EXPLOSION RISK OF COAL SUSPENDED PARTICLES 6
Arkhipov Vladimir, Patrakov Yuriy, Sechin Aleksandr, Usanina Anna

PHYSICAL MODEL AND BASES OF MATHEMATICAL MODELLING OF ABOVE-SURFACE GASIFICATION OF COAL 13
Arukov R. N., Subbotin A. N.

COMPOSITE SOLID FUEL: RESEARCH OF FORMATION PARAMETERS 18
Tabakaev Roman, Astafev Alexander, Kazakov Alexander, Zavorin Alexander

INFLUENCE OF RADIATION AND CONVECTIVE TYPES OF HEAT EXCHANGE ON THE NECESSARY DURATION OF TEMPERATURE MEASUREMENT BY THERMOCOUPLES 23
Atroshenko Yuliana K., Bychkova Alena A., Ozerova Irina P.

ABOUT INFLUENCE OF OPERATIONAL TEMPERATURES ON RELIABILITY OF OPERATION OF THE EQUIPMENT OF THERMAL POWER PLANTS 28
Atroshenko Yuliana K., Bychkova Alena A., Andyk Vladimir S.

NUMERICAL MODELING OF THERMOCAPILLARY DEFORMATION AND FILM BREAKDOWN IN A LOCALLY HEATED THIN HORIZONTAL VOLATILE LIQUID LAYER 33
Barakhovskaya Ella, Marchuk Igor

SCENARIO MODELING OF THERMAL INFLUENCE FROM FOREST FIRE FRONT ON A CONIFEROUS TREE TRUNK 39
Baranovskiy Nikolay V., Barakhnin Vladimir B., Andreeva Ksenia N.

THE STUDY OF DYNAMIC PROCESSES IN THE BOILER FURNACE WITH CIRCULATING FLUIDIZED BED 45
Gil A. V., Baturin D. A.

INFLUENCE OF GEOMETRIC CHARACTERISTICS OF CANAL ON PRECIPITATING CAPACITY OF SEPARATOR 49
Belyaev L. A., Zaitsev A. S., Kondakov A. A., Shevelev S. A., Matveev A. S.

HEAT TRANSFER DURING EVAPORATION OF CESIUM FROM GRAPHITE SURFACE IN AN ARGON ENVIRONMENT 54
Bespala Evgeny, Novoselov Ivan, Ushakov Ivan

THE INFLUENCE OF AGEING PROCESS ON THE BARIUM ALUMINATE PHASE COMPOSITION 59
Bezmaternykh A. O., Shvaljov Y. B., Kudiiarov V. N., Gubin V. E., Babihina M. N.

MODELING OF HIGH ENERGY IMPACT ON ICE IN TAKING INTO ACCOUNT THE TEMPERATURE 64
Bogomolov Gennady N., Glazyrin Victor P., Orlov Maxim Yu.

IDENTIFICATION OF SIZE AND CONCENTRATION OF SUBMICRON PARTICLES ON THE BASIS OF RAYLEIGH SCATTERING MODEL 69
Bondarchuk Ivan. S., Arkhipov Vladimir A., Zhukov Alexander S.

STUDY OF SOOT PARAMETERS IN THE PROCESS OF COMBUSTION OF CONDENSED SYSTEMS 73
Promahov Vladimir V., Zhukova Tatyana V., Ryzhikh Yuliya N.

DOUBLE CONTINUUM MODEL OF EVOLUTION OF PRECURSOR DROPLETS IN THE FLOW OF HEAT TRANSFER MEDIUM OF A PLASMA CHEMICAL REACTOR 79
Zhukov Ilya S., Bondarchuk Sergey S., Borisov Boris V.

HEAT MODE ELECTRONIC DEVICES AT ELEVATED EXTERNAL TEMPERATURES 83
Kravchenko Evgeny, Vershinina Kseniya, Boykova Tamara

MATHEMATICAL MODELLING OF PROCESS OF FLUORATION OF METAL TUNGSTEN 87
Brendakov Roman V., Shvab Alexander V., Brendakov Vladimir N.

STUDYING ONE OF THE PROBLEMS OF WAVE FILTRATION 90
Bubenchikov Michael, Mishcharina Elena, Libin Eduard

THE TEMPERATURE OF THE DIMETHYLHYDRAZINE DROPS MOVING IN THE ATMOSPHERE AFTER DEPRESSURIZATION OF THE FUEL TANK ROCKETS 95
Bulba Elena, Nemova Tatiana, Dyatchina Anna

NUMERICAL MODELLING OF HEAT TRANSFER DURING IMPACT OF A MOLTEN DROPLET ON A SURFACE 99
Cherepanov Roman O., Gerasimov Alexander V.

EFFECT OF THE REBURNING ZONE STOICHIOMETRY ON THE NOX CONCENTRATION AT THE THREE-STAGE COMBUSTION OF PULVERIZED COAL 103
Chernetskaya Nelya, Chernetskiy Mikhail, Zhuikov Andrey

INVESTIGATION OF TWO-PHASE FLOW IN SHORT HORIZONTAL MINI CHANNEL HEIGHT OF 1 MM 110
Ron'shin Fedor, Cheverda Vyacheslav

HEAT TRANSFER WHILE INCORPORATING METAL PARTICLES IN A MELTING SUPPORT PLATE 114
Mamontov Gennady Ya., Chuguev Danil N.

MATHEMATICAL MODEL OF HEAT TRANSFER IN MORPHOLOGICAL PART OF VEGETATION AT INFLUENCE BY THERMAL RADIATION FROM SURFACE FOREST FIRE FRONT 121
Baranovskiy Nikolay V., Demikhova Alena N.

MATHEMATICAL MODELLING OF THE FIXED-BED BIOMASS-COAL CO-GASIFICATION PROCESS 127
Donskoy Igor G.

JOINT ELECTRICITY AND HEAT PRODUCTION POSSIBILITIES ON THE BASIS OF THE DISTRICT TRIGENERATION PLANT 130
Ershov Vadim, Tarasova Ekaterina

EXPERIMENTAL OBSERVATION OF THE TEMPERATURE AT THE LIQUID/GAS INTERFACE 137
Velicanov V. D., Orlova E. G., Saigash A. S., Korneva O. S.

APPLICATION OF EVAPORATIVE COOLING FOR THE CONDENSATION OF WATER VAPORS FROM A FLUE GAS WASTE HEAT BOILERS CCP 142
Galashov Nikolay, Tsibulskiy Svyatoslav, Kiselev Alexander

THE DRAG REDUCTION OF MICROCHANNEL FLOW BY CONTRAST WETTABILITY 146
Gatapova Elizaveta Ya., Gluzdov Dmitry S.

THE BOILER FURNACE NUMERICAL MODELING OF BURNING NON-STANDARD FUEL 150
Gil A. V., Koksharev O. M., Lebed D. V., Starchenko A. V.

CONDITIONS AND CHARACTERISTICS OF COAL WATER SLURRY CONTAINING PETROCHEMICALS IGNITION BY HOT PARTICLE 154
Gal'chenko Nina K., Medvedev Valeriy V., Nigay Aleksandr G., Sivkov Dmitriy S.

THE ANALYSIS OF THE CHARACTERISTICS OF SWIRL CHAMBER FOR COOLING HOT WATER 158
Goldaev Sergey V., Tretyakova Kseniya B.

ESTIMATION OF THE SAFE ELECTRICAL LOAD ON SPIRAL BULBS HEATED BY CAPACITORS APPLIED TO MULTIPLE START-UPS OF OPEN GAS GENERATORS 163
Goldayev Sergey, Basalayev Sergey, Minkova Natalya

IGNITION OF WET WOOD PARTICLES IN THE STREAM OF HIGH-TEMPERATURE GASES 170
Syrodoy S. V., Gutareva N. Y., Bugaeva K. A.

THE EXPERIMENTAL STUDY OF THE EFFECT OF ADDING HIGH-MOLECULAR POLYMERS ON HEAT TRANSFER CHARACTERISTICS OF NANOFLUIDS 178
Guzei Dmitriy, Pryazhnikov Maxim, Minakov Andrey, Zhigarev Vladimir

NUMERICAL APPROACH TO THE EVALUATION OF ATMOSPHERIC POLLUTION BY COMBUSTION PRODUCTS OF SOLID FUELS 184
Gvozdyakov Dmitry, Gubin Vladimir, Shvab Svetlana, Tanishev Anton

THE EFFECT OF THE SUBSTRATE TEMPERATURE AND THE HEIGHT OF THE MINI-CHANNEL ON THE EVAPORATION OF LIQUID DROPLET BLOWN BY THE GAS FLOW 188
Isachenko Ekaterina, Orlik Evgeniy, Kabov Oleg

IMPROVING ENERGY EFFICIENCY CABLE PRODUCTION 193
Iashutina Olga, Vershinina Kseniya, Ivanova Evgeniya

NEUTRON AGE DETERMINATION IN FAST REACTOR MATERIALS USING THE GROUP METHOD 198
Kabanova Marina F., Yuzhakov Aleksey V., Kuzmin Arian V., Kobenko Yuriy V., Riabova Elena S.

ERRORS IN DETERMINING THE THERMAL CHARACTERISTICS OF LIQUIDS BY THE LASER-PULSE METHOD 206
Kats Mark D., Kats Ilija M.

MODELING OF THERMOHYDRAULIC PROCESES OF HEAT EXCHANGE EQUIPMENT 212
Savina Karina, Abramovskih Alexey, Haymin Vladimir, Lavrinenko Sergey

THE NUMERICAL SIMULATION APPLICATION FOR FIRE-TUBE BOILER HEATING SURFACE SAFETY EVALUATION 216
Khaustov Sergei A., Guk Olga V., Lisenkov Artem
INTERACTION BETWEEN HYDROGEN MOLECULES AND A CLOSED NANOTUBE 222
Tarasov Egor, Khudobina Yulia
STUDYING OF THE CONTACT ANGLE HYSTERESIS ON VARIOUS SURFACES 226
Kirichenko E. O., Gatapova E. Ya.
LEVITATION OF LIQUID MICRODROPLETS ABOVE A SOLID SURFACE SUBCOOLED TO THE LEIDENFROST TEMPERATURE 231
Kirichenko D. P., Zaitsev D. V., Kabov O. A.
EXPERIMENTAL DETERMINATION OF THERMAL CHARACTERISTICS OF MUNICIPAL SOLID WASTE 236
Bukhmirov Vyacheslav, Kolibaba Olga, Gabitov Ramil
EFFECT OF METAL ADDITIVES ON THE COMBUSTION CHARACTERISTICS OF HIGH-ENERGY MATERIALS 241
Korotkikh Alexander, Glotov Oleg, Sorokin Ivan
SOLVING EXCESS WATER PRODUCTION PROBLEMS IN PRODUCTIVE FORMATION 247
Kozyrev Ilya, Zyatikov Pavel, Deeva Vera
INVESTIGATION OF THE COMBUSTION STABILITY OF METHANE-AIR MIXTURE IN RECUPERATIVE BURNERS OF DIFFERENT GEOMETRIES 251
Krainov Alexey, Minkov Leonid, Moiseeva Ksenia
HIGH-SPEED INTERACTION ROD AND SEGMENTED PROJECTILES INTO MASSIVE OBSTACLES 256
Kramshonkov E.N., Krainov A.V., Shorohov P.V.
ANALYSIS OF INFLUENCE OF HEAT TRANSFER CONDITIONS ON THE UPPER COVER TO HEAT TRANSFER IN THERMOSYPHON 266
Krasnoshlykov Alexander
CONVECTION STUDY BY PIV METHOD WITHIN HORIZONTAL LIQUID LAYER EVAPORATING INTO INERT GAS FLOW 271
Kreta Aleksei, Lyulin Yuriy
THE INFLUENCE OF GAS-PHASE HYDROGENATION PARAMETERS ON THE PROCESSES OF ACTIVATION, SORPTION AND ACCUMULATION OF HYDROGEN IN THE POWDER OF PURE TITANIUM 276
Kudiiarov V.N., Babihina M.N., Gvozdyakov D.V.
FIELD RESEARCH OF FIRING STORED PEAT CRITICAL CONDITIONS 280
Kulesh Roman N., Orlova Ksenia Y.
CRITICAL HEAT FLUX IN NANOFLUIDS AT QUASI-STATIONARY AND STEPWISE HEAT GENERATION 285
Moiseev Mikhail, Zhukov Vladimir, Kuznetsov Denis
PENETRATION OF PARTICLES THROUGH MULTI-BARRIER SYSTEMS 289
Bubenchikov Michael, Bubenchikov Aleksey, Libin Eduard, Khudobina Yulia
APPLIED TECHNOLOGICAL DIRECTION OF POWER PLANT ASH AND SLAG WASTE MANAGEMENT WHEN KUZNETSK BITUMINOUS COAL IS BURNED 293
Lihach Snejana A., Grishchernko Maria V., Kulesh Roman N., Lebedev Vitali M., Orlova Ksenia Y.
THE INVESTIGATION OF STRUCTURE HETEROGENEOUS JOINT WELDS IN PIPELINES 297
Lyubimova Lyudmila, Fisenko Roman, Breus Sergei, Tabakaev Roman
EXPERIMENTAL RESEARCH OF INFLUENCE OF A RELATIVE PARTICLES POSITIONING IN A GAS STREAM ON CHARACTERISTICS OF THEIR AERODYNAMIC TRACES 303
Volkov Roman S., Osmolovskaya Maria M., Ozerova Irina P.
VERIFICATION OF CONJUGATE HEAT TRANSFER MODELS IN A CLOSED VOLUME WITH RADIATIVE HEAT SOURCE 307
Maksimov Vyacheslav I., Nagornova Tatiana A., Kurilenko Nikolai I.
DYNAMIC WETTING AND DEWETTING: COMPARISON OF EXPERIMENT WITH THEORIES 312
Orlova Evgeniya.G., Malyhina Ekaterina R., Feoktistov Dmitriy.V., Zhidkova Larisa.S.
DROPLET SPREADING PROCESS IMPACT ON IGNITION CHARACTERISTICS OF CONDENSED MATERIALS 318
Mamontov Gennady Ya., Taratushkina Galina V., Kurilenko Elena Yu.
MODELLING COAL COMBUSTION CONDITIONS AT ITS STORAGE IN A STACK 323
Masanik A.S., Vasilevskiy A.V.

PHYSICAL MODELING OF SUPERSONIC FLOW OVER BODIES WITH PLANE AND AXISYMMETRIC SHAPE ... 327
Maslov E.A., Faraponov V.V., Chupashev A.A., Matskevich V.V.
RADIANT HEATING OF LIQUID IN THERMAL ENTRY LENGHT ... 330
Medvedev G., Martynova A.
EXPERIMENTAL INVESTIGATION OF GAS HYDRATE PRODUCTION AT INJECTION OF LIQUID NITROGEN INTO WATER WITH BUBBLES OF FREON 134A ... 335
Meleshkin Anton V., Mezentsev Ivan V., Elistratov Dmitriy S.
BOILING OF BINARY ZEOTROPIC BLENDS IN THE PLATE HEAT EXCHANGER OF THE HEAT PUMP ... 338
Mezentseva Nadezhda N., Mezentsev Ivan V., Meleshkin Anton V.
THE RESEARCH OF THE BOILING CRISIS OF HYDROUS SOLUTION LITHIUM BROMIDE ... 342
Mironova Nina V., Morozov Vladimir S., Elistratov Sergey L., Ratkov Roman S.
CALCULATION OF INDUSTRIAL ENTERPRISE VENTILATION SYSTEM BY NETWORK INTEGRAL METHOD ... 347
Mihienkova Evgeniya I., Filimonov Sergey A., Neobyavlyayuschiy Pavel A., Boykov Dmitry V.
DYNAMICS OF EVAPORATION FRONT PROPAGATION IN FREON R21 WITH ADDITION OF NANOPARTICLES ... 353
Moiseev Mikhail, Kuznetsov Denis, Zhukov Vladimir
SURFACE CRYSTALLIZATION OF AQUEOUS SALT SOLUTION UNDER OVERHEATING AND OVERCOOLING ... 359
Nakoryakov Vladimir, Misyura Sergei, Morozov Vladimir
NON-ISOTHERMAL DESORPTION OF THIN LAYERS OF AQUEOUS SALT SOLUTIONS ... 362
Misyura Sergei, Morozov Vladimir
RESULTS OF STUDY OF SULFUR OXIDE REDUCTION DURING COMBUSTION OF COAL-WATER SLURRY FUEL THROUGH USE OF SULFUR CAPTURING AGENTS ... 365
Murko Vasiliy I., Karpenok Viktor I., Senchurova Yuliya A., Tailakov Oleg V., Bogomolov Alexander R., Khyamyalyainen Veniamin A.
STUDYING PERIODIC ROLLING WAVES ON THE SURFACE OF A FALLING LIQUID FILM ... 372
Nakoryakov Vladimir E., Ostapenko Vladimir V., Bartashevich Maria V.
SIMULATION OF TURBULENT NATURAL CONVECTION IN A LOCAL ZONE OF LARGESCALE AREA UNDER CONDITIONS OF THE RADIANT ENERGY SUPPLY ... 376
Nee Alexander, Valieva Liliya
THE OPPORTUNITY ANALYSES OF USING THERMOSYPHONS IN COOLING SYSTEMS OF POWER TRANSFORMERS ON THERMAL STATIONS ... 382
Nurpeiis Atlant, Nemova Tatiana
INFLUENCE OF THE TILT ANGLE OF LANCES ON THE FLUID FLOW INSIDE THE GASIFIER SHAFT ... 389
Openyshev P.V., Sheremet M.A.
DYNAMICS OF UF_6 DESUBLIMATION WITH THE INFLUENCE OF TANK GEOMETRY FOR VARIOUS COOLANT TEMPERATURE ... 392
Orlov Aleksey A., Tsimbalyuk Alexandr F., Malyugin Roman V., Glazunov Anatolij A.
RESEARCH OF ENERGY SAVING MODE OF BUILDING DISTRIBUTED CONTROL SYSTEM UNDER INFLUENCE OF MANY DISTURBING FACTORS ... 397
Morozov Maxim N., Ushakov Roman B.
EXPERIMENTAL RESEARCH OF THERMOPHYSICAL PROCESSES IN A CLOSED TWO-PHASE THERMOSYPHON ... 403
Feoktistov D.V., Vympin E.A., Nurpeiis A.E.
EVALUATION OF EFFECTIVENESS OF LIGNOSULFONATE APPLICATION FOR ORGANIC COAL-WATER FUEL RHEOLOGICAL PROPERTIES IMPROVEMENT ... 408
Osipov Vitaliy, Larionov Kirill, Matveev Alexander, Gromov Alexander
MATHEMATICAL MODELING OF THE HEAT TRANSFER PROCESS IN POROUS COOLING SYSTEMS UNDER INFLUENCE LOW ENERGY DISTURBANCES ... 412
Ovchinnikov Vyacheslav, Ruleva Eugenia, Yakimov Anatoliy
SUPERHEATER TUBE FLAT WALL STATIONARY TEMPERATURE FIELD ... 422
Parpiev A.T., Loginov V.S.
PERFORMANCE OF THE NUCLEAR POWER PLANT CONDENSING UNIT ... 428
Pavelyev Egor, Lavrinenko Sergey
SOLVING PROBLEM OF THERMAL CONDUCTION FOR PROVIDING STRENGTH OF ELECTRONIC UNITS ON THERMAL IMPACTS ... 433
Petrashchuk Svitlana, Kovtun Igor, Voznyak Andriy

AN EXPERIMENTAL PROCEDURE TO ESTIMATE LIFETIME OF WATER DROP WITH GRAPHITE INCLUSION UNDER INTENSIVE VAPORIZATION WITH EXPLOSIVE BREAKUP 438
Legros Jean C., Gumerov Vladislav M., Piskunov Maxim V.

NUMERICAL MODELING OF THE EFFECT OF THAWING OF SOIL IN THE AREA OF PLACING TANKS FOR STORAGE FUEL OF THERMAL POWER PLANTS AND BOILER 444
Polovnikov V.Yu., Ozhikenova V.Yu., Makhsutbek F.T.

DETECTION OF HEAT-STRESSED FURNACE WALL'S AREAS BASED ON RESULTS OF NUMERICAL SIMULATIONS OF THERMAL FLOWS 448
Popov V.V, Taylasheva T.S., Lebedev V.M.

SIMULATION OF ROOM CLIMATE OF PUBLIC BUILDINGS 452
Prorokova Maria, Bukhmirov Vyacheslav

MATHEMATICAL MODELING OF SALT LENS FORMATION PROCESS IN THE CATHODE ASSEMBLY OF ALUMINUM ELECTROLYSIS CELLS 455
Orlov Anton, Proshkin Alexander

THEORETICAL AND NUMERICAL CHARACTERIZATION OF THE THERMAL PHYSICAL PROPERTIES OF CARBON CERAMIC MATERIAL 461
Prosuntsov Pavel V., Taraskin Nikita Yu.

RESEARCH OF FERROMAGNETIC NANOPARTICLES TRANSPORT UNDER MAGNETIC FIELD 468
Pryazhnikov Maxim, Guzey Dmitriy, Minakov Andrey, Zhigarev Vladimir, Shebeleva Anna

THE DISTRIBUTION OF PARTICLES IN THE WALL REGION OF THE CYCLONE CHAMBER 472
Razva Aleksandr, Vasilevsky Michail, Pleshco Alice, Naumkin Aleksandr

THE LIQUID FILM FLOW WITH EVAPORATION: NUMERICAL MODELLING 477
Rezanova Ekaterina

ACHIEVEMENTS, PROBLEMS AND PROSPECTS OF THE SCIENTIFIC APPROACH AIMED FOR ENSURING RESOURCE-EFFICIENT TECHNOLOGIES IN THE CONTEXT OF PERSONNEL POLICY IN THE ENERGY SECTOR 481
Rostovtseva Veronika M., Nizkodubov Gavriil A.

ABOUT THE RESULTS OF THE DESTRUCTION OF THE MOLECULES OF LIQUID HYDROCARBONS IN THE FIELD OF ACOUSTIC CAVITATION 487
Rudenko Michael, Vasiltsov Mark, Gorbasenko Stepan, Lavrenyuk Tatiana

ANALYTICAL STUDY OF MICROWAVE HEATING OF THE COAL LAYER WITH MIXED BOUNDARY CONDITIONS OF I AND III TYPES 490
Salomatov Vladimir, Karelin Vadim, Paschenko Sergey, Salomatov Vasiliy

INVESTIGATION OF BOILING IN HIGH-POWER LED COOLING SYSTEM 499
Shatskiy Evgeniy E.

NUMERICAL SIMULATION OF TURBULENT FLOW OVER OF SEMI-CYLINDRICAL PROFILE AT ZERO ANGLE OF ATTACK 503
Shebelev Alexander, Gavrilov Andrey

NUMERICAL STUDY OF DEFORMATION DROPS OF WATER IN GAS STREAM 507
Shebeleva Anna, Chernetskiy Mikhail

MATHEMATICAL MODELLING OF THE PROCESS OF TUNGSTEN FLUORIDES REDUCTION BY HYDROGEN 512
Brendakov Roman, Shvab Alexander, Brendakov Vladimir

METHOD EXPERIMENTAL DEFINITION OF EFFICIENCY FUELBURN IN A GAS REACTORS WITH MINI CHANNELS 515
Slesareva Ekaterina, Elistratov Sergey, Ovchinnikov Valeriy

EFFECT OF OXIDIZING MEDIUM ON SYNTHESIS GAS CONTENT AT SOLID FUEL GASIFICATION 519
Korotkikh Alexander, Slyusarskiy Konstantin, Sorokin Ivan

MATHEMATICAL MODEL OF HEAT TRANSFER IN LAYERED STRUCTURE OF HUMAN SKIN INFLUENCED BY HEATED UP TO HIGH TEMPERATURES PARTICLE 523
Stuparenko Alexandr A., Solodkin Andrey S.

MATHEMATICAL MODELING OF NONSTATIONARY SEPARATION PROCESSES IN GAS CENTRIFUGE CASCADE FOR SEPARATION OF MULTICOMPONENT ISOTOPE MIXTURES 528
Orlov Alexey, Ushakov Anton, Sovach Victor

STUDYING THE DYNAMICS OF BREAKDOWN OF THIN HORIZONTAL LIQUID LAYERS WITH LOCAL HEATING 533
Spesivtsev Serafim, Lyulin Yuriy

FORMING THE COMPOSITION OF UNDERGROUND COAL GASIFICATION PRODUCTS IN THE SIMULATION OF VARIOUS HEAT AND MASS TRANSFER CONDITIONS IN THE COAL SEAM 538
Masanik A.S., Subbotin A.N., Tarasanov A.S.

INFLUENCE OF ABSORPTION OF THERMAL RADIATION IN THE SURFACE WATER FILM ON THE CHARACTERISTICS AND IGNITION CONDITIONS 545
Syrodoy Samen V., Gutareva Nadegda Yu., Taburchinov Roman I.

EXPERIMENTAL SETUP FOR STUDYING THE CAPTURE OF SULFUR AND NITROGEN OXIDES FROM FLUE GASES 551
Sysolyatin Andrey S., Kryukov Sergey V., Chemakin Maxim A., Dvorovenko Igor V.

SMALL DISTRIBUTED ENERGY RUSSIA: COMBINED HEAT AND POWER GENERATION 555
Astafev Alexander, Kazakov Alexander, Zavorin Alexander

EXPERIMENTAL STUDY OF TWO-PHASE THERMOSYPHON USING R-22 AS A WORKING FLUID 561
Theeb Maathe.A.

EXPERIMENTAL DETERMINATION OF HEAT TRANSFER COEFFICIENT UNDER FREE CONVECTION IN AN UNBOUNDED SPACE 570
Timoshenko Evgenya, Afanaseva Viktoria, Kravchenko Evgeny

THE EFFECT OF INCLINATION ANGLE ON CRITICAL HEAT FLUX IN A LOCALLY HEATED LIQUID FILM MOVING UNDER THE ACTION OF GAS FLOW IN A MINI-CHANNEL 574
Tkachenko Egor M., Zaitsev Dmitry V.

SIMULATIONS OF METHANE MOTION IN A VERTICAL COAL BED WHEN FILLED WITH WATER 578
Bubenchikov Aleksey, Tsyrenova Valentina, Tsydypov Sevan

COST PRICE OF PRODUCTS IN THE SYSTEM OF HEAT, REFRIGERATION AND ELECTRIC ENERGY PRODUCTION COMBINED AT THERMAL POWER PLANT 582
Tubolev Alexander, Romashova Olga, Belyaev Leonid

EXPERIMENTAL STUDY OF HEAT TRANSFER IN THE STEAM THERMAL SIPHON TUBE CHANNEL 586
Ulchiekov Mikhail A.

MATHEMATICAL MODELING OF HEAT TRANSFER IN THE SYSTEM "THE OBJECT OF HEATING – THERMAL INSULATION LAYER – THE EXTERNAL ENVIRONMENT" AND MICROCLIMATE PARAMETERS OF INDUSTRIAL PREMISES 590
Ulchiekov M. A.

EVALUATION OF THERMOSYPHON APPLICATION FOR COOLING THE MODULAR AUTOMATED CONTROL SYSTEMS 596
Ushakov Konstantin Yu., Peters Andrey N., Bogomolov Alexander R., Temnikova Elena Yu.

TWO-DIMENSIONAL PROBLEM OF TURBULENT NATURAL CONVECTION IN A SEMI-OPEN CAVITY WITH RADIANT HEATING OF INTERNAL BOUNDARIES 599
Nee Alexander, Valieva Liliya

IGNITION OF THE SOARING DROPLET SETS OF WASTE-DERIVED COAL-WATER SLURRY WITH PETROCHEMICALS 605
Valiullin Timur R., Dmitrienko Margarita A., Shevyrev Sergey A., Egorov Roman I.

ESTIMATES OF TWO-PHASE FLOW PARAMETERS IN THE CYCLONE CHAMBER 610
Razva Aleksandr, Vasilevsky Michail, Pleshco Alice, Asan Alibek

INTEGRAL CHARACTERISTICS OF STABILITY OF COAL-WATER SLURRIES AND COAL-WATER SLURRIES CONTAINING PETROCHEMICALS BY EVALUATION OF SEPARATION INTO LAYERS 616
Dmitrienko Margarita A., Nyashina Galina S., Vershinina Ksenia Yu., Lyrschikov Sergey Yu.

SOFTWARE PRODUCTS FOR THE NUMERIC EVALUATION OF OSCILLATION OF NOX IN THE PULVERIZED FUEL COMBUSTOR 623
Zueva E. A., Vizgavlyust N. V., Vorobyeva L. V., Starchenko A. V.

MODELING OF POLLUTANT DISPERSION IN ATMOSPHERE BASED ON SOFTWARE COMPLEX "SKAT" 627
Vorobiev Aleksander V., Dutova Ekaterina M., Latyshev Andrei S.

DETERMINATION OF NECESSARY AND SUFFICIENT OPERATIONAL CONDITIONS OF THE SLEEVE-TYPE BEARINGS IN AUTONOMOUS POWER PLANTS OF ORMAT TYPE UNDER CONDITIONS OF SIBERIA AND THE FAR EAST 631
Vysokomornaya Olga V., Vysokomorny Vladimir S., Beck Alexander S., Kosintsev Andrey G., Panin Vladimir F.

USING A HIGH-TEMPERATURE FLUE GASES IN THE TECHNOLOGY OF COMBUSTION NEUTRALIZATION OF WASTEWATER 636
Vysokomornaya Olga V., Zakharevich Yuliya S., Aykina Natalya N., Ushmaev Denis V.
IGNITION PARAMETERS OF DIESEL LAYERS UNDER CONDITIONS OF LOCAL HEATING 639
Zakharevich Arkadiy V., Zygin Mikhail S., Panchenko Pavel V., Vlasova Yekaterina A.
RESEARCH OF HEAT AND MASS TRANSFER PROCESSES IN THE NODES OF FREE-FLOW MICRO-HPP WITH THE USE OF 3D TECHNOLOGY 643
Yankovsky Stanislav, Kulupanov German, Andriyashchenko Vladimir
THE INFLUENCE OF GAS DIFFUSION ON THE DYNAMICS OF A SPHERICAL LAYER OF VISCOUS INCOMPRESSIBLE LIQUID AND HEAT AND MASS TRANSFER IN IT 646
Rezanova Ekaterina, Zakurdaeva Alla
RESEARCH OF HEAT RATES EFFECT ON THE PROCESS OF FUEL-BED GASIFICATION OF "BALAKHTINSKOE", "OSINNIKOVSKOE", "KRASNOGORSKOE" AND "BORODINSKOE" COAL DEPOSITS 651
Zenkov Andrey, Yankovsky Stanislav, Matveeva Anastasia, Lavrinenko Sergey, Gromov Alexander
MATHEMATICAL MODELING OF HEAT FRICTION CONTACT MASTER BELT WITH THE GUN MOUNT BARREL DURING THE PROCESS OF HIGH-SPEED MOTION 655
Zezulinsky Jaroslav, Guskov Anatolyi, Milevsky Konstantin
NUMERICAL ANALYSIS OF INFLUENCE OF ENGINEERING CONSTRUCTION IN THE AREA OF PLACING FUEL STORAGE TANKS ON THEIR HEAT LOSS 658
Polovnikov V. Yu., Makhsutbek F. T., Ozhikenova Zh. F., Zharova I. K.
DETERMINATION OF THE MINIMAL AMOUNT OF WATER FOR EFFECTIVE SUPPRESSION OF THE THERMAL DECOMPOSITION OF FOREST COMBUSTIBLE MATERIALS 663
Zhdanova Alena O., Nyashina Galina S., Osmolovskaya Maria M.
VERIFICATION OF MODEL OF CALCULATION OF INTRA-CHAMBER PARAMETERS IN HYBRID SOLID-PROPELLANT ROCKET ENGINES 667
Zhukov Ilya S., Borisov Boris V., Bondarchuk Sergey S., Zhukov Alexander S.
DYNAMIC REGIME OF IGNITION OF SOLID PROPELLANT 671
Zolotorev Nikolay, Kuznetsov Valeriy, Maslov Eugene
ANALYSIS OF FRACTIONATED FLY ASH AND SLAG AT KEMEROVO STATE DISTRICT POWER PLANT 674
Sysolyatin Andrey S., Zvingul Ivan A., Temnikova Elena Yu.
REGISTRATION OF TIMES OF PRESERVING THE LOWERED TEMPERATURES OF COMBUSTION PRODUCTS IN THE TRACE OF THE AEROSOL FLOW 678
Vojtkov Ivan S., Volkov Roman S., Chernova Galina A.
Author Index

edp sciences

MATEC Web of Conferences

Statement of Peer review

In submitting conference proceedings to MATEC Web of Conferences, I certify to the Publisher that I adhere to the **Policy on Publishing Integrity** of the journal in order to safeguard good scientific practice in publishing.

1. All articles have been subjected to peer review administered by the proceedings editors.
2. Reviews have been conducted by expert referees, who have been requested to provide unbiased and constructive comments aimed, whenever possible, at improving the work.
3. Proceedings editors have taken all reasonable steps to ensure the quality of the materials they publish and their decision to accept or reject a paper for publication has been based only on the merits of the work and the relevance to the journal.

Title, date and place of the conference

Heat and Mass Transfer in the System of Thermal Modes of Energy - Technical and Technological Equipment - 2016 („HMTTSC-2016")
April 19-21, 2016, Tomsk, 634050, Russia

Proceedings editor(s):

G.V. Kuznetsov, DcS
P.A. Strizhak, DcS
E.E. Bulba, PhD
A.O. Zhdanova, PhD

Date and editor's signature

July 07, 2016

MATEC Web of Conferences **72**, 01001 (2016) DOI: 10.1051/matecconf/20167201001
HMTTSC-2016

TEMPERATURE AND CONCENTRATION TRACES OF SPRAY FLOWS DURING MOTION IN A FLAME

Dmitry V. Antonov[1,*], *Alena* O. Zhdanova[1], and *Mark* D. Kats[1]

[1]National Research Tomsk Polytechnic University, 634050 Tomsk, Russia

Abstract. Heat and mass transfer models are developed on the base of experimental data and using Ansys Fluent software. These models allow prediction of the temperature and concentration traces of droplets. Transfer mechanisms of water droplets from different flames of flammable liquid (ethanol, kerosene и benzine) with temperature gases 450–850 K are analyzed. The paper considers aerosol flows with droplets sizes of 0.04–0.4 mm and concentration of $3.8\cdot10^{-5}$ –$10.3\cdot10^{-5}$ m^3 of droplets/m^3 of gas. The maximum gas temperature reduction in the trace of a moving liquid is ranged from 850 K to 600 K. The times of keeping the low temperature of the gas-vapor mixture in the droplets trace are from 13 s to 25 s relative to the initial gas temperature.

1 Introduction

At the present time gas-vapor-droplet technologies are used in various industries [1–3], for example, in firefighting, thermal treatment of liquids, ignition of liquid fuels. However, to talk about the broad and active use of high-temperature (over 1000 K) gas-vapor droplet technologies is quite difficult due to a rather limited information base (particularly, the reliable experimental).

Over the last yers researches are conducted [4, 5]. The main objective of various development and research are improving the efficiency of means and methods of atomization in high-temperature heat technology. Unfortunately, the fact is not taken account that the water droplets during motion in high-temperature gases reserve temperature and concentration traces which reduce the temperature in the combustion zone due to the evaporation of water droplets.

The goal of the present work is numerical analyses of heat and mass transfer processes at researches of temperature and concentration traces of spray flows during motion in a typical flame with using Ansys Fluent software.

* Corresponding author: dva14@tpu.ru

2 Physical model of heat transfer

When setting heat and mass transfer problems considered that initial water temperature equals T_{w0}=300 K, initial temperature gases equals T_{g0}=300 K. It was assumed that the thermal characteristics of the materials are not dependent on temperature. Fig. 1 shows the scheme of solution area for spray flow during motion in a flame.

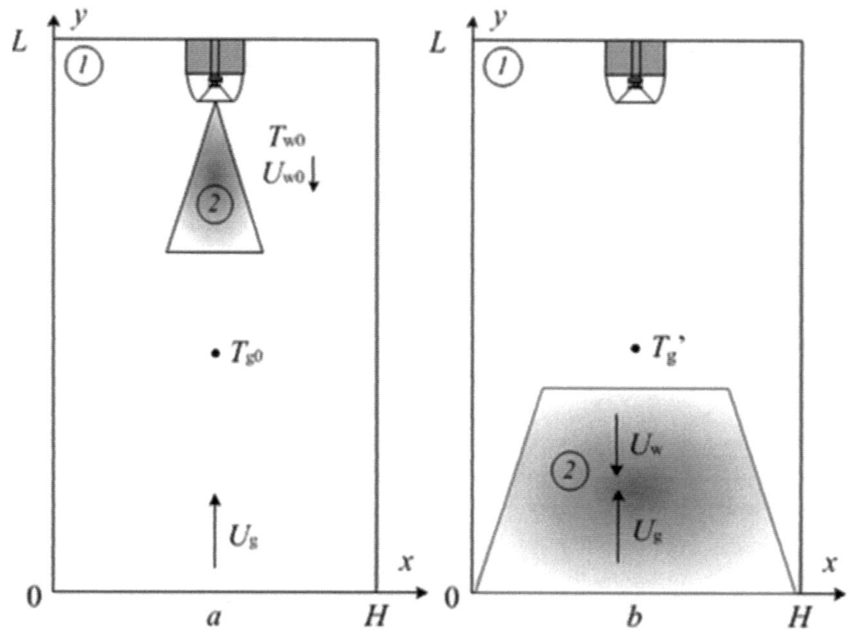

Fig. 1. Scheme of solution area for spray flow during motion in a flame: *1* – high-temperature of the gas-vapor mixture, *2* – spray flows.

3 Mathematical model and decision methods

Initial conditions (t=0): $T=T_{g0}$, $C_w=0$, $C_g=1$, $\psi=0$, $\omega=0$ at $0<x<H$, $0<y<L$; $T=T_{w0}$ at $0<r<R_d$, $0<\varphi<2\pi$.
Boundary conditions (t>0):

$$T = T_g, C_w = 0, C_g = 1, \frac{\partial \psi}{\partial y} = U_g \text{ at } y = 0, 0 < x < H; \tag{1}$$

$$T = T_a, \frac{\partial^2 C_w}{\partial y^2} = 0, \frac{\partial^2 C_g}{\partial y^2} = 0, \frac{\partial \psi}{\partial y} = -U_w \text{ at } y = L, 0 < x < H; \tag{2}$$

$$\frac{\partial T}{\partial t} = 0, C_w = 0, C_g = 0, \frac{\partial \psi}{\partial x} = 0 \text{ at } x = 0, 0 < y < L; \tag{3}$$

$$\frac{\partial T}{\partial t} = 0, C_w = 0, C_g = 0, \frac{\partial \psi}{\partial x} = 0 \text{ at } x = H, 0 < y < L. \tag{4}$$

Where t – time, s; T – temperature, K; T_{g0} – initial temperature of gases, K; T_{wo} – initial temperature of water, K; C_w – concentration of water; C_g - concentration of high-temperature gases; $\omega = \frac{\partial v}{\partial x} - \frac{\partial u}{\partial y}$ – rotation vector velocity, s^{-1}; ψ – flow function, m^2/s; x, y – coordinates of the Cartesian coordinate system, m; H, L – dimensions of solution area, m;

r, φ – coordinate of the spherical systems; R_d – radius of droplet, m; T_g – temperature of gases, K; U_g – velocity of high-temperature gases, m/s^2; U_w – velocity of spray flow, m/s^2; T_a – temperature of air, K.

In this formulation of the problem there are adiabatic horizontal and vertical walls with constant temperature $T_g \approx 1170$ K and $T_a \approx 298$ K. Velocity of sray flows and velocity gases equal $U_g = 1.5$ m/s^2 and $U_a = -5$ m/s^2.

Boundary conditions for liquid – gas were set taking into account vaporization:

$$R = R_1, 0 < \varphi < 2\pi, \lambda_1 \cdot \frac{\partial T_1}{\partial R} = \lambda_2 \cdot \frac{\partial T_2}{\partial R} - W_e \cdot Q_e, \rho_2 D_2 \frac{\partial C_w}{\partial r} = W_e, t > 0. \quad (5)$$

Where R_1 – droplet radius at the output of high-temperature gases, m; λ_1 – thermal conductivity of water, W/(m·K); T_1 – water temperature, K; λ_2 – thermal conductivity of water vapor, W/(m·K); T_2 – water vapor temperature, K; W_e – evaporation rate, kg/(m^2·s); Q_e – water vaporization heat, J/kg; ρ – water vapor density, kg/m^3; D_2 – diffusion coefficient, m^2/s.

The simulation used Ansys package. To solve this problem used time steps $\Delta\tau = 0.01$ s and grid $\Delta x = 0.01$ mm.

4 Results and discussion

Fig. 2 shows temperature fields in the channel after injection of aerosol for typical flame (ethanol, kerosene and benzine).

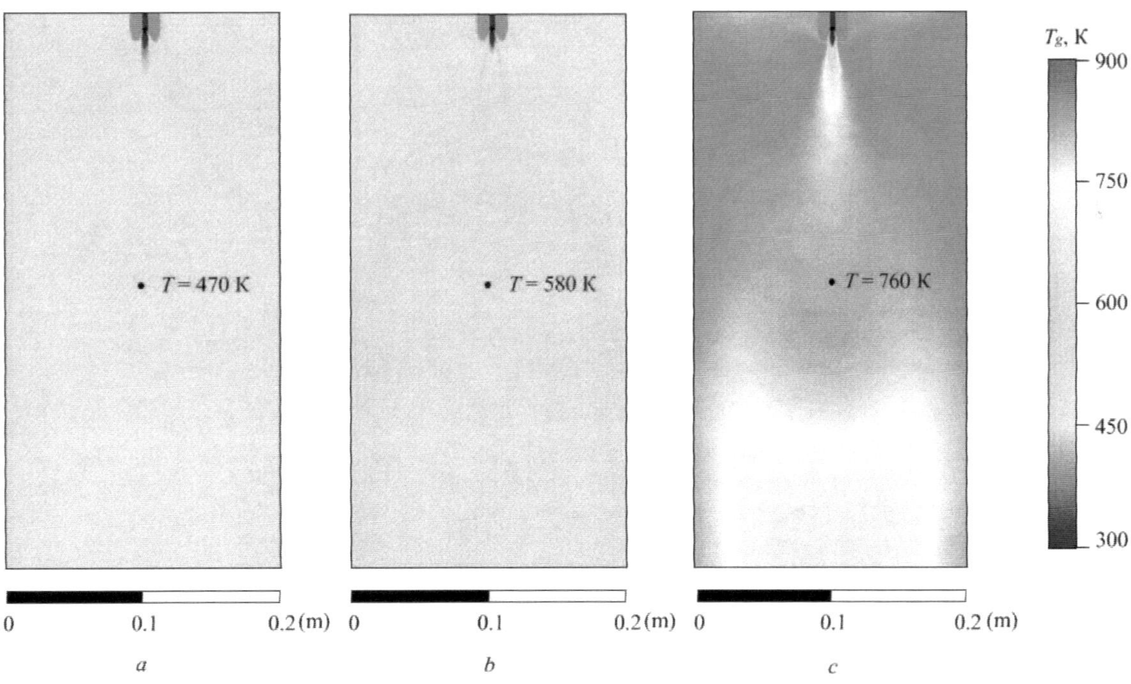

Fig. 2. Temperature fields in the channel after injection of aerosol: *a* – ethanol, *b* – kerosene, *c* – benzene.

Fig. 3 indicates a satisfactory correlation between the maximum temperatures (T_g') in the trace of the aerosol droplets obtained by experiment and when simulating. This allows us to make some conclusions. Firstly, the Ansys Fluent models can be used to predict of temperatures T_g' at different parameters (size, concentration, velocity, initial droplets temperature) of the aerosol and combustion products (temperature and velocity). Secondly, in the case of the aerosol, water evaporation is a determining mechanism for reducing the temperature in the trace.

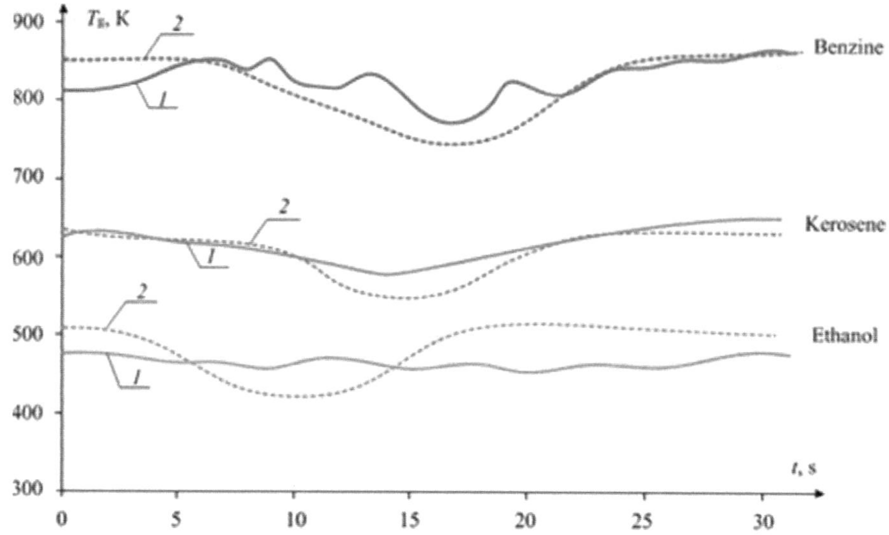

Fig. 3. Maximum temperature in the trace of water droplets: *1* – experimental data, *2* – simulation results.

Fig. 4 shows times of keeping temperature trace depending on droplet size for typical flame: ethanol, kerosene and benzine.

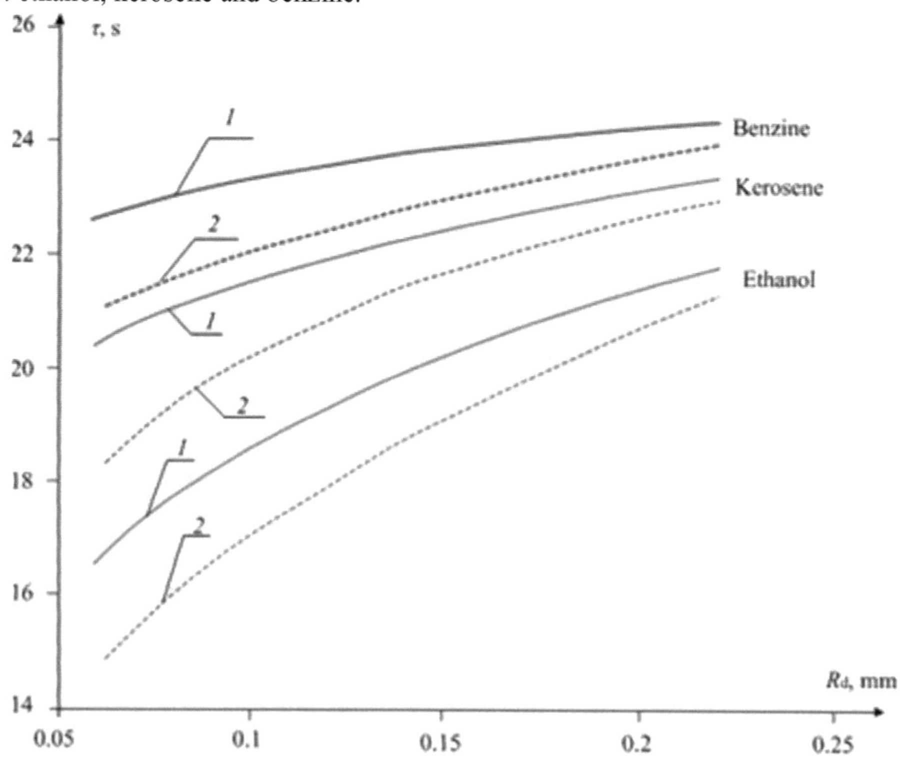

Fig. 4. Times of keeping temperature trace depending on droplet size for typical flame: *1* – experimental data [6], *2* – simulation results.

Dedicated on fig. 4 the experimental and theoretical values of the times of keeping temperature traces illustrate that with increasing deviation of the droplet size is minimized. For small R_d difference between the experimental and theoretical values of τ grows. This can be explained mainly by the fact that the experiments [6] showed rather intense deceleration and turn of small droplets. As consequence, the experimental times τ are longer than those obtained by the mathematical simulation.

5 Conclusion

According to the experimental results, physical and mathematical models of heat and mass transfer have been developed using the Ansys Fluent software. These models can be used in predicting gas temperatures, concentrations of vapor and gases in the trace of aerosol drip flows, as well as times of keeping low temperatures τ. Researches show satisfactory correlation of simulation results and experimental data.

Acknowledgments

The investigations with water aerosol were supported by the Russian Foundation for Basic Research (No. 15-38-20006). The work at the field of modeling of heat and mass transfer processes near boundary "liquid – gas" was supported by the scientific schools grant NSH-7538-2016.8.

References

1. J. E. Splittles, Y. D. Shikhmurzaev, Phys. Fluids **24** (2012)
2. P.A. Strizhak, J. Eng. Phys. Thermophys. **86**, 895 (2013)
3. A. Abbud-Madrid, D. Watson, J.T. McKinnon, Suppression and Detection Research and Applications Conference (2007)
4. Y.V. Lyulin, O.A. Kabov, Appl. Phys. Lett. **39**, 88 (2013)
5. R.V. Maltsev, A.K. Rebrov, J. Appl. Mech. Tech. Phys. **48**, 142 (2007)
6. D.V. Antonov, I.S. Voitkov, P.A. Strizhak, EPJ Web Conf. **110**, Article Number 010001 (2016)

METHOD OF COMPREHENSIVE ASSESSMENT OF EFFICIENCY OF DECREASE OF FIRE AND EXPLOSION RISK OF COAL SUSPENDED PARTICLES

Vladimir Arkhipov[1,*], *Yuriy* Patrakov[2], *Aleksandr* Sechin[3], *Andrey* Sechin[3], and *Anna* Usanina[2]

[1]National Research Tomsk State University, 634050 Tomsk, Russia
[2]Federal Research Center for Coal and Coal Chemistry of Siberian Branch of the Russian Academy of Sciences, 650610 Kemerovo, Russia
[3]National Research Tomsk Polytechnic University, 634050 Tomsk, Russia

Abstract. Methods and results of experimental study of explosive and wetting characteristics of coal dust are presented. Modified evaluation method of explosive characteristics of coal dust allows estimating critical concentrations of suspended dust at variation of temperature of ionizing ignition source corresponding to beginning of flame propagation by means of approximation method. Method of measuring the contact angle of particles of coal powder is based on visualization of the drop placed on the pressed coal.

1 Introduction

Coal dust is an inevitable element of practically all technological stages of production, transportation, preparation, and processing of coal. This determines the increased fire and explosion risk of enterprises of coal industry and energetics [1, 2]. Inflammation of the dust-air mixture is a multivariable process so accepted classifications and characteristics of coal dust are quite relative and does not take into account majority of factors that can ensure trouble-free operation of such enterprises.

One of the most effective ways to reduce the explosion of suspended coal particles is irrigation of mine face using drop-liquid aerosol. This reduces the risk of self-ignition of coal dust. It is necessary to conduct the complex experiments on concentration limits of flame propagation for dry and wetted coals from different deposits to evaluate the effect of the wettability of particular coal ranks on their explosiveness. In this regard, the development of new devices and methods for studying the wetting properties of the coal surface and critical conditions of the flame propagation in aerosuspension are still relevant since only a deep understanding of the combustion and coal particles wetting processes can give the effective technological decisions that reduces the likelihood of coal dust explosions.

In this work the methods of experimental study of fire and explosion dangerous and wetting characteristics of coal particles are presented.

[*] Corresponding author: leva@niipmm.tsu.ru

2 Experimental study of fire and explosion dangerous characteristics of coal suspended particles

Modified method of successive iteration is proposed to study the fire and explosion dangerous characteristics of coal dust. This method allows estimating the critical concentrations of suspended dust particles corresponding to the beginning of flame propagation at variation of temperature of ionizing ignition source.

Scheme of experimental setup for studying the fire and explosion dangerous characteristics of particles suspended in air is shown in fig. 1. The setup consists of test chamber *1*, round spray in coincidence with nozzle *2*, electromagnetic valve *3*, receiver *4*, manometer *5*, strain-gage pressure sensor *6*, valves *7*, *8*, *15*, *16*, thermocouple sensors *9*, *10*, strain gauge scanner recorder with oscilloscope *11*, multichannel recorder *12*, ignition source *13*, compressor *14*, forming unit of gas medium *17*.

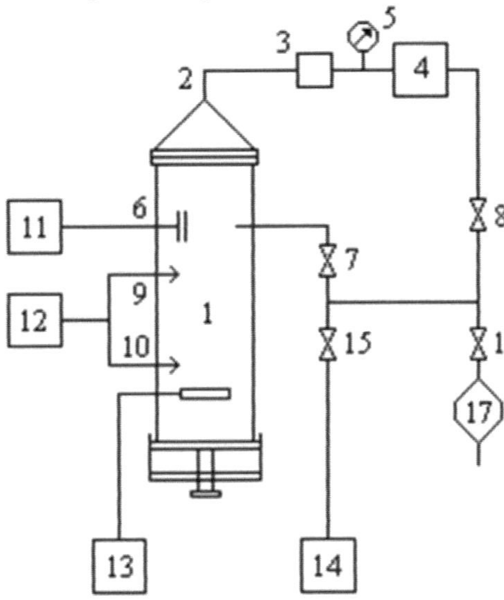

Fig. 1. Scheme of experimental setup for studying the critical conditions of flame propagation in cloud of suspended particles.

In previously sealed reactor chamber *1* with initial manometric pressure *p*=50.7 kPa in the receiver *4* and at driven recording units and initiating ignition source *13* the control experiment is conducted. The temperature of initiating ignition source *13* is 1050 °C. In future, the data obtained is used for comparison with results of work experiments.

Then, the substance studied weighing 0.05 gram is placed in the nozzle *2*, the air pressure *p*=50.7 kPa in the receiver is adjusted, the registration is beginning, and the setup is actuated.

If the flame spreads up to the top of the thermocouple and higher and there was the pressure increase that is different from blank experiment, the experiment is considered to be a success. Otherwise or in the case of absence of one of the indicators (flame spreading to the upper flange of the chamber *1* or the absence of pressure rise) the experiment is considered to be a failure.

Exposing the nozzle, it is checked whether all the substance take part in the spraying. If the nozzle remains non-atomized substance the experiment is repeated. At that the atomizing impulse value is increased by 10.1 kPa until the nozzle will not remain empty. If the nozzle is empty the experiment is repeated with the substance of previous mass, but atomizing impulse value is decreased on 10.1 kPa. This is the way to determine the optimal atomizing impulse for concrete substance of specified mass.

If the ignition and flame spread is not detected, the mass of substance is increased by 0.05 g. After positive result at given mass of the substance, the limiting conditions of flame spread increasing and decreasing the mass of the substance by 0.01 g is searched. This procedure remains until two dust concentrations that are different no more than on 0.01 g will be found. The first concentration corresponds to ignition occurring, the second one corresponds to ignition rejection. The presence or absence of ignition at these concentrations is examined by means of more than ten control experiments with the mass of the substance that does not give the ignition. In the case of ignition during the control experiments the mass of the substance is decreased by 0.01 g until the entire series of noncombustible suspended particles will be obtained.

After each experiment the bottom flange is removed by means of cutoff of the ignition source *13*, the reaction chamber *1* is cleaned of the settled dust and combustion products and purged with compressed air.

Defining the burning area of suspended dust particles at the change of temperature of initiating ignition source, the experiment is carried out for the above-mentioned technique. The ignition source temperature is reduced by 50 °C until the suspended dust particles ignite and burning spreads throughout the volume of the reaction chamber.

Testing of the method and experimental setup for studying the critical conditions of flame propagation in air-dust mixture is conducted using the benzoic acid. For particles with an equivalent diameter of d_p=(160÷200) μm, lower concentration limit of flame spread is equal to C = 36.7 g /m^3 at the initiating source temperature of 1100 °C. This is in good agreement with literature data [3]. Thus, the proposed experimental method is not only consistent with the standard one [4], but also applicable for expanding the fields of its application.

Effect of the initiating ignition source temperature was studied on the sample of Kuznetsk coal rank KO (0÷50 mm). From the results presented (fig. 2) it is seen that air-dust particles are more sensitive to initiation by the source with a high ignition temperature.

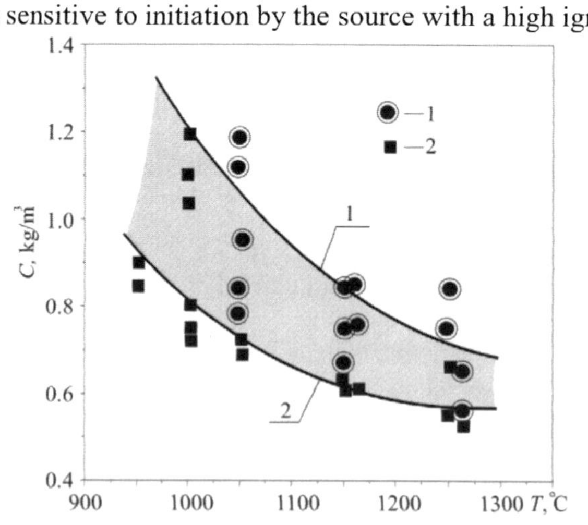

Fig. 2. Concentration dependence of coal dust burning field on ignition source temperature: 1 – burning with marked increase in pressure; 2 – burning without marked increase in pressure in testing chamber.

Analysis of the flame-front spread has been shown that the burning of coal dust up to some concentrations is characterized by weak growth of manometric pressure in the reaction chamber. In a similar way to burning of hydrocarbon blend with air [5-7] it is possible to suggest that for coal dust the regime of "coal" flame exists. The transition of the burning front to "hot" flames area (i.e. with a marked increase in pressure) occurs for the coal sample at increasing the concentration of air-dust mixture in (100÷150) g/m^3 and more, depending on the source ignition temperature. For example, the concentration of coal dust that characterizes the burning process of "cold" flame is approximately 750±50 g/m^3 at the temperature of initiating ignition source 1150 °C (the lower curve in fig. 2). But concentration characterizing the burning process in the "hot" flame regime is equal to 850±50 g/m^3 at the same temperature (upper curve, fig. 2).

In this case the mechanism of flame distribution on air-dust mixture is significantly different from the gas phase of "cold" flame of "ideal" systems and it can be simply interpreted as a "relay" transfer of the thermal pulse from one coal particle to another. And so it will be determined by larger number of parameters (dispersion of dust particles, oxygen concentration and thermal conductivity of gas phase, mechanism and rate of burning of the coal particle and others).

Thus, conducting studies on the standard method [4] it is impossible to change the temperature of the ignition source and this limits the understanding about the ignition process.

3 Experimental study of wetting characteristics of coal suspended particles

The study of the wettability of coal powder particles was conducted on the setup consisting of the matrix for pressing *1*, the sample pressed *2*, a device for droplets forming *3*, high-speed camera *4*, and drop *5* located on the coal powder sample pressed (fig. 3). The essence of the method for determining the coal powder wettability is to directly measure the contact angle of the drop located on the pressed coal particles briquette of cylindrical shape [8]. The problem of liquid penetration through the sample pressed is solved by means of preparing the briquette of specified highness and diameter, pressing at specified pressure in matrix of special form, and previous preparation of powder sample with specified particle size.

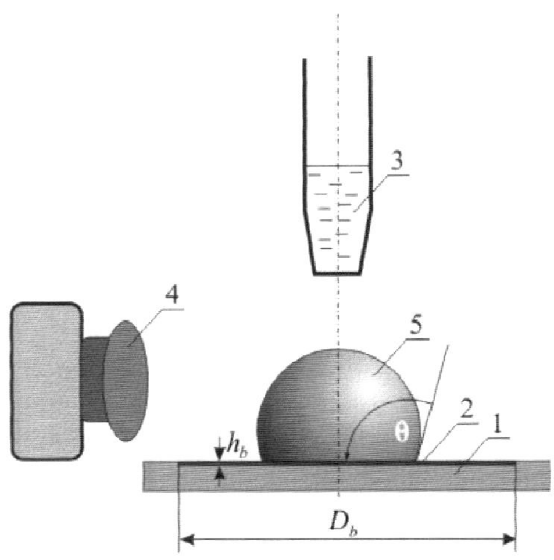

Fig. 3. Scheme of experimental setup for determining the wettability of coal powder.

It is necessary to provide the corresponding briquette diameter D_b since it is not known how liquid drop behaves on the surface of pressed briquette (it spreads in a thin liquid layer or forms the drop on it). The experiments for wetting case closed to full wetting performed for drop of silicon oil (PDMS-100) spreading on acrylic glass has shown that the maximum value of drop contact radius is equal to $r_k \cong 4D$. The briquette diameter is chosen from the condition $D_b \leq 2D$ that provides the location of the drop on the briquette surface and possibility of measuring the contact angle.

Evaluation experiments in choice of optimal height of the briquette h_b have shown that at value $h_b = 0.1D$ the uniformly height-distributed briquette can be obtained. For smaller values of the height h_b it is not enough the required weight of powder for the formation of tablets.

Experiments for coal samples have shown that more effective value of compacting pressure is equal to p>400 MPa. Choice of dispersivity of initial powder with maximum particle size no more than 100 μm and compacting pressure no more than 400 MPa allows decreasing the porosity of the sample significantly, obtaining the plane working surface and the stable drop on a plane surface.

Since the height of the briquette is low there is the possibility of the briquette destruction during their ejecting from the press mold. It is suggested to carry out the pressing of powder particles in the matrix to avoid the briquette destruction. The size of matrix corresponds to the required values of briquette diameter and height (Fig. 4). The contact angle measurement is performed on the briquette that is placed in the matrix for pressing.

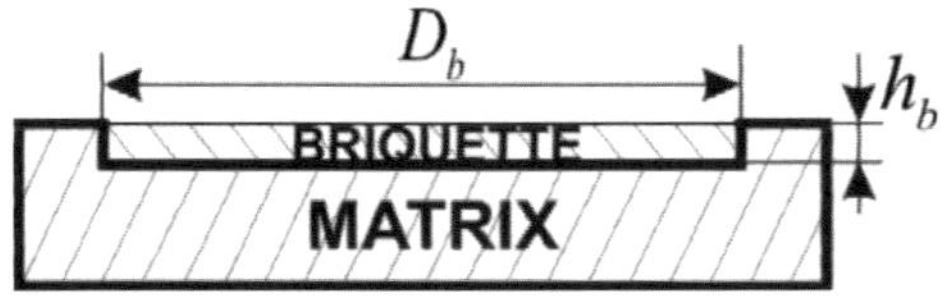

Fig. 4. Matrix for pressing of powder particles.

For low-viscosity liquids after their impact with the solid surface the decaying oscillations of contact angle and the drop height are observed. The less the impact velocity is, the lower the oscillation frequency is and the faster equilibrium drop shape is formed. Since it is difficult to provide the zero impact velocity at placing the liquid drop on a pressed sample it should be ensured the lowest impact velocity.

The experiments on liquid drop spreading on solid surface showed that average value of minimal impact velocity that can be provided is equal to 1 sm/s. It is shown that time of forming the equilibrium contact angle of the distilled water drop at u=1 sm/s is equal to t=200 ms. This value is lower limit of the time that corresponds to beginning of measurement of equilibrium contact angle. The choice of distilled water as an etalon liquid is described the following arguments. Firstly, it is more widely used liquid for the determining the powder wettability, and, secondly, for viscous liquid the value t will be more.

Necessity of decreasing the time of forming the equilibrium drop shape exists due to the fact that it is observed the evaporation process of the drop along with formation of equilibrium contact angle of the drop. The process of evaporation results in the change of gas phase (air is changed on saturated steam) and the contact angle value is changed.

In the work, the analysis of evaporation degree of the distilled water drop during their spreading on acrylic surface at room temperature was performed. The choice of this liquid is described the fact that it is one of volatile liquids. Calculation of the drop volume by means of processing the video for different time interval t_i was conducted to evaluate the evaporation degree:

$$V_k(t_i) = 2\pi \int_0^{r_k(t_i)} r h(r) dr,$$

where $V_k(t_i)$ is the drop volume in specific time t_i; $h(r)$ is the dependence of the drop height on radial coordinate for time moment $t=t_i$; $r_k(t_i)$ is the contact radius in specific time. Processing of the video sequence showed that marked evaporation of the drop (deviation of $V(t_i)$ from initial drop volume $V = \pi D^3/6$ on 0.5 %) begins at t_i~6 s. This time t can be considered as lower limit corresponding to the beginning of effect of physical and chemical processes, namely evaporation, on the drop. So the measurement of the contact angle must be performed in time interval t =(0.2 ÷ 6.0) s after placing the drop on the pressed sample.

Approbation of the proposed method for determining the wettability of coal powders was conducted for coal of different deposits. The results of determining the contact angle for different coal sample are presented in Table 1.

Table 1. Contact angle values θ for coal surface.

Sample	Contact angle, degree	δ, %
A1	90.1 ± 0.3	0.3
A2	53 ± 10	18.9
A3	61 ± 7	11.4
A4	67 ± 3	4.5
A5	60 ± 3	5.0
A6	87 ± 2	2.3
A7	91±7	7.7
A8	91±7	7.7
A9	80±2	2.5

The analysis of effect of chemical composition of coals on contact angle value showed that while the mass fraction of oxygen, nitrogen and sulfur β=O+N+S rises, the contact angle value is decreased (fig. 5).

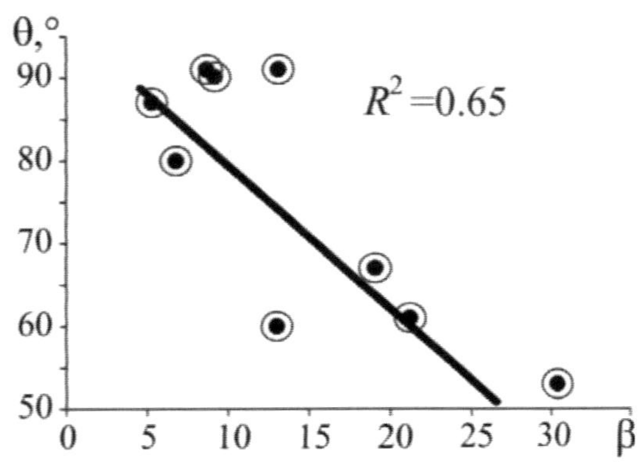

Fig. 5. Dependence of contact angle wettability of coal powders on parameter β.

The approximation curve is shown in fig. 5. The curve with coefficient of determination R^2=0.65 is obtained by means of least-squares method.

4 Conclusion

The modified method for studying the fire and explosion dangerous characteristics of coal mixtures and method for determining the contact angle of coal powders are presented.

• Experimental method for studying the fire and explosion dangerous characteristics provides the determination of critical conditions of flame spread in the conditions of uniform distribution dust cloud at varying the temperature of initiating ignition source. This allows conducting more accurate substance rating and so development of measures on fire and explosion risk.

• The possibility of occurrence of "cold" flame at ignition of air coal dust mixture and its transition in "hot" flame at increasing the dust concentration has been shown. This is important for theory of fire safety engineering.

• The method for studying the wetting characteristics of coal powders has some advantages. Direct measurement of contact angle excludes the error accumulation due to indirect determination. Plane surface of the sample excludes the error of determination of contact angle due to their inequality and roughness. Reducing the liquid penetration provides the stable drop on the sample surface that allows measuring its contact angle. The method excludes the contact angle determination error occurring due to the change of gas phase composition at the drop evaporation.

• The considered methods for studying the characteristics of coal can be used for development of new devices and technological decisions in mining works.

References

1. S.B. Romanchenko, Yu.F. Rudenko, V.N. Kosterenko, *Dust dynamics in coal mines* (Gornoe delo, Moscow, 2011)
2. V.V. Pomerancev, S.L. Shagalova, V.A. Reznik et al., *Spontaneous combustion and explosions of natural fuels dust* (Energiay, Leningrad, 1978)
3. A.N. Baratov , A.Ya. Korolchenko , N.G. Kravchuk et al., *Fire and explosion unsafety of substances and materials and their extinguishing agent* (Khimiya, Moscow, 1990)
4. GOST 12.1.044-89. OSSS, *Fire and explosion unsafety of substances and materials. Nomenclature of indicates and methods of their determination* (Izdatelstvo standartov, 1965).
5. V.Ya. Bonner, C.F. Tipper, Int. 10th Symp. Combust. **145** (1965)
6. J.A. Heand Schneider, C. Volanski, Revue Roumaine de Chimie **18**, 195 (1973)
7. J. A. Barnard, B. A. Harwood, Combust. Flame **21**, 354 (1973)
8. V.A. Arkhipov, D.Yu. Paleev, V.F. Trofimov, A.S. Usanina, Patent RF no. 2457464, IPC G01N 13/00. *Method for determining the wettability of powder materials* (Publ. 27.07.2012, Bull. no. 20)

PHYSICAL MODEL AND BASES OF MATHEMATICAL MODELLING OF ABOVE-SURFACE GASIFICATION OF COAL

R. N. Arukov[1], and *A.* N. Subbotin[1,*]

[1]National Research Tomsk Polytechnic University, 634050 Tomsk, Russia

Abstract. A mathematical model describing the heat and mass transfer processes of above-surface gasification of coal with high temperature steam is introduced. Numerical studies have allowed determining the temperature profiles according to the length of the sample and composition of gasification end products. It is found that such parameters as porosity of gasified fuel and oxidant flow have little influence on the composition of the synthesis gas. The main parameter is the temperature of the feed oxidant.

1 Introduction

Gasification of solid fuels is a universal method of its processing. The versatility of the solid fuel gasification techniques can be considered in three ways. Firstly, any solid fuel from organic waste to various anthracitic coal and black coal can be gasified, regardless of their chemical composition, the composition of the cindery part, sulfur impurities, size, moisture content and other properties.

Secondly, the important feature of the solid fuel gasification methods is their wide range of usage. Gas generator can handle with the largest chemical factories, producing millions of tons of ammonia or methanol per year, can supply fuel gas to major CHP plants and at the same time can provide gas to small autonomous power and chemical plants (for example, gas generating systems for cars), towns and villages, small chemical, engineering, or other factories.

2 Formulation of the problem

We consider the work area of the experimental stand [1], created for the high-temperature steam gasification of solid fossil fuels. Work area of the stand has a cylindrical shape, has a thermal insulation on the outside. We fill solid fuel, the water vapor is pumped through it, temperature of the vapor can reach 1473 K. Steam-gas mixture, which is formed at outlet from the cylindrical canister, is supplied to the plant where steam condensation occurs. The gas mixture remaining after the condensation of steam flows into the gas analyzer where its composition is determined. In this research, we propose a mathematical model of the

* Corresponding author: subbot@tpu.ru

process of the solid fuel conversion described above. On the basis of a mathematical model [2] the system of equations averaged over the radius of the working area and describing the researched process is recorded. Based on experimental studies conducted at the Department of the Kuzbass State Technical University [3], we deal with the following chemical reactions:

$$C + H_2O = CO + H_2 - 130.5 \text{ kJ / mol},$$
$$CO + H_2O = CO_2 + H2 + 169.3 \text{ kJ / mol},$$
$$C + O_2 = CO_2 + 395 \text{ kJ / mol},$$
$$2CO + O_2 = 2\,CO_2 + 571 \text{ kJ / mol}.$$

The task is to determine the composition of the gas end mixture, non-stationary temperature field in the sample, and outlet from it.
Averaging system of equations over the diameter of the cylindrical part of the working area [1] we obtain following system of equations for a description of the non-stationary mode of solid fuel conversion:

conservation of mass of coke

$$\rho_1 \cdot \frac{\partial \varphi_1}{\partial \tau} = - R_C , \quad R_C = R_{C1} + R_{C2} , \tag{1}$$

if $R_{C1} = s \cdot \rho_{4\cdot} \cdot \varphi_4 \cdot c_1 \cdot k_{s1} \cdot \varphi_{1\cdot} \cdot \exp(-E_{C1}/(R \cdot T_S))$,
$R_{C2} = s \cdot \rho_{4\cdot} \cdot \varphi_4 \cdot c_4 \cdot k_{s2} \cdot \varphi_{1\cdot} \cdot \exp(-E_{C2}/(R \cdot T_S))$
(we assume that in the process of conversion the mass of condensed water is equal to zero, then volume ratio $\varphi_2 = 0$),

the continuity of the gas phase

$$\frac{\partial\, \rho_4\varphi_4}{\partial \tau} + \nu \cdot \frac{\partial\, \rho_4\varphi_4}{\partial z} = R_C + R_4 . \tag{2}$$

where $R_4 = R^1_{CO} + R^2_{CO}$, $R^1_{CO} = k^1_{CO} \cdot c_2 \cdot c_5 \cdot \exp(-E^1_{CO}/(R \cdot T_4))$,
$R^2_{CO} = k^2_{CO}\left(c_1\, M / M_{51}\right)^{0.25}\left(c_2\, M / M_{52}\right) \cdot T_4^{-2.25} \cdot \exp\left(-E_{CO} / RT_4\right);$

the continuity of α – component of the gas phase

$$\frac{\partial(\rho_4\varphi_4 \cdot c_\alpha)}{\partial \tau} + \nu \cdot \frac{\partial(\rho_4\varphi_4 \cdot c_\alpha)}{\partial z} = \frac{\partial}{\partial z}(\rho_4\varphi_4 D_\alpha \frac{\partial c_\alpha}{\partial z}) + R_\alpha , \qquad \alpha = \overline{1.5} ; \tag{3}$$

energy conservation of the gas phase and coke

$$\rho_4 \cdot \varphi_4 \cdot c_\alpha \cdot (\frac{\partial T_4}{\partial \tau} + \nu \cdot \frac{\partial T_4}{\partial z}) = \frac{\partial}{\partial z}((\lambda_4 \cdot \varphi_4 + \lambda_R)\frac{\partial T_4}{\partial z}) - A_s \cdot (T_4 - T_S)$$
$$+ q^1_{CO} \cdot R^1_{CO} + q^2_{CO} \cdot R^2_{CO} - \frac{2}{r_0} \cdot \alpha_1 \cdot \varphi_4 \cdot (T_4 - T_S) \quad , \tag{4}$$

where $\lambda_R = \dfrac{16 \cdot \sigma \cdot T_4^3}{s}$;

$$\sum_{i=1}^{4} \rho_i \cdot \varphi_i \cdot c_{p_i} \cdot \frac{\partial T_S}{\partial \tau} = \frac{\partial}{\partial z}(\lambda_S \cdot \frac{\partial T_s}{\partial z}) + q_{C1} \cdot R_{C1} + q_{C2} \cdot R_{C2} + A_S \cdot (T_4 - T_S) +$$
$$\frac{2 \cdot r_1}{r_0^2} \cdot \alpha_3 \cdot (T_S - T_E) + q_R , \quad (5)$$

if $\lambda_S = \sum_{i=1}^{2} \lambda_i \cdot \varphi_i$, $\quad c_{p4} = \sum_{\alpha=1}^{4} c_{p_\alpha} \cdot c_\alpha$;

gas movement in the pores (in the form of a quadratic law of filtration) and the equation of state

$$v = \frac{k}{\mu + k \cdot \beta \cdot \rho_4 |V|} \cdot (\frac{\partial p}{\partial z}), \quad p = \frac{\rho_4 \cdot R \cdot T}{M_4}, \quad M_4 = 1 / \sum_{\alpha=1}^{4} \frac{c_\alpha}{M_{4\alpha}} . \quad (6)$$

The system of equations (1) - (6) can be solved with the boundary conditions:

$$\tau = 0: \quad \varphi_i = \varphi_{iH} , \ i = \overline{1.4}; \ c_\alpha = c_{\alpha H} , \ \alpha = \overline{1.5}, \ T = T_S = T_H , \quad p = p_H ; \quad (7)$$

$$z{=}0: \quad T_4 = T_{Pw} , \quad (\rho v)_w = \text{const}, \ \lambda_S \cdot \frac{\partial T_S}{\partial z} = \alpha_1 \cdot (T_S - T_{Pw}), \quad c_\alpha = c_{\alpha w} ,$$
$$z{=}h: \quad \partial c_\alpha / \partial z = \partial T_4 / \partial z = 0, \quad p = p_H , \ \lambda_S \frac{\partial T_S}{\partial z} = 0. \quad (8)$$

In equations (3) – (5) we enter a brief notation rates of chemical reactions. Here, v – the filtration rate; M_4 - Molecular weight of the gas phase; 4 index is attributed to the values characterizing the gas, and s – to the condensed phase; k – is a coefficient of permeability of the porous layer; $(\rho v)_H$ – the gas flow rate in the input section; T_S, T_4, T_E – is temperatures of the condensed phase, of the gas in the pores and of the external environment; $E_{L1}, E_{C1}, E_{C2}, k_{L1}, k_{C1}, k_{C2}$ – activation energy and pre-exponential factors of water evaporation reactions, heterogeneous reactions of coke oxidation; R - universal gas constant; $\lambda_{4\alpha}, c_{4\alpha}, M_{4\alpha}$ – thermal conductivity, mass concentration and molecular weight α-constituent, the parameter with an index $\alpha = 1$ corresponds to the oxygen (O_2), α =2 – CO, α =3 – CO_2, α =4 – H_2, α =5 – H_2O; p – is the pressure of the light-end products in the pores; $A_S = \alpha_1 \cdot s$ – is the volumetric heat transfer coefficient; s – is the specific surface area of pores; $q_R = \varepsilon_S \cdot \sigma \cdot T_S^4$ – is the radiative flux from the outer wall of the tube to the environment; ε_S - is the degree of emissivity; τ - is time; z – is the dimensional space coordinate directed along the tube axis; α_1, α_2, α_3 – is the heat transfer coefficients in the input, the output section and on the outer wall of the pipe; r_1, r_0 – inner and outer radii of the tube; $\rho_0, c_{P0}, \lambda_0, \varphi_i$ – is the density, heat capacity, thermal conductivity factor and the volume ratio: coke (i = 1), water (i = 2), fly ash (i = 3) and the gas phase (i = 4); $\lambda_4, \lambda_R, \lambda_S$ – is the coefficients of molecular and radiant thermal conductivity of the gas phase and the thermal conductivity of the condensed phase; σ –

Stefan - Boltzmann constant; μ – Dynamic viscosity coefficient; β – Constant in the quadratic law of filtration.

3 Analysis of the results of numerical calculations

Testing of the mathematical model was used for the future research. As the basis of the test of numerical calculations an experiment with a sample of "Berezovskaya" mine where the temperature in the layer of material is around 1273 K was taken [3].

Numerical calculations were performed for the following parameters: $\rho_1 = 1500$ kg/m^3, $\rho_2 = 1000$ kg/m^3, $\rho_3 = 1560$ kg/m^3, $\rho_4 = 1.25$ kg/m^3, $c_1 = 1630$ J/(kg·K), $c_2 = 4190$ J/(kg·K), $c_3 = 800$ J/(kg·K), $\lambda_1 = 0.4$ W/(m·K), $\lambda_2 = 0.61$ W/(m·K), $\lambda_3 = 0.8$ W/(m·K), $\lambda_4 = 0.0268$ W/(m·K)), $\lambda_{O_2} = 0.0268$ W/(m·K), $\lambda_{CO} = 0.025$ W/(m·K), $\lambda_{CO_2} = 0.038$ W/(m·K), $\lambda_{H_2} = 0.307$ W/(m·K), $\lambda_{H_2O} = 0.029$ W/(m·K), $\lambda_{N_2} = 0.025$ W/(m·K), $E_{S1} = 140$ kJ/mol, $q_{S1} = -1.0875 \cdot 10^7$ J/kg, $E_{S2} = 207$ kJ/mol, $q_{S2} = 9.64 \cdot 10^5$ J/kg, $E_{CO} = 106.5$ kJ/mol, $q_{CO} = -1.08 \cdot 10^6$ J/kg, $k_{S1} = 2.08 \cdot 10^4$ m/s, $k_{S2} = 306$ m/s, $S = 6 \cdot 10^5$ 1/m, $T_H = T_S = 293$ K, $p_H = 1.01 \cdot 10^5$ Pa, $\varphi_{1H} = 0.352$, $\varphi_{2H} = 0$, $\varphi_{3H} = 0$, $\varphi_{4H} = 0.648$.

The testing has shown that there is good agreement on the percentage of gasification products after warming up the sample. The temperature in the sample layer that was found numerically, also consistents with the results of laboratory tests. Thus, the proposed mathematical model can be used to analyze the composition of the gas in the study of organic raw materials conversion processes. During the tests the temperature and the concentrations of gas components relationships that were obtained by numerically solving the mathematical model (1) – (8), were different not more than 5% from the sample conversion process of "Berezovskaja" that was provided in the laboratory.

Thus, the proposed mathematical model can be used to analyze the composition of the gas in the study of organic raw materials conversion processes.

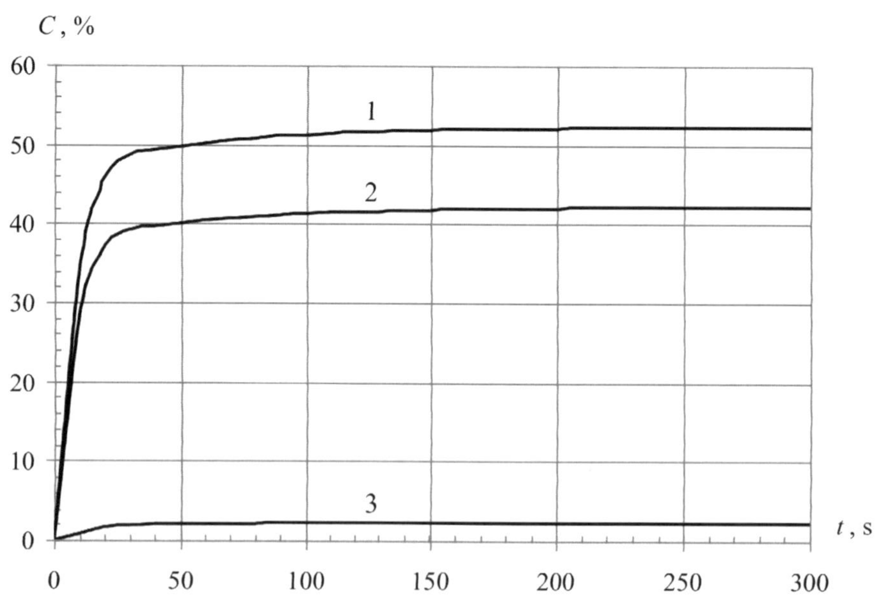

Fig. 1. Percentage of synthesis (water vapor) 873 K: line 1 – H_2, line 2 – CO, line 3 – CO_2.

Therefore, based on this model studies of the yield of synthesis gas depending on the temperature of the fed steam, porosity of solid gasified fuel and steam flow have been conducted.

Figures 1 and 2 show the volumes of gas components (in percent) formed by blowing coke with high temperature steam.

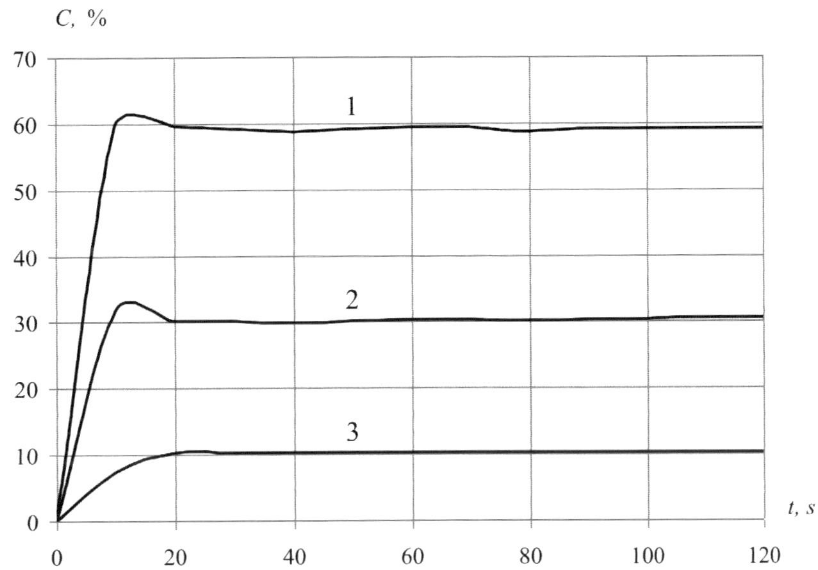

Fig. 2. Percentage of synthesis gas formed during the conversion of coke at a temperature of supplied oxidant (water vapor) 1473 K: line 1 – H_2, line 2 – CO, line 3 – CO_2.

Analyzing figures 1 and 2, it can be concluded that the increase in coke gasification temperature alters the composition of the synthesis gas. The proportion of hydrogen and carbon dioxide increases and carbon dioxide concentration decreases. The porosity and oxidant flow rate influence on the composition significantly weaker then the temperature, however, these relationships are not given.

4 Conclusions

On the basis of numerical calculations we can draw the following conclusion: a mathematical model describes the process of a solid fuel conversion with accuracy sufficient for practice, and it can be used to predict the composition of the synthesis gas in the gasification of solid fuels. Changing temperature of oxidant and its rate of supply and the fuel porosity influence on the composition of the resulting combustible gas.

References

1. G.V. Kuznetsov, A.N. Subbotin, Heat proc. Tech. **9** (2010) [in Russian]
2. R.N. Kulesh, A.S. Mazanik, A.N. Subbotin, Bulletin TPU. Power Eng. **4** (2014) [in Russian]
3. S.A. Sheviriev, A.R. Bogomolov, Power system **12** (2013) [in Russian]

COMPOSITE SOLID FUEL: RESEARCH OF FORMATION PARAMETERS

Roman Tabakaev[1,*], *Alexander* Astafev[1], *Alexander* Kazakov[1], and *Alexander* Zavorin[1]

[1]National Research Tomsk Polytechnic University, 634050 Tomsk, Russia

Abstract. Involving of local low-grade fuels resources in fuel and energy balance is actual question of research in the present. In this paper the possibility of processing low-grade fuel in the solid fuel composite was considered. The aim of the work is to define the optimal parameters for formation of the solid composite fuel. A result of researches determined that dextrin content in the binder allows to obtain solid composite fuel having the highest strength. The drying temperature for the various fuels was determined: for pellets production was 20-80 °C, for briquettes – 20-40 °C.

1 Introduction

One of the most important problems in modern energy is ensuring of energy safety countries in general and regions – particularly. Even in the Russian Federation, reached natural and fuel resources, more than 45 % of regions are dependent on the fuel delivery [1, 2]. It is also important that the fuel delivery is accompanied by transportation costs, which increased the fuel cost by several times. As a result, energy supply of such regions is provided by higher electricity tariffs and in conditions of constant depending on the integrity of traffic arteries.

The solution of this problem is involving of local organic raw materials, which often are low grade type of fuel. Such raw materials include peat, low-quality brown coal, biomass and household waste. The main problems of involving low grade raw materials to fuel and energy balance are its low thermotechnical characteristics and strength (brittleness and crumbliness). These problems lead to extraction difficult in the winter and high operational costs at traditional combustion in furnaces with grate firing: the need of raw materials drying, the high value of sifting, incomplete combustion and underburning, as a result the low boiler efficiency [3–5].

One of the most popular directions to low grade raw materials recycling for energy use is forming of fuel briquettes or pellets, called as solid composite fuel (SCF). The solid composite fuel is prepared from low grade raw materials by pressing [6] or products of its preliminary elevation, using the binder and forming equipment [7].

* Corresponding author: TabakaevRB@tpu.ru

The forming parameters (components rating and drying temperature) of solid composite fuel from thermal elevation products of low grade raw materials according to the technology [6] are researched in this work.

2 Material and methods

The research results for technology of fuel briquettes production from law grade raw materials, published in [8–10], and the aspiration to full using of thermal recycling products at the fuel briquettes production lead to necessary of binder addition on the stage of briquette forming.

As initial low grade fuel was researched the peat with following thermal and technical characteristics: moisture content (W_t^r) – 72.8 %, ash per dry mass (A^d) – 9.1 %, yield of volatile substances per fuel weight (V^{daf}) – 71.6 %, lowest calorific value (Q_i^r) – 3.1 MJ/kg.

The thermal recycling of peat was performed at a temperature 400 °C. The weight yield of thermal recycling products (relative the dried weight of raw materials): carbon residue – 43.4 %, pyrolysis condensate – 26.2 % (including tar – 6.3 %), gas – 30.4 %.

As it turns that a tar yield from researched peat is insufficient for based on its binder ability involving of all carbon residue to solid composite fuel. Additional of extra binder is necessary.

Dextrin was researched as one of the most accessible binders because costs for binders at production are very important factor, determined the SCF cost, and the binder process effects on their thermal and technical, strength characteristics.

Dextrin is polysaccharide, obtainable by thermal recycling potato or corn starch. It is applied mainly for adhesive means producing, as well as in food, light industry and foundry engineering.

The binder was obtained by dextrin addition to pyrolysis condensate, heated to a temperature of 50-70 °C. A binder with 5-%, 10-%, 15-%, 20-%, 30-% dextrin content (by weight) was researched. The carbon residue milled to particles of not more than 1 mm, after that mixed with binder. The ratios of carbon residue and binder was considered: (1 : 4); (1 : 3); (2 : 3); (1 : 1); (3 : 2). It is established that the binder is insufficient for forming at the using of ratios (1 : 1); (3 : 2), the mixture is too dry. The ratios (1 : 4); (1 : 3) is not allowed to form SCF due to lack of viscosity – the mixture is too liquid because a binder content is very high. The optimal ratio of carbon residue and binder in this case is (2 : 3). The water diluting is possible at low yield of pyrolysis condensate. As a result, SCF from contents in table 1 was obtained.

Using of 5% solution as a binder allows to involve all semi-coke, at the same time a homogeneous and sticky mixture was obtained. It is not resisted to forming and saved the received form. The 10- and 15% solutions using was shown the relevant or inclined to its results.

At the testing of more concentrated solutions of dextrin the forming mixture was obtained the inhomogeneous, consisting of the layers that have negative affect for SCF formation because of addition efforts to shaping.

Briquettes and pellets sizes for SCF production were chosen according by GOST 54248-2010 "Peat briquettes and pellets for heating purposes. Specifications": pellets of 20 mm in diameter and 20 mm of height, briquettes of 50 mm in diameter and 50 mm of height.

Table 1. Content of produced SCF.

SCF	Content
1	Carbon residue + tar (ratio 1:1) [4]
2	Carbon residue + binder (5-% dextrin content)
3	Carbon residue + binder (10-% dextrin content)
4	Carbon residue + binder (15-% dextrin content)
5	Carbon residue + binder (20-% dextrin content)
6	Carbon residue + binder (30-% dextrin content)

The drying of formed fuel (pellets) with different dextrin content was carried out at a temperature 20-40 °C. The results of mechanical compression tests for SCF according by GOST 21289-75 "Coal briquettes. Methods for the determination of mechanical strength" shows in figure 1: the highest mechanical compressive strength (P_{max}) was observed in solid composite fuel with dextrin content 5-10 %, the continued increase of dextrin share resulted to decrease of strength characteristics.

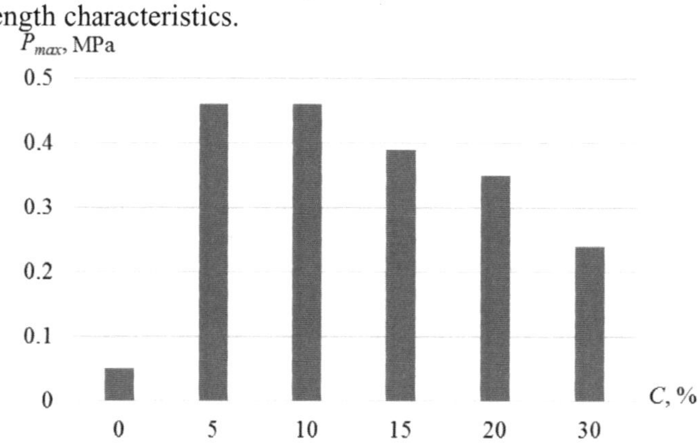

Fig. 1. The dependence of SCF (pellets) compressive strength (P_{max}) from dextrin content (C) in binder.

From practical and economical point of view [11] for SCF production is expedient to use a binder with minimal dextrin concentration, because a binder with 5 % its content has been used in further.

In example of pellets, made based on binder with 5% dextrin content in pyrolysis condensate, experimentally researched the drying temperature of SCF in the range of 20 to 140°C. The moisture losses during to drying of produced pellets shows in fig. 2a.

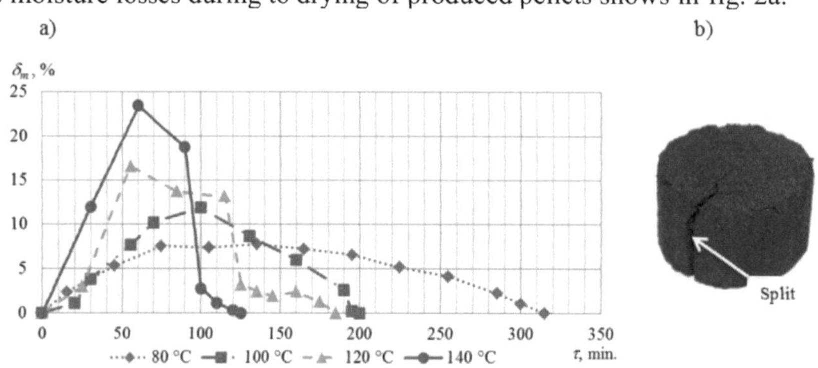

Fig. 2. The results of pellets drying: a) a moisture loss of pellet (δ_m) from time (τ) at the different dry temperature; b) pellet, dried at 140 °C.

Fig. 2a shows that moisture from SCF intensively evaporated at high drying temperature (100-140 °C). It leaded to emergence and development of surface porosity. The mechanical tests (table 2) show that all SCF examples have resistance to fracture at dropping. The results of compression test show that pellets, dried at 20-80 °C, had the highest strength. The increasing of drying temperature resulted to decrease pellets strength due to emergence pores on surface. The noticeable splits appeared at a temperature more than 120 °C (fig. 2b); it indicated to very high drying speed, sharply reduced the strength characteristics of pellet (table 2).

Table 2. The mechanical strength of the TCP depending on the drying temperature.

Type of SCF	Drying temperature, °C	Characteristics of mechanical strength	
		at dropping , %	*to compression, MPa*
Pellet	20-40	100	0.46
	80	100	0.46
	100	100	0.26
	120	100	0.26
	140	100	0.20

However, the increase of SCF size from pellets to briquettes contributed the changes to select of drying temperature. Drying of briquettes at a 80 °C lead to emergence of visible splits on their surface (fig. 3). The splits emergence explained by increase of briquettes diameter (as compared with pellets to 2.5 times), which lead to nonuniform heating.

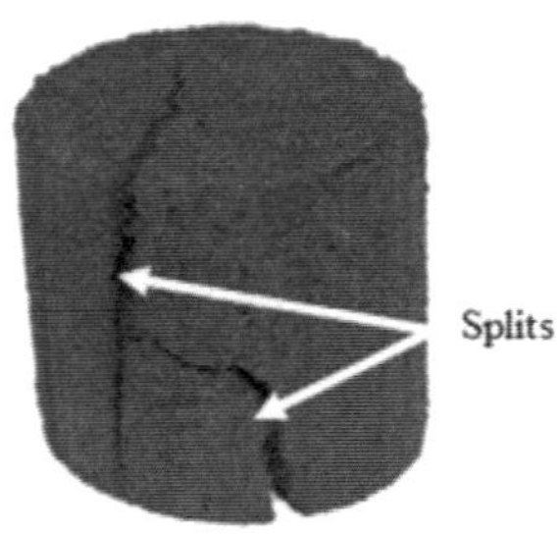

Fig. 3. Briquette (d=50 mm, h=50 mm) after drying at 80 °C.

Briquettes heated from outside surface to centre are unevenly due to low thermal conduction. A heated outside surface of briquette was solidified. In the process of continued drying the moisture inside briquette evaporated and vapor yield through an outside solidified furnace accompanied by pores formation. At the high drying speed intensity yield of evaporated moisture formed splits on surface of fuel briquette.

Briquettes with 5% dextrin content in forming mixture, dried at a temperature 20-40 °C have the following characteristics: moisture content (W_t^r) – 0 %, ash per dry mass (A^d) – 19.1 %, yield of volatile substances per fuel weight (V^{daf}) – 19.4 %, lowest calorific value (Q_i^r) – 21.1 MJ/kg, strength at dropping – 100 %.

3 Conclusions

The experiments of test dextrin in pyrolysis condensate as a binder in solution is allow to determine ratio of components for SCF forming and temperature of subsequent drying. The dextrin content 5-10 % in pyrolysis condensate at forming was ensured maximal fuel strength to compression. From practical and economical point of view is expedient to use a binder with minimal dextrin concentration (5 %). At the same time the ratio of semi-coke and binder based on dextrin was (2 : 3). The water diluting is necessary at low yield of pyrolysis condensate.

The drying temperature of SCF for pellets production was 20-80 °C, for briquettes – 20-40 °C. The higher drying temperature lead to forming of pores and splits on surface, reducing the mechanical strength of SCF. The fixed interval for pellets production is allows to producer choose the drying temperature in this range. The increased temperature up to 80 °C, allowing to reduce drying time, is preferred at high productivity. The lower temperature (20 °C) does not require of additional costs for drying, but increase the residence time of fuel at this production stage and requires the larger area for drying of SCF.

Acknowledgments

The reported study was partially supported by the Ministry of education and science of The Russian Federation, state order No. 13.948.2014/K.

References

1. Energy strategy of Russia for the period to 2030 (*confirm by the government edict of the Russian Federation No. 1715-p* 2009)
2. V.V. Lopatin, S.M. Martemyanov, Russ. Phys. J. **55**, 511 (2012)
3. O.S. Popel, B.F. Reutov, A.P. Antropov, Therm. Eng. **57**, 909 (2010)
4. S. A. Khaustov, A. S. Zavorin, K.V. Buvakov, N.A. Zakharushkin, MATEC Web Conf. **19**, 01020 (2014)
5. S.A. Khaustov, A.S. Zavorin, K.V. Buvakov, L.D. Kudryashova, A.V. Tshelkunova, EPJ Web Conf. **82**, 01041 (2015)
6. R. I. Muazu, Ju. A. Stegemann, Fuel Process. Technol. **133**, 137 (2015)
7. R.B. Tabakaev, A.V. Kazakov, A.S. Zavorin, Solid Fuel Chem. **49**, 267 (2015)
8. R.B. Tabakaev, A.V. Astafev, A.V. Kazakov, A.S. Zavorin, MATEC Web Conf. **23**, 01040 (2015)
9. R.B. Tabakaev, A.V. Astafev, A.V. Kazakov, A.S. Zavorin, M. Polsongkram, IOP Conf. Series: Mate. Sci. Eng. **93**, 012017 (2015)
10. R.B. Tabakaev, P.S. Gergelizhiu, A.V. Kazakov, A.S. Zavorin, IOP Conf. Series: Mate. Sci. Eng. **66**, 012052 (2014)
11. A.S. Zavorin, R.B. Tabakaev, P.Y. Novoseltsev, A.V. Astafev, MATEC Web Conf. **19**, 01015 (2014)

MATEC Web of Conferences 72, 01005 (2016) DOI: 10.1051/matecconf/20167201005
HMTTSC-2016

INFLUENCE OF RADIATION AND CONVECTIVE TYPES OF HEAT EXCHANGE ON THE NECESSARY DURATION OF TEMPERATURE MEASUREMENT BY THERMOCOUPLES

Yuliana K. Atroshenko[1,*], *Alena* A. Bychkova[1], and *Irina* P. Ozerova[1]

[1]National Research Tomsk Polytechnic University, 634050 Tomsk, Russia

Abstract. Results of research of influence of radiation heat exchange on duration of execution of measurement high темератур are given by thermocouples, influence of the free convection on results of temperature measurement is also considered by the thermocouples set in protective sleeves with a gap. Researches showed that radiation heat exchange leads to lowering of duration of heating of the thermocouple to 30% whereas the free convection in a protective sleeve practically doesn't exert impact on process of heating up of the sensor.

1 Introduction

Operation of the equipment on objects of such industries as power engineering, mechanical engineering, metallurgy and others, is often connected to operation in the conditions of high temperatures. Monitoring and control of technological processes on such productions is executed by the automatic systems of monitoring and control executing, first of all, measurements of parameters of processes and the equipment [1, 2].

Correct and fastest measurement of parameters of technological process (temperature, pressure, the expenditure, etc.) is a basis of safety, reliability, profitability and overall performance of the equipment [2, 3].

It is set [4,5] that the sensor mounting mode on the equipment has significant effect on an inertance of operation of primary transformers of temperature. However, heating up conditions also influence integral characteristics of process of heating.

It is known that with increase in temperature of the controlled environment heattransfer substantially amplifies a radiation component [6, 7]. This influence is more noticeable in the range of the taken temperatures exceeding 750 K. When reviewing mounting conditions of the sensor in a protective sleeve with filling of the free space with liquid materials influence of possible convection in liquid layers is possible.

* Corresponding author: julic55@tpu.ru

2 Physical model of heat transfer

In a figure 1, a –the installation diagram of the thermocouple I in a protective sleeve of II with filling of the free space with liquid III is shown. In such conditions influence of the free convection in liquid III layers is possible.

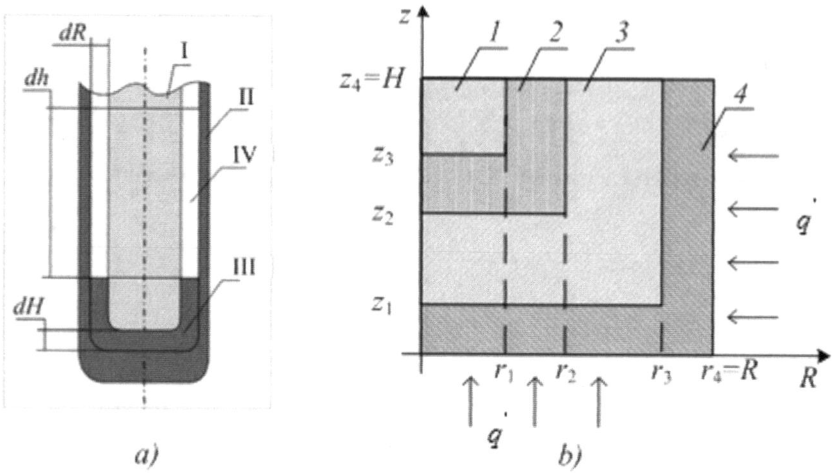

Fig. 1. The installation diagram of the thermocouple in a protective sleeve (a) and the diagram of area of the solution of the task (b): I – the thermocouple, II – a protective sleeve; III – liquid; IV – air; 1 – thermocouple junction; 2 – protective cap; 3 – aluminum oxide powder; 3 – protective cover.

In a figure 1, b – the diagram of area of the solution of the task of heat conduction for a case when on boundaries of area of the decision the heat flux of q is set is shown.

In case of the solution of the task the assumptions which aren't superimposing essential restrictions on a problem definition community are accepted:

1) heatphysical characteristics of elements of area of the decision don't depend on temperature;
2) the sensitive element has the correct cylindrical form.

The initial temperature of the systems shown in a figure 1 makes 20 °C.

The end of process of heating up was defined at the time of achievement by a thermocouple junction of temperature, other than an allowed error measured on value. Permissible deviation [8] for the thermocouple of K type makes ±1,5 in the range of temperatures from -40 to 375 °C; ±0,004·t in the range of temperatures from 375 to 1000 °C.

For area solutions of the task (fig. 1, *b*) are made the following sizes: H=5 mm; R=5 mm.

2 Mathematical model and decision methods

Process of heattransfer in the considered system (fig. 1) is described by system of differential equations:

$$c_1\rho_1\frac{\partial T_1}{\partial t}=\lambda_1\left(\frac{\partial^2 T_1}{\partial r^2}+\frac{1}{r}\frac{\partial T_1}{\partial r}+\frac{\partial^2 T_1}{\partial z^2}\right),\ t>0,\ 0<r<r_1,\ z_3<z<H; \qquad (1)$$

$$c_2\rho_2\frac{\partial T_2}{\partial t}=\lambda_2\left(\frac{\partial^2 T_2}{\partial r^2}+\frac{1}{r}\frac{\partial T_2}{\partial r}+\frac{\partial^2 T_2}{\partial z^2}\right),\ t>0,\ 0<r<r_2,\ z_2<z<z_3;\ r_1<r<r_2,\ z_3<z<H; \qquad (2)$$

$$c_3\rho_3\frac{\partial T_3}{\partial t}=\lambda_3\left(\frac{\partial^2 T_3}{\partial r^2}+\frac{1}{r}\frac{\partial T_3}{\partial r}+\frac{\partial^2 T_3}{\partial z^2}\right),\ t>0,\ 0<r<r_3,\ z_1<z<z_2;\ r_2<r<r_3,\ z_2<z<H; \tag{3}$$

$$c_4\rho_4\frac{\partial T_4}{\partial t}=\lambda_4\left(\frac{\partial^2 T_4}{\partial r^2}+\frac{1}{r}\frac{\partial T_4}{\partial r}+\frac{\partial^2 T_4}{\partial z^2}\right),\ t>0,\ 0<r<L,\ 0<z<z_1;\ r_2<r<r_3,\ z_2<z<H; \tag{4}$$

Where r – radial coordinate, m; z – axial coordinate, m; ρ – density, kg/ m^3; c – specific heat capacity, J / (kg · °C); λ – coefficient of heat conduction, W / (m · °C); indexes: 1 – thermocouple's junction, 2 – powder of an oxide of aluminum, 3 – a protective cover, 4 – air.

The equivalent heat flux on outline borders is defined as the amount of a convective heat flux of q_c and a heat flux caused by radiant heat exchange of q_r:

$$q_\Sigma = q_c + q_r\,. \tag{5}$$

Boundary conditions on inner and outline borders are given in the table 1.

Table 1. Boundary conditions of the task of heattransfer with the given external heat flux and boundary heat exchange.

Edge conditions	Decision area boundary	Edge conditions	Decision area boundary
$T_1(r_1,z)=T_2(r_1,z)$; $-\lambda_1\frac{\partial T_1}{\partial r}\Big\vert_{r=r_1}=-\lambda_2\frac{\partial T_2}{\partial r}\Big\vert_{r=r_1}$	($z_3<z<H$);	$T_1(r,z_3)=T_2(r,z_3)$; $-\lambda_1\frac{\partial T_1}{\partial z}\Big\vert_{z=z_3}=-\lambda_2\frac{\partial T_2}{\partial z}\Big\vert_{z=z_3}$	($0<r<r_1$);
$T_2(r_2,z)=T_3(r_2,z)$; $-\lambda_2\frac{\partial T_2}{\partial r}\Big\vert_{r=r_2}=-\lambda_3\frac{\partial T_3}{\partial r}\Big\vert_{r=r_2}$	($z_2<z<H$);	$T_2(r,z_2)=T_3(r,z_2)$; $-\lambda_2\frac{\partial T_2}{\partial z}\Big\vert_{z=z_2}=-\lambda_3\frac{\partial T_3}{\partial z}\Big\vert_{z=z_2}$	($0<r<r_2$);
$T_3(r_3,z)=T_4(r_3,z)$; $-\lambda_3\frac{\partial T_3}{\partial r}\Big\vert_{r=r_3}=-\lambda_4\frac{\partial T_4}{\partial r}\Big\vert_{r=r_3}$	($z_1<z<H$);	$T_3(r,z_1)=T_4(r,z_1)$; $-\lambda_3\frac{\partial T_3}{\partial z}\Big\vert_{z=z_1}=-\lambda_4\frac{\partial T_4}{\partial z}\Big\vert_{z=z_1}$	($0<r<r_3$);
$\frac{\partial T}{\partial z}=0$	($z=H$);	$\frac{\partial T}{\partial r}=0$	($r=0$);
$-\lambda_4\frac{\partial T_4}{\partial z}=q$	($z=0$);	$-\lambda_4\frac{\partial T_4}{\partial r}=q$	($r=R$);

The solution of the task is executed by method of finite differences on the basis of the four-dot implicit difference diagram [9]. The area of the solution of the task represents a grid 200×200 nodes with steps of 2,5·10^{-2} mm on axial and radial coordinates, a step on dt time = 0,001 sec [9].

3 Results and discussion

Simulation is executed taking into account heatphysical characteristics of elements of area of the solution of the task similar provided in operations [4, 5].

Distributions of temperatures on a sensitive element of the thermocouple in case of temperature measurement 1000 K taking into account and without radiation heat exchange are shown in a figure 2.

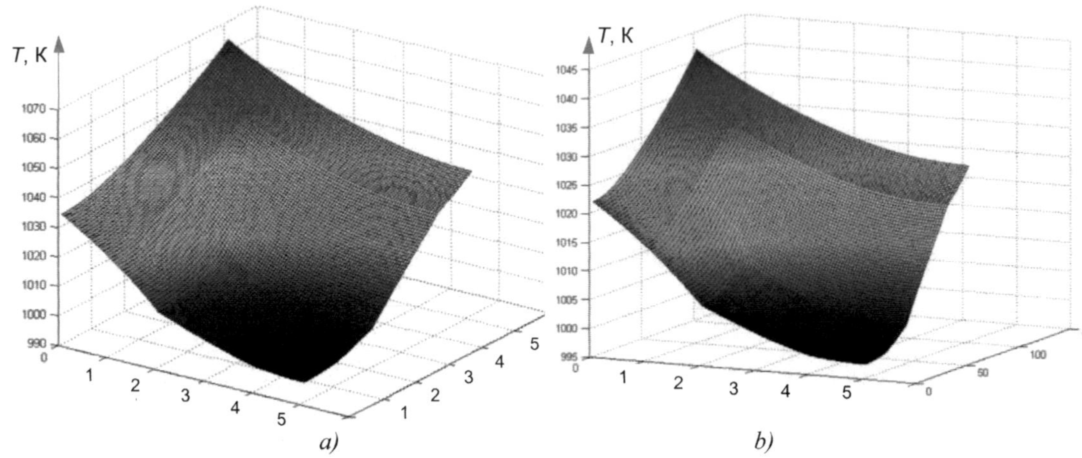

Fig. 2. Dependences of necessary duration of heating up of the sensor without (1) and taking into account (2) influence of radiant heat exchange.

It is visible that in identical conditions of measurement up to the temperature of 1000 To intensity of heating up of a sensitive element of the sensor taking into account radiation heat exchange (fig. 2, b) above

The dependences of minimum necessary times of heating of the TEP sensitive element received taking into account only convective and equivalent heat exchange are given in a figure 3.

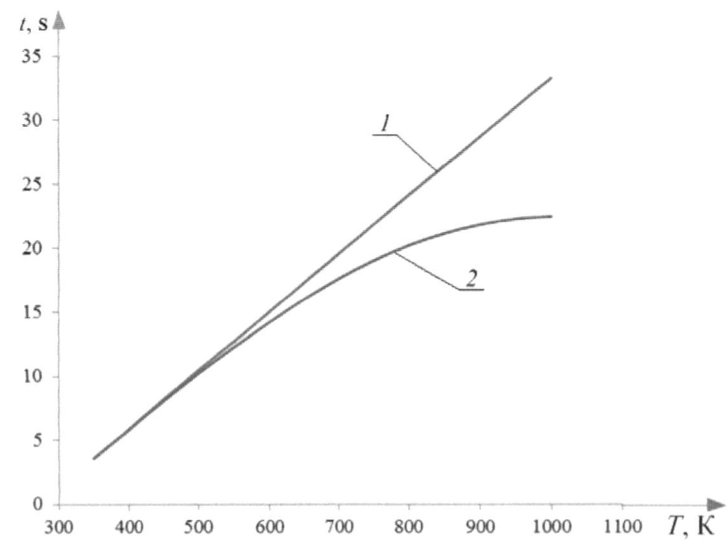

Fig. 3. Dependences of necessary duration of heating up of the sensor without (1) and taking into account (2) influence of radiant heat exchange.

It is visible that that influence of radiant heat exchange to 30% reduces the necessary duration of heating due to increase in the equivalent heat flux.

For an impact assessment of the free convection in liquid III layers (fig. 1, a) the equivalent coefficient of heat conduction was evaluated:

$$\lambda_Э = \lambda_{жс}\left[1+\frac{m\ (Gr_c \cdot Pr_c)^n}{Gr_c \cdot Pr_c + S}\right]. \qquad (6)$$

где m, n, S – the coefficients determined by the geometrical sizes and the form of the gap filled with liquid; Gr – Grashof number, Pr – Prandtle number.

Then the relative deviation of the equivalent coefficient of heat conduction from coefficient of heat conduction of liquid is defined by expression:

$$\delta\lambda = \frac{\lambda_Э - \lambda_{жс}}{\lambda_Э}\cdot 100\% = \frac{\dfrac{m\ (Gr_c \cdot Pr_c)^n}{Gr_c \cdot Pr_c + S}}{1+\dfrac{m\ (Gr_c \cdot Pr_c)^n}{Gr_c \cdot Pr_c + S}}\cdot 100\% = \frac{m\ (Gr_c \cdot Pr_c)^n \cdot 100\%}{Gr_c \cdot Pr_c + S + m\ (Gr_c \cdot Pr_c)^n} = 0{,}0021\ \%.. \qquad (7)$$

It is visible that the relative deviation of the equivalent coefficient of heat conduction taking into account influence of the free convection from coefficient of heat conduction of liquid isn't enough (less than 1%) that allows to draw a conclusion that the free convection in liquids (fillers of protective sleeves) can be neglected.

4 Conclusion

The conducted researches showed that influence of radiation heat exchange exerts the considerable impact on process of heat transfer in a sensitive element of the thermocouple in case of measurement of temperatures higher than 700 K, in particular, in case of temperature measurement 1000 K necessary time of heating up can be lowered to 30%. It is set that influence of the free convection in layers of the liquid filling a protective sleeve has no significant effect on process of heat-transfer in system "the protective sleeve – buffer material – thermocouple".

Acknowledgments

This research was supported by the Russian Foundation for Basic Research (No. 14-08-00057). The work at the field of modeling of heat and mass transfer processes near boundary "liquid – gas" was supported by the scientific schools grant NSH-7538-2016.8.

References

1. D. Matisková, M. Kotus, M. Balara, J. TEM **5**, 60 (2016)
2. M. Ranusawud, P. Limsuwan, K. Vatcharanukul, T. Somthong, Key Engineering Materials **613**, 133 (2014)
3. G. Bucur, O. Cangea, Petroleum - Gas University of Ploiesti Bulletin, Technical Series **66**, 1 (2014)
4. Y.K. Atroshenko and A.A. Bychkova, MATEC Web Conf. **23**, 01006 (2015)
5. Y.K. Atroshenko and A.V. Abramova, EPJ Web Conf. **110**, 01071 (2016)
6. M.A. Taymarov, Yu.V. Lavirko, E.E. Belyaeva, Proceedings of the higher educational institutions. Energy Sector Problems **7**, 69 (2015)
7. A.N. Makarov, Power Station **3**, 19 (2015)
8. GOST 6651–2009 *Thermal resistance of platinum, copper and nickel. General technical requirements and test methods*, Moscow (2009) [in Russian]
9. A.A. Samarskii, *The Theory of Difference Schemes* (Marcel Dekker, Inc., New York, 2001)

ABOUT INFLUENCE OF OPERATIONAL TEMPERATURES ON RELIABILITY OF OPERATION OF THE EQUIPMENT OF THERMAL POWER PLANTS

Yuliana K. Atroshenko[1,*], *Alena* A. Bychkova[1], and *Vladimir* S. Andyk[1]

[1]National Research Tomsk Polytechnic University, 634050 Tomsk, Russia

Abstract. The description of degree model of reliability of the heat power equipment in which as a factor of loading the working temperature of the equipment is used is provided, interval estimates of constant values of model are given. For a row of elements of constructions of boilers dependences of the relative increase in failure density are given by operation in the conditions of the increased temperatures for a long time.

1 Introduction

Reliability represents the complex index [1] characterizing properties of the equipment during the long time frame to save and reproduce working parameters in use. For the description of object properties from the point of view of its reliability indices are used [1]: faultlessness of operation, maintainability and other characteristics.

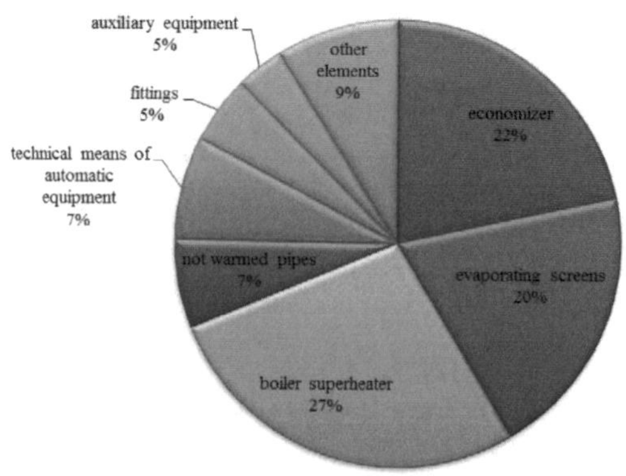

Fig. 1. The damaged elements of boiler aggregates.

Indices of reliability of the equipment substantially are defined by equipment operation modes (character and intensity of heat exchanging processes, burning, corrosion), depend on existence of deposits on surfaces and the processes proceeding in metals [1, 2]. In operation of boiler aggregates surfaces of heating (fig. 1) are most subject to damages.

* Corresponding author: julie55@tpu.ru

DOI: 10.1051/matecconf/20167201006

Pipes of furnace screens during an operating time are affected to radiative energy, corrosion influence of combustion products and other environments [2].

In the boiler installations working at supercritical parameters, pipes of radiation boiler superheaters are exposed to high-temperature corrosion which leads to changes in structure staly, to lowering of thermal stability and, finally, to thinning of pipes and appearance of bursts [2].

Thus, in use for extension of a resource of operation of the equipment and lowering of failure density special value has observance of the rules of operation including temperature condition.

2 Reliability model

Reliability of the equipment is characterized by numerical indices: time between failures, indices of readiness, maintainability, failure density, etc. Forecasting methods now in use of reliability are based on mathematical simulation of this parameter as functions from contributing factors. As temperature condition of operation exerts defining impact on reliability of the equipment, and corrupting, is generally caused, fatigue deformation, failure density can be defined by one-factor degree model [3]: $In(T)=V \cdot T^m$,

where V, m – constants, T – a loading factor (working temperature of the equipment, K). Constants of V and m – experimentally defined values, they depend on material of the equipment and other parameters. As the known numerical values of constants of V and m for specific type of the equipment it isn't enough (for some elements such data generally are absent), the ranges of their change were defined.

Interval values of constants of V and m were defined, proceeding from the given conditions: the average operating time of the equipment on a failure makes To_1 in case of normative value of working temperatures of the heat carrier T_1 (failure density in this case of In_1), in case of increase in temperature up to T_2 value the time between failures decreases, and failure density increases and makes In_1.

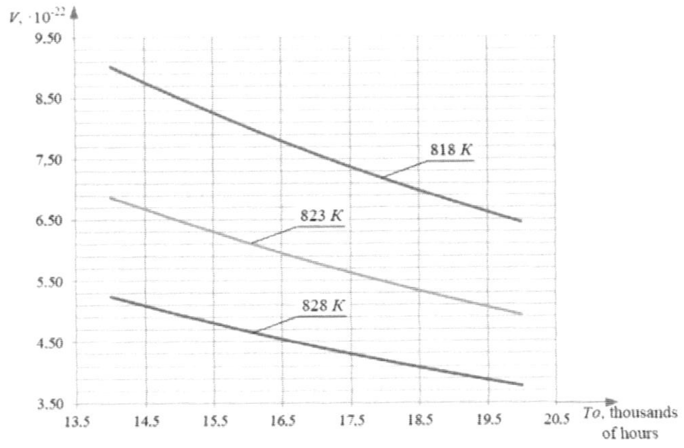

Fig. 2. Values of a constant of "V" for boiler superheaters boilers.

Based on the provided data [4, 5] it is possible to define value ranges of coefficients of V and m for different types of the equipment. Approximate values of a constant V in working conditions of the boiler superheater at temperatures of 818-828 K can be determined by the dependences shown in a figure 2. Values m in this case are in the range 5.79 … 5.87.

Support of the normative mode of exploitation of boiler superheaters is defined by quality of operation of systems of regulation, and also systems of thermal protection and locks.

3 Results and discussion

During operation of the equipment summary time of its operation in case of raised (in relation to project values) temperatures can reach several tens hours. In a figure 3 the relative change of failure density of the boiler superheater of the boiler depending on the summary duration of operation at the uprated temperatures is shown.

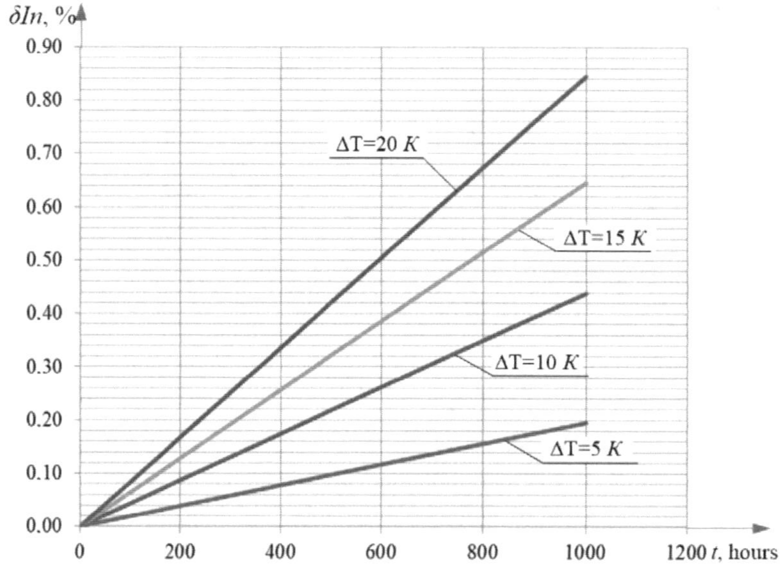

Fig. 3. Dependence of the relative increase in failure density of the boiler superheater by operation in the conditions of the increased temperature.

It is visible (fig. 3) that in the course of operation within 500 hours at a temperature exceeding estimated on 20 K change of failure density can reach 0.4 %, during operation of 1000 hours – to 0.85 %, etc. In this case authentic (exact) temperature measurement is important for implementation of its timely correction and/or actuating of protection in operation of the capital and ancillary equipment of thermal power plant as for the existing boiler installations having high rates of the worked-out park resource even little changes of failure density substantially increase probability of origin of alert conditions. It is connected to the fact that metal of elements of construction of a boiler in the course of the long operation in the conditions of high heat loads "accumulates" temperature fatigue, and sections of thinning of pipes appear. In such conditions even little relative change of failure density of the equipment can lead to alert conditions. Bursts in boiler superheaters are especially dangerous as until lowering of loading or break of a boiler (these processes can't instantaneous be executed) the depressurized section of the pipeline can significantly damage adjacent pipes of a packet and put the whole step of the boiler superheater out of action [2].

One of the major elements among ancillary equipment of boiler installations are smoke exhausters and forced draught fans. The average value of a time between failures for the forced draught fans is in the range of 23 thousand hours to 26 thousand hours, for smoke exhausters – in the range of 11 thousand hours to 14 thousand hours. It is connected to the fact that the smoke exhausters working with combustion gases with a temperature in the range from 423 K to 473 K are in "heavier" conditions in comparison with forced draught fans.

In a figure 4 the relative change of failure density the smoke exhausters and the forced draught fans boilers depending on the summary duration of operation is shown at the uprated temperatures of a work environment.

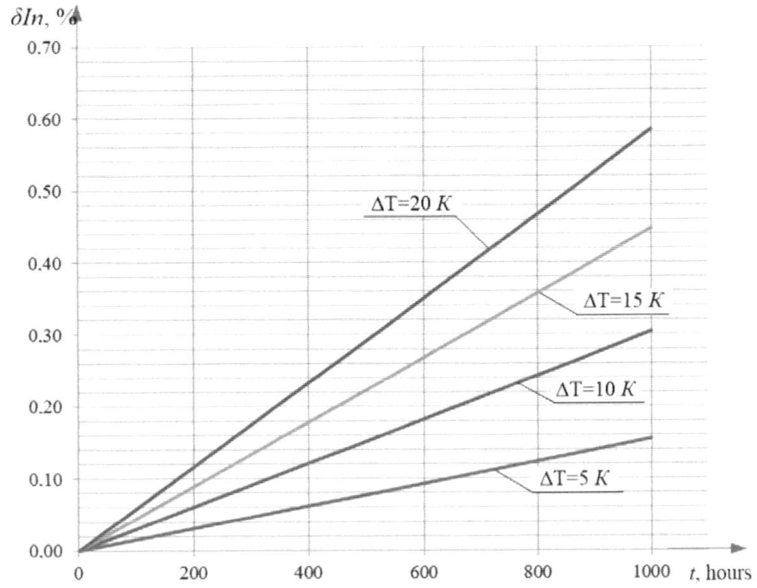

Fig. 4. Dependence of the relative increase in failure density the draft machines by operation in the conditions of the increased temperature.

It is visible (fig. 4) that in the course of operation within 600 hours at a temperature exceeding estimated on 20 K increase in failure density can reach 0.35 %. Therefore one of the main sources of failure the draft machines, is overheating of bearings. Insufficient quantity (or improper type) lubrications, the insufficient expenditure of the cooling water are basic reasons of origin of overheating of bearings. Timely detection of overheating of bearings will allow to prevent corrupting of the equipment.

4 Conclusion

The conducted researches showed that the summary operating time within 1000 hours of packets of the boiler superheater in the conditions of the increased temperatures leads to the relative increase in failure density up to 0.5 %. Such situations can arise owing to a high inertance of the thermoelectric transformers set in protective sleeves that leads to untimely response of systems of regulation to injection of cooling medium (feedwater, "own" condensate). Despite rather low values of increase in failure density (less than 1 %), growth of temperature of steam it is extremely unwanted since leads to the following consequences: excessive consumption of fuel, deterioration in a status of the equipment, stimulated actuating of systems of thermal protection.

The excessive consumption of fuel in case of not regulated increase in temperature of steam is caused by the fact that the management system fuel consumption of boiler installation connected to system of regulation of temperature of superheated steam will work out a signal on increase in fuel consumption for maintenance of the current temperature of steam.

Lowering of utilization properties of the equipment is connected to temperature dependences of indices of reliability of operation of elements of the equipment, caused by lowering of thermal stability of steel, accumulation of temperature fatigue of metals and other factors which in total can promote origin of alert conditions. The received values are defined, proceeding from the forecast of operation of the equipment in a zone of "not regulated" temperatures within 1000 hours whereas in practice in case of operation of the equipment on the prolonged park resource this value can reach 3–4 thousand hours and more. In this case the relative increase in failure density will make 1–2 %.

Acknowledgments

This work with intensive heat and mass transfer processes at the geterogenius system was supported by the Russian Foundation for Basic Research (No. 14-08-00057). The work at the field of modeling of heat and mass transfer processes near boundary "liquid – gas" was supported by the scientific schools grant NSH-7538-2016.8.

References

1. A.P. Lashitskij, G.P. Sutotskij, G.V. Vasilenko, V.M. Evtushenko, Zhongguo Gonglu Xuebao **11**, 4 (1998)
2. A.I. Andryushchenko, Reliability of the heat power equipment of TPP and NPP (Higher school Publ., Moscow, 1991) [in Russian]
3. V.V. Moskvichev, N.A. Makhutov, A.P. Chernyaev, Resistance to cracks and mechanical properties of constructional materials of technical systems (Science Publ., Novosibirskm, 2005) [in Russian]
4. GOST 28269-89 *Stationary boilers of big power. General technical requirements and test methods*, Moscow (2009) [in Russian]
5. SO 153-34.17.470-2003 *The instruction about an order of inspection and extension of service life of steam lines over a park resource*, Moscow (2003) [in Russian]

NUMERICAL MODELING OF THERMOCAPILLARY DEFORMATION AND FILM BREAKDOWN IN A LOCALLY HEATED THIN HORIZONTAL VOLATILE LIQUID LAYER

Ella Barakhovskaya[1,*], and *Igor* Marchuk[1,2]

[1]Novosibirsk State University, 630090 Novosiborsk, Russia
[2]Kutateladze Institute of Thermophysics SB RAS, 630090 Novosibirsk, Russia

Abstract. The problem of thermocapillary deformation and film breakdown in a thin horizontal layer of viscous incompressible liquid with a free surface is considered. The deformable liquid layer is locally heated. The problem of thermocapillary deformation of the locally heated horizontal liquid layer has been solved numerically for two-dimensional unsteady case. The lubrication approximation theory is used. Capillary pressure, viscosity and gravity are taken into account. Evaporating rate is supposed to be proportional to the temperature difference between the liquid and ambient. Heat transfer in the substrate is also simulated. The numerical algorithm for the joint solution of the energy equation and the evolution equation for the thickness of liquid layer has been developed. The model predicts the thermocapillary deformation of the liquid surface and the formation of dry spots. The dynamics of liquid surface, the dry spots formation and the velocity of the contact line have been calculated. The deformation of the free surface has been calculated for different values of the heating power and thickness of the liquid layer. The effect of surface tension coefficient and wetting contact angle on the velocity of the contact line motion has been analyzed. It has been obtained that the velocity of the contact line increases with the increase of the wetting contact angle value and of the surface tension coefficient.

1 Introduction

Heat transfer and film breakdown in thin liquid layer with local heating is an important challenge of thermal stabilization technique of electronic equipment [1-3]. The study of thermocapillary effects in the thin locally heated liquid layer is demanded problem in the field of technological processes as the thin films provide a high intensity of heat and mass transfer. But thin films are exposed to rupture [2,4]. The problem of thermocapillary breakdown of the locally heated horizontal liquid layer has been solved numerically in axisymmetric statement, fig.1. The thin layer approximation model is used [5]. Capillary pressure, viscosity and gravity are taken into account. Evaporating rate is supposed to be

[*] Corresponding author: ella94@bk.ru

proportional to the temperature difference between temperatures of the liquid surface and ambient. Heat transfer in the substrate is also simulated. Initially the liquid layer has flat surface and uniform temperature. The initial thickness of the liquid layer varies from 100μm to a few millimeters. The local heating of the liquid occurs from the substrate side. Deformations of the liquid surface are determined by the properties of the liquid, substrate and heater.

2 Problem statement

The problem considers a thin horizontal volatile liquid layer on the substrate with local heater, fig. 1. Liquid is viscous and incompressible. Heater is thin and uniform. Two-dimensional unsteady problem is considered. Geometry and system conditions are axisymmetric. Initially the liquid layer has flat surface and uniform constant temperature T_0. At the time $t=0$ the heater is activated and starts heating the substrate and liquid. Shear stress occurs on the liquid surface caused by the heterogeneity of its temperature. Thermocapillary flow and deformations of the liquid surface are developing. Gravity, surface tension, thermocapillary effect, viscosity, evaporation have a significant influence on the process.

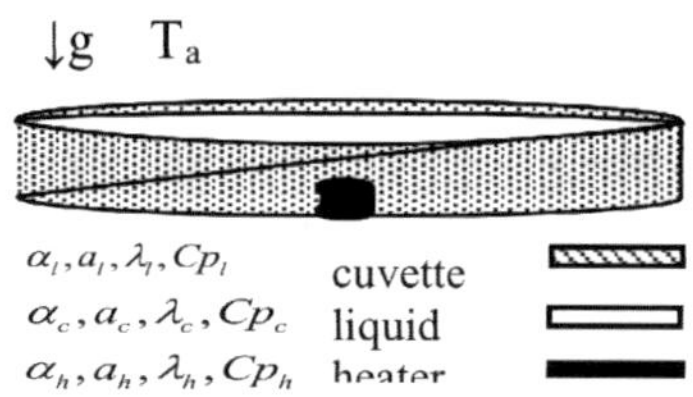

Fig. 1. Scheme of the process.

3 Mathematical formulation

3.1 Motion equation of the liquid

The mathematical formulation of the problem includes the Navier-Stokes equations, continuity and thermal conductivity. The system of equations in two-dimensional non-stationary case and in cylindrical coordinates [8]:

$$\begin{cases} \dfrac{\partial u}{\partial t} + u\dfrac{\partial u}{\partial r} + v\dfrac{\partial u}{\partial z} = F_r - \dfrac{1}{\rho}\dfrac{\partial p}{\partial r} + \nu\Delta(ru) \\ \dfrac{\partial v}{\partial t} + u\dfrac{\partial v}{\partial r} + v\dfrac{\partial v}{\partial z} = F_z - \dfrac{1}{\rho}\dfrac{\partial p}{\partial z} + \nu\Delta v \\ \dfrac{1}{r}\dfrac{\partial (ru)}{\partial r} + \dfrac{\partial v}{\partial z} = 0 \\ \rho C_p\left(\dfrac{\partial T}{\partial t} + u\dfrac{1}{r}\dfrac{\partial (rT)}{\partial r} + v\dfrac{\partial T}{\partial z}\right) = \dfrac{1}{r}\dfrac{\partial}{\partial r}\left(\lambda\left(r\dfrac{\partial T}{\partial r}\right)\right) + \dfrac{\partial^2 T}{\partial z^2} + Q \end{cases}, \quad (1)$$

The dimensional variables that are used into equations: u, v – the components of liquid velocity vector, $F_r = (g \cdot \tau)$, $F_z = (g \cdot n)$ - the components of the vector of external forces, τ - tangent vector along the surface, n – the normal vector to the surface, ρ - density, p -

capillary pressure, ν - kinematic viscosity, Q - bulk density of the heat sources, λ - coefficient of thermal conductivity; C_p - specific heat, T - temperature, t - time.

3.2 The evolution equation for liquid film thickness

Dynamics of thin films is well described by the evolution equation, which has been obtained by the system of equations (1). The evolution equation for liquid film thickness is obtained in the lubrication approximation theory [5,6]:

$$h_t + divq + J = 0 \tag{2}$$

h_t - velocity of the surface, $q = \frac{h^3}{3\mu} f + \frac{h^2}{2\mu}\tau$ - liquid flow rate along the r direction, $f = grad(\rho gh + \sigma H)$ - gradient of the modified pressure, $\tau = \sigma_T gradT$ - thermocapillary tangent stress, $J = J^*_{ev}(T_s(x) - T_a)$ - evaporation velocity, $T_s(x)$ - temperature of the surface, T_a - temperature of ambient, $J^*_{ev} = \alpha_{ev}/(r_{lv}\rho)$ - evaporation rate coefficient, α_{ev} - evaporation heat transfer coefficient, r_{lv} - latent heat of vaporization, μ - dynamic viscosity, ρ - density, g - gravity vector, h - film thickness, σ - surface tension, σ_T - linear coefficient, which determines the dependence of the surface tension on temperature, T - temperature, $H = \frac{h_{rr}}{(1+h_r^2)^{3/2}} + \frac{h_r}{r(1+h_r^2)^{1/2}}$ - double mean curvature.

Substituting the expression for the curvature, liquid flow rate along the surface and evaporation is obtained the equation in cylindrical coordinates:

$$h_t + \frac{1}{r}\frac{\partial}{\partial r}\left(r\left[\frac{h^3}{3\mu}\frac{\partial}{\partial r}(\rho gh + \sigma H) + \frac{h^2}{2\mu}\sigma_T\frac{\partial T}{\partial r} \right] \right) + \frac{\alpha_{ev}}{r_{lv}\rho}\Delta T = 0 \tag{3}$$

Equation (3) is a nonlinear differential equation of the first order in time and the fourth order in spatial variables relative to unknown function *h(t,r)*.

3.2.1 The boundary conditions

Boundary conditions for this problem are set naturally:
$h_r(t,0) = 0$ the condition of the axial symmetry is given in the center of the cuvette
$h_r(t,R) = 0$ the contact angle is given on the border of the cuvette
$q(t,0) = 0$ flow rate is equal to 0 in the center
$q(t,R) = 0$ the condition of impermeability of liquid through the walls

Also it is necessary to set boundary conditions in case of occurrence of film breakdown and formation of dry spots:
$h_r(t, r_{Dry\,Spot}) = \pm\tan\theta$
$q(t, r_{Dry\,Spot}) = 0$

3.3 The energy equation

The law of temperature distribution in the liquid layer is determined by the energy equation in cylindrical coordinates:

$$\rho C_p\left(\frac{\partial T}{\partial t}+u\frac{1}{r}\frac{\partial(rT_r)}{\partial r}+v\frac{\partial T_z}{\partial z}\right)=\frac{1}{r}\frac{\partial}{\partial r}\left(\lambda\left(r\frac{\partial T_r}{\partial r}\right)\right)+\frac{\partial^2 T_z}{\partial z^2}+Q, \qquad (4)$$

$Q=Q(t,r,z)=\begin{cases}Q=0, \text{ outside the heater}\\ Q=const>0, \text{ in the heater}\end{cases}$ - bulk density of the heat sources, λ - coefficient of thermal conductivity; C_p - specific heat; ρ - density of the liquid.

3.3.1 The boundary and initial conditions

The convective heat transfer coefficient is set on the substrate walls:

$$\lambda\frac{\partial T}{\partial n}\bigg|_W=\alpha(T-T_a),$$

where W indicates to conditions on solid surfaces.

In the condition of the top boundary the heat flux density is determined by the evaporation, fig.2.

$$\lambda\frac{\partial T}{\partial z}\bigg|_{z=h_w+h_l}=\alpha_{ev}(T-T_a), \qquad (5)$$

$\alpha_{ev}=J^*_{ev}r_{lv}\rho$ - the heat transfer coefficient for evaporation, r_{lv} - latent heat of vaporization, J^*_{ev} - evaporation rate coefficient. The coefficients for evaporation rates have been calculated using the data published in [7]: $J^*_{ev}=3.8\,\mu\text{m/s}\cdot\text{K}$, $\alpha_{ev}=9.72\,\text{kW/m}^2\text{K}$.

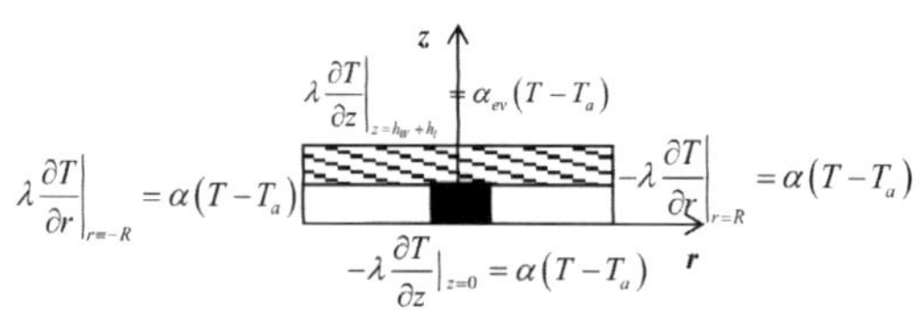

Fig. 2. Boundary conditions.

Initially the liquid layer has flat surface and uniform temperature:

$$h|_{t=0}(r,z)=h_0, \quad T|_{t=0}(r,z)=T_0=const \qquad (6)$$

3 Numerical modeling

For calculation it is used splitting into physical processes: distribution of heat and motion of liquid.

The evolution equation of the liquid layer thickness (3) is approximated at grid's nodes with finite volume method [5]. The grid with uniform steps in the space variables in the liquid and solid phases is set: $r_i=i\cdot dr$, $i=0,..,N$. The time step τ is also uniform. The value of the liquid layer thickness in the time t_k and i node: $h_i^k=h(r_i,t_k)$. A discrete analogue of eq. (3) is written for each volume V_i:

$$\frac{h_i^{k+1}-h_i^k}{\tau}+\frac{1}{r_i}\frac{q_{i+1/2}r_{i+1/2}-q_{i-1/2}r_{i-1/2}}{dr}+J_{ev}^*\Delta T_i=0 \quad (7)$$

System of nonlinear algebraic equations (7) that has been obtained by the approximation is solved with Newton's method. The Jacobians are calculated using numerical linearization [5]. The scheme has the second-order approximation for the spatial coordinates and the first-order in time.

The temperature in the liquid is calculated without allowance of the deformations. Since the heater is thin, it is considered that $Q=Q(t,r,z)$ is concentrated only in heater units. The difference analog of the energy equation (4) is solved by the method of fractional steps [9]:

$$\begin{cases}\dfrac{T_{ij}^{n+1/2}-T_{ij}^n}{dt}=\dfrac{\lambda}{2\rho Cp}\left(\dfrac{\left(1+\dfrac{dr}{2r}\right)T_{i+1,j}^{n+1/2}-2T_{ij}^{n+1/2}+\left(1-\dfrac{dr}{2r}\right)T_{i-1,j}^{n+1/2}}{dr^2}+\dfrac{T_{i,j+1}^n-2T_{ij}^n+T_{i,j-1}^n}{dz^2}\right)+\dfrac{q_{ij}^n}{2\rho Cp}\\[2ex] \dfrac{T_{ij}^{n+1}-T_{ij}^{n+1/2}}{dt}=\dfrac{\lambda}{2\rho Cp}\left(\dfrac{\left(1+\dfrac{dr}{2r}\right)T_{i+1,j}^{n+1/2}-2T_{ij}^{n+1/2}+\left(1-\dfrac{dr}{2r}\right)T_{i-1,j}^{n+1/2}}{dr^2}+\dfrac{T_{i,j+1}^{n+1}-2T_{ij}^{n+1}+T_{i,j-1}^{n+1}}{dz^2}\right)+\dfrac{q_{ij}^n}{2\rho Cp}\end{cases} \quad (8)$$

Boundary and initial conditions are set in the finite differences equations:

$j=1,..,M$

$T(j,0)=T(j,2),-$ symmetry

$T(j,N+1)=T(j,N-1)-$ adiabatic right wall

$i=1,..,N$

$T(0,i)=T(2,i)-\dfrac{\alpha}{\lambda}(T(1,i)-T_a)2dh$ -heat transfer coefficient is given on the free liquid surface

$T(M+1,i)=T(M-1,i)-\dfrac{\alpha_w}{\lambda_w}(T(M,i)-T_a)2dh_w$ -heat transfer coefficient is given on bottom wall

Initial condition: $T_{i,j}^0=0,\ i=0,..,N,\ j=1,..,M-1$.

Geometric parameters and parameters of the liquid, substrate and heater are specified. All this properties have a significant effect on the distribution of heat and evolution of liquid deformation. Calculations were performed for ethanol. The initial temperature of the liquid and ambient temperature is 20 ^{0}C.The ambient pressure is 1 Bar. The material of the cuvette is Teflon.

The resulting system of linear algebraic equations is solved at each step with the tridiagonal matrix algorithm. Calculations are performed successively. The time step for the evolution equation is done after the time step for the energy equation.

4 Numerical results

The deformation of the free surface has been calculated for different values of the heating power and thickness of the liquid layer. A dry spot is formed in a ring shape because of the axial symmetry, fig. 3. Surrounded by a dry ring liquid in the center is a drop, which tends to acquire an equilibrium shape, which is defined by Young–Laplace equation. Then the drop evaporates completely. The dynamics of the three phase contact line is calculated, fig. 3. It was obtained that the velocity of the contact line increases with the increase of the wetting contact angle value and of the surface tension coefficient.

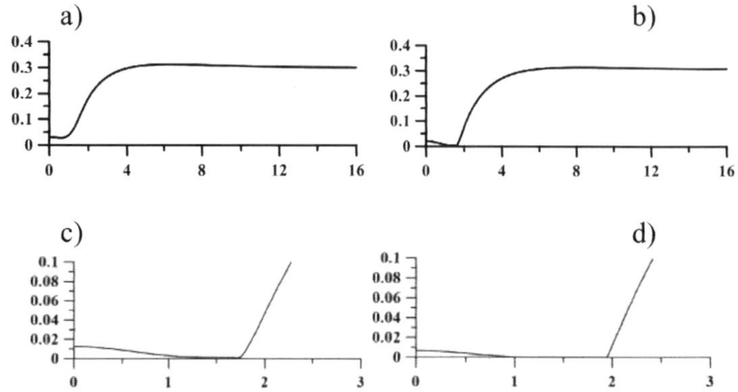

Fig. 3. Liquid layer breakdown and contact line evolution. Ethanol, $h_0 = 300\,\mu m$, $D_{cuvette} = 32\ mm, d_{heater} = 3.2\ mm, P = 0.3W$, heating time: a) 0.6 s, b) 0.9 s, c) 1.2 s, d) 1.5 s.

5 Conclusions

The numerical calculations with given model of thin horizontal liquid layer with local heating have shown that local heating of a horizontal liquid layer causes deformation of the liquid surface and thermocapillary flow. The model predicts the thermocapillary deformation of the liquid surface and the formation of dry spots with the intensive heating. Significant reduction of liquid film thickness occurs in the greatest value of the temperature gradient zone. Dry the spot is formed in the shape of a ring, but in the center is still liquid, which is surrounded by a dry spot. This liquid takes the form of droplets and ultimately evaporates completely. The velocity of the contact line increases with the increase of the wetting contact angle value and of the surface tension coefficient.

Acknowledgments

The study was financially supported by the Russian Science Foundation (Project 15-19-20049).

References

1. O.A. Kabov, E.A. Chinnov, High temp. **39**, 5 (2001)
2. O.A. Kabov, Yu.V. Lyulin, I.V. Marchuk, D.V. Zaitsev, International J. Heat Fluid Flow **28** (2007)
3. I. V. Marchuk, JET **18**, 3 (2009)
4. D.V. Zaitsev, D.A. Rodionov, O.A Kabov, Microgravity sci. technol. **19**, 3 (2007)
5. I.V. Marchuk, JET **24**, 4 (2015)
6. A. Oron, S.H. Davis, S.G. Bankoff, Rev. Mod. Phys. **69**, 3 (1997)
7. A.A. Fedorets, O.A. Kabov, I.V. Marchuk, TPL **37**, 2 (2011)
8. L.D. Landau, E.M. Lifshitz, *Hydrodynamics*, (1988)
9. N.N. Yanenko, *The Method of Fractional Steps*, (1971)

MATEC Web of Conferences **72**, 01008 (2016) DOI: 10.1051/matecconf/20167201008
HMTTSC-2016

SCENARIO MODELING OF THERMAL INFLUENCE FROM FOREST FIRE FRONT ON A CONIFEROUS TREE TRUNK

Nikolay V. Baranovskiy[1,*], *Vladimir* B. Barakhnin[2,3], and *Ksenia* N. Andreeva[1]

[1]National Research Tomsk Polytechnic University, 634050 Tomsk, Russia
[2]Novosibirsk State University, 630090 Novosiborsk, Russia
[3]Institute of Computational Technologies, 630090 Novosiborsk, Russia

Abstract. Scenario research results of heat transfer and tissue damage in three-layered tree trunk influenced by heat flux from forest fire are presented. The problem is solved in two-dimensional statement in polar coordinates. The typical range of influence parameters (heat flux from forest fire front, trunk radius, coniferous species, air temperature, duration of exposure and distance from fire line) is considered. Temperature distributions in different moments of time are obtained. Condition of tree damage by forest fire influence is under consideration in this research. Information summarized using tables with scenario and fire consequences results.

1 Introduction

Fire damages (thermal traumas) of trees in forest stand occur as a result of direct influence of forest fire more often. Fire damages involve either death of trees or their weakness. It leads to trees fall or their infection by harmful insects and mushrooms. In addition, thermal traumas cause anatomic changes in trees and not only weakness, but also strengthening of their growth and bearing [1-3].

As a result of fire influence trees get various damages which are shown in the form of [1-3]:
-trunk burns;
-burns of roots;
-crown burns.

Influence of surface forest fire on trees is often limited to an easy superficial burn or bark burn that can be painless for a tree if fire is not influence on cambium. Cambium is sensitive even to small rise of temperature. An external sign of cambium death is its russeting. According to literary data cambium is dead at heating to 57 °C. Dying off of the big sites of cambium in trunk or root circle can lead to drying of tree [1-3].

Phloem defeat at a root neck, around of a tree trunk and partially on a trunk circle is observed on forested territory where has passed surface fire of low intensity. Therefore the part of trees within next year will have a dying off of roots that will lead to mortality of

* Corresponding author: firedanger@tpu.ru

trees. It is established by results of the conducted researches, that there is phloem dying off at all types of forest fire. Consequently, possible restoration of forest stands on these sites is difficult to expect. Besides, mycelium presence on the considerable area of wood laying on a root and radical part of trunk of a pine (oak and birch) on all height of trunk will sharply reduce quality of wood. It leads to favorable conditions for intensive growth of various mushrooms and defeat of wood by insects [4].

Now the majority of papers is devoted to the analysis of an actual statistical material on forest fire incidents and their ecological consequences [5,6]. However, development of information-computation systems for geoecological monitoring of large forests at active forest fires on the basis of physically proved mathematical models of heat transfer in structure of the trees forming a forest stand may become one of perspective directions [7,8].

The purpose of the present work is scenario modeling of heat transfer in tree trunk structure of conifers at influence of radiation from forest fire front in flat statement and determination of conditions for thermal defeats formation.

2 Physical and Mathematical Statements

The mathematical model is developed taking into account following assumptions. Separately standing tree of coniferous breed is considered. For definiteness calculations is conducted for pine, fir and silver fir. The tree trunk is considered without presence of branches within this mathematical model. The trunk is considered as a three-layer structure (bark, inner bark and a core). Thermophysical properties of each layer material do not depend on temperature. Influence of moisture evaporation from external and trunk inside layers is neglected. Process of thermal decomposition of trunk wood on the basis of the kinetic scheme [9] is considered. Geometry of decision area is shown on fig. 1.

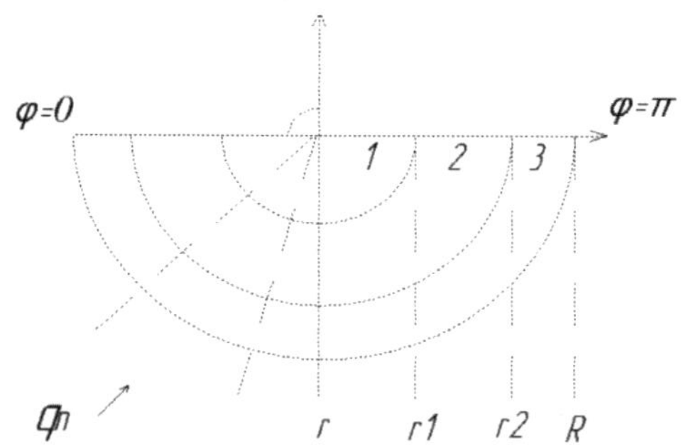

Fig. 1. Geometry of decision area: 1 – core; 2 – inner bark; 3 – bark.

Heat transfer processes in layered structure of tree trunk is described by the nonlinear non-stationary differential equations of heat conductivity with corresponding initial and boundary conditions:

$$\rho_1 c_1 \frac{\partial T_1}{\partial t} = \frac{\lambda_1}{r}\frac{\partial}{\partial r}\left(r\frac{\partial T_1}{\partial r}\right) + \frac{\lambda_1}{r^2}\frac{\partial^2 T_1}{\partial \varphi^2} - q_p k_1 \rho_1 \varphi_1 \exp\left(-\frac{E_1}{RT_1}\right),$$

$$\rho_2 c_2 \frac{\partial T_2}{\partial t} = \frac{\lambda_2}{r}\frac{\partial}{\partial r}\left(r\frac{\partial T_2}{\partial r}\right) + \frac{\lambda_2}{r^2}\frac{\partial^2 T_2}{\partial \varphi^2} - q_p k_1 \rho_2 \varphi_1 \exp\left(-\frac{E_1}{RT_2}\right),$$

$$\rho_3 c_3 \frac{\partial T_3}{\partial t} = \frac{\lambda_3}{r}\frac{\partial}{\partial r}\left(r\frac{\partial T_3}{\partial r}\right) + \frac{\lambda_3}{r^2}\frac{\partial^2 T_3}{\partial \varphi^2} - q_p k_1 \rho_3 \varphi_1 \exp\left(-\frac{E_1}{RT_3}\right),$$

$$T_i\big|_{t=0} = T_{i0},$$

$$r{=}R,\ \varphi \in (\varphi_1, \varphi_2),\ -\lambda_1 \frac{\partial T_1}{\partial r} = q_{ff},$$

$$r{=}0,\ -\lambda_1 \frac{\partial T_1}{\partial r} = 0,$$

$$\varphi{=}0,\ -\lambda_i \frac{\partial T_i}{\partial \varphi} = 0,$$

$$\varphi{=}\pi,\ -\lambda_i \frac{\partial T_i}{\partial \varphi} = 0,$$

$$r{=}R1,\ -\lambda_1 \frac{\partial T_1}{\partial r} = -\lambda_2 \frac{\partial T_2}{\partial r},\ T_1 = T_2,$$

$$r{=}R2,\ -\lambda_2 \frac{\partial T_2}{\partial r} = -\lambda_3 \frac{\partial T_3}{\partial r},\ T_2 = T_3,$$

The kinetic scheme of pyrolysis:

$$\rho_1 \frac{\partial \varphi_1}{\partial t} = -k_1 \rho_1 \varphi_1 \exp\left(-\frac{E_1}{RT_1}\right), \qquad \varphi_i\big|_{t=0} = \varphi_{i0},$$

$$\rho_2 \frac{\partial \varphi_1}{\partial t} = -k_1 \rho_2 \varphi_1 \exp\left(-\frac{E_1}{RT_2}\right), \qquad \varphi_i\big|_{t=0} = \varphi_{i0},$$

$$\rho_3 \frac{\partial \varphi_1}{\partial t} = -k_1 \rho_3 \varphi_1 \exp\left(-\frac{E_1}{RT_3}\right), \qquad \varphi_i\big|_{t=0} = \varphi_{i0},$$

$$\sum_{i=1}^{2} \varphi_i = 1,$$

$$\rho_1 \frac{\partial \varphi_1}{\partial t} = -k_1 \rho_1 \varphi_1 \exp\left(-\frac{E_1}{RT_1}\right),\ \varphi_i\big|_{t=0} = \varphi_{i0},$$

$$\rho_2 \frac{\partial \varphi_1}{\partial t} = -k_1 \rho_2 \varphi_1 \exp\left(-\frac{E_1}{RT_2}\right),\ \varphi_i\big|_{t=0} = \varphi_{i0},$$

$$\rho_3 \frac{\partial \varphi_1}{\partial t} = -k_1 \rho_3 \varphi_1 \exp\left(-\frac{E_1}{RT_3}\right),\ \varphi_i\big|_{t=0} = \varphi_{i0},$$

$$\sum_{i=1}^{2} \varphi_i = 1,$$

Where T_i, ρ_i, c_i, λ_i - temperature, density, thermal capacity and heat conductivity of tree trunk layers (1 - core, 2 – inner bark, 3 - bark). φ_i - volume fraction (1 - dry organic substance, 2 - gas). k_1, E_1 - thermokinetic constants of wood pyrolysis. q_p - thermal effect of pyrolysis. R - universal gas constant. r, φ - spatial coordinates. t - time coordinate. q_{ff} - radiant thermal flux from forest fire front.

The problem is solved by a method of finite differences. The multidimensional differential equations are solved by a locally-one-dimensional method. Marching method [10] is used for the decision of finite-difference analogues. Data of experimental burns, conducted by employees of the Institute of forest of the Siberian Branch of the Russian Academy of Science in vicinities of Krasnoyarsk are used for development of the basic scenarios of influence of radiant thermal flux from forest fire front. Dependence [11] is used for an estimation of thermal flux value from forest fire front:

$y = 326.37e^{-0.2791x}$.

Where x - distance to a fire front, m. y – thermal flux density, kW/m^2.

3 Results and Discussion

Numerical realization for mathematical model of heat transfer in layered structure of coniferous tree trunk is carried out using programming language of high level Object Pascal. IBM PC-compatible computer with Intel Core 4 Quad 3 GHz processor unit and 8 GB RAM was used for computing experiments. Using computation system based on multi-core processor with Hyper Threading technology for numerical realization of two-dimensional mathematical model allows to use physically proved resource effective algorithm for program realization. Various scenarios have been developed for an estimation of thermal influence on a tree trunk of coniferous breed. Typical distributions of temperature in layered structure of a tree trunk are presented to the certain moment of time on Fig. 2. Description of scenarios of fire influence is presented in table 1.

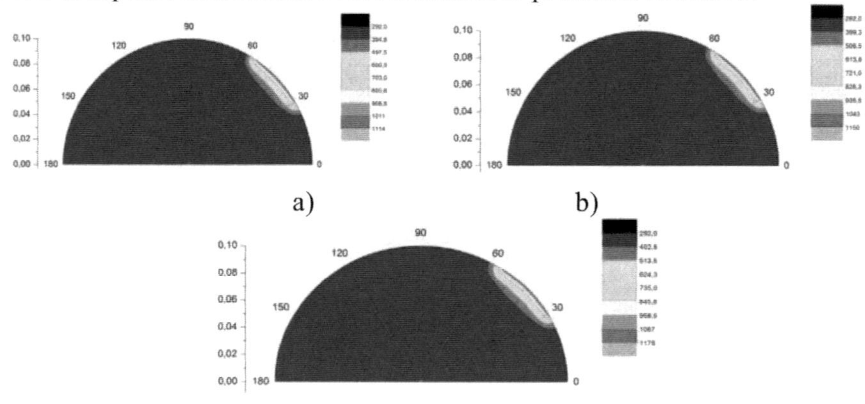

Fig. 2. Fields of temperature in layered structure of coniferous tree at the moment of time t=350 s: a - pine, b – silver fir, c – fir.

Table 1. The description of typical scenarios of influence of a radiant thermal stream from forest fire front on a coniferous tree trunk.

Season	Conifers	Distance to fire front, m	Core radius, m	Inner bark radius, m	Exposure time, s
Summer	Silver Fir	4	0.08	0.09	500
Summer	Pine	4	0.08	0.09	500
Summer	Fir	4	0.08	0.09	500

Short-term influence of radiation from forest fire front (few minutes) on coniferous tree trunk was accepted within the present scenario modeling. This assumption corresponds to the real scenario of forest fire behavior and characteristic velocity of forward forest fire front distribution in forest with coniferous trees.

The analysis of the obtained distributions of temperature shows, that eventually there is an intensive warming up, as bark layer and inner bark layer of coniferous tree trunk at influence of radiation from forest fire front. There is a thermal decomposition of wood and formation of thermal trauma site at reaching of critical temperatures in a tree trunk.

Typical data on conditions of thermal defeat of coniferous tree trunk in inner bark are cited in tables 1, 2. The analysis of tables allows to draw a conclusion, that longer influence of radiation from forest fire leads to thermal defeat of inner bark at smaller distances from a fire and density of thermal radiation. However, using thermophysical model is impossible to establish the fact of the subsequent tree mortality in forest stand. Temperature distributions presented on fig. 2 show, that the field of temperatures in tree trunk for various coniferous breeds has qualitative similarity. Appreciable distinctions of maximum temperatures are observed in bark layer. These distinctions decrease with spreading heat wave deep into a tree trunk. According to [12], there are distinctions in long-term prospect of dying off of trees of various breeds at rather identical influence of the raised temperatures. Besides, the range of influence time of the raised temperature in definite point of tree trunk and not only reaching of critical temperature are meant [13]. The given work has the big prospects of integration of results of numerical modeling (temperature fields in a tree trunk) with natural observations data for forest stands mortality in after fire period.

Table 2. Conditions of thermal defeat in tree trunk at influence time t=500 s. Surface forest fire of high intensity [11].

Conifers	Distan-ce, m	Heat flux, kW/m^2	Air tempera-ture, K	Trunk radius, m	Exposure time, s	Pyro-lysis	Thermal trauma
Pine	4	106.87	300	0.1	500	Yes	Yes
Silver fir	4	106.87	300	0.1	500	Yes	Yes
Fir	4	106.87	300	0.1	500	Yes	Yes

4 Conclusion

Scenario modeling results of heat transfer in coniferous tree trunk are resulted at influence of a radiant thermal flux from forest fire front in present paper. Conditions of thermal defeats occurrence in coniferous tree trunk are determined. Typical scenarios of forest fire danger (characteristic for summertime) were considered. Numerical modeling results for typical coniferous trees (fir, silver fir and pine) are obtained. Research of the heat transfer basic laws is carried out in layered structure of tree trunk at influence of surface forest fire of high intensity. Development of similar models opens prospects to creation of new generation information systems on the basis of the deterministic mathematical models of heat transfer in tree trunk. Working out of the complex deterministic-probabilistic approach for an estimation of forest fire ecological consequences with attraction of forest fire danger prognostic modeling results [14,15], mathematical models of forest fire influence on humans [16] and modern information technologies is possible [17,18].

Acknowledgments

The work was supported by the Russian President's grant (Scientific School project - 7538.2016.8).

References

1. G. M. Byram, Forest fire control and use **1** (1959)
2. W.F. Laurance, P. Delamonica, S.G. Laurance, H. Vasconcelos, T.E. Lovejoy, Nat. **404**, 836 (2000)
3. R.C.G. Mesquita, P. Delamonica, W.F. Laurance, Biol. Conserv. **91**, 129 (1999)
4. T.K. Kuryanova, A.D. Platonov, N.E. Kosichenko, Scientific journal of KubGAU **70**, 1 (2011)
5. J. Barlow, C.A. Peres, Philos. Trans. R. Soc., B **359**, 367 (2004)
6. C. Uhl, J.B. Kauffman, Ecol. **71**, 437 (1990)
7. N.V. Baranovskiy, V.B. Barakhnin, K.N. Andreeva, EPJ Web Conf. **110**, Article Number 01005 (2016)
8. N.V. Baranovskiy, K.N. Andreeva, Cloud Sci. **2**, 591 (2015) [In Russian)]
9. A. M. Grishin, *Mathematical modeling of forest fires and new methods of fighting them* (Russia, Tomsk: Publishing House of the Tomsk State University, 1997)
10. L. Vulkov, Lecture Notes in Computer Science **9045**, 434 (2015)
11. E.N. Valendik, I.V. Kosov, Siberian ecological journal **6**, 517 (2008) [In Russian]
12. T. Curt, W. Adra, L. Borgniet, For. Ecol. Manage. **258**, 2127 (2009)
13. T. Frejaville, T. Curt, C. Carcaillet, Front. Recent Dev. Plant Sci. **4** (2013)
14. E.P. Yankovich, N.V. Baranovskiy, 14th International Multidisciplinary Scientific Geoconference SGEM - 2014. GeoConference on Informatics, Geoinformatics and Remote Sensing **1**, 607 (2014)
15. E.P. Yankovich, N.V. Baranovskiy, K.S. Yankovich, 9th International Forum on Strategic Technology, IFOST 2014 (2014)
16. N.V. Baranovskiy, A.S. Solodkin, A.A. Stuparenko, EPJ Web Conf. **110**, Article Number 01075 (2016)
17. N.V. Baranovskiy, Cybernetics and Systems Analysis **51**, 471 (2015)
18. N.V. Baranovskiy, E.P. Yankovich, Journal of Automation and Information Sciences **47**, 11 (2015)

THE STUDY OF DYNAMIC PROCESSES IN THE BOILER FURNACE WITH CIRCULATING FLUIDIZED BED

A. V. Gil[1,*], and *D.* A. Baturin[1]

[1]National Research Tomsk Polytechnic University, 634050 Tomsk, Russia

Abstract. The paper presents a numerical simulation of the furnace with a circulating fluidized bed. Numerical study carried out on all volume of the combustion chamber. The results contours of particulate matter concentration and of velocities, as well as a graphical representation of changes in the concentration of particles on the bed height are shown. Simulation performed in Eulerian - Eulerian and Lagrange representation on a 3D model.

1 Introduction

Currently boilers with circulating fluidized bed (CFB) are widely used when low-quality coals are burnt. The quality of extracted coal with moisture and ash of more than 55 % causes difficulty of incineration by the traditional flare way (the slagging of furnace screens, flameout, etc.). Combustion in circulating fluidized bed has high efficiency and allows to achieve low emissions of harmful substances by binding of sulfur oxides and low-temperature combustion [1, 2].

However, this technology has several significant disadvantages associated with the mode of operation. Completeness of fuel combustion, the workload of the fluidized bed and gas-dynamic processes in the furnace are poorly understood, that lowers the reliability of boiler unit work (fig. 1) [2, 3].

The object of study is the U-shaped CFB boiler. The height of the furnace is 28 m. The boiler has 4 burners on the front for the coal fuel supplying into the furnace, and 6 slits on the back side to return unburned particles. The particles that are leaving the furnace are separated into 2 streams and enter the cyclones for separation (fig. 1).

2 Problem statement

The main method to study gas-dynamic processes, heat transfer and high-speed modes is mathematical modeling. The obtained information can be applied in the design of thermal plants, as well as in the evaluation of already operated units.

For multiphase flow, computational fluid dynamic models have been become more accepted in modeling gas-solids flow systems during the recent years. There are two

* Corresponding author: andgil@tpu.ru

approaches: Eulerian–Eulerian and Eulerian–Lagrangian models. In the Eulerian–Eulerian model, the solid phase is treated as a continuum, and an Eulerian framework is applied to describe the motion of the solids [4]. In most recent Eulerian–Eulerian model constitutive equations according to the kinetic theory of granular flow are incorporated [4-6]. Though physical characteristic of the solid particles such as the shape and size are included in the continuum representation through the empirical relations for the interfacial friction, these models do not recognize the discrete character of the solid phase [3].

Fig. 1. Sketch of the furnace: a) longitudinal section; b) cross section.

Baseline: project fuel – anthracite; fuel consumption BP = 9.74 kg/s; the volume of primary and secondary air 5.79/1.07; diameter of coal particles is 2.5 mm; the temperature from below 600 °C, in burners 300 °C; the air velocity below 1.5 m/s.

3 Mathematical model

At present, there is no generally accepted mathematical model of the numerical analysis for the MSC and therefore, the simulation results using the two methods are present in the work. Euler - Euler Method allows to trace gas-dynamic process in the furnaces of CFB boilers, but does not allow to consider the movement of individual particles of the fuel [7, 8]. Lagrange-Euler approach allows to investigate a large number of individual dispersed particles, but not quite correctly reflected the processes in the lower furnace with circulating fluidized bed [9, 10].

The computational grid consists of 137673 of cells that contains 27749 knots and has 272690 separate surfaces. The number of structural elements is sufficient to conduct the study because it provides acceptable accuracy of calculation while maintaining high speed of its holding. The research was carry out by using ANSYS Fluent 12.1.4.

4 Results and discussion

Different approaches designed more than 3000 iterations. Fig. 2 shows the concentration of substances in the furnace volume obtained by modeling by Euler method modeling.

Figure 2 shows that in the lower part of the furnace, a layer of 1.5 m height is formed, where the concentration of fuel reaches the maximum value. Then, by supplying air from below the particles are distributed throughout the volume of the combustion chamber. In the near wall region, the concentration of fuel is higher than in the central part of the furnace. Air has opposite values with respect to the fuel.

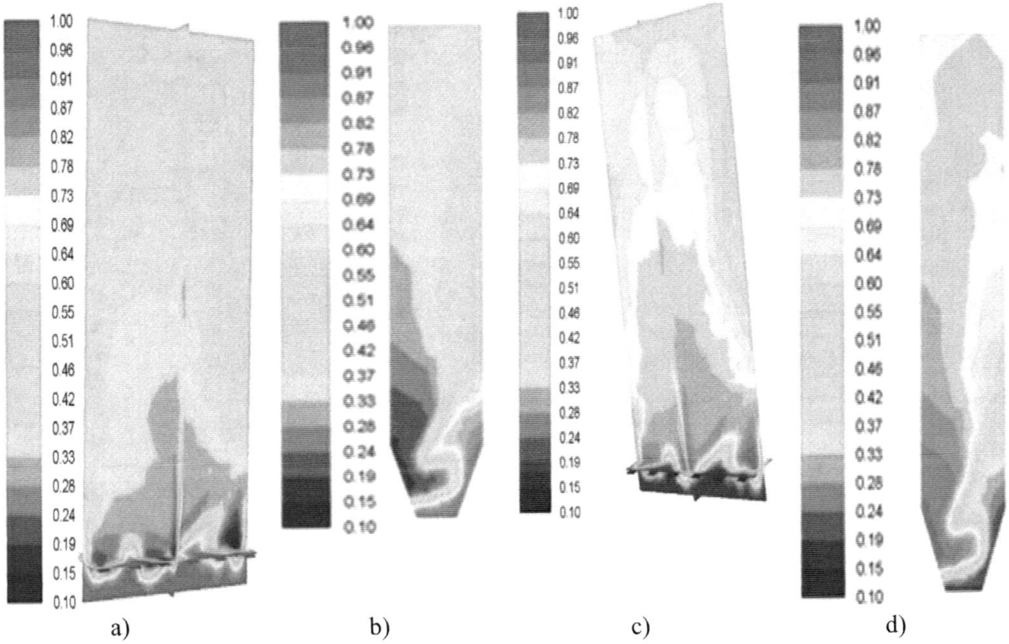

Fig. 2. Concentration of the estimated phases at the height of the furnace in vertical sections: a) solid particles in the section along the transverse axis; b) solid particles in the cross section along the longitudinal axis; c) air in the section along the transverse axis; d) air in the cross section along the longitudinal axis.

Figure 3 shows differences of the speed limit depending on the modeling method.

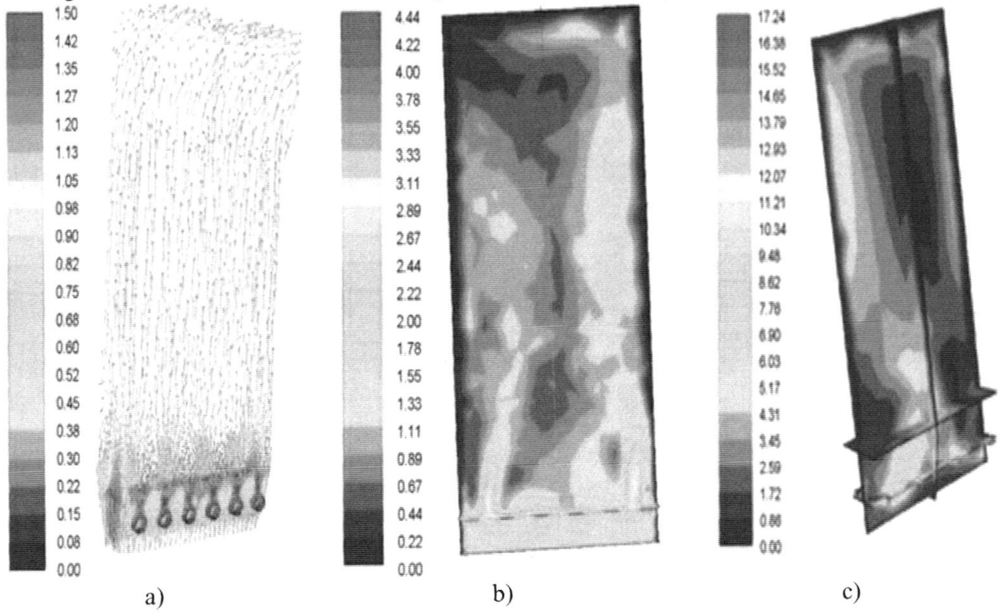

Fig. 3. Speed (m/s): a) cold venting; b) the Eulerian–Eulerian model; c) the Eulerian–Lagrangian model.

In the cold blowing (fig. 3, a) the speed limit in the combustion chamber is observed, the maximum speed corresponds to the air supply from below at 1.5 m/s. At the exit of the furnace, average speed is 0.85 m/s. In the simulation by the Euler method (fig. 3, b) the average speed in the lower part of the furnace volume is 2.5 m/s. Then with the height of the furnace the average speed drops to 1.5 m/s. Average speed in the simulation by the Lagrange method (fig. 3, b) in the region of the burners is 5 m/s, and at the height of the furnace chamber is 2-3 m/s.

5 Conclusions

Simulation using Eulerian–Eulerian method allows to observe the fuel layer and the subsequent distribution of fuel particles along the height of furnace volume. It is not possible to achieve such situation by the method of Lagrange–Euler.

There is some deference in the speed limit in the lower and upper parts of the combustion chamber using different modeling approaches and remains in the range of optimal values for furnaces with circulating fluidized bed.

Using modeling approaches allows collectively to assess gas-dynamic processes and the operation of boilers with circulating boiling layer under the condition of a rational analysis of simulation results using different mathematical approaches. Accordingly, for a better assessment of results and forecasting of operating processes in the boiler furnace with circulating fluidized bed is necessary to develop a hybrid mathematical model for circulating fluidized bed boilers.

References

1. P. Basu, *Combustion and gasification in fluidized beds* (CRC Press, FL) (2006)
2. Nan Zhanga, Bona Lua, Wei Wanga, Jinghai Li, Chem. Eng. J. **162** (2010)
3. A.V. Gil, A.Y. Gil, Eur.Phys. J.Web Conf. **82**, 01044 (2015)
4. Jun Xie, Wenqi Zhong, Baosheng Jin, Yingjuan Shao, Yaji Huang, Adv. Powder Technol. **24** (2013)
5. A.V. Gil, D. A., IOP Conf. Ser.: Mater. Sci. Eng. **93**, 012003 (2015)
6. A.V. Gil, D. A. Baturin, MATEC Web Conf. **23**, 01009 (2015)
7. Anusorn Chinsuwan, Animesh Dutta, Powder Technol. **193** (2009)
8. Mohammad Hadi Bordbar, Kari Myohanen, Timo Hyppanen Appl. Therm. Eng. **76** (2015)
9. G.A. Ryabov, O.M. Folomeev, D.A. Sankin, K.V. Khaneev, I.G. Bondarenko, D.A. Mel'nikov, Power Technology and Engineering **44** (2010)
10. A.V. Gil', A.S. Zavorin, A.V. Starchenko, S.V. Obukhov, Power Technology and Engineering **45** (2011)

INFLUENCE OF GEOMETRIC CHARACTERISTICS OF CANAL ON PRECIPITATING CAPACITY OF SEPARATOR

L. A. Belyaev[1], *A.* S. Zaitsev[1], *A.* A. Kondakov[1,*], *S.* A. Shevelev[1], and *A.* S. Matveev[1]

[1]National Research Tomsk Polytechnic University, 634050 Tomsk, Russia

Abstract. Designed mathematic model allows calculating trajectories of spherical non-deformable particles movement in the flow of viscous non-burnt liquid in the flat canal of complex geometry. According to the results of calculations graphs of dependence of precipitating particles share on their diameter and degree of curvature of the researched canal have been built.

1 Introduction

Economical efficiency and reliability of turbines work operating in the area below the curve of saturation mostly depends on moisture of vapor. Increase of final moisture leads to decrease of relative internal coefficient of efficiency and increased erosive wear of wheel space [1].

Different ways of vapor separation are used for reliable and effective work of turbine. One of the most popular ways is vapor separation in separators of louvered type.

The present work considers movement of two phase area consisting of non-burnt liquid and discrete phase in the form of separate spherical drops. Movement of the area occurs in the canal with smooth wave surface.

The purpose of the work is mathematic modeling of movement of drop moisture in the canal with set geometry and study of the process of precipitating at different canal curvature.

2 Mathematical model and study technique

Spherical non-deformable particle with known position, rate vector in the initial time and physical properties flies into the flat canal, principal scheme of which is depicted in picture 1. Then, the particle comes through the canal smoothly or under the influence of the area flow or actively interacts with solid surface and precipitates.

*Corresponding author: kondakov-alexandr@rambler.ru

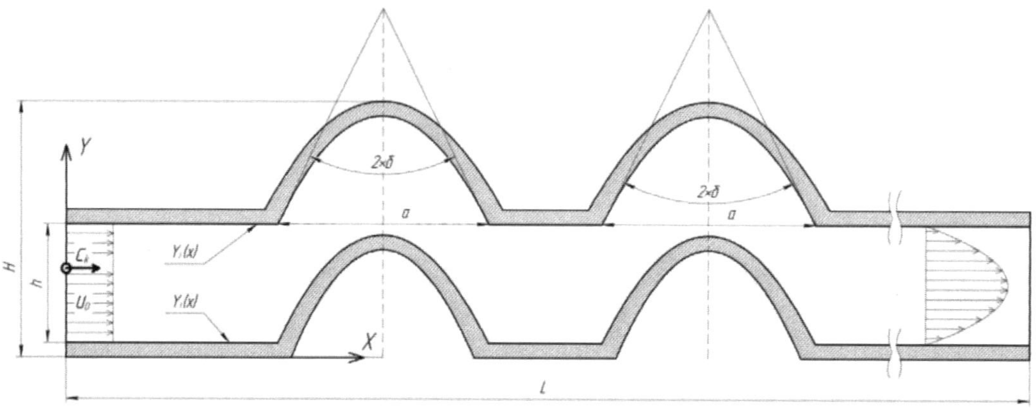

Fig. 1. Principal diagram of the researched canal.

Walls of the canal are set with functional dependences $Y_1(x)$ and Y_2 (x). Geometric characteristics of canal a, h, L are constant, and curvature of canal is determined by angle δ.

Let us consider flat movement in Descartes system of coordinates, where only aerodynamic resistance force influences a particle [2].

Equation of particle movement in projections on the axis of coordinates for the case has the following view [2]:

$$m\frac{dc_{kx}}{dt} = \rho F_{m.e.} c_x (u - c_{kx})\sqrt{(u - c_{kx})^2 + (v - c_{ky})^2};$$

$$m\frac{dc_{ky}}{dt} = \rho F_{m.e.} c_x (v - c_{ky})\sqrt{(u - c_{kx})^2 + (v - c_{ky})^2};$$

where

c_{kx}, c_{ky} - vector projections of particle rate on the axis of coordinates;

u, v - vector projection of flow rate on the axis of coordinates;

m – drop mass;

ρ – area density;

c_x - coefficient of hydro mechanical resistance of particles which depends on Reynolds number;

$F_{m.e.} = \frac{\pi d_e^2}{4}$ - area of midship cross section.

Flow area of continuum is described with the help of the system of differential equations of Navier-Stocks for flat stationary laminar flow [3], which consists of two equations of movement quantity and equation of continuity.

$$\begin{cases} \dfrac{\partial(\rho uu)}{\partial x}+\dfrac{\partial(\rho vu)}{\partial y}=\dfrac{\partial}{\partial x}\left(\mu\dfrac{\partial u}{\partial x}\right)+\dfrac{\partial}{\partial y}\left(\mu\dfrac{\partial u}{\partial y}\right)-\dfrac{\partial P}{\partial x}; \\ \dfrac{\partial(\rho uv)}{\partial x}+\dfrac{\partial(\rho vv)}{\partial y}=\dfrac{\partial}{\partial x}\left(\mu\dfrac{\partial v}{\partial x}\right)+\dfrac{\partial}{\partial y}\left(\mu\dfrac{\partial v}{\partial y}\right)-\dfrac{\partial P}{\partial y}; \\ \dfrac{\partial(\rho u)}{\partial x}+\dfrac{\partial(\rho v)}{\partial y}=0; \end{cases}$$

where

P- pressure;

μ- dynamic viscosity;

At the entrance of the canal rate vector is horizontal and has constant value. Conditions of sticking are set on the walls of the canal. We consider flow of liquid at the exit of the canal fully developed.

Border conditions for solution of the system:

1. Conditions at the entrance of the canal: $x = 0, 0 < y < H : u = const, v = 0.$
2. Conditions on the walls of the canal: $y = 0, y = H, 0 \le x \le L : u = v = 0.$
3. Conditions at the exit of the canal: $x = 0, 0 < x < H : \dfrac{\partial u}{\partial x} = \dfrac{\partial v}{\partial x} = 0.$

Control volume method is used for discretion of Navier-Stocks system in the present work. Using the method, calculated area is divided into a number of non-crossing control volumes in which hotspots can be found for calculation of designed value. Differential equation is integrated for each control volume. Power law of approximation of rate between net hotspots is used for calculation of received integrals.

SIMPLE Patankara - Spaulding algorithm is used for receiving rate areas and pressure [3].

As it can be seen from picture 1, flat canal has a complicated form that is why net chess with blocked areas in Descartes system of coordinates is used for calculation of flow area [4].

3 Results

Primary flow rate should be taken as u=0.3 m/s, with physical properties of air: $\mu=1.808\cdot10^{-5}$ Pa·s and ρ=1.205 kg/m^3 . Particle has the following dynamic viscosity and density: μ_k=0.001005 Pa·s and ρ_k=998 kg/m^3. Net chess is used in the calculation with dimensions 100×140 with permanent steps.

Walls of the canal are set as functional dependences so that geometric parameters a, h, L were constant at different angles δ.

Trajectories of particles movement in canal with diameter d_k=100 mkm and 200 mkm at primary position $y(0)$=0.05 m and $x(0)$=0 m are shown in picture 2. Particles with small diameter (small mass) can smoothly flow through canal without interacting with solid border and particles with big diameters bump with solid surface and precipitate.

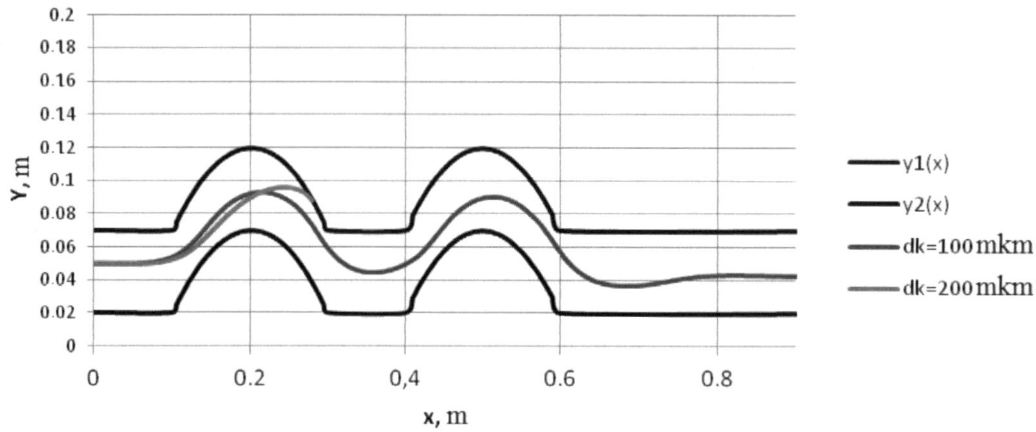

Fig. 2. Trajectory of particles movement with different masses.

In the given mathematic model influence of curvature curves of the researched canal on the particles precipitation percent of different diameter at equal border conditions of continuum flow and primary conditions for particle movement has been considered.

Share of precipitated particles in dependence of their diameter and canal curvature is shown in picture 3 (angle δ).

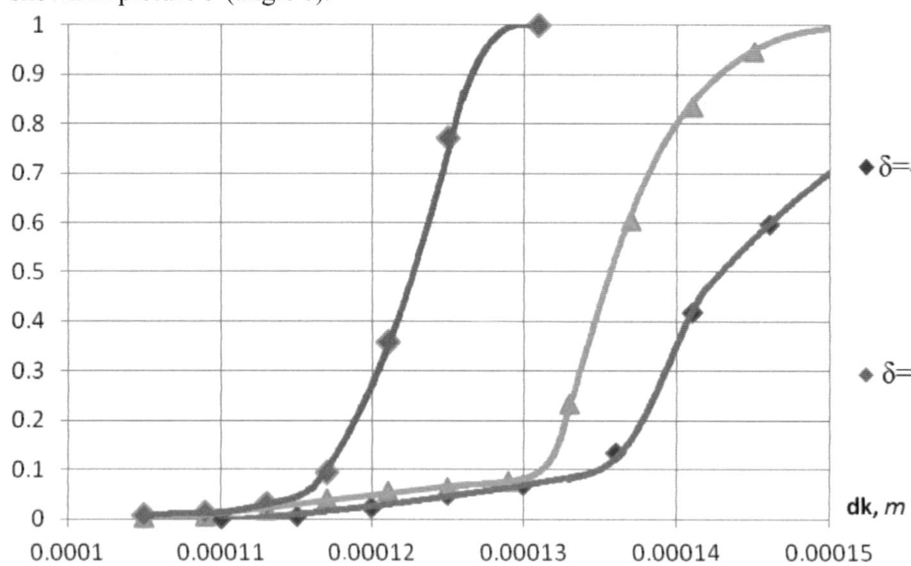

Fig. 3. Share of precipitating particles in dependence of precipitating drops on their diameter and canal curvature.

According to the received data of calculation we can conclude that increase of canal curvature leads to decrease of particles diameter where precipitation of all drops will occur. For the angle δ=42°, beginning with diameters d_k ≈130 mkm and more, almost all the particles will precipitate on the walls of the researched canal. For the angle δ=45° all the particles precipitate beginning with diameters d_k≈150 mkm.

4 Conclusion

Mathematic model of spherical particles movement in laminar flow of non-burnt liquid has been developed in the present work.

The result of calculations is receiving graphic dependence of precipitating drops on their diameter and canal curvature.

References

1. T. Kh. Margulova, *Atomic power stations* (1994)
2. M. E. Deich, *Gas dynamics of two-phase media* (1981)
3. S. V. Patankar, *Numerical Heat Transfer and Fluid Flow* (Taylor & Francis, 1980)
4. H.G. Versteeg, W. Malalasekera, *An introduction to computational fluid dynamics* (Pearson Education Limited, 2007)

HEAT TRANSFER DURING EVAPORATION OF CESIUM FROM GRAPHITE SURFACE IN AN ARGON ENVIRONMENT

Evgeny Bespala[1,*], *Ivan* Novoselov[1], and *Ivan* Ushakov[1]

[1]National Research Tomsk Polytechnic University, 634050 Tomsk, Russia

Abstract. The article focuses on discussion of problem of graphite radioactive waste formation and accumulation. It is shown that irradiated nuclear graphite being inalienable part of uranium-graphite reactor may contain fission and activation products. Much attention is given to the process of formation of radioactive cesium on the graphite element surface. It is described a process of plasma decontamination of irradiated graphite in inert argon atmosphere. Quasi-one mathematical model is offered, it describes heat transfer process in graphite-cesium-argon system. Article shows results of calculation of temperature field inside the unit cell. Authors determined the factors which influence on temperature change.

1 Introduction

By now all of the industrial uranium-graphite reactors (IUGR) are stopped in Russia. Upon that major part of them is either on the stage of shut down preparation or on the stage of shut down. Exception is IUGR EI-2, which was replaced on special radioactive waste parking site. However it is necessary to conduct operations connected with graphite stack disassembling and further radioactive waste burial for reactors of types of RBMK, EGP, AMB, AM (in Russia) reactors of types of Magnox, UNGG, AGR, HTR (not in Russia).

One of the most common radionuclides present in irradiated structural elements is ^{137}Cs. Thus, the specific activity of the nuclide in some samples of radiation sources can reach 3.2 TBq/kg [1]. In this case the radioactive cesium contamination observed in the materials that were in contact with nuclear fuel, since it is the product nuclear fission.

At the present time industry organizations offer different ways of processing radioactive waste (RW): electrooxidation (anodizing) in an oxygen-containing electrolyte [2], combustion in thermal oxidizer environment and pyrolysis [3], plasma treatment [4], thermochemical treatment [5], melting [6], vitrification [7]. Most of the known methods lead to increased volumes of waste, produced as by-product gases of reaction, which requires additional trapping means (bubblers, scrubbers or filters).

* Corresponding author: bespala_evgeny@mail.ru

In the case of using an inert gas as a transfer medium by thermal deactivation of radioactive waste the amount of secondary waste is reduced. With increasing gas temperature the rate of nuclides evaporation from the surface of materials is increased, which leads to increased productivity and the recovery rate of radionuclides. At the same time there is a transition from surface to volume processes. This is especially true during deactivation of graphite structural elements, the porosity of which is about 23 % [8]. Therefore, for the thermal deactivation of RW can be used argon low-temperature plasma.

In this case, simulations of heat transfer process in a high temperature argon flow for selecting of optimal regimes of graphite cleaning process by ^{137}Cs.

2 The methodology and calculations

Irradiated graphite constructional elements usually have surface contamination by radioactive cesium [9]. Heating of graphite-cesium surface by gas flow leads to a metal sublimation. Computational cell scheme is in the figure 1. There δ_{Cs} is layer thickness of metal cesium, δ_C is layer thickness of graphite surface. High temperature plasma flow consisting of noble gas (argon) is oriented along the graphite-cesium surface. The gas velocity at the entrance to computational cell with height of R above cesium surface is u_0. It must be noted that transversal and longitudinal dimensions of cell are much bigger than thickness of cesium layer δ_{Cs}. Graphite surface heats by high temperature flow contact and by bottom wall-graphite surface heat transfer. Authors used thermodynamic parameters of cesium as a mixture of its natural occurring isotopes.

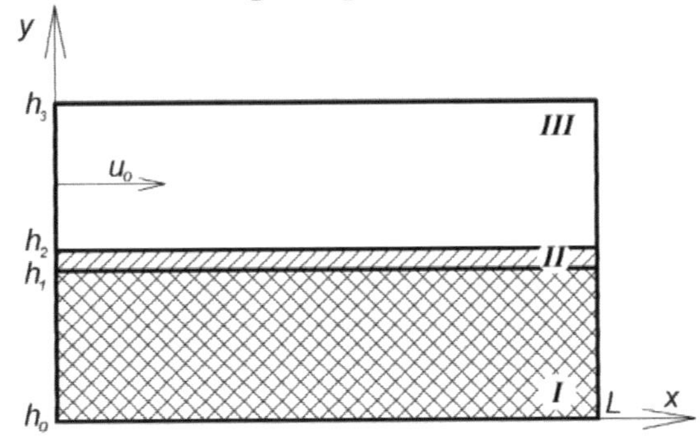

Fig. 1. Computational cell scheme: I – graphite, II – cesium, III – high temperature flow.

To simulate the heat transfer process we solved unsteady heat transfer equation in the Cartesian coordinate system. The direction of the x axis coincides with the direction of velocity vector u of high temperature gas flow.

$$\begin{cases} c_1\rho_1 \dfrac{\partial T}{\partial t} = \dfrac{\partial}{\partial y}(\lambda_1 \dfrac{\partial T}{\partial y}), \ h_2 < y < h_1 \\ c_2\rho_2 \dfrac{\partial T}{\partial t} = \dfrac{\partial}{\partial y}(\lambda_2 \dfrac{\partial T}{\partial y}), \ h_1 < y < h_0 \end{cases}, \quad (1)$$

where c_1, ρ_1, λ_1 – heat capacity, density and thermal conductivity of the metal cesium; c_2, ρ_2, λ_2 – heat capacity, density and thermal conductivity of graphite; h_0 – coordinate corresponding to the bottom surface of deactivating graphite; h_1 – the distance from the bottom to the top surface of graphite; h_2 – distance from the top surface to the surface of graphite in contact high temperature argon flow with cesium.

It was thought that the thermal and gas dynamic properties of argon dependent by temperature, but properties of graphite and cesium remain constant. On this basis the parameters of the plasma forming gas were chosen as follows:

$$u_{Ar} = u_0(\frac{T_{Ar}}{298}), \qquad (2)$$

$$\nu_{Ar} = 1.37 \cdot 10^{-5}(\frac{T_{Ar}}{298})^{1.75}, \qquad (3)$$

$$\lambda_{Ar} = 1.54 \cdot 10^{-3}(\frac{T_{Ar}^{1.5}}{T_{Ar} + 160}), \qquad (4)$$

$$\alpha_{Ar} = 2.04^{-5}(\frac{T_{Ar}}{298})^{1.75}, \qquad (5)$$

where ν_{Ar}, λ_{Ar}, α_{Ar} – kinematic viscosity, thermal conductivity, thermal diffusivity of argon at a temperature T_{Ar}, u_{Ar} – gas flow rate at the temperature $T_{Ar.}$

Heat balance equation on the surface of the metal cesium sublimation for the problem:

$$-\lambda_1 \frac{\partial T}{\partial y}\bigg|_{y=h_2(t)} = a(T_{Ar} - T_s), \qquad (6)$$

where a – heat transfer coefficient under convective heat transfer of a gas flow with a surface layer of metallic cesium, T_s,– cesium sublimation temperature.

From the equation 6 shows that the process of heat transfer at the boundary cesium-argon will be determined by the gas flow regime, since the amount of heat transfer coefficient a is uniquely associated with a Nusselt numbers and Reynolds numbers.

When solving equation 1 it was considered that the change in temperature between the layer of metallic cesium and graphite surface occurs by heat exchange, which characterized by a_1 at a constant coefficient of thermal conductivity of graphite. Changing of temperature of the external wall of graphite element is described by the equation 7:

$$-\lambda_2 \frac{\partial T}{\partial y}\bigg|_{y=h_0} = a_2(T_{h_1} - T_w), \qquad (7)$$

where a_2 – thermal conductivity of graphite, T_{h1} – the temperature on the boundary of cesium-graphite, T_w – temperature of the bottom surface of the graphite element.

We thought that at the initial time the temperature of the selected cells was the same in all areas and was T_0.

The numerical values of the parameters in the equation 2–7 take the following: c_1 = 273 J/(kg·K), c_2 = 710 J/(kg·K), ρ_1 = 1873 kg/m^3, ρ_2 = 1650 kg/m^3, λ_1 = 10 W/m·K, λ_2 = 150 W/m·K, μ = 137 g/mol, T_s, = 961 K, , $T_0 = T_w$ = 25 °C, u_0 = 10 m/s, R = 30 mm.

The initial data for modeling cesium evaporation process with a graphite surface was chosen to the initial temperature of the plasma gas T_{Ar} equal to 600 °C, 1000 °C, 1500 °C, 2000 °C. It was considered that the thickness of the graphite layer δ_C = 20 mm, and cesium layer δ_{Cs} was not more than 2 microns. The results of calculation of temperature changes on the surface of the cesium film shown in figure 2. The figure shows that the maximum temperature in the surface layer is lower than the initial temperature of the gas flow due to convective heat transfer between the argon and cesium as well as the heat transfer between the cesium and graphite. For example, cesium film temperature reaches 450 °C temperature with an initial weight average gas temperature plasma 600 °C, and at 2000 °C not exceeding 1950 °C. This is due to the fact that the lower boundary of the graphite sample maintained at a constant temperature equal to 25 °C. The time of heating the thin film of cesium to gas flow temperature does not exceed 8 minutes.

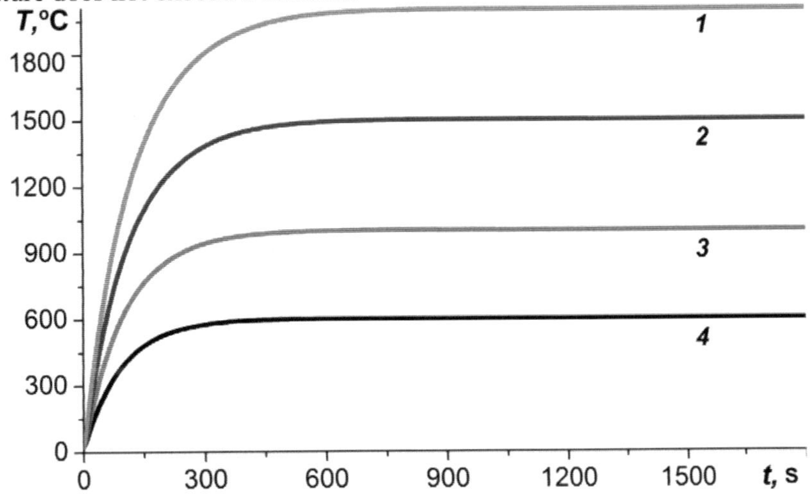

Fig. 2. Changing the cesium film temperature at different initial gas temperature of plasma: 1 – 2000 °C, 2 – 1500 °C, 3 – 1000 °C, 4 – 600 °C.

The velocity of the high temperature flow along the deactivated sample insignificantly effect on the heat transfer process (fig. 3).

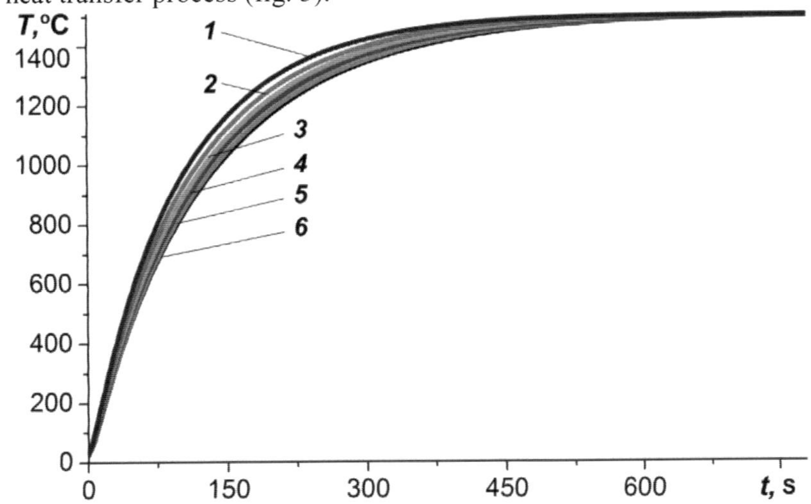

Fig. 3. Changing the cesium film temperature at different initial velocity of high temperature gas flow: 1 – 40 m/s, 2 – 20 m/s, 3 – 10 m/s, 4 – 5 m/s, 5 – 1 m/s, 6 – 0.1 m/s.

The graph shows that the release time of a thermodynamic system to a quasi-stationary mode varies from 7.5–8 minutes in the gas velocity range of 0.1–40 m/s. This is probably due to the fact that the under steady flow regime of high temperature flow does not occur significant change in the energy balance of the system. The change in the thermodynamic parameters of the system can be caused only by a change in thickness of cesium δ_{Cs}, which is negligible compared to an equivalent cell size.

3 Summary

The paper presents a mathematical model describing the heat transfer process in an argon-cesium-graphite system. The model was used to calculate the extraction rate of radioactive ^{137}Cs from the surface of graphite during plasma treatment of radioactive waste. It was shown that the significant factors influencing the degree of deactivation are weight average gas temperature of equilibrium plasma and velocity of gas flow.

Acknowledgments

The reported study was funded by RFBR according to the research project No. 16-38-00382 mol_a.

References

1. S.A. Dmitriev, A.S. Barinov, O.G. Batyuhnova, A.S. Volkov, M.I. Ozhovan, T,D, Scherbatova, *Technological basics of control systems of radioactive waste management* (NPO Radon, Moscow, 2007) [in Russian]
2. M.G. Golovachev, V.G. Lisovskih, Journal of Radioecological Problems and Conterminal Disciplines **14** (2010)
3. J.B. Mason, D. Bradbury, Nucl. Ener. J. **39** (2000)
4. V.F. Myshkin, V.A. Khan, V.G. Plekhanov, D.A. Izhoykin, E.V. Bespala, Rus. Phys. J. **57** (2015)
5. M.I. Ojovan, W.E. Lee, I.A. Sobolev, S.A. Dmitriev, O.K. Karlina, V.L. Klimov, C.N. Petrov, C.N. Semenov, J. Proc. Mech. Eng. **218** (2004)
6. R.K. Womack, JOM **51** (1999)
7. W.E. Lee, M.I. Ojovan, M.C. Stennett, N.C. Hyatt, J. Adv. Appl. Ceram. **108** (2006)
8. Yu.S. Virgiliev, I.P. Kalyagin, V,G, Makarchenko, Journal of Atomic Energy **36** (1974)
9. A.V. Bushuev, Yu.M. Verzilov, V.N. Zubarev, Journal of Atomic Energy **89** (2000)

THE INFLUENCE OF AGEING PROCESS ON THE BARIUM ALUMINATE PHASE COMPOSITION

A. O. Bezmaternykh[1,*], *Y.* B. Shvaljov[1], *V.* N. Kudiiarov[1], *V.* E. Gubin[1], and *M.* N. Babihina[1]

[1]National Research Tomsk Polytechnic University, 634050 Tomsk, Russia

Abstract. Barium aluminates were formed by the continuous precipitation method at different values of pH. The influence of ageing process and precipitation pH on the phase composition of the precipitate after firing was found. It was showed that there is growth of the $3BaOAl_2O_3$ concentration after ageing of precipitate during 7 days. Presumably, this process is associated with a decrease in the number of crystals, increasing their size in the ageing process of precipitate and chemical processes that occur during ageing: hydrolysis, hydration, dehydration.

1 Introduction

Dispenser cathodes are widely used in most electronic devices. Dispenser cathode is a cathode which is based on a porous sponge made of refractory metal. Tungsten powder is often used as the sponge material. The sponge pores are filled by the active substance to complete saturation of them. Mixture of barium oxide compounds with the acid or amphoteric oxides of some elements is the active substance. The operating principle of the cathode is based on the active substance interaction with metal of the sponge (reducing agent), thereby forming free barium, which activates the cathode.

Recently, technologies for obtaining chemically pure substances [1-3] found wide application. The classical method for preparing active substance is firing of barium carbonate with calcium carbonate and aluminum oxide (fig. 1a). However, this method has significant drawbacks. The products synthesized in this way are not sufficiently pure and have particle size heterogeneity.

In [4], the method for obtaining active substances was used by coprecipitation components, which formed barium-calcium aluminates and barium-calcium tungstates at subsequent firing of the precipitate. This technique has several advantages: rapidity of the reaction, the minimum amount of impurities. However, the disadvantage of this method is a batch process for the implementation of the process, leading to instability and poor reproducibility of the quality characteristics of the finished product.

* Corresponding author: Upenikova@yandex.ru

We propose to eliminate these drawbacks by the use of continuous precipitation technology at constant pH. This technique allows obtaining a substance with the desired final characteristics.

Furthermore, the precipitate final characteristics largely depend on precipitation parameters such as the concentration of the initial solutions, the value of pH, precipitation temperature, ageing time and temperature of the washing liquid.

In this paper we explored the effect of precipitation pH and ageing process on the formation of the final phase composition of the precipitate after firing.

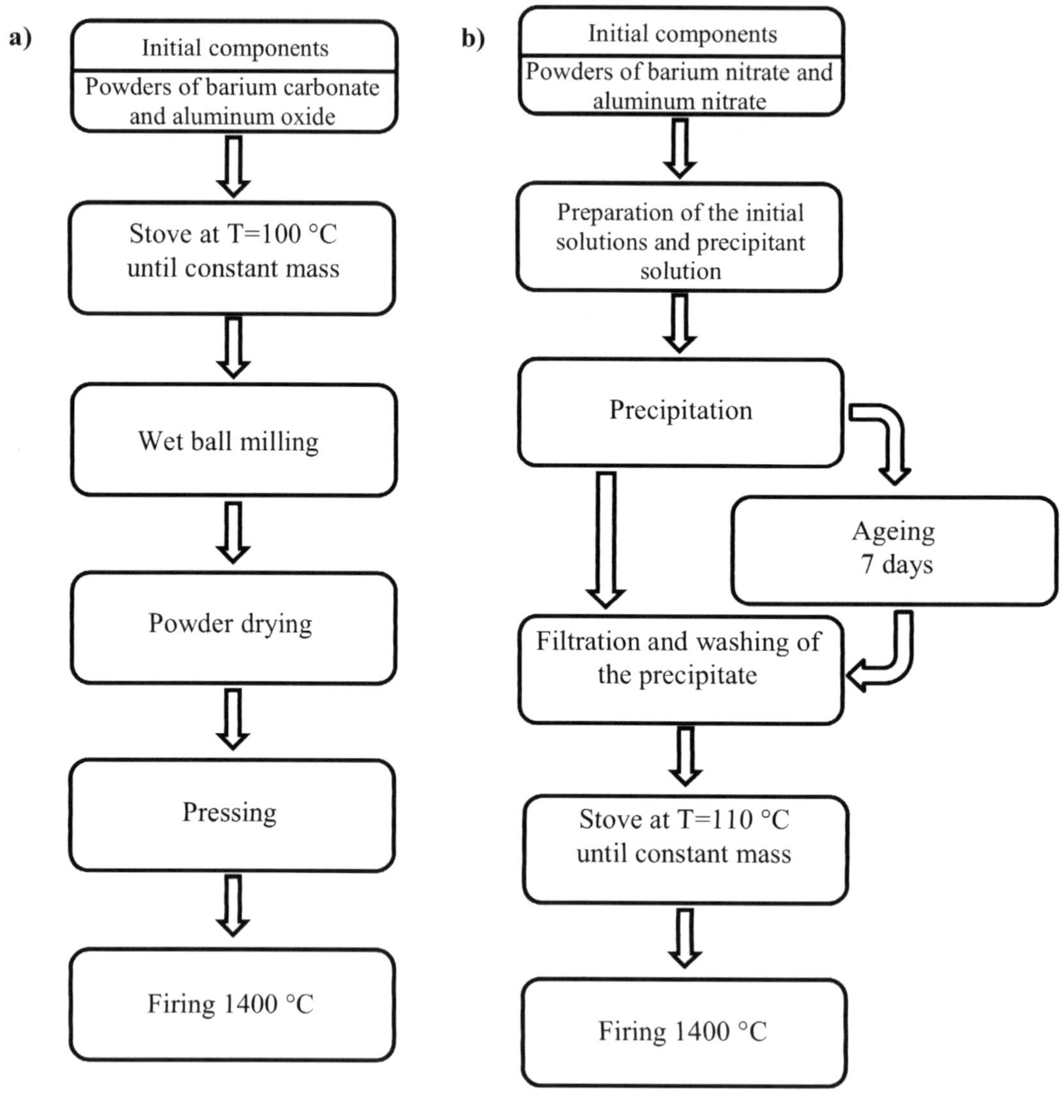

Fig. 1. The scheme of the barium aluminates synthesis a) classical technology b) precipitation method.

2 Materials and Methods

Initial solution of barium nitrate salts and aluminum was used for barium aluminates synthesis. Sodium carbonate solution acted as a precipitator. Data shown on the phase diagram [5] (fig. 2) was used for calculating the concentrations of the solutions.

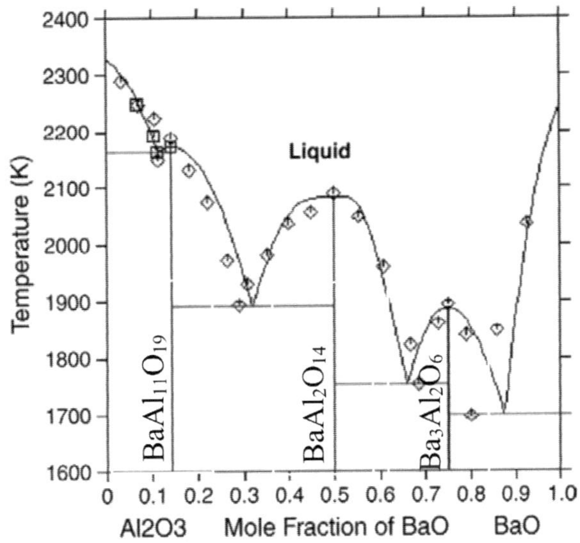

Fig. 2. Constitution diagram Al_2O_3 – BaO [5].

Phase composition of the precipitate after firing was obtained by the XRD method on the Shimadzu XRD-7000S diffractometer. The following parameters were used during the experiment: scan speed – 2° per minute; scanning step – 0.03°; range of angles – 20-50°; 1.5 seconds for information accumulation. The average crystallite size was determined using the Debye-Scherrer equation from the broadening of the diffraction peaks, taking into account the instrumental broadening and software PowderCell 2.4.

3 The Experimental Procedure

The preparation of the active substances was performed on the experimental facility which was previously successfully applied in the preparation of catalysts and sorbents of medical immunobiological drugs [6]. The scheme of experimental facility is shown in fig. 2. The construction of the experimental setup makes possible precipitation in periodic and continuous modes, at constant or variable volume, variable and constant pH, and different temperatures.

Solutions of aluminum nitrate and barium nitrate and sodium carbonate with given concentrations were supplied from tanks (1, 2) to the reactor (6) through the thermostat (5) by peristaltic pumps (3, 4). Precipitation was carried out at room temperature. Stirring was carried out using a mechanical stirrer (8). PH values were determined by pH-meter (9); electrodes are installed directly in the reactor. Adjustment and maintenance of the pH value was carried by changing the rate of sodium carbonate supplying. Formed precipitate was supplied into collector (11) by pump (10).

Precipitation was done at three values of pH: 6.5, 7.0, 7.5. Part of samples was kept in the mother liquor at room temperature during 7 days. The precipitate was washed to a negative reaction to an anion NO^{3-}. After washing in distilled water the precipitate was dried to constant weight at temperature of 110 °C. Then the process of firing was done for precipitate at temperature of 1400 °C during 2 hours. Brief scheme of the experiment is shown in figure 1b.

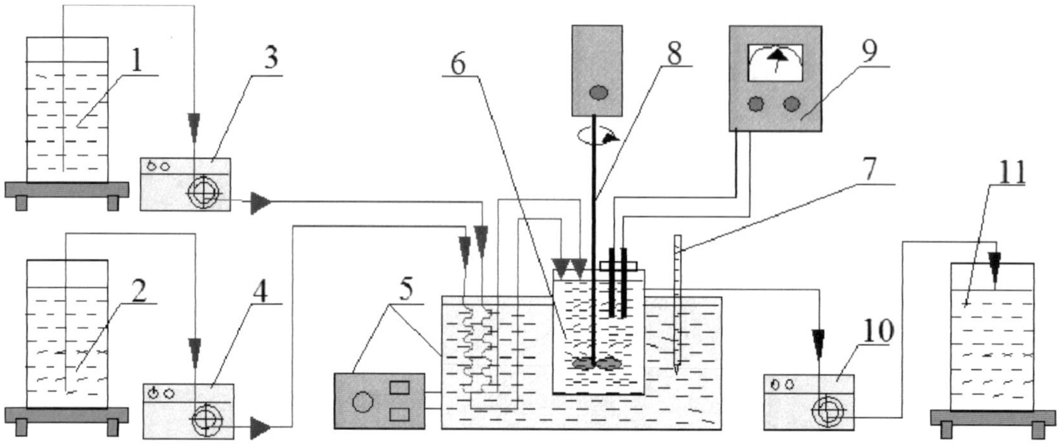

Fig. 3. The scheme of experimental facility: 1 – tank with a solution of barium nitrate and aluminum nitrate, 2 – tank with a solution of sodium carbonate, 3,4 – peristaltic pumps, 5 – thermostat, 6 – reactor, 7 – thermometer, 8 – mechanical stirrer, 9 – pH-meter, 10 – pump, 11 - collector.

4 Results and Discussion

The results of XRD analysis of the barium aluminate are shown in the Table 1.

Table 1. The results of the barium aluminate XRD analysis.

No.	Conditions		Phases	Phases volume content.%	Lattice parameters, A
	pH, values	Ageing time, days			
1	6.5	-	$BaAl_2O_4$	86.7	a=5.2263 c=8.7933
			$Ba_3Al_2O_6$	13.3	a=16.5199
2		7	$BaAl_2O_4$	57.9	a=5.2231 c=8.7939
			$Ba_3Al_2O_6$	42.1	a=16.5144
3	7.0	-	$BaAl_2O_4$	74.7	a=5.2243 c=8.7975
			$Ba_3Al_2O_6$	25.3	a=16.5041
4		7	$BaAl_2O_4$	38.8	a=5.2224 c=8.7944
			$Ba_3Al_2O_6$	61.2	a=16.5133
5	7.5	-	$BaAl_2O_4$	79.8	a=5.2206 c=8.7893
			$Ba_3Al_2O_6$	20.2	a=16.4984
6		7	$BaAl_2O_4$	17.7	a=5.2070 c=8.7879
			$Ba_3Al_2O_6$	82.3	a=16.4941

It should be noted that the mainly dominates monoaluminate barium in the materials obtained at a given pH and not subjected to aging. The change of pH does not influence the phase composition of the precipitate after firing.

Stage of ageing, which is allowed to proceed for 7 days, has a significant impact on the phase composition of the precipitate after firing. For example, at pH 6.5, 7.0 and 7.5, there is an increase of 3BaOAl2O3 content on 28.8 %, 25.3 % and 62.1 % respectively compared with precipitate after firing obtained by the same value of pH, but do not subjected to the ageing process.

It can be assumed that this process is related to physical and chemical phenomena occurring during ageing, such as the improvement of the structure of crystals, desorption of contaminants trapped in the precipitation process, as well as enlargement of the primary crystals.

Furthermore, there are possibly chemical processes in the precipitate when it stored under a layer of mother liquor. These chemical processes lead to changes in chemical and physical properties of the precipitate and in its chemical composition. Possible to assume that some processes such as hydrolysis, hydration and dehydration occur during the ageing processes of precipitate, which cause the increase the content of 3BaOAl2O6 in the material after firing.

5 Conclusion

Barium aluminates were formed by the continuous precipitation method at different values of pH. The influence of ageing process and precipitation pH on the phase composition of the precipitate after firing was found. It was showed that there is growth of the 3BaOAl2O3 concentration after ageing of precipitate during 7 days. Presumably, this process is associated with a decrease in the number of crystals, increasing their size in the ageing process of precipitate and chemical processes that occur during ageing: hydrolysis, hydration, dehydration.

Acknowledgements

This work was funded within the framework of realisation of Strategic Programme on National Research Tomsk Polytechnic University Competitiveness Enhancement in the Group of Top Level World Research and Academic Institutions.

References

1. C. Higashi, N.B. De Lima, J.R. Matos, C. Giovedi, C.C. Motta, Proc. IMOC (2005)
2. Q Wang, W. Liu, L. Dong, X. Zhu, X. Liu, J. Wang. Proc. IVEC (2015)
3. F.F. Sene, V.O. Santos, C.C. Motta, IVESC (2012)
4. A.A. Maklakov, E.P. Ostapchenko, J. Struct. Chem. **1** (1960)
5. X. Ye, W. Zhuang, C. Deng, W. Yuan, Z. Qiao, CALPHAD. **30** (2006)
6. L.D. Bystritskii, N.N. Dolzhenko, N.H. Stavitskaya, E.V. Stepanova, D.E. Sharygin, Y.B. Shvalev, RF Patent 2171678, A61K33/08. (2001)

MODELING OF HIGH ENERGY IMPACT ON ICE IN TAKING INTO ACCOUNT THE TEMPERATURE

Gennady N. Bogomolov[1,*], *Victor* P. Glazyrin[1], and *Maxim* Yu. Orlov[1]

[1]National Research Tomsk State University, 634050 Tomsk, Russia

Abstract. In this paper, the problems deep penetration of compact impactors into the ice, taking into account different temperatures were solved. The behavior of ice is described by the basic system equations of continuum mechanics, i.e., the equations of continuity, momentum and energy in the lagrangian approach. Medium are compressible, isotropic, no mass forces, internal sources of heat and thermal conductivity. Medium also includethe shockwave phenomena, as well as formation ''spall'' and ''shear'' damage. The stress tensor is divided into deviatoric and spherical components. Equation of statewas chosen in the form of Walsh. The components of the stress tensor deviator located on the elasticplastic flow model based on the equations of Prandtl-Reis associated with von Mises yield criterion. Initial impactor velocity was varied atfrom 50 to 325 m/s. Numerical simulation results showed the influence of temperature of the ice to the depth of penetration of the impactors.

1 Introduction

The urgency of research on highenergy impact on the ice in order to deepen scientific knowledge of the physics and mechanics of ice due to the intensive development is currently of the Arctic and the northern areas of Russia. In this case numerical simulatingis an effective tool to obtain detailed information about loading processes, significantly adding to the experimental results in a wide range of initial conditions.

Theoretical results on the investigation of the ice impulse load under a certain fixed temperature were obtained [1, 2]. However, in terrestrial conditions the ice can be kept at a temperature of 0 ^{0}C and up to −60 ^{0} C. Therefore, the mathematical model of the behavior of ice should be taken into account the dependence of the strength of physical and mechanical characteristics of the temperature.

In the present research the shock process of cylindrical and spherical impactors with an ice barrier was investigated. The subsonic range of initial velocity of impactors was studied.

*Corresponding author: glvp@list.ru

2 Mathematical model

The solution of the problem of phenomenological macroscopic model of continuum mechanics, which is based on elastoplastic behavior of the material. The basic system of equations is founded on the fundamental laws of conservation of mass, momentum and energy:

$$\frac{1}{\rho}\frac{d\rho}{dt} = -\left(\frac{\partial u}{\partial z} + \frac{\partial v}{\partial r} + \frac{v}{r}\right),$$

$$\rho\frac{du}{dt} = \frac{\partial S_{zz}}{\partial z} + \frac{\partial S_{rz}}{\partial r} + \frac{S_{rz}}{r} - \frac{\partial P}{\partial z},$$

$$\rho\frac{dv}{dt} = \frac{\partial S_{rr}}{\partial r} + \frac{\partial S_{rz}}{\partial z} + \frac{S_{rr} - S_{\theta\theta}}{r} - \frac{\partial P}{\partial r},$$

$$\frac{d\varepsilon}{dt} = \frac{P}{\rho^2}\frac{d\rho}{dt} + \frac{1}{\rho}\left[S_{zz}\frac{\partial u}{\partial z} + S_{rr}\frac{\partial v}{\partial r} + S_{\theta\theta}\frac{v}{r} + S_{rz}\left(\frac{\partial v}{\partial z} + \frac{\partial u}{\partial r}\right)\right].$$

$$\frac{DS_{zz}}{Dt} + \lambda\, S_{zz} = 2\mu\left(\frac{\partial u}{\partial z} + \frac{1}{3\rho}\frac{d\rho}{dt}\right),\quad \frac{DS_{rr}}{Dt} + \lambda\, S_{rr} = 2\mu\left(\frac{\partial v}{\partial r} + \frac{1}{3\rho}\frac{d\rho}{dt}\right),$$

$$\frac{dS_{\theta\theta}}{dt} + \lambda\, S_{\theta\theta} = 2\mu\left(\frac{v}{r} + \frac{1}{3\rho}\frac{d\rho}{dt}\right),\quad \frac{DS_{rz}}{Dt} + \lambda\, S_{rz} = \mu\left(\frac{\partial u}{\partial r} + \frac{\partial v}{\partial z}\right),$$

$$S_{rr}^2 + S_{zz}^2 + S_{rz}^2 + S_{rr}S_{zz} = \frac{1}{3}\sigma_m^2,\quad P = P(\rho,\varepsilon),\ \sigma_{ij} = S_{ij} - P\delta_{ij},$$

$$2\mu\left(e_{ij} - \frac{1}{3}e_{kk}\delta_{ij}\right) = \frac{DS_{ij}}{Dt} + \lambda S_{ij},$$

$$\frac{DS_{ij}}{Dt} = \frac{dS_{ij}}{dt} - S_{ik}w_{jk} - S_{jk}w_{ik}, \qquad \sigma_1 > \sigma_k$$

$$2e_{ij} = \nabla_j v_i + \nabla_i v_j;\quad 2w_{ij} = \nabla_j v_i - \nabla_i v_j;$$

$$3S_{ij}S_{ij} = 2\sigma_T^2.$$

$$A_p \geq A_P^*,\quad A_p = \int_0^{t_*} dA_p = \int_0^{t_*}\frac{\sigma_{ij}d\varepsilon_{ij}^p}{\rho},$$

where are ρ -density of the material; r, z - radial and axial coordinates; u, v - axial and radial components of the velocity vector; σ_{ij}, ε_{ij}, e_{ij} - tensor components of stress, strain and strain rate; ε - specific internal energy, δ_{ij} - Kronecker symbol , P - hydrostatic pressure; μ - shear modulus; σ_T - yield strength of the material; σ_k – ''spall'' damage; A_p - compressive shear plastic deformation; *D/Dt*– derivative of Yaumana. Yauman derivative take into account medium elements rotate while it is moving.

The strength of ice characteristics depend on the temperature as follows:

$$f_i = f_{i0}\left(1 + \frac{1}{f_{i0}}\frac{\partial f}{\partial t}\Delta T\right) = f_{i0} + a_i\Delta T,$$

where f_i - shear modulus, yield stress, tension spall and critical work of plastic deformation; f_{i0} - initial values of these characteristics. Constants a_i determined from experimental data [3, 4, 5] and the condition of the vanishing ice at zero temperature. The following values of strength characteristics used in the calculations:

μ_0 = 3.1 MPa; σ_{To} =1.85 MPa; σ_{ko} = 3.06 MPa; A_o=0.3 KJ/kg; a_1 = 0.035 MPa/degree; a_2= 0.030 MPa/degree; a_3 = 0.030 MPa/degree; a_4 = 0.02 KJ/ (kg×deg).

3 Numerical simulation

Physical formulation of the problems is following. The cylinder and the sphere of the same mass penetrate into ice barrier. Initial velocity impactors directed along its axis of symmetry normal to the plane of the barrier and was set equal to 50 to 325 m/s. The object of study is the ice cylinder. Cylinder height is 60 mm, but diameter is 45 mm.

There were 12 of calculation variants. In each variant the initial velocity of the impactor increased to 25 m/s. The cylinder has of dimensions (5×5) mm. Diameter of the spherical impactor is equal to 5.72 mm. Impactors weight are 0.77 g. The boundary conditions at the contact surfaces allow their sliding without friction. Material of impactors were steel U10A.

The initial conditions for problems of shock interaction of solids for the undisturbed state imply the vanishing stress component, pressure and internal energy. The numerical solution of the boundary value problem is carried out in a two-dimensional statement for the axial symmetry of themodified method Johnson G.R. It is known that this method is a lagrangian, but it allows solve the problem of deep penetration of heterogeneous structure, including modern safety structure. For this incorporated the algorithm erosion of triangulation elements and algorithm splitting nodes, the algorithm for constructing the free surface [6, 7]. The latest innovation should be considered smoothing algorithm for the contact surface [8].

For a plane and axial symmetry calculation program has been developed in a two-dimensional formulation that allows you to prepare online original data, including automatic partitioning of the computational domain. The computer program allows the application of five different equations of state for many materials of construction [8].

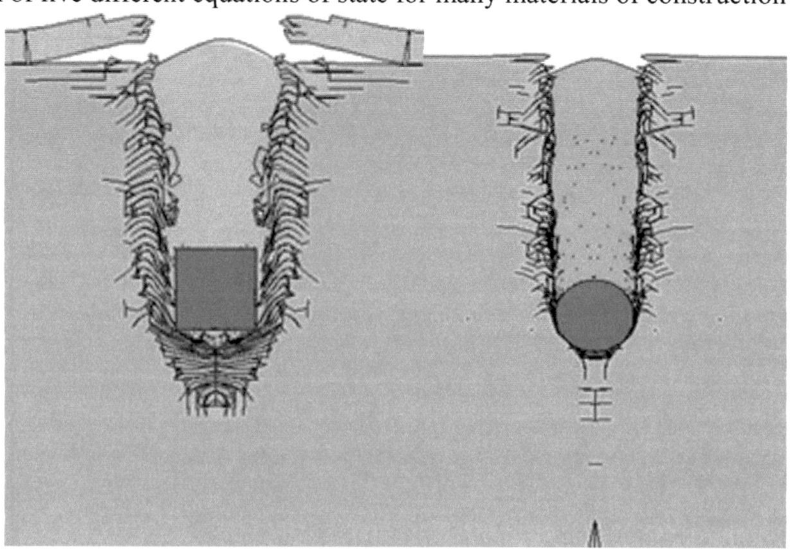

Fig. 1. Current configurations "Impactor – Ice", time 225 mks.

4 Results

The calculated configurations of "Impactor - Ice" are shown in figure 1. The initial velocity of the impactors are 225 m/s. It was found that the first destruction of the pockets formed in the ice in the contact zone "impactor - ice" at the begining process. The deeply penetrated impactors into the ice, the greater the area of destruction of ice was. The deformation of the cylindrical and spherical impactors occurs in the elastic field without failure impactors. The penetration of the cylindrical impactor accompanied by spall fracture front surface of the ice barriers. The depth of penetration of the spherical impactor was 15 % more than the depth of penetration of the cylindrical impactor. Obviously, this is explained streamline shape of the spherical impactor.

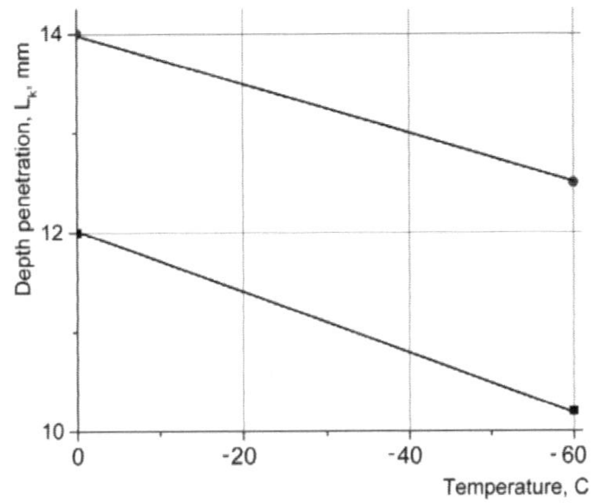

Fig. 2. Calculated dependence the depth of penetration of the impactors from ice temperature (● - spherical impactor, ■ - cylindrical impactor).

Figure 2 illustrates the dependence of penetration depth of impactors on the temperature. The upper line corresponds to the dependence of the spherical the impactor on the temperature and the bottom line corresponds to dependence of the cylindrical the impactor on the temperature.In this variant the initial velocity of impactors was equal to 175 m/s. The temperature changed from 0 to -60 ° C. It was found that with an increase in negative ice temperature depth penetration of impactors decreased in almost a linear relationship. It was found that the penetration of the cylindrical the impactor into ice decrease in depth of penetration was 16.3 %. At the same time the penetration of the spherical the impactor into ice decrease the depth of penetration was 15.2 %. In both cases, the impactors remained undamaged.

4 Conclusions

Thus the process penetration impactors into ice at low impact velocity have been simulated. Ice modeled by a single-phase medium with different negative temperature. Current and final parameters of the process, as well as identify areas destroyed by ice from the magistral crack and the first foci of destruction were predicted. The calculation results can be useful in the construction of approximate formulas for express analysis of the relevant bodies of shock interaction.

Acknowledgments

The work was supported by the RF Ministry of Education as part of the state task No. 2014 / 223 (project 1567).

References

1. V.P. Glazyrin, Ju.N. Orlov, Ju.N. Orlova, Izvestija vuzov. Fizika **52**, 7 (2009)
2. A. Combescure, Y. Cyuzel-Marmot, J. Fabis, Int. J. Solids Structures **48** (2011)
3. V.V. Bogorodskij, V.P. Gavrilo, Led. *Fizicheskie svojstva. Sovremennye metody gljaciologii* (1980)
4. K.S. Carney et. al., Int. J. Solids Structures **43** (2006)
5. R.G. Kraus et. al., Earth and Planetary Science Letters **289** (2010)
6. V.P. Glazyrin, M.Ju. Orlov, Ju.N. Orlov, Izvestija vuzov. Fizika **50**, 9 (2007)
7. A.V. Gerasimov, V.P. Glazyrin, V.N. *Barashkov i dr. Teoreticheskie i jeksperimental'nye issledovanija vysokoskorostnogo vzaimodejstvija tel,* (2007)
8. V.P. Glazyrin, Ju.N. Orlov, Research Journal of International Studies **11**, 42 (2015)

MATEC Web of Conferences **72**, 01014 (2016) DOI: 10.1051/matecconf/20167201014
HMTTSC-2016

IDENTIFICATION OF SIZE AND CONCENTRATION OF SUBMICRON PARTICLES ON THE BASIS OF RAYLEIGH SCATTERING MODEL

Ivan. S. Bondarchuk[1, *], *Vladimir* A. Arkhipov[1], and *Alexander* S. Zhukov[1]

[1]National Research Tomsk State University, 634050 Tomsk, Russia

Abstract. A method of identification of maximum size and concentration of submicron aerosol particles based on measurement of intensity attenuation of a parallel beam of probe optical radiation is described. Offered method makes it possible to determine both particle concentration and maximum particle size with controlled accuracy for aerosol media without any initial information about particle size distribution.

Particle size distribution of aerosol media (in the process of plasma chemical synthesis, preparation of fuel mixtures, analysis of combustion products etc.) laws of formation and its mass-geometry characteristics determine parameters and efficiency of many technology processes [1]. In this case the most accurate methods of study of such objects are non-contact optical methods. An approach making it possible to identify mass concentration and maximum size of condensed particles using the data of laser probing of aerosol formations is presented in this paper. According to offered method of measurement of optical density of dispersed medium measurements are performed in several points of spectral wavelength range: the boundaries of this range are selected on the basis of information about complex refractive index of particle material. When monochromatic radiation with the wavelength λ passes through a layer with the width l of uniformly distributed particles with the diameter D and mass concentration C_m its intensity is attenuated due to scattering and absorption by particles. Spectral transmittance T_λ, associated with Beer–Lambert–Bouguer law [2] is a quantitative characteristic of attenuation.

$$T_\lambda = I_\lambda / I_{0\lambda} = \exp(-\tau_\lambda) \tag{1}$$

where I_λ – radiation flux passed through the layer; $I_{0\lambda}$ – incident radiation flux.

Expression for spectral optical density τ_λ for monodisperse particles in the framework of assumptions of Mie theory has the following form [2]:

$$\tau_\lambda = \frac{1.5 C_m l}{\rho D} Q(\alpha, m) \tag{2}$$

* Corresponding author: ivanich_91@mail.ru

where ρ – particle material density;

Dependence $Q(\alpha,m)$ of dimensionless attenuation efficiency factor in equation (2) has a complex oscillatory character on diffraction parameter (Mie parameter) $\alpha=(\pi D)/\gamma$, from complex refractive index of particle material $m=n-i\chi$ (n – refractive index; χ – attenuation coefficient) and in common case is calculated using exact formulae of Mie theory.

In case of "small" particles if Rayleigh scattering condition is met $\alpha < 0.2$ attenuation efficiency factor is determined from the following analytical dependence [2]:

$$Q(\alpha,m)=F(m)\cdot\alpha\text{, where}\quad F(m)=\frac{24n\chi}{\left(2+n^2-\chi^2\right)^2+\left(2n\chi\right)^2} \tag{3}$$

After substitution of (3) into the expression for optical density (2) we have the following:

$$\tau_\lambda=\tau_\lambda^{\text{calc}}=\frac{1.5C_m l}{\rho D}\cdot F(m)\frac{\pi D}{\lambda}=\frac{1.5\pi C_m l}{\rho\lambda}F(m) \tag{4}$$

(4) implies that spectral density τ_λ in case of Rayleigh scattering does not depend on particle size D. Therefore, mass particle concentration can be determined from (4) using value of τ_λ:

$$C_m=\frac{\rho\lambda\tau_\lambda}{1.5\pi l F(m)} \tag{5}$$

In case of polydisperse particles, the expression for spectral optical density is determined as [2, 3]:

$$\tau_\lambda=\frac{1.5C_m l}{\rho}\cdot\frac{\int_0^\infty Q(\alpha,m)D^2 f(D)dD}{\int_0^\infty D^2 f(D)dD} \tag{6}$$

It should be noted that substitution of expression (3) for $Q(\alpha,m)$ in (6) as in case of Rayleigh scattering gives the same expression as (5) for mass concentration of particles.

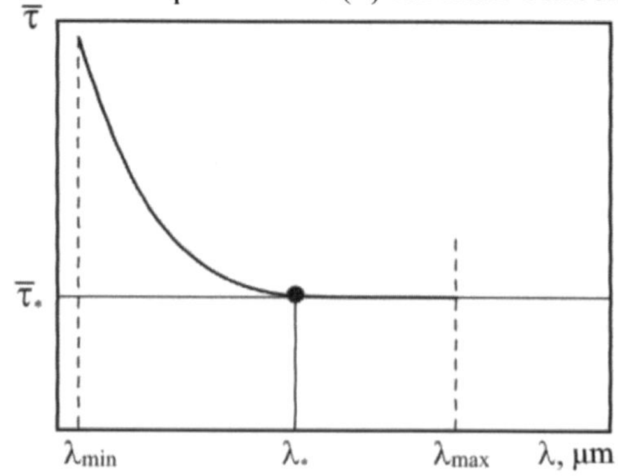

Fig. 1. Caption of the figure 1. Below the figure.

This method of C_m calculation from formula (5) using measured value of τ_λ is considered valid only for a limited range of probe radiation wavelength $\lambda \geq \lambda_*$.

To determine wavelength λ_*, limiting Rayleigh scattering region spectral optical density τ_λ^{meas} is measured in a certain wavelength $\lambda_{min} \le \lambda_* \le \lambda_{max}$ range and from the analysis of function $\bar{\tau}_\lambda = \bar{\tau}_\lambda^{meas} / \bar{\tau}_\lambda^{calc}$ limiting value λ^* is determined. Up to this value $\bar{\tau}_\lambda$ decreases monotonically (fig.1) and then $\bar{\tau}_\lambda = \bar{\tau}_{\lambda_*} = \text{const}$ for $\lambda > \lambda_*$. Particle mass concentration for region $\lambda > \lambda^*$ is determined using formula (5).

In order to calculate D_{max} value using relation given in [2]

$$D_{max} = \frac{\lambda_*}{\pi} \alpha_* (\lambda_*) \tag{7}$$

we need an algorithm of identification of boundary value of diffraction parameter α_*. Analysis of dependencies $Q(\alpha)$ (fig. 2) shows that α_* value limiting Rayleigh scattering region is determined by wavelength λ.

Calculated dependencies (for soot) are given in fig. 3. These charts were built on the basis of condition of 5 % and 10 % deviation $Q(\alpha)$ of Rayleigh approximation from the exact Mie theory solution.

Thus, there are certain values $\alpha^* = f(\lambda)$ for each wavelength which make Rayleigh approximation valid for $\alpha < \alpha_*$ with specified controlled accuracy. These dependences are well approximated by the following function

$$\alpha_* = a \cdot \exp(-b\lambda) \tag{8}$$

where $[\lambda]$ = μm, and constants a and b depend on optical constants of material of studied particles.

The following expression for D_m can be obtained from (7) and (8):

$$D_m = \frac{\lambda_*}{\pi} a \cdot \exp(-b\lambda) .$$

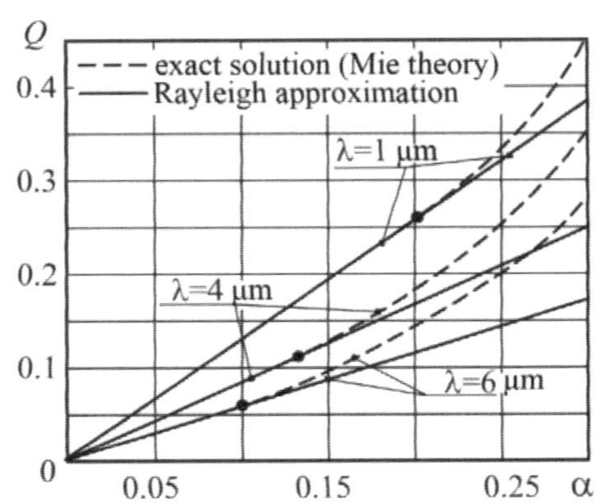

Fig. 2. Exact and approximated dependence of attenuation efficiency factor on diffraction parameter.

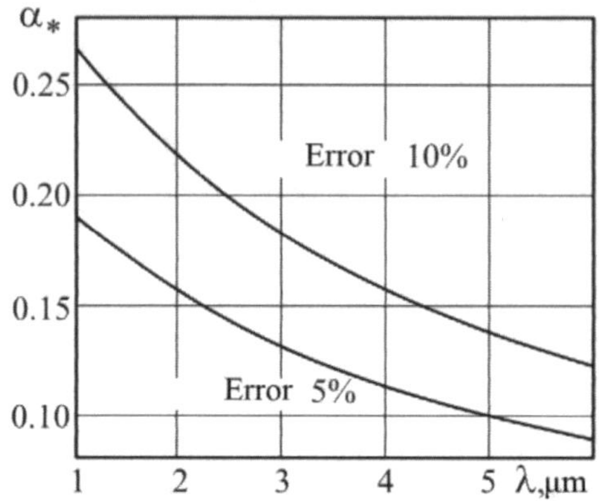

Fig. 3. $\alpha_* = f(\lambda)$ dependence for soot.

For $\lambda = (1 \div 6)$ μm dependence of refractive index n and attenuation coefficient χ on wavelength λ is as follows n=1.6-0.3 λ, χ = 0.6 λ [2]; approximation error comprises (2.5÷3.0) %. Final expression for maximum particle size D_m for strictly determined dependence $\alpha_* = 0.217 \cdot \exp(-0.155\lambda)$ is as follows:

$$D_m = \frac{0.217\lambda_*}{\pi} \cdot \exp(-0.155\lambda) \quad (9)$$

Thus, the offered method makes it possible to determine both particle concentration and maximum particle size with controlled accuracy for aerosol media without any initial information about particle size distribution.

Acknowledgments

This research was supported by the Ministry of Education and Science of the Russian Federation under the Federal Target Program "Research and development on priority directions of scientific-technological complex of Russia for 2014-2020", the agreement No. 14.578.21.0034, unique identifier PNI RFMEFI57814X0034.

References

1. I.S. Bondarchuk, I.A. Zhukov, B.V. Borisov, MATEC Web Conf. **23**, 01046 (2015)
2. V.A. Arkhipov, S.S. Bondarchuk, *Optical diagnostic methods of heterogeneous plasma combustion products: Study guide* (Tomsk State University, Tomsk, 2012) [in Russian]
3. V.A. Arkhipov, S.S. Bondarchuk, N.G. Kvesko, A.T. Roslyak, V.F. Trofimov, Atmos. Oceanic Opt. **17** (2004)

STUDY OF SOOT PARAMETERS IN THE PROCESS OF COMBUSTION OF CONDENSED SYSTEMS

Vladimir V. Promahov[1,*], *Tatyana* V. Zhukova[1], and *Yuliya* N. Ryzhikh[1]

[1]National Research Tomsk State University, 634050 Tomsk, Russia

Abstract. In the present information presents the results of studies of the effect of dispersion of ammonium perchlorate on the content and nature of the soot formed in the process of burning of butyl-rubber propellants.

Combustion products of condensed substances based on ammonium perchlorate (AP) contain soot-like substances. The laws of formation of these substances and their characteristics are not described in literature.

Results of study of influence of dispersity of ammonium perchlorate on the nature and content of the soot formed in the process of burning of butyl-rubber propellants (oxidizer excess ratio ≈0.43) with the following composition:

ammonium perchlorate – 73 %, butyl rubber – 24.8 %, quinol ether – 2.2 %.

Oxidizer dispersity was changed during the tests. AP of three types: fine fraction (less than 50μ), coarse fraction (200÷300) μ and bidisperse (mixture of the two previous fractions in the ratio of 1:1) was used. Samples were burned in the pressure range of (1÷8) MPa in nitrogen atmosphere.

Soot content was measured using two methods – laser diagnostics and direct collection of soot formed after burning up of a sample in a constant-pressure device.

A new approach was used for optical diagnostics. This approach makes it possible to determine both mass concentration and maximum size of condensed particles on the basis of the data of laser probing of combustion products.

According to offered method measurements of optical density $\lambda^{meas}(\lambda)$ of a dispersed medium are performed in several equally-spaced points of spectral wavelength range $\lambda=\lambda_{min}\div\lambda_{max}$. The boundaries of this range are selected on the basis of valid information about complex refractive index $m=n$-i·æ.

Particle mass concentration is determined for the range $\lambda\geq\lambda_*$; maximum particle size is determined using boundary wavelength λ_* and associated parameter α_* according to the following relations:

$$C_m = \frac{\tau^{meas}(\lambda)}{C \cdot K^{(p)}(\lambda)}, \quad (1)$$

[*] Corresponding author: bvborisov@tpu.ru

$$D_m = \frac{\alpha_* \lambda_*}{\pi}, \tag{2}$$

where $C = \text{const} = \frac{1.5l}{\rho_k}$, $K^{(p)}(\lambda)$ – coefficient in Rayleigh formula for attenuation efficiency factor

$Q(\alpha,m)$, $\alpha_*=f(\lambda_*)$, l – flow diameter, ρ_k – particle material density.

Boundary value λ_* (fig.1) is determined as abscissa of inflection point of the following function

$$\bar{\tau}(\lambda) = \frac{\tau^{meas}(\lambda)}{K^{(p)}(\lambda)}.$$

Expression for optical density $\tau(\lambda)$ for monodisperse particles in the framework of assumptions of Mie theory has the following form [1]:

$$\tau(\lambda) = \frac{1.5C_m l}{\rho_k D} Q(\alpha, m). \tag{3}$$

Dependences $n(\lambda)$, æ(λ) have a complex character and are determined experimentally.

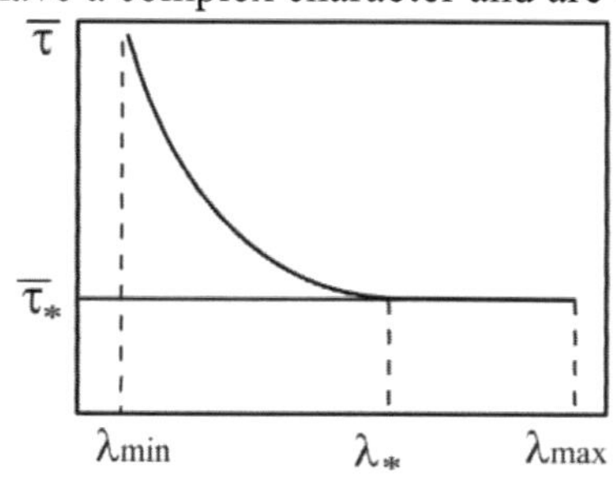

Fig. 1. Characteristic values of probe radiation wavelength.

Attenuation efficiency factor for $D << \lambda$ (Rayleigh approximation) is determined using the formula [1]:

$$Q^{(P)}(\lambda) = \frac{24n\,æ}{(n^2 - æ^2 + 2)^2 + (2n\,æ)^2}\alpha = K^{(P)}(\lambda)\cdot D. \tag{4}$$

Relations (4) and (3) imply the following:

$$\tau(\lambda) = \frac{1.5C_m l}{\rho_k} K^{(p)}(\lambda) = C \cdot K^{(p)}(\lambda) \cdot C_m. \tag{5}$$

As it can be seen from formula (5), $\tau(\lambda)$ does not depend on D and, therefore, concentration C_m can be determined from relation (1). Moreover, it can be shown that this method of determination of C_m can also be used for a polydisperse medium in case if Rayleigh approximation conditions are met for all particles [1]

$$\tau(\lambda) = \frac{1.5C_m l}{\rho_k} \cdot \frac{\int_0^\infty Q(\alpha, m) f(D) D^2 dD}{\int_0^\infty f(D) D^3 dD}. \tag{6}$$

where $f(D)$ – particle-size distribution function.

If Rayleigh approximation conditions are met for all particles, then from equations (6) and (4) we have:

$$\tau(\lambda) = \frac{1.5C_m l}{\rho_k} \cdot \frac{K^{(P)}(\lambda)\int_0^\infty D f(D) D^2 dD}{\int_0^\infty f(D) D^3 dD} = \frac{1.5C_m l}{\rho_k} K^{(P)}(\lambda). \quad (7)$$

Thus, expression (7) is identical to equation (5).

C_m calculation from formula (1) using measured values of optical density $\lambda^{meas}(\lambda)$ is considered valid only for a limited wavelength range $\lambda \geq \lambda_*$.

Wavelength λ_* boundary value is determined from the following relation

$$\bar{\tau}(\lambda) = \frac{\tau^{meas}(\lambda)}{K^{(p)}(\lambda)}.$$

The graph of function has two sections (fig.1) – monotonic decrease ($\lambda<\lambda_*$) and constant ($\lambda>\lambda_*$); the joint of these sections determines the wavelength boundary value λ_* of the area of Rayleigh approximation. Rayleigh approximation is not performed for area $\lambda<\lambda_*$ and calculation using relation (1) gives large error values. Concentration C_m for the range $\lambda \geq \lambda_*$ is determined using the following formula:

$$C_m = \frac{\bar{\tau}_*}{C} = \frac{\rho_k}{1.5l} \cdot \bar{\tau}_*.$$

A value $\alpha_* = f(\lambda)$, depending on wavelength λ is necessary to calculate maximum particle size D_m using expression (2). Illustration of this dependence for soot is given in fig.2. These charts were built on the basis of condition of 5 % and 10 % deviation of $Q(\lambda)$ from $Q(p)(\lambda)$ (fig.3). It should be noted that Rayleigh approximation condition is met for $\alpha < \alpha_*$ with specified controlled accuracy.

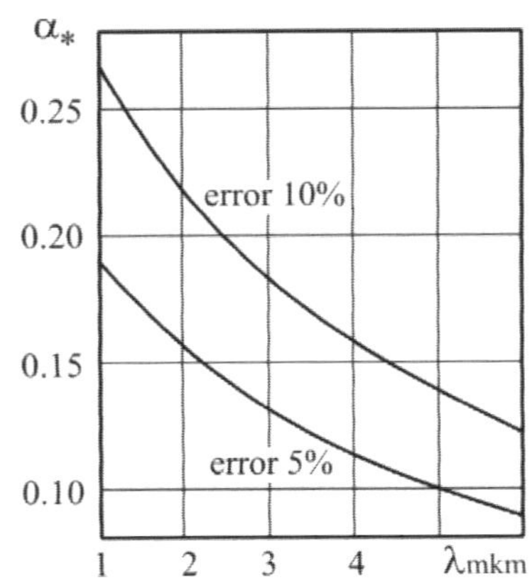

Fig. 2. Dependence of α_* on wavelength λ.

Dependence $\alpha_*(\lambda)$ has exponential character and can be approximated by the following equation:

$$\alpha_* = a \cdot \exp(-b\lambda). \quad (8)$$

Expression for D_m can be obtained from relations (8) and (2):

$$D_m = \frac{\lambda_*}{\pi} a \cdot \exp(-b\lambda).$$

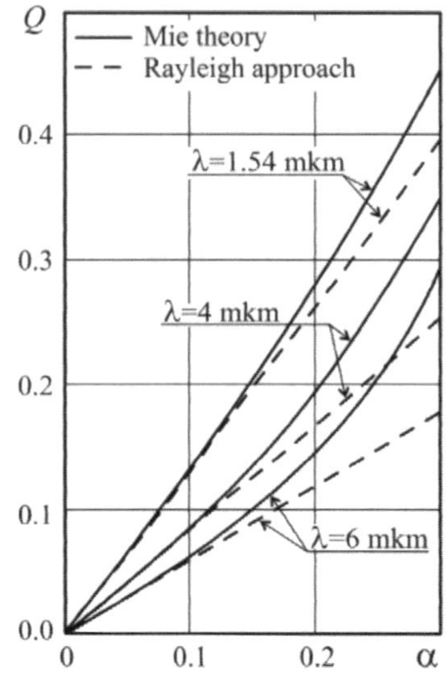

Fig. 3. Dependence of attenuation efficiency factor Q on diffraction parameter α.

According to the data from [2] for λ=(1÷6) μm the dependence of refractive index n and attenuation coefficient æ on wavelength λ is as follows: n = 1.6-0.3λ, $æ_* = 0.6\lambda^{0.6}$.

Approximation error in this case comprises (2.5÷3.0) %. Final expression for maximum particle size D_m for strictly determined dependence $\alpha_* = 0.217 \cdot \exp(-0.155\lambda)$ is as follows:

$$D_m = \frac{0.217\lambda_*}{\pi} \cdot \exp(-0.155\lambda).$$

This method makes it possible to determine both particle concentration and maximum particle size with controlled accuracy for various flames without any initial information about particle size distribution.

Soot samples were collected in the process of combustion of the system in a constant-pressure de-vice equipped with a quartz collector and an accessory for slow pressure drop preventing loss of solid combustion products. Unrestricted cylindrical propellant samples with the diameter of 10 mm and height of 70 mm were studied, their weight comprised (10÷12) g. Quartz collector after burning up and a slow pressure drop was placed into a dryer and was dried for 2-3 hours at 70-75 °C until a constant weight was reached. Soot content was determined using the following relation:

$$C = \frac{P_2 - P_1}{P_{samp}} \cdot 100\%,$$

where P_1 – weight of pure quartz collector, P_2 – weight of quartz collector with soot after drying,

P_{samp} – weight of studied sample.

Chemical analysis of soot [3] indicated that along with carbon particles it contains ammonium chloride. The quantity of the latter varies as oxidizer dispersity and sample combustion pressure change. Ammonium chloride content in a number of cases reaches 60 % of soot weight. Possible chemical interaction between chlorine and carbon and formation of so called insertion compounds is shown in [4].

The results obtained for composition of solid combustion products given in Table 1 indicate that increase of size of oxidizer crystals leads to increase of carbon content in solid combustion products and at the same time reduction of ammonium chloride content compared with systems containing oxidizer particles with sizes less that 50μ.

Comparison of the data obtained by optical diagnostic methods and selection of solid combustion products is given in Table 2. The data obtained by product selection represent only carbon content (w/o ammonium chloride).

The analysis performed has indicated that results obtained using different methods are comparable. Relatively low soot content in combustion products in case of use of laser diagnostics can likely be explained by deposition of particles on inner walls of the device containing burning sample.

Table 1. Soot composition.

AP size	carbon / NH_4Cl, wt % , at various pressures				
	1 MPa	2 MPa	3 MPa	6 MPa	8 MPa
Less than 50 μ	50/50	46/54	43/57	40/60	40/60
(200÷300) μ	67/33	–	65/35	–	60/40
bidisperse (1:1)	60/40	–	57/43	–	50/50

Table 2. Carbon content in combustion products.

AP size	Method	Carbon content, wt % , at various pressures				
		1 MPa	2 MPa	3 MPa	6 MPa	8 MPa
Less than 50 μ	selection	3.9	3.7	2.9	1.6	1.5
	optical	3.7	3.6	2.5	1.3	1.4
(200÷300) μ	selection	8.8	5.2	4.1	3.3	3.1
	optical	8.5	4.8	3.5	–	–
bidisperse (1:1)	selection	3.3	3.3	1.7	0.8	0.7
	optical	2.9	2.9	1.4	0.8	0.8

Thus, dispersity of oxidizer affects both total soot content in combustion products and chemical composition of solid combustion products. It is noted that increase of particle size leads to increase of carbon content in solid combustion products. In case of use of bidisperse oxidizer leads to reduction of soot content.

Acknowledgments

This research was supported by the Ministry of Education and Science of the Russian Federation under the Federal Target Program "Research and development on priority directions of scientific-technological complex of Russia for 2014-2020", the agreement No. 14.578.21.0034, unique identifier PNI RFMEFI57814X0034.

References

1. V.A. Arkhipov, S.S. Bondarchuk, *Optical diagnostic methods of heterogeneous plasma combustion products: Study guide* (Tomsk State University, Tomsk, 2012) [in Russian]

2. A.G. Blokh, *Heat exchange in steam boilers* (Energoatomizdat, Leningrad, 1984) [in Russian]
3. V.A. Klimova, *Basic methods of analysis of organic compounds* (Khimiya, Moscow, 1975) [in Russian]
4. Weygand-Hilgetag, *Experimental Methods in Organic Chemistry* (John Wiley & Sons Inc, 1972)
5. V.A. Arkhipov, S.S. Bondarchuk, *Formula of parameters communication of unimodal distributions of particles in the sizes with geometrical characteristics of probability density function* (Mechanics of fast flow processes. Tomsk: TSU, 1989) [in Russian]
6. V.A. Arkhipov, S.S. Bondarchuk, N.G. Kvesko, A.T. Roslyak, V.F. Trofimov, Athmospheric and Oceanic Optics **17**, 5 (2004)

MATEC Web of Conferences **72**, 01016 (2016) DOI: 10.1051/matecconf/20167201016
HMTTSC-2016

DOUBLE CONTINUUM MODEL OF EVOLUTION OF PRECURSOR DROPLETS IN THE FLOW OF HEAT TRANSFER MEDIUM OF A PLASMA CHEMICAL REACTOR

Ilya S. Zhukov[1], *Sergey* S. Bondarchuk[1], and *Boris* V. Borisov[2,*]

[1]Institute for Problems of Chemical & Energetic Technologies of the Siberian Branch of the Russian Academy of Sciences, 659322 Byisk, Russia
[2]National Research Tomsk Polytechnic University, 634050 Tomsk, Russia

Abstract. A double continuum physical and mathematical model of evolution of precursor in a vaporizing droplet of low-concentration solution of metal salt in the process of its heating in the flow of heat transfer medium of a plasma chemical reactor is represented in this paper. This model was developed to study the morphology of ceramic powder particles depending on solution characteristics and parameters of a high-temperature gaseous heat transfer medium which determine heat and mass transfer with a droplet medium.

Plasma chemical synthesis including thermo-chemical decomposition of liquid sprayed agents(precursors) in the flow of a high-temperature gaseous heat transfer medium [1,2] is one of the most promising technologies of production of ceramic powders with controlled parameters. The term “spray pyrolysis” is used for this technology in English [3].

Analysis of separate stages of plasma chemical synthesis nave indicated [1-4] that powder particle morphology is formed at the stage of droplet evaporation. Simultaneous solvent evaporation from the droplet surface and salt diffusion towards the center of the droplet is the basic process determining product structure. The relation between the rates of these processes not only determines the size of a salt "residue" particle formed in the process of solvent evaporation but also its structure – nut-type (with a relatively empty nucleus) or solid (without any cavities).

Physical and mathematical models of processes taking place in a plasma chemical reactor are necessary to provide controlled operation of the latter in order to produce particles with specified characteristics. At the same time, a number of well-known works are based on very significant simplifying assumptions in description of processes in the droplet of solution [1-4], particularly on approximate equality of solvent density and admixture density. This fact significantly reduces predictability of characteristics of synthesized metal oxide powders and their morphology (internal structure). The offered double continuum model of evolution of precursor droplet considers diffusion of an admixture with the density different from the density of the solvent.

* Corresponding author: bvborisov@tpu.ru

The basic assumption is the condition of conservation of volume of each component of the mixture in the process of its movement determined by the increase of mass fraction of salt at the boundary of precursor droplet due to solvent evaporation. This assumption leads to the following

$$\frac{\varrho^{s}}{\varrho^{so}}+\frac{\varrho^{m}}{\varrho^{mo}}=1\,, \qquad (1)$$

where ϱ^{s},ϱ^{m} – reduced densities (in the mixture) of salt and solvent;

From mass conservation equations for components

$$\frac{\partial\varrho^{s}}{\partial t}+\frac{\partial}{\partial r}\left(r^{2}\varrho^{s}u^{s}\right)=0,\frac{\partial\varrho^{m}}{\partial t}+\frac{\partial}{\partial r}\left(r^{2}\varrho^{m}u^{m}\right)=0$$

taking into account relation (1) we can obtain

$$\frac{\partial}{\partial r}\left(\varrho^{s}u^{s}+\varrho^{m}u^{m}\right)=0\,. \qquad (2)$$

where t – time, r – coordinate oriented from the centre of the droplet,
u^{s},u^{m} – velocities of mixture components.

Let us assume that salt velocity for the whole process is determined by the Fick's law using diffusion coefficient D and gradient of salt concentration c

$$u^{s}=D\frac{\partial c}{\partial r}\,.$$

Mass fraction c is linked with reduced densities of the mixture by the following relation

$$c=\varrho^{s}/\left(\varrho^{s}+\varrho^{m}\right).$$

Taking into account relation (2) for an elementary volume dV, bounded by surface S, laws of conservation of mass for mixture components can be represented as follows

$$\frac{\partial}{\partial t}\int_{V}\varrho^{s}dV+\int_{S}\varrho^{s}u^{s}dS+\int_{S}\varrho^{s}v dS=0\,,$$

$$\frac{\partial}{\partial t}\int_{V}\varrho^{m}dV+\int_{S}\varrho^{m}u^{m}dS+\int_{S}\varrho^{m}v dS=0\,,$$

$$u^{s}=D\frac{\partial c}{\partial x},\varrho^{m}u^{m}=\frac{\varrho^{mo}}{\varrho^{so}}\varrho^{s}u^{s}.$$

Convective mass transfer for a surface S moving with a velocity v^{*} is determined by droplet size reduction due to solvent evaporation.

Boundary conditions are defined on the basis of the physics of the process by the following relations.

$$u^{s}\Big|_{r=0}=u^{s}\Big|_{r=R}=0;\quad u^{m}\Big|_{r=0}=u^{m}\Big|_{r=R}=0.$$

At the main stage of the process when solvent evaporates under the influence of external heat flow it can be assumed that droplet surface temperature is equal to boiling temperature T_{vap} for given ambient pressure.

In the framework of the model of centrally symmetric reduced film [4, 5] (the radius of this film is determined by the condition of quasi-stationary heat exchange with the medium taking into account heat consumption for evaporation and vapor superheating up to the temperature of a heat transfer medium T_e) the relation for current rate of droplet radius reduction v^* can be written as follows

$$v^* = -\frac{dR}{dt} = \frac{\lambda_g}{\varrho^m R c_{vap}} \cdot \mathrm{Nu} \cdot \ln(1+\mathrm{B}), \quad u^s = D\frac{\partial c}{\partial r}.$$

where $B = c_{vap}\left(T_e - T_{vap}\right)/H_{vap}$ – Spalding number; $\mathrm{Nu} = 2 + 0.16\sqrt[3]{\mathrm{Re}_\Delta^2}$ – Nusselt number;

Re_Δ – Reynolds number of relative particle movement in a gas flow;

λ_g – gas heat transfer coefficient; c_{vap} – specific heat capacity of vapors; H_{vap} – heat of evaporation.

The following difference scheme can be used to solve these equations.

$$\frac{\bar{\varrho}_i^s \bar{V}_i - \varrho_i^s V_i}{\Delta t} = r_i^2 \varrho^s_{i-\frac{1}{2}} u^s_{i-\frac{1}{2}} - (r_i + \Delta r)^2 \varrho^s_{i+\frac{1}{2}} u^s_{i+\frac{1}{2}} + r_i^2 \varrho^s_{i-1} v_{i-\frac{1}{2}} - (r_i + \Delta r)^2 \varrho_i^s v_{i+\frac{1}{2}};$$

$$\frac{\bar{\varrho}_i^m \bar{V}_i - \varrho_i^m V_i}{\Delta t} = r_i^2 \varrho^m_{i-\frac{1}{2}} u^m_{i-\frac{1}{2}} - (r_i + \Delta r)^2 \varrho^m_{i+\frac{1}{2}} u^m_{i+\frac{1}{2}} + r_i^2 \varrho^m_{i-1} v_{i-\frac{1}{2}} - (r_i + \Delta r)^2 \varrho_i^m v_{i+\frac{1}{2}};$$

$$r_i = i \cdot r, V_i = \left(r_i^3 - r_{i-1}^3\right)/3, i = 0, 1, \ldots, N; \Delta r = R/N;$$

$$u^s_{\frac{1}{2}} = 0, u^s_{i-\frac{1}{2}} = D\frac{c_i - c_{i-1}}{\Delta r}, i = 2, 3, \ldots, N;$$

$$u^s_{i+\frac{1}{2}} = D\frac{c_{i+1} - c_i}{\Delta r}; i = 1, 2, \ldots, N-1; u^s_{N+\frac{1}{2}} = 0;$$

$$\varrho^m_{i\pm\frac{1}{2}} u^m_{i\pm\frac{1}{2}} = \frac{\varrho^{mo}}{\varrho^{so}} \varrho^s_{i\pm\frac{1}{2}} u^s_{i\pm\frac{1}{2}}, i = 1, 2, \ldots, N.$$

Overline over variables indicates their value at the next time layer. A stencil of the difference scheme is given in fig.1, where $f = (\rho^s; \rho^m; v; u^s; u^m)$.

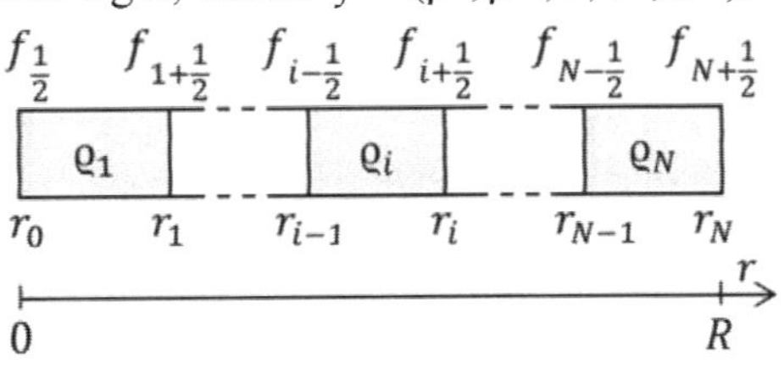

Fig. 1. Stencil of the difference scheme.

Thus, a double continuum model of precursor evolution for diffusion of admixture in the process of movement of vaporizing droplet medium along the loop of a plasma chemical reactor is formulated in this paper. A mass-conservative difference scheme is offered for numerical implementation of the model.

Acknowledgments

The reported study was funded by RFBR, according to the research project No. 16-38-60031 mol_a_dk.

References

1. A.L. Suris, *Plasma chemical processes and apparatus* (Khimiya, Moscow, 1989) [in Russian]
2. V.K. Larin, V.M. Kondakov, E.N. Malyi et al., Izv. vuzov. Non-ferrous metallurgy **5** (2003) [in Russian]
3. G.L. Messing, S.-C. Zhang, G.V. Jayanthi, J. American Ceramic Society **76**, 11 (1993)
4. I.S. Bondarchuk, I.A. Zhukov, B.V. Borisov, MATEC Web Conf. **23** (2015)
5. V.V. Pomerantsev, K.M. Arefyev, D.B. Akhmedov et al., *Fundamentals of practical combustion theory* (Energy, Moscow, 1986) [in Russian]

HEAT MODE ELECTRONIC DEVICES AT ELEVATED EXTERNAL TEMPERATURES

Evgeny Kravchenko[1,*], *Kseniya* Vershinina[1], and *Tamara* Boykova[1]

[1]National Research Tomsk Polytechnic University, 634050 Tomsk, Russia

Abstract. In this article we consider the numerical study of the effect increasing the ambient temperature of electronic device under external microwave heating.

1 Introduction

The problem of forecasting thermal conditions of technical equipment is the development of two promising methodologies related to development (PDfR - Probabilistic Design for Reliability [1]) and operation (DRM - Dynamic Reliability Management [2]) of modern electronic devices. The essence of these concepts is to identify the most significant factors affecting the service life of heat-loaded elements, followed by the development of mathematical models to predict the reliability of indicators based on the physics of bounce. As is known, the temperature is one of the key parameters. At the same time the thermal regime of technical facilities affects both the internal heat transfer and higher ambient temperature.

2 Methods of analysis of temperature fields in the presence of local heat sources

By modern methods of analysis of thermal simulation modes include devices with different software systems (eg, Flow Vision) [3]. There are also other approaches: a simplified mathematical model [4], the method of heating circuits (thermal resistance) [5], thermal monitoring [6], the finite element method [7], the finite difference methods [8], taking into account natural convection [9] radiation, together with the heat sink in the stationary and the cyclic operating conditions [10, 11].

* Corresponding author: kevatp@tpu.ru

3 Formulation of mathematical solving

Simulation of temperature field for the effects of microwave heating on the thermal regime of the system "polymer-semiconductor-composite" carried out in the range of operating temperatures in two dimensional axes.

In the two-dimensional formulation of the problem of heat transfer is reduced to the solution of non-stationary heat equation:

$$C(x,y)\rho(x,y)\frac{\partial T}{\partial t}=\frac{\partial}{\partial x}\left(\lambda(x,y)\frac{\partial T}{\partial x}\right)+\frac{\partial}{\partial y}\left(\lambda(x,y)\frac{\partial T}{\partial y}\right)+Q_{\text{heat}}(t,x,y)+Q_{\text{microwave}}(t,x,y), \quad (1)$$

where: C - specific heat; - The density; T - temperature; t - time; - coefficient of thermal conductivity; Q_{heat} - heat source; $Q_{microwave}$ - heat by microwave heating; x, y, z - coordinate.

When setting the initial conditions it was believed that at the initial time the temperature is evenly distributed:

$$T|_{t=0}=T_0(x,y), \quad (2)$$

where: T_0- the initial temperature.

The boundary conditions taken into account convective and radiative heat transfer:

$$x=0, y\in[0;L_y]: \quad -\lambda\frac{\partial T}{\partial x}=\alpha(T)(T_{\text{B}}-T)+\varepsilon_{pr}\sigma\left(T_{\text{B}}^4-T^4\right), \quad (3)$$

$$x=L_x, y\in[0;L_y]: \quad \lambda\frac{\partial T}{\partial x}=\alpha(T)(T_{\text{B}}-T)+\varepsilon_{pr}\sigma\left(T_{\text{B}}^4-T^4\right), \quad (4)$$

$$y=0, x\in[0;L_x]: \quad -\lambda\frac{\partial T}{\partial y}=\alpha(T)(T_{\text{B}}-T)+\varepsilon_{pr}\sigma\left(T_{\text{B}}^4-T^4\right), \quad (5)$$

$$y=L_y, x\in[0;L_x]: \quad \lambda\frac{\partial T}{\partial y}=\alpha(T)(T_{\text{B}}-T)+\varepsilon_{pr}\sigma\left(T_{\text{B}}^4-T^4\right). \quad (6)$$

wherein: α - the surface convective heat transfer coefficient of the EA member with the external environment; T_{B}- ambient temperature; - Stefan-Boltzmann constant; ε_{pr} - Reduced coefficient of the surface and the surrounding black environment.

Convective heat transfer coefficient depends on temperature and is determined for each point of the surface [9].

$$\alpha(T)=\left(1.42-1.4\cdot10^{-3}Taver\right)N\left(\frac{T-T_{\text{B}}}{L}\right)^{\frac{1}{4}}, \quad (7)$$

Powered emissivity products and the surrounding surface of the medium is given by [9].

$$\varepsilon_{\text{pr}}=\left(\frac{1}{\varepsilon}+\frac{1}{\varepsilon_{\text{aver}}}-1\right)^{-1}, \quad (8)$$

The intensity of the heat (W) with exposure to microwave [9]:

$$Q_{\text{microwave}}(t,x,y)=2\cdot\pi\cdot\varepsilon_0\cdot\varepsilon\cdot\text{tg}(\delta)\cdot\text{f}\cdot\text{E}^2. \quad (9)$$

4 The results of mathematical modelling

Temperature dependence of the maximum (curve 2, figure 1) and medium-temperature volume (curve 4, figure 1) without microwave heating at T_B = 320 K and under the influence of electromagnetic waves are presented in figure 1. Analysis of the results of numerical studies show that in the initial time interval (up to 40), there are significant differences in the absolute values of T_M and T_{aver}. However, at t = 100 s. to the differences in temperature estimates are not as large and ranges from 2 (3.4 curves in figure 1) to 5 K (curves 1.2 in figure 1).

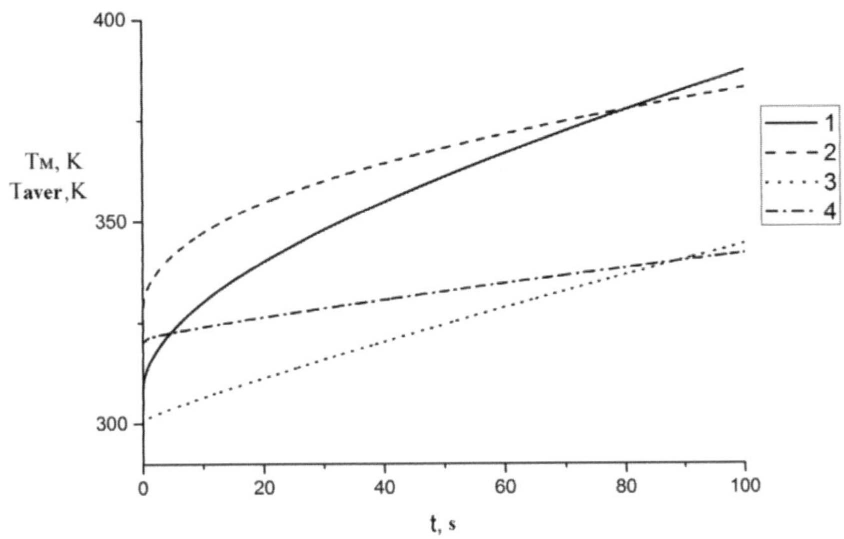

Fig. 1. Changes in the temperature characteristic of the system "polymer-semiconductor-composite' with time (1.3-T_M and T_{aver} at E = 1900 V/m, f = 7 GHz, T_B = 300 K; 2.4-T_M, at T_B = 320 K and without microwave heating).

5 Conclusion

It was found that at E = 1900 V/m in the frequency range of 3 - 7 GHz, the temperature increases from 10 to 25 K. The numerical study of the thermal regime of "polymer-semiconductor-composite" system has shown that the effect of microwave waves comparable to the increase in ambient temperature 20 K or increased local heat over 30 %.

Acknowledgments

This work was funded within the framework of realisation of Strategic Programme on National Research Tomsk Polytechnic University Competitiveness Enhancement in the Group of Top Level World Research and Academic Institutions (project VIU_ENIN_25_2016). The work at the field of modeling of heat and mass transfer processes near boundary between two layers was supported by the scientific schools grant NSH-7538-2016.8.

References

1. E. Suhir, Microelectronics Reliability **52** (2012)
2. Y. Wang, M. Enachescu, S.D. Cotofana, L. Fang, Microelectronics Reliability **52** (2012)
3. A.A. Aksenov, S.V. Zhluktov, N.F. Kudimov, E.E. Dream, M.D. Taran, O.N. Tretyakov, A.S. Shishaeva, Russian Academy of Sciences. Energy **2** (2013)
4. A. Borodin, A.A. Ivanov, J. Eng. Phys. **2** (2014)
5. E.C.W. Jong, J.A. Ferreira, P. Bauer, Power Electronics Letters IEEE **3** (2005)
6. A.B. Vlasov, Electrical Engineering **3** (2012)
7. S.G. Martyushev, I.V. Miroshnichenko, M.A. Sheremet, J. Eng. Phys. **1** (2014)
8. R.S. Volkov, O.V. Vysokomornaya, G.V. Kuznestov, P.A. Strizhak, J. Eng. Phys. Thermophys. **86**, 6 (2013)
9. G.V. Kuznetsov, E.V. Kravchenko, J. Eng. Phys. Thermophys. **88**, 6 (2015)
10. G.V. Kuznetsov, E.V. Kravchenko, EPJ Web Conf. **82** (2015)
11. M.D. Kats, I.M. Kats, J. Eng. Phys. Thermophys. **88**, 6 (2015)

DOI: 10.1051/matecconf/20167201018

MATHEMATICAL MODELLING OF PROCESS OF FLUORATION OF METAL TUNGSTEN

Roman V. Brendakov[1, *], *Alexander* V. Shvab[1], and *Vladimir* N. Brendakov[2]

[1]National Research Tomsk State University, 634030 Tomsk, Russia
[2]Seversk Technological Institute NRNU MEPhI, 636070 Seversk, Russia

Abstract. Wide use of products from metal tungsten in various branches of a national economy is caused by unique properties of this metal. By means of physical and mathematical modelling of process of receiving gaseous hexafluoride of tungsten it is possible to solve a problem of creation of import-substituting technology of processing of metal tungsten. The fluoride technology of processing of tungsten allows to put various coverings and to make products of tungsten of a various form which can't be received by traditional methods. Fluoride technology of processing of tungsten it is possible to present in the form sequences of two processes, process of fluorination of metal tungsten and process of restoration of fluorides of tungsten hydrogen.

1 Introduction

The process of fluorination of metal tungsten is carried out according to the chemical reaction given as

$$W + 3F_2 = WF_6. \quad (1)$$

To realize fluorination process, gaseous fluorine is passed over a layer of metal tungsten. Passing through the active zone of the reactor, fluorine interacts with tungsten powder to form tungsten hexafluoride (WF_6). Consequently, the concentration of fluorine in the gas stream gradually decreases due to the decreasing amount of fluorine and its dilution by tungsten hexafluoride formed. The reduction of concentration of fluorine in the two-component mix would slow down the reaction of fluorination. However, the stoichiometric coefficient in formula (1) suggests a reduction of the speed of the gas flow. In view of the limiting factors, which exert significant effect on the speed of mass-exchange processes and, hence, and on the efficiency of fluorination process, as such, the fluorine supply to the interaction surface, the speed of reaction of fluorination and the speed of removal of hexafluoride from the reaction zone.

* Corresponding author: brend989@gmail.com

2 The mathematical model

Suppose the limiting stage of the process of fluorination is the fluorine feed to the interaction surface; hence, the concentration of fluorine on the surface becomes equal to zero, and the equation of mass transfer can be written down as:

$$dM / d\tau = \alpha \cdot \beta_F \cdot C_F \cdot S,$$

where M – the mass of the fluorinated tungsten; α – stoichiometric coefficient of reaction (1); β_F – diffusion coefficient for fluorine; C_F – concentration of fluorine in the main gas stream; S – the area of fluorination of solid phase.

Another limiting case is when the limiting stage is the fluorination process as such; then the fluorine feed to the surface would exert practically no impact. Hence, the intensity of the fluorination process of can be written down in the form of speed of the reaction of fluorination:

$$dM / d\tau = \rho_w \cdot V \cdot S,$$

where ρ_w – the true density of metal tungsten and V – on the speed of fluorination reaction.

Suppose the stage of diffusive removal of tungsten hexafluoride from the fluorination surface is the limiting factor and the process of fluorination exerts no visible impact on the general speed of process, then the concentration of the key component (WF_6) on the surface of interaction will be almost equal to a unit. At the same time, the speed of the process of fluorination will be defined by the intensity of external mass transfer:

$$dM / d\tau = \beta_{WF} \cdot (1 - C_{WF}) \cdot S$$

where β_{WF} – diffusion coefficient of the key component of the mix; C_{WF} – concentration of tungsten hexafluoride in the main gas stream.

In case the diffusive and kinetic factors have an identical order of influence on the intensity of the process of fluorination, the concentration of the key component on the surface of interaction is less than the extreme concentration equal to a unit, and the concentration of fluorine on this surface is lower relative to that in the main stream. The concentrations of gas stream components on the surface of fluorination can be determined from the condition of equality between the general speed of fluorination process and the sum of speeds of individual processes:

$$dM / d\tau = \alpha \cdot \beta_F \cdot (C_F - C'_F) \cdot S = \beta_{WF} \cdot (C'_{WF} - C_{WF}) \cdot S = \rho_w \cdot V \cdot S, \quad (2)$$

where C'_F and C'_{WF} – concentration corresponding a mix component on the fluorination surface.

Thus, the mathematical model based on a system of differential equations in partial derivatives, which are similar to continuity equations and transfer equations derived for components of velocity vector, enthalpy and concentration of the key component of the mix, may simulate accurately the processes of producing gaseous tungsten hexafluoride and, given the boundary conditions, may also have numerical solution. Taking into account the balance ratios (2), it is sufficient to obtain algorithm for calculating the speed of reaction of fluorination of V in order to express the concentration of components in the mix of the fluorination surface, using the characteristics of the stream in the region under consideration, with iterations concluding the computational problem.

3 Results and discussion

The formula for calculating the speed of reaction of fluorination [1] is as follows:

$$V = K \cdot C'_{F}, \tag{3}$$

where K – a constant of fluorination reaction speed.

The latter constant will generally have the same value in certain situations. It is pertinent to cite herein the opinion of other workers: "A difference between the temperature on the reacting surface and the temperature of the gas stream will give the reaction speed of tungsten fluorination: the higher the difference of temperatures, the greater the reaction speed" [2]. The table of constants presented in [3] illustrates the dependence of constants of the reaction of fluorination on temperature in the reaction zone (T) on the one hand and the particle size distribution (δ) of metal tungsten powder. A regression analysis was made of the available experimental; on the base of results the following functional dependence was obtained for the constant of fluorination reaction speed:

$$K(\delta, T) = 180 \cdot \delta^{-0.073} \cdot e^{-3970/T}. \tag{4}$$

Using formulas (2), (3) and (4), a closed system of differential equations was obtained which form the basis of the mathematical model for addressing the process of fluorination of metal tungsten. For an assessment the operability of the model can be assessed from the results of test calculations presented in fig. 1.

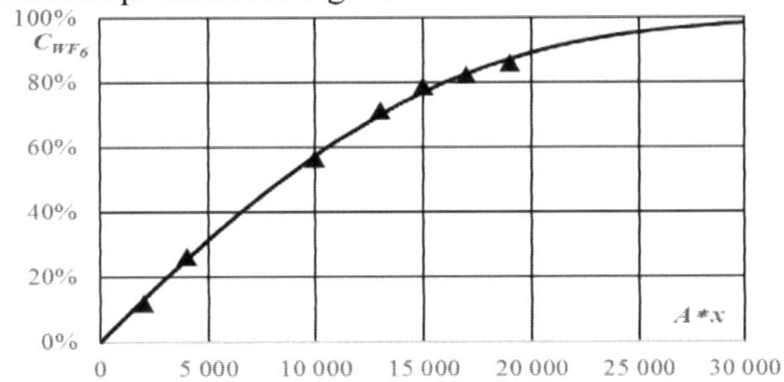

Fig. 1. Concentration of the key component of the mixture from the reaction zone length, *continuous line* – data calculated in the frame of model; ▲ - data reported in [4, 5].

A close match is evidently observed between the calculated and experimental data. This suggests that the proposed model can be applied for choosing optimum conditions of implementation of processes of fluorination of refractory metal powders.

References

1. Yu. M. Korolev, V. I. Stolyarov, *Restoration of fluorides of refractory metals hydrogen* (Moscow, Metallurgy, 1981)
2. Patent No. 2310608 - Way of receiving hexafluoride of tungsten. Authors: A. A. Galatian, A. G. Krupin, S. A. Kuzminykh, V. V. Lazarchuk, E. P. Marinenko, A. I. Mines., V. A. Khokhlov (Date of the publication 11/20/2007)
3. Yu. M. Korolev, *Int. Symposium on Inorganic fluorides: Chemistry and Technology* (TPU Publishing, Tomsk, 2014)
4. T.Sh. Agnokov, Yu.M. Korolev, M.F. Svidersky, and etc., Chemistry and technology of molybdenum and tungsten, Nalchik, KBGU **4** (1978)
5. T.Sh. Agnokov, Yu.M. Korolev, M.F. Svidersky, and etc., Chemistry and technology of molybdenum and tungsten, Nalchik, KBGU **5** (1979)

STUDYING ONE OF THE PROBLEMS OF WAVE FILTRATION

Michael Bubenchikov[1,2], *Elena* Mishcharina[2], and *Eduard* Libin[2,*]

[1]LLC "Gazprom Transgaz Tomsk", 634029 Tomsk, Russia
[2]National Research Tomsk State University, 634050 Tomsk, Russia

Abstract. Motions of particles through a permeable barrier are studied from the standpoint of Schrodinger wave dynamics. We consider time evolution of a substance concentration spot in the vicinity of one or two potential barriers. To solve the problem we developed a computer technology for integration of the non-stationary Schrödinger equation. The technology is based on symbol mapping and the matrix exponential method which was applied earlier to solve the classical heat equation. The developed theory is relevant to development of nanotechnology, in particular, to studies on permeability of nanoporous membranes.

1 Introduction

There is a serious problem in finding the coefficient of permeability for an ultra-thin layer of a nanoporous material. If to use methods of classical molecular dynamics [1-15], we risk getting a disproportionately large number of tests for starting molecules with different initial positions and velocities. In application of the Monte Carlo simulation, we are also confronted with a great volume of work associated with statistical processing of information. All this does not guarantee reliable values for the transmission and reflection coefficients. It is possible to introduce into consideration density of a medium being filtered through a nanoporous material. Then potential (elastic) interactions between the medium and the structure will be described by the wave equation. Additionally, if to note that the square of the wave function module is the density, integration of the nanoscopic Schrödinger equation allows finding the density distribution in front of, within and behind the barrier, and in the end, determining the transmission and reflection coefficients. Therefore, in recent years, there has been a number of works devoted to the use of this approach [16, 17], which in the future can be successfully applied to study permeability of materials composed of grapheme-like structures [18-24].

2 Problem statement

The well-known quantum mechanics [25-27] Schrödinger equation is as follows:

[*] Corresponding author: hudobina@mail2000.ru

$$\frac{\hbar}{i}\frac{\partial\psi}{\partial t}+U(x,y,z)\psi-\frac{\hbar^2}{2m}\Delta\psi=0. \quad (1)$$

With proper choice of scales for the variables x and t, it is also recorded in normalized form:

$$\frac{\partial\psi}{\partial t}=i\big(\Delta-U(x,y,z)\big)\psi. \quad (2)$$

It is not difficult to note that equation (2) is closely connected with the heat equation of the form:

$$\frac{\partial\theta}{\partial t}=\big(\Delta-U(x,y,z)\big)\theta. \quad (3)$$

The connection between the ψ and θ functions is that if we know any solution of equation (3) written in the form of: $\theta=\theta(x, y, z, t)$, the associated solution of equation (2) has the form: $\psi=\theta(x, y, z, it)$, which is a simple matter of replacing the real-time t by its imaginary value it. Additionally, the function ψ is always a complex value, even if the function θ is real. The square of the ψ function module in quantum mechanics is interpreted as the probability of particle localization in the elementary volume, and, therefore, the value of ψ is called the probability amplitude.

If to eliminate the time form equation (2), assuming that $\psi=\Psi e^{-ikt}$, for the function Ψ we obtain the stationary Schrödinger equation of the form:

$$\Delta\psi+\big(k^2-U(x,y,z)\big)\psi=0. \quad (4)$$

This equation does not contain the time and is managed more easily than equation (2) and it is used mainly for describing passages of particles through a potential barrier or for finding its own values and functions of the operator. The difficulty of analytical or numerical solutions of non-stationary equations (2) or (3) occurs because of the variable coefficient $U(x, y, z)$ present in these equations. Thus, the task is to choose an effective method of solving heat equation (3).

In the one-dimensional case, where the function ψ depends only on one spatial argument x, equations (2) and (3) can be written as:

$$\frac{\partial\psi}{\partial t}=i\left(\frac{\partial^2}{\partial x^2}-U(x)\right)\psi, \qquad \frac{\partial\theta}{\partial t}=\left(\frac{\partial^2}{\partial x^2}-U(x)\right)\theta. \quad (5)$$

At any arbitrarily chosen potential barrier $U(x)$, equations (5) can not be solved in an explicit analytic form. Therefore, to determine the general behavioural nature of the solutions, let us consider the case where there is no potential barrier, i.e. where $U(x)=0$ and equations (5) are of the form:

$$\frac{\partial\psi}{\partial t}=i\frac{\partial^2\psi}{\partial x^2}, \qquad \frac{\partial\theta}{\partial t}=\frac{\partial^2\theta}{\partial x^2}. \quad (6)$$

3 Calculations by matrix exponential method

The solution of the homogeneous heat equation of the form (6) can be formally written in the symbolic form, bearing in mind the expansion of the exponential differential operator in the Taylor series:

$$\theta(x,t) = e^{t\frac{\partial^2}{\partial x^2}}\theta(x,0) = \left[1 + t\frac{\partial^2}{\partial x^2} + \frac{t^2}{2!}\frac{\partial^4}{\partial x^4} + \frac{t^3}{3!}\frac{\partial^6}{\partial x^6} + \frac{t^4}{4!}\frac{\partial^8}{\partial x^8} + \cdots\right]\theta(x,0)\,. \quad (7)$$

Thus, for example, it is not difficult to verify directly that if $\theta(x, 0)$ is the polynomial in the variable x, then the row in formula (7) is cut off and we get the exact solution of heat equation (6). On the other hand, in case of numerical implementation of the solution, the differentiation operator with respect to variable x is usually replaced by its finite-difference approximation:

$$\frac{\partial^2\theta}{\partial x^2} \approx \frac{\theta_{n-1} - 2\theta_n + \theta_{n+1}}{h^2}\,. \quad (8)$$

Thus, the differentiation operator is the three-diagonal square matrix H_2 acting on the column vector consisting of numbers θ_n, n=1, 2, 3, …N. In this case, bearing in mind that θ is the column vectors, we can simply write solution (7) in the vector-matrix form:

$$\theta(x,t) = \mathrm{expm}(tH_2)\cdot\theta(x,0)\,. \quad (9)$$

The function of the matrix $f(\mathrm{A})=e^{\mathrm{A}}$ known as the matrix exponential is a standard program in a number of computer systems, and it can be used in solving more general equations (5) which take into account the effect of the potential barrier $U(x)$. It is only necessary to subtract the diagonal matrix $\mathrm{U}(x_i)$ from the differential operator H_2. Then we get:

$$\psi(x,t) = \mathrm{e}^{itH}\,\psi(x,0), \quad \theta(x,t) = \mathrm{e}^{itH}\,\theta(x,0), \quad \mathrm{H} = \mathrm{H}_2 - diag(U)\,. \quad (10)$$

Let us note the following properties of calculating the Schrödinger equation by the matrix exponential:

1. This method takes into account the presence of potential barriers in the Schrödinger equation, which is practically impossible to implement using analytical methods.
2. Time is included as a parameter directly in the matrix exponential, so formally, integration of the evolutionary coordinate is completed.
3. Calculating the matrix exponential H at N <300 takes a time of about 3-5 seconds. Thus, this method possesses high speed and accuracy.

Automatic execution of the zero boundary condition at the ends of the interval is due to the choice of the matrix H_2 as formula (8).

4 Calculation results

The figures below show the results of calculations in our program for some cases of potential barriers location in this area of solution. They present a diagram of probability amplitude distribution and its average value over time.

DOI: 10.1051/matecconf/20167201019

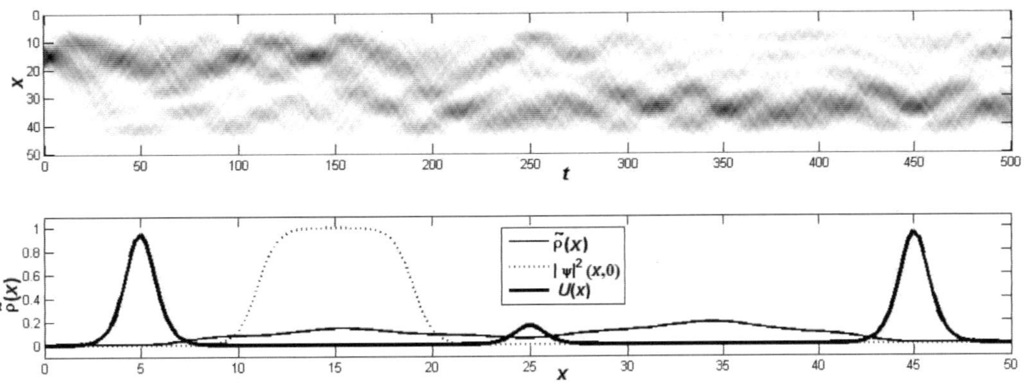

Fig. 1. Probability amplitude distribution and its average value over.

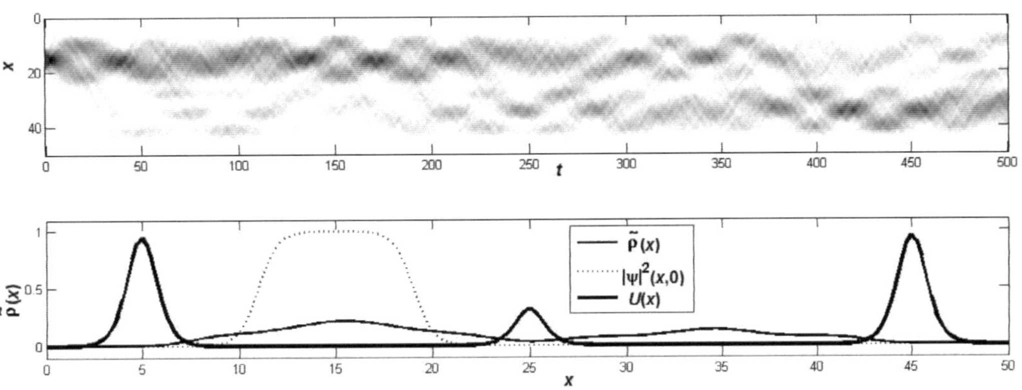

Fig. 2. Probability amplitude distribution and its average value over time.

The dots show the initial distribution of particle density, the solid line indicates the energy of the barriers. We examine permeability of the central barrier which has a significantly smaller size. The peripheral barriers only restrict movement of particles. Initially, the entire mass of particle was located to the left of the central barrier. As can be seen from the average density distribution (fig. 1), more than half the mass went over the barrier. In the second example (fig. 2) the transmission coefficient was lower. Thus, the application of the matrix exponential methods to solving the non-stationary Schrödinger equation showed the expected results when calculating the probability amplitude. This technology can be extended to two-dimensional problems which are of a greater scientific interest with respect to explanations of the absorption phenomenon in such problems.

Acknowledgments

This paper was written within the frame of Competitiveness Improvement Program of Tomsk State University and with financial support of the Russian Science Foundation, grant under Agreement No. 16-19-00089.

References

1. V.Ya. Rudyak, S.L. Krasnolutskii, D.A. Ivanov, J. Dok. Phys. **57** (2012)
2. V.Ya. Rudyak, A.A. Belkin, V.V. Egorov, D.A. Ivanov, Int. J. Multiph. **5** (2011)
3. V.Ya. Rudyak, S.L. Krasnolutskii, D.A. Ivanov, J. Microfl. Nanofl. **11** (2011)

4. V.Ya. Rudyak, S.N. Dubtsov, A.M. Baklanov, Tech. Phys. Let. **34** (2008)
5. V.Ya. Rudyak, S.L. Krasnolutskii, E.N. Ivashchenko, J. Eng. Phys. Thermoph. **81** (2008)
6. V.Ya. Rudyak, J. VDI-Berichte **1803** (2003)
7. V.Ya. Rudyak, A.A. Belkin, Tech. Phys. Lett. **29** (2003)
8. A.I. Potekaev, M.A. Bubenchikov, Russ. Phys. J. **54** (2011)
9. M.A. Bubenchikov, Russ. Phys. J. **54** (2011)
10. M.A. Bubenchikov, A.I. Potekaev, A.M. Bubenchikov, Russ. Phys. J. **56** (2013)
11. M.A. Bubenchikov, A.I. Potekaev, A.M. Bubenchikov, Russ. Phys. J. **56** (2013)
12. A.I. Potekaev, A.M. Bubenchikov, M.A. Bubenchikov, Russ. Phys. J. **55** (2013)
13. M.A. Bubenchikov, A.I. Potekaev, A.M. Bubenchikov, O.V. Usenko, A.V. Ukolov, IOP Conf. Series: Mat. Sc. Eng. **87**, 1 (2015)
14. A. M. Bubenchikov, M.A. Bubenchikov, O.V. Matvienko, E. A. Tarasov, O.V. Usenko, AIP Conf. Proc. **1698** (2016)
15. M. A. Bubenchikov, A. M. Bubenchikov, O. V. Usenko, V. A. Poteryaeva, S. Jambaa EPJ Web Conf. **110** (2016)
16. A.M. Bubenchikov, M.A. Bubenchikov, A.I. Potekaev, E.E. Libin, Yu.P. Khudobina, Russ. Phys. J. **57** (2014)
17. A.M. Bubenchikov, M.A. Bubenchikov, A.I. Potekaev, E.E. Libin, Yu.P. Khudobina, Rus. Phys. J. **58** (2015)
18. K.A. Krylova, Y.A. Baimova, S.V. Dmitriev, R.R. Mulyukov, Phys. Sol. St. **58** (2016)
19. A.V. Savin, E.A. Korznikova, S.V. Dmitriev, Phys. Sol. St. **57** (2015)
20. K.A. Bukreeva, R.I. Babicheva, S.V. Dmitriev, K. Zhou, R.R. Mulyukov, A.I. Potekaev, Russ. Phys. J. **57** (2014)
21. J.A. Baimova, B. Liu, S.V. Dmitriev, N. Srikanth, K. Zhou, Phys. Chem. Chem. Phys. **16** (2014)
22. E.A. Korznikova, S.V. Dmitriev, J. Phys. D: Appl. Phys. **47** (2014)
23. E.A. Korznikova, S.Y. Fomin,S .V. Dmitriev, Mat. Sc. Forum **845** (2016)
24. J.A. Baimova, S.V. Dmitriev, Mat. Sc. Forum **838** (2016)
25. H. Feshbach, Ph. M. Morse, *Methods of Theoretical Physics. Part II* (McGraw-Hill, New York, 1953)
26. L. D. Landau, E. M. Lifshitz: *Quantum Mechanics: Non-Relativistic Theory, Vol. 3* (Pergamon Press, Oxford, 1965)
27. J.W.S. Rayleigh, *The Theory of Sound, Vol.II* (Macmillan, London, 1878)

THE TEMPERATURE OF THE DIMETHYLHYDRAZINE DROPS MOVING IN THE ATMOSPHERE AFTER DEPRESSURIZATION OF THE FUEL TANK ROCKETS

Elena Bulba[1,*], *Tatiana* Nemova[2], and *Anna* Dyatchina[1]

[1]National Research Tomsk Polytechnic University, 634050 Tomsk, Russia
[2]Tomsk State Architecture and Civil Engineering, 634050 Tomsk, Russia

Abstract. This work includes the results of the numerical modeling of temperature changes process of the dimethylhydrazine (DMH) drops, taking into account the radial temperature gradient in the air after the depressurization of the fuel compartments rockets at high altitude. There is formulated a mathematical model describing the process of DMH drops thermal state modifying when it's moving to the Earth's surface. There is the evaluation of the influence of the characteristic size of heptyl drops on the temperature distribution. It's established that the temperatures of the small size droplets practically completely coincide with the distribution of temperature in the atmosphere at altitudes of up to 40 kilometers.

1 Introduction

The Liquid propellant rocket engines, most likely, will be the main engines for the heavy launch vehicles in the next 10-15 years. The main negative factor in the operation of such engines is their fuel - unsymmetrical dimethylhydrazine (heptyl) which may fall on the Earth's surface after the depressurization of the engine exhaust fuel tanks [1]. Until recently, there is not a valid opinion on the possibility of heptyl contamination of soil and water of the territories over which the exhaust engine is separated from the spacecraft, launched into outer space. There are attempts to modeling processes of atmospheric diffusion of heptyl vapors and droplets in the surface layers of the atmosphere. [2,3] But the problem of evaporation of heptyl in the atmosphere while driving to the ground in a production that takes into account all the major significant factors not solved so far. One of such factors is the thermal state of this liquid fuel droplets in the air (the surface temperature determines the rate of evaporation of heptyl [3]). It is known the solvings of the problems of the temperature field of this fuel droplets by assuming its homogeneity [2] and in the description of the temperature distribution in a drop using the heat equation in partial derivatives [3]. But these studies [3] were carried excluding the complex [4]

* Corresponding author: bulba@tpu.ru

distribution of temperature over the height of the atmospheric layer, which must have an important role in the formation of a drop temperature field and, consequently, the evaporation rate of heptyl.

2 Statement of the problem

The mathematical formulation of the problem of the formation of a small size spherical form drop temperature field is reduced to the solution of the heat equation for a sphere in the form of

$$\rho c\frac{\partial T}{\partial t}=\frac{1}{r^2}\frac{\partial}{\partial r}\left(\lambda r^2\frac{\partial T}{\partial r}\right), \tag{1}$$

with the initial condition

$$t=0, \quad T=T_0 \tag{2}$$

and with the boundary conditions

$$r=R_{к}: -\lambda\frac{\partial T}{\partial r}=\alpha(T^e-T),\ t>0,\ \alpha>0, \tag{3}$$

$$r=0 \quad \frac{\partial T}{\partial r}=0, \tag{4}$$

Where ρ - density of droplets, c - the specific heat capacity, λ - coefficient of thermal conductivity, r - radial coordinate, $R_{к}$ - outer radius of the droplets, T - temperature, T_0 - initial value of the drop temperature, T^e - ambient temperature, α - heat transfer coefficient.

The communication between ambient temperature and height [4] is presented in Table 1.

Table 1. Changing atmospheric temperature with height.

Outside Temperature, K	220	220	233	250	273	255	233	206
Height, m	10000	20000	30000	40000	50000	60000	70000	80000

The calculation of the drop temperature is necessary to determine the speed of the fall of v in the equation of its motion, taking into account the effect of air resistance. It is also necessary to know the velocity v to calculate the heat transfer coefficient α. The equation of motion of the spherical form drop is as follows:

$$\frac{dv}{dt}=g-\frac{3C_D\rho_{в}}{4d\rho_{г}}(v-v_{в})^2, \tag{5}$$

with the initial condition

$$t=0, \quad v=0, \tag{6}$$

where v - velocity of the fall of drops; $v_{в}$ - air velocity; $\rho_{в}$ - density of the air pollution; $\rho_{г}$ - the density of the liquid heptyl; d - droplet diameter; C_D - coefficient of resistance; g - acceleration of gravity.

The rate of air flow is assumed to be 0.

The drag coefficient of the particle C_D in the turbulent regime accepted constant value of 0.44. In general, any liquid droplet in flight by gravity force is deforming. In a process of deformation the various forms of a drop are possible (sphere, ellipsoid, "pancake", "pear" et al. [6, 7]). In experiments [6,7] was found that even during the time of relatively small flight distances of 1-2 meters the water, kerosene and acetone droplet goes through several cycles of such deformations changing its form. But with the growth speed of flight any liquid droplets are increasingly taking the form of a sphere. At the same time, less is the characteristic size of the drops, so the greater flight time it retains the shape of a sphere. That's why the spherical shape of the drop and the corresponding coefficient of resistance are adopted for the mathematical modeling. The heat transfer coefficient from the particles obtained values fall rate was calculated from the expressions

$$\mathrm{Nu} = 2 + 0,27\,\mathrm{Re}^{0.62}\,\mathrm{Pr}^{0.33}, \qquad (7)$$

$$\alpha = \frac{\lambda \mathrm{Nu}}{d}. \qquad (8)$$

There Nu, Re, Pr – are Nusselt number, Reynolds number, Prandtl number, respectively. Prandtl number was assumed to be 0.7, and the Reynolds number was calculated using the equation obtained from (4) the values of the velocities for the respective points in time. The kinematic viscosity coefficient is adopted in accordance with the tabulated data [4].

3 Results and discussion

The results of numerical solution of the problem (1) - (8) are shown in figure 1. There are presented the dependences of a heptyl drops temperature on a height of three specific sizes. It also includes a comparison with the results obtained previously in [2], excluding the radius droplet temperature gradient.

The numerical analysis shows that at small droplet sizes characteristic (curves 4, 6) there is a coincidence of a heptyl and air temperature over the entire height of incidence (curve 3). For large drops (curve 1), these temperatures are significantly different. Therefore, we can conclude that the temperature of characteristic small size droplets can be considered identical to the air temperature at the respective altitude. But for the larger heptyl drops, with greater thermal inertia and a high-speed fall, it is necessary to take into account the distribution of temperature drop by drop.

The comparisons with the previous [2] results (figure 1) shows that at small characteristic size of drops (curves 4, 6, and curves 5, 7, curve 3), their temperatures are almost identical. For the large heptyl droplets model (1) - (8) clarifies established in [2] the temperature distribution.

The theoretical investigation illustrates the thermal state of the heptyl drops driving in the atmosphere. After the depressurization of the fuel tanks of launch vehicles spacecraft airborne cloud descending to the surface of the Earth is changing. The larger drops, flying some distance, are spliting into smaller ones. The calculation results show that the larger is the droplet size , then the higher is the temperature and a possibly rate of evaporation at high altitudes. Small droplets also fall at a slower rate. Heptyl cloud, which may be imposed on long distance, is forming at high altitudes.

To derive the possibility of heptyl drops precipitation on the surface of the Earth is necessary to estimate the rate of evaporation of the fuel at the set value of the drops temperature at different heights. This can be done using the device [8.9]. But for modeling heptyl evaporation process it requires the data on the coefficients of evaporation [8] of the liquid fuel, which has not yet been established. Therefore, one of the tasks, the solution of which is necessary to conclude about the possibility of loss of heptyl in the Earth's surface, is to define the phase transformation characteristics of this fuel in the temperature range corresponding to the Earth's atmosphere.

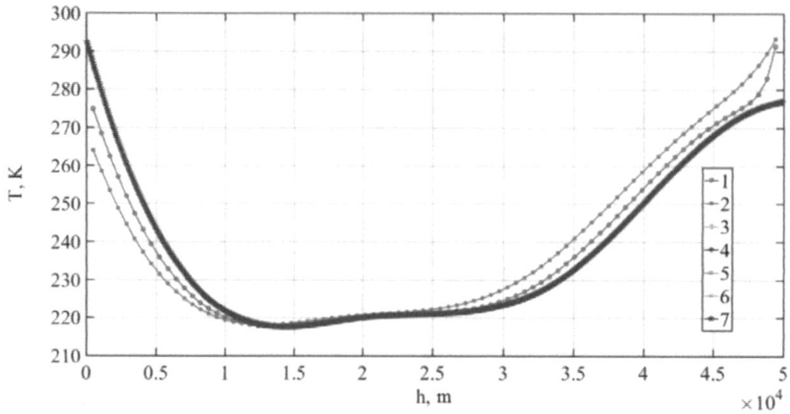

Fig. 1. The temperature of a heptyl spherical droplet with a characteristic size: 1 - d = 10 mm, taking into account the temperature gradient along the radius; 2 - d = 10 mm without radial temperature gradient; 3 - the temperature drop over the entire height; 4 - d = 1 mm, taking into account the radial temperature gradient; 5 - d = 1 mm without the temperature gradient along the radius; 6 - d = 1 mm, taking into account the temperature gradient along the radius; 7 - d = 1 mm without the temperature gradient along the radius.

4 Conclusion

The resulting numerical studies of the temperature distribution in height of falling drops of DMH based on the radius of the spherical particles of the temperature gradient for the large drops illustrate the effect of the characteristic size on the thermal processes occurring in the droplets. It is found that the temperature of the small size droplets almost completely coincide with the air temperatures over the entire height of heptyl incidence.

Acknowledgments

The work was supported by the Russian Federation President's grant (NSH project - 7538.2016.8).

References

1. E.L. Aleksandrov, *Ecological problems of the effects and risks of space engineering on the environment: A Reference Guide* (ANKIL, Moscow, 2000)
2. G.V. Kuznetsov, T.N. Nemov, E.E. Bulba, Izv.vuzov. Physics **6** (2006)
3. A.E. Dolotov, G.V. Kuznetsov, T.N. Nemova, Izv.vuzov. Physics **4** (2008)
4. Standard aktmosfera. Options. - M .: State Committee of the Council of Ministers of the USSR standards (1977)
5. V.I. Terehov, M.A. Pahomov, *Heat and mass transfer and hydrodynamics in a gas-drop flow* (Monograph NGTU. Novosiboirsk, 2008)
6. R.S. Volkov, G.V. Kuznetsov, P.A. Strizhak, Int. J. Heat Mass Transfer **85** (2015)
7. R.S. Volkov, O.V. Vysokomornaya, G.V. Kuznetsov, P.A. Strizhak, Tech. Phys. Lett. **41**, 2 (2015)
8. G.V. Kuznetsov, P.A. Strizhak, J. Eng. Thermophys. **17**, 3 (2008)
9. R.S. Volkov, G.V. Kuznetsov, P.A. Strizhak, Tech. Phys. Lett. **59**, 7 (2014)

NUMERICAL MODELLING OF HEAT TRANSFER DURING IMPACT OF A MOLTEN DROPLET ON A SURFACE

Roman O. Cherepanov[1,*], and *Alexander* V. Gerasimov[1]

[1]EDP National Research Tomsk State University, 634050 Tomsk, Russia

Abstract. SPH-based numerical technique for modelling of impact of molten drops on a surface with heat transfer and phase transitions effects is proposed. Computational algorithm uses SPH with procedure of restoring of particle consistence and variational approach to calculation of acceleration field. Also, boundary algorithm for free and contact surfaces in 3D setting are developed.

1 Introduction

In plasma coating techniques metal powder is heated to a molten state and the flow of drops is directed onto the surface of tool. Molten droplets spread out over the surface, cool and solidify during interaction. The particles solidify to form a coating deposit on the substrate's surface. It is important to have a clear understanding of the physics of process and to have a models of droplet impact on the surface of substrates for better control of the generation of splats and the structure of coating. Experiments are used to get a link between factors related to droplet impact and final morphology and quality of coating, but this is an expensive way. Numerical models are more preferable for prediction of coating quality and properties, but it need a good computational model, able to treat high deformation, liquid flow and solidification, heat transfer and contact interaction of droplet and surface. Important side is ability to make modeling in 3D, because of significant loose of information in simplified 2D-setting.

2 Computational Method

To calculate the parameters of splats, various approximation methods are used with different degrees of simplification. So, the representation of drops in the form of the equivalent cylinders provides a way to semi-analytical solution of the problem. Computational fluid dynamics or numerical method of solid mechanics lead to computer aided physical modeling in 3D setting.

[*] Corresponding author: rcherepanov82@gmail.com

In this paper we propose the use of smooth particle method (SPH) [1], adapted to the modeling of the deformation process of the liquid droplet with a phase transition at crystallization in a large range of speeds.

Basis of SPH approximation is equation

$$f^i \cong \int f(\vec{x})W(\vec{x}-\vec{x}^i,h)d\vec{x} \tag{1}$$

where h is smoothing parameter, which define a radius of fluency for points, $\vec{x}$ – is a space coordinate, W - smoothing function.

Spatial derivatives are defined via:

$$f^i_{,\alpha} = \frac{\partial f^i}{\partial x_\alpha} \cong \int f(\vec{x})W_{,\alpha}(\vec{x}-\vec{x}^i,h)d\vec{x} \tag{2}$$

Corresponding to (2) *particle approximation* is written as:

$$f^i_{,\alpha} = \frac{\partial f^i}{\partial x_\alpha} \cong \sum_k f^k W_{,\alpha}(\vec{x}^k-\vec{x}^i,h)\frac{m^k}{\rho^k} \tag{3}$$

where $\vec{x}^k, f^k, m^k, \rho^k$ – radius-vector, approximated function value, mass and density at k-th point.

Equation (3) have C^0-consistency [2] near surfaces and boundaries and at non-uniform particle distribution, and for restoring particle consistency special approaches needed. According [2], a test vector-function is introduced in form:

$$\Delta(\vec{x}) = (1, x_0, x_1, x_2) \tag{4}$$

Approximation of test function (1) and its derivatives (2) can be found with C^0-consistency, but due to the fact that this values are known, this approximation can be used to construct correction procedure and restore particle consistency. Let's define :

$$\Delta(\vec{x})_\alpha = \begin{cases} x_\alpha; \alpha > -1; \\ 1; \alpha = -1; \end{cases} \tag{5}$$

Matrix of approximations for this function at n-th point, found via uncorrected SPH approximation (3):

$$T_{\beta\alpha}(\vec{x}^n) = \sum_m \Delta_\alpha(\vec{x}^m - \vec{x}^n)W_{,\beta}(\vec{x}^m - \vec{x}^n, h)\frac{m^k}{\rho^k}; \tag{6}$$

Lets define correction matrix as

$$B^n_{\alpha\beta} = B_{\alpha\beta}(\vec{x}^n) = \left[T_{\alpha\beta}(\vec{x}^n)\right]^{-1}; \alpha, \beta = -1,0,1,2; \tag{7}$$

Now, corrected approximation of function *f(x)* value or its derivatives can be found, if we build full matrix of uncorrected approximations (3) and apply correction matrix (7):

$$f^i_{,\alpha} = \frac{\partial f^i}{\partial x_\alpha} \cong T^i_{\alpha\beta} \cdot \left\{\sum_k \frac{m^k}{\rho^k} f^k W_{,\beta}(\vec{x}^k - \vec{x}^i, h)\right\} \tag{8}$$

For calculation of flow of liquid and solid phases, approximation (8) applied to lagrangian of system:

$$L = \sum_k m^k \left(\frac{\vec{v}^k \vec{v}^k}{2} - u^k\right) \tag{9}$$

where $\vec{v}^k$ - is particle speed, u^k - is internal energy (per mass). In both phases (liquid and solid) internal energy change can be written in form:

$$\frac{du^k}{dt} = \sigma^k : \dot{\varepsilon}^k \tag{10}$$

This implies:

$$\frac{dL}{dt} = \sum_k m^k \left(v^k a^k - \sigma^k : \dot{\varepsilon}^k \right) \tag{11}$$

Since the relation between a tensor of rate of deformation and velocity of nodes is linear:

$$\dot{\varepsilon}^k = \frac{1}{2}\left(\left(\frac{\partial v_i^k}{\partial x_j} \right) + \left(\frac{\partial v_i^k}{\partial x_j} \right)^T \right) = Symm\_part\left(T_{j\beta}^k \sum_n v_i^n \frac{m^n}{\rho^n} W\left(\vec{x}^n - \vec{x}^k, h \right)_{,\beta} \right) \tag{12}$$

(11) leads to a relation between the accelerations of nodes and stress field:

$$a_\gamma^n = -\left(\sum_k \frac{m^k}{\rho^k} \sigma_{\gamma j}^k T_{j\beta}^k W_{,\beta}^{kn} \right) / \rho^n \tag{13}$$

For calculus of heat transfer (8) is applied to heat transfer equation [3]

$$\frac{dT}{dt} = \frac{-\left(\lambda T_{,j} \right)_{,j} + Q}{C_P \rho} \tag{14}$$

Heat production of liquid is caused by viscous stress and defined as:

$$Q_{liquid} = \sigma : \dot{\varepsilon} \tag{15}$$

Heat production of solid is caused by plastic strain:

$$Q_{solid} = \sigma : \dot{\varepsilon}^{pl} \tag{16}$$

where $\dot{\varepsilon}^{pl}$ - rate of plastic strain, found via Mizes yield condition. For calculation of phase transition during solidification, latent heat is introduced to equation (14) via additional heat source, which operates when material cooled to temperature of melting.

Boundary condition for heat transfer is heat insulation on free surfaces and ideal heat contact at droplet-surface interface. For impact problem we use a condition of zero normal stress at free surface and equal normal stress and equal speeds on contact surface.

In described method this boundary conditions does not need additional operations, such as "ghost particles" or special integration procedures for (2) to treat a boundary.

Smoothing length h was selected to provide sufficient count of neighbors for each particle for correction matrix to be well defined. Usually neighbor count was from 11 to 16.

Time step is defined via Courant–Friedrichs–Lewy condition: $\Delta t < \min\left[h / C \right]$, where C is the sound speed. At practice stability condition for heat transfer problem is more weak while droplet radius exceed 1 μm.

3 Results

A 38 μkm radius droplet of molten stainless steel is impacting with a velocity of 200 m/s on stainless steel substrate. Parameters of steel: density 7.87 g/cm3, E=200 GPa, G=70 GPa, Ypl=1.2 MPa, viscosity of molten steel 5.4 mPa*s, heat conductivity 28 Wt/m/K, heat capacity 0.56 kJ/kg, temperature of melting 1280 ^{0}C, latent heat 84 kJ/kg, initial temperature of droplet 1380 ^{0}C, initial temperature of substrate 0 ^{0}C.

Fig. 1. Impact of molten droplet on substrate. t=70 ns (a), t=140 ns (b), start of solidification at t=140 ns: solidificated points are marked with dark grey, solid- red and melted with lime (c).

As it shown at fig. 1, solidification of melted material starts at contact surface approximately 140 ns after contact and significant fragmentation of droplet due to high initial speed is observed. Energy conservation during calculation doesn't exceed 0.1 %, full change of momentum doesn't exceed 0.01 %.

Acknowledgments

This work was supported by the Russian Science Foundation (RSF) No. 16-19-10264.

References

1. L.B. Lucy, Astronomical J. **82** (1977)
2. M.B. Liu, G.R. Liu, Applied Numerical Mathematics **56**, 1 (2006)
3. M. Biot, J. Math. Mech. **15**, 2 (1966)

EFFECT OF THE REBURNING ZONE STOICHIOMETRY ON THE NO_X CONCENTRATION AT THE THREE-STAGE COMBUSTION OF PULVERIZED COAL

Nelya Chernetskaya[1,*], *Mikhail* Chernetskiy[1,2], and *Andrey* Zhuikov[1]

[1]Siberian Federal University, 660074 Krasnoyarsk, Russia
[2]Institute of Thermophysics SB RAS, 630090 Novosibirsk, Russia

Abstract. Numerical study of heat and mass transfer taking into account the combustion of coal particles in the furnace at the three-stage combustion of pulverized coal was performed. Analysis of the reburning zone stoichiometry on the concentration of nitrogen oxides at the furnace outlet was made. The values of excess air in the primary and reburning combustion zones, providing for the concentration of nitrogen oxides at the furnace outlet is not more than 350 mg/m^3 and unburned carbon not more than 1 % when burning coal with a high content of nitrogen were established.

1 Introduction

One of the main atmosphere polluters, greatly impacting on human health at the coal combustion in thermal power plants boilers are nitrogen oxides. Nitrogen oxides occurring during coal combustion for the mechanism of formation are divided into thermal, fast, and fuel. Thermal NO_x results from the reaction of atmospheric nitrogen with oxygen at temperatures above 1500 °C. Prompt NO_x is generated in the presence of hydrocarbon radicals, it prevail in fuels with high value of the hydrogen to carbon H:C ratio. Fuel NO_x contributes to the total quantity of nitrogen oxides formed during flaring coal. Fuel nitrogen contained in coal, partly comes out with the volatile with the remaining part appearing during the combustion of coke residue.

Applying technological methods to the combustion process allows significantly reduce the nitrogen oxides concentration in exhaust gases of boilers. Works [1-7] are known for the two most effective technological methods of NO_x reduction: applying low-NO_x burners (LNBs) and, two and three-stage combustion.

Three-stage combustion (reburning) is the most complex process, but can give a significant reduction in the nitrogen oxides concentration. As shown by laboratory [8-11], as well as a number of industrial tests, this scheme of combustion permits reduction the nitrogen oxides emissions by up to 50 %. The use of these methods can lead to increased unburned carbon-in-fly ash at the furnace outlet. Therefore, the reduction of unburned

*Corresponding author: micch@yandex.ru

carbon-in-fly ash at the furnace outlet, in addition to reducing NO_x concentrations, is one of the criteria of these methods.

Computational fluid dynamics methods are essential part of the work during implementation of reburning. Models describing the rate of coal burnout and the nitrogen oxides formation play a significant role in the accuracy of the results. The complexity of these processes explained, in particular due to their close relationship with the physical and chemical phenomena, taking place in the furnace [13]. Therefore, in spite of a number of works in this direction [4, 6, 12], the choice of the mathematical model, which would ensure correct results, remains relevant.

This paper presents the influence of reburning zone stoichiometric ratio on the NO_x concentration and unburned carbon-in-fly ash by mathematical modeling.

2 Problem statement and research methods

As research object selected combustion chamber unit at supercritical steam parameters with capacity of 660 MW with a steam capacity of 1830 t/h at a pressure of 28 MPa and temperature of 600 °C [14]. This boiler is a pilot project of Russian developers. Project fuel for this boiler is Kuznetskiy coal, the composition and characteristics of whose are presented in table 1.

Table 1. Composition and characteristics of coal.

Consist, %							Q^r_i, kcal / kg	V^{daf}, %	R_{90} %
W^r	A^r	C^r	S^r	H^r	N^r	O^r			
11.0	18.92	53.92	0.43	3.84	1.71	10.18	5000	41.0	18

The furnace with solid slag removal for the boiler has a prismatic shape (18.7×18.7 m) with corner tangential placement of the burners. Given the stringent requirements for NO_x emission and taking into account the high nitrogen content in the fuel, to achieve the minimum NO_x concentrations provides a furnace scheme with the implementation of reburning (fig. 1a). Burners are placed in three tiers. Each burner has two nozzles of coal-air mixture. The reburn burners are tangential location. OFA nozzles are arranged

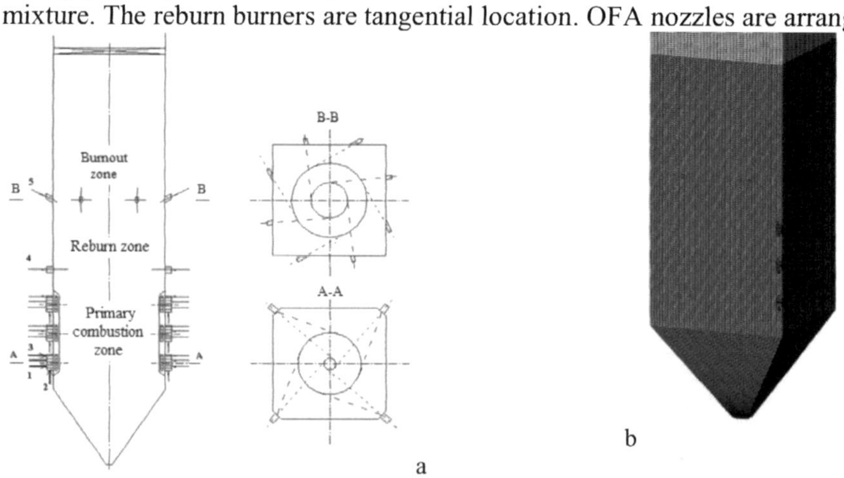

Fig. 1. The geometry of the combustion chamber. a) furnace scheme. 1 – coal-air mixture, 2 – secondary air, 3 – gas recirculation, 4 – reburn fuel, 5 – OFA б) computational mesh of the furnace.

tangentially on the furnace walls with the direction of twist opposite to the main torch. In the draft of the boiler, to the combustion chamber imposes restrictions on concentration of nitrogen oxides - not more than 350 mg/m^3.

The experimental studies [2, 7, 8] showed that one of the main parameters having a significant influence on the nitrogen oxides concentration at the outlet of the furnace with reburning technology is the reburning zone stoichiometry. When reducing the stoichiometric ratio in the reburning zone on NO_x concentration at the outlet of the furnace is reduced, the amount of unburned carbon in the fly ash is increasing.

The main aim of this research is the numerical investigations of stoichiometric ratio influence at the furnace outlet the regenerative combustion zone on the concentration of NO_x and unburned carbon- in-fly ash.

The computational were performed using the in-house finite-volume block-structured CFD package SigmaFlow. This CFD-package has been successfully used for solving the optimization process in the furnace [15, 16]. The complete combustion model considers the process as a flow of incompressible nonisothermal multicomponent reacting gas-particle mixture, using Euler- Lagrangean frames for the gas-phase and particles respectively. The field variables are described by Reynolds averaged equations for conservation of mass, momentum, enthalpy and mixture fraction of species concentration. The gaseous phase is treated as a single-phase fluid in which the effects of the dispersed reacting coal particles on the gaseous phase were accounted for by the PSI-Cell method. The RANS equations are closed with the standard high-Re-number k-ε turbulence model. More details about the model and numerics of SigmaFlow can be found in Chernetskiy et al. [17]. Calculation of volatile fuel components combustion is based on the use of global irreversible reactions between fuel and oxidant. To describe the reaction in turbulent flows with large mixing time used a hybrid model using the kinetic model and the eddy break up model to determine the reaction rate. The modeling of radiant energy transfer is conducted basing on the P1 approximation of spherical harmonics for a grey medium. The advantage of this method is easiness of its matching with methods of aerodynamics and heat transfer calculation realized in curvilinear meshes. The absorption coefficients were calculated using the weighted-sum-of-gray-gases model. The Lagrange method was used in the present work to describe of particle motion. Coal particle devolatilization as well as char combustion are approximated by one- component schemes using a kinetic-diffusion model. In the modelling of NO_x formation two mechanisms are taken into account: formation of thermal NO_x (according Zeldovich's model [18]), formation of prompt NO_x and formation of fuel NO_x, (Chen [19] and DeSoute [20]).

Conservation equations for gas phase are written down in a generalized conservation law in a control volume. For the volume finite-difference analog of equation is written down. For calculation of diffusion flow on the face of control volume centrally-difference scheme with second order precision is employed. At the approximation of convective terms Leonard's scheme is employed that is substantially minimizes the circuit viscosity. For connection velocity and pressure fields SIMPLE-C procedure is employed.

3 Results and discussion

Geometry and computational mesh of the combustion chamber are shown in figure 1. To perform the computational studies it is accepted that the main burner is supplied 80 % of the fuel, on the reburn zone of 20 %. Distribution of coal, air and gases recirculation main burners is uniform. The excess air coefficient at the exit from the active combustion zone (main burner) ~ 1.1, excess air coefficient in the recovery zone was varied in the range 0.9 – 1.05, and at the furnace outlet is 1.2.

According to the results of computational studies established that the tangential burner ensures the formation of a central vortex in the center of the furnace and secondary vortices along the walls. The gases flow through nozzles with the reverse twist flow, relative to the main burners, ensures good mixing of combustion products with coal-air flows.

Figure 2 shows temperature fields in the central section of the furnace at different excess air ratio in the reburn zone. For all variants, in the primary combustion zone, the temperature near the walls does not exceed 1000 °C, which is admissible for the Kuznetsky coal from the point of view of heating surfaces slagging. While decreasing the excess air ratio at the exit of the reburn zone, increases the amount of air supplied to the OFA nozzles for afterburning products of incomplete combustion. This leads to differences in the temperature field in the OFA zone. But in general, for all variants, there is a homogeneous temperature distribution in horizontal cross section and incident radiative fluxes along the perimeter of the combustion chamber, ensuring excellent mixing of the overfire air and the upward coal-gas-air flow and a good fuel burnout. The maximum combustion temperature reaches 1650 °C. At the outlet from the furnace chamber temperature are in the range 1140-1150 °C, which avoids slagging of screens.

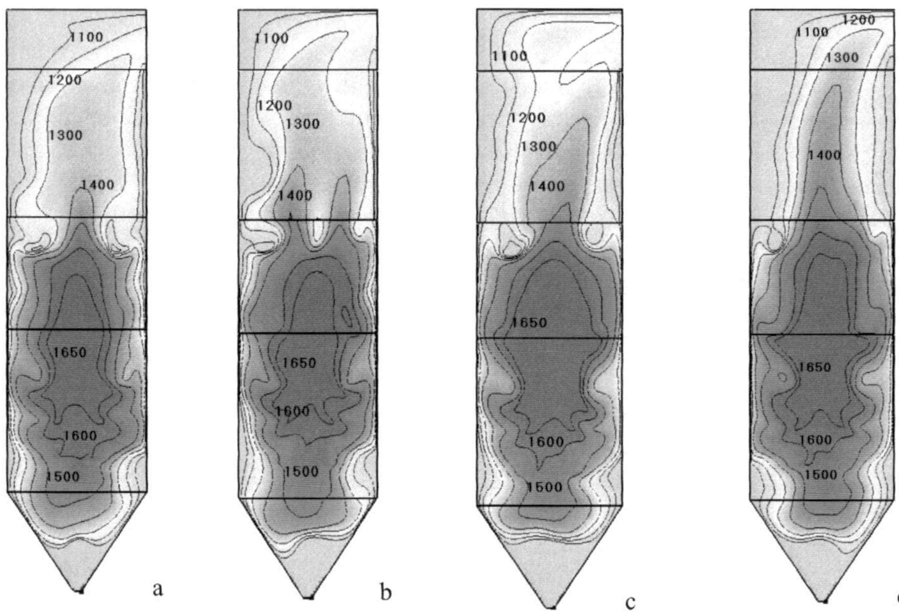

Fig. 2. Temperature field in the central section of the furnace at different reburning zone stoichiometry, °C. a) $\alpha = 0.9$; b) $\alpha = 0.98$; c) $\alpha = 1$; d) $\alpha = 1.05$.

Figure 3 shows the nitrogen oxides concentration in the central section of the furnace. From the drawings it is seen that all cases in the active combustion zone demonstrate the same distribution and nitrogen oxides concentrations. At the exit from the primary combustion zone the NO_x reading is ~ 550 mg/m^3 for all options due to same value of excess air at the exit from the primary combustion zone.

In the reburn zone while decreasing the air excess ratio there is a more effective interaction of incomplete combustion products with NO_x (fig. 3) formed in the main burner zone, and there is a recovery of them to the molecular nitrogen. To the greatest extent, the reduction of NO_x occurs when $\alpha = 0.9$, where at the level of the OFA nozzles the nitrogen oxides concentration reaches values of 130 mg/m^3. For the case of $\alpha = 1.05$ there is a significant NO_x concentration in the central zone (fig.3d) and the mean value NO_x concentration at the level of the OFA nozzles is more than 650 mg/m^3. At the exit of the combustion chamber for all considered variants is observed a uniform distribution of the nitrogen oxides concentration. The NO_x values for option with $\alpha = 0.9$ is 160 mg/m^3, whereas for the case of $\alpha = 1.05$ – 650 mg/m^3.

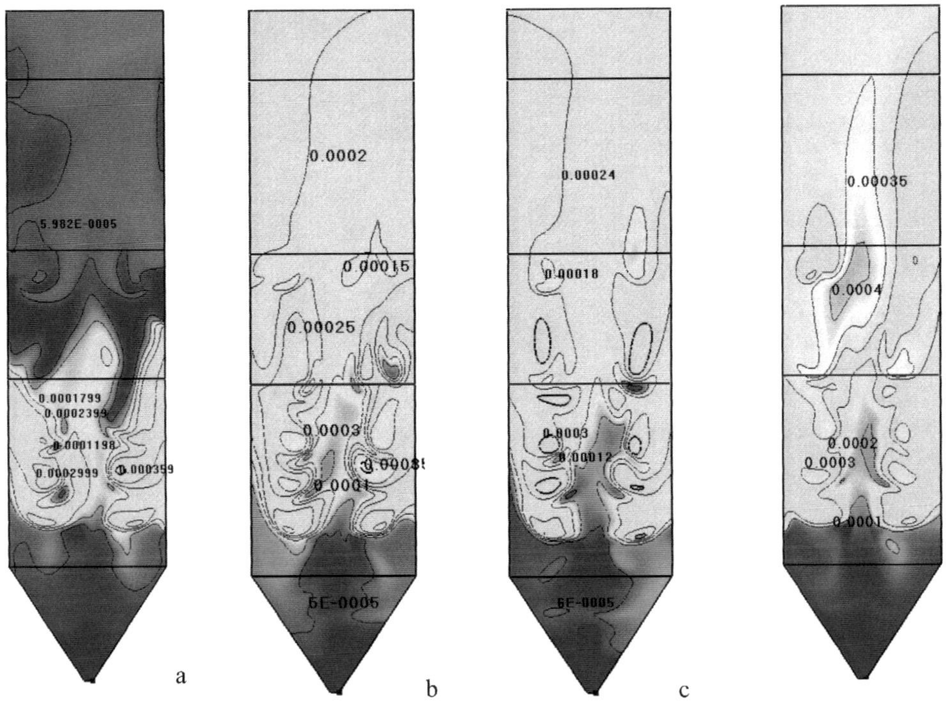

Fig. 3. Nitrogen oxides concentration in the central section of the furnace at different reburning zone stoichiometry, kg/kg. a) $\alpha = 0.9$; b) $\alpha = 0.98$; c) $\alpha = 1$; d) $\alpha = 1.05$.

Despite the existence of the reburning zone and the same coefficient of excess air at the outlet of the furnace there is a difference in unburned carbon-in-fly ash. Reducing the amount of air in reburning zone increases the number of incomplete combustion products at the reburning zone outlet. So, for the case of α= 1.05, unburned carbon-in-fly ash q4 is ~ 0.4 %, whereas for the case of α= 0.9 – q4 ~ 2 %. Figure 4 shows the dependence of the NO_x concentration and unburned carbon-in-fly ash at the furnace outlet from the excess air ratio at the reburninig zone.

From figure 5 we can see the decrease the nitrogen oxides concentration and increase the unburned carbon-in-fly ash at the furnace outlet when reducing the stoichiometric ratio in the reburn zone.

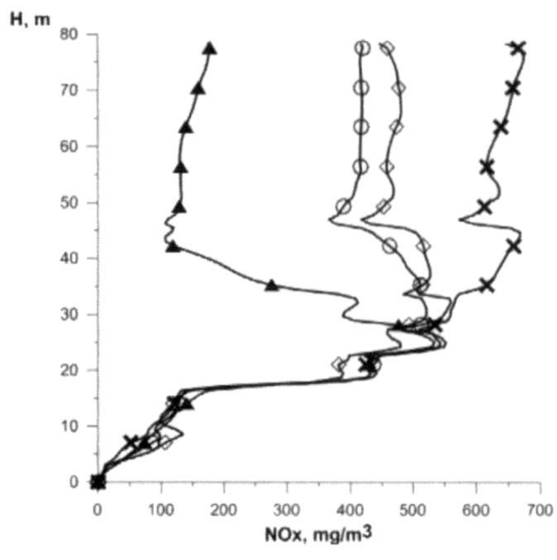

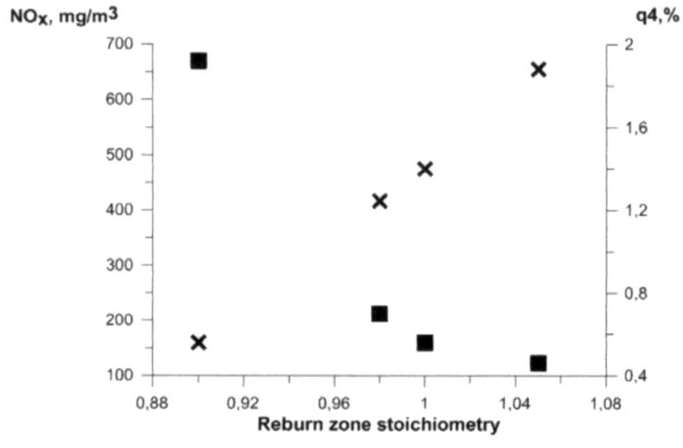

Fig. 4. NO_x concentration at the furnace height. ▲ – α = 0.9; ○ – α = 0.98; ◇ – α = 1.0; × – α = 1.05.

Fig. 5. The dependence of the NO_x concentration and unburned carbon at the outlet of the furnace from reburning zone stoichiometry. ■ – q4, × – NO_x.

Given the imposed project constraints to the furnace, the nitrogen oxides concentration is not more than 350 mg/m^3 and the value of unburned carbon-in-fly ash no more than 1 %, the value of α must be in the range of 0.94 – 0.98.

4 Conclusion

The performed numerical studies reveal a significant influence of stoichiometry in the reburning zone in the formation the nitrogen oxides concentration. It is shown that the change stoichiometric ratio in the range of 0.9 – 1.05 leads to changes in the NO_x at the furnace outlet on more than 4 times.

The dependencies of the NO_x concentration and unburned carbon-in-fly ash at the outlet of the furnace unit capacity of 660 MW while burning Kuznetsky coal from the stoichiometric ratio in the reburn zone have been defined. It is shown that to ensure the nitrogen oxides concentration at the furnace outlet at less than 350 mg/m^3 and unburned carbon at less than 1 %, the value of α must be in the range of 0.94 – 0.98.

Acknowledgments

The research on modelling combustion coal and associated heat and mass transfer phenomena was supported by RFBR, research project No. 15-38-20969.

References

1. J.C. Kramlich, W.P. Linak, Prog. Energy Combust. Sci. **20** (1994)
2. V.I. Babii, V.R. Kotler, S.P. Titov, T.M. Balashova, Topical issues of burning energy fuel (1980)
3. V.R. Kotler, *Reduction in nitrogen oxides emissions boilers of thermal power plants during solid fuel combustion* (Souzenergo, 1982)
4. H. Zhou, K.F. Cen, J.R. Fan, Energy **29**, 1 (2004)
5. C.T. Bowman, 24th Symposium (International) on Combustion, The Combustion Institute, Pittsburgh (1993)
6. S.C. Hill, L.D. Smoot, Prog. Energy and Combust. Sci. **26**, 4 (2000)

7. B. Staiger, S. Unterberger, R. Berger, K.R.G. Hein, Energy **30**, 8 (2005)
8. L.D. Smoot, S.C. Hill, H. Xu, Prog. Energy and Combust. Science **24**, 5 (1998)
9. B. Folsom, Fuel Energy Abstr. **4**, 227 (1997)
10. B.X. Shen, Q. Yao, X.C. Xu, Fuel Processing Technology **85**, 11 (2004)
11. H.P.Wan, C.S. Yang, B.R. Adams, S.L. Chen, Fuel **87**, 3 (2008)
12. W.P. Adamczyk, S. Werle, A. Ryfa, Applied Thermal Engineering **73** (2014)
13. Y.A. Anikin, I.S. Anufriev, E.Y. Shadrin, O.V. Sharypov, Thermophysics and Aeromechanics **21**, 6 (2014)
14. A.G. Tumanovskii, M.Yu. Altukhov, A.L. Shvarts, and other, Power Technology and Engineering **44** (2010)
15. M.Yu. Chernetskii, A.N. Alekhnovich, A.A. Dekterev, Thermal Engineering **59**, 8 (2012)
16. Maidanik, M.N., and other, Thermal Engineering **58**, 6 (2011)
17. M.Yu. Chernetskiy., A.A. Dekterev., A.P. Burdukov, K. Hanjalić, Fuel **135** (2014)
18. Y.B. Zeldovich, P.Y. Sadovnikov, D.A. Frank-Kamenetckiy, AS USSR **317** (1947)
19. W. Chen, PhD thesis of Brigham Young University (1994)
20. G.G. DeSoete, In: Fifteenth symposium (international) on combustion. The Combustion Institute (1975)

MATEC Web of Conferences 72, 01023 (2016) DOI: 10.1051/matecconf/20167201023
HMTTSC-2016

INVESTIGATION OF TWO-PHASE FLOW IN SHORT HORIZONTAL MINI CHANNEL HEIGHT OF 1 MM

Fedor Ron'shin[1,*], and *Vyacheslav* Cheverda[1]

[1]Kutateladze Institute of Thermophysics SB RAS, 630090 Novosibirsk, Russia

Abstract. The experiments with two-phase flow in the short horizontal rectangular minichannel with the height of 1 mm and width of 29 mm have been carried out using water and gas nitrogen. The five two-phase flow patterns have been recognized in the minichannel: churn, stratified, annular, bubble, and jet. These regimes are plotted on a graph and the boundaries between them determine precisely. The height of a horizontal minichannels has a significant role on boundaries between the flow regimes.

1 Introduction

Currently there is a rapid development of miniature heat exchangers. Studies show that the heat exchange systems with mini- and micro-dimensions turn out to be more energy efficient than macro-systems with a characteristic channel size larger than 1 mm [1]. With decreasing channel thick-ness the surface to volume ratio of a channel increases, which leads to high heat transfer rate in these systems. Such systems are becoming ever more in demand in the microelectronics, aerospace industry, transport and energy.

A considerable number of studies have been published in the field of two-phase flows in minichannels. An overview of publications on the two-phase flow in the channels of various configurations can be found in [2-4]. The analysis of the works shows that the structure of the two-phase flow is mainly influenced by the following parameters: geometry and dimensions of the channel [5, 6], parameters of the input area [7] and fluid properties, such as viscosity and surface tension [8, 9]. Many works describe round channels, although rectangular channels are more promising for the use in thermal stabilization systems. In rectangular channels the flow pattern is qualitatively consistent with regimes in the pipes, although the boundaries between the regimes are substantially different. There are two linear sizes, and depending on their relationship the process pattern significantly changes. Works [10-12] consider mechanisms, influencing the regimes forming in minichannels. The channel height is shown to have a significant influence on the mechanism of the two-phase flow formation.

The aim of present work is to investigate the regimes of two-phase patterns and boundaries between them in the short (length of 80 mm) horizontal minichannel with the

* Corresponding author: f.ronshin@gmail.com

height of 1 mm and width of 29 mm using the modern techniques and to compare them with the channel with cross-section of 0.3×30 mm^2.

2 Experimental results

The regimes of two-phase flow in the minichannel with cross-section of 1×29 mm have been experimentally studied using fluorescence technique (this technique is described in [13]) and a digital video camera. The flow regime map has been created for the investigated channel (fig. 1). The superficial gas U_{SG} and liquid U_{SL} velocities, determined as the volumetric gas or liquid flow rate divided by the channel cross-section, are used as the coordinates in the figure. The following two-phase flow patterns have been detected in the channel: churn, stratified, annular, bubble, and jet.

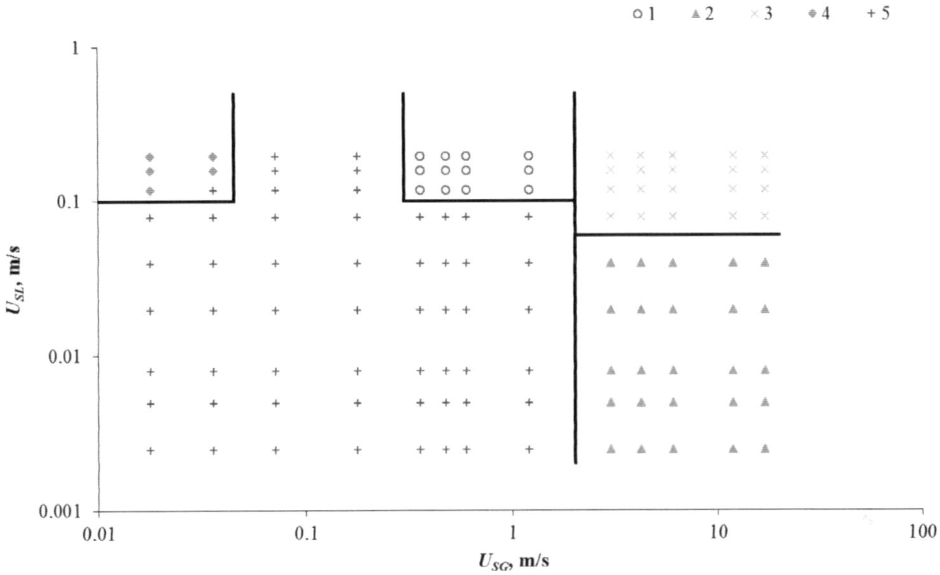

Fig. 1. Map of two-phase flow in the minichannel with cross section of 1×29 mm^2.Flow regimes: (1) churn; (2) stratified; (3) annular; (4) bubble; (5) jet.

In the range of low superficial velocities of liquid, gas moves in the central part of the channel, while the bulk of liquid moves along its periphery along lateral walls. Disturbances on the surface of the liquid are not observed. The stationary jet regime is observed at low superficial velocities of liquid and gas, when the gas flow occupies no more than a half of the channel cross section. The jet flow pattern is shown in figure 2c (U_{SL}=0.039 m/s, U_{SG} = 0.178 m/s). The liquid film moves along the bottom wall of the channel, the liquid completely fills the lateral sides of the channel. By increasing superficial liquid velocity, the liquid pulsations occurs in the lateral parts of the channel. By increasing the superficial gas velocity, when the gas begin to occupy larger volume than the liquid, the transition to the stratified flow regime takes place. In this regime, part of the liquid moves along the bottom channel wall in the form of a film together with the gas stream. The liquid fills the lateral sides of the channel. The stratified flow pattern is shown in figure 2a (U_{SL}=0.02 m/s, U_{SG}=16.95 m/s). In this case liquid film moves along the bottom wall of the minichannel and a liquid pulsation can be observed in the lateral parts of the channel. By increasing superficial liquid velocity, liquid emissions occur from the lateral sides of the channel and one forms the film on the upper wall of the channel. The transition to the annular flow regime takes place. The film in such flow regime moves along the channel together with the gas stream at the top and bottom walls of the channel. In this regime gas together with droplets forms the core of the flow in the center of microchannel. Gas occupies the most part of the microchannel cross-section and near with the lateral walls, the pulsating motion of liquid is observed. the annular flow pattern is sown in figure 2b (U_{SL}=0.195 m/s, U_{SG}=16.95 m/s). In this regime, one can see a liquid film moving on the window. The bubble flow regime is observed at high superficial liquid velocities and low superficial gas velocities. In such regime, liquid comprising many small gas bubbles moves along the minichannel. The size and amount of bubbles vary depending on liquid and gas flow rates, but the sizes of the bubble are always much smaller than the channel width. The bubble frequency increases with increasing superficial liquid and gas velocities. The breaks in the bridges begin to appear with increasing the superficial gas velocity, the transition to the churn flow regime takes place, figure 2d. This flow pattern is discussed in more details in [14]. The main characteristic of this flow regime is the existence of continuous and broken liquid bridges. The transition from the jet regime to the churn regime occurs with an increase of superficial velocity of the liquid and the transition is accompanied by the appearance of continuous liquid bridges, which are stable and the number of which in the minichannel was more than one. In the figure 2d one can see the broken bridges, which is not observed in the jet and stratified flows. The transition from the bubbly (continuous filled liquid bridges) to the churn regime was accompanied by the appearance of breaks in the bridges. Upon the transition to the annular flow regime, the continuous filled liquid bridges disappear; upon the passage to the bubbly regime, they become stable.

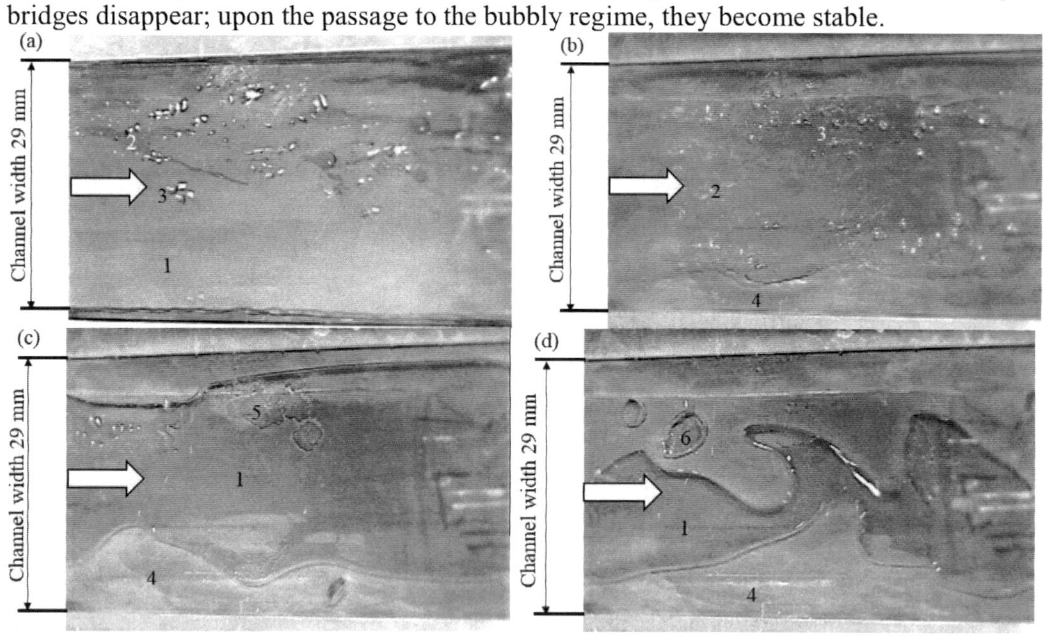

Fig. 2. Typical photographs of the regimes: (a) Stratified flow regime at U_{SL} = 0.02 m/s, U_{SG} = 16.95 m/s; (b) Annular flow regime at U_{SL} = 0.195 m/s, U_{SG} = 16.95 m/s; (c) Jet flow regime at U_{SL} = 0.039 m/s, U_{SG} = 0.178 m/s; (d) Churn flow regime at U_{SL} = 0.195 m/s, U_{SG} = 0.6 m/s; 1 – liquid film on the bottom channel wall, 2 – liquid film on the upper channel wall, 3 – liquid drops, 4 – the channel regions fully filled with liquid, 5 – unwetted zone on the bottom channel wall, 6 – gas bubbles.

DOI: 10.1051/matecconf/20167201023

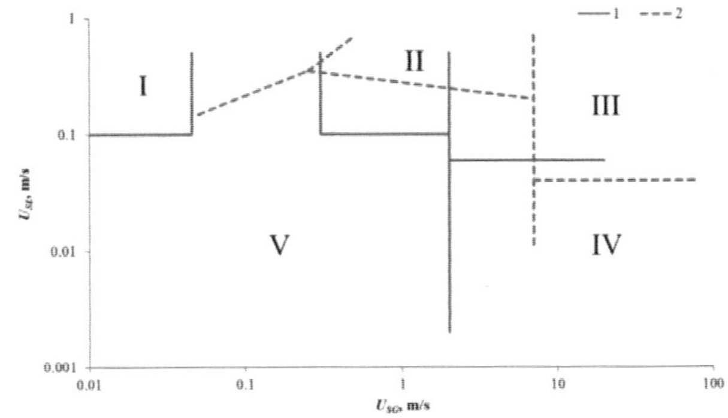

Fig. 3. Influence of the channel highs on the flow patterns for the channels with sections of 1×29 mm^2 (1) and 0.3×30 mm^2 (2). Flow patterns: (I) bubble, (II) churn, (III) annular, (IV) stratified, and (V) jet.

The comparison between two flow pattern map for the microchannels with sizes 1×29 mm^2 and 0.3×30 mm^2 is demonstrated in Figure 3. It is seen that with a decrease in the channel height the boundaries between the jet and the stratified modes, and the jet and the annular modes shift towards higher reduced velocities of the gas, significantly increasing the area of the jet mode. The boundary between the churn and the jet regimes is shifted towards higher reduced velocities of the fluid. The area of the bubble regime is shifted towards higher reduced velocities of the gas and liquid. The boundary between stratified and annular regimes does not change significantly.

Acknowledgments

We gratefully acknowledge the support from the Ministry of Education and Science of Russia (No. 14.604.21.0053 Project identifier RFMEFI60414X0053).

References

1. V.V. Zamashchikov, A.A. Korzhavin, E.A. Chinnov, Combust. Explos. Shock Waves **50**, 381 (2014)
2. N. Shao, A. Gavriilidis, P. Angeli, Chemical Engineering Science **64**, 2749 (2009)
3. E.A. Chinnov, O.A. Kabov, High Temp. **44**, 773 (2006)
4. E.V. Rebrov, Theor. Found. Chem. Eng. **44**, 355 (2010)
5. J.W. Coleman, S. Garimella, Int. J. Heat Mass Transfer **42**, 2869 (1999)
6. T.S. Zhao, Q.C. Bi, Int. J. Multiphase Flow **27**, 765 (2001)
7. V. Haverkamp, V. Hessel, H. Löwe et al., Chem. Eng. Technol. **29**, 1015 (2006)
8. S. Waelchli, P.R. von Rohr, Int. J. Multiphase Flow **32**, 791 (2006)
9. R. Pohorecki, P. Sobieszuk, K. Kula, W. Moniuk, M. Zielinski, P. Cyganski, P.Gawinski, Chem. Engng J. **135**, S185 (2008)
10. O.A. Kabov, E.A. Chinnov, V.Cheverda, Microgravity sci. technol. **XIX-3/4**, 44 (2007)
11. E.A. Chinnov, V.V. Guzanov, O.A. Kabov, Tech. Phys. Lett. **35**, 653 (2009)
12. E.A. Chinnov, F. Ron'shin, O.A. Kabov, Int. J. Multiphase Flow **80**, 57 (2016)
13. E.A. Chinnov, J. Heat Transfer **138**, 7 (2016)
14. E.A. Chinnov, F.V. Ron'shin, V.V. Guzanov, D.M. Markovich, O.A. Kabov, High Temp. **52**, 681 (2014)
15. E.A. Chinnov, F.V. Ron'shin, O.A. Kabov, Tech. Phys. Lett. **41**, 817 (2015)

DOI: 10.1051/matecconf/20167201024

HEAT TRANSFER WHILE INCORPORATING METAL PARTICLES IN A MELTING SUPPORT PLATE

Gennady Ya. Mamontov[1,*], and *Danil* N. Chuguev[2]

[1]National Research Tomsk Polytechnic University, 634050 Tomsk, Russia
[2]PJSC "LUKOIL"

Abstract. The work defines and solves the heat transfer problem for "particle – support plate" system heated to high temperatures by means of mathematical simulation methods. In the course of the formulated problem, the numerical studies were carried out under conditions of a high density of the particle material in comparison with the support plate material.

1 Introduction

Over the last few decades, the processes running if different hardening coatings are applied on the surface of mechanical engineering, petroleum chemistry and power engineering equipment structural units have been studied intensively [1]. In these circumstances, various approaches based predominantly on theoretical studies are used [1, 2]. But up to the present moment there are no published models that in a unique manner describe the high-temperature interaction mechanisms of a single particle and a support plate surface (that is an element of some equipment) under conditions of a high-temperature impact on the latter one (e.g. while plasma coating). It is hard if not impossible to visualize by experiment an interaction process of a fully or partially melted metal single droplet heated to high temperatures and a support plate (that can be also melted). However, it is known [3] (based on the results of the particle and support plate adhesion zone microstructure analysis) that even under conditions of the particle motion low speed at the impact moment (up to 20mps), in numerous cases, the latter are incorporated in a support plate.

One of the possible mechanisms of such incorporation can be a displacement by a particle of the melted metal boundary layer at its contact at the instant of encounter. Furthermore, the energy necessary for penetration of such support plate layer can be released also as a result of the particle metal crystallization. Traditionally, while plasma coating at the instant of encounter, the latter presents a heterogeneous structure – a droplet some part (20-25 %) of which is crystallized. Until recently, the mathematical models taking into consideration the aforementioned interaction processes of a particle and a support plate under conditions of hardening coating application (at high temperatures) while simulating the main characteristics of hardening plasma technologies of processing equipment structural elements were not in use.

* Corresponding author: gmamontov@tpu.ru

The work objective is a mathematical simulation of the heat transfer while incorporating a metal particle in a support plate that melts at a heating process by means of energy accumulated in the particle and a crystallization heat under conditions of low speed gravity sedimentation.

2 Problem definition and solution pattern

A disk-shaped particle positioned at the surface of a metal sheet is considered (fig. 1). The temperature of the metal sheet is lower than that of the particle the bottom part of which (10% from top to bottom) is crystallized. As a result of the particle heat conductivity, a thin layer of the support plate metal is heated to the melting point (or higher) at contact with the heating source (particle) and a thin melt sheet is formed. Under the action of gravity, the particle expels the melt and incorporates into the support plate, which is heated to a full particle crystallization by means of the phase transition heat release. The incorporation speed is determined by the support plate metal fusion penetration. The studied process is transient due to the energy limitedness accumulated in the particle.

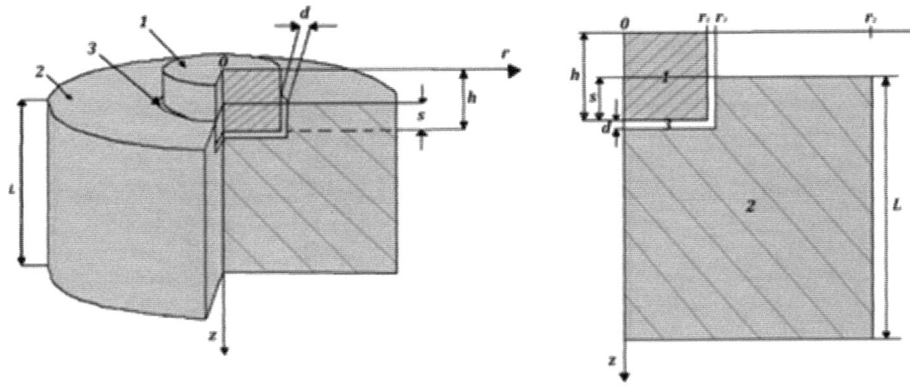

a) b)

Fig. 1. Overview diagram of the studied process (a). Geometric layout of the process under study (b). 1, 2, 3 – particle, support plate and melt area respectively.

While setting a problem it was supposed, in a similar way to [4], that the particle keeps a disk shape throughout the crystallization process; the maximum temperatures reached under conditions of the phase transition do not exceed the melted support plate metal vaporization point; the particle energy is sufficient to incorporate at depth corresponding to a melting point T_p; the particle advance speed to a depth of the support plate is determined by the isotherm $T=T_p$ advance speed on the particle symmetry axis.

To describe the heat transfer processes within the studied "particle – support plate" system, a mathematical model developed to solve condensed material firing problems was used as baseline through the in-situ heating of the single heated to high-temperature particle [5, 6].

The nonlinear differential equation system of transient heat conductivity, continuity, motion and energy for the "particle – support plate" system as part of the stated physical analog is as follows:

$$c_1\rho_1\frac{\partial T_1}{\partial t}=\lambda_1\left(\frac{\partial^2 T_1}{\partial r^2}+\frac{1}{r}\frac{\partial T_1}{\partial r}\right)+\lambda_1\frac{\partial^2 T_1}{\partial z^2}+\frac{Q_1W_1}{h} \quad (1)$$

$$c_2\rho_2\frac{\partial T_2}{\partial t}=\lambda_2\left(\frac{\partial^2 T_2}{\partial r^2}+\frac{1}{r}\frac{\partial T_2}{\partial r}\right)+\lambda_2\frac{\partial^2 T_2}{\partial z^2}+\frac{Q_2W_2}{h} \quad (2)$$

$$c_3\rho_3\frac{\partial T_3}{\partial t}=\lambda_3\left(\frac{\partial^2 T_3}{\partial r^2}+\frac{1}{r}\frac{\partial T_3}{\partial r}\right)+\lambda_3\frac{\partial^2 T_3}{\partial z^2} \tag{3}$$

$$\frac{\partial u}{\partial r}+\frac{u}{r}+\frac{\partial v}{\partial z}=0 \tag{4}$$

$$\frac{\partial u}{\partial t}+u\frac{\partial u}{\partial r}+v\frac{\partial u}{\partial z}=-\frac{1}{\rho_3}\frac{\partial p}{\partial r}+\mu(\frac{\partial^2 u}{\partial z^2}+\frac{1}{r}\frac{\partial u}{\partial r}-\frac{u}{r^2} \tag{5}$$

$$c_3\rho_3\left(\frac{\partial T_3}{\partial t}+u\frac{\partial T_3}{\partial r}+v\frac{\partial T_3}{\partial z}\right)=\lambda_3\frac{\partial^2 T_3}{\partial z^2}+\mu\left(\frac{\partial u}{\partial z}\right)^2 \tag{6}$$

$$\frac{\partial}{\partial r}(ru)+\frac{\partial}{\partial z}(rv)=0 \tag{7}$$

$$\frac{\partial v}{\partial t}+u\frac{\partial v}{\partial r}+v\frac{\partial v}{\partial z}=-\frac{1}{\rho_3}\frac{\partial p}{\partial z}+\frac{\mu}{\rho_3 r}\frac{\partial}{\partial r}\left(r\frac{\partial v}{\partial r}\right)+g \tag{8}$$

$$c_3\rho_3\left(\frac{\partial T_3}{\partial t}+u\frac{\partial T_3}{\partial r}+v\frac{\partial T_3}{\partial z}\right)=\lambda_3\frac{\partial^2 T_3}{\partial r^2}+\mu\left(\frac{\partial u}{\partial r}\right)^2 \tag{9}$$

The mass melting and crystallization rate is calculated using the formula:

$$W_i=\rho_i V_i, where\ i=1,2 \tag{10}$$

$$V_i(r,z,t)=\frac{\xi(r,z,t+\Delta t)-\xi(r,z,t)}{\Delta t} \tag{11}$$

Based on prime geometric relationships V_i can be viewed as:

$$V_i(r,z,t)=\left[\frac{T(r,0,t+\Delta t)-T_m}{T(r,0,t+\Delta t)-T(0,h,t+\Delta t)}-\frac{T(r,0,t)-T_m}{T(r,0,t)-T(0,h,t)}\right]\frac{h}{\Delta t} \tag{12}$$

$$s=\int_0^{t_k}V_2dt. \tag{13}$$

It is accepted that the particle, support plate and support plate melt area symmetry axes are in line. The particle size – ($0\le r\le r_1$ and $0\le z\le h$), support plate – ($0\le r\le r_2$ and $0\le z\le L$), melt area – ($0\le r\le r_1$, $h\le z\le h+d$) $\cup$ ($r_1\le r\le r_3$ and $h-s\le z\le h+d$).

The (1) – (9) system boundary conditions are written as:

$$z=0, 0\le r\le r_1:\ -\lambda_1\frac{\partial T_1}{\partial z}=\alpha\left(T_g-T_1\right)+\sigma\varepsilon_{em}\left(T_g^4-T_1^4\right) \tag{14}$$

$$r=r_1, 0\le z\le h-s:\ -\lambda_1\frac{\partial T_1}{\partial r}=\alpha\left(T_g-T_1\right)+\sigma\varepsilon_{em}\left(T_g^4-T_1^4\right) \tag{15}$$

$$r=r_1, h-s\le z\le h:\ -\lambda_1\frac{\partial T_1}{\partial r}=-\lambda_3\frac{\partial T_3}{\partial r}, T_1=T_3 \tag{16}$$

$$z=h, 0\le r\le r_1:\ -\lambda_1\frac{\partial T_1}{\partial z}=-\lambda_3\frac{\partial T_3}{\partial z}, T_1=T_3 \tag{17}$$

$$r=0, 0\le z\le L+h-s:\ -\lambda_i\frac{\partial T_i}{\partial r}=0, where\ i=1,2,3 \tag{18}$$

$$z = h - s, r_1 \leq r \leq r_2 : -\lambda_i \frac{\partial T_i}{\partial z} = \alpha(T_g - T_i) + \sigma \varepsilon_{em}(T_g^4 - T_i^4), where\ i = 2, 3; \tag{19}$$

$$r = r_3, h - s \leq z \leq h + d : -\lambda_3 \frac{\partial T_3}{\partial r} = -\lambda_2 \frac{\partial T_2}{\partial r}, T_3 = T_2 \tag{20}$$

$$z = h + d, 0 \leq r \leq r_3 : -\lambda_3 \frac{\partial T_3}{\partial z} = -\lambda_2 \frac{\partial T_2}{\partial z}, T_3 = T_2 \tag{21}$$

$$r = r_2, h - s \leq z \leq L + h - s : -\lambda_2 \frac{\partial T_2}{\partial r} = 0 \tag{22}$$

$$z = L + h - s, 0 \leq r \leq r_2 : -\lambda_2 \frac{\partial T_2}{\partial z} = 0 \tag{23}$$

$$r = 0,\ h \leq z \leq h + d : u = 0 \tag{24}$$

$$z = h,\ 0 \leq r \leq r_1 : u = 0 \tag{25}$$

$$r = r_1, h - s \leq z \leq h : \upsilon = 0 \tag{26}$$

$$r = r_3, h - s \leq z \leq h + d : \upsilon = 0 \tag{27}$$

$$z = h - s,\ r_1 \leq r \leq r_3 : u = 0 \tag{28}$$

$$z = h + d,\ 0 \leq r \leq r_3 : u = U_m \tag{29}$$

$$z = h,\ r_1 \leq r \leq r_3 : \upsilon = V_m \tag{30}$$

The (1) – (9) system initial conditions are as follows:

$$\text{where } t = 0: T_1(r, z) = T_p,\ T_2(r, z) = T_s,\ s = 0,\ d = 0 \tag{31}$$

The given emissivity factor ε_{em} of an object system where the radiant heat transfer is in progress, is calculated using the formula [11]:

$$\varepsilon_{em} = \frac{1}{\frac{1}{\varepsilon_g} + \frac{1}{\varepsilon_m} - 1}. \tag{32}$$

The key agreed notations are: T – temperature; λ – heat conductivity; c – heat capacity; ρ – density; r, z – cylindrical coordinates; t – time; Q, W – heat effect and phase transition rate; u, υ – melt rate coordinates; μ – dynamic viscosity; V – linear advance speed of melt and crystallization boundary; h, s, d, L – particle, support plate and hollow space typical dimensions.

The problem of a metal melt flow in a slot hollow space between the particle and a solid support plate part was solved as part of a thin passage model [7].

The stated equation system with the corresponding boundary conditions and additional relations is solved by means of a finite-difference method developed to solve the heat transfer nonlinear problems under conditions of active physical-chemical algorithm transformations [8, 9]. A nonlinear and unequally spaced difference mesh was used. Owing to the absence of experimental findings and solutions of even very simplified similar problems, the numerical simulation result reliability was assessed by means of numerical solutions based on condensing grid consecutive order.

3 Findings and discussions

The numerical studies within the limits of the formulated problem were carried under conditions of a higher particle material density as opposed to the support plate metal. With such ratio of densities, we meet the particle incorporation condition only due to gravitation work that is stronger than the lift force acting on the particle on the side of the melt. To demonstrate more clearly the studied process conformity, the density ρ_1/ρ_2 ratio was as great as practicable in the real world – 7.8.

Figure 2 displays the distinctive positions of typical isotherms while incorporating the particle in a melted support plate part. It can be seen that the melt sheet bed depth reaches 0.001 m. under the particle and near the flat side. The distinctive positions of typical isotherms and field velocities in a vertical and horizontal part of the hollow space (fig. 1) are presented in figure 3. The studied process conformity analysis was carried out.

It is determined that if the particle is considered as a metal droplet which is melted at the contact moment with the support plate, in the course of crystallization the metal layer, which is the front side adjacent to the particle, freezes at rapid-fire pace even under relatively low (200-250 K) heat overloads in relation to the support plate. On further motion, the particle is made as a disk the front side of which is solid and the remaining part is melted.

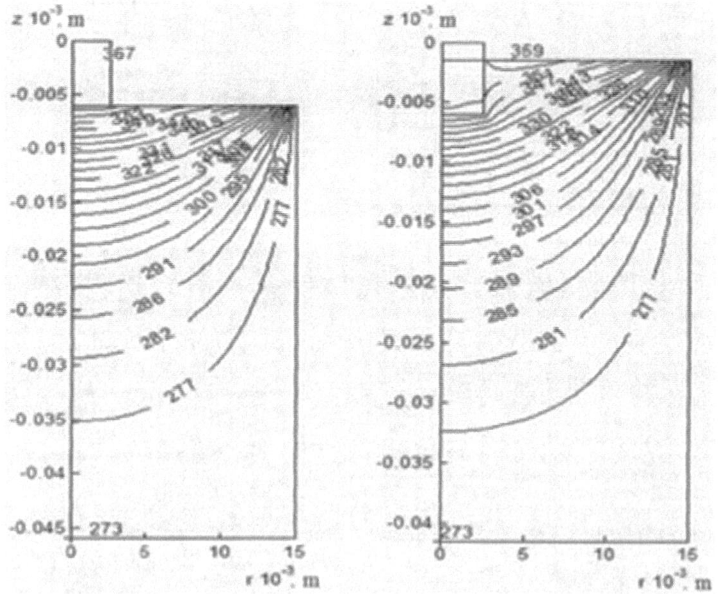

Fig. 2. The typical isotherm position while incorporating a metal particle (a – initial instant, b – finite time instant).

The numerical simulation results give grounds for conclusion that if the particle has a certain kinetic energy store in case of sedimentation on the support plate and is solid, it will incorporate in the support plate without any postheating due to the heat transition phase. In this case, the incorporation depth will be determined by the impact velocity and deformation (virtual) of the particle. Thus, the ratio of characteristic heat time, particle deformation, support plate fusion penetration and particle incorporation is essential.

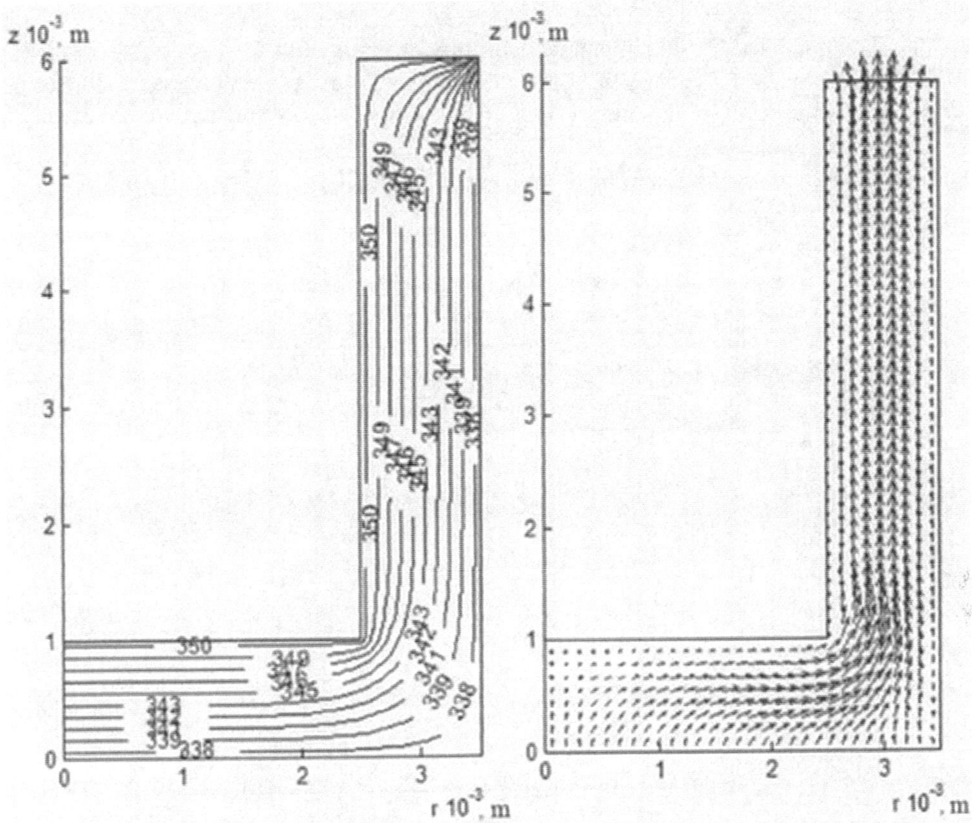

a) b)

Fig. 3. Typical isotherm position (a) and field velocities (b) in a vertical and horizontal passage. (Re = 100).

The numerical study results showed that the beneficial significant effect could be obtained by means of the support plate pre-heating to temperatures close to the material melting point.

4 Conclusion

As a result of the heat conductivity problem solution for "heated up to high temperature particle – melting support plate" system, the possibility to incorporate the particle in a material three-dimensional layer of the hardenable detail within the limits of the basic process characteristic change that corresponds to coating technologies, was determined. The formulated mathematical model and the heat transfer problem solution method can be used to select technological processes for hardening and protecting coating application on the surface of machine and device components.

Acknowledgments

The work was supported by the Russian President's grant (Scientific School project - 7538.2016.8).

References

1. Yu.A. Chumakov, A.G. Knyazeva, Combust. Explos. Shock Waves **48** (2012)
2. E.M. Ivanov, A.A. Uglov, Physics and chemistry of material processing **2** (1985)
3. A.M. Dubasov, V.V. Kudinov, M.X. Shorshov, Physics and chemistry of material processing **6** (1971)
4. G.V. Kuznetsov, G.Ya Mamontov, G.V. Taratushkina, Combust. Explos. Shock Waves **40**, 1 (2004)
5. G.V. Kuznetsov, G.Ya. Mamontov, G.V. Taratushkina, Khimicheskaya Fizika **23**, 3 (2004)
6. G.V. Kuznetsov, P.A. Strizhak, J. Eng. Thermophys. **17** (2008)
7. P. J. Roache, *Computational Fluid Dynamics, Hermosa Publishers, Albuquerque* (NM, USA, 1976)
8. V.L. Strakhov, A.N. Garashchenko, G.V. Kuznetsov, V.P. Rudzinskii, Combust. Explos. Shock Waves **37** (2001)
9. O.V. Vysokomornaya, G.V. Kuznetsov, P.A. Strizhak, Russ. J. Phys. Chem. B **5** (2011)

MATEC Web of Conferences **72**, 01025 (2016) DOI: 10.1051/matecconf/20167201025
HMTTSC-2016

MATHEMATICAL MODEL OF HEAT TRANSFER IN MORPHOLOGICAL PART OF VEGETATION AT INFLUENCE BY THERMAL RADIATION FROM SURFACE FOREST FIRE FRONT

Nikolay V. Baranovskiy[1,*], and *Alena* N. Demikhova[1]

[1]National Research Tomsk Polytechnic University, 634050 Tomsk, Russia

Abstract. Numerical research results of heat transfer in layered tree trunk influenced by heat flux from forest fire are presented. The problem is solved in two-dimensional statement in Cartesian system of co-ordinates. The typical range of influence parameters of heat flux from forest fire is considered. Temperature distributions in different moments of time are obtained. Condition of tree damage by forest fire influence is under consideration in this research.

1 Introduction

There are few growing plant layers in forest depending on biological features of the major plants, their age and certain physical-geographical conditions. Each layer - accurately enough delimited horizons of active plant parts concentration. Layers can be formed by plants of one or two and more types. Following layers can be distinguished in forest [1]:

- Forest stand. Crown layer is set of crones of the closed trees. Two crown layers can be marked out in moderate forests, but five crown layers can be differentiated in tropical forests. There is layer of very high trees towering over crown layer of forest stand in tropical rain-forests.
- Shrub layer. This layer consists of bushes and low trees.
- Grass or grass with shrubbery.
- Moss or/and lichen.
- Forest litter. This layer consists of vegetation residues on a soil surface.

Properties of forest fuel [2] have great value for forecasting of forest fire danger. So-called fuel models use for their description. Enough detailed characteristic of forest fuel types involved in process of surface forest fire is known. Plants and their residues of various level of decomposition are forest fuel in forest stand, which can burn at fires [3]. Practically all plants concern to combustible materials in forest biogeocenosis (live and dead, and also tree waste, fallen trees, litter, humus and peat horizons.

It is possible to consider three levels in forest fuel classification [4]:

* Corresponding author: firedanger@tpu.ru

- Classification of elementary parts of forest fuel complexes, including separate plants from surface cover, shrub and undergrowth, low bushes, morphological parts of trees (branches, snags, dead fallen wood etc).
- Classification of simple complexes (forest fuel layers) inside biogeocenosis.
- Classification of biogeocenosis as multiple forest fuel complexes.

Forest fuel segmentation of all layers into groups is considered in classification of N.P. Kurbatskiy [5, 6] not only using their place in biogeocenosis, but also and function carried out by them during fire. All forest fuel separated in next groups [5]: 1) burning conductors, 2) sustaining combustion, 3) inhibitory burning.

The purpose of the present work is mathematical modeling of heat transfer processes in separate morphological part of grass vegetation at influence of radiation from front of surface forest fire.

2 Physical and Mathematical Statements

The mathematical model is developed taking into account following suppositions. The separate morphological part of grass plant (for definiteness - caulis) is considered. The caulis of blade is considered only without presence of lamina. The caulis is considered as three-layer structure that consists of internal cavity, pulp and protective layer [7].

The basic assumptions are resulted:

1. Short-term influence of a radiant thermal flux from surface forest fire front (fire of low intensity) is considered during the period previous possible ignition of blade.
2. The three-layer system (structure) of blade caulis with effective thermophysical characteristics is considered.
3. It is supposed, that caulis thermophysical characteristics do not depend on temperature.
4. Thermal decomposition of caulis tissues under the influence of the raised temperatures is neglected.
5. Evaporation of cellular moisture from caulis tissues also is neglected.
6. Conduction is basic mechanism of heat transfer in caulis tissues.
7. The monolithic structure of caulis tissues is considered.

Geometry of decision area is described on fig. 1.

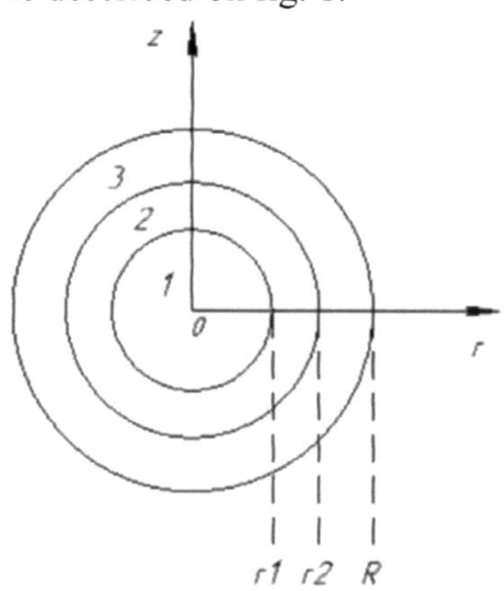

Fig. 1. The scheme of decision area.

Processes of heat transfer in layered structure of blade caulis till the moment of possible ignition are described by the non-stationary differential equations of heat conductivity with corresponding initial and boundary conditions:

$$\rho_1 c_1 \frac{\partial T_1}{\partial t} = \frac{\lambda_1}{r}\frac{\partial}{\partial r}\left(r\frac{\partial T_1}{\partial r}\right) + \lambda_1 \frac{\partial^2 T_1}{\partial z^2},$$

$$\rho_2 c_2 \frac{\partial T_2}{\partial t} = \frac{\lambda_2}{r}\frac{\partial}{\partial r}\left(r\frac{\partial T_2}{\partial r}\right) + \lambda_2 \frac{\partial^2 T_2}{\partial z^2},$$

$$\rho_3 c_3 \frac{\partial T_3}{\partial t} = \frac{\lambda_3}{r}\frac{\partial}{\partial r}\left(r\frac{\partial T_3}{\partial r}\right) + \lambda_3 \frac{\partial^2 T_3}{\partial z^2},$$

$$T_i\big|_{t=0} = T_{i0},$$

$$r=R,\ -\lambda_3 \frac{\partial T_3}{\partial r} = q_{ff},$$

$$r=0,\ -\lambda_1 \frac{\partial T_1}{\partial r} = 0,$$

$$z=H,\ -\lambda_i \frac{\partial T_i}{\partial z} = 0,$$

$$z=R1,\ -\lambda_i \frac{\partial T_i}{\partial z} = 0,$$

$$r=R1,\ -\lambda_1 \frac{\partial T_1}{\partial r} = -\lambda_2 \frac{\partial T_2}{\partial r},\ T_1 = T_2,$$

$$r=R2,\ -\lambda_2 \frac{\partial T_2}{\partial r} = -\lambda_3 \frac{\partial T_3}{\partial r},\ T_2 = T_3,$$

Where T_i, ρ_i, c_i, λ_i - temperature, density, thermal capacity and heat conductivity of layers of blade caulis (1 - air cavity, 2 - pulp, 3 - protective layern). r, z – cylindrical coordinates. t - time coordinate. q_{ff} - radiant thermal flux from forest fire front.

The problem is solved by a method of finite differences [8]. The multidimensional differential equations are solved by a locally-one-dimensional method [8].

3 Results and Discussion

Numerical realization of mathematical model for heat transfer in layered structure of caulis is carried out using programming language of high level Object Pascal. IBM PC compatible computer with Intel Pentium Core 2 Duo 3 GHz processor unit and 8 GB RAM was used for computing experiments. The one-dimensional mathematical model allows to develop resource effective algorithm for program realization. Verification of one-dimensional model is spent using flat statement in cylindrical coordinates in the present research. Typical temperature distribution in layered structure of caulis is presented for certain time moment on fig. 2.

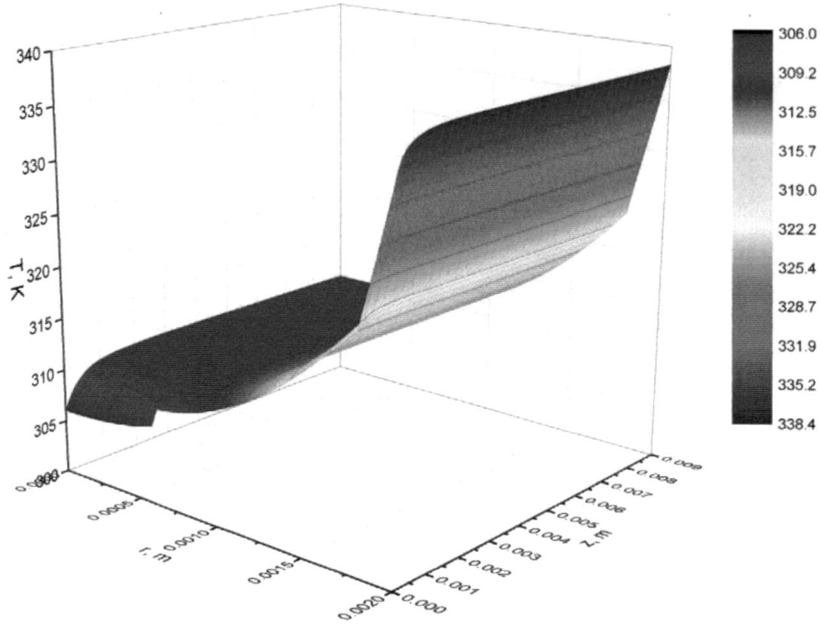

Fig. 2. Temperature distribution in caulis structure at influence of radiation from forest fire (q_{ff}=7 kW/m^2) at the moment of time t=3 s.

Dependences of temperature in protective layer and pulp of caulis from time are presented on figs. 3, 4.

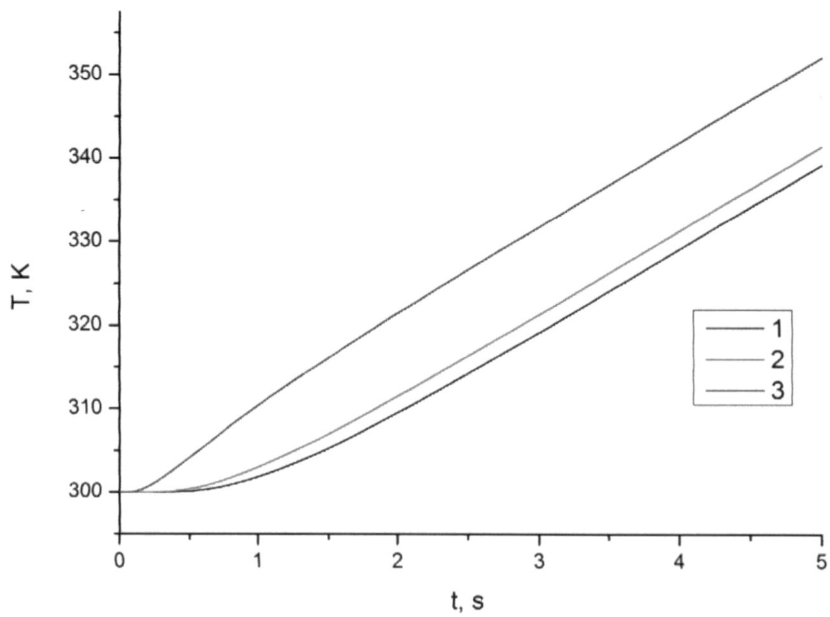

Fig. 3. Dependence of temperature on time in protective layer of caulis: 1 - r=0.00185 m, 2 - r=0.0019 m, 3 - r=0.00195 m.

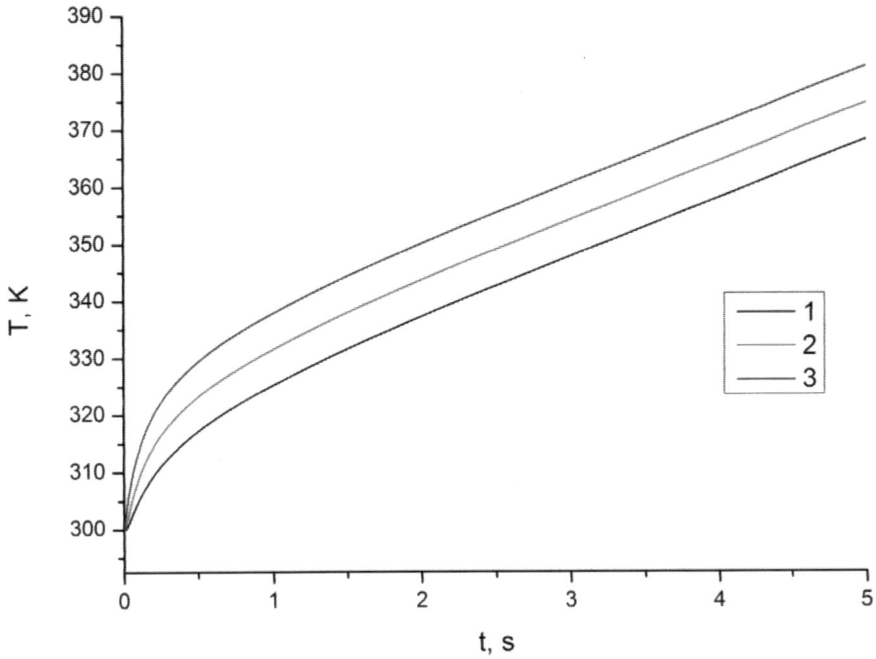

Fig. 4. Dependence of temperature on time in pulp of caulis: 1 - r=0.0008 m, 2 - r=0.001 m, 3 - r=0.0015 m.

The analysis of the obtained distributions allows making conclusion that there is an intensive warming up, both inside protective layer and pulp layer at influence of radiation from front of surface forest fire. Short-term influence of radiation from front of forest fire [9] (few seconds) on blade caulis was considered in the present work.

The preliminary analysis of heat transfer processes in layered structure of blade caulis shows, that it is necessary to use more perfect models for the full analysis of influence of radiation from forest fire front on morphological parts of plant. Such models should consider moisture evaporation from pulp and pyrolysis of dry organic substance [10-13]. However, using of thermophysical model for heat transfer simulation in caulis allows to make tentative estimations about time scales of ignition delay of plant at influence of a radiant thermal flux. Development of such mathematical models is necessary for creation of physically proved methods for an estimation of forest fire danger, pyrological estimations of surface cover in forest. Besides, such mathematical models can be used in information systems of biogeocenosis ecological monitoring.

4 Conclusion

The simple mathematical model of heat transfer in layered structure of blade caulis at influence of radiation from surface forest fire is presented in the current work. Results of the present research will allow to develop more perfect models for an estimation of forest fire danger and ecological consequences of forest fires. Furthermore, development of the complex deterministic-probabilistic approach for estimation of forest fires ecological consequences is possible taking into account prognostic modeling results of forest fire danger [14-15] and up-to-date information technologies [16-19].

Acknowledgements

Work was implemented under financial support of Russian Foundation for Basic Research. Grant N 16-41-700831.

References

1. V.I. Ufimtsev, O.A Kupriyanov, Strelnikov *Bulletin KemGU* **4**, 8 (2013)
2. G.M. Byram, *Combustion of forest fuels // Forest fire control and use* (Ed. K.P. Davis, 1959)
3. N.P. Kurbatskiy, *Questions of forest pyrology* (1972)
4. A.V. Volokitina, M.A. Sofronov, *Classification and mapping of vegetative combustible materials* (2002)
5. N.P. Kurbatskiy, *Questions of forest pyrology* (1972)
6. N.P. Kurbatskiy, *Technician and tactics of forest fire suppression* (1962)
7. I.I. Andreeva, L.S. Rodman, *Botany: studies for high schools* (2005)
8. L. Vulkov, Lecture Notes in Computer Science **9045**, 434 (2015)
9. N.V. Baranovskiy, Proceedings of SPIE **9680** (2015)
10. N.V. Baranovskiy, A.V. Zakharevich, D.S. Osotova, EPJ Web Conf. **82**, 01020 (2015)
11. D. Mohan, C.U. Jr. Pittman, P.H. Steele, Energy and Fuels **20**, 848 (2006)
12. G. Varhegyi, M.J. Antal, T. Szekely, P. Szabo, Energy and Fuels **3**, 329 (1989)
13. R. Capart, L. Khezami, A.K. Burnham, Thermochimica Acta **417**, 79 (2004)
14. E.P. Yankovich N.V. Baranovskiy, K.S. Yankovich, 9th International Forum on Strategic Technology, IFOST 2014, Paper 6991108 (2014)
15. G.V. Kuznetsov, N.V. Baranovsky, EPJ Web Conf, **76**, 01028 (2014)
16. R.H. Fraser, Z. Li, Remote Sensing of Environment **82**, 95 (2002)
17. Y.-H. Zhang, M. J. Wooster, O. Tutubalina, G. L. W. Perry, Remote Sensing and Environment **87**, 1 (2003)
18. N.V. Baranovskiy, Cybernetics and Systems Analysis **51**, 471 (2015)
19. N.V. Baranovskiy, E.P. Yankovich, Journal of Automation and Information Sciences **47**, 11 (2015)

MATHEMATICAL MODELLING OF THE FIXED-BED BIOMASS-COAL CO-GASIFICATION PROCESS

Igor G. Donskoy[1,*]

[1]Kutateladze Institute of Thermophysics, 630090 Novosibirsk, Russia

Abstract. The paper considers mathematical modelling of downdraft fixed-bed gasification process of the mixtures of woody biomass and coal. Biomass/coal ratio, biomass moisture content and air equivalence ratio are varying parameters. Boundaries of the efficient gasification regimes are estimated.

1 Introduction

Biomass (at first as forest products and agricultural wastes) has significant potential as a source for energy production [1]. However, energy use of biomass faces some technical difficulties: biomass combustion is usually not enough efficient and reliable process. Concerning this, other approaches in biomass conversion are developed, including co-conversion of biomass and coal.

There are different ways of thermochemical conversion besides combustion. Gasification is one of possible alternative technologies: it is conversion of solid fuel into product gas containing CO and H_2. The gas has better combustion characteristics and could be used in heat and electricity production [2, 3]. Co-gasification may have following advantages: providing stable thermal regimes for moist biomass gasification, and increasing of fuel mixture reactivity by biomass addition.

Presented work is devoted to investigation of the fixed-bed woody biomass and coal co-gasification process efficiency under co-current fuel and air flows condition (downdraft gasification process).

2 Model input parameters

Reactor concerned in model is vertical cylinder tube with following sizes: bed height is 0.5 m; bed diameter is 8 cm. Mean particle sizes: 5 mm for coal and 10 mm for biomass. Fuel feed rate changes from 1 kg/h (for coal only) to 2 kg/h (for biomass only). Main model variables are biomass mass portion in feed (0-100 %), air equivalence ratio (0.25-0.8) and initial biomass moisture content (10-40 %). Efficiency criteria are cold gas efficiency (the

* Corresponding author: donskoy.chem@mail.ru

degree of solid fuel heating value into gas heating value conversion) and fuel conversion degree (average for biomass-coal mixture).

Mathematical model developed in previous works [4, 5] was used, added by main features of co-gasification process (particles sizes difference and fuel reactivity). Mixtures of Azeysk brown coal and woody biomass are considered (fuels characteristics are listed in table). It is supposed that fuel reactivities in mixture do not change compared with individual fuel reactivities.

Table 1. Properties of coal and biomass.

% mass	W^r	A^d	C^{daf}	H^{daf}	O^{daf}	$(N+S)^{daf}$	V^{daf}
Coal	10	17	76.4	5.3	16.3	2.0	44
Biomass	10-40	1	47.2	5.7	47.0	0.1	85

3 Modelling results

Calculation results are presented at figs. 1 and 2. It is possible to estimate limits of biomass-coal mixture efficient conversion regimes area. In all cases maximum cold gas efficiency is reached at boundaries of parameters area. If initial biomass moisture content is 10 % or 20 %, efficient conversion of biomass without coal is possible. If moisture content is more than 20 %, it is necessary to maintain high air ratios to achieve high fuel conversion degree that leads to low cold gas efficiency.

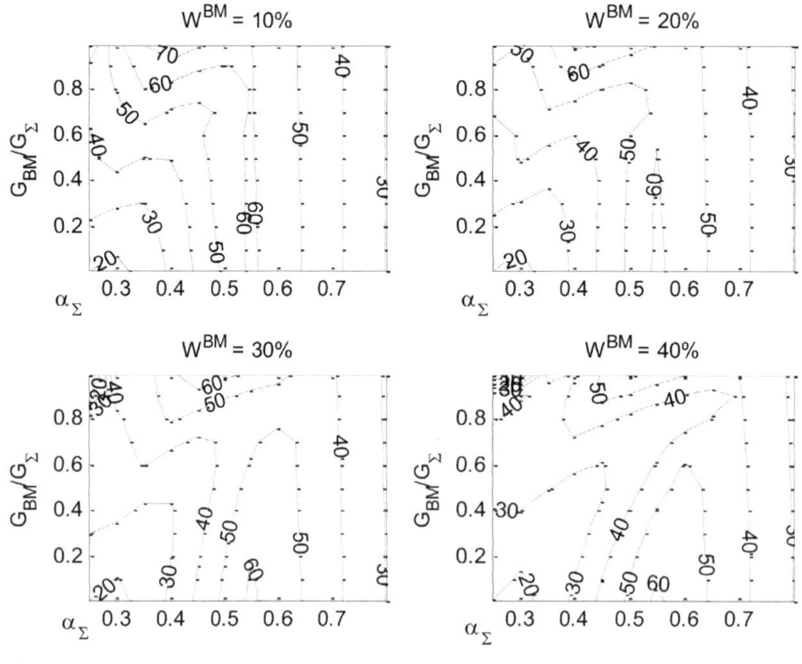

Fig. 1. Cold gas efficiency dependence on air ratio (α_Σ), biomass mass fraction (G_{BM}/G_Σ) and its initial moisture (W^{BM}).

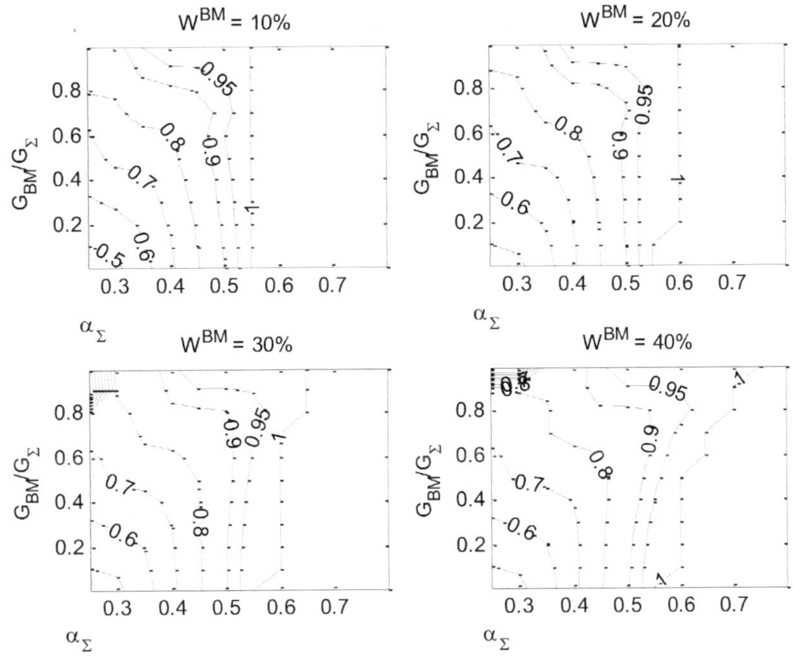

Fig. 2. Average fuel conversion degree dependence on air ratio (α_{Σ}), biomass mass fraction (G_{BM}/G_{Σ}) and its initial moisture (W^{BM}).

Lower limit of cold gas efficiency may be fixed at 60 % (allowable value in comparison with existing low-capacity gasifiers [6]). Then reasonable biomass fraction in fuels mixture will naturally decrease with moisture content: it is 50 % at 20 % of moisture in raw biomass; 15 % at 30 % moisture in raw biomass; and less than 5 % at 40 % of moisture in raw biomass.

Future work could be directed to investigation of efficient methods of heat recirculation for moist fuels gasification. Using this, it is possible to push boundaries of efficient biomass-coal co-gasification regimes.

This work considers thermal interaction of fuels in mixture only (e.g. by heat of chemical reactions). However, chemical interactions are possible that lead to changes in fuel reactivities. Development of mathematical models taking into account these features could be an object of future research.

Acknowledgments

This work was carried out at the Kutateladze Institute of Thermophysics SB RAS and financially supported by the Russian Science Foundation (project number 15-19-10025).

References

1. A. Demirbas, Prog. Energ. Comb. Sci. **30** (2004)
2. M. Asadullah, Renew. Sust. Energ. Rev. **29** (2014)
3. J. Rizkiana, G. Guan, W.B. Widayatno et al., Fuel **134** (2014)
4. I.G. Donskoi, A.V. Keiko, A.N. Kozlov et al., Therm. Eng. **60** (2013)
5. I.G. Donskoy, A.V. Keiko, A.N. Kozlov et al., Renew. Biores. **4**, 1 (2016)
6. D.A. Svishchev, A.N. Kozlov, I.G. Donskoy et al., Fuel **168** (2016)
7. A.H. Tchapda, S.V. Pisupari, Energies **7** (2014)

MATEC Web of Conferences 72, 01027 (2016) DOI: 10.1051/matecconf/20167201027
HMTTSC-2016

JOINT ELECTRICITY AND HEAT PRODUCTION POSSIBILITIES ON THE BASIS OF THE DISTRICT TRIGENERATION PLANT

Vadim Ershov[1,*], and *Ekaterina* Tarasova[1]

[1]National Research Tomsk Polytechnic University, 634050 Tomsk, Russia

Abstract. Renewable energy has become increasingly important in the last decade due to the limited reserves of fossil fuels and environmental issues related to CO2 emissions. The paper investigates the joint production possibility of electricity and heat in the Tomsk district Stepanovka, using modern trigeneration plant. The analysis was performed based on a comparison of the two alternatives for the area: the use of a diesel power plant (DPP) or gas-piston units (GPU) on biofuel. Technical and economic indicators for two different systems were calculated to evaluate the projects effectiveness.

1 Introduction

Reduction of natural hydrocarbon resources, rise in price of oil and gas production systems, electric and heat energy tariffs increase, environment pollution by emission of carbon dioxide and global warming [1,2] enhance the interest in the use of renewable energy sources. Personal power plants, operating on renewable energy, can become an alternative to traditional diesel power plants, operating on imported and expensive fuel, for rural areas of Tomsk region.

Energy resources consumption reduction and their efficient allocation can be provided by more efficient power plant with a combined production of heat / electricity / cold. [1]

The heat excess generated in the motor can be used for heat and cold generation for consumer needs. Usually, heat demand is dramatically reduced in summer, due to heating and ventilation necessity deficiency. However it's possible to dispose of the large volumes heat by means of cold power generation for manufacture process needs. This is possible by including compressor or absorption conditioners in the heat network diagram. This allows to increase the plant efficiency in summer time. [1, 2, 3]

The current state of the Stepanovka district communal infrastructure is characterized by a high grade of equipment deterioration; deterioration rate ranges from 48 % to 82 % for most objects. On the average, 68 % of utilities have served set time-limit. This results in excessive networks losses, the low equipment efficiency factor, excessive accident risk. This requires a higher cost of operation and maintenance of municipal infrastructure. Special attention should be given to following utilities problems:

* Corresponding author: kachalov@tpu.ru

• high level of fixed assets depreciation;
• engineering systems low reliability, especially electric power supply systems;
• insufficient replacement scale of electrical networks and modernization of boiler stations and DES equipment;
• insufficient network replacement scale and retrofitting.

According to the project the electrical energy production is planned to be covered by the diesel power plant and the biomass gas piston installation, the electric and heating systems will be reconstructed. Wastes from agriculture and timber industry may be used as a biofuel source. The exhaust steam will be provided for socially important facilities heating such as kindergarten, school, hospital and others to enhance efficiency.

The article deals with three major project aspects:
• required heat and electricity loads analysis and assessment;
• the technical and economic viability analysis;
• environmental impact assessment.

2 The district energy consumption volume assessment

It is necessary to evaluate the heat and electric power consumption volumes for power supply system optimization. The utilities consumption standard values for the hot water supply in premises are shown in the Table 1.

Table 1. Consumption standard values for the hot water supply.

No.	The premises accomplishment rate	The utilities consumption ratio (m^3 per month for 1 person)
1	Dwellings with central water supply, sewage and hot water supply	1.16
2	Dwellings with central water supply, hot water supply and without centralized sewerage	0.91
3	Dwellings with central water supply, sewage and hot water, equipped with sinks, kitchen sinks, showers	2.51
4	Dwellings with central water supply, sewage and hot water, equipped with seating baths, sinks and shower	3.02
5	Dwellings with central water supply, sewage and hot water, equipped with 1500-1700 mm baths, sink and shower	3.11

The utilities consumption ratio values for dwellings heating are shown in the Table 2.

Table 2. The utilities consumption ratio values for heating in residential and non-residential premises of the Tomsk region during the heating season.

Number of floors in a building	Gcal per 1 sq. meter of total floor area per month	
	Dwellings built before 1999	Dwellings built after 1999
1	0.0462	0.0194
2	0.0457	0.0175
3	0.0288	0.0177
4	0.0288	0.0155
5	0.0247	0.0155

3 The power plant configuration alternatives

1) Diesel power plant
2) Biofuel gas-piston installation
The annual saving of equivalent fuel consumption, obtained due to the combined development of energy by a mini-CHPP in comparison with a separate power generation by CPP and thermal energy by a local boiler station:

$$\Delta B = B_{eq} - B \text{ kg eq.t./year;} \tag{1}$$

Where $B = b_E^{mCHPP} E_{YEAR}^{mCHPP}$ - annual equivalent fuel consumption by mini-CHPP. B_{eq} – annual equivalent fuel consumption generated separately by CPP and thermal energy in a local boiler station including transport energy losses;

$b_E^{mCHPP} = \dfrac{0.123}{\eta_E^{ICE}}$ – equivalent fuel unit consumption of mini-CHPP, gas-piston internal combustion engine (ICE);

$E_{YEAR}^{mCHPP} = E_{CONS}$ - annual power generation of mini-CHPP, kW*h/year.

The fuel economy of the combined energy development is distributed in proportion to (types of the developed energy) relative shares:

$$\Delta B_E = \Delta B \beta_E \text{ kg eq.t./year;} \tag{2}$$

$$\Delta B_H = \Delta B \beta_H \text{ kg eq.t./year;} \tag{3}$$

Average annual unit consumption values of mini-CHPP equivalent fuel:

- for generated electric energy:

$$b_E^{mCHPP} = \frac{B_E^{HPP1} - \Delta B_E}{E_{CONS}(1+\alpha_{ON})} \text{ kg eq.t./kW*h;} \tag{4}$$

α_{ON} - power consumption coefficient for mini-CHPP own needs;

- for generated heat energy Q_H kW*h:

$$b_H^{mCHPP} = \frac{B_{BOIL} - \Delta B_H}{Q_H} \text{ kg eq.t./kW*h;} \tag{5}$$

Calculated average annual unit consumption values of mini-CHPP equivalent fuel:

- for generated electric energy:

$$B_E^{mCHPP} = Э_{GEN}^{mCHPP} \cdot b_E^{mCHPP} \text{ kg eq.t./year;} \tag{6}$$

- for generated heat energy:

$$B_H^{mCHPP} = Q_H \cdot b_H^{mCHPP} \text{ kg eq.t./year;} \tag{7}$$

The equivalent value of a total average annual consumption of conditional fuel used for mini-CHPP spent for production of electric and thermal energy:

$$B_{EQ} = B_E^{mCHPP} + B_H^{mCHPP} \text{ kg eq.t./year;} \tag{8}$$

The corrected shares of the equivalent fuel consumption used by mini-CHPP for generation of electric and thermal energy:

$$\gamma_E = \frac{B_E^{mCHPP}}{B_{EQ}} \tag{9a}$$

$$\gamma_H = \frac{B_H^{mCHPP}}{B_{EQ}} \tag{9b}$$

Total annual expenses for mini-CHPP:

$$E = E_{H\Sigma} + E'_{BOIL} + E_S + E_{ALL} + E_{AM} \text{ thousand rub;} \tag{10}$$

Where $E_{H\Sigma} = E_H^{REA} + E_H^{EST}$ - total annual expenditures for mini-CHPP biogas;

$E_H^{REA} = B_{BG}^{OVER} C_{BG}$ - total annual expenses for mini-CHPP biogas;

$E_H^{EST} = B_{BG}^{EST} C_{BG}$ - total annual expenses for power supply;

B_{BG}^{EST} - total consumption of mini-CHPP biogas for power supply;

C_{BG} - biogas costs, related to one ton of equivalent fuel;

E'_{BOIL} - expenses for additional heat power supply during winter, generated by boiler station;

$E_S = 12nE_E^{mCHPP}W$ - annual expenses for staff salary;

n - mini-CHPP staff coefficient, person/MW;

E_E^{nCHPP} - mini-CHPP electrical output capacity, MW;

W - corporate employee monthly salary, rub/month;

$E_{ALL} = 0.3E_S$ - salary allocation, rub;

Annual depreciation:

$$E_{AM} = p_{AM} C_\Sigma \tag{11}$$

Where p_{AM} - amortization allocations standard

Total mini-CHPP installed costs:

$$C_\Sigma = (C_{EQ} + C_{DE} + C_{SC} + C_{CS} + C_{JC}) \tag{12}$$

Where $K_{OБ}$ - mini-CHPP equipment costs, rub.;

$C_{DE} = (0.05 - 0.1)C_{SC}$ - design costs, rub.;

$C_{SC} = (0.15 - 0.2)C_{EQ}$ - construction and installation costs, rub.;

$C_{CS} = (0.03 - 0.05)C_{EQ}$ - commissioning costs, rub;

$C_{JC} = k_{CONN} E_E$ - power system connection costs, rub.;

k_{CONN} - unit connection costs, rub. /kW;

Production annual costs:

- electrical energy

$$E_E = \gamma_E E_\Sigma \text{ rub./year;} \quad (13)$$

-heat energy

$$E_H = \gamma_H E_\Sigma \text{ rub./year.} \quad (14)$$

Mini-CHPP electrical and heat energy prime costs (что это за термин?--себестоимость):

- electrical energy prime costs:

$$S_E = \frac{E_E}{E_{CONS}} \text{ rub./kW·h;} \quad (15)$$

- heat energy prime costs:

$$S_H = \frac{E_H}{Q_H^{mCHPP}} \text{ rub./kW·h; or } S_H = \frac{E_H}{1163 Q_H^{mCHPP}} \text{ rub./Gcal;} \quad (16)$$

Annual money inflow by means of energy saving using gas-piston mini-CHPP power supply system:

$$\Delta C_{YEAR} = C_{EL} + E_H + P_E; \quad (17)$$

Annual electric energy costs if electric network is used for power supply:

$$C_{EL} = E_{CONS} T_E \text{ rub./year;} \quad (18)$$

Where T_E - tariff on electric energy, consumed from central electric network, rub./kW*h;

Heat power supply costs using local boiling station:

$$C_H = Q_H T_H \text{ rub./year;} \quad (19)$$

Where T_H - local boiling station heat energy tariff, rub./kW·h.

Excess electric power selling, generated by mini-CHPP and provided to electric networks:

$$\Delta E_{YEAR} = E^{mCHPP} - E_{CONS} \text{ kW·h/year;} \quad (20)$$

Excess electric power selling annual profit:

$$P_E = \Delta E_{YEAR} T_E \text{ rub./year;} \quad (21)$$

Annual profit from construction, taking into account expenses of mini-CHPP and excess electric power sale, rub/year:

$$P_t = \Delta C_{YEAR} - \Delta H_{TC} - \Delta H_{PI} + A - E \ ; \qquad (22)$$

$\Delta H_{TC} = 0.02 C_{\Sigma}$ - property tax change due to investment, rub/year;

$\Delta H_{PI} = 0.2 \Delta C_{YEAR}$ - tax payments increment, rub/year;

$A = p_{AM} C_{\Sigma}$ rub/year – depreciation charges;

p_{AM} – depreciation charges coefficient;

Mini-CHPP construction net income:

$$NI = \sum_{0}^{n} \Delta C_{YEAR} - C_{\Sigma} \ ; \qquad (23)$$

Where $\sum_{0}^{n} \Delta C_{YEAR}$ - Money inflow during n mini-CHPP maintenance years, rub.

Net present value:

$$NPV = \sum_{t=0}^{n} \frac{Pt}{(1+g)^t} - \sum_{t=0}^{m} \frac{Kt}{(1+g)^t} \ ; \qquad (24)$$

Gas-piston biogas mini-CHPP construction payback period:

$$T_{ок} = \frac{C_{\Sigma}}{P_t} \ ; \qquad (25)$$

4 Environment emissions reduction

Annual gross atmosphere emissions were calculated for the operating diesel installations according to the technique [2]. As basic data we used:

- data from technical manufacturer documentation of diesel installation operational power for calculation of the maximum onetime emissions;
- calculations result of standard values of diesel engine annual fuel consumption for the evaluation of gross emissions in the atmosphere.

The average specific values of emissions indicators which were given in the technique [2] reflect the main toxicity parameters change regularities of diesels depending on a load and high-speed operating mode of the power plant, and also power and tip speed ratio of the diesel engine [2], [3]. Also, working process organization features of multi-cylinder powerful diesel installations are taken into account.

Emissions calculations are performed for the following harmful substances coming to the atmosphere with the exhaust gases of the diesel installations:

- carbon oxide (CO);
- nitrogen oxides (NOx) (in terms of NO2);
- hydrocarbons (CH)).

For stricter emissions accounting of gas-diesel installations it is expedient to be guided by ecological indicators certificates [2.3].

5 Conclusion

Energy resources consumption reduction and their efficient allocation can be provided by more efficient power plant with a combined production of heat / electricity.

Calculation results have shown that during the internal combustion engines work on pyrolysis gas there is a significant decrease in gross emissions in comparison with diesel installations. Calculated decrease has accounted for:

- ✓ CO by 42%;
- ✓ NOx by 50%;
- ✓ CH by 93%;
- ✓ C (soot) by 99,5%;
- ✓ SO_2 by 99,6%;
- ✓ CH_2O by 98%.

References

1. European Technology Platform on Renewable Heating and Cooling: 2020 – 2030 – 2050, Common Vision for the Renewable Heating & Cooling sector in Europe, Luxembourg: Publications Office of the European Union (2011)
2. H. Zinko et al., International Energy Agency (2008)
3. A. Nosrat, J. M. Pearce, Appl. Energy **88**, 9 (2011)

MATEC Web of Conferences **72**, 01028 (2016) DOI: 10.1051/matecconf/20167201028
HMTTSC-2016

EXPERIMENTAL OBSERVATION OF THE TEMPERATURE AT THE LIQUID/GAS INTERFACE

V. D. Velicanov[1], *E.* G. Orlova[1,*], *A.* S. Saigash[1], and *O.* S. Korneva[1]

[1]National Research Tomsk Polytechnic University, 634050 Tomsk, Russia

Abstract. Experimental studies on measurements of the temperature profile in the vicinity of the evaporating distilled water layer were carried out. The temperature was measured with a chromel-alumel thermocouple, when it moves under the action of the precision actuator directed perpendicular to the evaporating layer. According to the experimental results the fluctuation in temperature of the air and liquid was found in the immediate vicinity of the liquid/gas interface. However, temperature jumps near the interface was not observed.

1 Introduction

The main problem at modelling of processes involving the phase transitions is the determination of the boundary conditions prescribed on the thermodynamic variables, such as temperature and partial pressure on the liquid/gas interface. The computation of heat transfer at condensation or evaporation of liquid films is the example of these processes [1-7]. There is a nonequilibrium region in the vicinity of the interface due to the small mean free path of vapour molecules. It means that the temperature and vapour pressure are deflected from the surface temperature and the saturation pressure at this temperature [8]. That is, there is a possibility of an abrupt change of temperature and pressure [9] in the vicinity of the liquid/gas interface, even for Knudsen numbers close to zero. To date, the vapours generated above the liquid surface are considered as an ideal gas in the modelling of the evaporation or the condensation. In this case, the temperature and pressure gradients at the interface are determined by solving the problem on definition of the boundary conditions using Boltzmann equation.

The mathematical model of the liquid evaporation including the temperature gradient at the interface was developed in [9]. It was found that the evaporation rate of liquid is higher by three orders, if the temperature and jumps boundary conditions were used in accordance with the kinetic theory in comparison with using the "continuity" of the temperature and pressure [9].

The temperature field at the liquid/gas interface was studied experimentally during evaporation of the ultrapure water obtained by of Milli-Q purification system [10]. The constant thickness of the layer was kept by a dispenser device (syringe pump). The

* Corresponding author: fdv@tpu.ru

temperature was recorded by the thermocouples with a diameter of 150 μm, located on the movable part of the linear actuator. The thickness of the liquid layer and the surface temperature were kept constant and were equal to 1.75 mm and 31 °C, 52 °C, 92 °C, respectively. According to the results of the experiments the temperature gradient at the interface was recorded. The average value of the gradient was 1.3 ° C, and it is increased with increasing substrate temperature (the heater). For example, the gradients for temperatures of 31, 52 and 92 °C were 0.2, 1 and 3 °C, respectively.

The temperature gradient indicates the presence of thermal resistance on the liquid surface. If this resistance is present, it is necessary to consider the inequality of evaporation and condensation coefficients in the numerical study of these processes.

The aim of this work is to determine experimentally the existence of the temperature gradient at the liquid/gas interface under the condition of thin liquid layer evaporation into the atmosphere.

2 Experimental procedures

The studies were carried out on the experimental setup (fig. 1) with using the equipment of the shadow method [11, 12].

The temperatures of the upper and lower surfaces of the heater, liquid and environment were measured using six chromel-alumel thermocouples *6* (OMEGA 5TC-TT-K-40-36). Two by two thermocouples were located on the top and bottom surfaces of the metal plate, one in the liquid layer and one in the environment inside the box.

Signals from thermocouples, were transferred to the controller-ADC *7* (NI cDAQ-9171) further to the personal computer *12*, where the processing and displaying on the screen of data were implemented in software developed in LABVIEW Signal Express 2012. The substrate used was PTFE; the liquid was distilled water. The surface temperature of the substrate was kept constant equal to 56 °C.

Temperature measurements in the immediate vicinity of the liquid/gas interface were performed using the calibrated thermocouple K type (OMEGA 5TC-TT-K-40-36) with a thermojunction of 80 microns. The responses were tested with standard resistance thermometers ETS-100 with the error in the test temperature ranges of 0.05 °C. The estimation of measurement errors was carried out as described in [13]. According to results of calculation the error does not exceed 2 %.

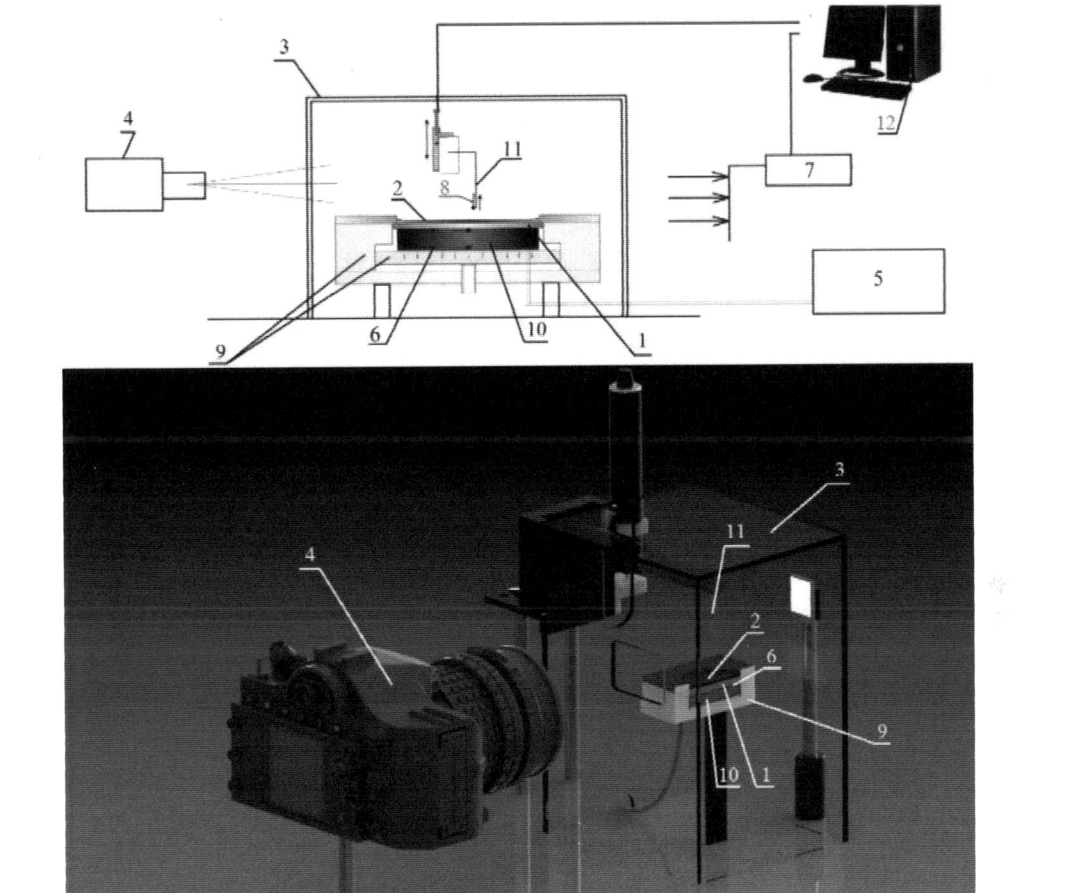

Fig. 1. Scheme of the experimental setup: 1 – substrate; 2 – liquid; 3 – transparent box; 4 – photographic camera; 5 – power supply; 6, 8 – thermocouples; 7 – analog-to-digital converter; 9 – thermal insulating wall; 10 – metal plate; 11 – gauge (needle of the actuator); 12 – personal computer.

The position of the micro-thermocouple was controlled by the precision actuator Zaber with increment of 1 mm. The travel speed of the movable part of the actuator was 100 μm/s.

Distilled water was dosed to the substrate after heating its surface up to a predetermined temperature of 56 °C. In this case the layer of air at a distance of 2 mm above the surface was heated to 55 °C (fig. 2). Next, the actuator was turned on and the thermocouple moved to the surface of liquid/gas interface.

3 Results and discussion

Figure 2 presents the typical experimental dependence of the temperature change from the position of thermocouple in the vertical plane relative to the liquid/gas interface.

The point 0.0 on the vertical axis corresponds to the liquid/gas interface. Negative values of this axis are the coordinates of the liquid phase (distilled water); the positive ones are the gas phase (air). Presented dependence (fig. 2) was plotted based on the results of five experiments as the arithmetic mean of the recorded data.

The fluctuation in temperature of the air and liquid in the immediate vicinity of the liquid/gas interface was found according to analyses of the dependence (fig. 2). What is more, no temperature jumps were observed on a distance of ± 80 microns from the interface (fig. 2). This distance corresponds to a thermojunction of a thermocouple. The fluctuation nature of the temperature can be explained as follows. When liquid is dozed on the surface the under layer, which contacts to the surface, warms up. The temperature gradient and the density gradient appear; they directed from the interface. It leads to liquid convection. The top colder layer of liquid moves down, and warmed layer rises to the interface. It is obvious that the circulation of liquid affects the movement of the air near the interface. And as a result, the fluctuations of the temperature appear. Under the normal conditions with atmospheric pressure this thickness is equal to 1 mm for water vapour [9].

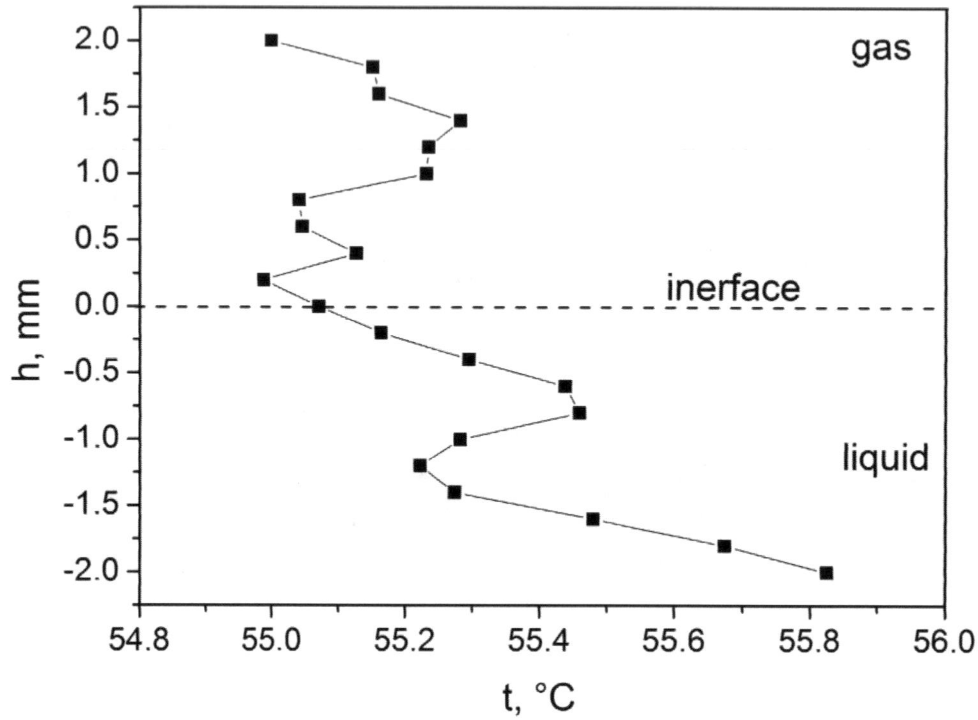

Fig. 2. Dependence of the temperature from the position of thermocouple in the vertical plane relative to the liquid/gas interface.

Therefore, for further solving the problem of determining the thermodynamic parameters at the interface and their influence on the process of evaporation and condensation it is necessary to conduct the experimental studies using thermocouple with a diameter of a thermojunction up to 10 μm.

According to results of the experiments the condensate layer was found to form on the thermocouple during its movement to the interface. To the best of our knowledge, it will affect the recorded parameters.

So, it can be concluded that it is necessary to use a thermocouple with horizontally ("telescoping") arranged electrodes. Such arrangement will correspond to the isotherms of the liquid/gas interface and will reduce the heat loss associated with the temperature gradient across the electrodes of thermocouples. Also it will reduce the thickness of the condensate layer forming on the surface of the thermocouple.

Acknowledgments

The reported study was supported by the grant of President of Russian Federation for the government support of young Russian scientists (MK-6810.2016.8).

References

1. S. Ya Misyura, Chemical Physics Letters **583**, 34 (2013)
2. V.P. Lebedev, V. V. Lemanov, S. Ya. Misyura, V. I. Terekhov, Fluid Dynamics **28**, 624 (1993)
3. K. Batischeva, E. Orlova, D. Feoktistov, MATEC Web Conf. **19**, 01001 (2014)
4. I. Afanasyev, E. Orlova, D. Feoktistov, EPJ Web Conf. **82**, 01054 (2015)
5. D.O. Glushkov, G.V. Kuznetsov, P.A. Strizhak, Thermal Science **19**, 1541 (2015)
6. A.S. Krasnoshlykov, V.I. Maksimov, EPJ Web Conf. **82**, 01046 (2015)
7. V.I. Maksimov, T.A. Nagornova, V.S. Chernyshev, MATEC Web Conf. **23**, 01051 (2015)
8. J. W. Cipolla Jr., H. Lang, S. K. Loyalka, J Chem Phys. **61**, 69 (1974)
9. E. Ya. Gatapova, I.A. Graur, F. Sharipov, O.A. Kabov, Int. J. Heat Mass Transfer **83**, 235 (2015)
10. E.Ya. Gatapova, R.A. Filipenko, Yu.V Lyulin, I.A. Graur, I.V. Marchuk, O. A. Kabov, Thermophysics and Aeromechanics **22**, 701 (2015)
11. G.V. Kuznetsov, D.V. Feoktistov, E.G. Orlova, J. Eng. Phys. Thermophys. **89**, 317 (2016)
12. D.V. Feoktistov, E.G. Orlova, A.G. Islamova, MATEC Web Conf. **23**, 01054 (2015)
13. N.A. Yaryshev, *Theoretical Fundamentals of Non-Stationary Temperature Measurements* (Energoatomizdat, Leningrad, 1990)

APPLICATION OF EVAPORATIVE COOLING FOR THE CONDENSATION OF WATER VAPORS FROM A FLUE GAS WASTE HEAT BOILERS CCP

Nikolay Galashov[1,*], *Svyatoslav* Tsibulskiy[1], and *Alexander* Kiselev[1]

[1]National Research Tomsk Polytechnic University, 634050 Tomsk, Russia

Abstract. The object of the study are boilers that burn organic fuel and the recovery boilers (RB) of the combined cycle plant (CCP), which are also working on the products of the combustion of hydrocarbon fuels. The purpose of research is to find technologies that increase efficiency of the thermal power plant (TPP) and technologies that reduce the environmental impact on the environment by burning fossil fuels. The paper deals with the technology of the boilers burning hydrocarbon fuel with condensation of water vapor from the exhaust flue gases. Considered the problems caused by using of this technology. Research shows that the main problem of this technology in the boilers is the lack of reliable methods of calculation of heat exchangers, condensers. Particular attention is paid to the application of this technology in the recovery boilers combined-cycle plants, which are currently gaining increasing use in the generation of electricity from the combustion of gas in power plants. It is shown that the application of technology of condensation of water vapor in RB CCP, the temperature decreases of exhaust gases from 100 to 40 °C, allows increasing the efficiency of the RB with 86.2 % to 99.5 %, i.e. at 12.3 %, and increase the efficiency of the CCP at 2.8 %.

1 Introduction

For many years, the design and operation of boilers that burn organic fuel, one of the main tasks was to prevent condensation of the vapor in the exhaust flue gases, as contained in the condensate, sulfur oxides, nitrogen and other harmful substances lead to metal corrosion of the heating surfaces and flues, and as the destruction of the inner wall of the chimney.

In order to prevent condensation of the vapor is necessary to maintain the temperature of the flue gases above the dew point, at which the combustion of natural gas, depending on the excess air, is about 40–60 °C, and the combustion of coal, depending on the chemical composition of the flue gas may be below or above this temperature. Taking into account the reducing the temperature exhaust gases at lower boiler loads for nominal mode it usually take for 70-110 °C gas and coal 140-160 °C. High temperature leads to a great loss of

* Corresponding author : gal@tpu.ru

heat from the exhaust gases, resulting in a gross efficiency of the boiler at the gas does not exceed 95 %, and coal – 90 %, based on the bottom heat of combustion of fuel. At the same time useless lost heat spent on evaporation of moisture, and oxides are all harmful substances out with the flue gases and pollute the environment.

To improve the efficiency of the boiler, the reliability of the gas ducts, exhaust and chimney, as well as reducing the emission of harmful substances, based on available technology cooling flue gases below the dew point by means of heat exchangers, condensers. This technology is currently recognized as one of the most resource- and energy-efficient and have been widely used in many countries around the world, as evidenced by the rapidly growing number of scientific publications [1-15] and commercial offers for the sale of boilers based on it. While this technology is increasingly used in small boilers running on gas, but there are a number of works on the application of it to the hot water and vapor boilers of high power, as well as in the recovery boilers (RB) combined cycle gas turbines (CCGT) [16].

To provide reliable condensing flue gas temperature should be below the dew point, i.e., when working with natural gas – to be below 44 °C (excess air ratio 2.5), wherein when calculating the boiler efficiency at higher heating value will be around 99 %, while in non-condensing boiler it will be 85 %, which reduces fuel consumption by 15 %.

The amount of harmful emissions (CO_2, NO_X) in the flue gas leaving boiler with condensation of water vapor in 5-7 times lower than in traditional gas-fired boilers. A coal-fired boilers have the effect of trapping of harmful substances in the condensation of water vapor from the flue gases even higher.

2 Application problem

The combustion of a hydrocarbon fuel, depending on the type of fuel in the flue gas is formed about 10-15 % moisture. To capture moisture from the flue gases and the contact surface applied-exchangers capacitors. Application for cooling flue gas condensing heat exchanger surface display type allows air or water flow in the circuit of stations useful in the process. The coefficients of heat transfer from the flue gas to the heating surfaces with a deep cooling, accompanied by condensation of water vapor, considerably higher coefficients of convective heat and substantially comparable to the heat exchange coefficients for the contact devices. The disadvantages of surface heat exchangers should include large size, which often does not allow to place them directly in the boiler; high cost of the metal, often stainless steel.

More often for dehumidifying contact apply flue gas heat exchangers in which the flue gas for cooling and condensation of the evaporator, which is applied in the cooling of the gases by reaction with droplets or jets of water film. These devices are much smaller in size and surface do not require expensive materials, standing. They have a high heat transfer rate substantially greater than the heat transfer coefficients of convective heat transfer. Their drawback is that the heated water is absorbed from the flue gas oxygen, carbon dioxide, sulfur oxides and becomes corrosive-aggressive properties, so for further use requires cleaning. For natural gas RB CCP condensate practically is not exposed to contamination.

Currently, there are a number of methods for the calculation of heat exchangers with condensation of water vapor from the gas, but the variety of types and designs devices requires further research to improve the methods of calculation of heat exchangers.

Significant problem when using condensing boilers of water vapor from the flue gas is to ensure reliable operation of the flues and chimney. The low temperature of the flue gas with a residual content of water vapor can lead to condensation in the ducts and chimney, and the destruction of equipment. Therefore, after the heat exchanger drained gases preheated stream of hot gases passing by the heat exchanger, which leads to underutilization of the heat of condensation in the boiler. At low flue gas temperatures can be for flues and flue pipes to apply new construction materials not subject to corrosion by the condensation of water vapor.

3 Prospects for the use of technology at CCP

Consider how is possible to increase the efficiency of RB and efficiency of CCP using this technology in the recovery boiler.

The efficiency of CCP can be determined by the equation [17, 18]

$$\eta_{CCP} = \eta_{GTP} + (1-\eta_{GTP})\eta_{RB}\eta_{STP}, \tag{1}$$

where η_{GTP} – efficiency of the gas turbine plant (GTP); η_{RB} – efficiency of the RB; η_{STP} – КПД steam turbine plant (STP).

For example, take the CCP that has $\eta_{GTP} = 0.36$ and $\eta_{STP} = 0.33$. In RB from GTP gases with excess air ratio flow $\alpha_{EX} = 2.7$ with temperature 600 °C, and from RB exhaust to the chimney at a temperature of 100 °C. Efficiency of RB determined by the equation

$$\eta_{RB} = \frac{h_{GTP} - h_{EX}^{100}}{h_{GTP} - h_{15}}, \tag{2}$$

where h_{GTP} – enthalpy of gas at the inlet RB, at the given parameters is equal to 650.4 kJ/kg; h_{EX}^{100} – enthalpy gas at the outlet RB, at the given parameters is equal to 102.4 kJ/kg; h_{15} = 15.4 kJ/kg – enthalpy gas at the temperature 15 °C, this is the temperature the air inlet to the compressor GTP. Substituting into the equation (2) the enthalpy can be obtain η_{RB} = 0.862, and from the equation (1) obtain the $\eta_{CCP} = 0.542$.

Now, assume that the output gases from the RB set-pin condenser heat exchanger in which the gases are cooled by water to a temperature $t_{EX} = 40$ °C, which is below the dew point $t_{DP} = 43$ °C. The moisture content at the inlet of the heat exchanger-condenser at a temperature of 100 °C and the excess air coefficient $\alpha = 2.7$ is $d_C = 56$ g/kg dry gases and the output will be 47 g/kg. Thus, condensation of water vapor flow rate with respect to entering of the RB gases will be $\Delta d_C = 9$ g/kg. The heat of condensation can be useful in CCP scheme used to heat water or air, and condensation can be used after removal of dissolved gases, such as additional water. In this case, efficiency of RB is determined by the equation

$$\eta_{RB}^{cond} = \frac{(h_{GTR} - h_{EX}) + (1-\Delta d_C)(h_{EX} - h_{EX}^{C}) + \Delta d_C r + d_C(h_C - h_{AW})}{h_{GTR} - h_{15}}, \tag{3}$$

where h_{EX}^{C} = 41.0 kJ/kg – enthalpy at t_{EX} = 40 °C and α_{EX} = 2.7; h_C – enthalpy of the condensate, at the temperature 40°C is equal to 167.6 kJ/kg, heat of condensation r = 2406 kJ/kg; h_{AW} = 41.9 kJ/kg –enthalpy of additional water at 10 °C. As a result equation (3), obtain the η_{RB}^{cond} = 0.995, according to the equation (1) obtain $\eta_{CCP} = 0.570$.

Thus, application of the technology of condensation of water vapor in the RB can improve the efficiency with 86.2 % to 99.5 %, i.e. 12.3 %, and the efficiency of the CCP to increase by 2.8 %.

4 Conclusion

Application of technology of boilers with condensation of water vapor from the flue gas is promising to improve the efficiency of facilities and protecting the environment.

- The main problem of the use of technologies of condensation heat from the flue gas in boilers is the lack of reliable methods of calculation of heat exchangers, condensers. This issue requires further research.
- Application of the technology of condensation of water vapor from the of exhaust gases in the RB CCP for example considered it possible to increase efficiency of RB from 86.2 % to 99.5 %, i.e. 12.3 %, and the efficiency of the CCP to increase by 2.8 %.

Acknowledgments

The work has been supported by the Russian Federal focused program "Research and development on priority orientation of science and technology in Russia 2014–2020", unique identifier 'Applied Research and Experimental Developments' project RFMEFI58114X0001.

References

1. M. Osakabe, Heat Transfer – Asian **3**, 12 (2001)
2. D.F. Che, Y.H. Liu, C.Y. Gao, Energy Convers. Manage. **45**, 16 (2004)
3. D.F. Che, Y.D. Da, Z.N. Zhuang, Heat Mass Transfer **41**, 6 (2005)
4. X. Shi, D. Che, B. Agnew, J. Gao, Int. J. Heat Mass Transfer **54**, 10 (2011)
5. K. Jeong, M. J. Kessen, H. Bilirgen, E.K. Levy, Int. J. Heat Mass Transfer **53**, 8 (2010)
6. K. Hwang, C.H. Song, K. Saito, S. Kawai, Appl. Therm. Eng. **30**, 8 (2010)
7. X.H. Ma, X.D. Zhou, Z. Lan, Y.M. Li, Y. Zhang, Int. J. Heat Mass Transfer **51**, 10 (2008)
8. E. Levy, H. Bilirgen, C. Samuelson, K. Jeong, M. Kessen, C. Whitcomb, International Technical Conference on Coal Utilization and Fuel Systems (2008)
9. D. Hazell, *Modeling and Optimization of Condensing Heat Exchangers for Cooling Boiler Flue Gas* (Ph.D. Dissertation, Lehigh University, 2011)
10. K. Jeong, *Condensation of Water Vapor and Sulfuric Acid in Boiler Flue Gas* (Ph.D. Dissertation, Lehigh University, 2008)
11. M. Lavigne, *Numerical Simulations of Condensing Heat Exchangers with Oxyfuel Flue Gas* (Thesis, Lehigh University, 2010)
12. E. Levy, H. Bilirgen, M. Kessen, D. Hazell, B. Carney, Am. Soc. Mech. Eng. **1** (2011)
13. D.D. Hazell, *Modeling and Optimization of Condensing Heat Exchangers for Cooling Boiler Flue Gas* (Ph.D. Dissertation, Lehigh University, 2011)
14. P. Frank, *Fundamentals of Heat and Mass Transfer* (John Wiley&Sons, Inc., Hoboken, 2007)
15. M. Lavigne, *Numerical Simulations of Condensing Heat Exchangers with Oxyfuel Flue Gas* (Thesis, Lehigh University, 2010)
16. I.V. Budakov, V.D.Burov, New in the Russian power industry **11**, 11 (2011)
17. A.D. Truhny, *Combined cycle gas turbine power plants* (Publishing House MEI, 2013)
18. R. Kehlofer, B. Rukes, F. Hannemann, F. Stirnimann, *Combined-Cycle Gas & Steam Turbine Power Plants* (PennWell Corporation, 2009)

MATEC Web of Conferences **72**, 01030 (2016) DOI: 10.1051/matecconf/20167201030
HMTTSC-2016

THE DRAG REDUCTION OF MICROCHANNEL FLOW BY CONTRAST WETTABILITY

Elizaveta Ya. Gatapova[1,*], and *Dmitry* S. Gluzdov[1,2]

[1]Kutateladze Institute of Thermophysics SB RAS, 630090 Novosibirsk, Russia
[2]Novosibirsk State University, 630090 Novosibirsk, Russia

Abstract. The model is developed and the boundary condition is approximated at the lower boundary of the channel. It is shown that a hydrophobic surface can significantly reduce the hydraulic resistance due to the occurrence of bubble layer. Moreover, at the slip ratio equal to 1 mm, the water flow rate in the channel with the height of 1 mm increases by 2.5 times.

1 Introduction

The endeavor of manufacturers to increase the processors speed requires the reduction of their sizes that inevitably leads to problems associated with heat transfer. Contemporary liquid cooling systems are effectively used in combination with the microchannel structure. At that, the smaller the microchannel thickness, the more efficient is heat removal. However, the reduction of the microchannel thickness leads to a strong increase in the hydraulic resistance that substantially prevents heat transfer and results in the necessity to maintain high pressure drops at the microchannel inlets and outlets [1, 2]. This, in turn, affects the size of the entire cooling system.

It is clear that the further improvement of advanced liquid cooling systems is associated with the problem of reducing the hydraulic resistance in microchannels. One of the previously proposed solutions to this problem is the application of additional microstructure on the surface of the microchannels [3].The essence of this solution lies in the fact that the additional microstructure of the microchannel contains gas, which does not hinder the liquid when contacting with the applied surface, unlike the microchannel walls, to which liquid is adhered as consistent no-slip boundary condition [4].

This technique may allow reducing the hydraulic resistance of up to 40% [5]. However, it has drawbacks related to the high cost of improved microchannel manufacturing, as well as to the problem of keeping gas within the microstructure [6], maintaining flow stability [7], and preventing rupture of a liquid film [8,9].

Another solution to the problem of high hydraulic resistance in a microchannel is proposed in [10]. It was revealed that on the microchannel surface coated with a hydrophobic material, a bubble layer is formed. We proposed to use this effect to ensure liquid slippage at the surface of local microchannel sections that will allow reducing

[*] Corresponding author: egatapova@gmail.com

hydraulic resistance. The application of contrast hydrophobic coating can enhance heat transfer and improve the liquid pumping conditions in general.

The results obtained previously in [10], support further research prospects in this area. The aim of this work is to study the flow characteristics in a microchannel with applied contrast hydrophobic coating through a numerical solution of the two-dimensional system of Navier-Stokes equations with a singularity in the boundary condition at the lower boundary.

2 Formulation of a problem

A tentative diagram of the problem is shown in fig. 1. The lower wall of the microchannel with specified length and height is coated with contrast hydrophobic surface to form bubble layer. The distribution of velocity and pressure at the inlet to the microchannel, as well as all the parameters of the flowing liquid, such as viscosity and density, are known. The liquid is considered incompressible. Water served a working liquid.

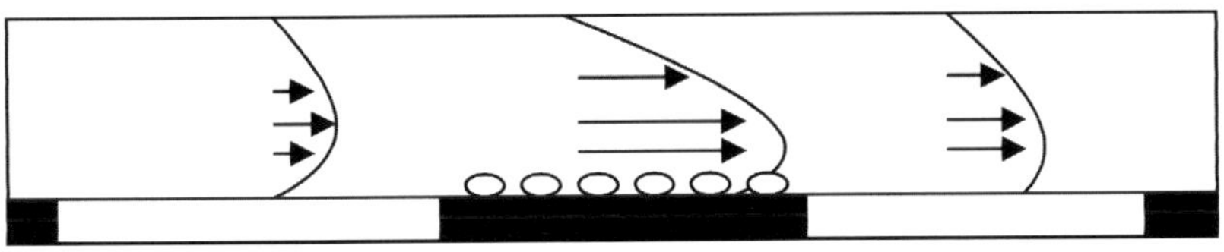

Fig. 1. Visualization of the problem.

It is required to find velocity and pressure fields in the microchannel through numerical solution of the Navier-Stokes equations.

To obtain a complete system of the Navier-Stokes equations (1), we use at the lower boundary the conventional Navier condition with the feature in the β function, which takes its maximum value on the surfaces with a hydrophobic coating and becomes zero on a conventional microchannel surface. On the boundary between the regular and hydrophobic surfaces, the β function is smoothed.

$$\begin{cases} \dfrac{\partial u}{\partial t} + u\dfrac{\partial u}{\partial x} + v\dfrac{\partial u}{\partial y} = -\dfrac{1}{\rho}\dfrac{\partial P}{\partial x} + \eta\left(\dfrac{\partial^2 u}{\partial x^2} + \dfrac{\partial^2 u}{\partial y^2}\right) + g\sin\alpha \\ \dfrac{\partial v}{\partial t} + u\dfrac{\partial v}{\partial x} + v\dfrac{\partial v}{\partial y} = -\dfrac{1}{\rho}\dfrac{\partial P}{\partial y} + \eta\left(\dfrac{\partial^2 v}{\partial x^2} + \dfrac{\partial^2 v}{\partial y^2}\right) + g\cos\alpha \\ \dfrac{\partial u}{\partial x} + \dfrac{\partial u}{\partial y} = 0 \\ u = \beta(x)\dfrac{\partial u}{\partial y} \end{cases} \quad (1)$$

Here u and v - are the horizontal and vertical velocity components, respectively; η - is the kinematic viscosity.

The boundary condition for the horizontal velocity component on the upper wall corresponds to the conventional no-slip boundary condition, while the condition on the lower wall is approximated by formula (2) using the technique shown below. For the vertical velocity we use the impermeability condition. In other cases, where the boundary condition is needed in the numerical scheme, we use the so-called "soft conditions", i.e. the first derivative of a sought quantity is made equal to zero. Function $\beta(x)$ (2) has the following configuration (fig. 3).

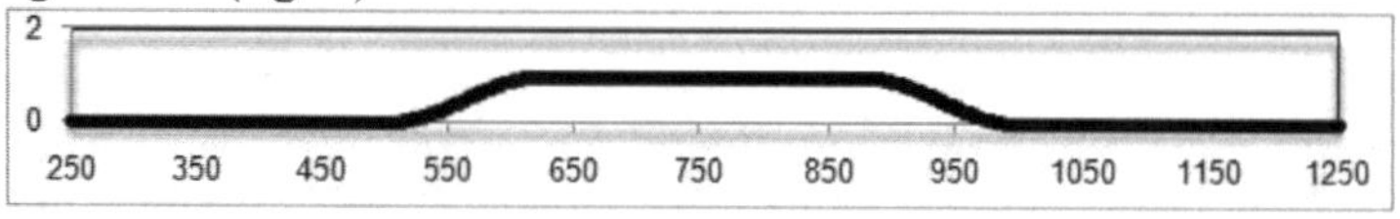

Fig. 2. Approximation of $\beta(x)$ function.

$$\beta = \begin{cases} 0, x \in [l_i; l_{i+1}] \\ \beta_{\max} \left[\dfrac{1 + \sin\left(\pi \dfrac{x - l_{i+1}}{\Delta l} - \dfrac{\pi}{2} \right)}{2} \right], x \in [l_{i+1}; l_{i+1} + \Delta l] \\ 1, x \in [l_{i+1} + \Delta l; l_{i+2} - \Delta l] \\ \beta_{\max} \left[\dfrac{1 - \sin\left(\pi \dfrac{x - l_{i+2} + \Delta l}{\Delta l} - \dfrac{\pi}{2} \right)}{2} \right], x \in [l_{i+2} - \Delta l; l_{i+2}] \end{cases} \quad (2)$$

where $\beta_{\max}$ - is the maximum value of the slippage coefficient, l - is the length of the section with hydrophobic coating, Δl - half of the section, on which the function is smoothed, i – is the number of a section with hydrophobic coating.

3 Results

The problem is solved for the Poisson flow of the water at the section with the bubble layer, pressure gradient $\partial P/\partial x$ = 0.1 atm/m, dynamic viscosity μ = 0.001 kg/m*s, the characteristic bubble size δ = 1 μm, and the slippage length β=1 mm. We have revealed that the flow rate has increased 2.5 times in comparison with that for the section without bubble layer (fig.3).

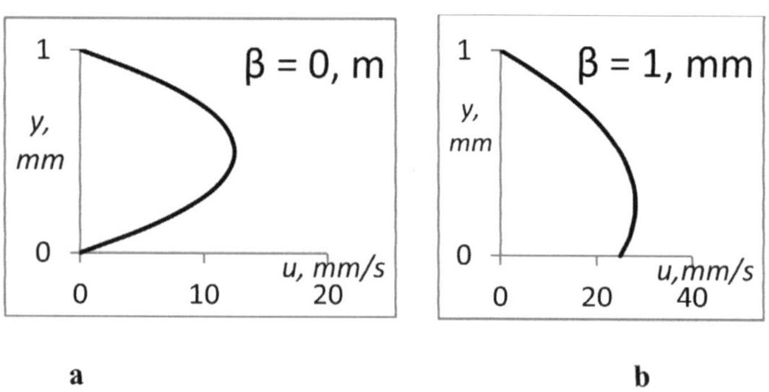

Fig. 3. The velocity profile: a - section with the conventional surface; b - section with the hydrophobic surface.

4 Conclusions

A model was developed and the boundary condition was approximated at the lower boundary of the channel. It is shown that a hydrophobic surface can significantly reduce the hydraulic resistance due to the occurrence of bubble layer. Moreover, at the slip ratio equal to 1 mm, the water flow rate in the channel with the height of 1 mm increases by 2.5 times.

Acknowledgments

The work was supported by the Russian Science Foundation (project No. 14-19-01755).

References

1. T.M. Squires and S.R. Quake, Rev. Mod. Phys. **77**, 977 (2005)
2. H.A. Stone, A.D. Stroock, A. Ajdari, Annual Reviews of Fluid Mechanics **36** (2004)
3. J. Ou, B. Perot, and J. P. Rothstein, Phys. Fluids **16** (2004)
4. E. Karatay, A.S. Haase, C.W. Visser, C. Sun, D. Lohse, P.A. Tsai, R.G.H. Lammertnik, Proc. Nat. Acad. Sci. **110**, 8422 (2013).
5. J.P. Rothstein, Annual Reviews of Fluid Mechanics **42**, 89 (2010)
6. D. Byun, J. Kim, H.S. Ko, H.C. Park, Phys. Fluids **20**, 113601 (2008)
7. V.S. Ajaev, E.Ya. Gatapova, O.A. Kabov, Phys. Fluids **25**, 122102 (2013)
8. D.V. Zaitsev, O.A. Kabov, V.V. Cheverda, N.S. Bufetov, High Temperature **42**, 3 (2004)
9. D.V. Zaitsev, D.P. Kirichenko, O.A. Kabov, Tech. Phys. Lett. **41**, 6 (2015)
10. E.Ya. Gatapova, V.S. Ajaev, O.A. Kabov, JETP Lett. **101**, 3 (2015)

THE BOILER FURNACE NUMERICAL MODELING OF BURNING NON-STANDARD FUEL

A. V. Gil[1,*], *O.* M. Koksharev[1], *D.* V. Lebed[2], and *A.* V. Starchenko[3]

[1]National Research Tomsk Polytechnic University, 634050 Tomsk, Russia
[2]OJSC DGK, Vladivostok, Russia
[3]National Research Tomsk State University, 634050 Tomsk, Russia

Abstract. The aim of paper is to study the combustion processes of organizing non-standard fuel burning using a numerical approach in a boiler unit BKZ–210–140 with steam capacity of 210 t/h. The Euler-Lagrange aerothermochemical technique of describing physical and chemical processes in the furnace volume by the FIRE 3D application package on the base was used. Special attention is paid to the numerical modeling results conclusion of the aerodynamics processes, heat exchange, and combustion in the furnace volume.

1 Introduction

Often in the course of thermal power plants continuous operation standard solid fuel exceeded its storage. In such event boiler operation at the off-design coal is prohibited and therefore it is necessary to test the boiler for possibility of non-standard fuel burning [1].

This paper considers the possibility of organizing non-standard fuel burning using a numerical approach [2, 3] in a boiler unit with steam capacity of 210 t/h, which are common in regions of Siberia, the Far East and Primorski Krai.

2 Problem statement

The research object is the vertical water-tube boiler unit E-210-140 with single drum, natural circulation, double-pass configuration and dry-ash removal. Combustion chamber is equipped with four straight-flow burners with tangential layout, has volume of 1093 m^3 and a cross-sectional size of 7424 × 7808 mm. The standard fuel is Chihezsky brown coal.

As an alternate fuel adopted Pavlovsky brown coal. Table 1 shows the basic burning characteristics of the design and alternate coal [4].

* Corresponding author: andgil@tpu.ru

Table 1. Burning characteristics of the design and alternate coal.

	Unit of measure	Chihezsky brown coal	Pavlovsky brown coal
Brand	–	1B	1B
Humidity	%	43.0	41.5
Ash	%	12.5	19.6
The sulfur content	%	0.2	0.4
The carbon content	%	30.3	25.7
The hydrogen content	%	2.5	2.3
The nitrogen content	%	0.4	0.3
The oxygen content	%	11.2	10.2
Lower heating value of the fuel	MJ/kg (kkal/kg)	10.72 (2560)	9.13 (2180)

Comparing the two types of coal, it can be noted that mineral content of the Pavlovsky brown coal is higher and carbon content is lower. Therefore, the non-standard fuel value is lower that increase its consumer in comparison to the standard fuel.

3 Mathematical model

In this study, the Euler-Lagrange aerothermochemical technique of describing processes by the FIRE 3D application package on the base was used [5, 6]. Turbulent gas properties are calculated using the "k-ε" two-parameter turbulence model. Radiative heat transfer is represented within the P1 method approximation of spherical harmonics.

Tasks are solved numerically on non-staggered grids in the Cartesian coordinate system using the relaxation method. Discretization of initial differential equations is performed by using the finite volume method [7]. Diffusion flow on the facets of the control volume are approximated by a central difference scheme, which has the second order of accuracy. To determine the convective flows Van Leer monotonic upwind scheme of the second-order was used.

4 Results and discussion

Figure 1 shows the formation of dead area close to the aerodynamic nose, topping, back and front walls. The velocity in these areas does not exceed 1.5 m/s. With the rising of gas circuit, velocity is aligned across the width and depth of the furnace and reach the average value of 10 m/s. In the horizontal cross section along the axis of the burners, maximum speeds occur at the outlet of embrasures - more than 16 m/s. With the approaching the axis of combustion chamber the velocity is reduced and has a value of about 10 m/s. When flows encounter occurs, there are the formation of vortex in vertical planes in above and below regions next to the zone of burners.

Figure 2 shows the pattern of temperature distribution in different sections. Analyzing these results, three zones can be highlighted. The first one is the dry-bottom hopper zone: the temperature in this area is about 1300 K. The second one is the zone from the dry-bottom hopper beginning to the aerodynamic nose: the temperature here is about 700 K. The third one begins from aerodynamic nose and continues above it: the temperature range around 1400 K.

Temperature distribution fields in furnace volume is chaotic. There are local zones of high temperatures - the top right corner of furnace.

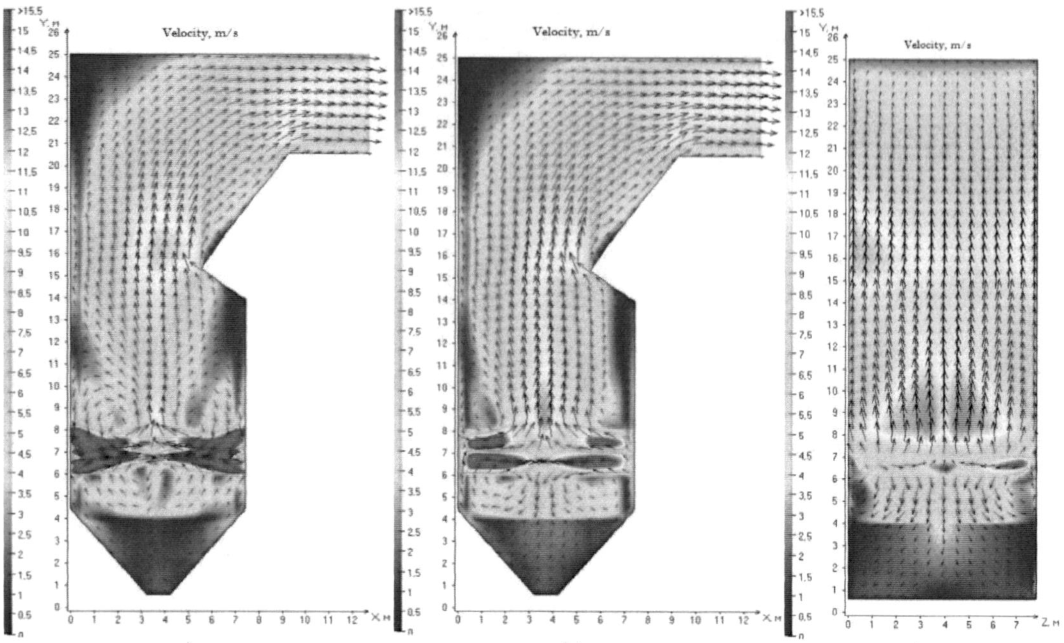

a) b) c)

Fig. 1. The aerodynamic flow pattern in a longitudinal cross-section: a - at a distance of 1 meter from the side wall; b - along the transverse axis of the furnace; c - along the longitudinal axis of the furnace.

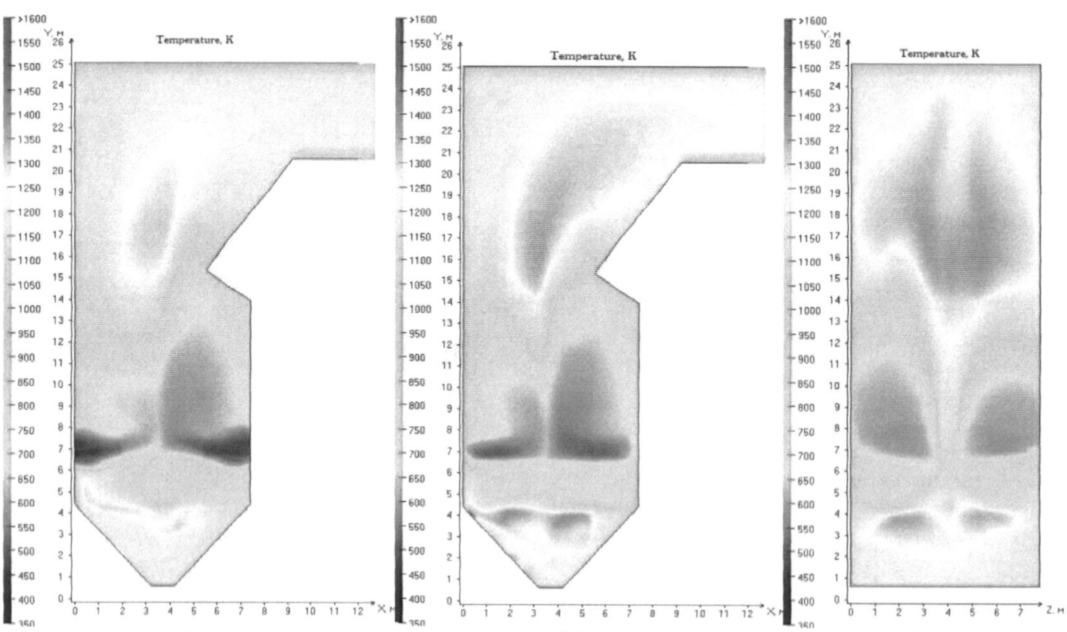

a) b) c)

Fig. 2. The temperature distribution in the longitudinal cross-section: a - at a distance of 1 meter from the side wall; b - along the transverse axis of the furnace; c - along the longitudinal axis of the furnace.

Figure 3 shows a graph of the average and maximum temperatures in the horizontal cross-section of the furnace.

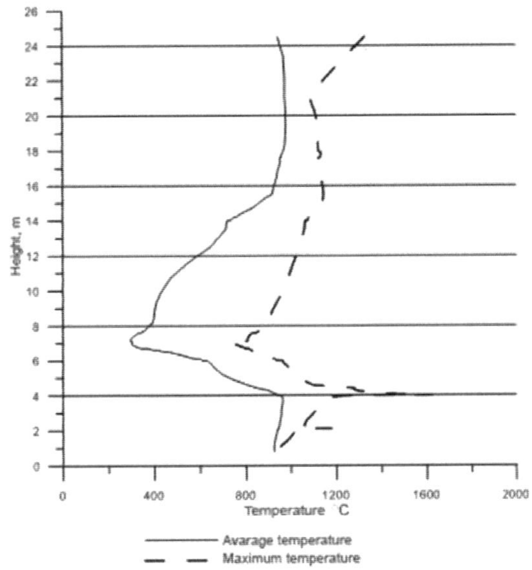

Fig. 3. Height variation of the maximum mean integral temperature values in the horizontal cross-section of the furnace.

From this graph it is seen that the temperature in the bottom of the furnace is from 950 to 1000 °C; then it gradually decreases – at the level of the burners the temperature is about 300 °C; then increases again – at the furnace exit it is 1000 °C.

5 Conclusions

The resulting furnace exit gas temperature is slightly different from the calculated by normative method [4] (~ 30 °C), what indicates the adequacy of accepted models and approaches.

Overall, the results of mathematical modeling prove the possibility of Pavlovsky brown coal burning in researched boiler furnace. The picture of the particles distribution in the furnace volume uniquely corresponds to its aerodynamic structure and confirms the vortex nature of their movement. In addition, the temperature distribution in the longitudinal cross-section picture shows that screen surfaces work in less loaded mode and less prone to slagging while combusting of the Pavlovsky brown coal.

References

1. A.V. Zavorin, A.V. Gil, P.S. Khaustov, R.B. Tabakaev, D.A. Buslov, IOP Conf. Ser.: Mater. Sci. Eng. **66**, 012038 (2014)
2. Z.F. Tian, P.J. Witt, M.P. Schwarz, W. Yang, Energy Fuels **26**, 2 (2012)
3. S. Belosevic, M. Sijercic, S. Oka, D. Tucakovic Int. J. Heat Mass Transfer **49** (2006)
4. Thermal Computation of Boilers. Normative Method, 3rd Edition, (NPO TsKTI, St. Petersburg, 1998)
5. A.V. Gil, A.V. Starchenko, Thermophysics and Aeromechanics **19**, 3 (2012)
6. A.V. Gil, A.Y. Gil, EPJ Web Conf. **82**, 01044 (2015)
7. N.V. Vizgavljust, A.V. Starchenko, A.V. Gil, T.S. Taylasheva, EPJ Web Conf. **82**, 01044 (2015)

CONDITIONS AND CHARACTERISTICS OF COAL WATER SLURRY CONTAINING PETROCHEMICALS IGNITION BY HOT PARTICLE

Nina K. Gal'chenko[1], *Valeriy* V. Medvedev[2], *Aleksandr* G. Nigay[2,*], and *Dmitriy* S. Sivkov[2]

[1]Institute of Strength Physics and Materials Science SB RAS, 634055 Tomsk, Russia
[2]National Research Tomsk Polytechnic University, 634050 Tomsk, Russia

Abstract. Characteristics of CWS containing petrochemicals ignition by single hot particle in disk shape established experimentally. The main components of fuel composition are coal 45 %, water 45 %, used engine oil 10 %. The main parameters of local heat sources are: material – steel, diameter 10–12 mm, high 4–8 mm. As a result of experimental study is established limit (minimum) conditions for ignition of CWS containing petrochemicals by hot particle and the impact of local heat source parameters on the main process characteristic – ignition delay time.

1 Introduction

The urgency of investigating of ignition characteristics of new fuels based on coal suspensions associated with the major scientific problem, which includes two tasks: recycling of solid and liquid combustible waste of coal and oil refineries, low-grade solid fuel, used oils; power generation by burning such fuel compositions.

Last years fuels based on the coal-water slurry used for power generation in countries with dynamically developing fuel and energy complex (China, India, USA, Germany). Coal-water slurry (CWS) has advantages in comparison with typical solid and liquid fuels for thermal power plants: high environmental characteristics, fire safety in preparation for burning, economical operation of power plants.

The technological process of CWS burning similar to the processes that take place in the furnaces of coal-fired boilers. However, there are some differences. The main features of the combustion process for CWS are: low-temperature surface activation reaction at the fuel ignition stage, the growth of the specific reaction surface in the main combustion zone, due to the intensification of the interaction reaction of carbon with water vapor.

Last years various characteristics of CWS research numerically [1–3] and experimentally [4–6]: rheological characteristics, methods of preparation, transport and storage, combustion parameters. Developments of furnaces construction for efficient combustion of CWS, conducted by the authors [7, 8], have a great importance for the practice.

* Corresponding author: nigayalexandr@mail.ru

Highly promising direction in the development of CWS subjects is to create new composite liquid fuels on their basis (CWS containing petrochemicals) that consists of three main components: coal, water and combustible petrochemicals (motor, turbine, transformer, transmission oils) [9–12]. Results of combustion processes research of such fuels let us develop the theoretical bases of recycling technology for a large group of industrial facilities to energy production.

The purpose of present work is the experimental study of limit (minimum) conditions for ignition of CWS containing petrochemicals by hot particle and establishing the impact of local heat source parameters on the main process characteristic – ignition delay time.

2 Experimental methods

Investigation of the conditions and characteristics of ignition of CWS containing petrochemicals by single hot particles carried out using an experimental setup. Its scheme is shown in fig. 1.

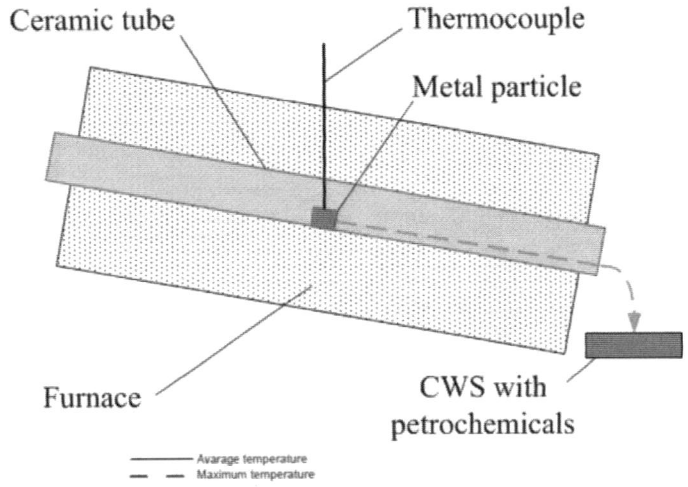

Fig. 1. Scheme of an experimental setup.

The main part of the experimental setup is a ceramic tube with diameter 50 mm and length 350 mm. The tube is heated to a predetermined temperature before the experiment. After temperature stabilization the metal particle in the disk shape placed in the central part of the ceramic tube. Particle heating takes 5–7 minutes. As a result of short-term mechanical impact hot particle by sliding along the ceramic tube deposited on the surface of CWS containing petrochemicals. The interaction process of local heat source with fuel is fixed by high-speed videocamera Phantom V411 (the frame size 800x600 pixels, the frequency 1000 frames per second).

Steel particles with diameter 10–12 mm and high 4–8 mm is used in experiments. Its initial temperature is varied in the range of 950–1100 K. Flame appearance in the vicinity of the particle corresponds to the ignition moment. Ignition is considered stable if the flame was observed more than 3 s after combustion initiation. About 5–8 approaches carried out under identical conditions to determine the ignition delay time.

CWS containing petrochemicals contains two combustible components: brown coal from the "Borodinskiy" section of the Krasnoyarsk region and used engine oil. Main characteristics of fuel components shown on Table 1 and 2.

Table 1. The results of technical and elemental analysis of brown coal from the "Borodinskiy" section of the Krasnoyarsk region.

W^a, %	A^d, %	V^{daf}, %	Q^a_s, MJ/kg	C^{daf}, %	H^{daf}, %	N^{daf}, %	S^{daf}, %	O^{daf}, %
4.11	4.12	47.63	22.91	73.25	6.516	0.79	0.435	18.99

Table 2. The main characteristics of petrochemical (used engine oil).

Density at 293 K, kg/m^3	Humidity, %	Ash, %	Flash temperature, K	Ignition temperature, K	Heat of combustion, MJ/kg
871	0.28	0.78	405	491	44

The mass fraction of components in CWS containing petrochemicals are coal 45 %, water 45 %, used engine oil 10 %.

3 Results and discussion

Fig. 2 shows dependences of ignition delay time of CWS containing petrochemicals on initial temperature of hot steel particle.

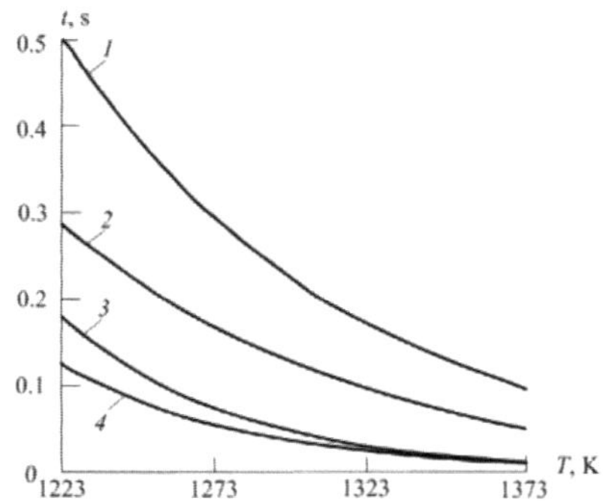

Fig. 2. Dependences of CWS containing petrochemicals ignition delay time on initial temperature of hot particle: *1* – *d*=10 mm, *h*=4 mm; *2* – *d*=10 mm, *h*=6 mm; *3* – *d*=12 mm, *h*=5 mm; *4* – *d*=12 mm, *h*=8 mm.

Curves *1–4* (fig. 2) shows the exponential dependence of the ignition delay time of fuel composition on hot particle initial temperature. Obtained result agrees well with the main provisions of the chemical kinetics [13].

The higher initial temperature of steel particle the lower ignition delay time of CWS containing petrochemicals. The minimum duration of the induction period corresponds to particles with a maximum heat content (*d*=12 mm, *h*=8 mm). Heat content of particle decreases when its size reducing. It leads to an increasing of ignition delay time (fig. 2).

Video frames of ignition are presented in fig. 3 during interaction process of hot particle with CWS containing petrochemicals.

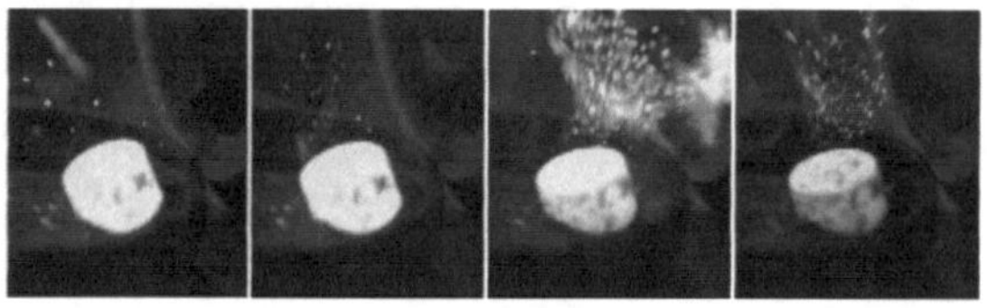

Fig. 3. Video frames of CWS containing petrochemicals ignition by hot particle.

It is shown (fig. 3) that at the initial moment products of flammable liquid evaporation ignites in the vicinity of the particle. Similar phase transitions occur during heating of liquid droplets on metal substrates [14, 15]. As a result of fuel near surface layer heating happens enhancement of combustible liquid evaporation. Volatiles stream contains small coal particles that combustion in oxidant medium.

4 Conclusion

Ignition of CWS containing petrochemicals by hot steel particle is experimentally investigated. The limit conditions (temperature higher 950 K) for fuel composition ignition by local heat source is established. The dependence of the main process characteristics (ignition delay time) on the initial particle temperature is exponential.

Acknowledgments

This work was supported by the grant of the President of the Russian Federation MK-6491.2016.8.

References

1. S. Tavangar, S.H. Hashemabadi, A. Saberimoghadam, Fuel Process. Technol. **132**, 153 (2015)
2. S. Belošević, I. Tomanović, V. Beljanski, D Tucaković, T. Živanović, Appl. Therm. Eng. **74**, 102 (2015)
3. G.V. Kuznetsov, V.V. Salomatov, S.V. Syrodoy, Combust., Explos. Shock Waves **51**, 4 (2015)

A. Kijo-Kleczkowska, Fuel **90**, 2 (2011)

4. E.G. Gorlov, Solid Fuel Chem. **38**, 6 (2004)
5. I.I. Lishtvan, P.L. Falyushin, E.A. Smolyachkova, S.I. Kovrik, Solid Fuel Chem. **43**, 1 (2009)
6. S.E. Pashchenko, V.V. Salomatov, S.V. Alekseyenko, *Method for Hydrocarbon Fuel Combustion*. Patent for invention RUS 2407947 03.08.2009 (2009) [in Russian]
7. A.Z. Zhuk, B.V. Kleymenov, A.E. Sheindlin, Therm. Eng. **58**, 1154 (2011)
8. K.Yu. Vershinina, D.O. Glushkov, P.A. Strizhak, G.V. Kuznetsov, Solid Fuel Chem. **50** (2016)
9. D.O. Glushkov, P.A. Strizhak, K.Yu. Vershinina, Applied Therm. Eng. **96**, 536 (2016)
10. D.O. Glushkov, D.P. Shabardin, P.A. Strizhak, K.Yu Vershinina, Fuel Process. Technol. **143**, 60 (2016)
11. D.O. Glushkov, S.V. Syrodoy, A.V. Zhakharevich, P.A. Strizhak, Fuel Process. Technol. **148**, 224 (2016)
12. D.A. Frank-Kamenetskii, *Diffusion and Heat Transfer in Chemical Kinetics* (Nauka, Moscow, 1987) [in Russian]
13. G.V. Kuznetsov, D.V. Feoktistov, E.G. Orlova, Thermophysics and Aeromechanics **23**, 17 (2016)
14. G.V. Kuznetsov, D.V. Feoktistov, E.G. Orlova, J. Eng. Phys. Thermophys. **89**, 317 (2016)

THE ANALYSIS OF THE CHARACTERISTICS OF SWIRL CHAMBER FOR COOLING HOT WATER

Sergey V. Goldaev[1], *Kseniya* B. Tretyakova[1,*]

[1]National Research Tomsk Polytechnic University, 634050 Tomsk, Russia

Abstract. The alternative way of cooling water circulating in the swirl chamber (SC) was considered. The method for calculating SC was improved and automated. The analysis of the main characteristics of the SC was performed.

1 Introduction

In many industries there is a problem of cooling the circulating water to the desired temperature. Artificial ponds, as well as cooling towers, are used as coolers. While operating the cooling towers there is an uneven water distribution in the cross section; liquid carryover and hazardous substances from the cooling towers, that lead to a negative impact on the environment; freezing of the elements of the cooling towers and breakdown take place in winter. The application of swirl chamber (SC) with a spray device was proposed in order to improve the efficiency of the cooling process of circulating water in industrial plants, [1].

2 Description of the swirl chamber

Hot water flows to SC (fig. 1) through the nozzle 1, gets to the spray gun 2, mounted on the shaft, where it is distributed to disks with different diameters.

[*]Corresponding author: tretyakova-kseniya@mail.ru

DOI: 10.1051/matecconf/20167201033

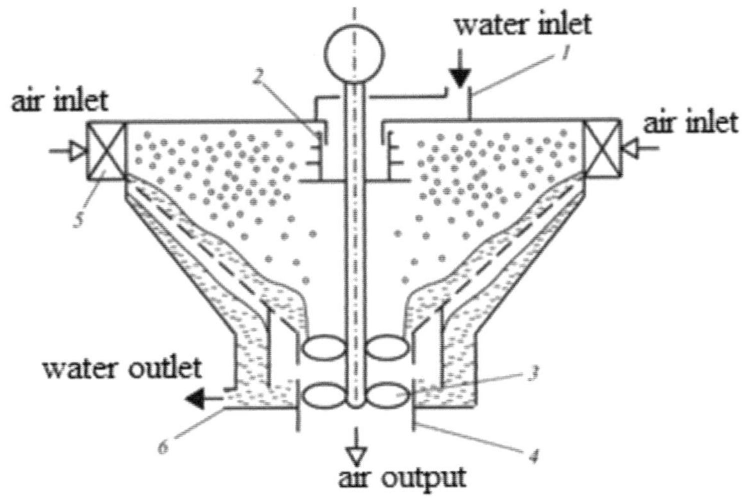

Fig. 1. Scheme of SC disk spray gun:
1 - nozzle for hot water inflow; 2 - disk spray gun; 3 - blades; 4 - nozzle for air outflow;
5 - tangentially blade swirler; 6 - nozzle for cooled water outflow.

When the shaft rotates, the liquid come off the disks of the spray as drops, it throws in different directions, forming a wide spray pattern. Rotating at the same angular velocity, the disks create polydisperse drop spray, providing uniform filling of the working area of the machine, eliminating stagnant zones and increasing the efficiency of heat and mass transfer process [1].

The blades 3 are mounted on the shaft. They are located in the nozzle 4 for the air outflow in several rows at a distance from each other, that lets the the working area of VC increase. Rotating on the shaft blades create a low pressure in the nozzle 4, that provides the air draft into the chamber through the tangentially blade swirler 5. The air gets rotation motion, moving toward the center of the machine, involving the drops. This interaction leads to the formation of the fine rotating drop layer in the working area. The cooled water is derived from the SC through the nozzle 6 [1].

3 Results and discussion

The short-cut method, that provides sufficient accuracy of calculation, was developed.

The formula [1] describes the calculation of the temperature of the drops cooling.

As the dimensions of some parameters are unknown, the formula is replaced for the simplified formula, found from the equation of heat transfer between air and water.

$$T_W = T - (T - T_{W0})e^{-B_k \tau}, \quad (1)$$

Where $B_k = 6\alpha / (c_L \rho_L d_k)$, τ – the contact time of the maximum diameter drops and the air, s, at $\tau = 0$, $T_L = T_{k0}$.

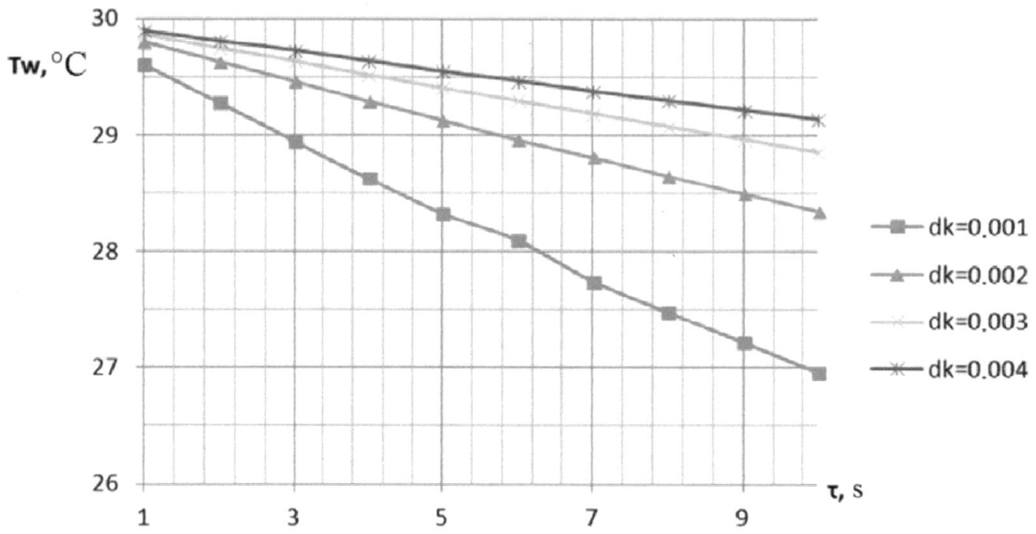

Fig. 2. The dependence of the temperature of the cooling drops and the drop diameter.

Figure 2 shows dependence of the temperature of the cooling drops and the contact time. The graph shows that the smaller the diameter of the drops, the more effective the cooling.

The diameter of the output nozzle *m* was found by the known values of air mass flow G_m in kg/s, its density ρ_G at a given temperature, number of drive turns *n* according to the formula for calculating the axial fan [2]:

$$D_d = 4\left(G_m / n\rho_G\right)^{1/3}, \tag{2}$$

In order to obtain the minimum air flow resistance the speed of the air, outflowing from the SC equals the speed at the entrance of the machine, m/s:

$$W_{en} = 4G_m / \pi D_d^2 \rho_G , \tag{3}$$

Hydraulic resistance of the device was determined from the expression $\Delta p = \xi_{en} \dfrac{\rho_G W^2{}_{en}}{2}$, where $\xi_{en} = 5.5(L_m / G_m)^{0.4}$ coefficient of resistance chamber

Useful capacitance of the drive, W:

$$D_d = 4\left(G_m / n\rho_G\right)^{1/3}, \tag{4}$$

Figure 3 shows that the flow resistance increases with increasing the flow rate.

The analogue of Glaser for assessing the effectiveness of cooling was introduced to the methodology of the calculation.

$$\chi = Q / N, \tag{5}$$

Where $Q = \alpha \cdot \Delta T \cdot S_k$,- heat load, W; α – coefficient of heat transfer from the water to the air, W/(m²κ); ΔT – the difference between the water and air temperatures, κ; S_k – the drop area, m^2

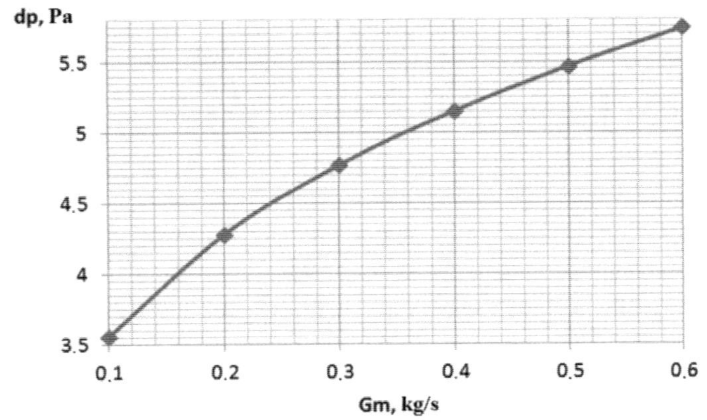

Fig. 3. Dependence of the unit hydraulic resistance from the mass air flow.

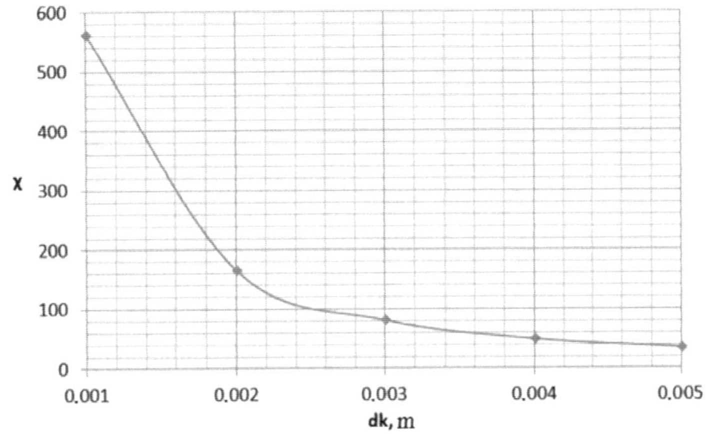

Fig. 4. Dependence of the Glaser criterion analogue from the droplet size.

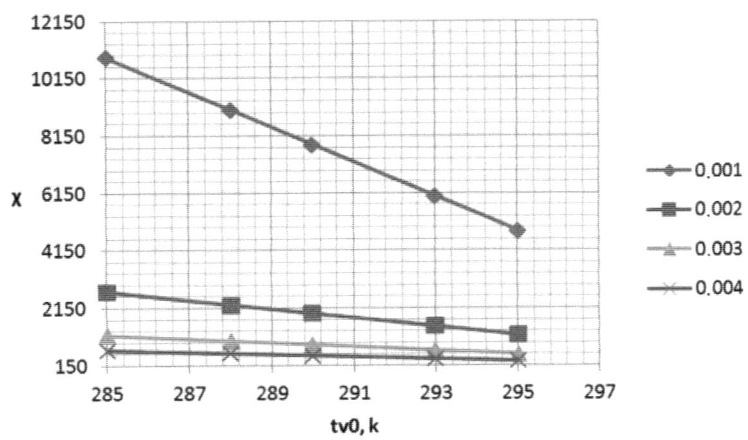

Fig. 5. Dependence of Glaser criterion analogue from the air temperature at the entrance to the block.

Figures 4 – 6 show the dependence of the Glaser criterion analogue from the droplet size. Its highest value of 560 was obtained for the drops of 1 mm diameter. With the increasing of the droplet size the value of this criterion decreases and with the diameter of 2 mm it equals 160, with a diameter of 3 mm - equals 80.

The comparative analysis has shown that the spraying cooling towers and cooling towers with regular packing are the most effective, but their disadvantage is in the presence of droplets ablation, minimizing the creation of swirling air flow.

The effectiveness of the cooling water was estimated by the parameter $E_T = 1 - T_L / T_{L0}$, and the proportion of evaporated water was estimated by the parameter $E_m = (1 - L_m / L_{m0}) \cdot 100\%$ [7]

Cooling of water flow rate slows down in the SC with air flow decreasing, as the driving force for heat transfer tends to zero [7].

With the increase in equilibrium moisture content the water evaporation rate decreases with increasing initial air temperature. In addition, increase of the air velocity entering the SC influences on the reduction of the percentage of water vapor, as the water drops are in the work area less time at the same volume [7].

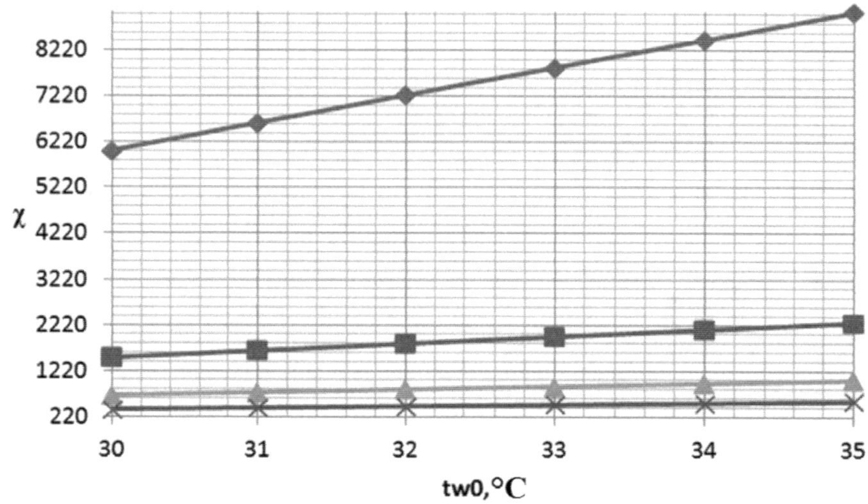

Fig. 6. Dependence of Glaser criterion analogue from the water temperature at the entrance to the block.

5 Conclusion

Approximate method for calculating the size of the SC in Turbo Pascal was implemented, the calculation results satisfactorily meet the data of the mathematical modeling [1].

References

1. A.V. Dmitriev, O.S. Dmitriev, A.N. Nikolaev, Ind. Power **8** (2013)
2. V.M. Cherkassy, *pumps, fans, kompressory* (Energoatomisdat, Moscow, 1984) [in Russian]
3. D.G. Pazhi, V.S. Galustov, *Basic techniques of atomization of liquids* (Chemistry, Moscow, 1984) [in Russian]
4. Y.I. Arefiev, N.V. *Spiridonov Proceedings of meetings and conferences on hydraulic engineering. Hydro – aerodynamics studies and design of cooling of thermal and nuclear power plants* (Energoatomnzdat., Leningrad, 1985) [in Russian]
5. A. H. Lefebvre, Hemisphere Publishing (1989)
6. B.C. Ponomarenko, Chem. Ind. **7** (1996)
7. A.V. Dmitriev, O.S. Makusheva, N.A. Nikolaev, Chem. Pet. Eng. **7** (2011)

MATEC Web of Conferences **72**, 01034 (2016) DOI: 10.1051/matecconf/20167201034
HMTTSC-2016

ESTIMATION OF THE SAFE ELECTRICAL LOAD ON SPIRAL BULBS HEATED BY CAPACITORS APPLIED TO MULTIPLE START-UPS OF OPEN GAS GENERATORS

Sergey Goldayev[1], *Sergey* Basalayev[2,*], and *Natalya* Minkova[2]

[1]National Research Tomsk Polytechnic University, 634050 Tomsk, Russia
[2]National Research Tomsk State University, 634050 Tomsk, Russia

Abstract. A method was developed for the automated calculation of the safe electrical load on spiral bulbs warmed by capacitors for the multiple start-ups of open gas generators under water. An energy criterion was proposed that would, for spiral bulbs with known dimensions and thermal properties, allow the selection of the type of pulsed power supply providing wire heating required for dibasic solid fuel (DSF) ignition, and maintaining its operative state.

1 Introduction

For the enabling of multiple start-ups and stops of underwater open gas generators (OG) based on dibasic solid fuels (DSF) [1], a method [2] was proposed, according to which the block fuel 1 is located in an upright position (figure 1).

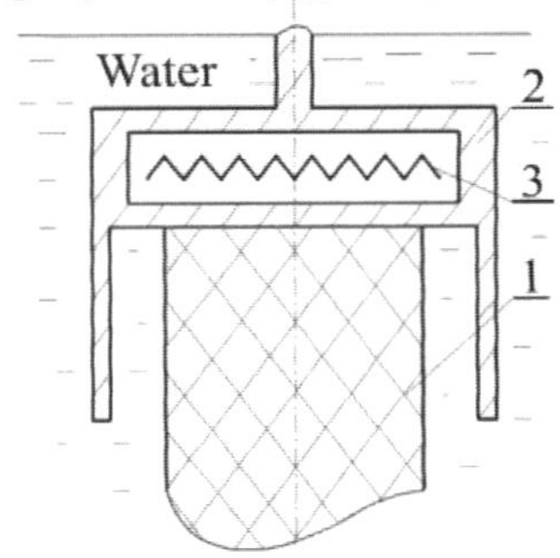

Fig. 1. Arrangement scheme of the block and the heat-resistance glass in the ignition.

Localization of the intended combustion zone at the upper end is achieved by using heat-resistant glass (HG) 2, at the bottom of which a CH 3 alloy with high electrical resistivity is placed (Table 1). The HG covers the sidewalls of the top part of the block with

* Corresponding author: tarm@niipmm.tsu.ru

liquid that can completely fill the margin formed between the walls and the block.

Table 1. Thermal properties of alloys with high electrical resistance.

Alloy	ρ_P, kg/m^3	c_P, J/kg·K	$\rho_0 \cdot 10^8$, Ohm·m	T_{pl}, °K	$\gamma \cdot 10^4$, 1/K	$(\rho_P\, c_P/\rho_0)$ 10^{12}
Constantan	8900	410.33	49	1533	0.10	7.45
Nickel	8900	468.84	7.24	1426	0.54	57.6
Nichrome	8400	460.57	110	1733	4.0	3.86

Burning is maintained due to the forced movement down of the HG, and the extinction of DSF occurs after the HG's retraction upward.

An experimental check confirmed [3] that the proposed method of DSF burning in an aquatic environment provides for stable ignition, stable combustion and reliable burning interruption.

2 The task

To ensure an emergency start-up of the OG, the energy of a charged capacitor may be used [4], for example, the capacitor included in the construction of the explosive device PC-2000m [5]. Its characteristics are shown in Table 2. The stored energy of the capacitor was estimated by the formula $E_0 = 0,5C_0U_0^2$.

Table 2. Parameters of pulsed power sources.

No.	Brand	Capacity C_0, uF	Voltage U_0, V	Energy E_0, J	Mass, kg
3	CSM –3	4	1600	5	1.6
4	PC – 2000 m	50	1200	35	5.2

3 Approximate methods

The experience of condenser blasting machine design was used to calculate the heating portion of the ignition of DSF, located near the contact zone of CH with the bottom of the HG [5]. It was thought that the discharge of a capacitor with the size of C_0 at the initial voltage of U_0, occurs to an external circuit having only active resistance. Then the force change of the current over the period of time t is described by the following expression [5]

$$I = (U_0 / R_h)\exp\left[-t/(R_h \cdot C_0)\right], \tag{1}$$

where R_h is the electrical resistance of the wire.

The heat flux density generated by CH is represented as a function exponentially decreasing with time [4]

$$q_h = q_{h0}\exp(-2t / R_h \cdot C_0), \tag{2}$$

where $q_{h0} = U_0^2 / (R_h \cdot S_h)$ is the maximum heat flux created by CH at the moment of the circuit closing; S_h is a heating square, covered by CH.

The approximate method of calculating the characteristics of the ignition described below is based on a criterion - the achievement of the surface of the "temperature flashpoint" sample equal to $T_{bc} \approx 500$ K.

For the dependence of the surface temperature of the DSF sample on the time at the accident inert heating stage, an analytical solution with the following form was obtained [4]:

$$\Theta_{sp}(\tau) = \Theta_c + \frac{q}{n\sigma}\left[\frac{a_2\Psi_2\left(a_2\sqrt{z}\right) - a_1\Psi_1\left(a_1\sqrt{z}\right)}{a_2 - a_1} - e^{(-\sigma\tau)} + \frac{(a_2+a_1)}{\sqrt{\pi}}\int_0^{\tau}\frac{e^{[\sigma(z-\tau)]}}{\sqrt{z}}dz\right], \quad (3)$$

where similar complexes are marked in such a way for brevity

$$\Psi_i\left(a_i\sqrt{\tau}\right) = \exp\left(a_i\sqrt{\tau}\right)^2 \operatorname{erfc}\left(a_i\sqrt{\tau}\right) \quad . \quad (4)$$

Here $\operatorname{erfc}(z) = 1 - \operatorname{erf}(z)$ is an additional integral of probability.

The dimensionless parameters which are part of (3) are calculated in that way:

$$a_1 = n(1-\delta_1)/2,\ a_2 = n(1+\delta_1)/2,\ \delta_1 = \sqrt{1-4n\,\mathrm{Bi}_c},\ 1-4n\,\mathrm{Bi}_c > 0$$

$$\sigma = \frac{2t_r}{R_h C_0};\ \mathrm{Bi}_c = \frac{\alpha_{ef} l_m}{\lambda_k},\ q = \frac{q_h l_m}{\lambda_k T_m},\quad n = \frac{c_h \cdot \rho_h \cdot d_h + c_s \cdot \rho_s \cdot d_s}{c_k \cdot \rho_k \cdot l_m} .$$

Here $t_r = 2.3R_{el} \cdot C_0$ is the time of discharge of the capacitor [6]; physical properties of DSF, spiral bulbs and heat-resistant glass: ρ_k is density, kg/m^3; c_k is the specific heat capacity of the mass, J/(kg·K); λ_k, a_k are thermal conductivity coefficients, V/m·K and thermal diffusivity, m^2/s; c_h are the specific heat capacity of the mass, ρ_h is density and d_h is the diameter of the wire; c_s is the specific heat capacity of the mass, ρ_s is density and δ_s is the thickness of the HG's bottom; α_{ef} is the effective heat transfer coefficient.

The scale coordinates and density of the heat flux are:

$$l_m = \sqrt{a_k t_r},\ q_m = \lambda_k \cdot T_0 / l_m .$$

The calculation of the probability integral was carried out according to the interpolation formula [6] which has one coefficient

$$\operatorname{erf}(z) = \sqrt{1 - \exp\left(-1.26z^2\right)} .$$

The costs of heating the bottom part of the HG are calculated with the help of the n parameter in the DSF ignition scheme shown in figure 1.

$$q_{HC} = \frac{1}{t_r}\int_0^{t_r} q_H dt = \frac{U_0^2}{4.6R_h \cdot S_h}.$$

Let us note that q_{HC} indirectly depends on C_0 through t_r and affects the parameter σ, reducing it with the increasing of C_0. Another option in data processing is the use of the initial value of the specific electrical energy equal to the ratio of $A_0 = E_0 / S_h$.

4 The results of the ignition time

An analysis was conducted for CH from nichrome wires of different diameters. Table 3 shows the calculated data of t_z depending on A_0 for the N type of DSF.

Table 3. Ignition time at varying the heat load on CH in the presence of the screen.

A_0, mJ/m^2		0.25	0.50	0.75	1.0	1.25
1)	t_z, ms	∞	0.2	0.14	0.10	0.065
2)		0.3	0.13	0.095	0.086	0.85
3)		0.3	0.075	0.03	0.02	0.018

The length of the wire was 140 mm and the diameter for variants 1 and 2 was 0.1 mm, and for variant 3 was 0.25 mm. For fixed U_0 = 1600 V capacitance the values ranged from 4 to 40 uF. The thermal load A_0 varied from 0.06 to 0.69 mJ/m^2, and t_z decreased from 9.1 to 8.3 ms.

Installing HG onto the end part of the block leads to the appearance of a screen between the surface of DSF and CH of 1 mm thickness. The varying of U_0 from 950 to 2000 V increased E_0 from 22.6 to 100 J, the corresponding thermal load increased from 0.3 to 1.34 mJ/m^2, and t_z decreased from 347 to 60 ms (variant 1). During the discharge the surface temperature of the block reached only 396 K when C_0 = 5 uF and U_0 = 1600 V.

In the next series of calculations C_0 changed from 10 to 80 uF, and the voltage was U_0= 1600. The results are shown in the form of the 2nd variant. From the structure of dependence (2) for the heat flux density generated by CH it follows that its amplitude is defined by the initial voltage U_0. The capacitance value affects the development of the process of heating the sample in time - for large C_0 (σ parameter is less) the temperature rise is slower and the ignition time increases.

The change in d_p is affected through the n and q_{HC} parameters. A wire 2.5 times as thick was used in the 3rd variant and as a result, the n parameter value became 2.5 times larger. However, $R_h \sim 1/d_p^2$ and therefore q_{HC} increased to 6.25, which resulted in a decrease of t_z.

Thus, the analysis showed that in the presence of a thin steel screen between CH and the end surface of a block from DSF, sources that allow the production of an energy density of more than 30 mJ/m^2 in the section of DSF heating are necessary for the realization of this process. In particular, a PC – 2000 m explosive device can be used.

5 Estimation of the safe electrical load on spiral bulbs heated by capacitors

To ensure re-ignition of the block of the same CH it is necessary that its electrical resistance changes little and that contact with the connecting wires is maintained.

The conductor may break after its heating to a temperature comparable to the melting temperature T_{pl} of the corresponding metal with high resistivity shown in Table 1. The results of such an analysis for the variant of CH heating from an autonomous power supply of direct currents are shown in the work [7].

Within the thermo-technical approach of evaluating the load on the conductors of nichrome or another alloy, the maximum current strength is determined that can be passed through the wire heater surrounded by air, so that the maximum temperature does not exceed 1273 K [8].

In another source [9], it is necessary to calculate the strength of the current going through thin nichrome wire of 0.1 mm in diameter and 2 m in length, placed horizontally in the air, having a temperature of 293 K, at which its destruction will occur, if it is known that nichrome has T_{pl} = 1733 K.

There is such a task in [10]. An electric detonator in the form of a cylinder of 0.1 mm in diameter and 5 mm in length is in air whose temperature is T_0=303 K, the intensity of cooling heating wire is characterized by the heat transfer coefficient $\alpha = 10$ W/(m^2·K)). The thermal properties of the detonator are: thermal conductivity $\lambda_h = 20$ W/(m·K), $a_h = 5\cdot10^{-5}$ m^2/s, the melting temperature of the detonator's metal is 1173 K, and the electrical resistance $R_h = 0.2$ Ohm. The task is to determine the time after which the detonator would explode when passing through it a constant current of 3 A, neglecting the radiation and escape of the heat to the attachment at the ends of the detonator. The value of t = 7.62 ms is given in the answer.

Let us add to the results of the DSF ignition analysis for spiral bulbs heated by capacitors an estimation of the time required to reach the temperature limit T_{pl}.

Let us sum up the formulations given in [7] due to the reducing current (1).

Let us write the heat balance of the area of wire based on the energy conservation equation.

When an electrical current passes through a conductor of cylindrical shape of d_p in diameter and l_p in length, heat is generated which can be calculated by the formula of Joule [9]:

$$Q = I^2 R_p = U_0^2 \cdot \exp\left(-2t / R_p C_0\right) \sigma_p / \left(\rho_{el}\, l_p\right), \tag{5}$$

where σ_p is the cross-sectional square of the conductor.

Cooling of the conductor is carried out by free convection and radiation, and c can be taken into account by an appropriate heat transfer coefficient [8]. Let us use the effective value of α equal to 2000 W/(m^{2}K) [11], which corresponds to the intensity of the heat exchange between wire of nickel alloy and chromium, heated when passing through an electrical current and DSF embedded in the surface layer. Estimation of the Bio number $\mathrm{Bi} = \alpha d_p / 2\lambda$ shows that for thick wires in the task it is $\mathrm{Bi} = \alpha d_p / 2\lambda = 2\cdot10^3 \cdot 0.5\cdot10^{-3} / 2\cdot11.6 = 0.043$. Consequently the thermally thin body model is applicable [9].

The heat balance equation for the wire has the following form

$$c_p \rho_p V_p dT_p / dt = Q - \alpha\left(T_p - T_w\right) S_p, \tag{6}$$

where c_p is the specific heat capacity of the mass of the conductor; ρ_p is its density; V_p S_p is the volume and the square of the surface of the conductor.

Having substituted into the differential equation (6) the expression for heat evolution (5), we obtain a linear inhomogeneous ordinary first-order differential equation:

$$dT_p / dt + A\left(T_p - T_w\right) = B \exp\left(-Dt\right), \tag{7}$$

where

$$A = \frac{\alpha S_p}{\left(c_p \cdot \rho_p \cdot V_p\right)}; \quad B = \frac{U_0^2}{\left(c_p \cdot \rho_p \cdot V_p \cdot R_{el}\right)}; \quad D = \frac{2}{R_p \cdot C_0}.$$

The temperature of the wire is equal to the temperature of the wall of the heat-resistant glass at the initial moment of time: at $t = 0$ $T_p = T_w$.

Using the expression for the general integral (7) from [12] and determining the integration constant using initial conditions, we obtain

$$T_p = T_w + B\left[\exp(-Dt) - \exp(-At)\right]/(A - D). \quad (8)$$

The described algorithm for calculating the desired temperature was employed using a PC.

The following results were obtained in the course of the parametric analysis. The temperature of the nichrome wire $d_p = 0.2 \cdot 10^{-3}$ m, $l_p = 0.12$ m in length, reached a maximum value $T_p = 650$ K by the time moment $t = 0.08$ ms after using the capacitor subversive machine CSM–3 with $U_0 = 1600$ V and the capacity $C_0 = 4$ uF. A twofold reduction f d_p leaving the other parameters unchanged led to the wire after the heating time t = 0.24 ms reaching the maximum temperature of 1697 K, which exceeds the melting temperature of nichrome. Using a pulsed source of PC–2000m with higher energy (E_0 = 50 J) contributes to the intense increase in the temperature of the wire, which at the time t = 0.06 ms had already been exceeding T_{pl}.

Analysis of the obtained results showed that in cases when E_0 is less than the energy required to heat the wire to the melting temperature $E_p = c_p \rho_p V_p \left(T_{pl} - T_w\right)$, then the wire maintains a functional state.

For example, heating the wire of d_p = 0.2 mm and l_p = 0.2 m up to T_{pl} requires 34.3 J of energy. The PC–2000m pulsed source has 36.0 J of energy. By the time moment t = 0.51 ms the temperature of the wire reaches 1693 K. The choice of wire of d_p = 0.5 mm and the previous length requires 128.5 J of energy to heat it to the temperature T_{pl} .The use of a PC–2000m pulsed source allows heating of the wire to 694.6 K.

Another option for CH conservation in the functional state is to switch it off after its heating, for example, up to 1000 K. In practice, this can be realized with the aid of a technical solution [13], where the shunt resistor is set parallel to the heating wire and a fuse strip is placed successively with this juncture. When using an electro-initiator via a current impulse of an amplitude from 2 A and duration of 30 ms there occurs a fail-proof ignition of the working charge with a probability of not less than 0.999, and the subsequent fail-proof shutdown of the electro-initiator with the same probability.

6 Conclusion

A method was developed for the automated calculation of the safe electrical load on spiral bulbs heated by capacitors for multiple start-ups of an open underwater gas generator. An energy criterion has been proposed which allows the selection of a type of pulsed power supply that enables the heating of wire required for DSF ignition, and maintaining the functional state of spiral bulbs with known dimensions and thermal properties.

Acknowledgments

The reported study was funded by RFBR according to the research project No. 16-03-00630a.

References

1. V.D. Barsukov, S.V. Goldaev, *Teoriya, eksperiment, tehnicheskie predlozheniya* (Izd-vo Tom. un-ta, Tomsk, 2003) [in Russian]
2. V.D Barsukov, S.V Goldaev, N.P, Minkova, S.N. Polenchyuk, Patent 2345277 RF Sposob szhiganiya unitarnogo tverdogo topliva v zhidkoy srede, Opebl. v BI No.3 ot 27.01.2009
3. V.D. Barsukov, S.A. Basalaev, S.V. Goldaev, N.P. Minkova, Khim. Fiz. Mezoskopiya **17**, 2 (2015)
4. V.D. Barsukov, S.V. Goldaev, N.P. Minkova, N.A. Babushkin, Vestnik TSU: Matematika i mehanica **3** (2015) [in Russian]
5. M.M. Graevskiy, *Spravochnik po elektricheskomu vzryvaniyu zaryadov* (Randevu, Moscow, 2000) [in Russian]
6. A.N. Reznikov, L.A. Reznikov, *Teplovie protsessy v tehnologicheskih sistemah* (Mashinostroenie, Moscow, 1990) [in Russian]
7. S.V. Goldaev, N.A. Babushkin, S.A. Basalaev, EPJ Web Conf. **23**, Article Number 01014 (2015)
8. K.D. Voskresenskiy, *Sbornik raschetov i zadach po teploperedache* (Gosud. energetich. izd-vo, Leningrad, 1959) [in Russian]
9. F.F. Tsvetkov, R.V. Kerimov, V.I. Velichko, *Zadachnik po teplomassoobmenu* (Izdatelskiy dom MEI, Moscow, 2008) [in Russian]
10. F. Kreit, U. Black, *Osnovy teploperedachi* (Mir, Moscow, 1983) [in Russian]
11. E.B. Volkov, T.A. Syritsyn, G.Y. Mazing, *Statika i dinamika raketnyh dvigatelnyh ustanovok* (Mashinostroenie, Moscow, 1978) [in Russian]
12. I.N. Bronschtein, K.A. Semendyaev, *Spravochnik po matematike* (Nauka, Moscow, 1986) [in Russian]
13. P.A. Voronov, O.L. Ignatov, Y. M. Sidorkin, A.V. Sokolov, A.A. Uskov, Patent 2156946 RF Elektroinitsiator s samootklyucheniem ot istochnika toka MKI 7 F42 b 3/12 /– Zayavl 12.02. 1999 Opubl. 27.09.2000 [in Russian]

MATEC Web of Conferences **72**, 01035 (2016) DOI: 10.1051/matecconf/20167201035
HMTTSC-2016

IGNITION OF WET WOOD PARTICLES IN THE STREAM OF HIGH-TEMPERATURE GASES

S. V. Syrodoy[1,*], *N.* Y. Gutareva[1], and *K.* A. Bugaeva[1]

[1]National Research Tomsk Polytechnic University, 634050 Tomsk, Russia

Abstract. The problem of wet wood ignition of particles in a high-temperature gas has been solved numerically in the framework of a quasi-frontal model of evaporation. The basic characteristics of the ignition has been shown in the article. The influence of the different parameters on the ignition conditions has been shown in this work.

1 Introduction

Formation of a large amount of wood waste [1, 2], which are formed at preparation and processing of wood, is the most urgent issue [3, 4] of wood industry for today. This waste includes slab, bark, sawdust, shavings, wood dust, wood chips. Storage and warehousing of them are almost impossible for a number of reasons, main of which is high fire hazard and explosion of timber [5, 6]. For this reason, the most promising and cost-effective way to dispose of "secondary" timber is its implementation (as fuel) to the overall balance of thermal and power generation [7-10]. It should be added that the use of wood waste as a raw material for the boiler units has a number of significant advantages [11-15] over traditional (oil, gas, coal) fuels. Firstly it is a significant expansion of the resource base of modern boilers [11] and the possibility of autonomous power supply [12] distant (from the main settlements) areas (places of lumbering or extraction of oil). The important advantage is the fact that the wood is the only renewable fuel [13] while reserves of the conventional fuels (oil and coal) are limited. It should also be noted environmental friendliness of wood [14]. We can see almost no sulfur in its composition [15], so there is practically no sulfur oxide in burning products, which generated during the combustion of wood. CO_2 concentration can be minimized through better choice of the design characteristics of the flue boilers devices [16]. It also worth noting that 0.5 % to 3 % of ash is formed by burning the timber [17], while for some coals, this figure may reach 50 % [18]. Accordingly, it can be said that the creation of energy efficiency and perspective from ecology point of view boiler units working on wood waste, is one of the most perspective trends in modern energy technology [19].

However it is known that the wood (in most cases) is a highly-watered fuel [20, 21]. Accordingly, an essential stage of thermal treatment of the particle to the fire is the process of evaporation.

It should be noted that the mathematical modeling of phase transformation (e.g. evaporation), is one of the most difficult problem. This is due to the fact that in conditions

* Corresponding author: ssyrodoy@yandex.ru

of high heat moisture evaporation occurs in a narrow (much less than the linear size of the particle) constantly moving area ("front" of evaporation). The latter fact is complicated by high water evaporation endothermic effect (up to 2.5 MJ / kg).

By now quite a large number of [22-25] models of the process dehydration of highly-watered porous particles have been designed. However, most of them are based on some assumptions, which can significantly simplify the procedure of modeling. For example, the temperature equilibrium [22] on the partition border of the system "water-vapor" or application method of moisture-potential [23]. Models of [24, 25] "drying" of wet timber are based on the concept of volume phase transition of the distribution in space. The most complete evaporation process describes the model [26], which is based on the idea that a phase transition occurs in the front, on the border of the system "water-vapor". And the temperature depends on the rate of evaporation. Numerical modeling of the dehydration of the fuel particle has been realized by using the method of time step accumulation [27].

However, it is worth noting that this approach is difficult realized in a high-temperature heating at the decision the problem of ignition (processes of evaporation, pyrolysis and ignition occur together), and can lead to significant inaccuracies of calculations. Accordingly, the mathematical modeling of heat and mass transfer processes in the ignition highly-watered fuels (such as wet wood), in the conditions of high temperature heating at the joint main of basic thermal treatment processes (water evaporation, pyrolysis of the organic part of the fuel), is an urgent (unresolved to date) problem.

The purpose is the mathematical modeling of the ignition of damp wood particles in high-temperature heating in the gaseous environment.

2 Formulation of the problem

The model of the ignition process has been accepted. At the initial time ($\tau = 0$) spherical wet wood particle gets to a stream of high-temperature environment. The process of fast water evaporation begins as a result of intensive heating. The front of evaporation moves away from the surface into the interior of the particle. A porous layer with a high thermal resistance is formed. Further heating initiates thermal decomposition of wood, and volatile substances are allocated. Vapor-gas mixture is formed near the surface. It ignites at critical values of temperature and concentrations. The period of time from the start of the heat exposure to the moment of ignition is the ignition delay time (t_{ign}).

The mathematical formulation of the problem, which corresponds to the above physical model, has been formulated as a system of differential equations in partial derivatives:

– the energy equation for the wood particle:

$$Cp_w(T)\rho_w(T)\frac{\partial T_1}{\partial t} = \frac{1}{r^2}\frac{\partial}{\partial r}\left[r^2 \cdot \lambda_w \cdot \frac{\partial T_1}{\partial r}\right] \pm \sum_{i=1}^{n} Q_i \cdot W_i - \frac{Q_e \cdot W_e}{h_{sp}} \quad (1)$$

$$0 < r < r_0$$

Where: $\sum_{i=1}^{n} Q_i \cdot W_i = Q_{td} \cdot W_{td} + Q_c \cdot W_c$

– the energy equation for the gas environment:

$$Cp_g \rho_g \frac{\partial T_2}{\partial t} = \frac{\lambda_g}{r^2}\frac{\partial}{\partial r}\left[r^2 \cdot \frac{\partial T_2}{\partial r}\right] + Q_b \cdot W_b \quad (2)$$

$$r_0 < r < r_g$$

– the equation of diffusion of pyrolysis products in the gas phase:

$$\rho_g \frac{\partial C_1}{\partial t} = \frac{D_1 \cdot \rho_g}{r^2} \frac{\partial}{\partial r}\left[r^2 \cdot \frac{\partial C_1}{\partial r} \right] - W_b \tag{3}$$

$$r_0 < r < r_g$$

– the equation of diffusion of water vapor in the environment:

$$\rho_g \frac{\partial C_2}{\partial t} = \frac{D_2 \cdot \rho_g}{r^2} \frac{\partial}{\partial r}\left[r^2 \cdot \frac{\partial C_2}{\partial r} \right] \tag{4}$$

$$r_0 < r < r_g$$

Mass rate of evaporation (W_e kg / (m^2 · s)) has been calculated as in [28]:

$$W_e = W_0 \cdot \exp\left(\frac{Q_e \cdot \mu \cdot (T_e - T_f)}{R \cdot T_f \cdot T_e} \right). \tag{5}$$

The coordinate evaporation border has been found from the expression:

$$r_F = r_0 - \int_0^{\tau} u_F d\tau \tag{6}$$

Where $u_f = W_e/\rho_{wt}$ — linear rate of advance of the evaporation front, m/s.

Upon reaching the surface of the particle conditions $T_s \geq T_{std}$ (T_s — surface temperature, T_{std} — temperature of thermal decomposition start) decomposition of the organic part of the fuel with the release of volatiles starts in the particle. The process is described by the thermal decomposition of chemical kinetics:

$$\frac{\partial \eta(r,t)}{\partial t} = \left[1 - \eta(r,t)\right] \cdot k_{td} \cdot \exp\left(- \frac{E_{td}}{R \cdot T(r,t)} \right) \tag{7}$$

Pyrolysis rate has been calculated using the expression:

$$W_{td} = \left[1 - \eta(r,t)\right] \cdot k_{td} \cdot \rho(t) \cdot \exp\left(- \frac{E_{td}}{R \cdot T(r,t)} \right) \tag{8}$$

In the “near-wall” area of the particle this reaction occurs: $CH_4 + O_2 = CO_2 + H_2O + 55546$ kJ/kg. The rate of it has been calculated from the mathematical expression of the Arrhenius law:

$$W_b = (1 - C_1 - C_2) \cdot C_1 \cdot k_b \cdot \rho_g \cdot \exp\left[- \frac{E_b}{R \cdot T(r,t)} \right] \tag{9}$$

Oxidant concentration has been determined from the material balance of the system “oxidizer — water vapor — volatile”:

$$C_1 + C_2 + C_3 = 1 \tag{10}$$

It is assumed that part of the combustible solid residue is mainly composed of carbon. Accordingly, the reaction occurs at the surface of the wood particle $C + O_2 = CO_2 + 18560$ kJ / kg. The rate of this reaction has been calculated from the expression:

$$W_C = k_C \cdot \rho_W \cdot \exp\left[-\frac{E_C}{R \cdot T(r,t)}\right] \tag{11}$$

The system of equations (1) — (11) has been solved with the boundary conditions:

$$0 < r < r_0,\ T(r,0) = T_0,\ r_0 < r < r_g,\ T(r,0) = T_e,,\ \eta(r,0) = \eta_0.$$

$$\frac{\partial T(0,t)}{\partial r} = 0 \tag{12}$$

$$\frac{\partial T^2(r_g,t)}{\partial r^2} = 0 \tag{13}$$

$$\frac{\partial C_1^2(r_g,t)}{\partial r^2} = 0 \tag{14}$$

$$\frac{\partial C_1(r_0,t)}{\partial r} = \frac{W_{td} \cdot r_0}{\rho_g \cdot D_1} \tag{15}$$

$$\frac{\partial C_2(r_0,t)}{\partial r} = \frac{W_e \cdot r_0}{\rho_g \cdot D_1} \tag{16}$$

$$\frac{\partial C_2^2(r_g,t)}{\partial r^2} = 0 \tag{17}$$

The boundary condition of the fourth kind is performed at the interface of the system "wood particle — environment":

$$\lambda_W \left.\frac{\partial T_1}{\partial r}\right|_{r=r_0-0} = \lambda_g \left.\frac{\partial T_2}{\partial r}\right|_{r=r_0+0} \tag{18}$$

$$T(r_0;t) = T(r_0;t)$$

The designations: T_0 — the initial temperature of the particle, K; T_e — ambient temperature, K; η_0— the initial value of the parameter of burnout; $\sum_{i=1}^{n} Q_i \cdot W_i$ — the complex, which takes into account the total thermal effect of the process of pyrolisis and ignition of the organic part of fuel, W/m^3; $Q_b \cdot W_b$ — the complex, which takes into account the thermal effect of thermochemical interaction the pyrolysis products with the oxidant, W/m^3; $Q_e \cdot W_e$ — the complex, which takes into account the thermal effect of water evaporation, W/m^2; $Q_{td} \cdot W_{td}$— the complex, which takes into account the thermal effects in evaporation of water, W/m^3; $Q_c \cdot W_c$— the complex, which takes into account thermal effect of thermochemical interaction the organic part of the fuel with the oxidant, W/m^3; k_{td} — pre-exponential factor of thermal decomposition reaction, 1/s; k_b — pre-exponential factor of the reaction of combustible components, 1/c; C_1 — concentration of combustible components; C_2 — concentration of water vapor; E_{td} — the activation energy of the thermal decomposition reaction, J/(mole·K); E_b — the activation energy of the oxidation reaction of the combustible components, J/(mole·K); η(r,t) — burnout parameter; T_f — water temperature corresponding to the freezing point, K; μ – the molar mass of water, kg/mole; R — universal gas constant, J/(mole·K); W_0 — mass rate of water evaporation at the temperature T_f kg/(m²·s); Q_e — the thermal effect of water evaporation, J/kg; Q_c — the thermal effect of the carbon ignition reaction, J/kg; Q_{td} — thermal effect of thermal decomposition reaction, J/kg; Q_b — thermal effect of volatile ignition reaction, J/kg; T_e — temperature of the boundary of evaporation, K; λ_w — thermal conductivity coefficient of wood, W/(m·K); λ_g — thermal conductivity coefficient of external environment, W/(m·K); Cp_w — heat capacity of wood, J/(kg·K); Cp_g — heat capacity of external environment, J/(kg·K); ρ_w — wood density, kg/m3; ρ_g — environment density, kg/m3; r_0 — particle radius, m; r_g — outer radius of gas area, m; h_{sp} — spreading parameter of evaporation front.

The thermal conductivity, heat capacity and density of the wet and "dehydrated" parts of fuel have been calculated based on the volume fractions of the major components:

$$\begin{gathered}\lambda_1 = \varphi_4\lambda_4+\varphi_5\lambda_5 \quad \lambda_2 = \varphi_5\lambda_5+\varphi_6\lambda_6 \\ Cp_1 = \varphi_4 C_4+\varphi_5 C_5 \quad Cp_2 = \varphi_5 C_5+\varphi_6 C_6 \\ \rho_1 = \varphi_4\rho_4+\varphi_5\rho_5 \quad \rho_2 = \varphi_5\rho_5+\varphi_6\rho_6\end{gathered} \tag{19}$$

Where: φ — volume fraction of the fuel component: 4 — water; 5 — wood; 6 — water vapor.

3 Results

The figure 1 shows the time of ignition delay of wood particles, depending on the ambient temperature. The analysis of the graphs shows that ambient temperature rise reduces the ignition delay time. It is worth noting the non-linear character of the graphs (1)-(3). It demonstrates the essential influence of the complex co-occurring physical and chemical processes during the thermal treatment. So after analyzing the dependences $\tau_{ign}(T)$ of particles diameter $\delta = 0.5\cdot10^{-3}$ m - $\delta = 5\cdot10^{-3}$ m it is possible to note significant differences of ignition delay times at ambient temperature T_e =800 K. The last suggests a fairly high degree of influence on the process of evaporating moisture ignition.

Figure 2 shows the dependence of the delay time of ignition of wood particles with a diameter $\delta=1\cdot10^{-3}$ m on a fraction of moisture in the structure of fuel. The analysis of the dependences shows that the increase of fuel moisture leads to significant ignition delay (almost in 2 times). This is because high water content in the wood heat removal is greatly increased due to evaporation of moisture and thus complete "dehydration" of the particle occurs much longer.

Figure 3 shows the temperature distribution in the system of "fuel particles — gas environment" at the moment of ignition. The analysis of dependency T(r) indicates that the ignition of wood particles under conditions of relatively low ambient temperature (Te <1000K) takes place only after completely removing moisture. In other words, it requires deeper heating of the main fuel layer. Meanwhile in the case of high-temperature heating (Te>1000K) wood particles can ignite before their complete dehydration. It is possible to note the characteristic bend at the graph 1 (figure 3). It shows the interface of the system "wet particle — dry wood." It can be concluded that the subsequent burning of such flooded fuel will significantly differ from the oxidation of dry wood particles.

DOI: 10.1051/matecconf/20167201035

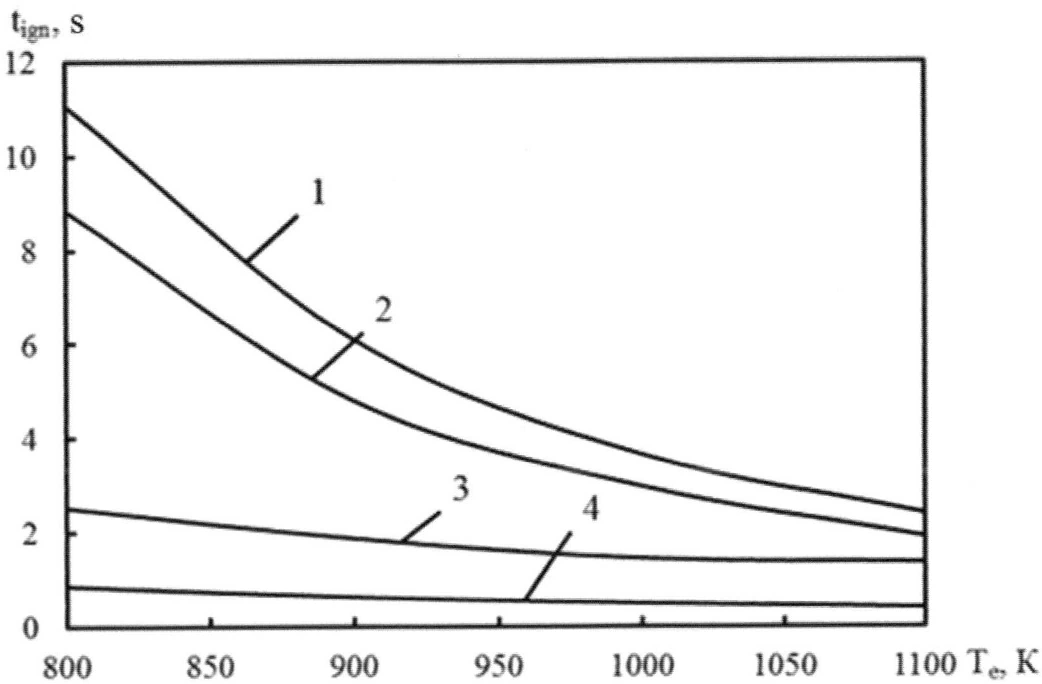

Fig. 1. The dependence of the ignition delay times of wood particles (humidity 10%) the ambient temperature at a rate (δ) particle: 1 - $\delta=5\cdot10^{-3}$m; 2 - $\delta=3\cdot10^{-3}$m; 3- $\delta=1\cdot10^{-3}$m; 4- $\delta=0.5\cdot10^{-3}$m.

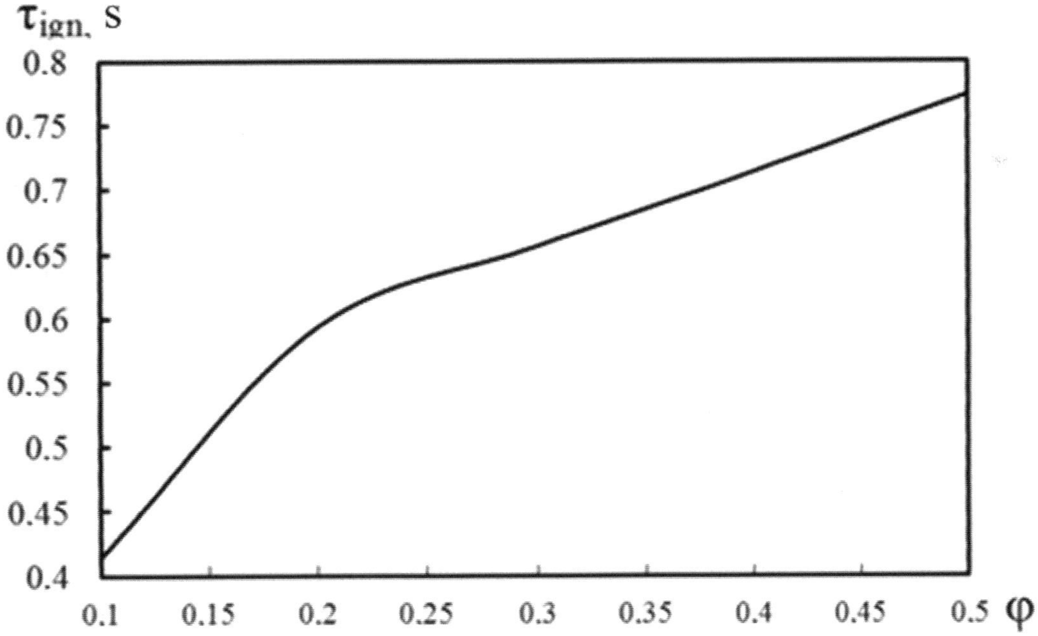

Fig. 2. The dependence of the ignition delay times of wood particles on a fraction of moisture in its structure.

4 Conclusion

Delay times of the ignition of wet wood particles have been set with varying sizes ranging from $\delta = 0.5\cdot10^{-3}$m to $\delta = 5\cdot10^{-3}$m in different heat transfer conditions based on the results of mathematical modeling. It is established that the co-occurrence of the complex of thermal treatment processes (slow heating, water evaporation, thermal decomposition of the organic part of the fuel, gas-phase ignition of volatiles) has a significant impact on characteristics and conditions of the ignition.

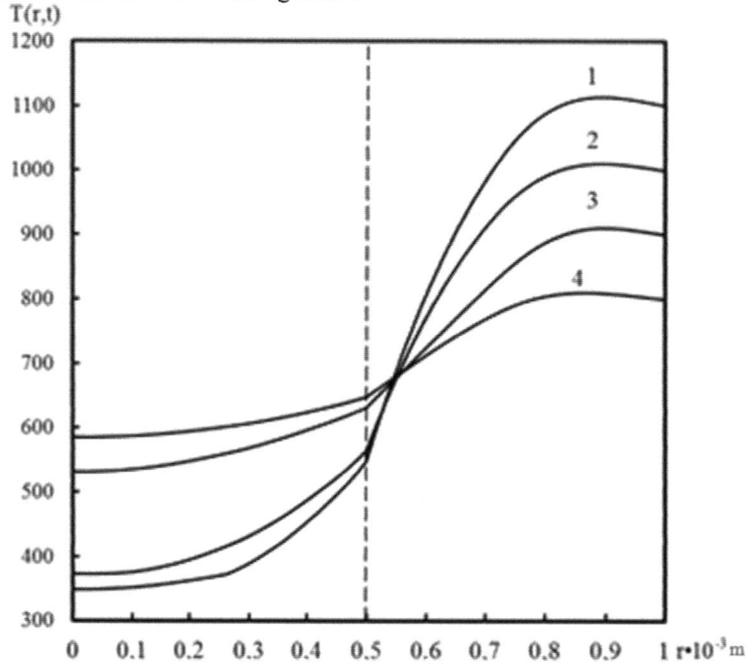

Fig. 3. The temperature distribution along the radius in the system "wood particle — gas environment" diameter $\delta=1\cdot10^{-3}$ m at ambient temperatures: 1 — T_c=1100 K; 2 — T_c=1000 K; 3 — T_c=900 K; 4 — T_c=800 K.

According to the results of the numerical experiment, two different modes of ignition of wood particles have been installed — a high-temperature (Te>1000 K) and low temperature (Te≤1000 K). The ignition occurs only after complete dehydration of the main layer of fuel, in the case of ignition of wood particles in a medium with an oxidant temperature Te <1000 K. Upon heating wood particles in the medium with ET> 1000 K, ignition can occur before the complete removal of moisture. It is obvious that the subsequent combustion under continuing evaporation will substantially differ from the oxidation of the same (size and composition of the component), but dry particles.

Acknowledgments

The work at the field of modeling of heat and mass transfer processes near boundary between two layers was supported by the scientific schools grant NSH-7538-2016.8.

References

1. E. Ratajczak, G. Bidzińska, A. Szostak., M. Herbeć, Resour., Conserv. Recycl. **97** (2015)
2. N.S.L. Srivastava, S.L Narnaware, J.P. Makwana, S.N. Singh, S. Vahora, Renewable Energy **68** (2014)
3. J. F. Eshun, J. Potting, R. Leemans, J. Cleaner Prod. **26** (2012)

4. E. T. Acquah, C. Whyte, Office of Sustainable Development, USAID Africa Bureau **139** (1998)
5. M. van Blijderveen, E. A. Bramer, G. Brem, Waste management **32**, 9 (2012)
6. G. Hehar, O. Fasina, S. Adhikari, J. Fulton, Fuel Processing Technology **127** (2014)
7. H. Athari, S. Soltani, M. A. Rosen, S. M. S. Mahmoudi, T. Morosuk, Renewable Energy **92** (2016)
8. A.T. Thattai, V. Oldenbroek, L. Schoenmakers, T. Woudstra, P.V. Aravind, Appl. Energy **168** (2016)
9. P. C. Kuo, W. Wu, Energy Convers. Manage. **111** (2016)
10. P. C. Kuo, W. Wu, Chem. Eng. Sci. **142** (2016)
11. A. Favero, E. Massetti, Resource and Energy Economics **36**, 1 (2014)
12. A.M. Hendricks, J.E. Wagner, T.A. Volk, D.H. Newman, Biomass and Bioenergy **88** (2016)
13. G.M.J. Herbert, A.U. Krishnan, Renewable Sustainable Energy Rev. **59** (2016)
14. M.C. Heller, G.A. Keoleian, M.K. Mann, T.A. Volk, Renewable Energy **29**, 7 (2004)
15. C.A. Balaras, K. Droutsa, E. Dascalaki, S. Kontoyiannidis, Energy and buildings **37**, 5 (2005)
16. C. Yu, L. Li, L. Zeng, Z. Tang, Powder Technol. **254** (2014)
17. D.E. Priyanto, S. Ueno, N. Sato, H. Kasai, T. Tanoue, H. Fukushima, Fuel **174** (2016)
18. J. He, M. Tan, R. Zhu, Y. Zhao, P. Hong, Y. Jiang, Fuel **172** (2016)
19. C. Li, C. Gillum, K. Toupin, B. Donaldson, Energy Convers. Manage. **103** (2015)
20. Y. Ding, C. Wang, S. Lu, Energy Convers. Manage. **98** (2015)
21. W.C.R. Chan, M. Kelbon, B.B. Krieger, Fuel **64**, 11 (1985)
22. S.L. Mitchell, M. Vynnycky, Journal of Computational and Applied Mathematics **300** (2016)
23. F.R.G.B. Silva, G. Gonçalves, M.K. Lenzi, E.K. Lenzi, Int. J. Heat Mass Transfer **63** (2013)
24. V. Zadin, H. Kasemägi, V. Valdna, S. Vigonski, M. Veske, A. Aabloo, Applied Mathematics and Computation **267** (2015)
25. M.V. Belleghem, H.J. Steeman, M. Steeman, A. Janssens, M.D. Paepe, Building and Environment **45**, 11 (2010)
26. S.V. Syrodoy, G.V. Kuznetsov, V.V. Salomatov, Thermal Engineering **62**, 10 (2015)
27. S.V. Syrodoi, G.V. Kuznetsov, V.V. Salomatov, Solid Fuel Chem. **49**, 6 (2015)
28. S.V. Syrodoy, V.V. Salomatov, N.Y. Gutareva, EPJ Web Conf. **76**, Article Number 01018 (2014)
29. I.I. Markov, About the mechanism of evaporation at the interface of two media / I.I. Mark, A.A. Khashchenko, O.V. Evening, the Russian Defense Ministry, SKGTU, North-Caucasian branch of Technological Sciences. Collection of Scientific Works, A series "Physic-chemical, Stavropol (2004)

THE EXPERIMENTAL STUDY OF THE EFFECT OF ADDING HIGH-MOLECULAR POLYMERS ON HEAT TRANSFER CHARACTERISTICS OF NANOFLUIDS

Dmitriy Guzei[1,*], *Maxim* Pryazhnikov[1], *Andrey* Minakov[1,2], and *Vladimir* Zhigarev[1]

[1]Siberian Federal University, 660041 Krasnoyarsk, Russia
[2]Institute of Thermophysics SB RAS, 630090 Novosibirsk, Russia

Abstract. The experimental study of influence of adding high-molecular weight polymers on heat transfer characteristics of water-based nanofluids with Al2O3 nanoparticles. Volumetric concentration of particles was very low, 10-3 %. The volumetric particle concentration was equal to 2%. We investigated the dependence of heat transfer coefficient and pressure drop on the concentration of additives high polymers.

1 Introduction

Intensification of heat transfer and the related problems of experimental and theoretical studies are currently assuming an important and rapidly developing field of the teaching of heat transfer. In all machines, equipment and technologies there is a need of intensive heat removal, which uses a different kind of heat transfer equipment. One way to solve the problem of heat transfer enhancement of the process is the use of liquids with a dash of nanoparticles of various compositions, called "nanofluids".
The term "nanofluids" was first proposed in [1] to paper to the two-phase system consisting of a base fluid and nanoparticles high thermal conductivity material. Typical base fluids are water, organic liquids (ethylene glycol, oil), polymeric solutions. The material for the nanoparticles are metals, metal oxides, carbon nanotubes.

The aim of this study was to: Experimental study ways to reduce the friction in the turbulent heat transfer nanofluids using polymeric additives for the creation of nanofluids with desired transport properties and rheology.

2 Description of the experimental setup

The diagram of the installation to study the heat transfer coefficient is shown in fig 1. The installation is a closed loop with a circulating coolant. The working fluid is pumped through the heated test section to the heat exchanger, where heat is removed by the thermostat. The

*Corresponding author: gudimas@yandex.ru

flow rate of the working fluid in the loop is controlled by controlling the pump power by means of laboratory transformer. Power input to the pump is measured through Omix meter.

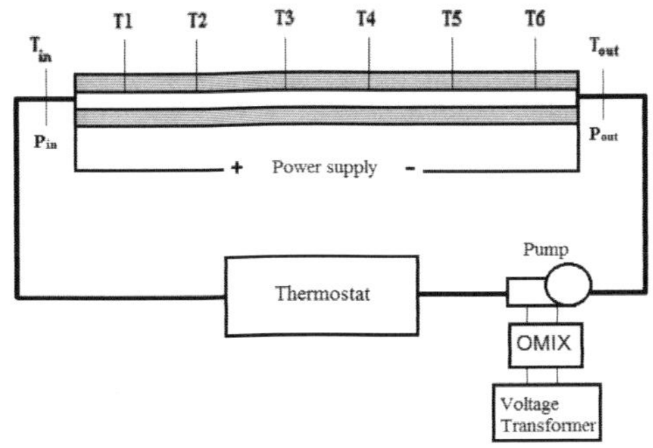

Fig. 1. A schematic diagram of the measuring system.

The heated test section is a stainless steel tube 6 mm in diameter, 1m long with wall thickness of 0.5 mm. The tube was heated by supplying electric current directly to its wall. This heating technique allows obtaining a constant wall heat flux boundary condition. In addition, this heating technique is universal and easily applicable to tubes of any cross section. The tube is insulated by multi-layer insulation. The heating power is controlled by a transformer. Six chromel-copel thermocouples fixed on the tube wall at the equal distance from each other were used to measure the local temperature of the tube. Temperature measurements were carried out by TPM-200 meters. In addition, the temperatures at the inlet and outlet of the heated test section were measured using thermocouples. At that, the thermocouple designed to measure the fluid temperature at the outlet of the loop was located at a considerable distance from the end of the heated test section to ensure uniformity of the fluid temperature in the metering point. To determine the temperature of the liquid at the outlet the liquid must be very good to mix. To do this, at the exit of the heated channel equipped with a special mixer. The mixer is a cylindrical chamber with a diameter 3 cm and 5 cm in length, filled with polypropylene rings (diameter of 5 mm and a length of 7mm). The mixer and loop section between the heater and the fluid temperature metering point was also insulated.

In all experiments, electric power supplied heater was measured. This electric power compared with the heat capacity transferred to liquid. The installation was well heat insulated. Therefore, the difference between the thermal and electrical capacities not exceed 3-5%.

The pressure drop measurements were carried out using a differential manometer OWEN PD200. The designed experimental setup was tested based on known empirical data for heat transfer of pure water. The water flow rate ranged from 0.65 to 2 l/min that corresponds to the range of Reynolds number from 2300 to 8000.

3 The testing of the experimental setup

Figure 2 shows the comparison of the experimental data on the average Nusselt number depending on the Reynolds number with the known empirical correlation [2]:

$$Nu = 0.021 \cdot \mathrm{Re}^{0.8}\,\mathrm{Pr}^{0.43},$$

Where $Nu = \alpha d / \lambda$ – is the Nusselt number, $\alpha = GC_p(T_{out} - T_{in})S^{-1}(T_w - \bar{T})^{-1}$ – is the average heat-transfer coefficient; G – mass flow rate, C_p – is the specific heat of the fluid; S – is the area of the lateral surface of the channel; T_{out}, T_{in} – are fluid temperatures at the channel inlet and outlet; $\bar{T} = (T_{in} + T_{out})/2$ – is the average fluid temperature; T_w – is the arithmetic mean of the channel wall temperature, obtained by averaging of the values of six thermocouples; Pr – is the Prandtl number; $\mathrm{Re} = \rho U d / \mu$ – is the Reynolds number; λ and μ – are fluid thermal conductivity and viscosity coefficients, respectively; U – is the superficial velocity; and d – is the tube diameter. This values calculated at average temperature $\bar{T} = (T_{in} + T_{out})/2$.

As is obvious from the plots presented in fig. 2, experimental data obtained in the turbulent flow regime, are in good agreement with the empirical correlation [2]. The discrepancies do not exceed 5% that is comparable to the accuracy of the correlation.

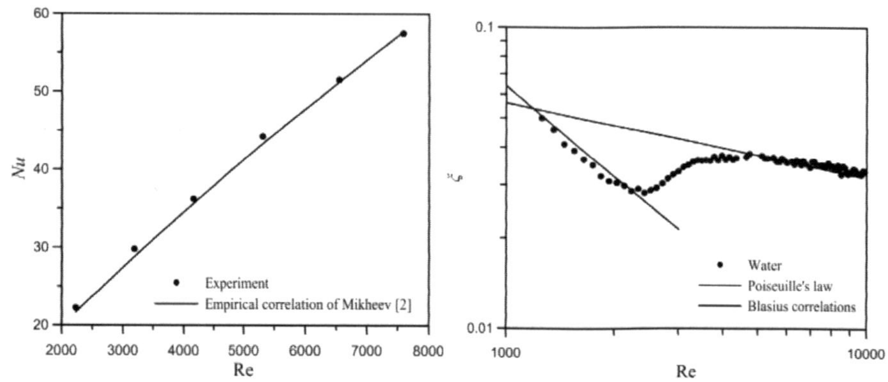

Fig. 2. Heat transfer coefficient (a) and friction factor (b) for pure water versus Reynolds number.

In addition to the heat transfer of pure water, pressure drop was measured as well. Figure 2b shows the measured correlation between friction factor and Reynolds number for pure water. The friction factor was calculated by known formula:

$$\zeta = \frac{2d\Delta P}{\rho U^2 L}.$$

Where U – is the superficial velocity; d – is the diameter; L – is the test section length; and ΔP – is the measured pressure drop.

For comparison, the chart shows also calculations by Poiseuille's theoretical relationship $\zeta = 64/\mathrm{Re}$ for laminar flow and Blasius correlation $\zeta = 0.316\,\mathrm{Re}^{-0.25}$ for turbulent flow. It is obvious that within the range of Reynolds numbers from 2300 to 3000 we observe flow transition from laminar to turbulent regime. The measured pressure drop values are consistent with the theoretical values with an accuracy of 5%.

Initial conditions (t=0): $T=T_{g0}$, $C_w=0$, $C_g=1$, $\psi=0$, $\omega=0$ at $0<x<H$, $0<y<L$; $T=T_{w0}$ at $0<r<R_d$, $0<\varphi<2\pi$.

Boundary conditions (t>0):

$$T = T_g,\ C_w = 0,\ C_g = 1,\ \frac{\partial \psi}{\partial y} = U_g \ \text{at} \ y = 0,\ 0 < x < H; \tag{1}$$

$$T = T_a,\ \frac{\partial^2 C_w}{\partial y^2} = 0,\ \frac{\partial^2 C_g}{\partial y^2} = 0,\ \frac{\partial \psi}{\partial y} = -U_w \ \text{at} \ y = L,\ 0 < x < H; \tag{2}$$

$$\frac{\partial T}{\partial t}=0, C_w=0, C_g=0, \frac{\partial \psi}{\partial x}=0 \text{ at } x=0, 0<y<L; \quad (3)$$

$$\frac{\partial T}{\partial t}=0, C_w=0, C_g=0, \frac{\partial \psi}{\partial x}=0 \text{ at } x=H, 0<y<L. \quad (4)$$

Where t – time, s; T – temperature, K; T_{g0} – initial temperature of gases, K; T_{wo} – initial temperature of water, K; C_w – concentration of water; C_g - concentration of high-temperature gases; $\omega=\frac{\partial v}{\partial x}-\frac{\partial u}{\partial y}$ – rotation vector velocity, s^{-1}; ψ – flow function, m^2/s; x, y – coordinates of the Cartesian coordinate system, m; H, L – dimensions of solution area, m; r, φ – coordinate of the spherical systems; R_d – radius of droplet, m; T_g – temperature of gases, K; U_g – velocity of high-temperature gases, m/s^2; U_w – velocity of spray flow, m/s^2; T_a – temperature of air, K.

In this formulation of the problem there are adiabatic horizontal and vertical walls with constant temperature $T_g \approx 1170\text{ K}$ and $T_a \approx 298\text{ K}$. Velocity of sray flows and velocity gases equal $U_g=1.5\text{ m/s}^2$ and $U_a=-5\text{ m/s}^2$.

Boundary conditions for liquid – gas were set taking into account vaporization:

$$R=R_1, 0<\varphi<2\pi, \lambda_1\cdot\frac{\partial T_1}{\partial R}=\lambda_2\cdot\frac{\partial T_2}{\partial R}-W_e\cdot Q_e, \rho_2 D_2\frac{\partial C_w}{\partial r}=W_e, t>0. \quad (5)$$

Where R_1 – droplet radius at the output of high-temperature gases, m; λ_1 – thermal conductivity of water, W/(m·K); T_1 – water temperature, K; λ_2 – thermal conductivity of water vapor, W/(m·K); T_2 – water vapor temperature, K; W_e – evaporation rate, kg/(m^2·s); Q_e – water vaporization heat, J/kg; ρ – water vapor density, kg/m^3; D_2 – diffusion coefficient, m^2/s.

The simulation used Ansys package. To solve this problem used time steps $\Delta\tau = 0.01$ s and grid $\Delta x = 0.01$ mm.

4 The results of the experiments

Numerical and experimental investigation of the influence of additives of high molecular polymers on the characteristics of turbulent heat transfer of nanofluids was conducted.. The Reynolds number is varied from 2000 to 10000. Distilled water was used as a base fluid. The polymers used range of acrylic polymers with a molecular weight of from 1 to 20MDa anionic charge and from 3 to 40 %. As nanoparticles was used a powder of aluminum oxide. The diameter of the nanoparticles ranged from 10 to 100 nm. The volumetric particle concentration was equal to 2%. The obtained suspension was subjected to hour ultrasonic bath "Sapphire".

The experiments were measured and the average value of the local heat transfer coefficient at the channel walls, and the pressure drop at the inlet and outlet of the channel at different concentrations and polymer nanoparticles.

It has been shown that the addition of polymer does not reduce the value of the coefficient of heat transfer at turbulent forced convection of nanofluid (see fig. 3).

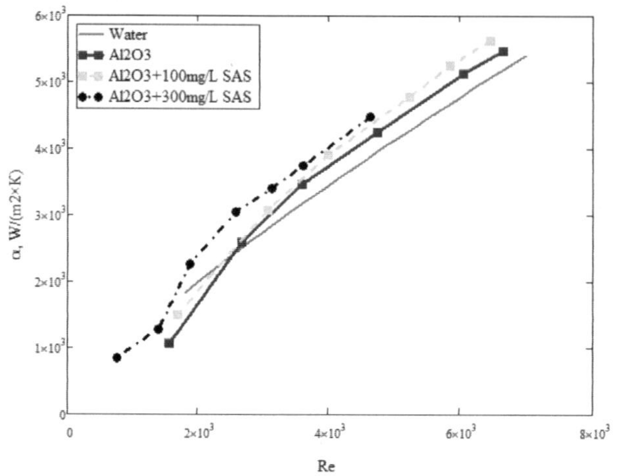

Fig. 3. The dependence of the average heat transfer coefficient on the Reynolds number and the surfactant concentration (anionic polyacrylamide flocculant, $M = 20 \cdot 10^6–22 \cdot 10^6$ anionic charge - 40%).

The measured values of the pressure drop in the channel for turbulent flow of nanofluids with additives of polymers (fig. 4). It is experimentally shown that the addition of high molecular weight polymers can reduce friction during turbulent flow of nanofluids. So in particular, it is shown that adding polyacrylamide in the nanofluid reduces the pressure drop in the channel in the turbulent flow regime by approximately 60% compared to the nanofluid without polymer. With the increase of polymer concentration, the pressure drop decreases monotonically and goes to the next level after that does not depend on the concentration of the polymer. This reduction occurs only in turbulent flow regime, which suggests that the reason for the decline is due to the Toms effect. It was found that for lower pressure drops in turbulent flow of nanofluids are needed in some times higher concentration of polymer than needed for pure liquid. Thus, in particular, to reduce the pressure drop by 67 % for pure water, the concentration of polyacrylamide of about 100 ml/L, in nanofluids the concentration of polymer for the same reduction of friction should be at 250-300 mg/L.

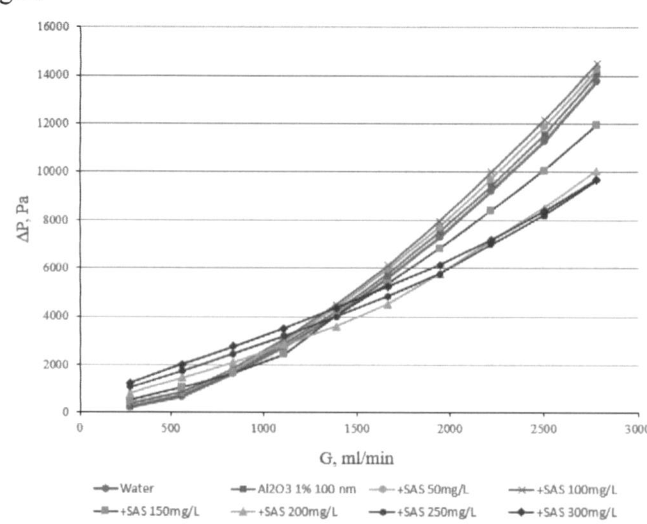

Fig. 4. The differential pressure dependence of nanofluids based on Al_2O_3 particles on the flow rate and different concentrations of surfactants. (anionic polyacrylamide flocculant, $M = 20 \cdot 10^6–22 \cdot 10^6$ anionic charge – 40 %).

This is, adsorption of the molecules of the polymer nanoparticles. A study of the molecular weight of the polymer to reduce the value of the pressure when pumping nanofluids. The results are shown in fig. 5.

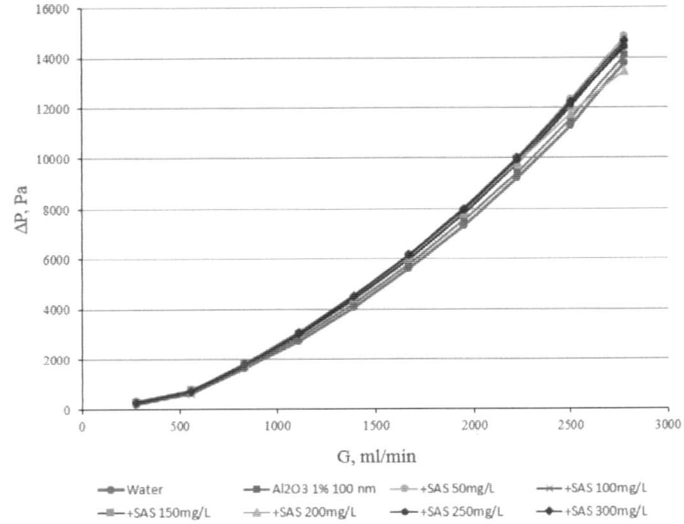

Fig. 5. The differential pressure dependence of nanofluids based on Al_2O_3 particles on the flow rate and different concentrations of surfactants. (anionic polyacrylamide flocculant, $M = 1 \cdot 10^6$ anionic charge – 3 %).

The influence of molecular weight and anionic charge of the polymer to reduce the value of friction pumping nanofluids in the turbulent regime. So in particular, it was shown that the additive of polyacrylamide with a molecular weight of 1 MDA and a concentration of 300 mg/L for nanofluids with particles of aluminum oxide almost does not reduce friction. While the same additive of polyacrylamide with a molecular mass of 20 MDA reduces friction by about 50 %. Thus it is shown that for maximum effect to reduce friction it is necessary to apply a polymer with a higher molecular weight.

Acknowledgments

This work was financially supported by the Russian Foundation for Basic Research (Project No. 16-38-00569).

References

1. S.U.S. Choi, ASME FED **66** (1995)
2. F.F. Tsvetkov, B.A. Grigoriev, *Heat and Mass Transfer* (Power Engineering Institute, Moscow, 2005)

NUMERICAL APPROACH TO THE EVALUATION OF ATMOSPHERIC POLLUTION BY COMBUSTION PRODUCTS OF SOLID FUELS

Dmitry Gvozdyakov[1,*], *Vladimir* Gubin[1], *Svetlana* Shvab[1], and *Anton* Tanishev[1]

[1]National Research Tomsk Polytechnic University, 634050 Tomsk, Russia

Abstract. This research presents the results of mathematical modeling of acid rain formation and its fallout in the areas adjacent to the site of the energy complex. Four thermal periods that are typical for regions of sharply continental climate has been considered. The influence of convective parameters on the formation mechanism of atmospheric precipitation has been estimated. The contribution of solar energy flux to the research process has been analyzed.

1 Introduction

Nowadays one of the main sources of emission combustion products of solid fuels in atmospheric is chemical and energy complexes [1]. Every year emissions of toxic compounds, i.e. SO_3, in environment are increase. [2]. The interaction (condensation on the surface of microscopic particles) of the sulfuric anhydride with the components of air can lead to formation of acid rain [1]. Researches [3] are shown that acid formation cause considerable damage not only in the acid rain, actually. According to previous research [4] – with the increase (or decreasing) in the speed of air masses can change the mass flow rate of condensation SO_3 on the droplet surface. Experience to solve problems [5–6] shows the ability of numerical approach in the processes of transformation and distribution in airspace of combustion of natural fuel thermal power plants produce.

The purpose of this research is the numerical approach to the study of the process of formation and distribution of the sedimentation of acid rain in the atmosphere adjacent to the thermal power plant, on the surface of condensation nuclei, with the main significant factors: unsteady heat and mass transfer; phase transformations; solar radiation; the speed of movement of air masses; temperature; drops of flight dynamics.

We used an iterative algorithm developed for solving nonlinear heat conduction problems in a local heat exposure [7] method of finite differences.

According to [8], the influence of non-sphericity coefficient drops characterized non-sphericity G, as determined ratio of the surface area of a real drop in the area of similar size sphere.

* Corresponding author: Dim2003@tpu.ru

2 Mathematical problem

Heat equations drops and movement within the accepted physical model of the test process is as follows:

$$\left.\begin{aligned} \rho_1 \cdot C_P \cdot \frac{\partial T_1}{\partial t} = \lambda\left(\frac{\partial^2 T_1}{\partial r^2} + \frac{2}{r}\cdot\frac{\partial T_1}{\partial r}\right) \\ \frac{dU_X}{dt} = \frac{3\cdot G\cdot C_D\cdot \rho_1\cdot(U_X - U_0)\cdot(U_X - U_0)}{8\cdot\pi\cdot\delta\cdot\rho_2} \\ \frac{dV_Y}{dt} = g + \frac{3\cdot G\cdot C_D\cdot \rho_1\cdot(V_Y - V_0)\cdot(V_Y - V_0)}{8\cdot\pi\cdot\delta\cdot\rho_2} \end{aligned}\right\} \quad (1)$$

where $\rho_{1,2}$ – density, kg/m^3; C_P – isobaric heat capacity, $J/(kg{\cdot}K)$; λ – coefficient of thermal conductivity, $W/(m{\cdot}K)$; T – temperature, K; t – time, s; V_X, V_Y – drop velocity in the direction of the axes x and y, respectively, m/s; C_D – drag coefficient; g – acceleration of gravity, m/s^2; U_0, V_0 – velocity components ambient, m/s.

The initial and boundary conditions are:

$$t = 0: \quad T = T_0, \quad 0 \le r \le \delta;$$

$$r = 0: \quad -\lambda_1 \cdot \frac{\partial T_1}{\partial r} + Q\cdot\left(W_K^3 + W_K^4\right) - P = -\lambda_2 \cdot \frac{\partial T_2}{\partial r}, \quad T_1 = T_2; \quad r = \frac{\delta}{2}: \quad \frac{\partial T_1}{\partial r} = 0,$$

where T_1, T_2 – the temperature drops and a gaseous medium, respectively, K; W_k^3 W_k^4 – mass condensation rate, $kg/(m^2{\cdot}s)$; Q – the heat of reaction, J/kg; δ and r – the characteristic dimensions of the drops, m; P – density of solar energy flux, W/m^2.

Two thermal periods are analyzed, typical for regions with sharply continental climate (i.e., Siberia). The value of the solar energy flux density was assumed to be the value of the solar constant. Below in fig. 1 illustrates the results of numerical studies in the form of temperature drop of the surface from a distance (blue dashed lines correspond to the temperature of the air).

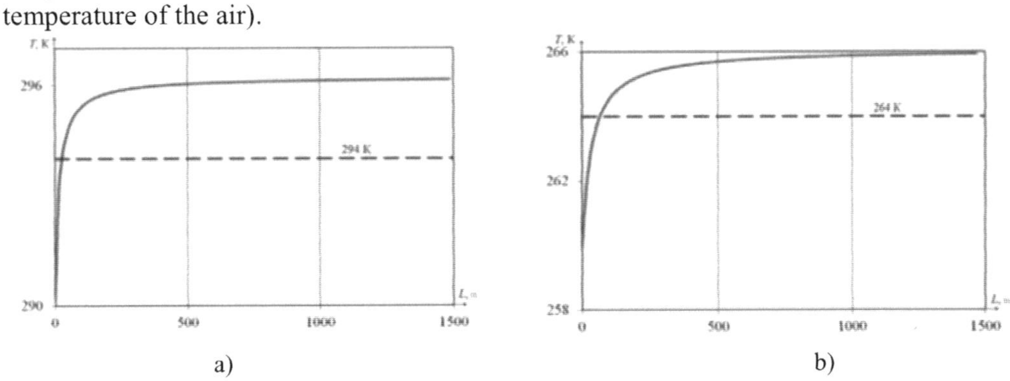

Fig. 1. Temperature fields of drop at the time 1800 *s* at the wind speed of 5 *m/s*: a) in the summer; b) in the winter.

Secondly, the influence of the wind speed on warming formed drops is substantially. The difference between the temperature of sulfuric acid droplets formed surfaces in the airspace, adjacent to the thermal power plant, and the environment is 0.5 ... 1.2 K. Numerical evaluation of the impact of the speed of movement of air masses (fig. 1) is consistent with studies [8, 9]. Below in fig. 2 are shown the results of mathematical modeling.

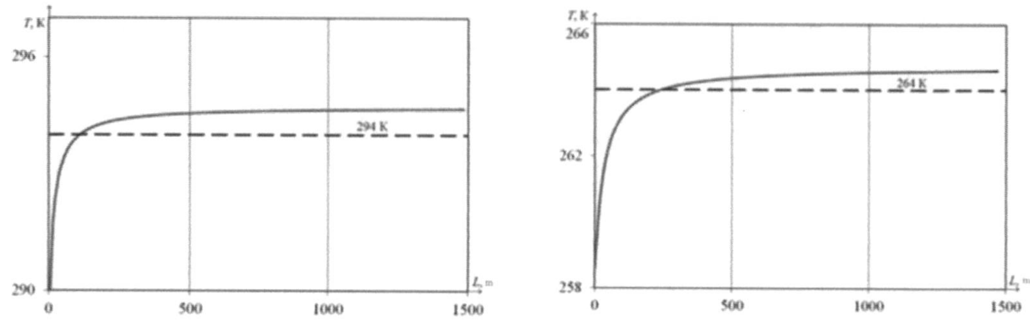

a) b)

Fig. 2. Temperature fields of drop at the time 1800 *s* when at the wind speed of 10 *m/s*: a) in the summer; b) in the winter.

Table 1 shows a comparison of the results of research [9] and the present study. Comparison of the numerical values of mathematical modeling of the formation of acid formation on the surface of condensation nuclei is performed at the wind speed of 5 m/s and ambient temperature of 294 K.

Table 1. Dimensions drops of sulfuric acid with different mechanisms of formation of acid rain.

t, s	[9]	Condensation of SO_3 & H_2O with considering *P*
	$\delta_{ch} \cdot 10^3$, *m*	$\delta_{ch} \cdot 10^3$, *m*
0	δ_0=5.0·10^6, *m*	
600	0.960	0.923
1800	1.616	1.553

Here δ_0 is the size of condensation nuclei in the initial time, *m*; δ_{ch} – is particle size formed at time *t*.

3 Conclusion

Evaluation of the numerical parameters of acid precipitation leads to the conclusion that co-condensation of vapor of sulfuric anhydride and steam based on the solar energy flux significantly (up 4 %) slows the process of formation of sedimentary deposits. It should be noted that after 600 s precipitation with a typical size of 1 mm are formed.

Acknowledgments

The research was realized with financial support of Minobrnauki of Russia in framework of FTP "Research and development in prior direction of scientific-technological complex of Russia in 2014-2020 years", unique R&D identifier RFMEFI58114X0001.

References

1. Y. Izrael, I.Nazarov, A.Pressman, *Kislotnye dozhdi* (Acid rain. Gidrometeoizdat. Leningrad, 1989)
2. E. Volkov, E. Gavrilov, F. Dugih, *Flue pipe TPP and NPP* (Energoatomizdat. Moscow, 1987)
3. D. Gvozdyakov, V. Gubin, Butlerov Communications **37**, 10 (2013)

4. M.Thomas, Impact of air pollution on plants. Geneva (1962)
5. D. Gvozdyakov, V. Gubin, Chem. Pet. Eng. **51** (2015)
6. D. Gvozdyakov, V. Gubin, EPG Web Conf. **82**, Article Number 01026 (2015)
7. G. Kuznetsov, M. Sheremet, Russ. Microelectron. **37**, 2 (2008)
8. V. Terekhov, M. Pakhomov, *Heat and mass transfer and hydrodynamics in gas-drop flows* (NSTU, Novosibirsk, 2008)
9. D. Gvozdyakov, V. Gubin, Chem. Pet. Eng. **3** (2015)

THE EFFECT OF THE SUBSTRATE TEMPERATURE AND THE HEIGHT OF THE MINI-CHANNEL ON THE EVAPORATION OF LIQUID DROPLET BLOWN BY THE GAS FLOW

Ekaterina Isachenko[1,2], *Evgeniy* Orlik[1,3,*], and *Oleg* Kabov[1,4]

[1]Kutateladze Institute of Thermophysics SB RAS, 630090 Novosibirsk, Russia
[2]Novosibirsk State University, 630090 Novosibirsk, Russia
[3]Novosibirsk State Agricultural University, 630039 Novosibirsk, Russia
[4]National Research Tomsk Polytechnic University, 634050 Tomsk, Russia

Abstract. The experimental setup has been designed and fabricated to study the evaporation of liquid droplets, blown by the gas flow in a mini-channel with a height of 6 and 9 mm. Substrates were removable, and the surface temperature of the substrate was maintained at a constant level. The shadow method was the main method of measurement. A series of experiments with a water drop of 100 μl volume settled on a polished stainless substrate were carried out. As a result the dependences for the droplet evaporation rate were obtained in the range of temperatures 25-70 °C and Reynolds numbers of gas 0-2000.

1 Introduction

Drop is a fairly simple physical object, including complex physical processes and interactions. Evaporation of liquid droplets on solid surfaces is a key point in many processes such as: painting, coating, air humidification, irrigation in agriculture, cooling of hot surfaces, etc. The evaporation of droplets has been widely studied numerically and experimentally, but mostly in motionless air. The studies show that the contact angle and its hysteresis have a significant impact on droplet evaporation [1, 2, 3]. Authors of [4] studied the effect of gravity on the drop spreading over a solid surface. In work [5] it is shown that Marangoni effect (flow driven by surface tension gradient) modifies the drop interface shape and impacts on evaporation. The influence of gas flow on droplet evaporation is known, but all these studies were conducted in the open air with insufficiently controlled conditions [6]. This paper presents the experimental results for evaporation of the droplet under gas action, placed on a heated substrate in a mini-channel.

* Corresponding author: orlik.evgeniy@gmail.com

2 The experimental setup and conditions

The scheme of experimental setup is shown in figure 1. It includes a channel with height varying from 3 mm to 20 mm, removable substrates, the temperature regulation system of the substrate, and the air supply system. The side walls and the upper cover were equipped with optical windows for visualization. The aim of this work is to study the effect of the substrate temperature and the Reynolds number of the air flow on the drop evaporation rate in a channel with height of 6 and 9 mm. The working fluid was ultrapure water, obtained through a Milli-Q system. The drop was placed into the channel using a syringe. The initial drop volume was about 100 μl. The experiments were conducted on the stainless steel substrate with a polished surface. The Reynolds number of the gas flow varied from 0 to 2000. The gas temperature was maintained equal to 25 °C. Heating of the substrate was carried out by Peltier elements contacting with the substrate. The substrate temperature during the experiment was regulated with a precision of 0.20 °C with the help of the controller PR-59. The experiments were conducted for three temperature values 25.0, 50.0 and 70.0 °C.

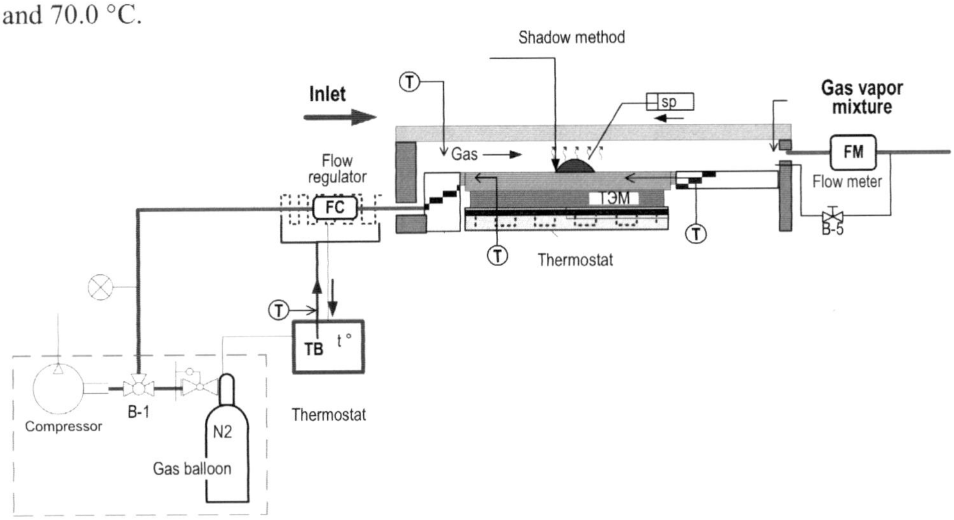

Fig. 1. Scheme of the setup.

To register the contour of the sessile drop on the solid substrate the shadow method was used. Its principle is based on the lighting of the physical object by a parallel light beam, whose shadow is recorded by the camera, as shown in figure 2. Optical equipment allows obtaining images with a resolution of 6 μm/pixel (see figure 3). The obtained images were processed by different methods with the help of software (the Drop Shape Analysis by KRÜSS), including on the basis of the solution to the Young-Laplace equation [7]. The most suitable for measuring the contact wetting angle in the experiments was the tangential method, since the drop has a slightly asymmetric shape due to the air flow impact. The principle of the tangential method is based on the adjustment of the contour according to the equation of conical sectors. Then at the point of intersection of the contour line and the baseline, the derivative of this equation equal to the contact angle is taken.

DOI: 10.1051/matecconf/20167201038

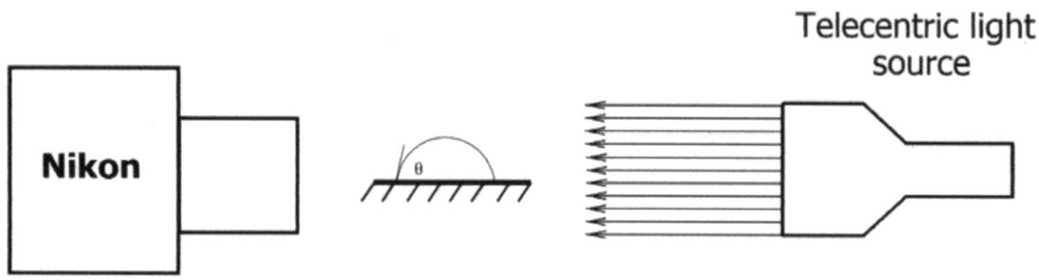

Fig. 2. Shadow method.

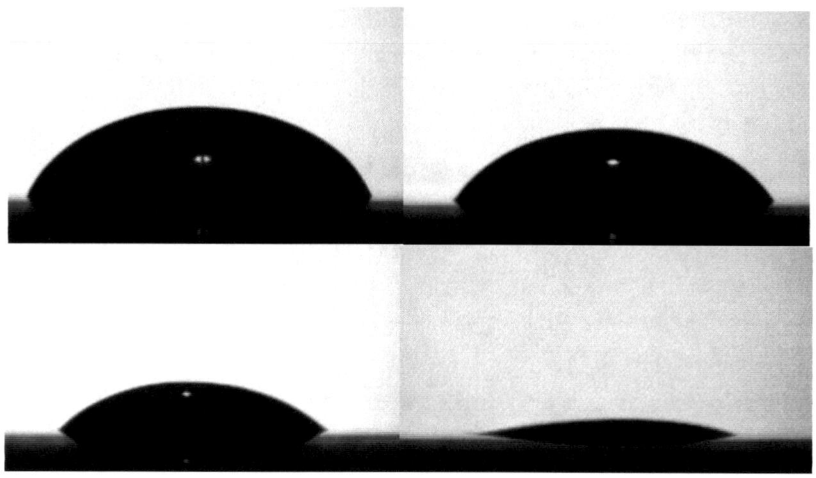

Fig. 3. Photo of the drop using the shadow method. The interval between frame is 2 minutes. T = 70 °C, Re = 1140, channel 9 mm.

3 Results

The first series of experimental data was obtained at constant substrate temperature of 25 °C and at different air flow velocities in the channel of 6 mm and 9 mm height (see figure 4). It is seen that the channel height does not affect the nature of the curve. In the 6 mm channel, the drop lifetime is significantly less than in the 9 mm channel. This may be explained by the fact that at the same Reynolds numbers, the gas velocity in the 6 mm channel is higher than in the 9 mm channel. The superficial velocity in the 6 mm channel at maximum Re number = 2000 corresponds to 5 m/s.

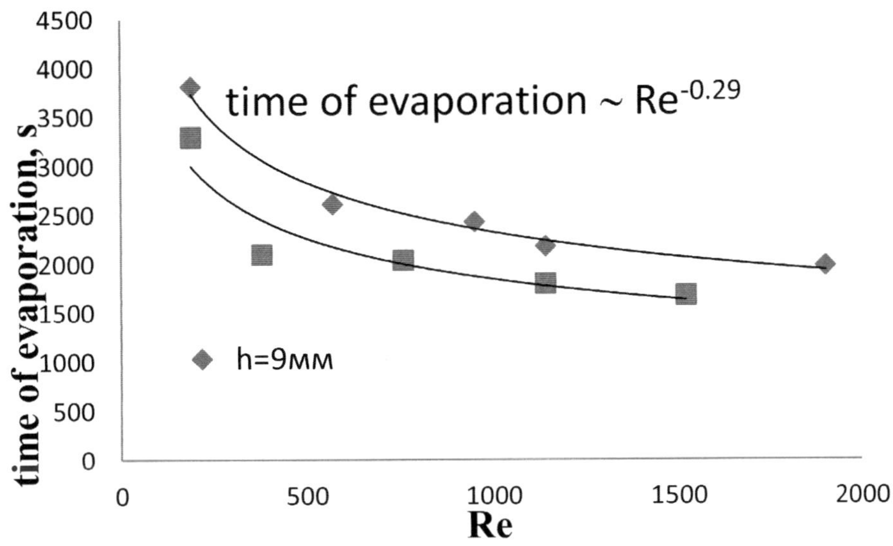

Fig. 4. Drop lifetime depending on the Reynolds number. $T_{substrate}$ = 25 °C.

Analyzing the results of the experiment shown in figure 5, it may be concluded that in the range Re = 1140-1900 the evaporation rate is determined by the substrate temperature and less depends on the height of the channel and the Reynolds numbers of the gas.

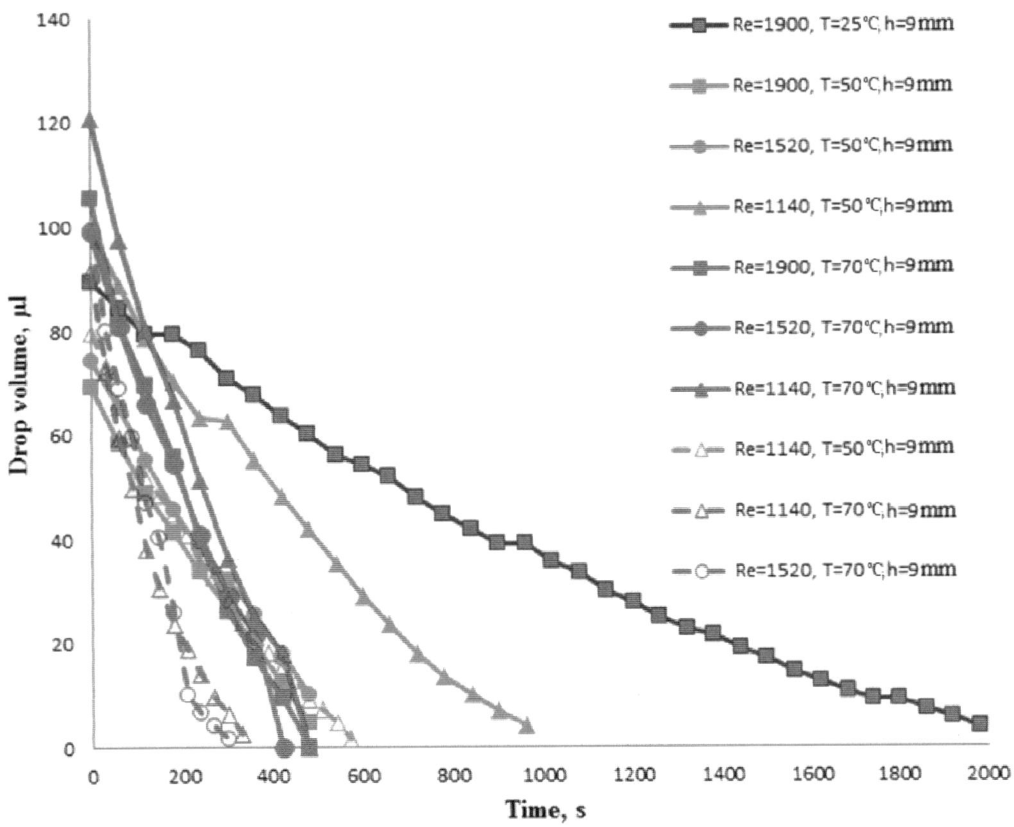

Fig. 5. The effect of temperature and height of the channel on the evaporation of droplets for different Reynolds numbers of the gas.

Figure 6 shows the results of the evaporation rate Q, related to the drop base. Despite the fact that there is some scatter in the data, it is visible that the evaporation rate is weakly dependent on time. The height of the channel affects the rate of evaporation, as well as in the case with the experiment without the substrate heating. It can be noted that before the disappearance of the drop, i.e. its complete evaporation, some curves show intensification of evaporation, which may be explained by the fact that the drop becomes almost flat (see. figure 3). Since the interval between frames is large enough (30 seconds), then further analysis is required, and new measurements performed with a smaller time interval between frames.

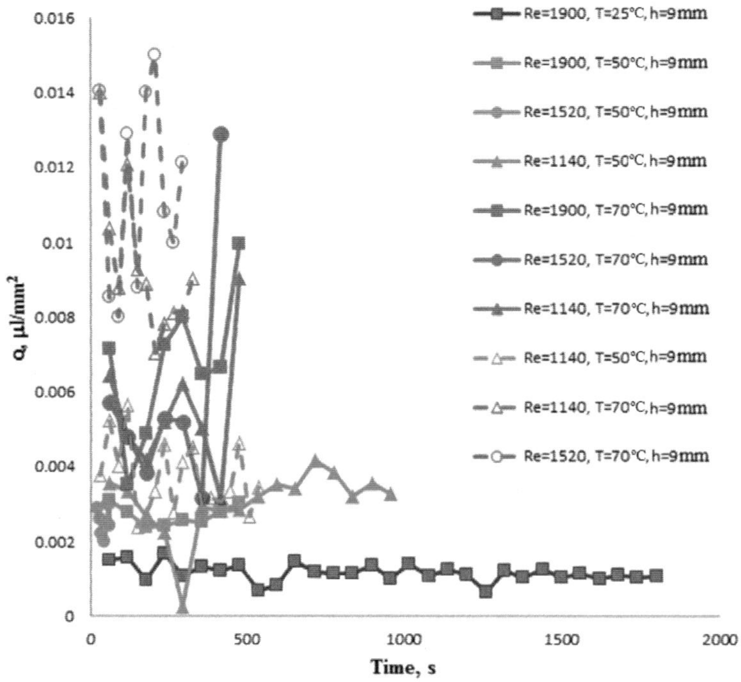

Fig. 6. Evaporation rate of water droplet via time.

Acknowledgments

The authors express their deep gratitude for the support of this work to the Russian Science Foundation (project No. 14-19-01755).

References

1. O.A. Kabov, D.V. Zaitsev, Dokl. Phys. **58**, 7 (2013)
2. D.V. Zaitsev, D.P. Kirichenko, O.A. Kabov, Tech. Phys. Lett. **41**, 6 (2015)
3. A.A. Semenov, D.V. Feoktistov, D.V. Zaitsev, G.V. Kuznetsov, O.A. Kabov, Thermophysics and Aeromechanics **22**, 6 (2015)
4. E.Ya. Gatapova, A.A. Semenov, D.V. Zaitsev, O.A. Kabov, Colloids Surf. **441** (2014)
5. Y. Tsoumpas, S. Dehaeck, A. Rednikov, P. Colinet, Langmuir **31**, 49 (2015)
6. L. Bin, R. Bennacer, A. Bouvet, Appl. Therm. Eng. **31** (2011)
7. I.V. Marchuk, V.V. Cheverda, P.A. Strizhak, O.A. Kabov, Thermophysics and Aeromechanics **22** (2015)

MATEC Web of Conferences **72**, 01039 (2016) DOI: 10.1051/matecconf/20167201039
HMTTSC-2016

IMPROVING ENERGY EFFICIENCY CABLE PRODUCTION

Olga Iashutina[1,*], *Kseniya* Vershinina[1], and *Evgeniya* Ivanova[1]

[1]National Research Tomsk Polytechnic University, 634050 Tomsk, Russia

Abstract. During the energy calculation is made at different temperatures of the heating surface. The influence of the speed of pulling on the cost of the finished products of cable products. The interrelation of speed broaching and temperature of the heating surface.

1 Introduction

The manufacturing process of cables consists of several stages: drawing, cure, drawing, extruding, winding on a spool. The most energy-consuming is the cure. This physical-chemical process takes quite a long time intervals (hundreds of seconds) occurs at a low pulling rate and high temperature heat insulating product. It is also necessary to monitor the temperature of the surface of the insulating layer, which should not exceed the temperature of the beginning of the thermal decomposition of the material sheath.

Each plant produced dozens of cable products. They are classified into solid and stranded, power, installation, monitoring, mine cables, control cables, etc.

Table 1 shows the prices for power cables unarmored several manufacturing plants.

Table 1. Cost of power unarmored cables.

Manufacturer	Brand Cable Products	Cost, ruble/meter
"Content", Tomsk	VVG 3x1.5	20.90
"Roscable", Tomsk	VVG 3x1.5	17.85
"ETM", Tomsk	VVG 3x1.5	22.70

The purpose of this work to reduce energy consumption and improve resource efficiency of production of cable products. To achieve this goal have been resolved following tasks:

1. Determination of the minimum energy associated with a decrease in temperature of the heating surface.
2. Determination of the effect of speed on the energy consumption of production broach.

* Corresponding author: osy1@tpu.ru

2 Problem statements

When setting objectives assumed that the cable is passed through a specialized camera with different heating temperature (543 K, 593 K). It is considered [1], that the output of such a chamber is a product characterized by a high degree of polymerization of the insulating sheath over the entire thickness [1]. The term "complete polymerization" [2] means the end of the chemical process in the insulating layer to the condition $\varphi \approx 1$ (φ - the degree of completion of the main chemical reaction).

As a first approximation considered when modeling the system shown in fig. 1.

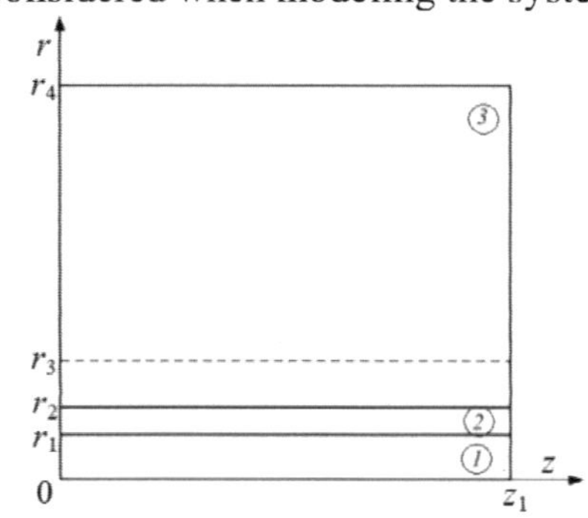

Fig. 1. Scheme for solving the problem of heat transfer area at $0 \le t \le t_p$: *1*-cable core, 2 - cable shell, 3 - the air heating chamber.

It was believed that the cable consists of a core 1 and shell 2. A product with an initial temperature T_0 and constant speed w_c moves through the heating chamber. The shell is heated at a significantly higher temperature 3. Vault chamber temperature (initial temperature of the air in the chamber) T_v taken much more T_0. As a result, overheating of the cable insulating layer is polymerized. Weight completion determined by the degree of polymerization of the insulating material and φ describes the quality of the final product. Time of completion of the polymerization ($\varphi \approx 1$ through the entire thickness ($r_1 < r < r_2$) shell) t_p is the primary integral characteristic of the process.

When setting objectives into account convective and radiative heat transfer mechanisms.

Selecting a cylindrical coordinate system for modeling due to the fact that the cable products are often multi-layered elongated along the axis of symmetry of the long cylinders [3]. To consider setting an axially symmetric (fig. 1).

3 Mathematical model

The system of differential equations of unsteady heat transfer in a partitioned system "heating chamber - air - insulation shell - core cable" (fig. 1), corresponding to the physical formulation of the problem, has the form [4, 5].

Heat conduction equation for the cable core ($0<r<r_1$, $0<z<z_1$):

$$\rho_1 C_1 \frac{\partial T_1}{\partial t} = \lambda_1 \left(\frac{\partial^2 T_1}{\partial r^2} + \frac{1}{r}\frac{\partial T_1}{\partial r} + \frac{\partial^2 T_1}{\partial z^2} \right) \tag{1}$$

The energy equation for insulating cable shell $r_1<r<r_2$, $0<z<z_1$):

$$\rho_2 C_2 \frac{\partial T_2}{\partial t} = \lambda_2 \left(\frac{\partial^2 T_2}{\partial r^2} + \frac{1}{r}\frac{\partial T_2}{\partial r} + \frac{\partial^2 T_2}{\partial z^2} \right) + q_2 \rho_2 \frac{d\varphi_2}{dt}, \tag{2}$$

when $$\frac{d\varphi_2}{dt} = (1-\varphi_2)k_2^0 \exp\left(-\frac{E_{a2}}{R_t T_2}\right). \quad (3)$$

The energy equation for air in the heating chamber ($r_2<r<r_4$, $0<z<z_1$):

$$\rho_3 C_3 \frac{\partial T_3}{\partial t} = \lambda_3\left(\frac{\partial^2 T_3}{\partial r^2} + \frac{1}{r}\frac{\partial T_3}{\partial r} + \frac{\partial^2 T_3}{\partial z^2}\right). \quad (4)$$

Initial (t=0) conditions $T_1=T_0$ at $0\le r<r_1$, $0\le z\le z_1$;$T_2=T_0$ and $\varphi=\varphi_0$ at $r_1\le r\le r_2$, $0\le z\le z_1$; $T_3=T_v$ at $r_2<r<r_4$, $0\le z\le z_1$.

The boundary conditions at $0<t\le t_p$:

$z=0,\ z=z_1,\ 0\le r<r_1$ $$\frac{\partial T_1}{\partial z} = 0;$$

$z=0,\ z=z_1,\ r_1\le r\le r_2$ $$\frac{\partial T_2}{\partial z} = 0;$$

$z=0,\ z=z_1,\ r_3\le r\le r_4$ $$\frac{\partial T_3}{\partial z} = 0;$$

$r=0,\ 0\le z\le z_1$ $$\frac{\partial T_1}{\partial r} = 0;$$

$r=r_1,\ 0\le z\le z_1$ $$-\lambda_1\frac{\partial T_1}{\partial r} = -\lambda_2\frac{\partial T_2}{\partial r},\ T_1=T_2;$$

$r=r_2,\ 0\le z\le z_1$ $$-\lambda_2\frac{\partial T_2}{\partial r} = -\lambda_3\frac{\partial T}{\partial r},\ T_2=T_3;$$

$r=r_4,\ 0\le z\le z_1$ $T=T_v$.

Here ρ – density, kg/m^3; C – heat capacity, J/(kg·K); T – temperature, K; t – time, c; λ – thermal conductivity, W/(m·K); r, z – cylindrical coordinate system, m; q_2 – thermal effect of the polymerization reaction, J/kg; φ – polymerization degree; k_2^0 – pre-exponential factor of the chemical reaction, c^{-1}; E_{a2} – the activation energy of a chemical reaction, J/mol; R_t – universal gas constant, J/(mol·K); ν – kinematical viscosity, m^2/s; g – acceleration of free fall, m/s^2; T_0 – the initial temperature of the conductor and the cable shell, K; T_v – primary air temperature in the chamber, K; indexes "1", "2", "3" correspond to the core, cable sheath and air in the chamber.

The system of nonlinear nonstationary differential equations (1) - (4) with appropriate boundary conditions is solved by finite difference method [6]. Difference analogues of differential equations (1) - (4) are solved locally one-dimensional method. For solving nonlinear equations, the method of iterations. To assess the reliability of the results of numerical modeling were used algorithms based on the verification of the conservativeness of difference schemes used [7, 8, 9].

4 Results and discussion

Consider energy as the temperature of the heating surface of the Ths1 = 543 K and Ths2 = 593 K speed and pulling cable products υ = 1 m / s.

Electricity tariff of 5 rubles per 1 kW / h at the cable company.

Power the air chamber Vac1 = 27.84 kW, at Ths1 = 543 K.
1 day (8 hours) energy consumption:

$$\Sigma\ \mathrm{con1} = T\cdot V\cdot C = 8\cdot 27.84\cdot 5 = 1113.48 \text{ rubles.} \quad (5)$$

1 month (take 21 working days in a month):

$$\Sigma\ \mathrm{con1m} = T\cdot V\cdot C = 8\cdot 27.84\cdot 5\cdot 21 = 23383.08 \text{ rubles.} \quad (6)$$

Over the year (in 2015 - 247 days):

$$\Sigma\ \mathrm{con1y} = T\cdot V\cdot C = 8\cdot 27.84\cdot 5\cdot 247 = 275059.2 \text{ rubles.} \quad (7)$$

Power the air chamber Vac2 = 32.94 kW, at Ths1 = 593 K.
1 day (8 hours) energy consumption:

$$\Sigma\ \mathrm{con2} = T\cdot V\cdot C = 8\cdot 32.94\cdot 5 = 1317.56 \text{ rubles.} \quad (8)$$

1 month (take 21 working days in a month):

$$\Sigma\ \mathrm{con2m} = T\cdot V\cdot C = 8\cdot 32.94\cdot 5\cdot 21 = 27668.76 \text{ rubles.} \quad (9)$$

Over the year (in 2015 - 247 days):

$$\Sigma\ \mathrm{con2y} = T\cdot V\cdot C = 8\cdot 32.94\cdot 5\cdot 247 = 325447.2 \text{ rubles} \quad (10)$$

Here C - the tariff for electricity, rubles; Vi - power consumption, kW; T - the number of hours of consumption, hours.

By reducing the temperature of the heating surface on the Δ Ths = 50 K savings per year will be:

$$\mathrm{Syear} = \Sigma\ \mathrm{con2y} - \Sigma\ \mathrm{con1y} = 325447.2 - 275059.2 = 50.388 \text{ rubles.} \quad (11)$$

Determine the impact of the speed of the cable pulling on energy products.

Consider υbr1 = 0.5 m / s and υbr2 = 1 m / s at a temperature of the heating surface of Ths = 543 K.

When υbr1 = 0.5 m / s for 1 day (8 hours) to manufacture the cable product length l = 14400 m with υbr2 = 1 m / s – l = 28.800 m.

Energy consumption at the heating surface Ths = 543 K for 1 day will be:

$$\Sigma\ \mathrm{con1} = T\cdot V\cdot C = 8\cdot 27.84\cdot 5 = 1113.48 \text{ rubles} \quad (12)$$

By increasing the speed of pulling, the cost of the finished product is greatly reduced. In this cable polymerization product is uneven, that is the degree of polymerization not all the points of the polymerization layer is close to unity, which in turn reduces the quality of the finished product and have a negative impact on its physical and chemical properties.

5 Conclusion

For the normal process of polymerization of cable products required to comply with several conditions: a small speed broaching and the temperature of the heating surface, which does not exceed the thermal decomposition temperature of the beginning of the shell material of the cable. To improve the energy efficiency of production of cables should greatly reduce the temperature of the heating surface. In turn, with a decrease in the surface temperature of the insulating layer to reduce the feeding speed is assumed, which will increase the cost of the finished products. It is therefore necessary to find a balance between the temperature of the heating surface and the advance speed.

Acknowledgments

This work was funded within the framework of realisation of Strategic Programme on National Research Tomsk Polytechnic University Competitiveness Enhancement in the Group of Top Level World Research and Academic Institutions (project VIU_ENIN_25_2016). The work at the field of modeling of heat and mass transfer processes near boundary between two layers was supported by the scientific schools grant NSH-7538-2016.8.

References

1. F.F. Koshelev, A.E. Kornev, N.S. Klimov, *General Technology of Rubber* (Chemistry, Moscow, 1968)
2. I.I. Tugov, *Chemistry and Physics of Polymers* (Chemistry, Moscow, 1989)
3. Cable Control. Power Cables. Part 1. Designator Catalog LLC ''Kamsky Cable''
4. P.J. Roache, *Computational Fluid Dynamics*, (Hermosa Publishers, 1982)
5. Y. Jaluria, Natural Convection: Heat and Mass Transfer, Pergamon Press, Oxford (1980)
6. A.A. Samarskii, *The Theory of Difference Schemes*, (Marcel Dekker, Inc., 2001)
7. E.V. Ivanova, G.V. Kuznetsov, P.A. Strizhak, EPJ Web Conf. **76**, 01024 (2014)
8. E.V. Ivanova, G.V. Kuznetsov, P.A. Strizhak, Chem. Pet. Eng. **51**, 3 (2015)
9. D. Veenendaal, P. Block, Eng. Struct. **75** (2014)

NEUTRON AGE DETERMINATION IN FAST REACTOR MATERIALS USING THE GROUP METHOD

Marina F. Kabanova[1,*], *Aleksey* V. Yuzhakov[1], *Arian* V. Kuzmin[1], *Yuriy* V. Kobenko[1], and *Elena* S. Riabova[2]

[1]National Research Tomsk Polytechnic University, 634050 Tomsk, Russia
[2]Samara State Social-Pedagogical University, 443099 Samara, Russia

Abstract. The article deals with the methods of identifying fast neutron age in sodium (*Na*) and uranium-238 (*238U*); describes the model of advanced and effective fast neutron nuclear reactors (FN), where *Na* is a coolant while *238U* is involved in the fuel cycle in large quantities; justifies the choice of the group method for calculating the neutron age value in the substances mentioned above that can show the accuracy of the used constants for *Na* and estimate various versions of multilevel description of neutron moderation in *238U* – the most powerful resonance absorber of the neutron reactor active zone.

1 Introduction

The idea of the fast reactor was first put forward by E. Fermi during 1940s and independently from him by or fellow Russian physicist A. Leipunskii. However, only by the 50s of our century it will have become possible to hope that there will be a programme of changing into these effective but expensive nuclear reactors. The fuel for them can be a more widespread *238U* and *232Th* and the coolant – sodium which vital neutronic characteristics are analysed in this paper [1].

Fast neutron ageing is connected with their migration in the moderating process. This characteristic is important while calculating the active zone of the atomic reactor as it determines the slowing down neutron leakage. The reconciliation of the theoretical and experimental methods of identifying neutron age shows the choice validity of the nuclear-physical constants. The nuclear-physical constants are meant mostly for engineer-physicists for reactor development, operation and protection. These characteristics are constantly renewed while obtaining new experimental and theoretical results and are generalized in the relevant reference books and library database.

* Corresponding author: mak16@tpu.ru

2 The peculiarities of the analysed elements

In the research works dedicated to age calculation in graphite [2] and in sodium [3] the group constants were used that helped to get the results that go in accordance with either direct measurement or with the calculation based on the previously established group consonants [5]. The peculiarity of graphite, the first nuclear reactors moderator, is the absence of resonance while moderating from 2 *MeV* till indic resonance *E=1.46 eV*. The peculiarity of neutron moderation in sodium structure *238U* is resonance absorption (fig. 1).

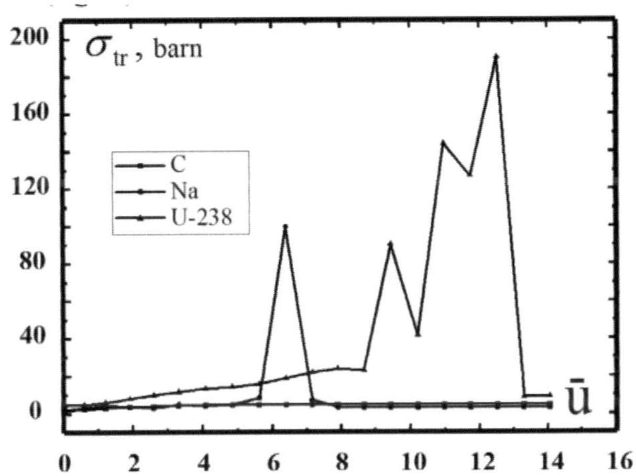

Fig. 1. Microscopic transport cross-section of carbon, sodium and uranium *238U* depending on average lethargy value in groups.

Basic energy characteristics 28-group constants are demonstrated in tab. 1, where E_n shows group energy range while $\bar{u}$ corresponds to the lethargy average value in the group.

Table. 1. Neutron energy group structuring .

Group	E_n	$\bar{u}$	Group	E_n	$\bar{u}$
-1	14.5–14.0 *MeV*	-1.96361	**13**	**4.65–2.15 *keV***	**6.37713**
0	14.0–10.5 *MeV*	-1.81238	14	2.15–1.0 *keV*	7.14665
1	10.5–6.5 *MeV*	-1.44692	15	1.0–0.46 *keV*	7.91219
2	6.5–4.0 *MeV*	-0.96508	16	465–215 *eV*	8.67971
3	4.0–2.5 *MeV*	-0.48551	17	215–100 *eV*	9.44923
4	2.5–1.4 *MeV*	0.02532	18	100–46.5 *eV*	10.21478
5	1.4–0.8 *MeV*	0.59784	19	46.5–21.5 *eV*	10.98230
6	0.8–0.4 *MeV*	1.20397	20	21.5–10.0 *eV*	11.75182
7	0.4–0.2 *MeV*	1.89712	21	10.0–4.65 *eV*	12.51736
8	0.2–0.1 *MeV*	2.59027	22	4.65–2.15 *eV*	13.28488
9	100.0–46.5 *keV*	3.30702	23	2.15–1.0 *eV*	14.05440
10	46.5–21.5 *keV*	4.07454	24	1.0–0.465 *eV*	14.81995
11	21.5–10.0 *keV*	4.84406	25	0.465–0.215 *eV*	15.58747
12	10.0–4.65 *keV*	5.60961	T	0.0253 *eV*	18.18561

In this table as well as in the following ones energy group with a resonant scattering *Na* .is shown. Using a part of mean group crossing from the paperwork [5] that are important for calculating neutron age till any detectable value where the following group constants are demonstrated: σ_c - microscopic cross-section of radioactive neutron capture;

σ_{in} - inelastic scattering cross-section with account of reaction cross-section $(n,2n)$ and $(n,3n)$; $\sigma_s \equiv \sigma_e -$ μ - average cosine of the scattering angle, calculated without including the group of neutron after scattering; ζ - average lethargy increment under elastic scattering (average logarithmic decrement). General type of some mentioned characteristics dependence is shown in fig. 2.

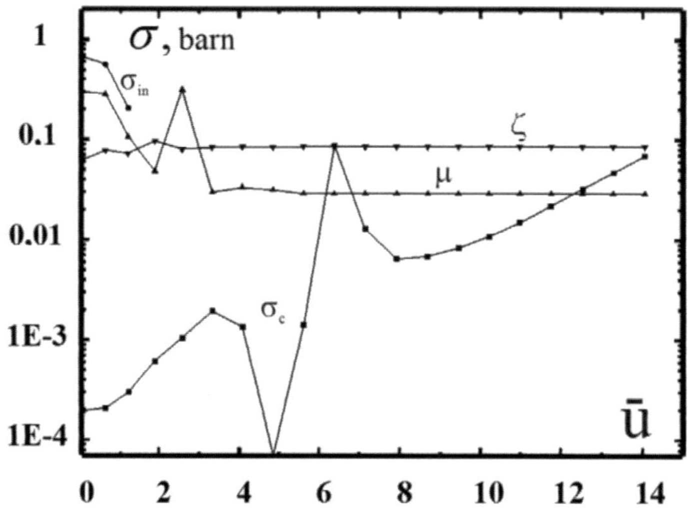

Fig. 2. Sodium characteristics in interval 4-23 groups of the 28-group constants.

The data were completed with the calculation of effective microscopic transport cross-section that is calculated using the formula

$$\sigma_{tr}(u) = \sigma_c(u) + \sigma_{in}(u) + \sigma_s(u)\cdot\left(1\text{-}\mu(u)\right),$$

and with the neutron removal cross-section from the group due to elastic neutron scattering with the sodium nucleus [6]

$$\sigma_R^{(i)} = \left(\zeta\sigma_s\right)^{(i)} / \Delta u,$$

where Δu – the group width per lethargy unit.

According to the group theoretic method [5, 6] neutron age in i-group is calculated using the following formula

$$\tau_i = \frac{D_i}{N\cdot\left(\sigma_a^{(i)} + \sigma_R^{(i)}\right)}, \quad (1),$$

where $N = 0.0254 \cdot 1024\,sm^{-3}$ – nuclear sodium concentration with the density 0.971 g / sm^3, and the absorption cross-section is calculated using the formula:

$$\sigma_a^{(i)} = \sigma_c^{(i)} + \sigma_{in}^{(i)}, \quad (2)$$

The sun of cross-sections in (1) $\sigma_a^{(i)} + \sigma_R^{(i)}$ is usually called functional cross-section or cross-section removal. The neutron diffusion code in the group is equated using the formula:

$$D_i = \frac{1}{3 \cdot \sum_{tr}{}^{(i)}} = \frac{1}{3 \cdot N\sigma_{tr}{}^{(i)}}, \quad (3)$$

Thus, formulas (1)-(3) make it possible to calculate the age in each energy group. Since we are interested in the area of moderation, that was modeled in experiments in age determination [2], meaning lethargy diapason $u = 0...14.125$ which corresponds to the neutron moderation length from $E_0 = 2$ *MeV* till indium resonance energy, the calculation data of age determination in groups shown in tab. 2 cover only this energy diapason.

Table 2. Age group calculation data .

Group	E_n	D_i, $barn^{-1}$	$\sigma_a^{(i)} + \sigma_R^{(i)}$, $barn$	τ_i, sm^2
4	2.5–1.4 *MeV*	6.05768	0.23359	1020.99
5	1.4–0.8 *MeV*	4.60722	0.44722	405.58
6	0.8–0.4 *MeV*	3.25962	0.45277	283.44
7	0.4–0.2 *MeV*	3.65336	0.53004	271.36
8	0.2–0.1 *MeV*	5.84937	0.38259	601.92
9	100.0–46.5 *keV*	2.65205	0.56604	184.46
10	46.5–21.5 *keV*	3.33791	0.44936	292.45
11	46.5–21.5 *keV*	2.8407	0.52674	212.32
12	10.0–4.65 *keV*	1.63994	0.91845	70.30
13	**4.65–2.15 *keV***	**0.13553**	**11.09442**	**0.48**
14	2.15–1,0 *keV*	1.88764	0.80852	91.92
15	1.0–0.465 *keV*	4.00631	0.38119	413.77
16	465–215 *eV*	4.22677	0.35935	463.08
17	215–100 *eV*	4.21301	0.36458	454.95
18	100–46.5 *eV*	4.20977	0.36686	451.78
19	46.5–21.5 *eV*	4.20397	0.36854	449.10
20	21.5–10.0 *eV*	4.19456	0.37828	436.56
21	10.0–4.65 *eV*	4.18067	0.38856	423.60
22	4.65–2.15 *eV*	4.16052	0.40114	408.34
23	2.15–1.0 *eV*	4.13131	0.42618	381.65

As constantly used moderation borders in experiments are modeled from the average fission neutron energy spectrum equal to *2 MeV* till indium resonance $E_{In} = 1.46$ *eV*, i.e. they are only part of 4 and 23 groups, it is important to correct the age determination in them using a simple interpolation. As in diapason of group 4 with the energy 2.5…1.4 *MeV* neutron age while moderating in this energy interval is calculated $\tau_4 =$ 1022.99 sm^2, it is obvious that in interaval 2.0…1.4 *MeV* the age equals 556.9 sm^2, and in the last 24 group – 229.3 sm^2.
On composing all group elements including the correction in 4 and 23 groups we have the following age value

$$\sum \tau_i^{2\,MeV...1.46\,eV} = 6702.6\, sm^2.$$

There are no direct experiments in age determination in sodium that is why the results of the numerical experiment [7] were used to check the data. The above mentioned study shows that for the same moderation diapason the age value was 6616.6 sm^2. Thus, fractional accuracy of our result is 1.3 %.

4 Age determination in 238U

The difficulty in getting reliable data in uranium-238 is connected with the presence of resonance peaks in the area of moderation for the transport cross-section. Fig.1 shows strong dependence of transport cross-section of uranium-238 on energy. Basic group energy characteristics [5] are shown in tab. 1.

The study [5] has reliable data concerning nuclear-physical characteristics; they are used for the reactor design and shielding. For uranium-238 there are two constant libraries that were received before 1977: ABBN-MICRO and ABBN-78. In ABBN-MICRO the cross-sections for plutonium-239 and uranium-238 have some experiment-calculation discrepancy that lead to the development of ABBN-76 where cross-section for the mentioned nuclides was improved (fig. 3). Due to this fact, ABBN-78 library is recommended for design calculation.

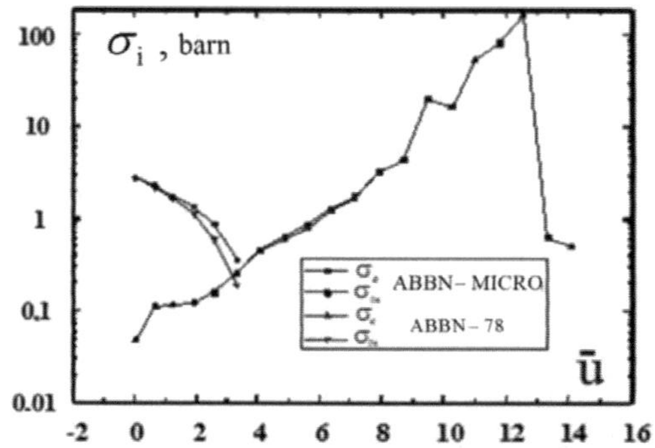

Fig. 3. Cross-section improvement for radiative capture and inelastic scattering *238U* in ABBN-78.

Group data take average logarithmic decrement $\bar{\zeta}$ from energy into account. This decrement is easily seen in the regions $E > 1\ keV$ (fig. 4). It is also necessary to take into account the fact that average scattering angle cosine μ in the low-energy regions is permanent $(\mu = 2/2A)$ while in the high-energy regions it decreases.

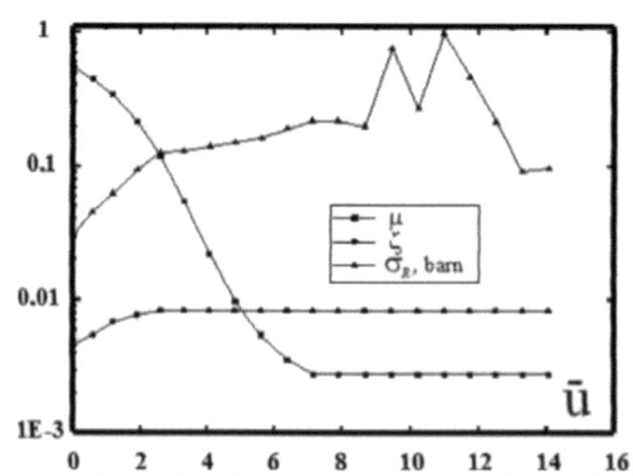

Fig. 4. The average scattering angle cosine dependence, average logarithmic decrement and cross-section removal for *238U*.

According to the group theoretic method [6] neutron age in i-group is calculated in the same way as for *Na* from formulas (1-3). For the age calculation a part of the average group cross-sections (groups 4-23) was taken from the study [5].

Thus, formulas (1), (2) allow us to calculate the age in each energy group. As we are interested in the moderation area that was modeled in experiments in age determination in graphite [2]: lethargy diapason u=0…14.125 that corresponds to neutron moderation length from $E_0 = 2$ *MeV* till indium resonance energy $E_{In} = 1.46$ *eV*, the calculation data of neutron age in uranium-238 in groups shown in tab. 3 cover only this energy diapason

As the required borders are only part of 4 and 23 groups, it is necessary to correct the age determination in them using a simple interpolation. As in diapason of group 4 with the energy 2.5…1.4 *MeV* neutron age while moderating in this energy interval is calculated $\tau_I = 11.496$ sm^2, it is obvious that in interaval 2.0…1.4 *MeV* the age equals 6.271 sm^2, and in the last 24 group – 16.123 sm^2.

Table 3. Calculation data of group age in the constant libraries.

Group	E_n	ABBN-MICRO		ABBN-78	
		D_i, sm	τ_i, sm^2	D_i, sm	τ_i, sm^2
4	2.5–1.4 *MeV*	1.54717	11.49645	1.54717	11.49645
5	1.4–0.8 *MeV*	1.40881	**12.01036**	1.4299	**12.56411**
6	0.8–0.4 *MeV*	1.1632	**12.48323**	1.18274	**13.37243**
7	0.4–0.2 *MeV*	0.87193	**11.7176**	0.8955	**13.91574**
8	0.2–0.1 *MeV*	0.68496	**12.17116**	0.70537	**16.71442**
9	100.0–46.5 *keV*	0.59139	**16.54923**	0.60009	**21.76872**
10	46.5–21.5 *keV*	0.53488	**18.82203**	0.53545	**19.29159**
11	46.5–21.5 *keV*	0.49128	**12.97186**	0.49311	**13.94287**
12	10.0–4.65 *keV*	0.4461	**8.9531**	0.44825	**9.69591**
13	4.65–2.15 *keV*	0.37314	**5.25108**	0.37452	**5.5282**
14	2.15–1.0 *keV*	0.3184	**3.28065**	0.32028	**3.52325**
15	1.0–0.465 *keV*	0.29782	1.77742	0.29782	1.77742
16	465–215 *eV*	0.30669	1.36464	0.30669	1.36464
17	215–100 *eV*	0.07804	0.07832	0.07804	0.07832
18	100–46.5 *eV*	0.16928	0.2121	0.16928	0.2121
19	46.5–21.5 *eV*	0.04886	0.01872	0.04886	0.01872
20	21.5–10.0 *eV*	0.0555	0.01387	0.0555	0.01387
21	10.0–4.65 *eV*	0.03702	0.00459	0.03702	0.00459
22	4.65–2.15 *eV*	0.77678	22.39832	0.77678	22.39832
23	2.15–1.0 *eV*	0.75346	26.87146	0.75346	26.87146

Due to age calculation data in two libraries it is clear that there is a discrepancy in groups 5-14 (fig. 4).

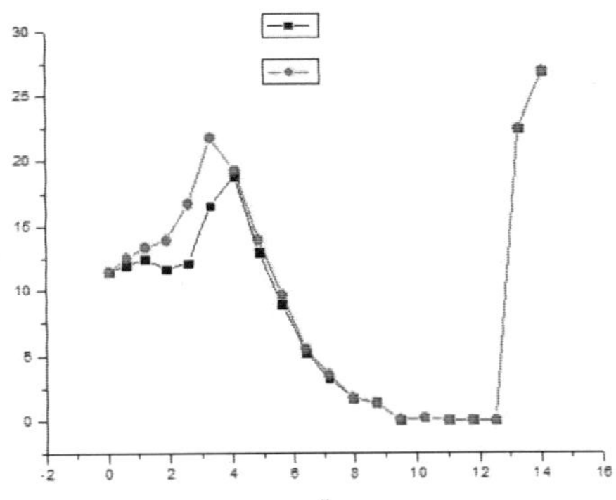

Fig. 5. The average scattering angle cosine dependence, average logarithmic decrement and cross-section for *238U*.

On composing all group elements including the correction in 4 and 23 groups we have the following age value in ABBN-MICRO:

$$\sum \tau_i^{2\,MeV...1.46\,eV} = 161.43\ sm^2.$$

In ABBN-78:

$$\sum \tau_i^{2\,MeV...1.46\,eV} = 177.53\ sm^2.$$

The calculated data differ by 9 %. It is impossible to judge the accuracy of the results as there are no experimental data. In [8] there is the age value in uranium-238 that was calculated by IPPE even before 1969. According to the analytic results of IPPE the neutron age in uranium-238 at moderation till indium resonance energy equals 288.6 sm^2 which means that the fractional accuracy of our results is 38.5 %.

4 Conclusions

Group method is the most detailed and demonstrative way that takes the following things into account: the character of effective microscopic cross-section of radioactive capture σ_c, inelastic scattering σ_{in}, average cross-section angle cosine μ and average lethargy increment under elastic scattering (fig. 2, fig. 3, fig. 4).

1. The existence of cross-section resonance for *Na* (in our case the elastic scattering cross-section) multiplies the chances of moderating neutron removal from the group due to resonance self-shielding (see the highlighted part of table 2).
2. The reproducibility precision of the age calculating results using the data of the studies [5] and [7] shows that quite an accurate constant determination was obtained by 1964
3. This paper deals with the neutron age calculation in uranium-238 at moderation till indium resonance using group method. There is also a comparative analysis of the two constant systems. Due to the correction of the cross-section radiative capture and inelastic scattering in ABBN-MICRO the neutron age results in uranium-238 are completely different.
4. Currently ABBN-78 library tends to be the most accurate constant system. However, the absence of experimental data on neutron age in uranium-238 does not allow to judge about the accuracy of nuclear-physical constants of ABBN-78 library.
5. The group method that was used in this research provided precise data for *Na* and data divergence for uranium-238 which means that it is important to look for other methods or experimental ways for calculating the age of uranium-238.

References

1. E.O. Adamov, L.A. Bolshov. I.H. Ganev, *The white book of nuclear power engineering* (MINATOM, 2001) [in Russian]
2. A.V. Kuzmin, Tomsk Polytechnic University Journal **323**, 2 (2013)
3. A.V. Kuzmin, *The basis of neutron transfer* (Laboratory course, Tomsk) (2010) [in Russian]
4. I.V. Gordeev, D.A. Kardashev, A.V. Malyshev, *Nuclear-Physical Constants (Handbook)* (Atomizdat, Moscow, 1963)
5. L. P. Abagyan, N.O. Bazazyants, I.I. Bondarenko, M.N. Nikolaev, Group constants for calculating nuclear reactors (Atomizdat, Moscow, 1964)

6. G.G. Bartolomey, G.A. Bat, V.D. Bajbakov, M.S. Alkhutov, *The basis of the theoretical and practical methods of nuclear power reactors calculation* (Moscow, 1989)
7. L.V. Marchenko, J.A. Sergeev, *The squared moderation length calculation for various substances in 18 and 26-group approximation and their comparison with experimental data* (Moscow, 1969)
8. A.V. Kuzmin, *Experimental and calculated determination of neutron age for various substances* (Tomsk, 2011)

ERRORS IN DETERMINING THE THERMAL CHARACTERISTICS OF LIQUIDS BY THE LASER-PULSE METHOD

Mark D. Kats [1,*], and *Ilija* M. Kats [1]

[1]National Research Tomsk Polytechnic University, 634050 Tomsk, Russia

Abstract. The analysis of test conditions of determining the thermal diffusivity of liquids by the laser pulse method are analyzed. Numerical solution of the nonstationary temperature pattern of three-layer sample when exposed to the surface of the thermal pulse of high power and low time span problem was found. The limits of the range of possible changes in the value of the heat transfer rate to the heating surface, corresponding to the conditions of reliable determination of the thermal diffusivity of a typical organic liquid ethanol without changing its phase state was determined.

1 Introduction

Organic liquids are increasingly used in various fields of science and technology. It used as solvents[1], coolant-cutting fluids [2] and high temperature fluids [3]. Every year, new organic liquids are developing for special purpose of machinery and technology [4–5]. Thermal processes in the technical fluids and its thermal characteristics plays an important part in reliability control and normal operation of technical systems.

For these reasons, determination of the thermal characteristics with high accuracy becomes more actual task, especially in the development of new products and technical systems.

To investigate the thermal properties of such materials, it is best to use highly efficient procedures, for example, the laser pulse – laser flash (LFA) method [6, 7]. It is based on the absorption of a pulse of radiant energy in a thin layer by the frontal "hot" surface of the sample and the approximate solution of the one-dimensional heat conduction problem for an infinite plate. From the time-temperature dependence of the reverse ("cold") surface of the sample, one can determine the maximum heating temperature T_{max} and the time τ_{05} for half the maximum heating temperature of the cold surface of the sample to be attained.

The thermal diffusivity a of the material being investigated when using LFA method, is calculated as

$$a = 0.1388 \cdot L^2 / \tau_{0.5}, \quad (1)$$

where L is the thickness of the sample.

* Corresponding author: riki@tpu.ru

If thermal diffusivity is known, the specific heat capacity c, and the thermal conductivity λ of the material can be calculated by formulas from [6,7]. The practical interest is the use of LFA method to determine the thermal diffusivity of organic liquids.

For materials that are solid determination of their thermal characteristics is not straightforward. For liquids this task becomes complicated by the necessity of creation of the measuring cell providing intensive heating of a uniform and plane-parallel layer of liquid of the known thickness. For this reason, the task of determination of the thermal characteristics of the liquids is divided to two tasks. The first task is providing the conditions of heating corresponding to the fundamental principles of LFA method (small duration and high intensity of heating) without liquid boiling up. The second task is by selecting the measuring cell layer thicknesses required to provide reliable determination of thermal characteristics, the temperature rise to "cold" surface layer of high thermal conductivity material used to make the measuring cell [8–11].

The design of the measuring cell is mainly determined by the equipment used for research thermal characteristics of materials. For example, measurement cell of laser flash analyzer LFA- 457 is a thin-walled metal cylinder (with a cover), which is placed under investigation liquid. The laser pulse is applied to the bottom of the measuring cell, temperature registration is carried out infrared sensor on the opposite side [12].

Exposure to high heat fluxes q = $10^7...10^9$ W/m^2 should lead to an increase in surface temperature (T_p) of the substance. Heating of the measuring cell surface by a laser with rate of heat transfer $q=10^7...10^9$ W/m^2 increase in surface temperature (T_p) of the test fluid.

The range of T_s is limited by the organic liquids initial boiling point, which for the majority liquids is 350....390 K [13]. This condition requires a limitation in the rate of heat transfer during experiment. During the heating of the measuring cell by a laser pulse, the temperature of the liquid surface may excess initial boiling point. This fact significantly reduces the accuracy of the thermal diffusivity of the test substance. For this reason, the magnitude of the pulsed rate of heat transfer must be limited to values of the temperature in any point of the layer of liquid below its boiling point.

The purpose of this work is analysis of the conditions determining the thermal characteristics of organic liquids by a laser pulse method.

2 Statement of the problem

To solve the problem of determining the thermal characteristics of organic liquids was used mathematical model from [14 –16] and one-dimensional Thermal Conductivity method for a three-layer infinite plate with constant rate of heat transfer q, pulse duration τ_{imp} and the absence of heat exchange with the environment.

In case of a problem definition it was supposed that the transverse size of the measuring cell is many times more liquid layer thickness can not be taken into account a possible two-dimensional heat-transfer process and heating due to the heat on the side walls of the cylindrical body.

Also, it was supposed, that thermal characteristics of the materials of the measuring cell and organic liquid do not depend on temperature. This assumption is justified for many materials in real-rated range of possible changes in temperature during the experiments even at high values of the heat flux to the on-heated surface of the sample due to the short pulse duration.

The problem is solved in a Cartesian coordinate system. Solutions represent a three-layer plate total thickness L_3 (fig. 1). Surface segment x=0 was heated by the energy pulse q. The pulse duration is τ_{imp}.

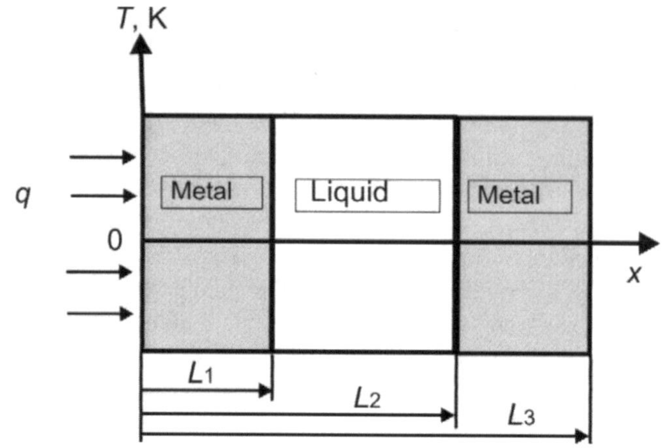

Fig. 1. Scheme of the area of solution.

The mathematical model consists of the three nonstationary heat equation

$$c_1\rho_1 \frac{\partial T_1}{\partial t} = \lambda_1 \left(\frac{\partial^2 T_1}{\partial x^2} \right), \quad 0 < x < L_1 \tag{2}$$

$$c_2\rho_2 \frac{\partial T_2}{\partial t} = \lambda_2 \left(\frac{\partial^2 T_2}{\partial x^2} \right), \quad L_1 < x < L_2 \tag{3}$$

$$c_1\rho_1 \frac{\partial T_3}{\partial t} = \lambda_1 \left(\frac{\partial^2 T_3}{\partial x^2} \right), \quad L_2 < x < L_3 \tag{4}$$

with boundary conditions:

$$x = 0: \quad -\lambda_1 \frac{\partial T_1}{\partial x} = q, \quad \tau_{imp} > t > 0; \tag{5}$$

$$x = 0: \quad -\lambda_1 \frac{\partial T_1}{\partial x} = 0, \quad t > \tau_{imp}; \tag{6}$$

$$x = L_1: \quad T_1 = T_2; \; -\lambda_1 \left(\frac{\partial T_1}{\partial x} \right) = -\lambda_2 \left(\frac{\partial T_2}{\partial x} \right) \tag{7}$$

$$x = L_2: \quad T_2 = T_3; \; -\lambda_2 \left(\frac{\partial T_2}{\partial x} \right) = -\lambda_1 \left(\frac{\partial T_3}{\partial x} \right) \tag{8}$$

$$x = L_3: \quad \lambda_1 \frac{\partial T_3}{\partial x} = 0, \quad t > 0; \tag{9}$$

$$t = 0: \quad T_1 = T_2 = T_3 = T_0, \tag{10}$$

where ρ is the density of the material; q is a rate of heat transfer at the surface; T, T_0 are the actual and initial temperature; t is the time; subscripts 1, 2 are referred relate to the material of the measuring cell and the liquid.

3 Solution method

The above nonlinear boundary value problem was solved by the method of finite differences [17] using an iterational algorithm [18, 19], developed for the numerical solution of high-temperature heat-transfer problems in regions with local energy sources.

Problem (2)–(10) belongs to the category of singularly perturbed problems [20]. This problem is more complicated those problems in [14-16]. The main difficulty of this task is the necessity of solving a system of three heat equations with a significant (hundreds of times) difference of materials heat diffusion of each layer.

In this case, to reduce the heat retention of the process, the metal plates (1) and (3) heat diffusion must be maximized. The heat diffusion of the organic liquids do not exceed 0.65 W/(m·K) [13]. Therefore, the numerical solution of heat diffusion problem for three-layer infinite plate with significantly different coefficients of thermal conductivity in a high-intensity pulsed heating is independent problem. Its solution is largely due to the choice of grid parameters. Therefore, the difference grid time steps chosen similarly to [11, 12] from 10^{-5} to 10^{-12} s.

The values of heat transfer rate and pulse duration taken in accordance with the actual possible characteristics of the laser source of identification of thermal properties of materials research apparatus. These conditions comply with the parameters of the laser source of modern quantum optical generator [21]: heat transfer rate $q = 10^7 \ldots 10^8$ W/m^2, pulse duration τ_{imp}=0.1…2 ms and emission wavelength 1.06 mkm.

4 Results and discussion

In [14–17] was found, that in determining thermal characteristics of solid materials by laser pulse method there is methodological errors caused adopted in formulating the problem [6, 7] assumptions. The errors in determining the thermal characteristics of material can be defined [14-16] by comparison of the actual values of thermal diffusivity (a) with the material obtained by numerical simulation of the implementation conditions of experiments on determination of thermal characteristics materials by this method. The errors of liquids thermal conductivity are determined by the method specified in [14–16].

The verification of the reliability of estimates numerical simulation was made temperature pattern of yogurt calculation (layer thickness of 1 mm, $\lambda = 0.267$ W/(m·K), $c = 1600$ J/(kg·K),) [11], located in the copper measuring cell with wall thickness of 0.5 mm ($\lambda = 110$ W/(m·K), $c = 380$ J/(kg·K), $\rho = 8530$ kg/m^3) [11], rate of heat transfer was $q = 2.4 \cdot 10^7$ W/m^2, a pulse duration was 1.5 ms, T_0 = 273 K.

Grid time steps was not above $1 \cdot 10^{-6}$ c. The initial step in the coordinate x = $5 \cdot 10^{-7}$ m, and in the areas near the borders of the materials chosen irregular (the algorithm of geometric series) [22]: $x_{i+1} = q \cdot x_i$, where q - coefficient of variation. At the boundary surface the step of simulation was decreased, and after crossing boundary surface it was increased. The coefficient of variation has been assumed constant (q= 0.02).

The difference between calculation and experience-based data [11] was about 0.2 %. This is confirms the accuracy of the numerical simulation results.

The thermal characteristics was calculated for the organic liquids in table 1 [13]. It was supposed, that the measuring cell was made from iron with wall thickness 0.5 mm ($\lambda = 20$ W/(m·K), $c = 551$ J/(kg·K), $\rho = 7894$ kg/m^3) [23]. The parameters of the laser heating and the initial temperature selected like in the same problem in [11].

Table 1. Thermal characteristics of organic liquids [13].

Liquid	C, J/(кg·K)	λ, W/(m·K)	ρ, кg/m^3	a, m^2/s	$T_к$, K
Water	4183	0.60	998.2	$1.4\cdot10^{-7}$	373
Methylbenzen	1630	0.139	885.5	$9.6\cdot10^{-8}$	383.6
Ethanol	1909	0.172	812.1	$1.1\cdot10^{-7}$	351.4

The disadvantages of the method [6, 7] is the assumption of an infinite magnitude of the heat transfer rate of the laser radiation to the heated surface. In practice, this value cannot be infinite. On the other hand, heating of the liquid by the laser with heat transfer rate $10^7...10^8$ W/m^2 can lead to its boiling. This fact can greatly increase the error in determining the thermal diffusivity. For example, the boiling point T_k of Methylbenzene is 383.6 K, for Ethanol its 351.4 K [13]. The last condition requires limitation of the heat transfer rate in the experiment at the borders of measuring cell – liquid.

At the fig. 2 is shown the temperature distribution T(x) in the Methylbenzene layer thickness 1 mm with a pulse duration 1.5 ms, the magnitude of the heat transfer rate $2.4\cdot10^7$ W/m^2, depending on the duration of the heating period

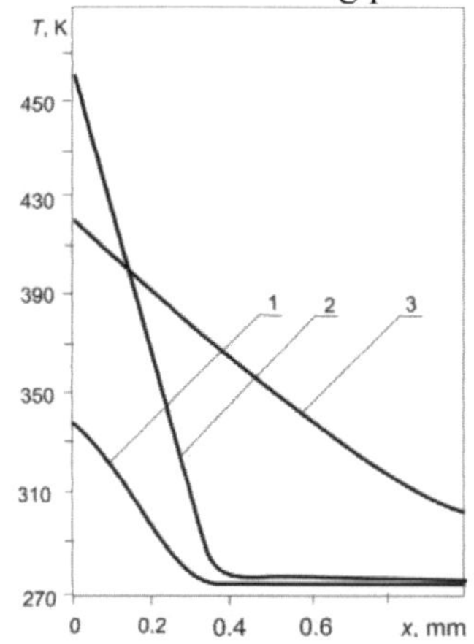

Fig. 2. The temperature distribution T(x) in the Methylbenzene layer thickness 1 mm with a different pulse durations: 1 – 1.5 ms; 2 – 100 ms; 3 – 1 s.

From fig. 2 it is seen that at the time after pulse ending on the borders of the materials L_1 the liquid temperature varies and reaches its maximum 457 K at t = 100 ms. This fact can be explain, that in conditions of low heat diffusion of the organic liquids heat energy accumulated in the material of the steel wall of the measuring cell after the laser pulse, and flows into the liquid, respectively, the temperature of a thin boundary layer increases. Then, the heat flow to the deeper layers of the liquid and the temperature is decreasing.

Simulations have shown necessity of substantial reduction of limit values (maximum) heat flows to determine liquids heat diffusion with thermal conductivity by the pulse method. In cases, when the wall thickness of the measuring cell is 0.5 mm, a liquid layer is 1 mm and the length of the pulse laser heating is $\tau_{imp} = 1.5$ ms, the maximum allowable heat flow is as follows: water $1.35\cdot10^8$ W/m^2; Methylbenzene – $5.84\cdot10^7$ W/m^2; for ethanol – $4.5\cdot10^7$ W/m^2.

The numerical results of heat diffusion of the liquids at a value of heat flux $2.4\cdot10^7$ W/m^2 and $\tau_{imp} = 1.5$ ms are given in table 2.

Calculations show the possibility of determining heat diffusion of organic liquids by typical LFA method without its boiling with minimal errors.

Table 2. The numerical results and reference values of the heat diffusion of three types of liquids.

Liquid	*a* (lookup value), m^2/s	*a* (calculated value), m^2/s	Definition error, %
Water	$1.4 \cdot 10^{-7}$	$1.28 \cdot 10^{-7}$	0.34
Methylbenzene	$9.6 \cdot 10^{-8}$	$9.5 \cdot 10^{-7}$	0.9
Ethanol	$1.1 \cdot 10^{-7}$	$1.11 \cdot 10^{-7}$	0.28

5 Conclusion

In this wore was shown possibility of determining the heat diffusion of organic liquids by the laser pulse method. Also, the parameters of the laser pulse, in which such determination is possible was defined. Analysis of systematic errors in the method and comparison of literature data with the results showed that the error is less than 1 %.

References

1. A. Weisberger, A. Proskauer, J. Riddick, *The organic solvents. Physical properties and cleaning methods* (Publishing House of Foreign. Lit-ry, Moscow, 1958)
2. L.V. Khudobin, *Technology and engineering application of cutting fluids for machining* (TNT, Moscow, 2016)
3. Y.M. Babikov, D.S. Tales, *Organic and silicone heat transfer fluids* (Energia, Moscow, 1975)
4. V.A. Suchilin, I.E. Gribut, S.A. Golikov, Electr. information. compl. chem. **7**, 4 (2011)
5. P. Wasserscheid, W. Keim, Angew. Chem. Int. Ed. **39**, 21 (2000)
6. W.J. Parker, R.J. Jenkins, C.P Butler, J. Appl. Phys. **32**, 9 (1961)
7. M. Akoshima, T. Baba, Int. J. Thermophys. **27**, 4 (2006)
8. Y. Tada, M. Harada, M. Tarigani, W. Equchi, J. Rev. Sci. Instrum. **49**, 9 (1978)
9. N.S. Srinivasan, X.G. Xiao, S. Seetharaman, J. Appl. Phys. **75** (1994)
10. J. Blumm, A. Lindemann, S. Min, Thermochim. Acta **455** (2007)
11. E.P. Oliviera, Z.E. Silva, Cr.K. Silva, 20th Intern. Congress of Mechan. Engin. (2009).
12. *The thermal diffusivity and conductivity analyzer LFA 457 McroFlash* https://www.netzsch-thermal-analysis.com/ru/produkty- reshenija/ temperaturoprovodnost -i-teploprovodnost/lfa-457-microflash/
13. N.B.Vargaftik V.P.Filippov, A.A.Tarizmanov, E.E. Totskiy, *Handbook of heat-conductor fluids and gases* (Energoatomisdat, Moscow, 1990)
14. G.V. Kuznetsov, M.D. Katz, Meas. tehn. **55**, 4 (2012)
15. G.V. Kuznetsov, M.D. Katz, Meas. tehn. **53**, 6 (2010)
16. G.V. Kuznetsov, M.D. Katz, Meas. tehn. **52**, 4 (2009)
17. A.A. Samarskiy *Theory of difference schemes* (Nauka, Moscow, 1983)
18. O.V. Vysokomornaya, G.V.Kuznetsov, P.A. Stizhak, Russ. J. Phys. Chem. B **5**, 4 (2011)
19. D.O. Glushkov, G.V. Kuznetsov, P.A. Stizhak, Russ. J. Phys. Chem. B **5**, 6 (2011)
20. K. Chang, F. House, *Nonlinear singularly perturbed boundary value problems.* (Mir, Moscow, 1988)
21. Ed.G.A.Abilsiinov, *Industrial lasers. Ref. Benefit 1. T.* (Mashaneering, Moscow, 1991)
22. A.A. Samarskiy, Y.I. Popov, *Difference methods for solving problems of gas dynamics* (Nauka, Moscow, 1992)
23. V.V. Chirkin, *Thermal properties of materials of nuclear engineering: Manual* (Atomizdat, Moscow, 1968)

MATEC Web of Conferences **72**, 01042 (2016) DOI: 10.1051/matecconf/20167201042
HMTTSC-2016

MODELING OF THERMOHYDRAULIC PROCESES OF HEAT EXCHANGE EQUIPMENT

Karina Savina[1], *Alexey* Abramovskih[1], *Vladimir* Haymin[1], and *Sergey* Lavrinenko[1,*]

[1]National Research Tomsk Polytechnic University, 634050 Tomsk, Russia

Abstract. You should leave 8 mm of space above the abstract and 10 mm after the abstract. The heading Abstract should be typed in bold 9-point Arial. The body of the abstract should be typed in normal 9-point Times in a single paragraph, immediately following the heading. The text should be set to 1 line spacing. The abstract should be centred across the page, indented 17 mm from the left and right page margins and justified. It should not normally exceed 200 words.

1 Introduction

As it is known [1], the development of heat exchangers for mobile power stations and modular units is an actual problem of modern heat engineering, housing, communal service and many other coupled industries. Special attention is given to projects of energy efficient equipment in the regions with an autonomous power supply system. It should be noted, that in the Russia a project to create mobile floating nuclear power plants of small power and reduced-enrichment uranium fuel possibility [2] was propsed by State Corporation for Atomic Energy "Rosatom" [3-5].

Operating experience of block-modular facilities produce thermal and electrical energy, it leads to the conclusion that the technical and economic indicators depends on operating conditions.

The research results showed that it is necessary to carry out modeling, taking into account the specific conditions of the operation of power unit in order to determine the efficiency of heat exchange equipment.

2 Numerical analysis

Based on the foregoing numerical analysis of the effectiveness of the degasser was carried out, using the FLOW SIMULATION Solid Works package block for calculating. This unit makes possible to perform 3-dimensional finite element stationary and non-stationary calculations for real geometry considering the nonlinear properties of solids and fluids. The accepted calculation model considers both mechanisms of internal heat transfer and heat transfer from the degasser case to the environment.

* Corresponding author: serg86@tpu.ru

The software package FLOW SIMULATION provides the possibility to simulate a convective fluid flow in closed and open areas with various fluids and pressures.

The complex tasks related to the issues of hydrodynamics and heat exchange in the SolidWorks Flow Simulation is solved by a system of differential equations of motion, continuity, energy, the thermal conductivity in the channel walls [6, 7].

The main stages of the development of computer models in SolidWorks Flow Simulation are [1]:

a) the creation of a 3D model;

b) the construction of a grid computational domain;

c) the imposition of boundary conditions;

d) the calculation and visualization of the derived temperature field, pressure, etc.

3 Computational models

As the initial data a 3-D model of the degasser's structure was used. In the first stage, the area of the degasser with the adjacent space in the room are highlighted, the boundary conditions are imposed, as well as the direction of gravity is given for calculating the thermogravitational convection of ambient air (fig. 1).

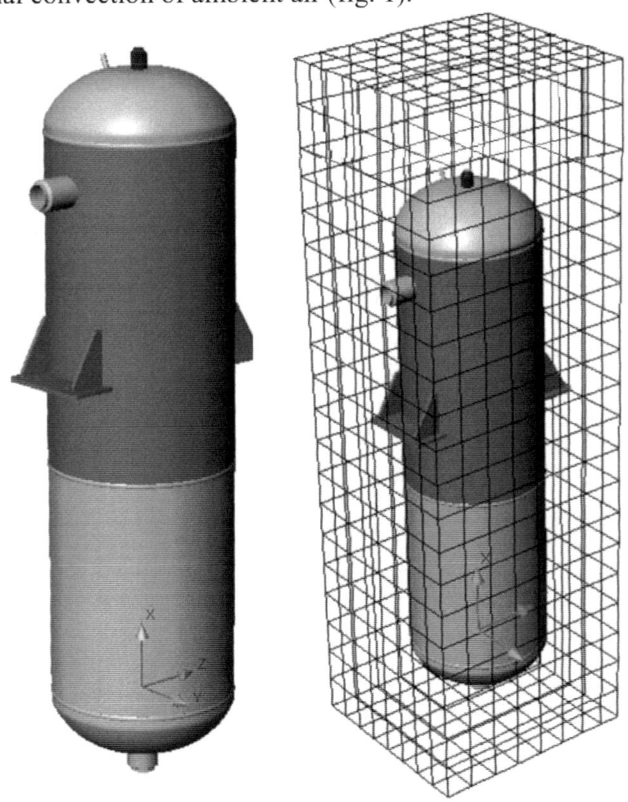

Fig. 1. Three-dimensional structure of the degasser and the basic grid of the model.

Computational grid is built on the basis of computing resources by using its local refinement in the parietal layers, narrow channels and thin walls.

In the modeling, degassers design elements and fluids are assigned to the properties of the respective materials.

The material's thermal conductivity of the degasser's body (which made of stainless steel) in the calculation model was set by using table (Tab. 1) data.

Table 1. The thermal conductivity of the heat exchange degasser's surface.

Temperature, K	Coefficient of thermal conductivity, W/(m^{2}K)
293	15
593	19.8
903	26.5
1103	27.8

In the path of the inner degasser's surface, which is directly contacting with the circulating water at 348 K, the heat transfer coefficient was taken to be equal 800 W/(m^{2}K), and the paths, that are contacting with the gaseous medium - 10 W/(m^{2}K). On lattices level there is a mixed contact with the water-gas medium, so the heat transfer coefficient on these surfaces is assumed to be equal 200 W/(m^{2}K).

4 Results of calculation

Studying of circulating air flows in the surrounded has made it possible to obtain the distribution of the external heat transfer coefficient, heat flux density and temperature of structure elements (fig. 2).

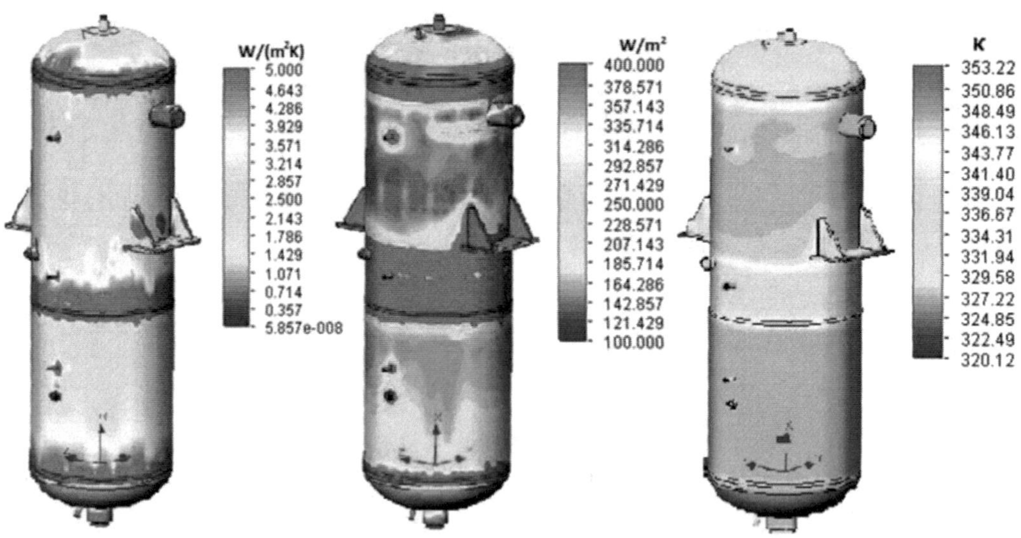

Fig. 2. Estimated distribution of heat transfer coefficient (W/m^{2}K), the heat flux (W/m^2) and the temperature (K) on the surface of the structural elements.

As a result of the three-dimensional thermal-hydraulic calculation degasser's case, with using the calculating block FLOW SIMULATION packet SolidWorks, was found, that the heat transfer coefficients at the outer surface is unevenly distributed, and changes in the range from 2 to 5 W/(m^{2}K), the heat flux may vary from 100 to 400 W/m^2, and the maximum temperature of the base plate's bottom does not exceed 348 K.

5 Conclusion

The general principles laid down in the code and its flexible structure make it possible to mode the heat transfer processes in the units of different configurations and sizes. The proposed approach allows a fast and sufficiently accurate analysis of the efficiency of heat exchange equipment in various operating conditions, including those for mobile small and medium capacity power units.

Using SolidWorks package allows modeling of hydraulic processes in a variety of elements, such as pumps, accumulators, valves of various designs, and so on, which makes it possible to accurately carry out strength calculations.

References

1. A.G. Bagoutdinova, Y.D. Zolotonosov, O.V. Shemelova, Bulletin of Kazan Technological University **14** (2014)
2. V.A. Ruppel, Yu.S. Tretyakova, S.V. Lavrinenko, et al. SMART GRIDS 2015. MATEC Web Conf. **37**, Article Number 01059 (2015)
3. M.S. Khvostova, Secure-ness in the technosphere **2** (2012)
4. V.V. Zamukov, A.E. Baburkin, A.V. Doroshchenko, S.V. Belchenko. Shipbuilding **2** (2007)
5. V.V. Zamukov, A.E. Baburkin, A.V. Doroshchenko, V.S. Stepanov. Probl. At. Sci. Technol., Ser.: Safety Assurance of NPP **24** (2009)
6. A.A. Alyamovsky, A.A. Sobachkin, E.V. Odintsov. *Computer modeling in engineering practice* (BKHV-Peterburg Publ., St. Petersburg, 2008)
7. K. Larionov, S. Lavrinenko, V. Gubin, et al. XVth International Conference Linguistic and Cultural Studies: Traditions and Innovations (LKTI 2015), Procedia Social and Behavioral Sciences **206** (2015)

 DOI: 10.1051/matecconf/20167201043

THE NUMERICAL SIMULATION APPLICATION FOR FIRE-TUBE BOILER HEATING SURFACE SAFETY EVALUATION

Sergei A. Khaustov[1,*], *Olga* V. Guk[1], and *Artem* Lisenkov[1]

[1]National Research Tomsk Polytechnic University, 634050 Tomsk, Russia

Abstract. The numerical simulation is applied for fire-tube boiler heating surface safety estimation. Thermal processes in an inflatable fire-tube furnace during its emergency operation were simulated using the finite volume method with Euler approximation and the implicit pressure based algorithm. Study results reproduce failures connected with increasing of impasse aerodynamic resistance. The method of these failures prediction is suggested. Simulation has shown that entering the amount of coolant into combustion volume results in burner fan incapability to overcome the impasse resistance of the furnace. The simulation results are visually confirmed during the inspection of emergency boilers.

1 Introduction

Fire-tube boilers have a number of advantages over water-tube, but at the same time requiring more careful attitude in the design, manufacture, installation and operation [1]. The design of fire-tube boilers and their essential parts must ensure reliability and safety of operation. The main indicators are reliability, retentive durability, maintainability. Reliability is provided during the design, construction, operation and maintenance. Mistakes made at any stage, may negate the material resources, labour costs and time spent on all other stages.

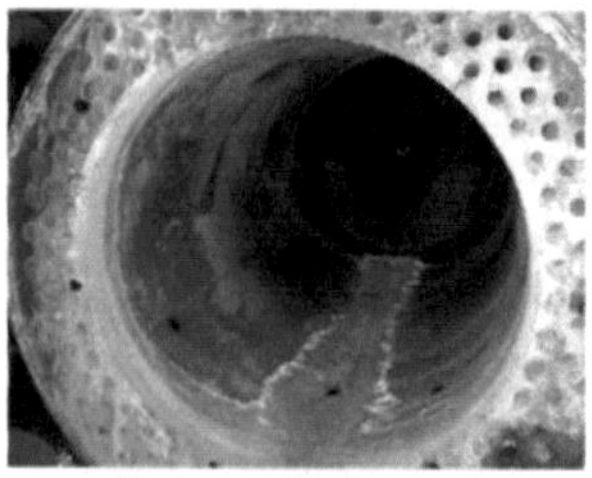

Fig. 1. The example of fire-tube boiler state after several years of operation.

In the case of feasibility studies, fire-tube boilers are used for non-project fuel burning such as waste oil, crude oil, and other sub-standard fuels [2]. After a certain service fire-tube boilers installed in some domestic and autonomous sources of heat were prone to

* Corresponding author: khaustovsa@tpu.ru

corrosion. Sometimes this led to the emergency mode: the volume of the combustion chamber was littered with water stains, leakages were observed in the welds (fig. 1), there also were cracks in the tube plates and fire-tube surface bulges. In addition, elements of charred insulation were observed at the boiler cover.

Engineering design of fire-tube boiler has no unified regulatory framework [3]. Due to the lack of a uniform methodology for the construction of fire-tube boilers the numerical simulation method (NSM) is increasingly used. Benefits of NSM are expressed in the possibility of obtaining almost realistic picture of the combustion flow in the boiler, depending on the design and operation mode, which gives grounds to refute or confirm the structural decisions [4, 5]. This method is used to investigate failures by simulating the conditions present during the incident. By NSM the further operational reliability of the boiler equipment in a certain period can be evaluated and set the necessary actions for the extension of the whole complex devices life [4 - 6].

2 NSM of heating surfaces safety review

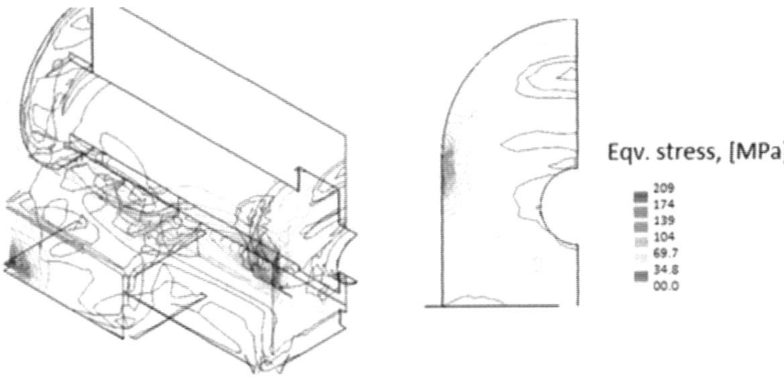

Fig. 2. The stress field of plating and the rear wall [7].

Computational and experimental analysis was carried out of the strength of the fire-tube boiler design by numerical simulation using the finite element method (FEM) in order to determine the need to consider the value of thermal stresses in fire-tube boiler design [7].

Table 1. Distribution of potential and kinetic energy of deformation [7].

	f_{01}=21 Hz		f_{02}=33.3 Hz		f_{03}=45.6 Hz	
	$E_{\text{п}}$	$E_{\text{к}}$	$E_{\text{п}}$	$E_{\text{к}}$	$E_{\text{п}}$	$E_{\text{к}}$
Fire tube	0.8	19.2	38.6	9.0	0.6	0.8
Back recirculation chamber and inspection orifice	0.6	7.0	34.4	73.4	1.0	1.1
Boiler shell, composed of housing head	46.8	25.6	9.1	0.6	95.0	95.6
Back head	7.2	2.4	11.5	16.4	0.8	1.0
Front head with front recirculation chamber	0.2	16.0	2.0	0.3	1.2	0.7
Lugs	44.4	29.8	4.4	0.3	1.4	0.8

On the basis of the numerical model fire-tube boiler various parameters strength diagnostics were estimated. Membrane bending stress and strain energy distribution were considered as key criteria for identifying weakness in the design of the boiler.

Based on the research results presented (fig. 2 and table 1) it can be concluded that the boiler shell is an element with the highest concentrations of maximum stress and strain energy. A critical area is the design of the rear wall (fig. 2). Maximum stresses in the structure appear in the boiler drum near the junction of the rear wall (209 MPa). But it does not exceed the yield stress (215 MPa) for this element of material P265GH [8] (russian analogue 20K steel).

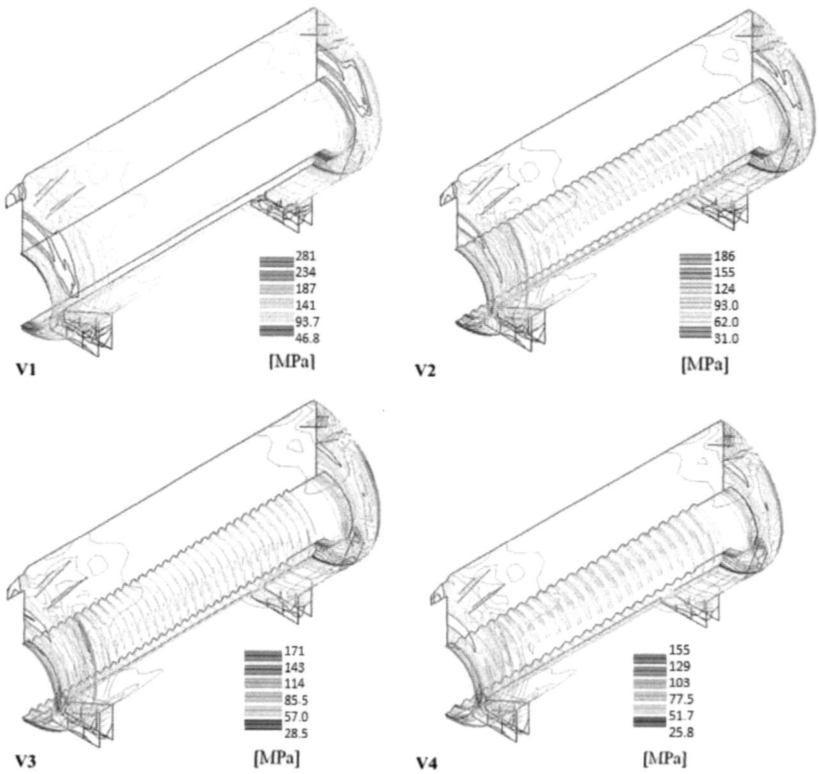

Fig. 3. The stress field models [9].

In [9] the influence of the form of the furnace tube of fire-tube boiler thermal deformation by NMS and FEM is estimated. Four fire-tube furnaces were investigated (fig. 3):

- V1 with smooth shape of the combustion chamber;
- V2 corrugations with a step equal to 151 mm and a depth of - 50 mm;
- V3 increments - 151 mm and depth - 75 mm;
- V4 c spacing of 200 mm and a depth of 75 mm.

In the model, the maximum stress V1 equal to 281 MPa, are located at the junction of the furnace and the boiler casing. The maximum value is higher stresses than the yield strength of the material components, which is 211 MPa for the material P295GH fire tube (analogue - 18K steel) and 187 MPa for the material P265GH shell [8]. The maximum stress in the elements of the models with a corrugated furnace takes place in the model V2, and the value is 186 MPa, which is comparable to the yield strength of the material. V4 model has the smallest value of stress equal to 155 MPa.

Table 2. Maximal deformations and stresses in the boiler substructures [9].

Boiler model	Pressure loads			Thermal loads		
	Maximal deformation [mm]	Maximal stress [MPa]		Maximal deformation [mm]	Maximal stress [MPa]	
		Furnace tube	Boiler body		Furnace tube	Boiler body
V1	4.19	70	136	7.91	240	240
V2	5.66	85	135	7.37	134	134
V3	9.77	105	135	7.30	80	80
V4	10.50	111	135	7.30	84	84

Based on the results (table 2) got by authors [9] the maximal stress caused by pressure in the boiler with a smooth furnace design is almost twice lower than in the corrugated furnace at higher stresses caused by the thermal loads. This suggests that the corrugated combustors have greater flexibility during thermal loading.

Based on the review it can be concluded that the use of numerical simulation of combustion processes in the fire tube boilers is necessary to assess the reliability of the heating surfaces. It is used to identify critical areas and shortcomings in the design of the boiler unit; discover the causes of unstable operation; reproduce the output process of equipment failure; determine the source of an emergency. But at the same time authors failed to find any information about NSM of emergency modes connected with increasing of impasse aerodynamic resistance for example caused by the one of following issues: wrong burner, wherein its design does adjust the correct flame geometry, and entering the coolant into combustion volume.

3 Simulating the Emergency Mode

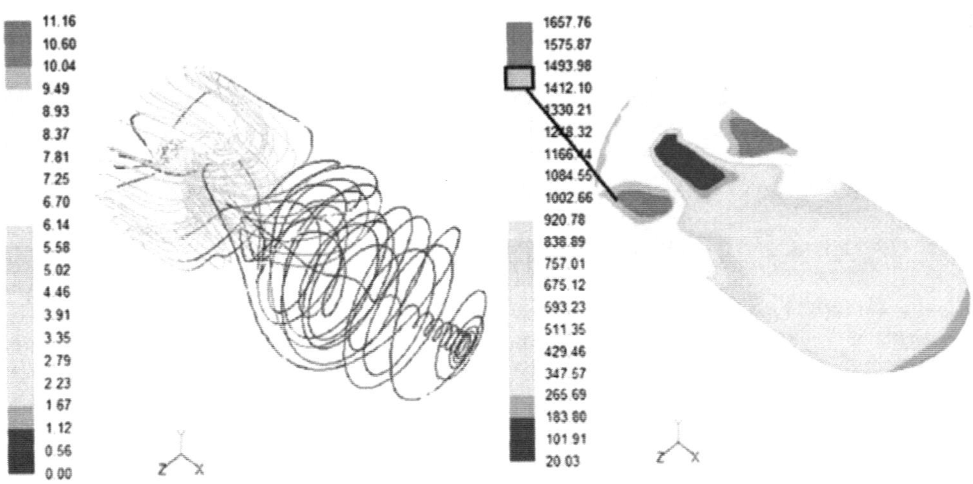

Fig. 3. Boiler in emergency mode: pathlines (m/s) and mean temperature contours (°C).

The aim of this paper is to simulate thermal processes in an inflatable fire-tube furnace during the emergency mode connected with increasing of impasse aerodynamic resistance in order to diagnose and predict complex defects of boiler heating surfaces. Authors discuss the following failures: wrong burner, which does not allow adjusting the correct flame geometry, and an increasing of impasse aerodynamic resistance caused by entering the coolant into combustion volume. Both failures result in burner incapability to overcome the resistance of the furnace.

To simulate this emergency mode, the furnace plating considered to have leakages in the welds allowing water (coolant) getting into combustion volume. Water entering flaming furnace volume through the cracks almost instantly vaporises. According to the saturated steam table, vaporizing water 1672-fold expands resulting in the fan incapability to overcome the impasse aerodynamic resistance.

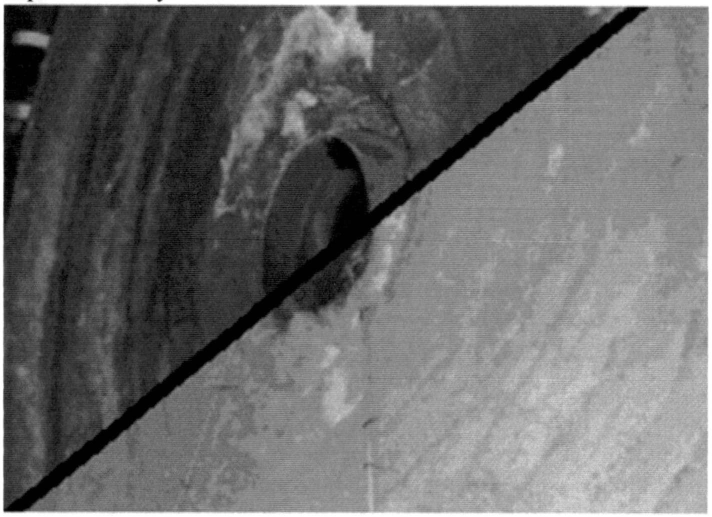

Fig. 4. Burned up cover insulation (pictured left) against the initial state (pictured right).

NSM of emergency mode indicates the sophisticated vortex motion within the impasse. High backpressure causes the formation of a stagnant zone filled by vaporised leaked water (fig. 4). Inertia of steam leakages provokes formation of asymmetric vortex twisting in furnace impasse [10, 11]. Such highly twisted aerodynamics caused by high impasse gas-dynamic resistance intensifies burning and increases the temperatures in the immediate vicinity of the burner [10]. As a result, the combustion takes place near the boiler front (fig. 4) and leads to high cover platting thermal stresses: the flame before blowing into the convection pipes comes into contact the insulation [10]. In this way insulation is burned during the boiler operation (fig. 4).

4 Discussions and Conclusion

Using the numerical simulation method, a processes of boiler emergency operation with fan incapability to overcome the fire tube resistance are visually represented. Authors of paper [10] describe an impasse aerodynamic resistance increasing due to wrong burner selection, this paper detects the same symptoms occurring, if the amount of coolant enters the combustion volume.

The simulation results are visually confirmed during the inspection of emergency boilers. Charred insulation of the boiler cover (fig. 4) visually demonstrates an excessive backpressure at the fire tube impasse. Together with the fact that the heat absorption of fire tube is reduced, the exhaust gases temperature increases, resulting in boiler efficiency reducing down to 89 %. Thermal losses connected with heat transfer to the environment through the boiler insulation also rise. After insulation cover is burned up (fig. 4) boiler efficiency will fall sharply. Furthermore, excessive thermal loads lead to burnouts of the tube sheet including weld ruptures caused by thermal expansion.

Analysis of the obtained temperature contour concludes that aerodynamic resistance enlargement due to wrong burner selection or the coolant entering the combustion volume can be diagnosed at early stages by controlling the boiler cover temperature.

Acknowledgments

The reported study was partially supported by the Ministry of education and science of The Russian Federation, state order No. 13.948.2014/K.

References

1. *Historical Advancement of Fire-Side Heating Surface in Fire-Tube Boilers, American Boiler Manufacturers Association* (Arlington, Va., USA, 2004)
2. S.V. Dolgov, E.K. Savchenko, S.A. Khaustov, R.B. Tabakaev, A.S. Zavorin, EPJ Web Conf. **110**, Article Number 01074 (2016)
3. S.A. Khaustov, A.S. Zavorin, K.V. Buvakov, N.A. Zakharushkin, MATEC Web Conf. **19**, Article Number 01020 (2014)
4. R. A. Visloguzov, R. B. Tabakaev, A. S. Zavorin, S. V. Dolgov, K. I. Klochko, International Conference on Mechanical Engineering, Automation and Control Systems (TPU, Tomsk, Russia), 6986943 (2014)
5. R.B. Tabakaev, P.S. Gergelizhiu, A.V. Kazakov, A.S. Zavorin, IOP Conf. Series: Mate. Sci. Eng. **66**, Article Number 012052 (2014)
6. S. A. Khaustov, E. I. Maksimova, International Conference on Mechanical Engineering, Automation and Control Systems (TPU, Tomsk), 7414903 (2015)
7. B. Gaćeša, V. Milošević-Mitić, T. Maneski, D. Kozak, J. Sertić, Technical Gazette **18**, 1 (2011)
8. *Flat products made of steels for pressure purposes. Non-alloy and alloy steels with specified elevated temperature properties* (European standards, EN 10028-2: 2003)
9. B. Gaćeša, T. Maneski, V. Milošević-Mitić, M. Nesterović, A. Petrović, Thermal science **18**, 1 (2014)
10. A.S. Zavorin, S.A. Khaustov, N.A. Zaharushkin, IOP Conf. Series: Mate. Sci. Eng. **66**, 1, Article Number 012029 (2014)
11. S.A. Khaustov, A.S. Zavorin, K.V. Buvakov, L.D. Kudryashova, A.V. Tshelkunova, EPJ Web Conf. **82**, Article Number 01041 (2015)

INTERACTION BETWEEN HYDROGEN MOLECULES AND A CLOSED NANOTUBE

Egor Tarasov[1], and *Yulia* Khudobina[1,*]

[1]National Research Tomsk Polytechnic University, 634050 Tomsk, Russia

Abstract. The present paper assesses the state of low-energy molecules in the vicinity of the crystal structure of a closed nanotube. It is proposed to use the continuum description of the impact energy from carbon graphene structures. In this approach, a closed tube consists of an open part and two fullerene hemispheres. Calculations revealed that adsorption of gas molecules by surface crystals is not the case of capturing by a part of the nano-object surface, but that of involvement of molecules in a complex orbital motion around the particle.

1 Introduction

Leaving aside the problem of gas filtration through open nanotubes, which has been the subject for a significant amount of research [1-6], in this paper, we consider a closed nanotube and associated sorption motions of molecules. The results of the comparative analysis of methane sorption by three different types of carbon nano-porous media (nanotubes, carbon fibre and silicon carbide) are given in the work [7]. In general, grapheme-like structures cause considerable interest [8-14] and for the purpose of studying them, classical approaches of molecular [15-28] and wave dynamics [29-31] are applied.

A nanotube can be regarded as a closed and twisted graphene plate. The hexagonal graphene grid is composed of regular hexagons, so that, if in the centre of each hexagonal element we place a point an join it to the vertices, we get a set of equilateral triangles with sides of 0.142 nm. The positions of nodes in the hexagonal grid are controlled by C-C bonds, the same as in the crystalline structure of diamond. But there are large tubes containing thousands of nodes in the crystalline lattice, as well as systems of nanotubes, including about a million carbon atoms. In such cases, it is advisable to use the continual approach.

2 Continual tube model

In order to realize the continual approach, the LJ-potential preliminarily used here must be modified in such a way that, when performing continual integration over the surface of the 2D-material, we could obtain a convergent value of the integral action. The above mentioned modification may be the following [8]:

* Corresponding author: hudobina@mail2000.ru

$$dU(\rho)=4q\varepsilon\left(\frac{\sigma}{\rho}\right)\text{th}\left[\left(\frac{\sigma}{\rho}\right)^{11}-\left(\frac{\sigma}{\rho}\right)^{5}\right]ds\,, \quad (1)$$

here ρ is the distance between the centre of the test molecule and the centre of the unit area on the surface of the 2D-material, q is density of energy distribution on the surface, ds is the elementary area on the tube surface. For a flat or a cylindrical grapheme structure it is easy to find that $q \approx 28$ nm^{-2}. Naturally, introducing cylindrical coordinates and integrating (1) over the side surface of the open nanotube, we obtain:

$$U(r,z)=4\varepsilon aq\int_0^{2\pi}\int_0^{h}\left(\frac{\sigma}{\rho}\right)\text{th}\left[\left(\frac{\sigma}{\rho}\right)^{11}-\left(\frac{\sigma}{\rho}\right)^{5}\right]d\varphi' dz'\,. \quad (2)$$

Moreover $\rho=\sqrt{r^2+a^2-2ar\cos\varphi'+(z-z')^2}$, a is the radius of the nanotube and h is its length. Double integration in (2) can be performed numerically, for each of the integrals using in series the trapezoidal rule.

When integration in (2) is completed, the test molecule velocity can be found from the energy integral:

$$\frac{mv^2}{2}+U=\frac{mv_0^2}{2}+U_0\,. \quad (3)$$

Or as a result of a numerical solution of the fundamental equation of dynamics:

$$m\frac{d\vec{v}}{dt}=-\text{grad}U\,. \quad (4)$$

To integrate this equation it is quite useful to apply the standard version of the Runge-Kutta scheme of the fourth-order accuracy.

3 Calculation results

Globally, in interactions of molecules with nanoscale structures, and even macroscopic bodies, it is possible to identify the mode of rapid interaction and the sorption mode. All the results presented here relate to the latter case. Molecules having high energy either move through an open tube, or get reflected. Therefore, we consider interactions of low-energy molecules with tubes. Let us place the origin of the Cartesian coordinate system on the tube axis in the very middle of it, so that its left edge is determined by the coordinate $z = -5$ nm and the right one – by $z = 5$ nm. We run the hydrogen molecule in the plane $z_0 = 4$, i.e. almost to the edge of the tube, from the position $x_0 = 2.8r$, $y_0 = 0$. In this case, its primary velocity components are: $u_0 = 125$ m/s, $v_0 = 37.5$ m/s, $w_0 = 0$.

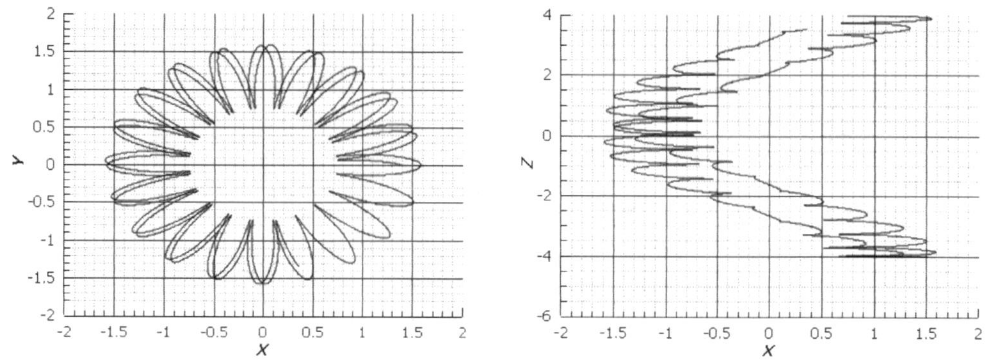

Fig. 1. Projected paths of hydrogen molecule in *xy* and *xz* plane for tube of $l = 10$ nm, $r = 0.5$ nm.

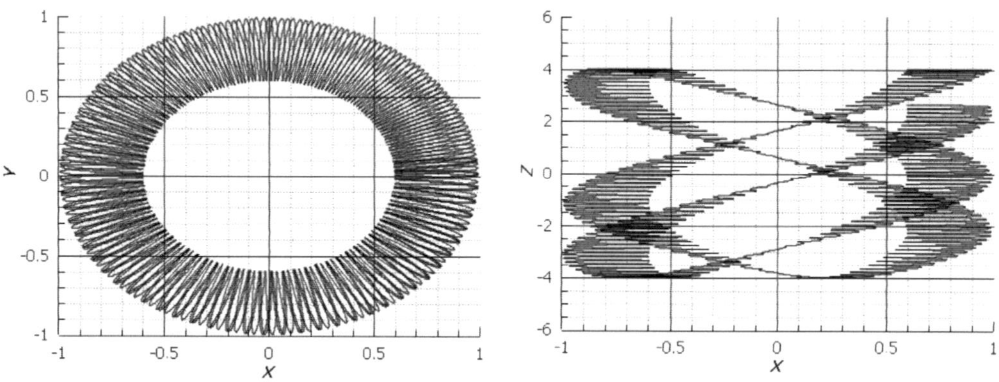

Fig. 2. Projected paths of hydrogen molecule in *xy* and *xz* plane for tube of $l = 10$ nm, $r = 0.35$ nm.

Since the smoothed tube is an ideal object and the total energy of the molecule remains constant, the trajectories presented above turned out to be ideal too. The movement in the *xy* plane is defined by two modes of vibration: the basic mode is associated with circular motion and the radar frequency is determined by the forces of attraction and repulsion. Another vibration is realized along the length of the tube. The frequency of this oscillation is caused by the asymmetric influences of the tube's parts. If at any time point the molecule is at the right end of the tube, the long-range attractive forces directed to the mass centre of the tube make it return to the left end, etc. Fluctuations of a higher frequency presented in the *xy* plane are generated by the same type of movements during which the molecule moves from the repulsion zone to the attraction area and vice versa.

Thus, in the nature of motion it is possible to identify one frequency associated with the revolution of the molecule around the axis of the tube, another one - with longitudinal movements (along the axis of the tube) and the third one – with radial vibrations of the molecule.

The performed calculations allow evaluating the sorption amount, knowing which, in turn, it is possible to find the associated mass composed of low-energy molecules. The suggested numerical model also allows finding the limiting energy, exceeding which provides the molecule with a way out of the gravitational field of the nanoparticle. From the values of maximum energy in the axes "particle energy – molecule energy" it is easy to construct a sorption curve separating the two global modes of interaction between molecules and particles. Calculations show that adsorption of gas molecules by surface crystals is not the case of holding them on the fragment of the surface, but that of involving them in a complex orbital motion around the particle.

Acknowledgments

This paper was written within the frame of Competitiveness Improvement Program of Tomsk State University.

References

1. D. Mattia, H. Leese, K.P. Lee, J. Membrane Sc. **475** (2015)
2. A. Surapathi, H. Chen, E. Marand, J.K. Johnson, Z. Sedlakova, J. Membrane Sc. **429** (2013)
3. D. Zhao, J. Ren, H. Li, X. Li, M. Deng, J. Membrane Sc. **467** (2014)
4. A.L. Ahmad, Z.A. Jawad, S.C. Low, S.H.S. Zein, J. Membrane Sc. **451** (2014)
5. N. Gilani, J. Towfighi, A. Rashidi, T. Mohammadi, M.R. Omidkhah, A. Sadeghian J. Appl. Surf. Sc. **270** (2013)
6. Y. Zhao, B.T. Jung, L. Ansaloni, W.S. Winston Ho, J. Membrane Sc. **459** (2014)
7. L. Liu, D. Nicholson, S.K. Bhatia, J. Ch. Eng. Sc. **121** (2015)
8. K.A. Krylova, Y.A. Baimova, S.V. Dmitriev, R.R. Mulyukov, Phys. Sol. St. **58** (2016)
9. A.V. Savin, E.A. Korznikova, S.V. Dmitriev, Phys. Sol. St. **57** (2015)
10. K.A. Bukreeva, R.I. Babicheva, S.V. Dmitriev, K. Zhou, R.R. Mulyukov, A.I. Potekaev, Russ. Phys. J. **57** (2014)
11. J.A. Baimova, B. Liu, S.V. Dmitriev, N. Srikanth, K. Zhou, Phys. Chem. Chem. Phys. **16** (2014)
12. E.A. Korznikova, S.V. Dmitriev, J. of Phys. D: Appl. Phys. **47** (2014)
13. E.A. Korznikova, S.Y. Fomin, S.V. Dmitriev, Mat. Sc. Forum **845** (2016)
14. J.A. Baimova, S.V. Dmitriev, Mat. Sc. Forum **838-839** (2016)
15. A.I. Potekaev, M.A. Bubenchikov, Russ. Phys. J. **54** (2011)
16. M.A. Bubenchikov, Russ. Phys. J. **54** (2011)
17. M.A. Bubenchikov, A.I. Potekaev, A.M. Bubenchikov, Russ. Phys. J. **56** (2013)
18. M.A. Bubenchikov, A.I. Potekaev, A.M. Bubenchikov, Russ. Phys. J. **56** (2013)
19. A.I. Potekaev, A.M. Bubenchikov, M.A. Bubenchikov, Russ. Phys. J. **55** (2013)
20. M.A. Bubenchikov, A.I. Potekaev, A.M. Bubenchikov, O.V. Usenko, A.V. Ukolov, IOP Conf. Series: Mat. Sc. Eng. **87**, 1 (2015)
21. A. M. Bubenchikov, M.A. Bubenchikov, O.V. Matvienko, E. A. Tarasov, O.V. Usenko, AIP Conf. Proc. **1698** (2016)
22. V.Ya. Rudyak, S.L. Krasnolutskii, D.A. Ivanov, J. Dok. Phys. **57** (2012)
23. V.Ya. Rudyak, A.A. Belkin, V.V. Egorov, D.A. Ivanov, Int. J. Multiphys. **5** (2011)
24. V.Ya. Rudyak, S.L. Krasnolutskii, D.A. Ivanov, J. Microfl. Nanofl. **11** (2011)
25. V.Ya. Rudyak, S.N. Dubtsov, A.M. Baklanov, Tech. Phys. Lett. **34** (2008)
26. V.Ya. Rudyak, S.L. Krasnolutskii, E.N. Ivashchenko, J. Eng. Phys. Thermoph. **81** (2008)
27. V.Ya. Rudyak, J. VDI-Berichte **1803** (2003)
28. V.Ya. Rudyak, A.A. Belkin, Tech. Phys. Lett. **29** (2003)
29. A.M. Bubenchikov, M.A. Bubenchikov, A.I. Potekaev, E.E. Libin, Yu.P. Khudobina, Russ. Phys. J. **57** (2014)
30. A.M. Bubenchikov, M.A. Bubenchikov, A.I. Potekaev, E.E. Libin, Yu.P. Khudobina, Rus. Phys. J. **58** (2015)

30\. M. A. Bubenchikov, A. M. Bubenchikov, O. V. Usenko, V. A. Poteryaeva, S. Jambaa, EPJ Web Conf. **110** (2016)

MATEC Web of Conferences **72**, 01045 (2016) DOI: 10.1051/matecconf/20167201045
HMTTSC-2016

STUDYING OF THE CONTACT ANGLE HYSTERESIS ON VARIOUS SURFACES

E. O. Kirichenko[1,2], and *E.* Ya. Gatapova[1,*]

[1]Kutateladze Institute of Thermophysics SB RAS, 630090 Novosibirsk, Russia
[2]Novosibirsk State University (NSU), 630090 Novosibirsk, Russia

Abstract. This paper is devoted to investigation of the contact angle hysteresis on various surfaces. It was carried out by two different methods: measuring the advancing and the receding contact angles and measuring the contact angles at water droplet evaporation under isothermal conditions. Data obtained using two methods have been compared. The influence of the contact angle hysteresis on the mode of the drop evaporation has been shown.

1 Introduction

In connection with rapid development of electronic technology there is a need in creation of new highly efficient methods of heat transfer enhancement [1]. Significant improvement in the performance of heat exchange devices may be reached by using surfaces with nano-coating [2]. Studies of the properties of these surfaces, such as the contact angle hysteresis [3,4] play an important role in the choice of their application [5,6].

In this paper, we study eight identical glass substrates with coatings applied using different technologies. The following coatings were considered: substrate manufactured using the silanization technology (absorption of trichloroctadecylsilane by silicon oxide); applying the photoresists of different types on a substrate, and applying a Teflon layer at various speeds of rotation. The resulting surfaces have different properties.

On these substrates, the contact angle and the contact angle hysteresis were measured by different methods. The measurements were carried out using the Drop Shape Analyzer from Kruss.

2 Experimental setup and research methods

The Drop Shape Analyzer (Kruss company) allows measuring the contact angle by the sessile method or the captive bubble method. The measuring range is 1-180°, and the resolution is ± 0.1°. For the analysis of the drop shape 6 methods are used (polynomial, length-height, the conical segment, the circumference, Yuong-Laplace equation, and the distance between the points). The liquid is dosed automatically. High-precision systems for dosing and positioning the liquid droplets in combination with the system of recording,

* Corresponding author: egatapova@gmail.com

analysis of the video (images) and calculation of surface energy and surface tension provide for measurements in an automatic mode. Digital camera has recording speed of 61 - 311 frame/sec, and max/min frames resolution: 780 x 580 /780 x 60 px. Optical system has sevenfold automatic zoom and coverage 3.7x2.7 - 23.2x17.2 mm. Photo of the experimental setup is presented in figure 1.

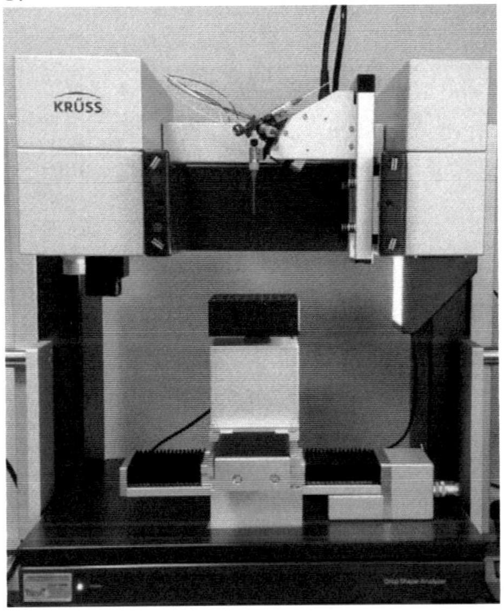

Fig. 1. A device for measuring the contact angle DSA100-E KRUSS.

The contact angle hysteresis was studied by two methods. The first method is measuring the advancing and the receding contact angles. The working fluid is ultraclean deionized water (MilliQ). Using the syringe pump, a drop is dispensed on the surface with constant rate of 0.2 ml/min, and at this time the advancing contact angle is determined. Then, at reaching the maximum volume of 5 μl, the syringe pump intakes the liquid with a constant rate of 0.2 ml/min. At that time the receding contact angle is measured. The drop volume and the liquid flow rate are chosen based on the properties of the liquid. The difference between the values of the advancing and receding angles is the hysteresis. The second method of studying the hysteresis of the angle is isothermal evaporation. A drop of 5 μl volume gets down to the surface and evaporates into the atmosphere. The contact angle is measured in the course of time.

3 The results of research

3.1. The method of advancing and receding contact angle

It is experimentally shown that the contact angle and the hysteresis are dependent on the technology of coating application. Figure 2 shows the dependence of the advancing/receding contact angle on the speed of the contact line on the surface, obtained by silanization. On the left, there are values of the receding contact angle, and on the right, values of the advancing contact angle. The hysteresis of the contact angle is determined in the vicinity of zero velocity of the contact line, and on this surface it is 10° - 20°.

 DOI: 10.1051/matecconf/20167201045

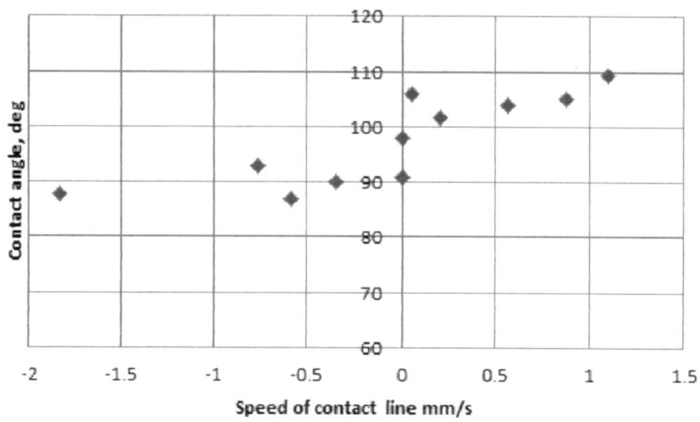

Fig. 2. The dependence of the advancing/receding contact angle on the speed of the contact line. The surface is Silane OTS.

Figure 3 shows the dependence of the advancing/receding contact angle on the speed of the contact line on the surface, obtained by application of photoresist Noa65. The value of contact angle hysteresis is about 30° - 40°. There is no negative velocity, since during the liquid intake the contact line remains fixed and does not move throughout the process.

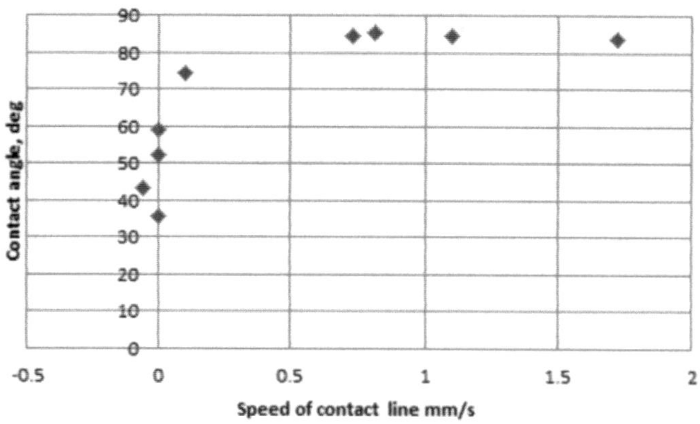

Fig. 3. The dependence of advancing/receding contact angle on the speed of the contact line. The surface is Noa65.

Fig. 4a and 4b present typical photographs of contact angle measurements using DSA 100 of the resting water droplet of 5 μl volume. The contact angle on the surface of the Silane OTS is 97° (fig. 4a), and on the surface of Noa65, it is 81° (fig. 4b).

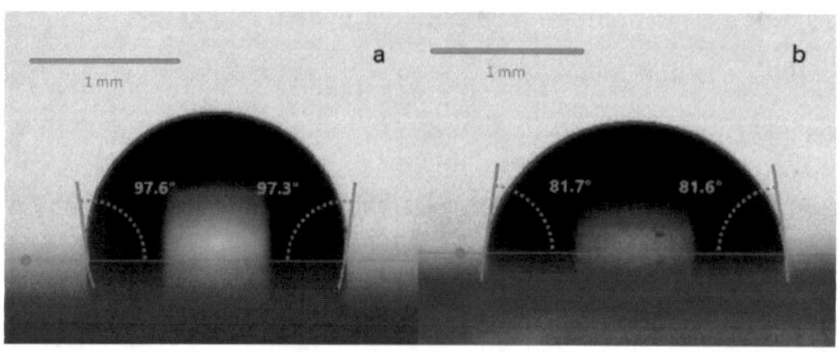

Fig. 4. Contact angle (a) OTS Silane surface b) Noa65 surface.

3.2. The method of isothermal droplet evaporation

Using the isothermal evaporation method the drops with an initial volume of 5 μl were investigated on the earlier described surfaces. While evaporating a liquid drop demonstrates one of three modes: the contact line moves, and the contact angle remains constant throughout the evaporation process (depinning); the contact line is fixed, and the contact angle continuously decreases (pinning); or a mixed mode. Figure 4 shows the dependence of the contact angle on the speed of the contact line on the surface of the Silane OTS.

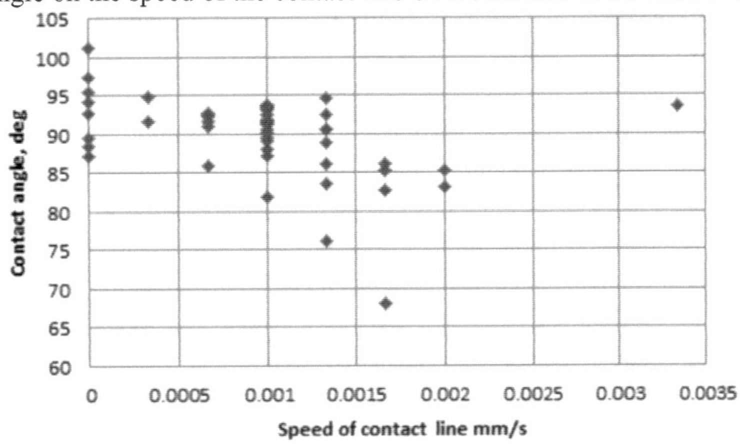

Fig. 5. The dependence of the contact angle on the speed of contact line. The surface is Silane OTS.

To compare the results with the method of advancing/receding contact angle the contact angle hysteresis is also determined in the vicinity of zero velocity of the contact line, and is 10° - 20°, which is consistent with the results, obtained by the first method. The average rate of change of the droplet contact radius, equal to the ratio of the initial contact radius to the drop lifetime, on the surface is 0.00085 mm/s. Relative to this speed, the speed values shown in figure 5, are rather large.

On the Noa65 surface the hysteresis is about 50° (fig. 6). Thus, the difference in determining the hysteresis by two methods is about 10°.

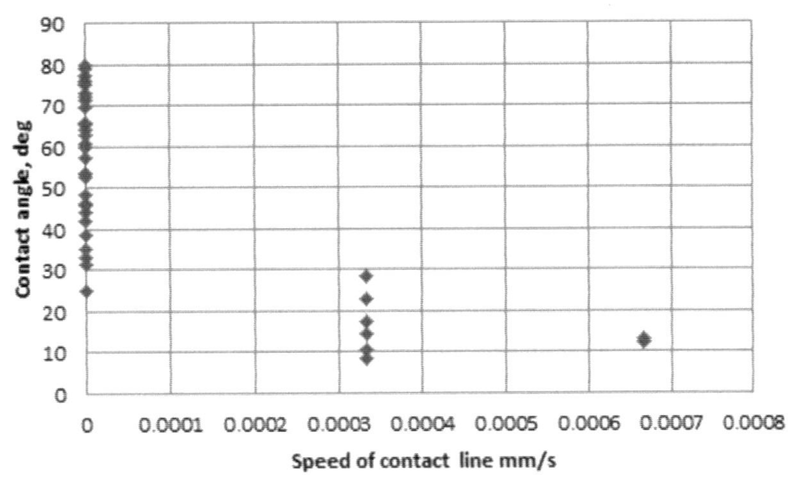

Fig. 6. The dependence of the contact angle on the speed of contact line. The surface is Noa65.

On these surfaces, the droplet demonstrates different modes of evaporation (fig. 7). On the surface obtained by silanization, the contact angle remains constant for almost all the time of drop evaporation. Whereas on the surface with applied photoresist, the contact line remains fixed, which leads to a permanent change in the contact angle.

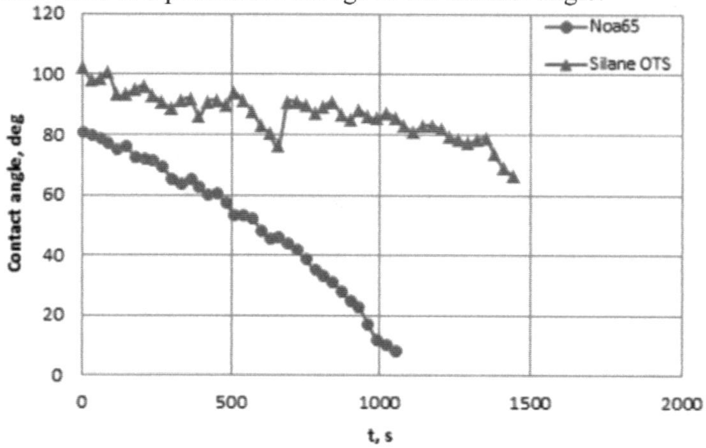

Fig. 7. The dependence of the contact angle on the time of evaporation.

The contact angle in figure 7 for both surfaces is changed in accordance with the measured the contact angle hysteresis (fig. 5, 6).

4. Conclusions

The investigation has been carried out to determine the contact angle hysteresis on various surfaces. The hysteresis was studied by two methods: by using standard procedures for DSA100, as well as at evaporation of water drops in the atmosphere under normal conditions. Two common types of coatings were studied: Silane and NOA 65. It has been found out that on the NOA 65 surface the difference in determining the hysteresis by two methods is about 10°, and on the Silane substrate the measurements coincide. The surface obtained by applying a photoresist NOA 65 can be used to create a flow with a constant contact radius. The surface obtained through silanization can be applied to organize flows with a constant contact angle.

Acknowledgments

The work was supported by the Russian Science Foundation (project No. 16-19-10675).

References

1. A. Bar-Cohen, K. Matin, N. Jankowski, D. Sharar, J. Electron. Packag. **137**, 1 (2014)
2. N. Miljkovic, R. Enright, Y. Nam, K. Lopez, N. Dou, J. Sack, E.N. Wang, Nano Lett. **13**, 179 (2013)
3. Jr. R.E. Johnson, R.H. Dettre, The journal of physical chemistry **68**, 1 (1964)
4. O.A. Kabov, D.V. Zaitsev, Dokl. Phys. **58**, 7 (2013)
5. A. Safonov, V. Sulyaeva, N. Timoshenko, E. Gatapova, O. Kabov, E. Kirichenko, A. Semenov, MATEC Web Conf. **37**, Article Number 01047 (2015)
6. E.Ya. Gatapova, A.A. Semenov, D.V. Zaitsev, O.A. Kabov, Colloids Surf., A **441** (2014)

LEVITATION OF LIQUID MICRODROPLETS ABOVE A SOLID SURFACE SUBCOOLED TO THE LEIDENFROST TEMPERATURE

D. P. Kirichenko[1,2,*], *D.* V. Zaitsev[1,3], and *O.* A. Kabov[1,3]

[1]Kutateladze Institute of Thermophysics SB RAS, 630090 Novosibirsk, Russia
[2]Novosibirsk State University, 630090 Novosibirsk, Russia
[3]National Research Tomsk Polytechnic University, 634050 Tomsk, Russia

Abstract. Evaporation of liquid microdroplets that fall on a solid surface with the temperature of below the Leidenfrost temperature is studied. It has been found out that sufficiently small liquid droplets of about 10 microns can suspend at some distance from the surface (levitate) and do not reach the surface; at that, the rate of droplet evaporation is reduced by an order as compared to microdroplets, which touch the surface. It is determined that in contrast to microdroplets, which touch the surface, the specific evaporation rate of levitating droplets is constant in time.

1 Introduction

Among the effective solutions for removing high heat flux densities, there are the systems of spray cooling [1] as well as the systems using stratification of a two-phase flow in a microchannel [2, 3]. Under the certain conditions, heat in these systems is removed due to intensive evaporation of liquid microdroplets. However, due to the Leidenfrost effect [4] the liquid microdroplets cannot reach the heating surface [5], and this may reduce the efficiency of apparatus cooling and cause its failure. The behavior of liquid microdroplets falling onto a solid surface, subcooled to the Leidenfrost temperature (about 200°C for water), is studied insufficiently. The phenomenon of droplet levitation at relatively low temperatures [6.7] can have a significant effect on operation efficiency of the two-phase cooling systems.

2 Experimental setup and research methods

The experimental setup is shown in fig. 1. The working section is made of stainless steel with an embedded copper core of 10x10 mm^2. The surface is heated using the nichrome tape, where the current from a power source is supplied. The temperature of the substrate surface was controlled by thermocouples mounted in the copper core. The surface temperature of the copper core was 135-155 °C. The experiment was carried out using the working section open to the atmosphere at the air temperature of 25-27 °C. Liquid

* Corresponding author: dmitriy_kirichenko@mail.ru

microdroplets with the size of about 10-100 microns were fed to the working surface from a nozzle mounted at some centimeters above the heating area. Deionized nanofiltered distilled water at the room temperature was used as the working liquid. According to images obtained via FASTCAM SA1.1 high-speed camera (5600 f/s with resolution of 1024x1024 pixels), the diameter and height of droplet levitating were measured. Optical resolution of the system was about 1 μm/pixel.

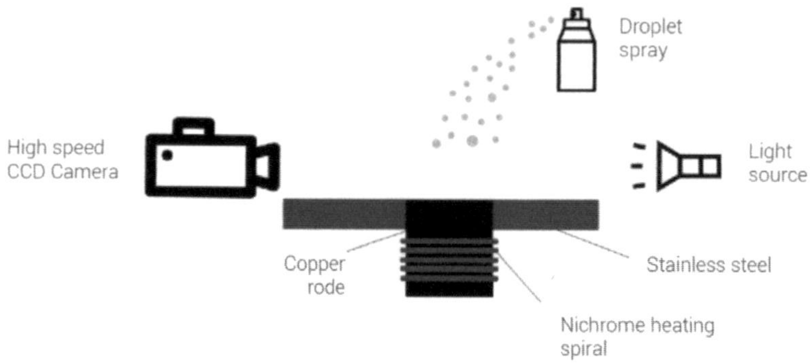

Fig. 1. Scheme of experimental section. (*Fastcam, copper core, heating, nichrome spiral, stainless steel, light source, droplet spray*).

3 Investigation results

It has been found out that microdroplets with the size of about 10 μm, falling on a heated surface with relatively low speed, do not touch it, but "suspend" over a solid surface at some distance; at that, the rate of droplet evaporation reduces by an order as compared to the microdroplets, which touch the surface. A photograph of levitating droplet and its reflection from the substrate is shown in fig. 2. Due to evaporation, the droplet size decreases with time. It should be noted that the droplet shape is not changed at evaporation and remains perfectly spherical, in contrast to levitation of large drops at the temperatures above the Leidenfrost temperature.

Dependence of a change in the radius of evaporating droplet on time is shown in fig. 3 for different surface temperatures and initial droplet sizes. Data were approximated with sufficient accuracy by linear relationships. It can be seen that at the same substrate temperature for the droplets with different initial sizes, inclination of lines is close. With an increase in the substrate temperature, the slope of the line increases.

Fig. 2. Microdroplet (R=50.5 μm), levitating at height h=4.2 μm. Substrate temperature – 155 °C.

DOI: 10.1051/matecconf/20167201046

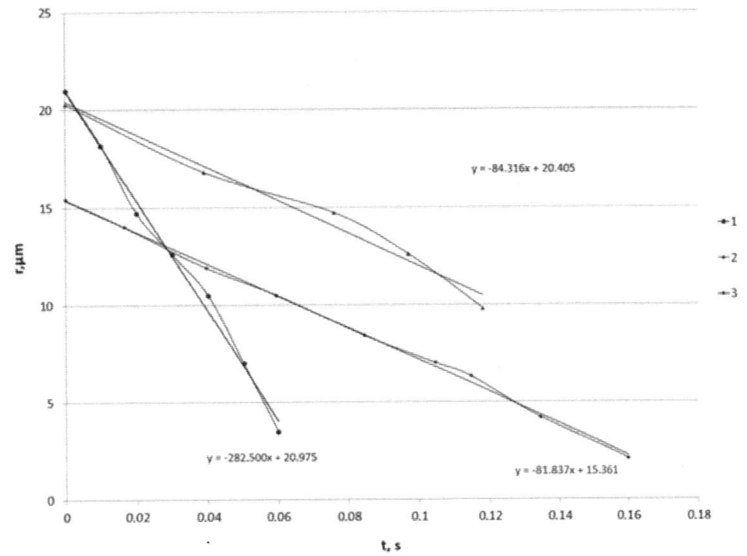

Fig. 3. Change in the radius of levitating droplet vs. time. 1 – substrate temperature – 155 °C; 2,3 – substrate temperature T=135 °C.

Dependences of specific evaporation rate of a droplet (a loss of the droplet weight per a time unit per a surface area unit) on time, plotted by approximating straight lines shown in fig. 3 are presented in fig. 4. It can be seen that the specific evaporation rate does not change during droplet evaporation, but increases with an increase in temperature. In the case, when the droplet touches the heated surface, the specific evaporation rate is not constant; it increases drastically in the last moments of droplet existence [8, 9].

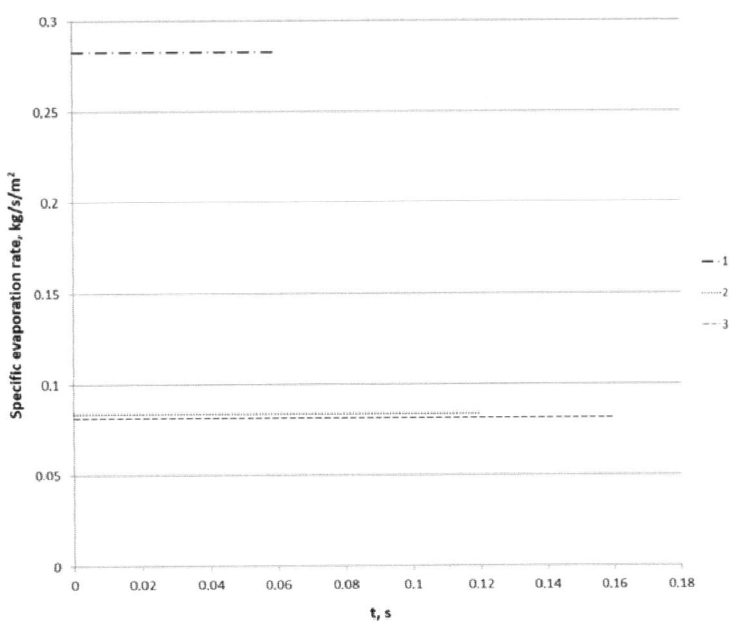

Fig. 4. Specific rate of levitating droplet evaporation. 1 – substrate temperature – 155 °C; 2,3 – substrate temperature – 135 °C.

The height of droplet levitation, h, was calculated as a half distance between the real droplet and its reflection (fig. 2). The diagram of dependence between dimensionless value $h^*=h/R$ and droplet radius R is shown in fig. 5. The obtained results were compared qualitatively with generalized results of [5] for levitation of water microdroplets over the surface, heated to the temperature of 400 °C. It is evident that the exponents of approximating dependences are close, but dependence of [5] is significantly higher.

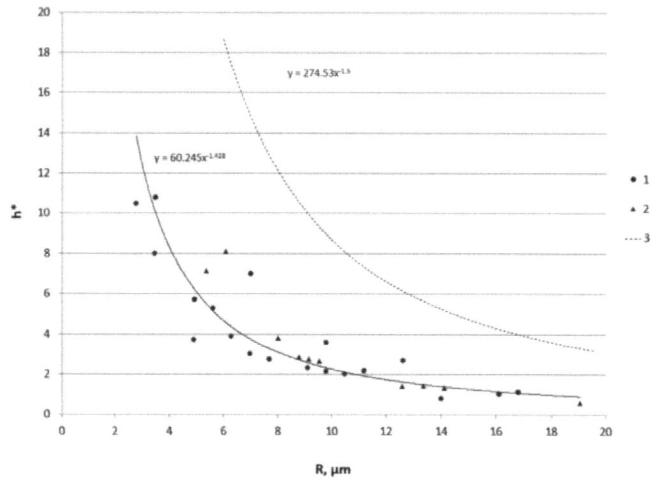

Fig. 5. Dimensionless height of droplet levitation vs. its radius. 1 – substrate temperature 135 °C; 2 – 155 °C; 3 – substrate temperature 400 °C [5]. Solid line – generalization of data 1 and 2.

4. Conclusions

Interaction of the falling liquid microdroplets with the solid surface, whose temperature is lower than the Leidenfrost temperature, was studied experimentally. It has been found out that:

- Microdrolets with the size of about 10 μm, falling on the heated surface with relatively low velocity, do not touch it, but "suspend" over the surface at some distance.
- The time of levitating droplet evaporation becomes an order higher in comparison with the microdroplets, which touch the surface.
- The specific evaporation rate of a levitating droplet is constant in time in contrast to the droplets contacting the surface, whose specific evaporation rate increases drastically with time.

The law of variation of the droplet levitation height vs. its radius is qualitatively similar to the law described in [5] for the substrate with the temperature higher than the Leidenfrost temperature.

Acknowledgments

The work was supported by the Russian Science Foundation (project No. 14-39-00003).

References

1. J. Kim, Int. J. Heat Fluid Flow **28**, 4 (2007)
2. O.A. Kabov, Y.V. Lyulin, I.V. Marchuk, D.V. Zaitsev, Int. J. Heat Fluid Flow **28**, 1 (2007)
3. D.V. Zaitsev, D.A. Rodionov, O.A. Kabov, Tech. Phys. Lett. **35**, 7 (2009)

4. M. Shirota, M.A.J.V. Limbeek, C. Sun, A. Prosperetti, D. Lohse, Phys. Rev. Lett. **116**, 6 (2016)
5. F. Celestini, T. Frisch, Y. Pomeau, Phys. Rev. Lett. **109**, 3 (2012)
6. A.A. Fedorets, I.V. Marchuk, O.A. Kabov, Tech. Phys. Lett. **37**, 2 (2011)
7. A.A. Fedorets, I.V. Marchuk, P.A. Stryzhak, O.A. Kabov, Thermophysics and Aeromechanics **22**, 4 (2015)
8. E.Y. Gatapova, A.A. Semenov, D.V. Zaitsev, O.A. Kabov, Colloids Surf., A **441** (2014)
9. A.A. Semenov, D.V. Feoktistov, D.V. Zaitsev, G. V. Kuznetsov, O.A. Kabov, Thermophysics and Aeromechanics **22**, 6 (2015)

DOI: 10.1051/matecconf/20167201047

EXPERIMENTAL DETERMINATION OF THERMAL CHARACTERISTICS OF MUNICIPAL SOLID WASTE

Vyacheslav Bukhmirov[1,*], *Olga* Kolibaba[1], and *Ramil* Gabitov[1]

[1]Ivanovo State Power Engineering University, 153003 Ivanovo, Russia

Abstract. The article is devoted to the experimental study of effective thermal characteristics of municipal solid waste (MSW) wet layer in the drying process. The research is based on the zone method. Series of laboratory experiments was performed. The drying kinetics and temperature curves were obtained on the base of which the kinetic coefficients were determined. We constructed a graph of the kinetic coefficients of the material moisture. The research results can be used for determination the temperature and moisture content fields in the MSW layer during its processing.

1 Introduction

In world practice an overwhelming amount of municipal solid waste (MSW) are taken to landfills. Increased cost of MSW burial and delivering them to the grave sites, a permanent environmental hazards connected with the disposal of large volumes of waste, as well as the difficulty of selection and arrangement of new landfills are stimulating the industrial waste processing as a way takes into account the most requirements of the economy, ecology and resources. Industrial processing provides an integrated approach to addressing disposal, recycling and disposal of waste. It makes it possible to save land resources, to produce of waste new products (heat and electricity), to solve environmental problems. The gradual transition from landfill to industrial processing is the main trend of the solution of solid waste problem in the world.

MSW is a heterogeneous mixture of complex morphological structure, including: food and vegetable waste, paper and textile components, plastic, leather, rubber, wood, ferrous and nonferrous metals, glass, stones, bones.

When considering the entire complex of problems associated with the collection, transportation, disposal and recycling of solid waste in the first place the question of the composition and the properties of this material. If for decision the problem on the collection and transportation of MSW the information on their moisture content and density is sufficient, then for the choice of disposal and utilization method and technology it is necessary to obtain complete information about the morphology and elemental

* Corresponding author: buhmirov@tot.ispu.ru

composition, including thermal properties of MSW. The morphological structure of MSW is given in Table 1. [1]

Table 1. Morphological composition of MSW in different climatic zones, % by weight.

Component	Climatic zone		
	medium	south	north
Food waste	35…45	40..49	32….39
Paper	32….35	22….30	26..35
Wood	1…2	1…2	2…5
Black scrap metal	3….4	2….3	3….4
Nonferrous scrap metal	0.5….1.5	0.5….1.5	0.5….1.5
Textile	3….5	3….5	4….6
Bones	1…2	1…2	1…2
Glass	2….3	2….3	4….6
Leather, rubber	0.5…1	1	2…3
Stone, stucco	0.5…1	1	1…3
Plastic	3….4	3….6	3….4
Other	1…2	3….4	1…2
Screenings (less than 15 mm)	5….7	6….8	4….6

In world practice for the MSW recycling and disposal are using thermal, chemical, biological and physico-chemical methods. The content in the MSW to 60 - 70% organic (combustible) faction prefers their processing energy efficient thermal methods, where the energy for processing itself can be removed from MSW while producing the new fuel and on the basis of electricity and heat.

2 Statement of the Problem

Thermal processing of solid waste is most often carried out in shaft furnaces (thermal reactors) in which MSW layer sequentially through drying processes and pyrolysis with subsequent gasification of solid carbonaceous residue. Gaseous fuel formed in the processing is removed to the consumer for use in thermal technological process.

Solid waste can be attributed to particulate material, which in the presence of moisture internal require high energy consumption for the drying process. In order to select a rational regime of the reactor it is necessary to determine the temperature field of solid waste layer, drying time raw material and energy costs in the process. Knowledge of these parameters is possible to calculate the speed of charging, the height of the drying zone of the thermal reactor and, therefore, to design it properly.

Determination of MSW layer temperature fields as a multi-porous body, makes it necessary to determine the dependence of its effective heat-physical properties on the moisture content and temperature.

Currently, there are many different methods and techniques for determination of the effective thermal characteristics of the porous bodies [2,3]. As for the solid waste, in literature there are only scattered data obtained based on experimental studies of thermal properties of some individual MSW components [4,5,6].

The purpus of this research is an experimental study of thermal properties of the MSW layer.

3 Experiment and processing

As shown in Table 1, the base components of MSW with the greatest percent in the morphological structure are: food waste, paper (cardboard) and textile, which in the form of moisture due to the material related to the capillary-porous colloidal bodies. To identify the physical picture of the heat and mass transfer in colloidal capillary-porous bodies you need to know the dependence of the thermal coefficients on the physical - chemical properties and moisture content, which is the main factor that determines the thermal properties of moist bodies.

In general, heat and mass transfer process is characterized by variability of physical parameters: temperature and moisture content. In order to determine the material temperature and moisture content dependences of the transfer coefficients we put a series of experiments. In this connection, we used a zonal method which ensures the accuracy of determining the physical characteristics values, acceptable in technical calculations.

Figure 1 shows the experimental setup. The experimental procedure was as follows. The test material (MSW layer sample) was placed in two thin-walled copper cylinder with a diameter of 16 mm and a length of 110 mm. One cylinder is suspended for weighing device, the other – is used for fixing the material temperature change using thermocouples connected to the inverter MBA8. Then cylinders were placed in an insulated pipe, purged with a hot heating medium at a speed of 8 m / s, excluding the external diffusion resistance to vapor transport from the open ends of the sample into the environment. Temperature change over time of the material along the length of the sample and the mass loss are fixed. Thus, the drying kinetics curves and the temperature curves were obtained, from which the effective thermal conductivity coefficient was calculated.

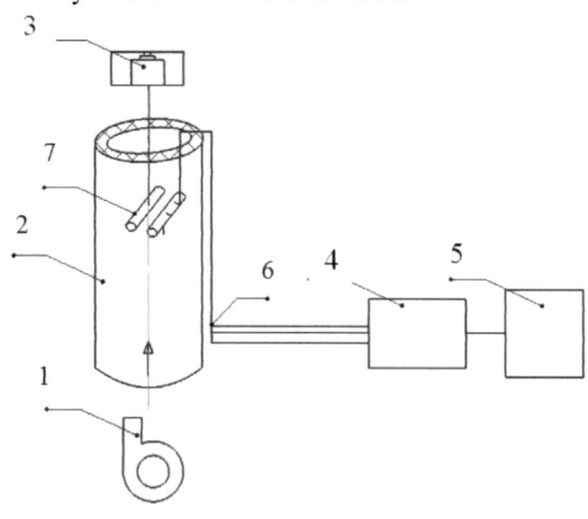

Fig. 1. Schematic of the experimental setup. 1. Air heater; 2. Insulated pipe; 3. Electronic balance; 4. Unit MBA-8; 5. PC; 6. Thermo-couples TCA (c); 7. Samples.

In accordance with the zonal method, the interval of moisture concentration variation in the solid phase $(U_S \div U_E)$ and the temperature $(T_S \div T_E)$ was divided into n zones, each of which assumes a constant thermal conductivity coefficient.

To determine the effective thermal conductivity the zonal method was supplemented by removing the temperature curve. Each temperature curve was divided into n zones corresponding to of kinetics curve zones. For each of the zones T_{S_i} and T_{E_i} were measured and the thermal balance was made:

$$q_i = \frac{m_c \cdot c_c \cdot \Delta T_i + m_M \cdot c_W \cdot \Delta T_i + m_c \cdot \Delta U_i \cdot r}{\Delta \tau_i} \qquad (1)$$

Then, considering that the quantity of heat required to heat up the mass of material, mass of moisture contained in it and evaporate the moisture is supplied by heat conduction through the side surface of the sample, we determined the effective thermal conductivity of moist material in the drying process by the formula:

$$\lambda_i = \frac{q_i \cdot \delta}{F \cdot (T_H - T_{M_i})} \tag{2}$$

In the equations (1), (2) is indicated: m_c - mass of dry substance in the sample, kg; c_c - heat capacity of dry substance, J/(kg·K); $\Delta T_i = T_{Ei} - T_{Si}$ - temperature change of the material over time $\Delta\tau_i$,° C; m_M - the mass of moisture in a sample to the time τ_i, kg; c_W - water heat capacity, J/(kg·K); δ – radius of the sample, m; F - the lateral surface of the sample, m^2; $T_H - T_{M_i}$ – the temperature difference between the heating medium and the average temperature of the material in the *i*-th time interval, ° C.

Effective thermal diffusivity was determined by the well-known formula:

$$a_i = \frac{\lambda_i}{c_i \cdot \rho_i} \tag{3}$$

Specific heat of moist body linearly dependent on the moisture content, was determined on the principle of additive:

$$c_i = c_c(1 - \bar{U}_i) + c_W \cdot \bar{U}_i \tag{3}$$

The density of the material to be dried was determined by the expression:

$$\rho_i = \rho_c(1 - U_i) + \rho_W \cdot U_i \tag{4}$$

where in - ρ_W, ρ_c the density of water and dry material respectively.

4 Results and Discussion

The experimental results are presented graphically in fig. 2 - 4.

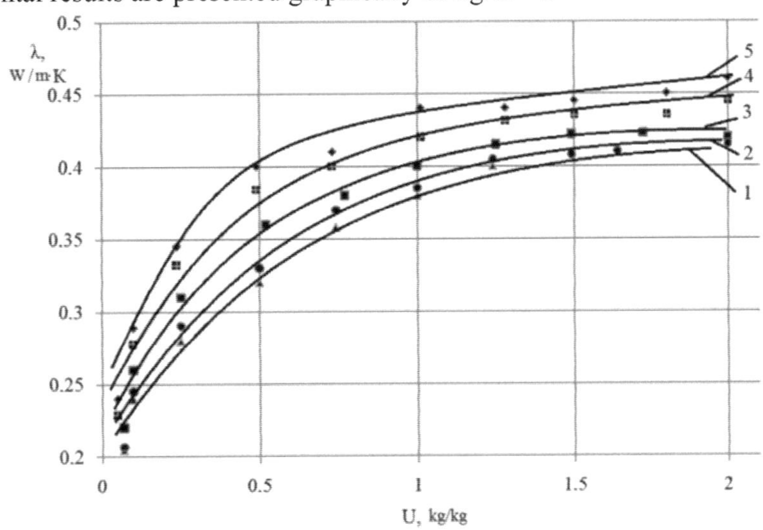

Fig. 2. Variation of the effective thermal conductivity of MSW during process at temperatures of the drying agent (1– 132°C, 2– 146°C, 3– 157°C, 4– 175°C, 5– 192°C).

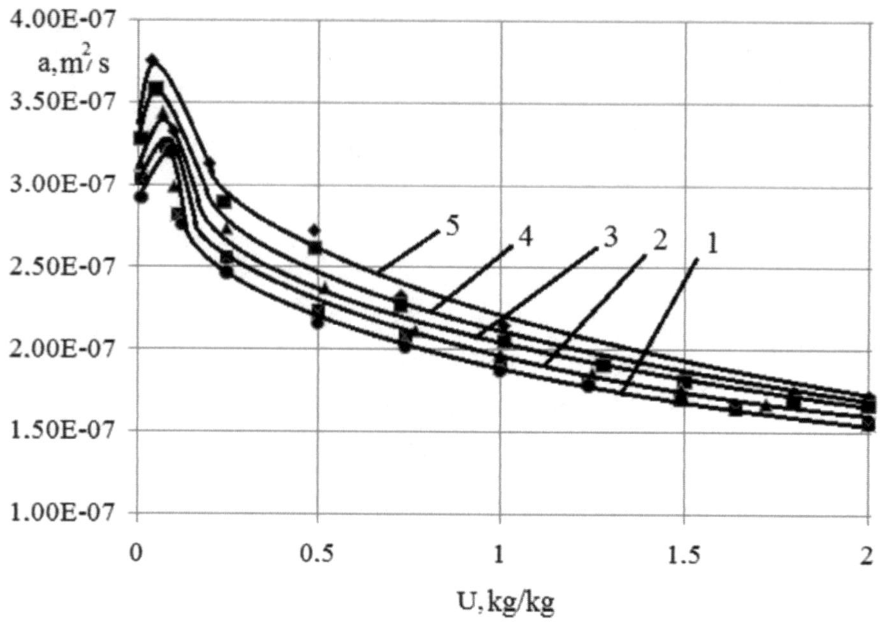

Fig. 3. Variation of the effective thermal diffusivity of MSW during process at temperatures of the drying agent (1– 132°C, 2– 146°C, 3– 157°C, 4– 175°C, 5– 192°C).

5 Conclusion

As a result of the experiments drying kinetics curves and temperature curves were obtained. On the base of the curves effective thermal conductivity and thermal diffusivity of MSW were calculated.

The results of this research can be used to calculate the temperature field and moisture concentrations in the MSW layer during its processing in the thermal reactor.

Acknowledgments

The work within the framework of the state contract No. 2461. Development of research method and process of manufacture of gaseous fuel and energy based on thermal processing of organic raw materials.

References

1. *Thermal utilization of solid household waste* (SRI concept Stromkompozit, Krasnoyarsk, 2006)
2. L.L. Vasiliev, S.A. Tanaeva, *Thermal properties of porous materials* (Science and technology, Minsk, 1971)
3. G.N. Dulnev, Y.P. Zarichnyak, *The thermal conductivity of mixtures and composite materials* (Energy, Leningrad, 1974) [in Russian]
4. A.S. Ginzburg, M.A. Gromov, T.I. Krasovskaia, *Thermophysical characteristics of food products* (Moskow, Food Industry, 1980) [in Russian]
5. S.A. Ilyin, V.M. Fokin, ASTU Messenger **2** (2006)
6. A.V. Mitrofanov, J. Vestnik ISPU **2** (2015)

EFFECT OF METAL ADDITIVES ON THE COMBUSTION CHARACTERISTICS OF HIGH-ENERGY MATERIALS

Alexander Korotkikh[1, 2, *], *Oleg* Glotov[3], and *Ivan* Sorokin[1]

[1]National Research Tomsk Polytechnic University, 634050 Tomsk, Russia
[2]National Research Tomsk State University, 634050 Tomsk, Russia
[3]Institute of Chemical Kinetics and Combustion SB RAS, 630090 Novosibirsk, Russia

Abstract. Thermodynamic calculation of combustion parameters and equilibrium composition of HEMs combustion products showed, that at the increase of aluminum powder dispersity the specific impulse and combustion temperature of solid propellants are reduced due to the decrease of the mass fraction of active aluminum in particles. Partial or complete replacement of aluminum by metal powder (B, Mg, AlB_2, Al\Mg alloy, Fe, Ti and Zr) in HEMs composition leads to the reduce of the specific impulse and combustion temperature. Replacement of aluminum powder by boron and magnesium in HEM reduces the mass fraction of condensed products in the combustion chamber of solid rocket motor. So, for compositions HEMs with boron and aluminum boride the mass fraction in chamber is reduced by 24 and 36 %, respectively, with respect to the composition HEMs with Al powder. But the mass fraction of CCPs in the nozzle exit increases by 13 % for HEMs with aluminum boride due to the formation of boron oxide in the condensed combustion products. Partial replacement of 2 wt. % aluminum powder by iron and copper additives in HEM leads to the reduce of CCPs mass fraction in chamber by 4–10 % depending on the aluminum powder dispersity duo to these metals are not formed condensed products at the HEMs combustion in chamber.

1 Introduction

The modern high-energy materials (HEMs) contains an oxidizer (ammonium perchlorate or nitrate), fuel-binder (inert or active polymeric rubbers), nitramines [1–3] and metal fuel (aluminum powder) [4–6], which allows to increase the basic power characteristics of solid propellants at the burning in the combustion chamber. The main characteristics of combustion in relation to the propulsion systems are the dependence of the burning rate versus pressure, combustion temperature of solid propellant, specific impulse [7], which is equal to the increase in the value of thrust implemented with the combustion of unit propellant mass, as well as chemical and particle-size distribution of condensed combustion products (CCPs).

* Corresponding author: korotkikh@tpu.ru

One way to the increase of HEMs combustion characteristic is the use of combustion catalysts or metal powders and their oxides as a component of solid propellants. The aim of paper is decision possible of metal additives in HEMs with partial replacement of aluminum powder in order of the increase of the combustion parameters propellants and the reduce of the condensed combustion products amount with using a thermodynamic calculation program Terra [8].

2 Combustion parameters

Thermodynamic calculation of combustion parameters and equilibrium composition of the combustion products was carried out for model composition of HEMs [9], which contains ammonium perchlorate of 64.6 wt. %, butadiene rubber of 19.7 wt. % and aluminum powder of 15.7 wt. %. We considered various aluminum powders with different particle size. Mass fraction of active aluminum in composition of powders (brands of ASD and Alex nanopowder) amount from 0.85 to 0.99, depending on their dispersity. The mean particle diameters of aluminum powders of different grades are shown in Table 1. In the capacity of additives were selected such metal powders as: boron, magnesium, titanium, iron, copper, zirconium, aluminum boride and alumino-magnesium Al/Mg alloy. We suggested, that the metal additive is introduced into the model composition of HEM by partial replacement of 2 wt. % aluminum powder.

Table 1. The mean particle diameters of aluminum powder.

Powder brand	d_{10}, μm	d_{20}, μm	d_{30}, μm	d_{32}, μm	d_{43}, μm
ASD-1	10.4	12.0	13.6	17.4	21.9
ASD-4	1.23	1.66	2.28	4.34	7.34
ASD-6	0.85	1.17	1.60	3.01	4.72
ASD-10	0.72	0.90	1.09	1.60	2.11
Alex	0.12	0.13	0.14	0.16	0.18

In the calculations we considered the initial oxidation of aluminum particles. In accordance with the results of our study the active (metal) aluminum Al content in powder of different parties and storage periods, which determined by the chemical analysis methods, we have adopted the average values of active aluminum content (Table 2). The remaining mass fraction of powder ($1-m_{Al}$) is equal aluminum oxide Al_2O_3. For example, Alex aluminum nanopowder is a mixture Al and Al_2O_3 in the weight ratio of 85/15. In the case of powder ASD-10 chemical analyzes are not carried out, and the quantity of active aluminum was calculated by extrapolation of the linear empirical approximation of Al from the value of the specific area of particles surface S_{sp} for powder ASD-1, ASD-4, ASD-6. In this evaluation for ASD-10 we took $S_{sp} = 10000$ cm^2/g. The initial oxidation of the metal additives Ti, Fe, Zr, B, Cu, AlB_2 and Al/Mg alloys we considered equal to zero. The weight ratio of the components in the alloy is assumed equal to Al/B = 0.55/0.45 and Al/Mg = 0.50/0.50.

Table 2. Mass fraction of aluminum metal and its oxide in various brands of aluminum powder.

Component	Mass fraction of powder components, wt. %				
	ASD-1	ASD-4	ASD-6	ASD-10	Alex
Al	0.995	0.980	0.964	0.947	0.850
Al_2O_3	0.005	0.020	0.036	0.053	0.150

In calculation we used the scheme of solid rocket motor, which shown in fig. 1. The pressures ratio of $p_c/p_a = 4.0/0.1$ MPa was used for gas-dynamic degrees of combustion products flow expansion. The index "c" is marked in the combustion chamber, and the index "a" – at the nozzle exit.

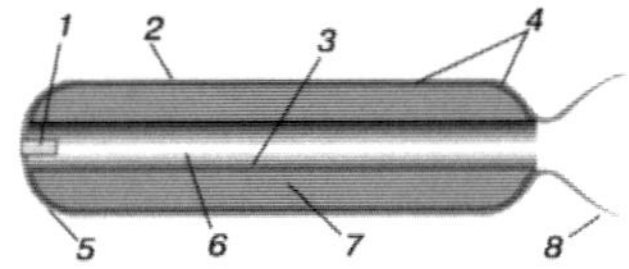

Fig. 1. The design of solid rocket motor: 1 – the igniter; 2 – a motor body; 3 – the burning surface; 4 – an insulation; 5 – a forward head; 6 – a central channel; 7 – the solid propellant charge; 8 – the nozzle.

The main calculated combustion parameters of HEMs such as the enthalpy I, the combustion temperature T_{ad}, the equilibrium adiabatic exponent k, relative molecular mass of the gas combustion products μ, the coefficient of excess oxidizing elements α, the flow rate of gas products from the nozzle w and specific impulse J are shown in Table 3.

Table 3. Calculated combustion parameters of HEMs with metal additives.

Metal powder	I, kJ/kg	T_{ad}, K	k	μ, g/mol	α	w, m/s	J, m/s
B*	-1841	2215	1.21	19.09	0.335	2228	2482
Mg*	-1841	2603	1.17	18.40	0.422	2289	2509
Al*	-1841	2638	1.18	17.11	0.406	2351	2590
ASD-1	-1854	2636	1.18	17.10	0.407	2349	2588
ASD-4	-1893	2631	1.18	17.07	0.408	2346	2582
ASD-6	-1937	2624	1.18	17.04	0.410	2341	2575
ASD-10	-1979	2617	1.18	17.02	0.412	2336	2568
Alex	-2228	2561	1.19	16.95	0.422	2299	2519
Alex+B	-2179	2366	1.20	18.04	0.409	2285	2517
ASD-4+B	-1886	2408	1.20	18.21	0.397	2329	2571
AlB_2	-1841	2353	1.15	18.97	0.363	2303	2558
Al\Mg (alloy)	-1841	2567	1.17	18.12	0.414	2325	2552
Alex+Cu	-2179	2463	1.20	17.48	0.426	2244	2456
ASD-4+Cu	-1886	2564	1.19	17.32	0.414	2302	2523
Alex+Fe	-2179	2477	1.20	17.53	0.424	2244	2455
ASD-4+Fe	-1886	2576	1.19	17.36	0.412	2303	2523
Alex+Ti	-2179	2530	1.19	17.01	0.422	2278	2496
ASD-4+Ti	-1886	2608	1.18	17.08	0.410	2325	2552
Alex+Zr	-2179	2540	1.19	17.01	0.425	2275	2489
ASD-4+Zr	-1886	2621	1.19	16.99	0.413	2328	2553

*Note: where and in Table 4 B, Mg and Al are not contained metal oxides.

With the increase of aluminum powder dispersity for a number of ASD-4→ASD-6→ASD-10→Alex the specific impulse J and combustion temperature T_{ad} of propellants are reduced due to the decrease of the mass fraction of active aluminum in the starting powders. Partial replacement of 2 wt. % aluminum by metal additive (B, Mg, AlB_2, Al\Mg alloy, Fe, Ti and Zr) in the composition of HEMs the values of J and T_{ad} are reduced.

3 Parameters of condensed combustion products

Table 4 shows the calculated parameters of condensed combustion products at the combustion of HEMs: mass fractions m_c and m_a, the phase composition of condensed particles contained in the combustion products at the pressures $p_c = 4.0$ MPa and $p_a = 0.1$ MPa, respectively.

Table 4. The calculated parameters of CCPs.

Metal powder	m_c, m_a	Phase composition and mass fraction of condensed particles
B*	0.17175 0.37286	$[B_2O_3]_c = 0.01904$; $[BN]_c = 0.13587$; $[B_4C]_c = 0.01684$ $[B_2O_3]_a = 0.19266$; $[BN]_a = 0.13643$; $[B_4C]_a = 0.02746$;$[C]_a = 0.01631$
Mg*	0.19665 0.25803	$[MgO]_c = 0.19665$ $[MgO]_a = 0.25803$
Al*	0.25272 0.29605	$[Al_2O_3]_c = 0.25272$ $[Al_2O_3]_a = 0.28825$; $[C]_a = 0.00780$
ASD-1	0.25325 0.29575	$[Al_2O_3]_c = 0.25325$ $[Al_2O_3]_a = 0.28823$; $[C]_a = 0.00752$
ASD-4	0.25478 0.29460	$[Al_2O_3]_c = 0.25478$ $[Al_2O_3]_a = 0.28798$; $[C]_a = 0.00662$
ASD-6	0.25642 0.29283	$[Al_2O_3]_к = 0.25642$ $[Al_2O_3]_a = 0.28735$; $[C]_a = 0.00548$
ASD-10	0.25787 0.29062	$[Al_2O_3]_c = 0.25787$ $[Al_2O_3]_a = 0.28638$; $[C]_a = 0.00424$
Alex	0.26253 0.27569	$[Al_2O_3]_c = 0.26253$ $[Al_2O_3]_a = 0.27569$
Alex+B	0.20036 0.29253	$[Al_2O_3]_c = 0.20036$ $[Al_2O_3]_a = 0.24039$; $[B_2O_3]_a = 0.02685$; $[BN]_a = 0.02529$
ASD-4+B	0.19089 0.29821	$[Al_2O_3]_c = 0.19089$ $[Al_2O_3]_a = 0.25594$; $[B_2O_3]_a = 0.00045$; $[BN]_a = 0.04182$
AlB_2	0.16798 0.33444	$[Al_2O_3]_c = 0.07914$; $[BN]_c = 0.08884$ $[Al_2O_3]_a = 0.12644$; $[BN]_a = 0.13171$; $[B_2O_3]_a = 0.07629$
Al\Mg alloy	0.20855 0.26815	$[MgAl_2O_4]_c = 0.20611$; $[MgO]_c = 0.00244$ $[MgAl_2O_4]_a = 0.20695$; $[MgO]_a = 0.06120$
Alex+Cu	0.23721 0.25903	$[Al_2O_3]_c = 0.23721$ $[Al_2O_3]_a = 0.24057$; $[Cu]_a = 0.01846$
ASD-4+Cu	0.24307 0.27369	$[Al_2O_3]_c = 0.24307$ $[Al_2O_3]_a = 0.25641$; $[Cu]_a = 0.01728$
Alex+Fe	0.23744 0.24058	$[Al_2O_3]_c = 0.23744$ $[Al_2O_3]_a = 0.24058$
ASD-4+Fe	0.24355 0.25641	$[Al_2O_3]_c = 0.24355$ $[Al_2O_3]_a = 0.25641$
Alex+Ti	0.26037 0.27392	$[Al_2O_3]_c = 0.23207$; $[Ti_3O_5]_c = 0.02830$ $[Al_2O_3]_a = 0.24057$; $[TiO_2]_a = 0.03335$
ASD-4+ Ti	0.25617 0.28128	$[Al_2O_3]_c = 0.23146$; $[Ti_3O_5]_c = 0.02471$ $[Al_2O_3]_a = 0.25541$; $[Ti_2O_3]_a = 0.02442$; $[C]_a = 0.00145$
Alex+Zr	0.26008 0.26759	$[Al_2O_3]_c = 0.23317$; $[ZrO_2]_c = 0.02691$ $[Al_2O_3]_a = 0.24057$; $[ZrO_2]_a = 0.02702$
ASD-4+Zr	0.25954 0.28274	$[Al_2O_3]_c = 0.23283$; $[ZrO_2]_c = 0.02671$ $[Al_2O_3]_a = 0.25555$; $[ZrO_2]_a = 0.02660$; $[C]_a = 0.00059$

For a number of ASD-4→ASD-6→ASD-10→Alex as a result of the assumption of the presence of aluminum oxide in initial aluminum powder, the mass fraction of condensed particles in the combustion products is increased.

It should be noted, that during the HEM combustion with boron in the chamber the quantity of condensed boron oxide is small and boron nitride in the combustion products dominate. Due to the nature of boron and magnesium additives (the ratio of molar masses of elements and oxides, the stoichiometric coefficient of oxidation reactions), a partial replacement of aluminum by boron and magnesium (in the mechanical mixture and alloy) reduces mass fraction m_c of CCPs in chamber. So, for compositions HEMs with boron and aluminum boride AlB_2 mass fraction m_c is reduced by 24 and 36 %, respectively, relative to the composition HEMs with active aluminum powder.

For compositions HEMs with iron and copper additives also have the reduce of m_c at 4–10 % depending on the aluminum powder dispersity. During HEMs combustion in the chamber these compounds are not formed in condensed product. As a result, the combustion products contain less alumina Al_2O_3 and the mass fraction of a condensed phase is decreased. Accordingly, HEMs compositions with iron and copper additives the reduction effect of mass fraction m_c is more pronounced in the case of HEM with Alex aluminum nanopowder, than HEM with microsized aluminum powder ASD-4, due to a larger content of alumina Al_2O_3 in Alex. Additives of titanium and zirconium in the case of HEMs with ASD-4, in opposite, Ti_3O_5 and ZrO_2 oxides make an additional contribution to the condensed products and increase mass fraction m_c.

Analyzing the data calculations pertaining to the nozzle exit, we note an increase of mass fraction of CCPs m_a of 13% for HEMs with aluminum boride AlB_2 due to the formation of boron oxide B_2O_3 in the combustion products. In the case of HEMs compositions with Alex and ASD-4 metal additives reduce the mass fraction of CCPs at the nozzle exit.

4 Conclusion

1. Thermodynamic calculation of HEMs combustion parameters and equilibrium composition of the combustion products showed, that at the increase of aluminum powder dispersity for a number of ASD-4→ASD-6→ASD-10→Alex the specific impulse and combustion temperature of solid propellants are reduced due to the decrease of the mass fraction of active aluminum in powder particles. Partial or complete replacement of aluminum by metal powder (B, Mg, AlB_2, Al\Mg alloy, Fe, Ti and Zr) in HEMs composition leads to the reduce of the specific impulse and combustion temperature.

2. Replacement of aluminum powder by boron and magnesium (in the mechanical mixture and alloy) in HEMs reduces the mass fraction of condensed products in the combustion chamber of solid rocket motor. So, for compositions HEMs with boron and aluminum boride the mass fraction in chamber is reduced by 24 and 36 %, respectively, with respect to the composition HEMs with active aluminum powder. But the mass fraction of CCPs in the nozzle exit increases by 13 % for HEMs with aluminum boride due to the formation of boron oxide in condensed combustion products.

3. Partial replacement of 2 wt. % aluminum powder by iron and copper additives in HEMs leads to the reduce of CCPs mass fraction in the combustion chamber by 4–10 % depending on the aluminum powder dispersity duo to these metals are not formed condensed product during the HEMs burning in the combustion chamber.

4. Additives of titanium and zirconium in HEMs with ASD-4 make an additional contribution to formation of titanium and zirconium oxides in the condensed products and increase the CCPs mass fraction in the combustion chamber.

Acknowledgments

The reported study was funded by RFBR according to the research project No. 16-03-00630 a "Study of bimetal powders effect on the characteristics of ignition, combustion and agglomeration of high-energy materials".

References

1. K. Takahashi, Sh. Oide, T. Kuwahara, Propellants, Explos., Pyrotech. **38** (2013)
2. M.W. Beckstead, K. Puduppakkama, P. Thakreb, V. Yang, Prog. Energy Combust. Sci. **33** (2007)
3. M. Kohga, K. Okamoto, Combust. Flame **158** (2011)
4. V.A. Arkhipov, A.G. Korotkikh, V.T. Kuznetsov, A.A. Razdobreev, I.A. Evseenko, Russ. J. Phys. Chem. B **5**, 4 (2011)
5. O.G. Glotov, Combustion, Explosion, and Shock Waves **42**, 4 (2006)
6. V.A. Arkhipov, S.S. Bondarchuk, A.G. Korotkikh, V.T. Kuznetsov, A.A. Gromov, S.A. Volkov, L.N. Revyagin, Combustion, Explosion, and Shock Waves **48**, 5 (2012)
7. V.A. Arkhipov, A.B. Kiskin, V.E. Zarko, A.G. Korotkikh, Combustion, Explosion, and Shock Waves **50**, 5 (2014)
8. B.G. Trusov, Eng. J.: Science and Innovation **1** (2012)
9. A.G. Korotkikh, V.A. Arkhipov, O.G. Glotov, K.V. Slyusarskiy, IOP Conf. Series: Mate. Sci. Eng. **93** (2015)

MATEC Web of Conferences **72**, 01049 (2016) DOI: 10.1051/matecconf/20167201049
HMTTSC-2016

SOLVING EXCESS WATER PRODUCTION PROBLEMS IN PRODUCTIVE FORMATION

Ilya Kozyrev[1], *Pavel* Zyatikov[1], and *Vera* Deeva[1,*]

[1]National Research Tomsk Polytechnic University, 634050 Tomsk, Russia

Abstract. One of the important developments of the Russian Federation national economy is a petroleum resource. Water shut off techniques are used in the oilfields to avoid the massive water production. We describe a technology for solving excess water production problems focusing on the new gel-based fluid which can be effectively applied for water shutoff. We study the effect of the gel-based fluid solution experimentally to show the feasibility of its treatment the in the near wellbore region to solve the excess water production problem.

1 Introduction

Petroleum resources are one of the important developments of the Russian national economy. However, the oil production in Russia fell down caused by the deterioration in reserve structure and quality, the high watercut of the oil production, the low productivity of the mature oil fields, and the final stage of the development of the basic production zone. The deterioration of oil reserves connects with water flooding of the productive oilfields. A greater understanding of water formation leads to an improved perception of the scientific progress and the technological development of the oil and gas industry. Excess water production is a very serious problem during waterflooding of reservoir formations. To solve the problem, many kinds of technologies have been developed. The main advanced approaches include squeezing shutoff fluids, which may be either resin-based fluids, or gel-based fluids [1].

The purpose of this paper is first to analyze the behavior of horizontal well of the oilfields then to present a new gel-based fluid which can be effectively applied for water shutoff in the near wellbore region. The results of the experimental study highlight the feasibility of the gel treatment to solve the excess water production problem.

2 Control of excessive water production

The analysis of the structure of oil reserves shows that importance of using a new technique and approach to improve the production efficiency and reduce costs of the mature fields. Thus, the important way to stimulate oil production and to increase conformance is to bring into production certain of the horizontal well. At the same time, the water problem occurs in horizontal well which could be solved by applying water-shutoff technologies such as

*Corresponding author: veradee@mail.ru

gel-based fluids. The injection of water-shutoff gels reduces water production and help address the problem when a high-permeability zone with a flow barrier in horizontal well. Figure 1 demonstrates the fractures and faults intersecting an aquifer in horizontal well [2].

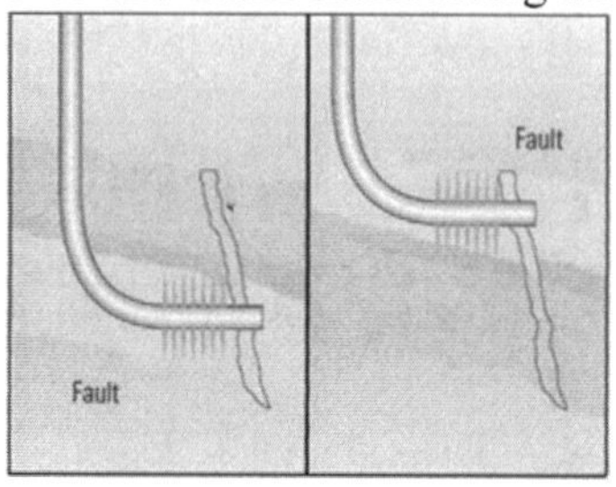

Fig. 1. Fractures or faults from a water layer (horizontal well).

Figure 2 shows the fluid distribution in reservoir containing water, oil and gas before production or injection begins [2].

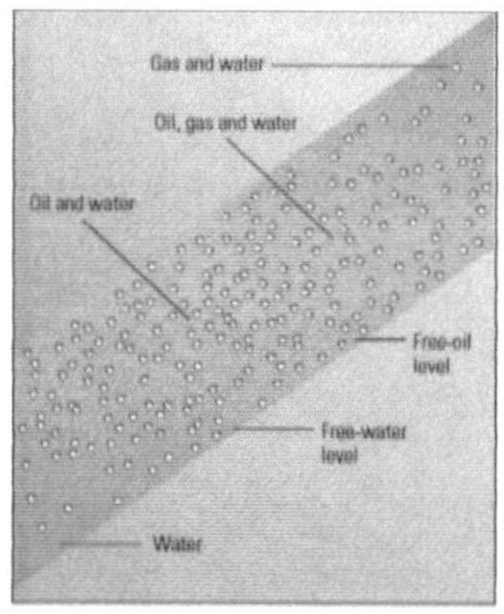

Fig. 2. Typical fluid distribution before production or injection begins.

Furthermore, major drawbacks of the most gel-based fluids are the limited penetration in the matrix and so selectively flow in the fractures, the heat sensibility, toxic property and high cost. Thus, there is a need to investigate effective and reliable gel-based fluids to become an impermeable barrier with a high penetration.

A laboratory study was conducted to characterize watershutoff gels that are injected in the horizontal well into fractures (or other high permeability anomalies) that are in direct contact with production wells. Laboratory experiments estimate the gelling kinetics and the dicalcium silicate ($CaCl_2$) rheological properties offering attractive possibilities of obtaining the physical-chemical properties and the gel waterproofing capacity.

Experiments are conducted in strong accordance with standard technique of the surface chemistry and physics of oil formation [3]. Recently, methods of mathematical statistics [4] have been used as supporting materials for calculation and report. We determine the dependence of the gel time on initial reagent concentration and the temperature being nonlinear. The results are confirmed by the law of mass action, so called Guldberg and Waage expression, and Arrhenius equation:

$$\frac{\partial C}{\partial t} = z_0 \exp\left(\frac{-E_a}{RT}\right) C_1^n C_2^l \quad (1)$$

where $W_\varphi - C$ is the gel concentration from a stock solution; %; z_0 is the pre-exponential factor (or simply the prefactor), s^{-1}; R is the universal gas constant; E_a is the activation energy, j/k/mol; T is the temperature, in kelvin; C_1 is the hydrochloric acid concentration (HCl), unit fraction, C_2 is the concentration of the dicalcium silicate ($CaCl_2$), unit fraction, t is the gel time, sec.

The most important of the thermodynamic characteristics during the gel formation is the activation energy (E_a). Overall from this series of experiments (table 1) we find the value of the energy 40 j/k/mol using the least square method for data interpretation.

Under initial condition the equation (1) gives us the follow solution: of $n = 1.15$, $l = 0.51$, $z_0 = 805235.8$ s^{-1}. It could therefore seem that the influence of the salinity of water in the concentration range of 1 to 1.20 g/sm^3 on the gel time is linear (figure 3). In this paper we control the salinity of water through the control of the concentration of the solution of $CaCl_2$ and NaCl.

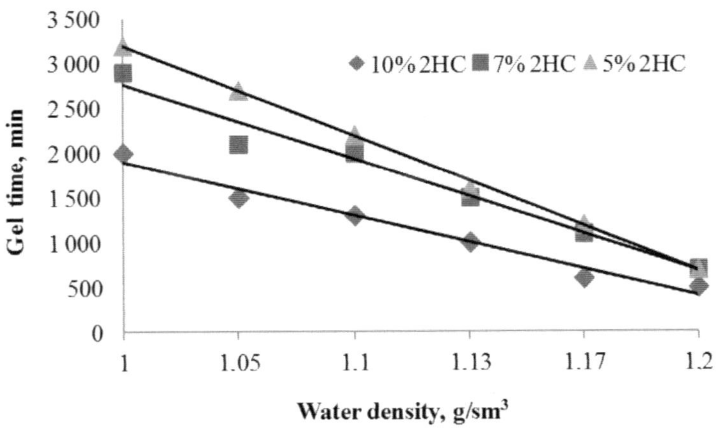

Fig. 3. The dependence of the gel time on the water density (in the temperature 22°C and in the concentration of HCl 8%).

During this laboratory study, we determine the selectivity of the water-shutoff gels which make evidence of the fact that the competent gel is formed in the water-wet core, by comparison the weak gel is formed in the oil-saturated core. We assume that it is connected with the limited adhesive properties of the gels in particular oil-saturated core. Table 1 summarizes the results of experiments and highlights a good gel waterproofing capacity.

Table 1. The characteristic of the near wellbore region.

Properties	**Variables**
Reservoir types	Fractured and faulted formation
Reservoir temperature, ^{0}C	32
Formation pressure, MPa	7.8
Length of core, m	3.5 10^{-2}
Diameter , m	2.8–10 $^{-2}$
Porosity, %	18
Absolute permeability (under overburden stress 3.4 MPa), μm	0.08
Flow rate, m^3/min	5–10^{-6}
Water relative permeability under P-T conditions (before water shutoff treatment), k_{before}, μm^2	0.025
Pressure differential of core (after water shutoff treatment), MPa/m	12.45
Water relative permeability (after water shutoff treatment), k_{after}, μm^2	0.007

3 Discussion of Results

Thuswise most promising for the application of this gel-based fluid is to minimize the excess water in the near wellbore region. The injection of a water-shutoff gel could reduce water production without affecting oil production from the formation.

It should be mentioned that LLC "Burneftegas" effectively applies this gel for water shutoff hydraulical fracturing in an injector well.

Moreover, in one producing well of Sorovskoe field the water–handling capacity to cope with up to 90 % during choke-back. After squeezing shutoff fluid based on silicon coating backfill materials AKOR-BN 102, the ratio of the produced fluid is reduce more than 40 %, at the same level of the production rate (around 10 m^3/per day) [5].

Similar results were obtained by JSCo "Tatneftprom". After squeezing shutoff fluid based on polymeric composition "MPS REIS-H" in the Ivninskoe field the production rate is increased 8-fold while the average water cut is reduced by 20–28 % [6].

4 Conclusion

What is clear about petroleum is that it will continue to play a large role in our lives. For mature fields, as a complex subject in the oil and gas industries, several types of oil production and distribution problems are posed: selection of crude oil, water flooding, particle coarsening [7] and others. Excessive water production and has a serious economic and environmental impact thus How to control water-cut of mature oilfields is always a challenging task for field operators. This paper considered new gel that can impart large permeability reduction factors for water flow in productive formation. Our results imply that ineffective gel treatment is responsible for attaining smaller water-shutoff values in the field than in the laboratory.

References

1. J. Elphick and R. Seright, Petroleum Network Education Conference's 3rd Annual International Conference on Reservoir Conformance Profile Modification, Water and Gas Shutoff, Houston, Texas, USA, August (1997)
2. B. Bailey, J. Tyrie, J. Elphick, et al., Oilfield Rev. **12**, 1 (2000)
3. B.M. Kurochkin, *Instructions to proceed insulation works with the application of water-swellable polymer* (SPA "Drilling equipment", USSRVT, Moscow, 2002)
4. S.M. Slobodyan, Meas. Tech. **49**, 1 (2006)
5. V.M. Stroganov, *Project development and application of the silicon coating backfill materials* (LLC SPF "Nitpo", Krasnodar, 2009)
6. I.P. Novikov, A.S. Primachenko, V.M. Lapshina, Oil-gas exposition **2**, 41 (2015)
7. M.V. Vasilevsky, P.N. Zyatikov, A.Y. Burykin, V.S. Deeva, IOP Conf. Ser.: Earth Environ. Sci. **27**, 012020 (2015)

INVESTIGATION OF THE COMBUSTION STABILITY OF METHANE-AIR MIXTURE IN RECUPERATIVE BURNERS OF DIFFERENT GEOMETRIES

Alexey Krainov[1], *Leonid* Minkov[1], and *Ksenia* Moiseeva[1,*]

[1]National Research Tomsk State University, 634050 Tomsk, Russia

Abstract. The results of numerical investigations of 5.5% methane-air combustion stability in heat recuperative burners (counter flow burner, U-shape burner and Swiss-roll burner) are presented in this paper. The investigation is carried out with the use of commercial CFD package Ansys-Fluent. The boundaries of combustion stability of 5.5% methane-air mixture depending on the gas flow rate at the inlet of the burners are determined.

The problem of stable combustion of lean gas mixture is of practical importance at the designing energy-efficient and environmentally-friendly burners. The recuperative heat transfer mechanism can be used to initiate and sustain a stable combustion of lean gas mixture [1]. Combustion stability in the recuperative burners is provided by preheating the inflowing cold gas mixture due to heat transfer through the internal partition or through the wall of burner. Among heat recovery devices the counter flow burners [1, 2], U-shaped burner [1, 3] and Swiss-roll burner [1] can be distinguished. A configuration of burner has significant influence on the combustion stability of gas. The stable combustion of lean methane-air mixture in the burners of equal channel length but different configurations is achieved under the different conditions.

The investigation results of combustion stability of 5.5% methane-air mixture in the three recuperative heat exchangers above mentioned are represented in this work. Physical and mathematical formulations of the problem are based on the works [2, 3] with taking into account the variety of burner's geometry. The investigation is carried out by means of the commercial code Ansys-Fluent. The aim of this work is to determine the boundaries of combustion stability of 5.5% methane-air mixture depending on the gas flow rate at the inlet of the burners. The schemes of flow in the counter flow burner, U-shaped burner, Swiss-roll burner are depicted in figs. 1-3 correspondingly. General scheme of the gas flow can be described as follows. Cold methane-air mixture having a velocity u_{vh}, a temperature T_{vh}, a concentration of fuel species $Y_{vh,CH4}$ is fed in the inlet channel, the cross-sectional size of which d is greater than a thickness of the internal channel walls.

* Corresponding author: Moiseeva_KM@t-sk.ru

The simplest scheme of the gas flow is that corresponding to the counter flow burner, fig.1. Gas is supposed to be fed with the same flow rate in the two parallel channels connected with a partition wall. The gas flow direction in one channel of the counter flow burner is opposite to that in another channel. In the U-shaped burner the mixture flow passes the inlet channel up to the burner elbow after that it changes the flow direction and flows out through the outlet channel, fig.2. In the Swiss-roll burner having a twisted configuration the mixture flow passes successively through several channels and elbows as shown in fig.3

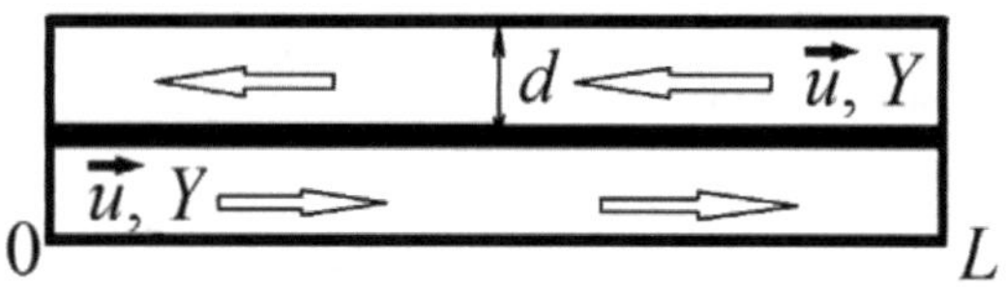

Fig. 1. The counter-flow burner.

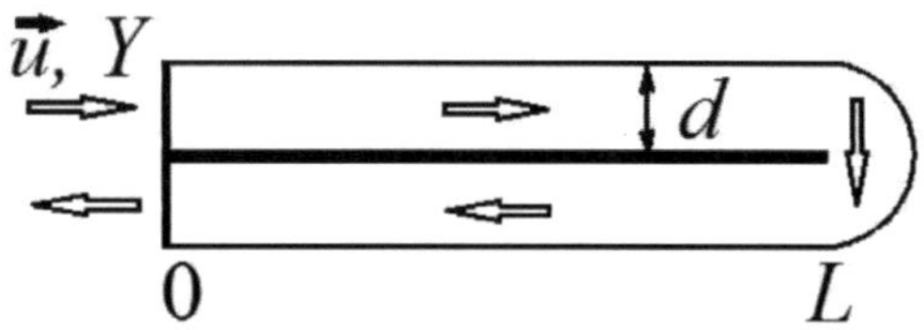

Fig. 2. The U-shape burner.

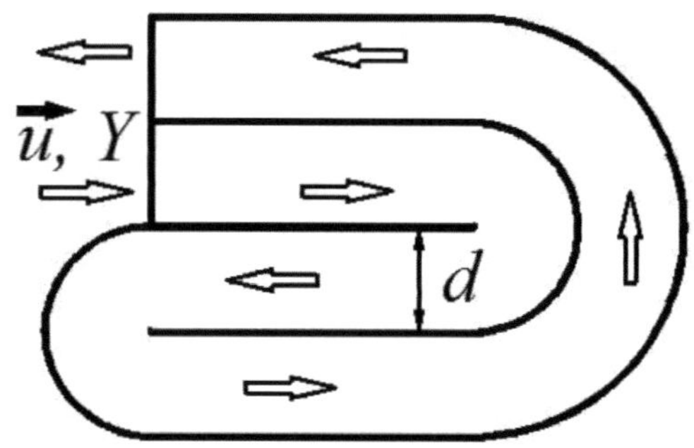

Fig. 3. The Swiss-roll burner.

The internal walls of a burner are supposed to be uniformly preheated to high temperature at the initial time. The external walls are thermal insulated. A perfect contact between internal walls and gas is supposed to take place.

In formulating the mathematical problem the following assumptions are made: the laminar flow of an incompressible viscous gas is considered; D diffusion coefficient, and λ gas thermal conductivity, as well as density and gas flow rate – they all depend on the temperature; the chemical process is supposed to be determined by a one-stage irreversible chemical reaction depending on the temperature according to the Arrhenius law.

The mathematical formulation of the problem is as follows.

The energy equation for the mixture is

$$\frac{\partial}{\partial t}(\rho E)+\nabla\cdot\left(\vec{u}\left(\rho E+p\right)\right)=\nabla\cdot(\lambda\nabla T)-\sum_{j=1}^{5}h_j^0 R_j,\ x,y\in \text{I, II, III}. \quad (1)$$

The energy equation for the dividing wall is

$$\frac{\partial}{\partial t}(\rho_1 E_1)=\lambda_1\nabla^2 T_1,\ x,y\in \text{IV}. \quad (2)$$

The mass-fraction equations for the species is

$$\frac{\partial(\rho Y_i)}{\partial t}+\nabla\cdot\left(\rho\,\vec{u}\,Y_i-\tilde{D}_m\,\nabla Y_i\right)=R_i,\ i=1..4,\ \ x,y\in \text{I, II, III}. \quad (3)$$

The continuity equation for the mixture is

$$\frac{\partial\rho}{\partial t}+\nabla\cdot(\rho\vec{u})=0. \quad (4)$$

The momentum equation for the mixture is

$$\frac{\partial(\rho\,\vec{u})}{\partial t}+\nabla\cdot(\rho\,\vec{u}\,\vec{u})+\nabla p=\nabla\cdot\boldsymbol{\tau}\,. \tag{5}$$

The equation of state for the mixture is

$$p=\rho RT\sum_{j=1}^{5}\frac{Y_i}{M_i}\,. \tag{6}$$

Where $\boldsymbol{\tau}=\mu\left(\nabla\vec{u}+\nabla\vec{u}^T\right)$ is the stress tensor, $E=\sum_{j=1}^{5}Y_j h_j+\frac{\vec{u}^2}{2}$ is the total energy of the mixture, $h_j=c_j\left(T-T_{vh}\right)$ is the enthalpy of j-species, h_j^0 – is the enthalpy of formation of j-species at the temperature of T_{vh}, $E_1=c_1\left(T_1-T_{1,vh}\right)$ is the enthalpy of the dividing wall, R_j is the rate of formation of j-species.

The coefficient of thermal conductivity and the coefficient of diffusion for the mixture are assumed to be functions of temperature [3]: $\lambda=\lambda_{st}\left(\frac{T}{T_{vh}}\right)^s$, $\tilde{D}_m=D_{st}\rho_{st}\left(\frac{T}{T_{vh}}\right)^s$. The rate of formation of i-species is determined by formula $R_i=\frac{\nu_i\,M_i}{\nu_{CH4}\,M_{CH4}}k_0\,\rho\,Y_{CH4}\,\exp\left(-\frac{E}{RT}\right)$. Where i =1 – CH_4, i = 2 – O_2, i = 3 – CO_2, i =4 – H_2O, i = 5 – N_2.

The boundary conditions at the inlet of the burner are as follows: the temperature of the mixture T_{vh} = 300 K, the velocity profile is supposed to be uniform and the velocity value $u_{x,vh}$ is set to variable, the mass fraction of species $Y_{vh,CH4}=0.035$, $Y_{vh,O2}=0.23$, $Y_{vh,CO2}=0$, $Y_{vh,H2O}=0$.

The boundary conditions at the outer wall of the burner are as follows:
non-slip condition is

$$\vec{u}=0\,, \tag{7}$$

zero heat flux condition is

$$\frac{\partial T}{\partial n}=0\,, \tag{8}$$

zero species flux condition is

$$\frac{\partial Y_i}{\partial n}=0\,. \tag{9}$$

The boundary conditions at the inner wall of the burner are as follows:
non-slip condition is

$$\vec{u}=0\,, \tag{10}$$

zero species flux condition is

$$\frac{\partial Y_i}{\partial n} = 0. \quad (11)$$

perfect thermal contact is

$$-\lambda_1 \frac{\partial T_1}{\partial n} = -\lambda \frac{\partial T}{\partial n}, \; T_1 = T. \quad (12)$$

Free outflow boundary conditions are used at the outlet of the burner.

The calculations were performed at the thermophysical and kinetic parameters of mixture corresponding to those of work [3]. A length and breadth of channels are set equal to $d = 6 \cdot 10^{-3}$ *m*, $d_1 = 2 \cdot 10^{-4}$ *m*, $L = 5 \cdot 10^{-2}$ *m*. Thermophysical parameters of internal walls are set equal to $c_1 = 687$ *J/(kg·K)*, $\lambda_1 = 30$ *W/(m·K)*, $\rho_1 = 7500$ kg/m^3. The initial temperature of internal walls is equal to 1700 *K*.

The cross component of mixture velocity $u_{y,vh}$ is assumed to be zero, whereas the longitudinal component of mixture velocity $u_{x,vh}$ varies from 0.01 *m/s* to 0.45 *m/s* in order to determine the maximum value of mixture flow rate, at which the stable combustion is possible.

The calculations were performed using SIMPLE method of the second-order accuracy. The time step was chosen equal to 10^{-5} *s*, the effect of the boundary layer was taken into account by mesh refinement near the wall, and the maximum step in space was $5 \cdot 10^{-6}$ *m*. Conservation of mass, momentum and energy balance was controlled at the computations. It was supposed that the solutions converged to the steady-state ones if residuals of equations were less than 10^{-5}.

The investigation results are represented in fig. 4-6. The figures shows the results obtained for limit values of the ongitudinal component of mixture velocity at the inlet $u_{x,vh}$. Little increasing in the mixture velocity at the inlet of burner leads to the removal of the combustion front the burner. According to results obtained the boundary of stable combustion for 5.5% methane-air mixture in the case of equal length channel burners of different types depends on the mixture velocity at the inlet of burners. Stable combustion is possible in the counter flow burner when the limit value of mixture velocity at the inlet $u_{x,vh}$ does not exceed 0.145 *m/s*. Changing the burner's configuration from the counter flow one to the U-shaped one (fig.5) enhances the range of the mixture velocity at the inlet of burner $u_{x,vh}$ up to 0.21 *m/s*. The twisted channel yields the widest range of the mixture combustion stability (fig.6) $u_{x,vh} = 0.3$ *m/s*.

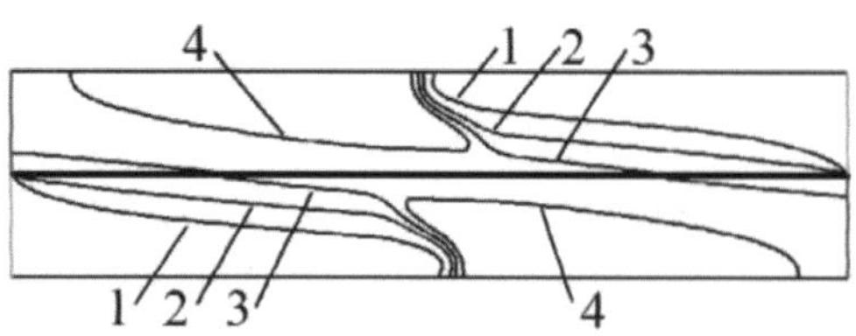

Fig. 4. $U_{x,vh} = 0.145$ *m/s*. 1 – 700 *K*, 2 – 1100 *K*, 3 – 1500 *K*, 4 – 1900 *K*.

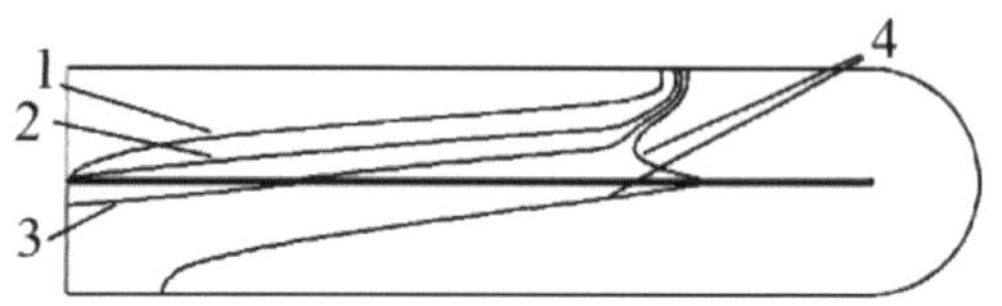

Fig. 5. $U_{x,vh} = 0.21$ *m/s*. 1 – 700 *K*, 2 – 1100 *K*, 3 – 1500 *K*, 4 – 1900 *K*.

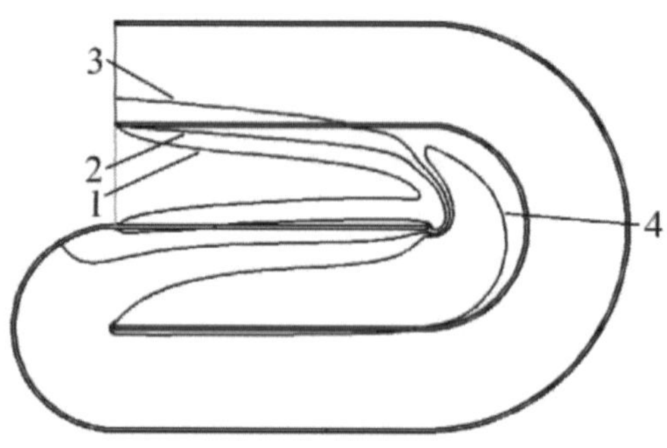

Fig. 6. $U_{x,vh} = 0.3$ *m/s*. 1 – 700 *K*, 2 – 1100 *K*, 3 – 1500 *K*, 4 – 1900 *K*.

The results obtained are in accordance with those of [2-3]. Indeed, the stable mixture combustion in the counter flow burner is possible when the combustion front in the upper channel is situated left of that in the lower channel in the case of right-left feeding for the upper channel and inversely for the lower channel. With that two high temperature zones overlapping each other are developed in the channels. Computations show that the combustion front is established only in the upper part of channel in the case of U-shaped burner that is in accordance with results of [3].

Acknowledgments

Research is funded by the Russian Science Foundation grant (project no. No. 16-33-60091 mol_a_dk).

References

1. Y. Ju, K. Maruta, Prog. Energy Combust. Sci. **37** (2011)
2. R. V. Fursenko, S. S. Minaev, Comb., Expl. Shock Waves **41** (2005)
3 A. Yu. Krainov, K. M. Moiseeva, Comb., Expl. Shock Waves **52** (2016)
4. S. Patankar, *Numerical heat transfer and fluid flow* (Hemisphere Publishing Corporation, New York, 1980)

MATEC Web of Conferences 72, 01051 (2016) DOI: 10.1051/matecconf/20167201051
HMTTSC-2016

HIGH-SPEED INTERACTION ROD AND SEGMENTED PROJECTILES INTO MASSIVE OBSTACLES

E.N. Kramshonkov[1,*], *A*.V. Krainov[1], and *P*.V. Shorohov[1]

[1]National Research Tomsk Polytechnic University, 634050 Tomsk, Russia

Abstract. In this paper we consider issues related to numerical modelling of processes of high-speed impact effect of a consistent series of damaging elements on the massive barrier. Research is carried out by comparing the finite speed of penetration of this series at a speed of penetration of the elongated one-piece (rod) of the missile of the same mass and diameter. Examines the range of initial impact speeds from 1 to 12 km/s taking into account such physical phenomena as fracture, melting and evaporation of the interacting bodies. The analysis showed that almost in all range of impact speeds segmented drummer is better solid elongated.

1 Introduction

High speed shock has been the most effective way to defeat the objects of different nature. Effectiveness of protective structures against impact devoted a lot of research. Constant search for effective means of destruction and the protection of space vehicles (missile defence, protection from dangerous space objects, etc.). In this regard, it seems necessary to study the impact in a wide range of impact speeds based on such physical phenomena as erosion, melting and vaporization of the interacting bodies – projectiles and obstacles. A special place in a number of lesion is segmented drums, the undoubted merit of which is their greater efficiency, in some situations, compared to traditional means of impact. Similar studies to destroy dangerous space objects in the form of a sphere from fragile material presented in [1], where the means of destruction used series (tandem) damaging elements, following each other.

2 Problem statement and initial data

Physical formulation of the problem, the results of which are analyzed in this paper is formulated as follows (see fig. 1). A series of cylindrical projectiles with initial velocity , interact with the massive obstacle. The drummers are arranged along a single axis at a certain distance from each other. The speed of the shock is directed along the symmetry

* Corresponding author: lux_veritatis@mail.ru

axis and the normal to the barrier. The problem is considered in the axis symmetric formulation. In this paper the results of solving two variants of the problem high-speed interaction of an elongated cylindrical and segmented projectiles with the massive obstacle in the range of impact velocities from 1 to 12 km/s. The problem is solved under the following initial data. An elongated firing pin is a cylinder of diameter 50 mm and length 200 mm; segmented – four compact segment with a length of 50 mm, with a line break between segments is 50 mm. The material of the drummers in both cases – hard steel. Barrier – mild steel, 250 mm thick.

For the mathematical description of the behavior of interacting bodies the elasto-plastic model of mechanics of a continuous medium taking into account the changes of porosity of deformed materials [2]. Modeling of different aggregate States of substances is carried out by means of interpolation wide-range equation of state describing the solid, liquid and gaseous phase [3]. When calculating the destruction is taken into account the concept of their formation as by type of job and shift, and in the implementation of their joint action [4]. After the destruction, the material is modeled by the environment in the form of sand or dust is not able to resist tensile loads, but perceiving compressive and shear loads in compression. Damage to the material as shear is typically the first stage of disruption, followed by a discontinuity at the isolation type already provisioned environment. The final stage of the calculation of damage is the realization of numerical algorithm of the separation of the body into separate pieces. The point of this algorithm is as follows. In some areas totally destroyed material, where the condition of a separated fracture, new surfaces, which divides the body into separate pieces. These fragments generally move (fly) with obviously different average speed [5-10].

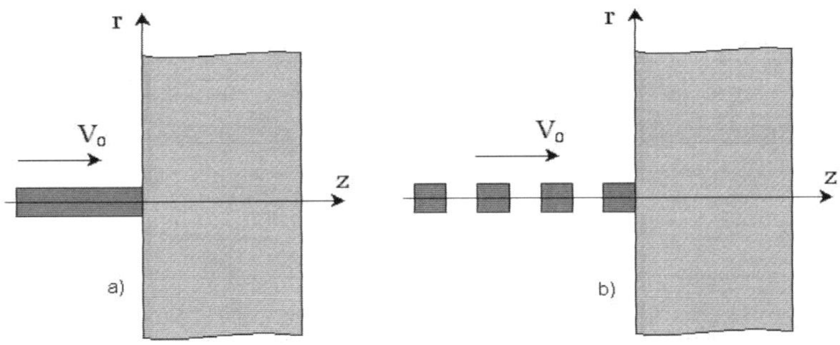

Fig. 1. The initial configuration of the penetration rod (a) and segmented (b) the drummer in the barrier.

As a research tool this study uses the method of numerical simulation based on 2-step method of second order accuracy in the Euler space variables [2, 9-13]. When calculating penetration segmented drummer applied the algorithm combining the contacting segments after actuation of interacting with the barrier segment to the size of a few computational cells [14-19]. The introduction of such algorithm is equivalent to the procedure of changing boundary conditions on the contact surface with a perfect slip condition at the rigid couplings [20-21]. Note that the use of such algorithm is possible only under the condition that all segments are made of the same material [22-25].

3 Analysis of the results of solving the problem

Below, as an example, it is considered two variants of the problem of the penetration rod and segmented impactor at impact velocities of 2 km/s and 12 km/s, as well as the analysis of the efficiency of a segmented impactor compared to the rod element. At high impact velocities, more than 8 km/s, the efficiency of the segmented impactor decreases due to the evaporation of the front segments.In fig. 2 shows the initial stage of penetration of impactors at a speed of 2 km/s at the time moment t = 40 microseconds. In the upper half-plane symmetry arrows indicate field of mass velocities, where the maximum length of arrows corresponds to the maximum value of the mass speed Vmax. Maximum speed in this case is the initial velocity of the kick. In the lower half-plane represented by intervals of relative pressure P/Pmax, the scale intervals of the pressures shown below the figure. Numerical values of Pmax are given in the caption to fig. 2; here shows the numerical values of Vk – velocity of the contact boundary (the speed of penetration of impactors), Hk is the depth of the crater.

Common to both options penetration is the maximum pressure in the contact border. This pattern is characteristic for the active phase of the penetration process when the average speed of the drummer or his segment than the velocity of the contact boundary. At the initial stage of penetration, one-piece cylinder has the advantage over the segmented firing pin, as the rate of penetration Vk, and the height Hk of the crater (fig. 2). However, along the length of the plastically deformed portion is not segmented drummer has more potential for penetration with higher speed Vk. This fact is confirmed by fig. 3, which shows the distribution of the mass velocity along the symmetry axis for both options penetration.

In fig. 4 shows a configuration of interacting bodies after the complete operation of both drummers, rod and segmented. In the lower half plane of symmetry designated area of devastation projectiles and obstacles. The right side of fig. 4 shows the designations of the various types of damage to the material: SHEAR – the material is destroyed on the shift type; a SPALL – material destroyed by the isolation type; the characters "break down" correspond to different degrees of atomization of the ruined material to the relative densities of 0.5 and 0.1. In the figure caption added numerical values of the Vt – speed of a back surface barrier on the symmetry axis.

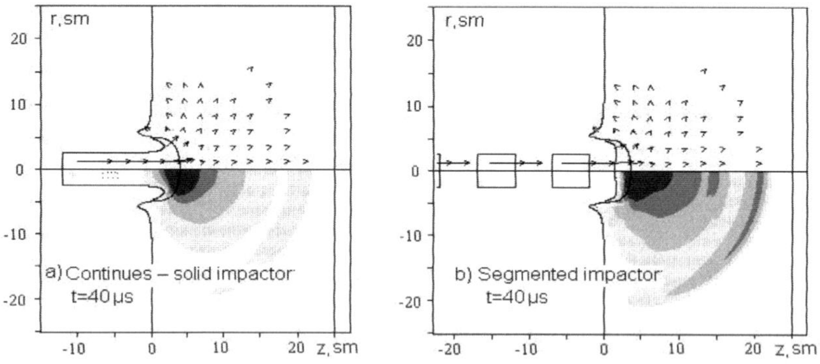

(a) $Pmax$ = 6.8 GPa, Vk = 0.9 km/s, Hk = 3.8 sm; (b) $Pmax$ = 2.8 GPa, Vk = 0.7 km/s, Hk = 3.5 sm
The scale of relative pressure intervals $P/Pmax$:

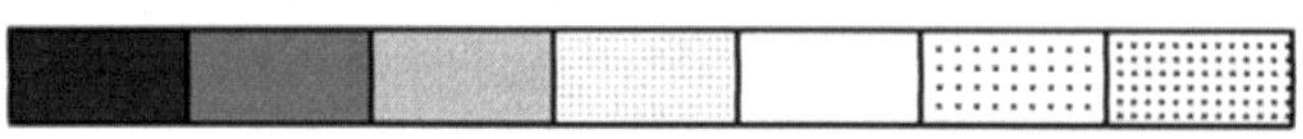

1 0.75 0.5 0.25 0.1 -0.05 -0.1

Fig. 2. The initial stage of penetration of impactor at an impact speed of 2 km/s.

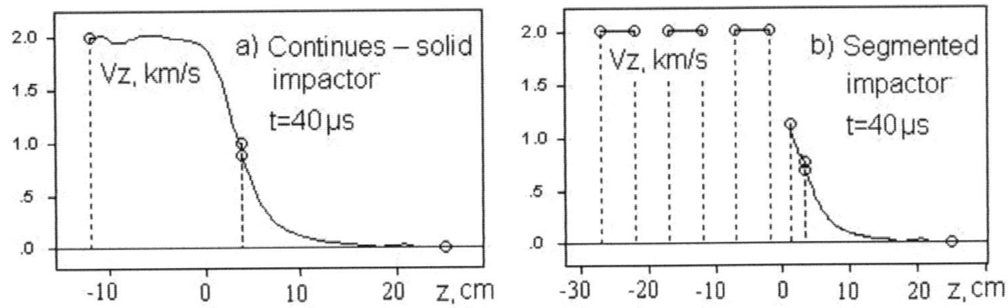

Fig. 3. The distribution of mass velocity along the symmetry axis.

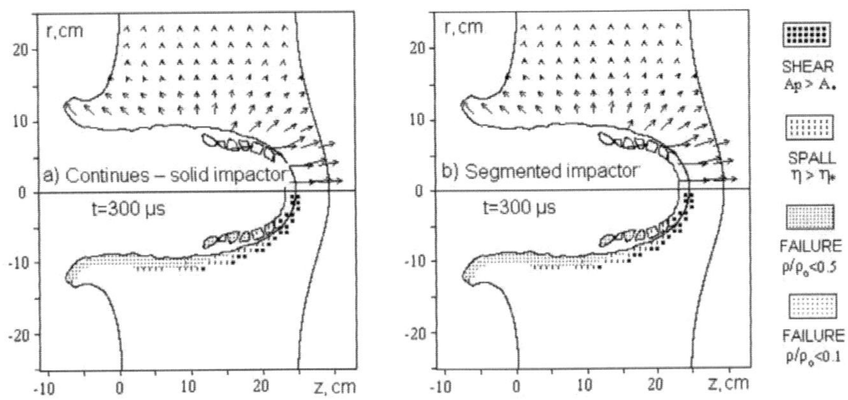

Fig. 4. The area of destruction after activation drummers: (a) *Vk* = 0.49 km/s, *Hk* = 24.7 cm, *Vt* = 0.32 km/s; (b) *Vk* = 0.84 km/s, *Hk* = 24.1 cm, *Vt* = 0.44 km/s.

On the numerical values of the quantities *Vk* and *Vt* shows that by time t = 300 μs occurs higher efficiency segmented drummer compared to the rod element. The depth of the crater *Hk* in both cases, the shock is almost aligned, in a period of 300 μs, this value was more core element. In fig. 5 shows the velocity dependence of the contact boundary of *Vk* and the back side of the obstacles *Vt* from the process time. For a truss element (fig. 5,a) is the characteristic stationary phase of penetration with almost constant speed *Vk* = 900 m/s, which continues in the period from 30 μs to 150 μs. Saw-tooth character of the curve *Vk* in fig. 5 b reflects the processes of penetration of 4 elements segmented drummer with some pauses in time.

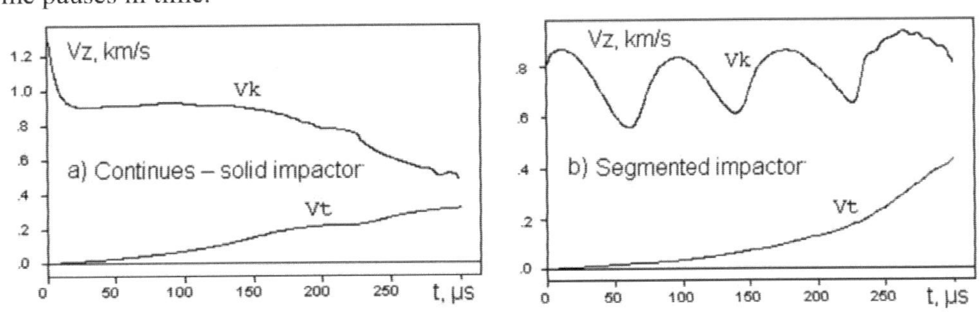

Fig. 5. The dependence of the velocity of the contact boundary of *Vk* and the back side of the obstacles *Vt* from the process time: (a) for the truss element; (b) for segmented drummer.

The final picture breaking through barriers is illustrated in figure 6. Here it has formed a final speed breakthrough *Vk*, which is the average speed of the fragments of the hammer and broken through the field of obstacles. Numerical values confirm his speed for greater efficiency segmented drummer compared to the rod element.

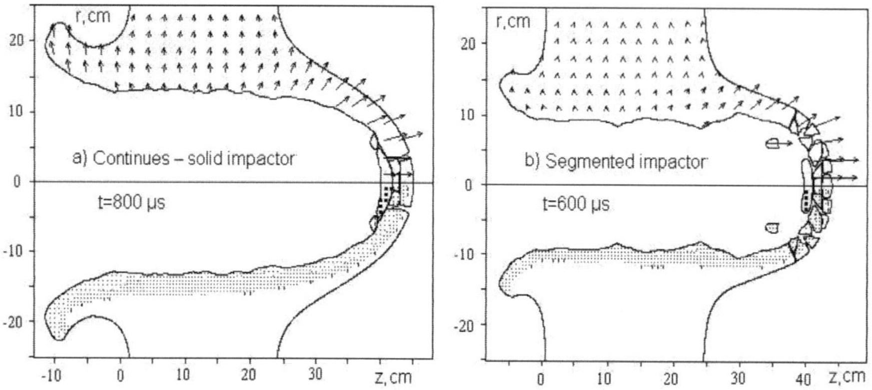

Fig. 6. The final stage of impactors penetration: a) *Vk* = 300 km/s; b) *Vk* = 510 m/s.

In fig. 7 shows the initial stage of penetration of impactors in the velocity of the shock interaction, is equal to 12 km/s. Here we have introduced the additional symbol regions of the state of the material, and below the figure summarizes the decoding of these States: MELTING – the internal energy of the medium is higher than energy of melting; P.GAS – internal energy of the material is higher than the energy of sublimation; GAS – substance is evaporated in the state. In this embodiment, the shock also still has the advantage in the intensity of penetration of the core element, as the rate of penetration *Vk*, and the depth of the crater *Hk*. The entire volume of the crater the core element are filled with the vaporized material of the impactor, and the temperature on the surface of the crater reaches 150000°C. The distribution of pressure and mass velocity along the symmetry axis are illustrated in figure 8. Given the fact that after unloading of the material is its internal energy decreases, the sublimated material may move either in the evaporated state or in the molten material. Therefore, the shock-compressed material is evaporated, its internal energy after unloading must remain higher than the energy of sublimation.

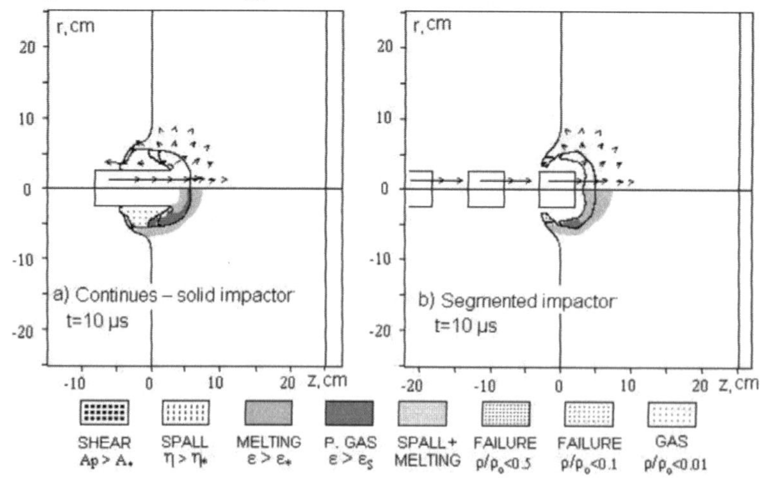

Fig. 7. The initial stage of penetration of impactor at an impact speed of 12 km/s: (a) *P*max = 201 GPA, *Vk* = 5.7 km/s, *Hk* = 5.8 cm; (b) *P*max = 97 GPA, *Vk* = 2.9 km/s, *Hk* = 4.9 cm.

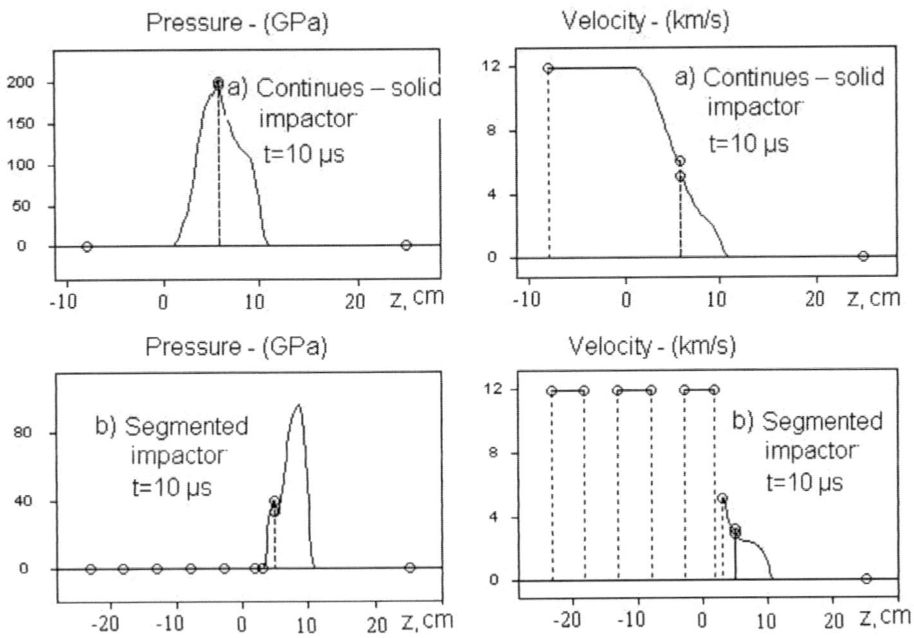

Fig. 8. The distribution of pressure and mass velocity along the symmetry axis.

Further evaporation of the materials the drums are presented in fig. 9. This stage of the process of penetration like an explosion inside the crater, vaporized substance drummers begins to scatter in the opposite direction with velocities of order 10 km/s (fig. 9, a). The active period of the penetration of the truss element is more intense than in segmented drummer. By the time 30 ISS truss element forms a deeper crater; more intensively scatters a cloud of vaporized material of the impactor; the region of high pressures (above 100 HPa) is shifted to the rear side of obstacles. However, by this time the penetration speed segmented drummer Vk higher than that of the core element, and the rear segment has not yet started any direct penetration into the barrier. The rear element of the segmented impactor (fig. 9, b) penetrates through a gaseous cloud formed by evaporation from the anterior segments.

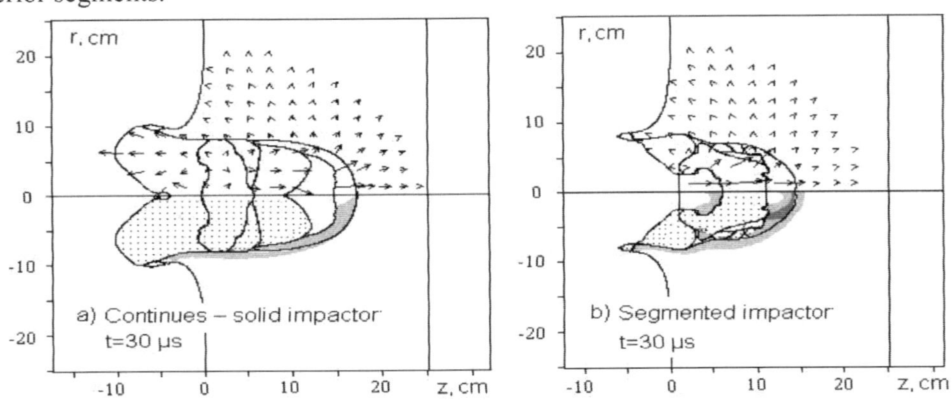

Fig. 9. Evaporation and expansion of the material of the drummer: (a) *P*max = 147 GPa, *V*max = 10.2 km/s, *Vk* = 4.5 km/s, *Hk* = 17.1 cm; (b) *P*max = 202 GPa, *V*max = 20.9 km/s, *Vk* = 6.0 km/s, *Hk* = 14.3 cm.

However, he is deformed and even partially transferred in the molten state, and the cloud of gas re it accelerates to velocities of order 20 km/s (fig. 10, b). This situation significantly reduces the efficiency of penetration of this type of firing pin under high impact velocities.

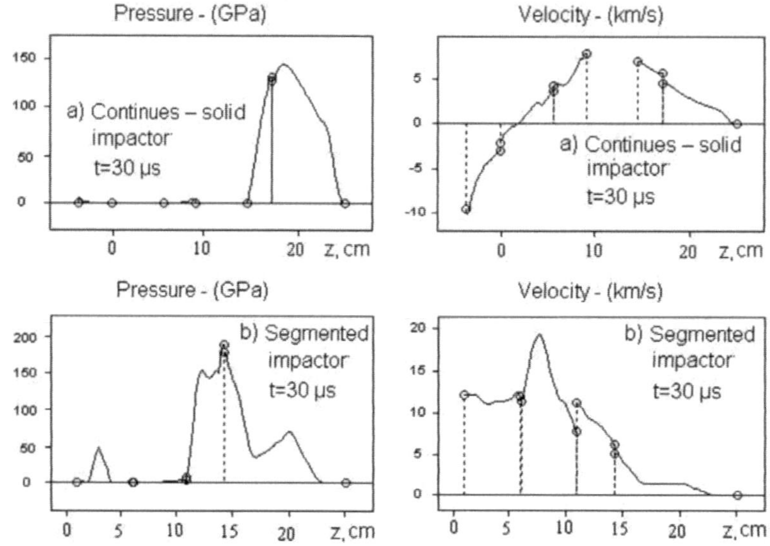

Fig. 10. The distribution of pressure and mass velocity along the symmetry axis.

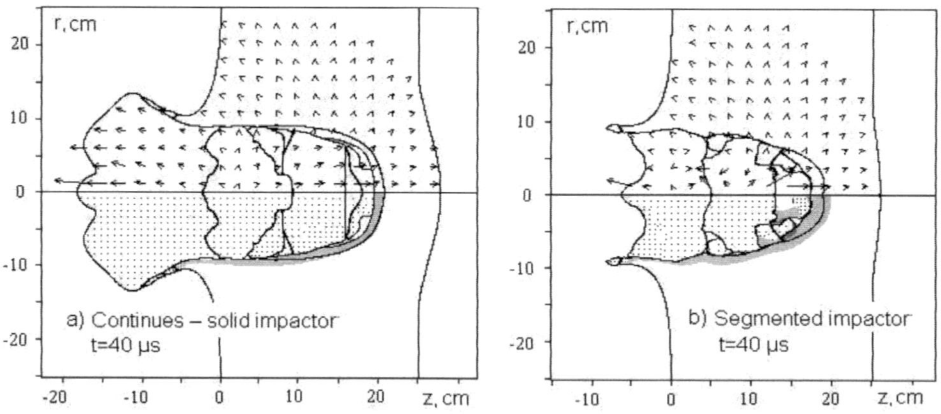

Fig. 11. The scattering of evaporated materials drummers and their penetration into the barriers: (a) Pmax = 45 GPa, Vmax = 18.0 km/s, Vk = 2.7 km/s, Hk = 20.7 cm; (b) Pmax = 84 GPA, Vmax = 20.0 km/s, Vk = 2.9 km/s, Hk = 18.9 cm.

To the time moment t = 40 μs active phase of penetration of the rod element is moved to the stage of inertial motion (fig. 11, a). Speed back side barriers was higher than the speed of the contact border (fig. 12, a). This fact promotes the formation of separated region of destruction in the zone of penetration of a target (see fig. 13, a). For a segmented impactor (fig. 11, b) a process of active influence had not ended; the rear segment of drummer, located in the solid-molten state continues to penetrate the barrier, and the rate of penetration Vk higher than that of the core element and above the speed of the back side of the obstacles (fig. 12, b).

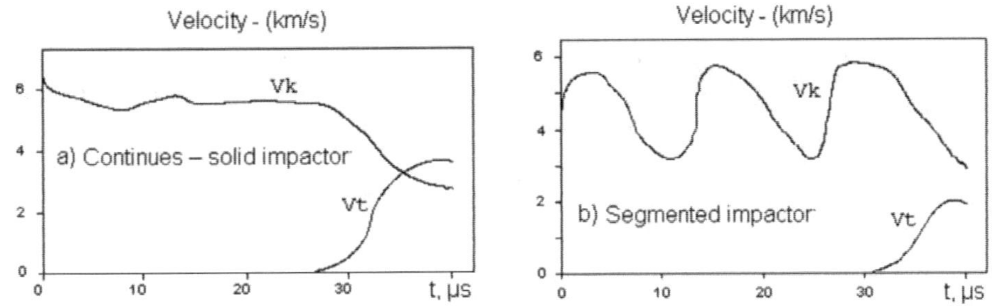

Fig. 12. The dependence of the velocity of the contact boundary of *Vk* and the back side of the obstacles *Vt* from the process time: (a) for the truss element; (b) for segmented drummer.

The final stage of penetration of impactors shown in fig. 13, illustrates the condition of the punched sections of obstacles and provides a quantitative assessment of the penetration velocity Vk. The segmented drummer this rate is almost two times higher than that of the core element. However, the total mass supergranny fragments from the core more than the segmented drummer, but they fly at a slower speed.

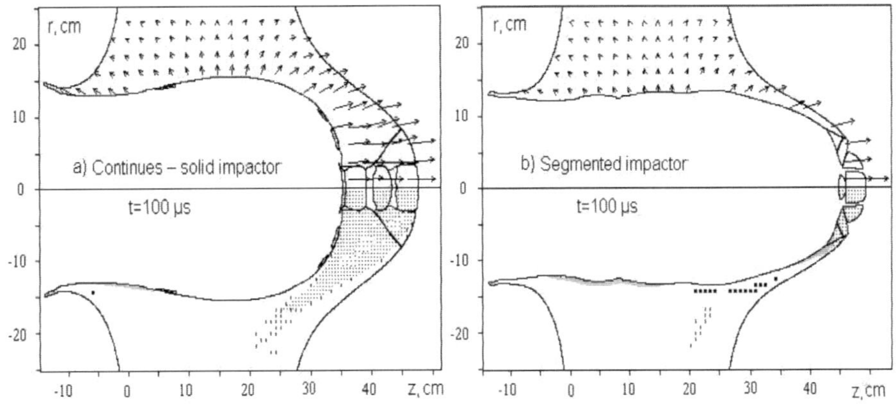

Fig. 13. The final stage of penetration of impactors: a) *Vk* = 2.4 km/s; b) *Vk* = 4.1 km/s.

End speed through breakthrough *Vk* is the main quantitative characteristic of the efficiency of penetration of the considered types of projectiles into a given barrier. As a result of numerical calculations in all range of impact velocities from 1 to 12 km/s the obtained value of this quantity. From these data, calculate the coefficient of resistance $K=(V_0-V_K)/V_0$ of barriers for solid rod and segmented *Kc*, *Ks*. The degree of efficiency of a segmented impactor compared to the elongated cylinder is calculated by the formula Kef = (*Kc* - *Ks*)/*Kc*.

The final curves the final velocity of the penetration for solid drummer (curve 1) and segmented (curve 2) depending on the impact velocity is shown in fig. 14, there is shown a curve of the degree of efficiency (curve 3). The main conclusion from these results is that segmented the drummer is more effective compared with extended rod for the whole range of impact velocities. The efficiency of the segmented impactor increases with increasing velocity of impact in the range of impact velocities from 2 km/s to 8 km/s. At ultra high speeds, more then 10 km/s, the value of Kef is not significantly reduced. This is due to the evaporation of the front segments and the dispersal of the vaporized substance in the opposite direction. At high impact velocities (> 6 km/s) formula for Kef only partially describes the effectiveness of both types of drummers. In this region, the hydrodynamic law of penetration, and percussionists spread over the surface of the holes, so the final speed of penetration, in this case, it is possible to speak as about the average speed of the front portions of the drummer and obstacles.

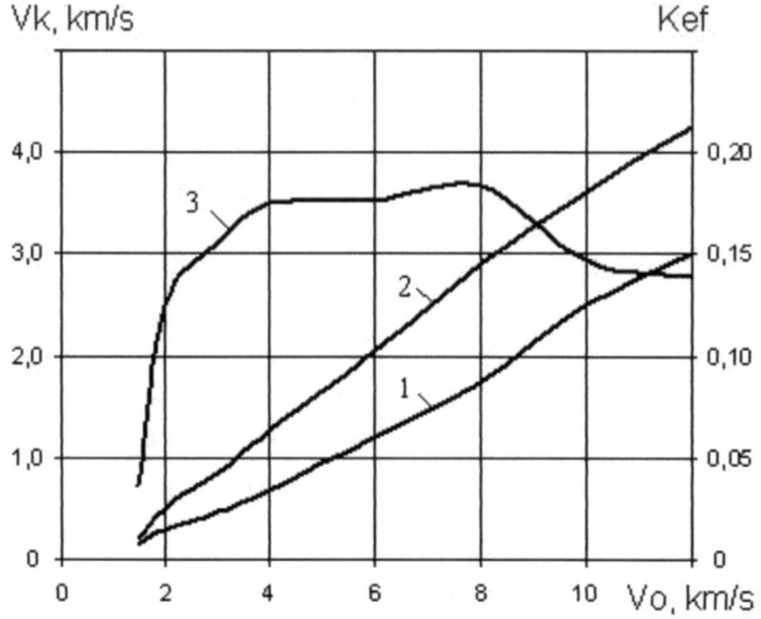

Fig. 14. Dependences of velocities breakthrough *Vk* (curves 1 and 2 for core and segmented drummer) and effectiveness Kef (curve 3) from the initial velocity of impact V_0.

References

1. V.G. Trushkov, E.N. Kramshonkov, E.A. Jagnitsina, J. News of high schools. Phys. **194** (2007)
2. V.G. Trushkov, *Numerical investigation of high-speed interaction objects of large deformation medium in the Euler space variables* (2007)
3. A.V. Zhukov, J. Def. Solid Mech. **70** (1987)
4. V.F. Tolkachyov, V.G. Trushkov, J. Chem. Phys. **2** (1993)
5. L.V. Efremova, A.I. Korneev, V. G. Trushkov, J. Phys. Comb. Expl. **2** (1987)
6. C.V. Razorenov, G.I. Kanel, E.N. Kramshonkov, K. Baumung, J. Phys. Comb. Expl. **5** (2002)
7. A. V. Kraynov, G. V. Kuznetsov, Web Conf. **76** (2014)
8. A.V. Kraynov, E.N. Pashkov, A.V. Ponomarev, J. Adv. Mat. Res. **1040** (2014)
9. G.V. Kuznetsov, A.V. Krainov, J. Appl. Mech. Tech. Phys. **5** (2001)
10. G.V. Kuznetsov, A.V. Krainov, J. High Temp. **2** (2003)
11. A.V. Kraynov, V.A. Arhipov, EPJ Web Conf. **110** (2016)
12. A.V. Kraynov, E.N. Pashkov, R.E. Lushnikov, V.A. Arkhipov, J. Key Eng. Mat. **685** (2016)
13. E.N. Kramshonkov, A.V. Kraynov, P.V. Shorokhov, EPJ Web Conf. **110** (2016)
14. H. Park, H. Kim, Int. J. Imp. Eng. **37** (2010)
15. P.M. Booker, J.D. Cargile, B.L. Kistler, V.La Saponara, Int. J. Imp. Eng. **36** (2009)
16. H. Wu, Q. Fang, Y. Peng, Z.M. Gong, X.Z. Kong, Int. J. Imp. Eng. **76** (2015)
17. Y.Y. Zhang, L. Chen, Int. J. Imp. Eng. **36** (2009)

18. T.T. Garfield, W.D. Richins, T.K. Larson, C.P. Pantelides, J.E. Blakeley, ICM'11 **10** (2011)
19. R.T. Moura, A.H. Clausen, E. Fagerholt, M. Alves, M. Langseth, Int. J. Imp. Eng. **37** (2010)
20. Hua Jiang, Xiaowo Wang, Shuanhai He, Mat. & Des. **39** (2012)
21. J.J. Rimoli, B. Talamini, J.J. Wetzel, K.P. Dharmasena, R. Radovitzky, H.N.G. Wadley, Int. J. Imp. Eng. **38** (2011)
22. T. Børvik, L. Olovsson, A.G. Hanssen, K.P. Dharmasena, H. Hansson, H.N.G. Wadley, J. Mech. Phys. Sol. **5** (2011)
23. C.J. Yungwirth, J. O'Connor, A. Zakraysek, V.S. Deshpande, H.N.G Wadley, J. Amer. Cer. Soc. **94** (2011)
24. Hamid Gheadnia, Ozdes Cermik, Dan B. Marghitu, Int. J. Imp. Eng. **86** (2015)
25. M. Scapin, L. Peroni, M. Peroni, Int. J. Imp. Eng. **40** (2012)

MATEC Web of Conferences **72**, 01052 (2016) DOI: 10.1051/matecconf/20167201052
HMTTSC-2016

ANALYSIS OF INFLUENCE OF HEAT TRANSFER CONDITIONS ON THE UPPER COVER TO HEAT TRANSFER IN THERMOSYPHON

Alexander Krasnoshlykov[1,*]

[1]National Research Tomsk Polytechnic University, 634050 Tomsk, Russia

Abstract. Numerical analysis of thermal conditions of a two-phase closed thermosyphon using the software package ANSYS FLUENT has been carried out. Temperature distributions at various heat transfer conditions on upper cover and different heat flow on bottom cover of thermosyphon have been obtained.

1 Introduction

Lithium-ion batteries play an important role as an source energy in a variety of applications (motor transport, aviation and space equipment) and have prospects for development through the use of less toxic materials, low weight, high energy and power density [1, 2]. However, a major obstacle to wide use of batteries of this type is the flammability and low performance at increasing temperature due to thermal runaway [1, 2].

Thermal runaway of the battery is the effect which occurs when the electrolyte temperature increases and the current flowing through the battery can rise to values approaching the short-circuit current [2]. Thermal runaway is able to affect the safety of the use of vehicles (e.g., aircraft [1]). Thermal runaway occurs at a combination of factors: the electrolyte temperature rise due to insufficient battery heat exchange with the environment; low level of electrolyte in the battery cell; aging active electrode mass [2].

Therefore, to maintain the temperature of the battery pack in a safe range for it is necessary to intensify the heat transfer from the outer surfaces.

Currently, system regulation thermal mode of batteries on the basis of the closed two-phase thermosiphons does not apply [3, 4]. Nowadays studies of thermal conditions thermosyphons at work with heat fluxes and the geometrical dimensions of the batteries used in vehicles (eg cars, aircrafts) has not been carried out.

It should be noted that the heat transfer characteristics of the two-phase closed thermosyphon affects a large number of factors [4]. One such factor is the heat transfer coefficient on the top cover of the thermosyphon. Experimental study of the regularities of heat transfer in thermosyphons often is problematic due to the multi-dimensionality of the factor space.

* Corresponding author: krasnas@tpu.ru

Mathematical modeling is the most rational method of investigation of thermal and hydrodynamic processes in thermosyphon.

There are mathematical models of physical processes occurring in the thermosyphon [5, 6]. But these models describe the work processes of heat transfer systems with a fairly significant assumptions on the mechanism of heat transfer. For example, in [5] accepted a model of the boundary layer flow in the vapor channel, and [6] discusses flow regimes at low speeds blowing vapor of the refrigerant from the heated surface and not taken into account a number of processes taking place in the vapor and liquid phases of the working fluid.

The purpose of work is mathematical modeling of heat transfer in two-phase closed thermosyphon at various conditions heat transfer on the top cover.

Accepted following scheme of the investigated process of heat transfer. The heat released during the technical elements of the system (e.g., battery) is supplied to the surface of the bottom cover of the thermosyphon (fig.1), as a result of the thermal conductivity of the coolant temperature increases. Near the surface of evaporation increases vapor pressure (P), vapor is moves upward and condenses on the inner surface of the top cover with the release of latent heat. The film of condensate flows down on the vertical walls of thermosyphon to heating zone.

2 Formulation of problem

Chart of the closed thermosyphon rectangular cross section considered in figure1.

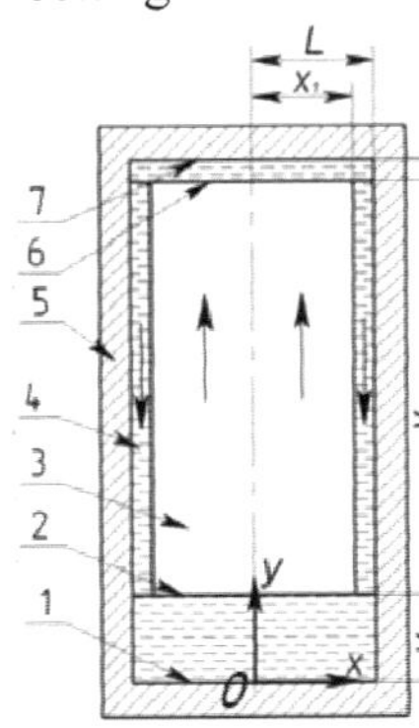

Fig. 1. Schematic diagram of the thermosyphon: 1 – bottom cover; 2 – surface evaporation; 3 – vapor flow; 4 – liquid film; 5 – vertical wall; 6 – condensing surface; 7 – top cover.

The equations of continuity, momentum and energy to be solved for the steam and the liquid film in the study area are of the form:

$$\frac{\partial \rho}{\partial t}+\frac{\partial(\rho u)}{\partial x}+\frac{\partial(\rho \upsilon)}{\partial y}=0; \tag{1}$$

$$\frac{\partial \rho_1 u_1}{\partial t}+\frac{\partial(\rho_1 u_1 u_1)}{\partial x}+\frac{\partial(\rho_1 \upsilon_1 u_1)}{\partial y}=\rho_1 g_x-\frac{\partial P_1}{\partial x}+\frac{\partial}{\partial x}\left(\mu_1\frac{\partial u_1}{\partial x}\right)+\frac{\partial}{\partial y}\left(\mu_1\frac{\partial u_1}{\partial y}\right); \tag{2}$$

$$\frac{\partial \rho_1 \upsilon_1}{\partial t}+\frac{\partial(\rho_1 u_1 \upsilon_1)}{\partial x}+\frac{\partial(\rho_1 \upsilon_1 \upsilon_1)}{\partial y}=-\frac{\partial P_1}{\partial y}+\frac{\partial}{\partial x}\left(\mu_1\frac{\partial \upsilon_1}{\partial x}\right)+\frac{\partial}{\partial y}\left(\mu_1\frac{\partial \upsilon_1}{\partial y}\right); \tag{3}$$

$$\frac{\partial \rho_2 u_2}{\partial t}+\frac{\partial(\rho_2 u_2 u_2)}{\partial x}+\frac{\partial(\rho_2 \upsilon_2 u_2)}{\partial y}=\rho_2 g_x-\frac{\partial P_2}{\partial x}+\frac{\partial}{\partial x}\left(\mu_2\frac{\partial u_2}{\partial x}\right)+\frac{\partial}{\partial y}\left(\mu_2\frac{\partial u_2}{\partial y}\right); \tag{4}$$

$$\frac{\partial \rho_2 \upsilon_2}{\partial t} + \frac{\partial(\rho_2 u_2 \upsilon_2)}{\partial x} + \frac{\partial(\rho_2 \upsilon_2 \upsilon_2)}{\partial y} = \frac{\partial P_2}{\partial y} + \frac{\partial}{\partial x}\left(\mu_2 \frac{\partial \upsilon_2}{\partial x}\right) + \frac{\partial}{\partial y}\left(\mu_2 \frac{\partial \upsilon_2}{\partial y}\right); \tag{5}$$

$$\rho_1 C_{p1}\left(\frac{\partial T_1}{\partial t} + u_1 \frac{\partial T_1}{\partial x} + \upsilon_1 \frac{\partial T_1}{\partial y}\right) = \lambda_1\left(\frac{\partial^2 T_1}{\partial x^2} + \frac{\partial^2 T_1}{\partial y^2}\right); \tag{6}$$

$$\rho_2 C_{p2}\left(\frac{\partial T_2}{\partial t} + u_2 \frac{\partial T_2}{\partial x} + \upsilon_2 \frac{\partial T_2}{\partial y}\right) = \lambda_2\left(\frac{\partial^2 T_2}{\partial x^2} + \frac{\partial^2 T_2}{\partial y^2}\right). \tag{7}$$

Where u, υ – the velocity components in the projection on the axis x, y, respectively; P – pressure; T – temperature; ρ – density; x, y – Cartesian coordinates; t – time; Cp – heat capacity; g – acceleration of gravity; λ – coefficient of thermal conductivity; μ – dynamic viscosity; indices 1, 2 – properties of the liquid and vapor.

The initial conditions for the system of equations (1-7) are given in the form:

$u(x,y)=0$; $T(x,y)=T_0$; $P(x,y)=P_0$.

The boundary conditions for the equations (1-7) have the form:

$$x = 0, 0 < y < H, \quad \frac{\partial T_2}{\partial x} = 0; \quad \frac{\partial u_2}{\partial x} = 0, \ \frac{\partial \upsilon_2}{\partial x} = 0,$$

$$x = L, \ 0 \le y \le H, \quad \lambda_1 \frac{\partial T_1}{\partial x} = 0; \quad u_1 = \upsilon_1 = 0,$$

$$x = x_1, y_1 \le y \le y_1 + y_2, \quad \begin{cases} T_1 = T_2 \\ \lambda_1 \dfrac{\partial T_1}{\partial x} = \lambda_2 \dfrac{\partial T_2}{\partial x}; \end{cases} \quad \begin{cases} u_1 = u_2 = 0 \\ \upsilon_2 = 0 \end{cases}, \ \frac{\partial \upsilon_1}{\partial x} = 0,$$

$$y = y_1, \ 0 \le x \le x_1, \quad \begin{cases} T_1 = T_2 \\ \lambda_1 \dfrac{\partial T_1}{\partial y} = \lambda_2 \dfrac{\partial T_2}{\partial y} - Q_e w_e - \upsilon_2 C_p \rho (T_2 - T_0), \end{cases} \quad \upsilon_2 = \frac{w_e}{\rho_2}, \ \upsilon_1 = \frac{w_e}{\rho_1},$$

$$y = y_1 + y_2, \ 0 \le x \le x_1, \quad \begin{cases} T_2 = T_1 \\ \lambda_2 \dfrac{\partial T_2}{\partial y} = \lambda_1 \dfrac{\partial T_1}{\partial y} + Q_c w_c + \upsilon_2 C_p \rho (T_1 - T_0), \end{cases} \quad \upsilon_1 = \frac{w_c}{\rho_1}, \ \upsilon_2 = 0,$$

$$y = 0, \ 0 \le x \le L, \ \lambda_1 \frac{\partial T_1}{\partial y} = q_h, \ u_1 = \upsilon_1 = 0,$$

where α – heat transfer coefficient, W/(m²·K).

Mathematical modeling of the task carried out in the package ANSYS FLUENT software. Mass rate of evaporation and condensation were calculated according to the formula [7]:

$$w_c = \beta \sqrt{\frac{M}{2\pi R T_H}} (P_2 - P_H),$$

where P_H – saturation pressure; T_H – saturation temperature; R – universal gas constant; M – molar weight; β – accommodation coefficient.

DOI: 10.1051/matecconf/20167201052

3 Results and Discussion

Numerical simulation of heat transfer in a two-phase closed thermosyphon rectangular cross section (figure 1) with the following geometric dimensions: transverse dimension L= 0.05 m; the longitudinal dimension H=0.3 m has been carried out. To investigate the effect of heat transfer conditions on the performance of the thermosyphon have been taken the following variations the heat transfer coefficient (α) on the top cover: 9.6 W/(m²·K); 15.6 W/(m²·K); 25.6 W/(m²·K). The heat flux density on the lower cover of thermosyphon in section y=0: $1.1\cdot10^4$ W/m^2; $1.7\cdot10^4$ W/m^2; $2.3\cdot10^4$ W/m^2. Water was considered as a working liquid.

The obtained distribution of temperature show that at q_h=$1.1\cdot10^4$ W/m^2 (fig. 2a) and the heat transfer coefficient α=9.6 W/(m^2·K) (curve 1) the difference T between the lower and upper covers of thermosyphon is about 6 °K. At changing α the top cover in the range of 15.6 W/(m^2·K) (curve 2) to 25.6 W/(m^2·K) (curve 3) the temperature difference (ΔT) in the test area is greater than α = 9.6 W/(m^2·K) and is 7.5 K and 8 K, due to the intense outflow of heat on the top cover of the thermosyphon.

Figures 2 and 3 show that increasing the heat flow density in the range of $1.1\cdot10^4$ W/m^2 to $2.3\cdot10^4$ W/m^2 leads to an increase ΔT by about 2°K. At q_h=$1.7\cdot10^4$ W/m^2 on the bottom cover of the study device (fig. 2b) temperature difference is 4°K, 5.5°K, 6°K, according to variants the heat transfer coefficients on the top cover, and at q_h=$2.3\cdot10^4$ W/m^2 (fig. 3) – 2°K, 3.5°K, 4°K.

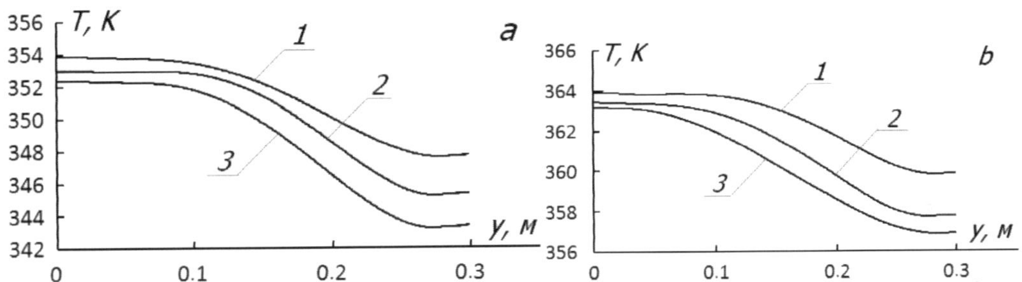

Fig. 2. The temperature distribution in the direction y: **a** – q_h=$1.1\cdot10^4$ W/m^2; **b** –q_h=$1.7\cdot10^4$ W/m^2. 1 –α=9.6 W/(m^2·K); 2 – α=15.6 W/(m^2·K); 3 – α=25.6 W/(m^2·K).

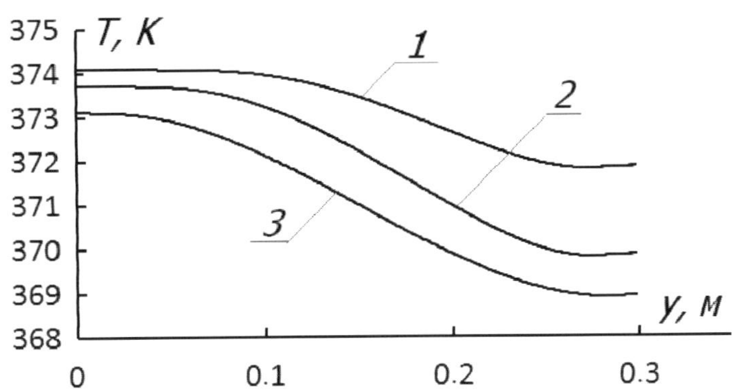

Fig. 3. The temperature distribution in the direction y at q_h=$2.3\cdot10^4$ W/m^2:1 – α=9.6 W/(m^2·K); 2 – α=15.6 W/(m^2·K); 3 – α=25.6 W/(m^2·K).

4 Conclusion

The dependence of the temperature difference between the upper and lower covers of a closed two-phase thermosyphon from different conditions of intensity heat transfer at the upper cover have been obtained. It is shown that an increase α in range 9.6–25.6 W/(m^2·K) leads to an increase ΔT to 2°K.

Acknowledgments

The work was held within the research state assignment "Science" No. 13.1339.2014/K (Code of Federal Target Scientific and Technical Program 2.1410.2014).

References

1. T. Song, Y. Li, J. Song, Z. Zhang, Proc. Eng. **80** (2014)
2. D.A. Khrustalyov, *Rechargeable batteries* (Emerald, Moscow, 2003)
3. J.L.G. Oliveira, C. Tecchio, K.V. Paiva, M.B.H. Mantelli, R. Gandolfi, L.G.S. Ribeiro, Appl. Th. Eng. **79** (2015)
4. M.K. Bezrodnyi, I.L. Pioro, T.S. Kostyuk, *Transfer processes in the two-phase thermosyphon systems* (Kiev, 2005)
5. G.V. Kuznetsov, A.E. Sitnikov, High Temp. (2002)
6. G.V. Kuznetsov, M.A. Al-Ani, M.A. Sheremet, J. Eng. Thermoph. (2011)
7. Ansys Help. FLUENT Theory Guide

MATEC Web of Conferences **72**, 01053 (2016) DOI: 10.1051/matecconf/20167201053
HMTTSC-2016

CONVECTION STUDY BY PIV METHOD WITHIN HORIZONTAL LIQUID LAYER EVAPORATING INTO INERT GAS FLOW

Aleksei Kreta[1, 2, *], and *Yuriy Lyulin*[2, 3]

[1]Novosibirsk State University, Novosibirsk, Russia
[2]Kutateladze Institute of Thermophysics SB RAS, Novosibirsk, Russia
[3]Tomsk Polytechnic University, Tomsk, Russia

Abstract. The paper is devoted to the experimental study of convection in a horizontal evaporating liquid layer (ethanol) of limited size under the action of gas flow (air). The two-dimensional velocity field in the liquid layer is obtained using the PIV method. The existence of a vortex convective flow within a liquid layer directed towards the gas flow has been revealed.

1 Introduction

Studying heat and mass transfer through the interface between liquid and gas is one of the most important problems nowadays. In the energy devices, the considered processes are implemented in mini- and microchannels of the apparatus such as heat pipes, film evaporators, two-phase cooling systems, etc. The intensive evaporating from the liquid layer surface under the gas flow causes various convective flows in the liquid layer. The shape and structure of the convection is quite complicated and can have a significant impact on the evaporation rate and hence the efficiency of heat exchangers. The convection flows within the evaporating liquid layer into inert gas, are actively studied experimentally [1-4] and theoretically [4-7] in the present time. Nevertheless works in this area is extremely insufficient for comprehensive understanding of the physics of processes related to heat and mass transfer through the interface between liquid and gas, as well as convection in the liquid layer.

The aim of this work is to experimentally study the structure of convective flows and to measure two-dimensional velocity field distribution in the liquid layer under the action of the gas flow using PIV method.

2 Experimental setup

The experimental studies were conducted on the setup shown in fig. 1. The setup consists of the following main components: test section, gas and liquid loops, data acquisition system,

* Corresponding author: alexy.slav@yandex.ru

thermal stabilization system, and optical techniques. Pure gas flows from the compressor to the test section. Gas mass flow rate at the inlet of the channel in the test section is set by means of mass flow controller. The liquid is injected into the test section by a precision syringe pump. The working liquid evaporates under the gas flow and the vapor-gas mixture flows to the outlet of the gas channel in test section. The vapor-gas mixture flow at the outlet is measured by a mass flow meter. The liquid temperature in the test section is determined by the temperature of the substrate base, which controlled with the help of thermoelectric module. The difference between the temperature of the gas and liquid does not exceed 0.1 ^{0}C. The position of the interface level is controlled with an accuracy of 10 micrometers using a Schlieren technique and the syringe pump [1]. Recording of the convection flows in the liquid layer was carried out by the PIV technique.

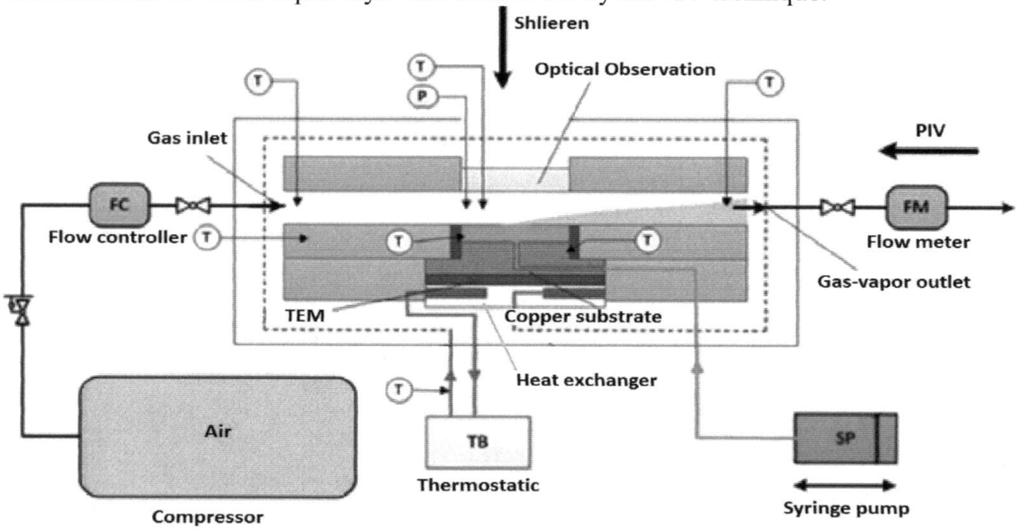

Fig. 1. Schematic diagram of the experimental setup.

The test section consists of several interconnected blocks. These blocks form a rectangular gas channel with a height of 3 mm and a width of 40 mm, as well as a rectangular 40x40 mm2 liquid chamber. A stainless steel plate with a square cut-out at the center is installed between the gas channel and the chamber with a liquid. The walls of the liquid chamber are made of transparent Plexiglas that gives the possibility to use PIV technique. The height of the liquid in the chamber varies from 1 to 10 mm and can be adjusted by means of a liquid substrate and special plates regulating the height. The cover of the test section is provided with transparent optical window installed above the liquid chamber.

2.1 PIV measurements

Recording of the convective structures in the liquid layer under action of the gas flow is carried out using the Particle Image Velocimetry (PIV) method. The schematic diagram of the method is shown in fig. 2. The method allows to determine the shape, velocity of the vortex structures in the liquid layer depending on various system parameters such as gas flow velocity, liquid temperature, the height of the liquid layer, etc. Detailed description of PIV method is given in [8].

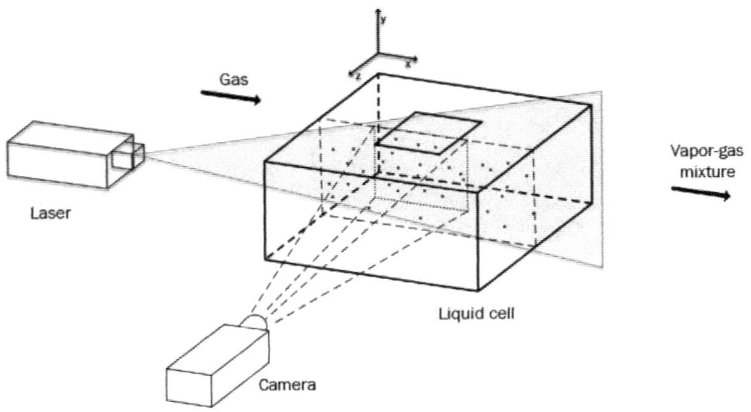

Fig. 2. Schematic diagram of the PIV system.

Instantaneous two-component velocity field is measured in the central cross section of the liquid layer directed towards the gas flow. In the beginning of the experiment, the particles (tracers) with a size of 5 microns are added to the liquid. The size and density of the tracers are chosen in such a way as to avoid the influence of the particles on the liquid flow pattern. A pulsed laser creates a thin (300 μm) light knife and illuminates small particles. The particles positions at the time of two consecutive laser flashes are recorded on two frames of a digital camera. A digital camera is located completely perpendicular to the light sheet. The flow velocity is determined by calculating a displacement of the particles during the time between the laser flashes. The determination of the particles displacement is based on application of correlation methods to the images of a tracer particle using a regular decomposition into elementary regions. The data processing is carried out using the "ActualFlow" software program. The program implements two main methods for image processing: the standard and the iterative cross-correlation methods [9]. The standard cross-correlation method is based on the employment of fast Fourier transform using the correlation theorem. However, for a more accurate calculation of the velocity vector is used an iterative cross-correlation method for calculation of two-component velocity field. The basic idea of the method consists in multiple processing of one and the same area using the result of processing at the previous iteration to estimate the processing parameters for the next iteration. As a consequence of this approach, it is possible to obtain a significant increase in the accuracy of the obtained data [9]. The cross-correlation method is used as the primary method to calculate two-component velocity field.

3 Results

The experiments were conducted at atmospheric pressure in a test section, and a stationary 3 mm thick liquid layer. Ethanol was used as the working liquid. The evaporation surface area was 100 mm^2 with the corresponding 10x10 mm cut-out in the plate. The gas flow rate ranged from 100 to 1000 ml/min with increment of 100 ml/min that corresponds to gas average velocity from 0.0139 m/s (Re=2.8) to 0.138 m/s (Re=28), respectively. The temperature of the "liquid-gas" system was maintained constant at a level of 20°C with a deviation less than 0.1oC.

Fig. 3 represents a two-component velocity field for a minimal (100 ml/min) and maximal (1000 ml/min) gas flow rates. The calculation of the average velocity of convective structures is conducted as follows. First, the area with a liquid vortex is determined visually. Next, a rectangular area with this vortex is selected in the "ActualFlow" program. The program displays a table with the coordinates and velocity projections along the axes x and y for each particle in the selected area. Particles with extremely low velocity vectors are removed from the table. Removed vectors are located near borders of the measuring area and outside of the liquid convective vortex. The velocity of each particle is determined by the formula:

$$V=\sqrt{(V_x)^2+(V_y)^2}$$

where, V_x and V_y – are x and y components of the velocity vector. The velocity of the convective vortex is calculated as the average value of the particle velocities in the measuring area.

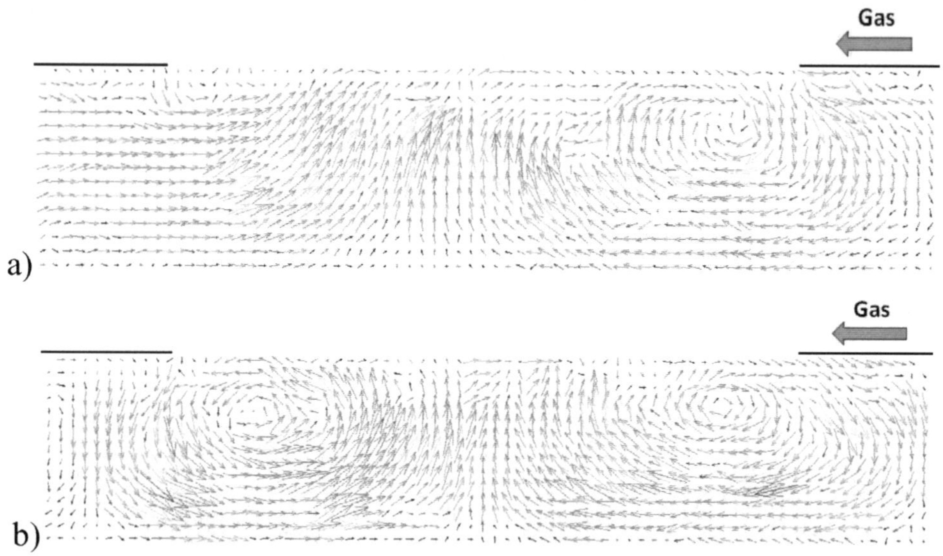

Fig. 3. Two-component velocity field. Liquid - ethanol. The layer thickness is 3 mm; a) gas flow rate is 100 ml/min (Vgas=0.0139 m/s, Re=2,8), b) gas flow rate is 1000 ml/min (Vgas=0.138 m/s, Re =28).

Fig. 3 (a) demonstrates the convective vortex at the initial section of the contact between liquid and gas. The direction of the vortex motion is opposite to the direction of the gas flow. The average velocity of the vortex is 0.18 m/s. Direction of a liquid flow opposite to the gas flow is due to the dominant influence of thermocapillary forces caused by the intensive evaporation at the interface. The gas flow cools the liquid surface causing a temperature gradient in the liquid layer that forces liquid to move against the gas flow. Fig. 3 (b) presents a two-component velocity field at a gas flow rate of 1000 ml/min. It is found the formation of two symmetrical and inverse convective vortexes within the liquid layer at the maximum gas flow rate. The first vortex at the initial area of the gas-liquid contact is moving in counter-current direction of the gas flow. Governing factor of the convective vortex is the thermocapillary effect as in the previous case (the gas flow rate of 100 ml/min). The second vortex moves in the same direction as the gas flow. This liquid movement occurs due to the action of the gas shear stress and the effect of the first vortex structure. The velocities of the first and second vortex structures are equal to 0.2412 and 0.2416 m/s, respectively.

4 Conclusion

Convection study by PIV method in a horizontal liquid layer evaporating into the gas flow has been performed. Effect of gas flow on the convective flows in the liquid layer has been considered. It is shown that the increase of the gas flow velocity induces the number of convective vortex in horizontal evaporating liquid layer and growth of its velocities. For minimum gas flow rate (100 ml/min), only one convective structure moving against the gas flow was observed. When increasing the gas flow rate to 1000 ml/min, another convective structure appears and moves along with the gas flow. Governing factor of the convective vortex moving in counter-current direction of the gas flow is the thermocapillary effect due to the intensive evaporation and the temperature decreasing near the attack border of the liquid.

Acknowledgments

The work was financially supported by the Russian Ministry of Education and Science (Project identifier RFMEFI61314X0011).

References

1. O.A. Kabov. Int. J. of Em. Multidisc. Liq. Sc. **2** (2010)
2. Yu.V. Lyulin, D.V. Feoktistov, I.A. Afanasiev, E.S. Chachilo, O.A. Kabov, G.V. Kuznetsov, Let. Tech. Phys. **41** (2015)
3. Yu.V. Lyulin, O.A. Kabov, Int. J. Heat Mass Transf. **70** (2014)
4. B. Scheid, J. Margerit, C.S. Iorio, L. Joannes, M. Heraud, P. Queeckers, P.C. Dauby, P. Colinet, Exp. Liq. **52** (2012)
5. O.N. Goncharova, O.A. Kabov, Int. J. Heat Mass Transf. **53** (2010)
6. O.N. Goncharova, O.A. Kabov, Int. J. Heat Mass Transf. **53** (2010)
7. H. Machrafi, C. S. Iorio, P.C. Dauby, Int. Phen. Heat Transf. **2** (2014)
8. H. Machrafi, C.S. Iorio, P.C. Dauby, Int. Phen. Heat Transf. **2** (2014)
9. C. Brossard, J. Monnier, P. Barricau, Opt. Diagn. Fl. (2009)

MATEC Web of Conferences 72, 01054 (2016) DOI: 10.1051/matecconf/20167201054
HMTTSC-2016

THE INFLUENCE OF GAS-PHASE HYDROGENATION PARAMETERS ON THE PROCESSES OF ACTIVATION, SORPTION AND ACCUMULATION OF HYDROGEN IN THE POWDER OF PURE TITANIUM

*V.*N. Kudiiarov[1, *], *M.*N. Babihina[1], and *D.*V. Gvozdyakov[1]

[1]National Research Tomsk Polytechnic University, 634050 Tomsk, Russia

Abstract. The technique for hydrogenation of powder hydrogen-storage materials based on gas-phase hydrogenation method has been developed in this work. The using of developed technique allow prepare pure titanium powder samples with hydrogen concentration in range from 0.5 to 4 wt.%. It was showed, that pressure increasing from 0.5 to 2 atm at constant temperature 400 °C leads to decreasing of powder activation time from 140 to 15 s. Temperature increasing on 50 °C and more degrees leads to significant decreasing of powder activation time. Pressure increasing from 0.5 to 2 atm leads to increasing of hydrogen sorption rate and concentration in 1.27 and 1.19 times respectively. Temperature increasing from 400 to 550 °C leads to increasing of hydrogen sorption rate in 1.3 times, but decreasing of hydrogen concentration in 1.45 times, which associated with decreasing of hydrogen saturation limit. Optimal values of temperature (400 °C) and pressure (2 atm) for pure titanium powder hydrogenation have been determined.

1 Introduction

One of the important tasks in the hydrogen energy is a task to develop effective hydrogen-storage materials (HSM). The selection of the optimal composition of the HSM and the synthesis parameters is carried out on the basis of data on the interaction of hydrogen with obtained materials. HSM should have high rates of hydrogen adsorption and desorption in the operating conditions, low activation time of adsorption and desorption, cycling stability [1]. Gas-phase hydrogenation is used for evaluation all these properties of synthesized HSM. Hydrogenation parameters such as temperature and pressure play an essential role in the hydrogen interaction with materials [2].

In this work technique for hydrogenation of powder HSM has been developed. Pure titanium powder was changed as a model material as soon as titanium is an important element of many HSM. The influence of hydrogenation parameters on the activation time,

* Corresponding author: viktor.kudiiarov@gmail.com

the rate of hydrogen sorption and absolute hydrogen concentration in the powder was investigated.

2 Materials and Methods

Pure titanium powder was used for investigation and it was put into cuprum capsule with quartz wool for isolation from wall of reaction chamber. Experimental chamber was put into automated complex Gas Reaction Controller LPB by Advanced Materials Research [3]. Then chamber was pumped out and heated with heating rate 6 °C/min for needed temperature. The temperature diapason was from 400 to 550 °C with step 50 °C. Chamber was filled by hydrogen for needed pressure after heating. Pressure diapason was from 0.5 to 2 atm with step 0.5 atm.

Titanium powder was kept in hydrogen medium at constant temperature and pressure until full hydrogenation and then slowly cooled with rate 2 °C/min in the hydrogen medium. Hydrogen concentration in powder was measured by volumetric method in the automated complex Gas Reaction Controller and with the help of hydrogen analyzer RHEN602 by LECO.

3 Results and Discussion

Hydrogen sorption curves by titanium powder at temperature 400 °C and pressure 0.5 atm (a) and at temperature 550 °C and pressure 2 atm (b) are shown on the figure 1.

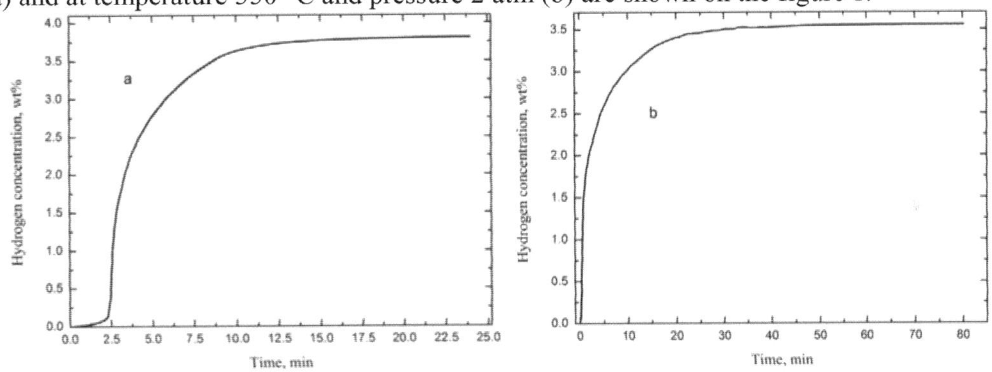

Fig. 1. Hydrogen sorption curves by titanium powder at temperature 400 °C and pressure 0.5 atm (a) and at temperature 550 °C and pressure 2 atm (b).

It is seen from figure 1a that at 400 °C there is a plateau-like section in a sorption curve that is associated with the activation of titanium powder. Than with pressure increasing the time of activation is reduced (figure 2a, the solid line) and at 2 atm it becomes negligible. Powder activation apparently occurs in the first few seconds of the hydrogen inlet into the chamber at hydrogenation at high temperatures (including 550 °C), since there is not the plateau-like section in the sorption curves (figure 1b).

Hydrogen sorption rates have been calculated based on the sorption curves depending on the pressure at 400 ° C (figure 2a, the dotted line) and depending on the temperature at a constant pressure of 2 atm. (figure 2b).

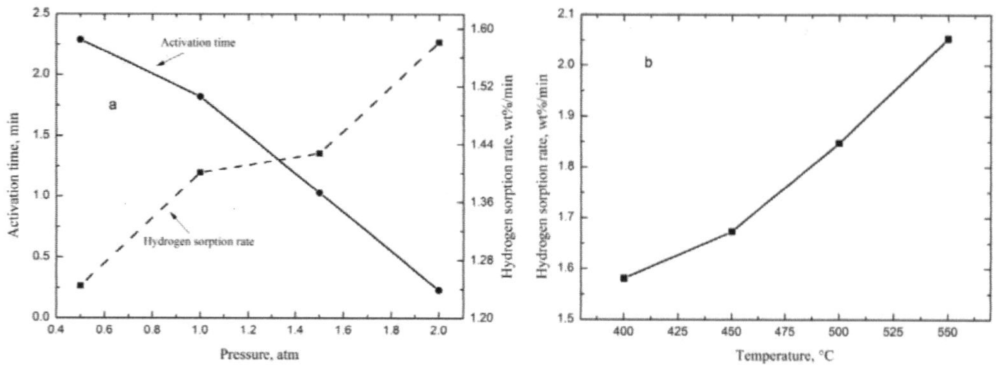

Fig. 2. The dependence of powder activation time (the solid line) and hydrogen sorption rate (the dotted line) on the pressure at constant temperature 400 °C (a) and dependence of hydrogen sorption rate on temperature at constant pressure 2 atm (b).

The opposite effect is observed for the dependence of the sorption rate on pressure and temperature. Pressure and temperature increasing leads to the hydrogen sorption rate increasing. However, if the pressure increases, a slight increase in the absolute hydrogen concentration is observed in the material after the hydrogenation, but the temperature increasing leads to the significant reduction of hydrogen concentration (Table 1), which is associated with a reduction of titanium saturation limit with temperature increasing [4].

Table 1. Hydrogen concentration in titanium powder after hydrogenation at different parameters.

Temperature, °C	Pressure, atm.	Hydrogen concentration, wt.%
400	0.5	3.82
400	1	3.87
400	1.5	3.93
400	2	4.01
450	2	3.92
500	2	3.78
550	2	3.51

The optimal parameters of hydrogenation (temperature and pressure) are necessary to determine when working with titanium-based HSM for balance between the rate of hydrogen sorption and absolute concentration. Temperature of 400 ° C and a pressure of 2 atm are optimal for the pure titanium powder.

Preparation of the experimental samples of titanium powder with different hydrogen concentrations in the range of 0.5 to 4 wt.% have interest for different research. For this reason hydrogenation was performed manually at a constant temperature of 400 ° C and hydrogen manual admitted into the chamber. Experimental samples of pure titanium powder with concentrations ranging from 0.5 to 4 wt.% were prepared By varying the amount of hydrogen in the reaction chamber.

4 Conclusion

The technique for hydrogenation of powder hydrogen-storage materials based on gas-phase hydrogenation method has been developed in this work. The pure titanium powder was changed for investigation as a model material. The using of developed technique allow prepare samples with hydrogen concentration in range from 0.5 to 4 wt.%. In additional, pressure and temperature influence on powder activation time, hydrogen sorption rate and hydrogen concentration was investigated. It was showed, that pressure increasing from 0.5 to 2 atm at constant temperature 400 °C leads to decreasing of powder activation time from 140 to 15 s. Temperature increasing on 50 and more degrees leads to significant decreasing of powder activation time.

Pressure increasing from 0.5 to 2 atm leads to increasing of hydrogen sorption rate and concentration in 1.27 and 1.19 times respectively. Temperature increasing from 400 to 550 °C leads to increasing of hydrogen sorption rate in 1.3 times, but decreasing of hydrogen concentration in 1.45 times, which associated with decreasing of hydrogen saturation limit.

Thereby, hydrogenation parameters play important role at working with titanium-based hydrogen-storage materials and it optimal values should be determined for balance between hydrogen sorption rate and hydrogen concentration. Temperature 400 °C and pressure 2 atm are for pure titanium powder such parameters.

Acknowledgements

This work was funded within the framework of realisation of Strategic Programme on National Research Tomsk Polytechnic University Competitiveness Enhancement in the Group of Top Level World Research and Academic Institutions.

References

1. B.A. Kolachev, P.E. Shalin, A.A. Ilin, *Hydrogen-storage alloys* (Metallurgy, Moscow, 1995)
2. L. Schlapbach, A. Züttel, Nat. **414** (2001)
3. V.N. Kudiiarov, A.M. Lider, S.Y. Harchenko, Adv. Mater. Res. **880** (2014)
4. I.P. Chernov, Yu. P. Cherdantsev, Yu. I. Tyurin, *Methods for metal-hydrogen systems investigation* (Energoatomisdat, Tomsk, 2004)

FIELD RESEARCH OF FIRING STORED PEAT CRITICAL CONDITIONS

Roman N. Kulesh[1, *], and *Ksenia* Y. Orlova[1]

[1]National Research Tomsk Polytechnic University, 634050 Tomsk, Russia

Abstract. This paper presents field research: influence of different factors on peat massif firing with local outside heat source. Minimum temperatures of defined size heat source which fires the peat under the wide range of moisture value and heat-exchange conditions with terrestrial environment were found. Besides, peat massif moisture values whereby no firing is undergoing within the range of the researched temperatures of ignition source were defined.

1 Introduction

Today, world's peat reserves are estimated at 186 billion tons. Great reserves and their constant renewal make peat one of the most important and strategically valuable feedstock, especially taking in consideration finiteness and exhaustibility of traditional fossil fuels such as oil, natural gas and coal [1].

Peat processing technologies and its intake in different economical industries are well known and widely proven. Prevention of peat firing during drying and storage is one of the distinctive features of the peat industry [2]. Whole technology of peat consumption from digging to ultimate product could be shown as a scheme: drainage of the peat deposit, digging, storage of the produced peat, transportation, long term warehouse storage, peat consumption as a fuel. There is a list of operations involved mostly in peat digging and preparation for usage process which are characterized as special fire risk operations due to peat humidity loss. Peat is easily self-ignite and combustible material, which might be inflamed with a relatively small ignition source [3, 4]. Often peat self-ignition is observed during its storage [5].

The main reason of stored peat firing is presence of an outside heat source. The mechanism of this process is defined firstly by appearance of separate fires (electric, mechanic or thermal sparks) at the incipient stage which fall on peat surface. In this case, if the thermal source has a sufficient energy, peat firing and deepening of the combustion source might occur. When this happens, firefighting becomes very problematical and in some cases even impossible.

Literature observation [6-15] showed that peat inflammability problem naturally and while stored is an important but understudied issue. It is possible that this might be the reason which stands on a way of practical time effective peat fire prevention and firefighting problem solution. This fact once again highlights the importance of this

* Corresponding author: ronikul@tpu.ru

research paper. Heat and mass transfer processes in inhomogeneous medium numerical modeling [6, 7] might be used for the purpose of the problem solvation. In order to do that one should know truth peat and peat pyrolytic decomposition products density and thermophysical properties. Unfortunately, literature observation did not allow to find mentioned above density and thermophysical properties truth value.

2 Physical statement of the problem

Peat is usually stored as defined size briquettes (peat stratums), for that reason the research handles the problem of peat briquette firing. Peat with density of 370 kg/m^3 is placed into a cylindrical cuvette with punched side surface. Cylindrical heat source with radius r_0 is placed on top of a peat stratum. The goal is to define firing parameters. The center of the heat source base is considered as an origin of coordinates. X and Y axis oriented along the top surface, and axis Z is vertical to the top peat surface.

The experimental variables of peat firing are firing fall time and minimum ignition source temperature, sufficient to initiate an ignition. During the experiment several thermocouples are places at different depth z under each heat source base. They trace temperature of certain peat stratum spots during the time of the experiment.

3 Characterization of peat firing parameters

It's well known, moisture has a great influence on ignition and combustion processes. Therefore, hot body ignition of peat samples with different moisture value experiments, were conducted. Research results of heat exchange processes taken place in peat layer when it had been fired with local heat source are shown on fig.1. Peat firing fall time (t_3) with different moisture value was obtained. All experiments were held with minimum reserved energy of an ignition source which ensures peat firing. Knowing onset temperature and sizes of an ignition source it is possible to find onset heat source energy. For that reason onset and current ignition source temperature and additionally temperature of certain peat stratum spots were measured during experiments. Presented characteristic curves show onset temperature is in accord with maximum (start) temperature, curve 1. Other curves on figure 1 show change in temperature of certain peat stratum spots with determined seats.

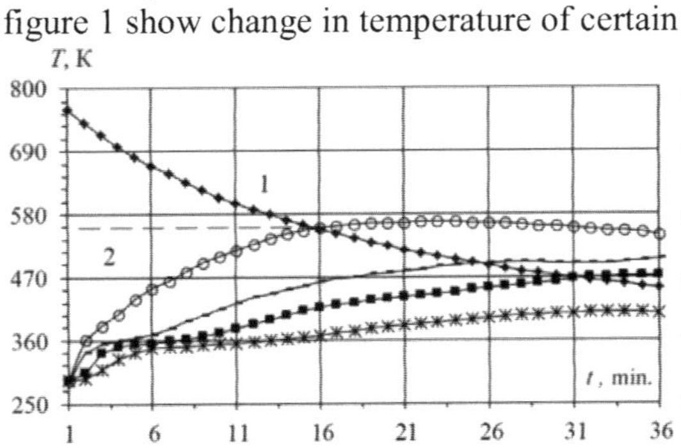

Fig. 1a. Change in temperature inside peat massif with moisture 10%.

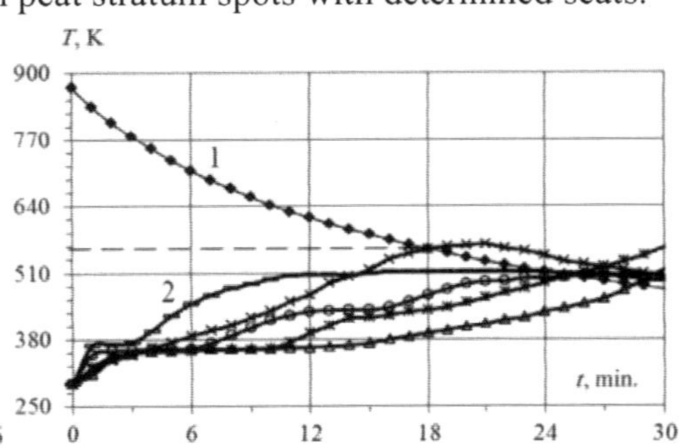

Fig. 1b. Change in temperature inside peat massif with moisture 20%.

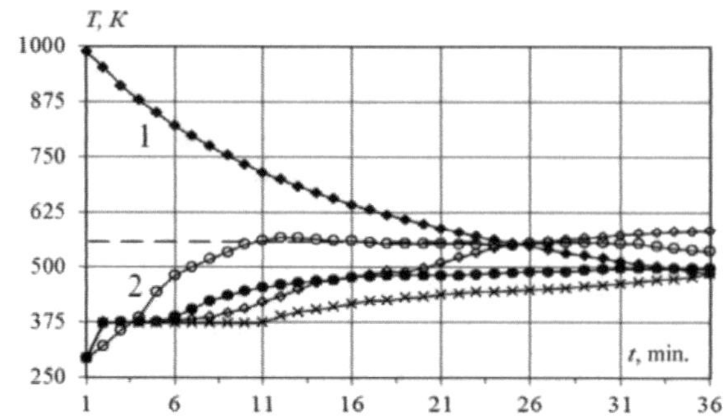

Fig. 1c. Change in temperature inside peat massif with moisture 30%.

Due to the high curves density there are only several results of the experiment with typical temperature changes in relatively remote spots shown on figures above. Dashed line characterizes ignition temperature, and dashed line and curve 1 intersection point characterizes ignition time. Figure 1 functional connections show that the higher peat moisture the longer the ignition time. The moment the ignition source touches peat surface ignition temperature starts dropping monotonely (curves 1). Upon that ignition source directs its energy as to peat layer, through thermal conductivity and radiation, as to terrestrial environment (near the ground air). This process leads to rise of peat bed massif temperature which tuned in by thermocouples inside the massif. The process stays in progress till the moment when ignition source temperature at any peat field research fixed spots becomes equal. Temperature recorded at the moment is peat ignition temperature. Next stage of the experiment shows further source temperature dropping.

It is necessary to notice that peat ignition moment at any spot on the surface local temperature rise is recorded. This fact is an additional proof of peat firing presence.

Obtained temperatures lets trace the development of combustion. Figure 1b shows combustion source traveling from one controlled area to another where temperature rises locally at 21 minutes in spot where combustion was absent till now. Further on combustion source moves to another controlled area. This results with temperature dropping in area of interests.

From functional connections above follows that the higher the peat moisture the longer the moisture vaporation process (vaporation takes place during first 3-9 minutes of the experiment). Moisture vaporation cools off reaction area intensively and results in its reduction in sizes due to oxygen deficit for combustion reaction from one side, and due to vaporation energy consumption. Whereas evaporated moisture filtration in peat massif pores leads to fullest peat layer exposure to heat (fig. 1c). This effect, shown by ΔT curves dispersion decrease, characterizes temperatures inside peat massif (fig. 3a – $\Delta T \approx 200$ K, fig. 3c – $\Delta T \approx 100$ K).

The set of experiments with thermal insulated heat source were done to evaluate influence of ignition source and terrestrial environment heat exchange. For that reason ignition source was covered with mullite thermal isolation layer. In these conditions it's fair to assume that the ignition source transfer its energy inside peat massif only. Thermal losses through thermal isolation in this case are neglected. Experiment results with outside heat exchange conditions changes (ignition source thermal isolation) are shown on figure 2. Thermal isolation temperature was recorded in order to evaluate the ignition source thermal losses through isolation (curve 3 on figure 2).

The set of experiments is characterized by more gradual ignition source temperature decrease (as a result of thermal losses to terrestrial environment minimization) and ignition fall time decrease. This property most likely is a result of some peat volume located under the ignition source quick preheating, and as consequence, intensive gasification (pyrolysis) of that and neighboring peat volume. Formed pyrolysis products mixed with water vapor serve to decrease thermal energy ejection by formed convection currents from peat massif. Those currents form additional thermal isolation layer which consist of hot gasification products and water vapor. This helps to decrease ignition source and terrestrial environment heat exchange and decrease its cool down time. Given effect is also prevent oxygen diving into peat massif which also helps to decrease ignition fall time (ignition time equals to 52 minutes).

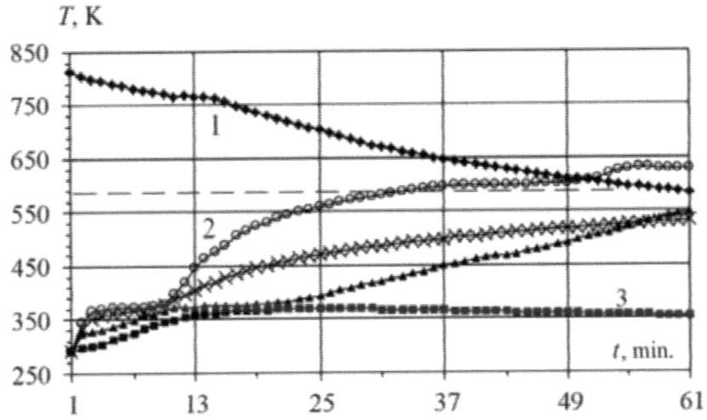

Fig. 2. Temperature change in peat massif with 20% moisture in case of thermal insulated ignition source.

All graphs show time interval which characterizes according the field research goal peat massif ignition only. Therefore the further combustion source spread is not presented. However it is important to notice that in all given experiments full peat sample burnout was detected, i.e. there were no cases of local burnout with further combustion reaction ending. The result allows concluding that local heat sources with some reserve energy might become a reason for a big fire.

It is determined that researched peat with 35-40% moisture does not light up because a burned out layer is quickly formed under the ignition source, and energy transferred through this thermal insulated layer is spent for moisture evaporation only. Water vapors create convection currents and with that force out oxygen necessary for chemical reaction and stop it from diving under the combustion source. By the time necessary mass transfer conditions for chemical transitions with terrestrial environment are set, the ignition source already does not have enough energy to initiate the combustion.

It is important that obtained results allow to make quantitative estimations taking in consideration all researched parameters and by that to be the base for further application in mathematical model and base for minimization peat ignition and its isolation recommendations development.

4 Obtained results analysis

Figure 3 shows generalized results of experiments with different ignition source and terrestrial environment heat exchange conditions. Base on quoted results analysis it is fair to state that 10% peat moisture increase leads to 5-7 minutes increase in ignition fall time (figure 3a). Besides, it has been elucidated that the increase in peat density will also result with an increase in ignition fall time and decrease in flame core size, i.e. as this takes place, oxygen delivery to combustion zone slows down. Conclusions of theoretical work prove the statement, stemming from experiments results [8].

The reason of experimentally found ignition fall time range spreading is most likely due to local different peat characteristics under the heat source and experimental uncertainty. In addition onset ignition source temperature is determined. What is more, besides good ignition source and working agent contact (cold state) were provided it is possible to have some minor inaccuracy. It happens because of orientation to decrease operation time value which might lead to only partial contact of heat source and peat surface.

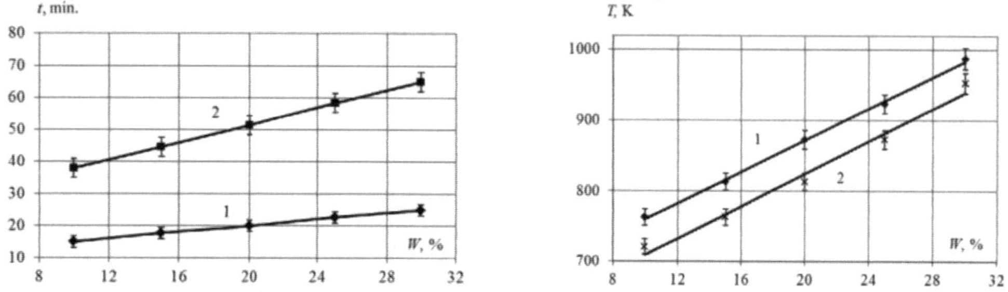

a) b)

Fig. 3. The ignition time and onset ignition temperature functional connection with peat moisture: 1 – without insulation; 2 – with thermal insulation.

5 Conclusion

It has been established that 10% peat moisture decrease with constant ignition source sizes will need 100-110 K ignition source temperature increase. During experiments with a thermal insulated ignition when peat moisture increases 10% ignition source temperature decrease gap will be 80-90 K.

The obtained research revealed that there is no ignition happening if peat moisture decrease up to 35-40% even if ignition source will be heated up to furnaces' maximum temperature – 1173 K. Hence hot ignition source of determined sizes could not become a reason for ignition of peat massif with moisture 35-40%.

During set of experiments with thermal insulated ignition source the ignition fall time increases 2-2.5 times with some onset heat source temperature decrease.

Experimentally obtained data and its comparison to other authors' published results [4-6] prove that peat, as a feedstock with typical characteristics, is a good choice for field research. This statement allows to apply main field research results to majority of peat from other deposits.

References

1. http://www.peatsociety.org/peatlands-and-peat/global-peat-resources-country
2. A.M. Grishin, A. S. Yakimov, JEPT **81**, 1 (2008)
3. A.E. Afanasyev, N.V. Churaev, *Nedra* (1992)
4. V.A. Nikiforof, *Peat deposits development and peat mechanical processing* (1979)
5. M.V. Popov, A. M. Shabarov, A. I. Gushchin, *Milled peat application in power industry* (1974)
6. A.N. Subbotin, *Doctoral dissertation* (2011)
7. X. Huang, G. Rein, H. Chen, ISC'30, **35** (2015)
8. A.N. Subbotin, R.N. Kulesh, Fire-safety (2009)
9. D.K. Thompson, B.M. Wotton, J.M. Waddington, Int. J. Wildland Fire **24** (2015)
10. A.N. Subbotin, Heat mass trans. MMF, **4** (2000)
11. A.N. Subbotin, IFJ **76** (2003)
12. A.N. Subbotin, Fire-expl. Saf. **16** (2007)
13. G. Rein, J. Garcia, A. Simeoni, V. Tihay, L. Ferrat, WIT Trans. Ec. Env. **119** (2008)
14. A. Usup, Y. Hashimoto, H. Takahashi, and H. Hayasaka, Trop. **14** (2004)
15. V. T. Kuznetsov, E. L. Loboda, Comb., Expl., Shock Waves **46** (2010)

CRITICAL HEAT FLUX IN NANOFLUIDS AT QUASI-STATIONARY AND STEPWISE HEAT GENERATION

Mikhail Moiseev[1, *], *Vladimir* Zhukov[1], and *Denis* Kuznetsov[1]

[1]Kutateladze Institute of Thermophysics SB RAS, Novosibirsk, Russia

Abstract. In this paper results of an experimental study on critical heat flux in nanofluid at quasi-stationary and stepwise heat generation are presented. Freon R21 with addition of 0.0077 vol.% of SiO_2 nanoparticles was used as test fluid. Boiling curves, critical heat fluxes and temperatures of boiling initiation were obtained for pure fluid and for nanofluid. It was shown that the addition of nanoparticles didn't affect heat transfer at pool boiling, but critical heat fluxes at quasi-stationary and stepwise heat generation were increased.

1 Introduction

Boiling is an effective way to remove heat from the surface that is widely used in various technical applications. The increasing requirements for efficiency and miniaturization of equipment give an impetus to the search of methods to intensify heat transfer during boiling and to increase critical heat flux (CHF) [1, 2]. In recent decades, one of the promising trends in enhancing the boiling characteristics is the use of nanofluids. The addition of nanoparticles can significantly increase both the heat transfer coefficient and CHF [3]. However, the obtained results are often contradictory and depend on many factors, so the practical use of nanofluids needs further systematic research. In addition, there is a necessity in the experiments to be carried out in nanofluids under non-stationary heat release and to simulate different emergency processes (depressurization of devices, removal of rods in nuclear power plants, etc.). At present there are only few studies on the topic. The available works demonstrate an increase in non-stationary CHF when adding nanoparticles [4, 5].

This work experimentally investigates boiling of Freon R21 with addition of 0.0077 vol.% of SiO_2 nanoparticles at quasi- and non-stationary heat release.

2 Description of the experiment

The experiments were carried out in a large volume of Freon R21 with addition of 0.0077 vol. % of SiO_2 particles with a size of 20-25 nm. The horizontal cylindrical pipe of stainless steel with a length of 50 mm, an outer diameter of 3 mm and a wall thickness of 0.5 mm

* Corresponding author: moiseevmikhail@gmail.com

was used as a working section. Heating of the section was performed by a controllable constant current source. The temperature of the wall and liquid in bulk under quasi-stationary conditions was determined by thin-film platinum resistance thermometers HEL-700. At non-stationary heat release the wall temperature was determined from the numerical solution of the heat equation. The experiments were conducted in the range of reduced pressures P/P_{cr} = 0.037–0.056 (P = 0.19–0.29 MPa). Detailed description of the setup is given in [6].

Experiments were carried out at two regimes of heat release. For the first regime, which we call "quasi-stationary heat release", the heat flux was increased by small amount step by step, and each time, stabilization of wall temperature was waited for. For the second regime, which we call "stepwise heat release", the heater was stepwise supplied with a constant current of 420 A. In this case the fluid that initially was at saturation temperature can reach high level of metastability, and development of boiling crisis changes drastically.

3 Results of the experiment

Fig. 1 shows the boiling curves in pure fluid and in nanofluid during the quasi-stationary heat release. The addition of the nanoparticles almost did not influence the efficiency of heat transfer. However, overheating of the wall, in which there is a transition from the region of single-phase convection to boiling incipience, decreased from 51 K to 25 K.

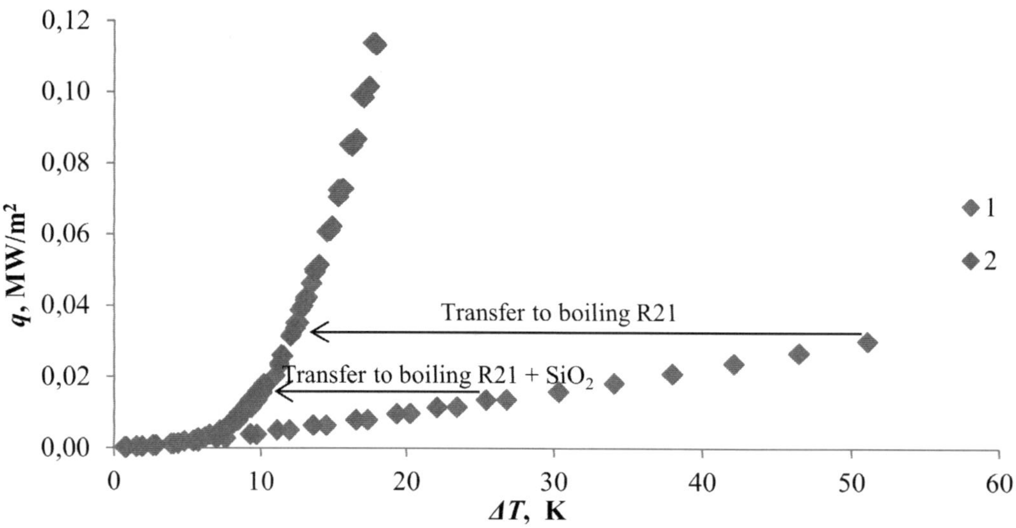

Fig. 1. Boiling curves. 1 – R21; 2 – R21 + 0.0077% SiO_2. P/P_{cr} = 0.037.

Experiments were carried out to determine CHF. In the pre-crisis region the electric current through the heater was gradually increased by a small value, and each time, stabilization of the wall temperature was waited for. The achievement of the CHF was determined visually by vapor film forming. The results of the experiments are presented in fig. 2. Experimental data for pure fluid and nanofluid are well described by Kutateladze formula:

$$q_{\rm cr} = kr\sqrt{\rho_{\rm v}}[g\sigma(\rho_{\rm l} - \rho_{\rm v})]^{1/4}, \quad (1)$$

where k is empirical coefficient, r – latent heat of evaporation, g - gravitational acceleration, σ – surface tension, ρ_v – vapor density, and ρ_l – liquid density.

Empirical coefficients were found from experimental data using least squares method. For the nanofluid k = 0.152, and for pure Freon R21 k = 0.137. Thus, the presence of

0.0077 vol. % of SiO_2 nanoparticles has led to the increase in the value of the first CHF by 11%.

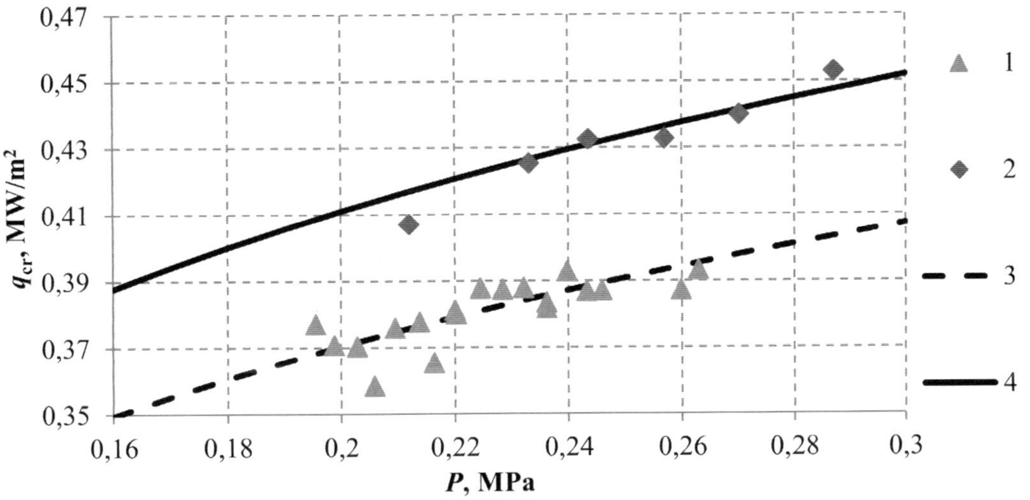

Fig. 2. The dependence of the critical density of the heat flux on pressure. Experimental data: 1 – Freon R21; 2 – Freon R21 + SiO_2; Calculation by Kutateladze $q_{cr} = kr\sqrt{\rho_v}[g\sigma(\rho_l - \rho_v)]^{1/4}$: 3 – $k = 0.137$; 4 – $k = 0.152$.

In the experiments under non-stationary heat release, the heater was stepwise supplied with a constant current of 420 A. The rate of the wall heating in this case was 2180 K/s, and overheating relative to the saturation temperature reached 70 K. An increase in the wall temperature results in an increase in the density of the heat flux, transferred to the liquid by thermal conductivity, since for the duration of an experiment (about 30 ms) the convection has no time to develop. The interface of a steam bubble growing at the nucleation site becomes hydrodynamically unstable when a certain wall overheat is reached. This leads to the initiation of evaporation front propagation along the heater (in the layer of metastable liquid) and the onset of the heat transfer crisis.

Fig. 3 shows the dependence of the heat flux density q, transmitted by the wall to the fluid, on the temperature of the wall overheating ΔT. The value of q was obtained from the numerical solution of the heat equation. Vertical lines mark the minimum temperatures of the wall, at which the evaporation front propagation was recorded. In pure Freon R21, the corresponding heat flux density amounted to 0.146 MW/m^2 (3 times less than CHF at quasi-stationary heat release). In nanofluid R21 + SiO_2, it is equal to 0.184 MW/m^2. Thus, the addition of SiO_2 nanoparticles has led to a 26% increase in CHF at stepwise heat release.

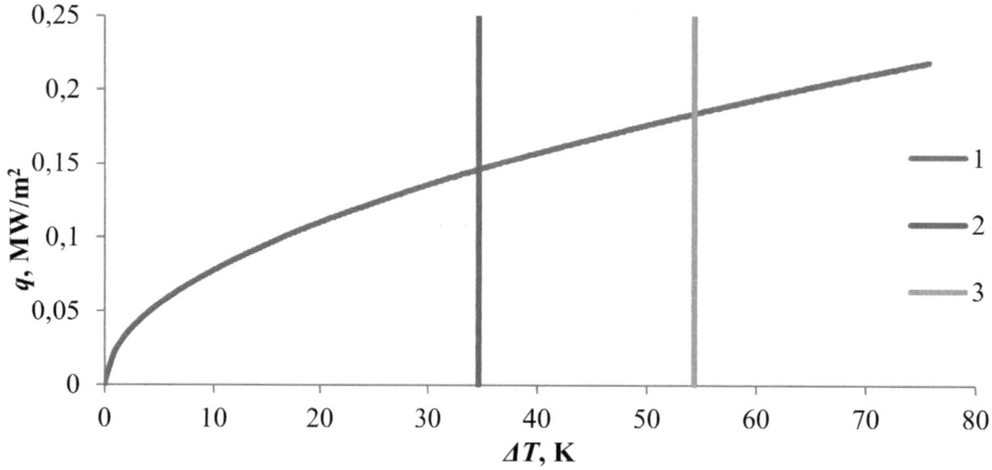

Fig. 3. 1 – heat flux at solid-liquid interface; 2 – front initiation in Freon R21; 3 – front initiation in nanofluid R21+0.0077 vol.% SiO_2. $P/P_{cr} = 0.037$.

4 Conclusions

Adding 0.0077 vol. % of SiO_2 nanoparticles in Freon R21 has not noticeably influenced the heat transfer coefficient during boiling, but has led to a decrease in boiling-up temperature. CHF in nanofluid under quasi-stationary heat release increased by 11% compared to pure fluid.

CHF at stepwise heat release was 3 times lower than CHF in quasi-stationary regime. Addition of nanoparticles provided 26% increase in CHF under non-stationary heat release. Wall overheat corresponding to boiling initiation also increased from 35 to 55 K.

Acknowledgment

The work was financially supported by RFBR (project No. 13-08-00178).

References

1. S.Y. Misyura, Exp. Therm. Fluid Sci. **75**, 43 (2016)
2. V.E. Nakoryakov, and S.Y. Misyura, JET **25**(1), 24 (2016)
3. D. Ciloglua, and A. Bolukbasib, App. Therm. Eng. **84**(5), 45 (2015)
4. V.I. Sharma, J. Buongiorno, T. McKrell, and L. Hu, Int. J. Heat Mass Transfer **63**, 425 (2013)
5. L. Fan, J. Li, D. Li, L. Zhang, Z. Yu, and K. Cen, Int. J. Therm. Sci. **91,** 83 (2015)
6. A.N. Pavlenko, A.A. Tairov, V.E. Zhukov, A.A. Levin, and M.I. Moiseev, JET **23(3)**, 173 (2014)

MATEC Web of Conferences **72**, 01057 (2016) DOI: 10.1051/matecconf/20167201057
HMTTSC-2016

PENETRATION OF PARTICLES THROUGH MULTI-BARRIER SYSTEMS

Michael Bubenchikov[1, *], *Aleksey* Bubenchikov[2], *Eduard* Libin[2], and *Yulia* Khudobina[2]

[1]"Gazprom Transgaz Tomsk", 634029 Tomsk, Russia
[2] National Research Tomsk State University, 634050 Tomsk, Russia

Abstract. The paper gives a theoretical justification for the effect of increasing permeability of ultra-thin membranes by altering their internal material structure, namely through the process of creating one or two energy voids within the permeable layer. An effective computing technology for solving the stationary Schrödinger equation is also described.

1 Introduction

Molecular dynamics is widely used in studies of physical processes in nanoporous structures [1-15], as well as in studying mechanical behaviour of nanoporous crystals [18-22]. However, for problems associated with passage of molecules through a barrier, wave dynamics appears to be more effective [23, 24]. In multi-barrier systems, in comparison with the case of a monolayer, particles are accumulated in a potential well, and thereby particle concentration in front of the barrier and behind it is increased. Since all this occurs in the wave mode of mass concentration and in the centre of the well there is the maximum amplitude of oscillations, the system with energy wells show generation of more intense transmitted waves, which drives a great mass over the barrier. Material particles: electrons, protons and atoms are associated with the de Broglie wave. If the wave frequency coincides with the frequency of one of the energy wells enclosed between two barriers, the system shows resonance phenomena which lead to an increase in particle concentration in the whole near-barrier area, in particular, behind the barrier, which means that there is an increase in the particle transmission coefficient. Since the authors were only interested in the final result concerning the integral permeability of the membrane instead of wave dynamics as such, all the conclusions reached in the work are based on the results of solving the stationary Schrödinger equation. The results of such studies may be interesting from the standpoint of technologies for creating effective functional layers with atomic selective permeability, as well as gas separation membranes for separation of helium from natural gas and others.

* Corresponding author: michael121@mail.ru

2 The method for solving the stationary Schrödinger equation

In problems of particle penetration through an energy barrier the following equation is used [25, 26]:

$$\frac{d^2\psi}{dx^2}+m\left[\lambda^2-U(x)\right]\psi=0. \tag{1}$$

This equation is commonly called the one-dimensional stationary Schrödinger equation and it describes the quantum-mechanical motion of a particle in one dimension. Here, the function $U(x)$ is the potential, λ^2 is the energy of the particle, and m is its mass. The desired function ψ is such that the square of its module determines the density of the probability of particle localization at the point x. Equation (1) is written here in the dimensionless normalized form, when the potential $U(x)$ and the energy of the particle λ^2 are attributed to the dimensional value of the potential barrier U_0, so that the maximum value of the potential $U(x)$ is assumed to be equal to 1. The mass of the particle m in equation (1) is also a dimensionless quantity referring to a certain standard value in accordance with the scales of the phenomenon under study. Clearly, equation (1) possesses not only a quantum sense. It is applicable to many other problems of theoretical physics which may consist in determining the resonant frequencies of acoustic or electromagnetic waves within an inhomogeneous cavity [27]. Equation (1) plays an important role in solutions of nonlinear partial differential equations where the method of inverse scattering is used [28]. In these problems it is necessary to restore the potential $U(x)$ when the function $\psi(x)$ is known.

The general solution of equation (1) can be obtained as a linear combination of two linearly independent solutions $w_1(x)$ and $w_2(x)$ which can be found using the traditional Runge-Kutta method by comparing the obtained result with the known exact solutions. The accuracy of calculations is controlled by the value of the Wronskian determinant $W=w_1(x)w_2'(x)-w_2(x)w_1'(x)$, as well as by the accuracy of following the equality $|B|^2+|A|^2=1$. The first item on the left side of this equation is the reflection coefficient and the second - the transmission coefficient. To meet these criteria in the numerical method of the Runge-Kutta type, one has to use a step of calculation which is too small and this greatly increases the computational time. In this connection, in this paper we propose to use a non-standard operation matrix analysis which is available in modern mathematical packages such as MatLab [29] or SciLab. This operation allows finding an orthonormal basis (null-space) for rectangular matrices.

For example, let the matrix **M** consist of 8 rows and 10 columns. Then, by carrying out the command **B**=null(**M**) we get the rectangular matrix **B** which has 10 rows and two columns each of which is orthogonal to all the rows of the matrix **M**. Additionally, the result is obtained with almost a maximum machining precision. In this method, equation (1) must be written in the finite difference form:

$$\psi_{n-1}+\left[-2+h^2m\left(\lambda^2-U_n\right)\right]\psi_n+\psi_{n+1}=0. \tag{2}$$

Next for equation (2) we construct a corresponding three-diagonal matrix, eliminate its last two lines and get the matrix **H** as a result. Calculating the null-space of the matrix **H**: **w**=null(**H**) gives a numerical representation of the fundamental solutions $w_1(x)$ and $w_2(x)$ of equation (1) in the form of two vector columns. The accuracy of the solution by this method is independent of the parameter λ and does not require a small step, which permits reliable calculation of both the function $\psi(x)$ and the transmission coefficient $D(m,\lambda)=|A|^2$ for particles passing through potential barriers of complex shapes.

2 Examples of calculations

The typical view of the graph $|\psi|^2(x)$ obtained by the numerical method is shown in fig.1, where the shape of the barrier is marked by a dotted line.

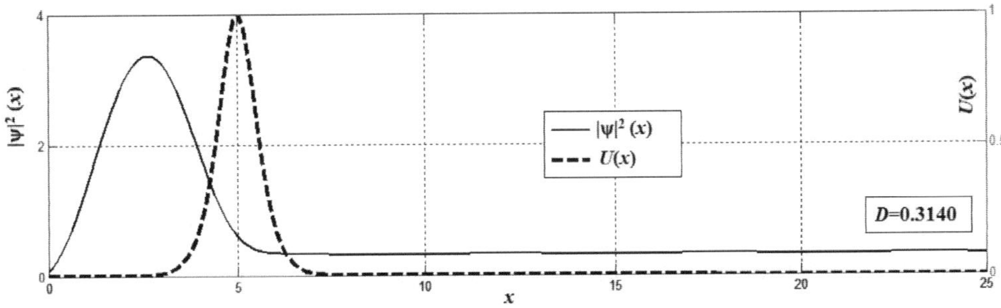

Fig. 1. Particle of parameters: m=1, λ=0.5 passing single barrier.

The transmission coefficient D in this case is of a little value equal to 0.3140. It would seem that addition of new barriers should reduce the general transmission coefficient. But, as is clear from fig.2, an increase in the number of barriers, by contrast, leads to an increase in the transmission coefficient. This is due to the fact that the density of probabilities $|\psi|^2$ in wells reaches its maximum values, which facilitates penetration of particles through the barriers.

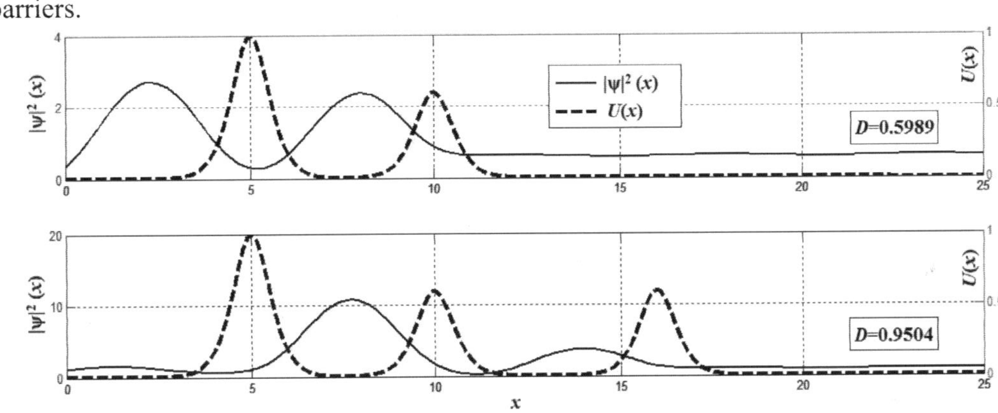

Fig. 2. Effect of adding new barriers on distribution of density of probabilities $|\psi|^2$ and particle transmission coefficient $D(m, \lambda)$.

Here we deal with a resonance effect of interaction between bound particles in potential wells and an oncoming particle. This phenomenon is observed only for certain values of the parameter λ and it is clearly seen on graphs of the function $D(\lambda)$ for a given shape of the potential barrier.

Calculations revealed that the effect of increasing permeability of a composite membrane refers only to certain particle energy. Changing the energy of particles falling onto the membrane, we inevitably get reduction of permeability for the same barrier system. Therefore, in practical problems of gas separation it is necessary to consider the Maxwell distribution of particle according to energies.

Acknowledgments

This paper was written within the frame of Competitiveness Improvement Program of Tomsk State University and with financial support of the Russian Science Foundation, grant under Agreement # 16-19-00089.

References

1. V.Ya. Rudyak, S.L. Krasnolutskii, D.A. Ivanov, J. Dok. Phys. **57** (2012)
2. V.Ya. Rudyak, A.A. Belkin, V.V. Egorov, D.A. Ivanov, Int. J. Multiph. **5** (2011)
3. V.Ya. Rudyak, S.L. Krasnolutskii, D.A. Ivanov, J. Microfl. Nanofl. **11** (2011)
4. V.Ya. Rudyak, S.N. Dubtsov, A.M. Baklanov, Tech. Phys. Lett. **34** (2008)
5. V.Ya. Rudyak, S.L. Krasnolutskii, E.N. Ivashchenko, J. Eng. Phys. Thermophs. **81** (2008)
6. V.Ya. Rudyak, J. VDI-Berichte **1803** (2003)
7. V.Ya. Rudyak, A.A. Belkin, Tech. Phys. Lett. **29** (2003)
8. A.I. Potekaev, M.A. Bubenchikov, Russ. Phys. J. **54** (2011)
9. M.A. Bubenchikov, Russ. Phys. J. **54** (2011)
10. M.A. Bubenchikov, A.I. Potekaev, A.M. Bubenchikov, Russ. Phys. J. **56** (2013)
11. M.A. Bubenchikov, A.I. Potekaev, A.M. Bubenchikov, Russ. Phys. J. **56** (2013)
12. A.I. Potekaev, A.M. Bubenchikov, M.A. Bubenchikov, Russ. Phys. J. **55** (2013)
13. M.A. Bubenchikov, A.I. Potekaev, A.M. Bubenchikov, O.V. Usenko, A.V. Ukolov, IOP Conf. Series: Mat. Sc. Eng. **87**, 1 (2015)
14. A. M. Bubenchikov, M.A. Bubenchikov, O.V. Matvienko, E. A. Tarasov, O.V. Usenko, AIP Conf. Proc. **1698** (2016)
15. M. A. Bubenchikov, A. M. Bubenchikov, O. V. Usenko, V. A. Poteryaeva, S. Jambaa, EPJ Web Conf. **110** (2016)
16. K.A. Krylova, Y.A. Baimova, S.V. Dmitriev, R.R. Mulyukov, Phys. Sol. St. **58** (2016)
17. A.V. Savin, E.A. Korznikova, S.V. Dmitriev, Phys. Sol. St. **57** (2015)
18. K.A. Bukreeva, R.I. Babicheva, S.V. Dmitriev, K. Zhou, R.R. Mulyukov, A.I. Potekaev, Russ. Phys. J. **57** (2014)
19. J.A. Baimova, B. Liu, S.V. Dmitriev, N. Srikanth, K. Zhou, Phys. Chem. Chem. Phys. **16** (2014)
20. E.A. Korznikova, S.V. Dmitriev, J. of Phys. D: Appl. Phys. **47** (2014)
21. E.A. Korznikova, S.Y. Fomin, S.V. Dmitriev, Mat. Sc. Forum **845** (2016)
22. J.A. Baimova, S.V. Dmitriev, Mat. Sc. Forum **838-839** (2016)
23. A.M. Bubenchikov, M.A. Bubenchikov, A.I. Potekaev, E.E. Libin, Yu.P. Khudobina, Russ. Phys. J. **57** (2014)
24. A.M. Bubenchikov, M.A. Bubenchikov, A.I. Potekaev, E.E. Libin, Yu.P. Khudobina, Rus. Phys. J. **58** (2015)
25. H. Feshbach, Ph. M. Morse, *Methods of Theoretical Physics. Part II* (McGraw-Hill, New York, 1953)
26. L. D. Landau, E. M. Lifshitz, *Quantum Mechanics: Non-Relativistic Theory, Vol. 3* (Pergamon Press, Oxford, 1965)
27. L. M. Brekhovskikh, *Waves in Layered Media* (Science, Moscow, 1973) [in Russian]
28. Alwyn Scoot, *Nonlinear Science. Emergence and dynamics of coherent structures* (University of Arizona, Oxford University Press, 2003)
29. S. R. Otto, J. R. Denier, *An Introduction to Programming and Numerical Methods in Matlab* (Springer-Verlag, London, 2005)

APPLIED TECHNOLOGICAL DIRECTION OF POWER PLANT ASH AND SLAG WASTE MANAGEMENT WHEN KUZNETSK BITUMINOUS COAL IS BURNED

Snejana A.Lihach[1,*], *Maria* V. Grishchernko[1], *Roman* N. Kulesh[1], *Vitali* M. Lebedev[2], and *Ksenia* Y. Orlova[1]

[1]Tomsk Polytechnic University, 634050 Tomsk, Russia
[2]Omsk State Transport University, 644046 Omsk, Russia

Abstract. Currently a lot of power plants have a problem with storage of coal combustion solid by-products (ash and slag). Holding capacity of existing power plants available ash dumps were enlarged and modernized repeatedly. Many plants have two or even three of them. Today new ash dump construction is economically inconvenient due to need to assign new plots of land and their inconveniently big distance from a plant, which increase ash and slag transportation expenses. The goal of our research work is to find promising directions for ash and slag waste mass utilization based on Kuznetsk bituminous coals experimental data on ultimate composition and properties. The experimental research of ash, slag and their mixture samples from ash dumps brought us to conclusion that the most promising direction for these materials application in large quantities is construction industry including road construction. Be-sides, we lined up some other directions for ash, slag, and ash and slag mixture possible application. These directions might not provide mass utilization but they are promising from a point of view of the researched waste properties.

1 Introduction

Constantly developing modern technologies and growing cities demand increase of electric and thermal energy supply. Power industry plants are quite capable to provide those in full-scale working with high force. Over the years power plants increase their output. Fuel volumes grow considerably, and as consequence waste volume grow as well.

2 Review

Power plants waste is an ash and slag mixture. It forms during several processes: when fuel is burned in steam boilers; when smokestack exhaust gases pass through fly-ash collectors; and when water necessary to awash the mixture into com-mon hydraulic removal channel

* Corresponding author: snejka.lix@yandex.ru

asses through. Then formed pulp from those channels with the use of pumps comes into an ash dump. Ash dump is a quite large piece of land with a cup-like hollow and an extended edge around the perimeter – border levee. Ash dump bottom and border levee are usually made of medium fine gravel and areneceous filler mixture.

According to Energy Forecasting Agency yearly output of ash and slag waste in Russia is about 30 million tons, and over 1,5 billion tons are ac-cumulated up to date. That said ash dumps capacity of 115 coal burning plants out of 172 is almost exhausted [1].

Ash and slag mixture flow to ash dump after it is being completely filled up (with mound formation) is totally unacceptable because there is a possibility of dusting during ash dump drying process. This might have a negative impact on objects within ash dump controlled access area: residential complexes, water bodies and other [2]. That is why ash and slag mixture disposal is so up to date problem for Russian thermal power plants nowadays.

Land assign for new ash dumps is rarely supported by City Halls on grounds of existing plots of land assignment (for residential construction, for agriculture use etc.), and also on grounds of water body locations being there. If ash dump will be located on such a land, it would be impossible to keep up of controlled access area, i.e. minimum distance from ash dump border levee to residential and nature objects. The distance is a subject to Russia Land Code. Based on the above we think that the best solution to the problem is ash and slag use in all possible ways with their removal from ash dumps. Therefore ash dump available capacity will increase.

There are a number of ways to dispose ash and slag mixture: as a compounding material in concrete, brick, ceramic production, as a filler for worked-out coal open cut during its restoration, as a material during road construction, as an ice covering material in winter time, as a material for agriculture use [2, 3]. However the most technologically reasonable and fast way to utilize ash and slag mixture – its application in road construction industry [4].

For possibility to use ash and slag waste in construction they must be certified as a construction material. Only after the certification it will classified as a raw material for industrial applications and could be called ash and slag materials (ASM). In order to get that done ash and slag waste go through sedimentation process, separation from fly-ash light fraction if any and drying. Then necessary analysis of ash and slag mixture characteristics, their compatibility to standards and of execution permission documentation takes place.

3 Research description

In the presented research we used ASM from Kuznetsk bituminous coal combustion as a sampling material to determine its applicability in road construction. During the research we studied initial fuel and ash and slag material characteristics found in laboratory setting (table 1).

While analysing table 1 we came to conclusion that researched ash in mostly identical to natural materials used in construction, for instance sand with SiO as a base, low calcium oxide content and spiked with aluminium oxides. The ash composition makes it a promising material from aluminium sulfate and corundum production point of view which used in waste water treatment and several areas of chemical industry. Besides it makes the researched ash also a promising material in terms of aluminosilicate microspheres production. Those are used in oil and gas, chemical, construction and machine-building industries [5, 6].

Table 1. Experimental research results of initial coal and ash after its combustion.

No.	Descriptors		Range
1	Grade of fuel		1CC
2	As-received basis	W^p, %	9.0
3		A^p, %	18.2
4		$S^p_к + S^p_{op}$, %	0.3
5		C^p, %	61.5
6		H^p, %	3.7
7		N^p, %	1.5
8		O^p, %	5.8
9	Lower calorific value, ccal/kg		5 700
10	Ash on the dry basis, %		20.0
11	Maximum humidity, %		--
12	Maximum ash content, %		25.0
13	Maximum sulfur content, %		--
14	Hygroscopic moisture, %		1.6
15	Normalized moisture, %*10^3kg/ccal		1.58
16	Normalized ash content, %*103kg/ccal		3.20
17	Dry ash free volatile-matter content , %		30.0
18	Ash fusion point, °C	Initial deformation	1 100-1 500
19		Weaking initiation	1 240-1 500
20		Fluid initiation	1 280-1 500
21	Ash chemical composition on sulfates absent basis	SiO_2, %	60.0
22		Al_2O_3, %	20.8
23		TiO_2, %	0.83
24		Fe_2O_3, %	10.4
25		CaO, %	2.9
26		MgO, %	1.4
27		K_2O, %	2.61
28		Na_2O, %	0.6

ASM is classified as a "virtually non-hazardous waste" which is the fifth class based on negative environmental dimension. It also falls into the first class based on effective specific activity of natural radioactive nuclides, A_{eff} (effective specific activity of natural radioactive nuclides is three times lower than maximum). It means that the re-searched ASM could be used in construction and road structures without limitations.

Requirements for ash and dump materials used in road construction are set according to regulatory documents [7]. In compliance with these documents cold resistance is the main criteria for ash and slag materials evaluation. Such-like international research use ASM samples mechanical characteristics in the induration process as the main applicability criteria [8].

Cold resistance is determined in laboratory setting and characterized by specific frost heave $K_{frost\ heave}$. It is ratio of vertical strain blow up in the process of sample freezing to its initial height ex-pressed as a percentage. According to the con-ducted experiments specific frost heave of carbonaceous ash and slag materials is no more than 3%. Therefore these materials will be intended to be described as a slightly rebound [7]. Such ash and slag mixtures might be used for earth roadbed embankment erection and roadbed surfacing with almost no limitations.

4 Conclusion

In conclusion we would like to note that problem of Russian power plant ash dump total capacity decrease in one of the most essential topics at the moment. If no action would be taken to solve the problem, ash dumps would have been filled out completely within next 5–10 years. Whereas characteristics of the researched ash and slag material allow to use them in construction industry for road pavement building without road bed quality deviation, and in different areas of chemical industry.

References

1. I.S. Kojuhovsky, Y.K. Celikovsky, Energ. **6** (2011)
2. J. Hot, M. Sow, C. Tribout, M. Cyr, Constr. Build. Mat. **110** (2016)
3. N. Nordin, M.M.A.B. Abdullah, M.F.M. Tahir, A.V. Sandu, K. Hussin, Int. J. Cons. Sc. **7** (2016)
4. M. Pasetto, N. Baldo, Constr. Build. Mat. **108** (2016)
5. A.S. Inozemtsev, E.V. Korolev, Ind. Civil Eng. **10** (2013)
6. Z.R. Ismagilov, N.V. Shikina, N.V. Zhuravleva, R.R. Potokina, N.A. Rudina, V.A. Ushakov, T.N. Teryaeva, Sol. Fuel Chem. **49** (2015)
7. *Engineering instructions for application of fly-ashes and ash and slag mixtures from various solid fuel combustion for earth roadbed, road base and road surface building* (ISCS 185-75.)
8. V.N. Castaldelli, J.C.B. Moraes, J.L. Akasaki,J.L.P. Melges, j. Monzó, M.V. Borrachero, L. Soriano, J. Payá, M.M. Tashima, Fuel **174** (2016)

MATEC Web of Conferences 72, 01059 (2016) DOI: 10.1051/matecconf/20167201059
HMTTSC-2016

THE INVESTIGATION OF STRUCTURE HETEROGENEOUS JOINT WELDS IN PIPELINES

Lyudmila Lyubimova[1] , *Roman* Fisenko[1] , *Sergei* Breus[1], and *Roman* Tabakaev[1,*]

[1]National Research Tomsk Polytechnic University, 634050 Tomsk, Russia

Abstract. Welding joints of dissimilar steels don't withstand design life. One of the important causes of premature destructions can be the acceleration of steel structural degradation due to cyclic mechanical and thermal gradients. Two zones of tube from steel 12H18N9T, exhibiting the structural instability at early stages of the decomposition of a supersaturated solid austenite solution, were subjected to investigation. Methods of x-ray spectral and structure analysis, micro hardnessmetry were applied for the research. Made the following conclusions, inside and outside tube wall surfaces of hazardous zones in welding joint have different technological and resource characteristics. The microhardness very sensitive to changes of metal structure and can be regarded as integral characteristic of strength and ductility. The welding processes are responsible for the further fibering of tube wall structure, they impact to the characteristics of hot-resistance and long-term strength due to development of ring cracks in the welding joint of pipeline. The monitoring of microhardness and structural phase conversions can be used for control by changes of mechanical properties in result of post welding and reductive heat treatment of welding joints.

1 Introduction

The main structural materials for the manufacture of superheaters for boiler, working on supercritical pressure, in Russia, including the welding joints, are austenitic steels 12H18N9T (Russian designation is 12X18H9T) and pearlitic steels 12H1MF (Russian designation is 12X1MF). Calculated on resource of work about 300 thousand hours, they are destroyed much earlier due to various reasons [1, 2]. The typical state for welding joints is the presence of chemical, structural, mechanical and geometrical heterogeneities, which are strain concentrators [3]. One of the important causes of premature destructions can be the acceleration of steel structural degradation due to cyclic mechanical and thermal gradients [4-6]. The structural conversions have a particular negative effect on welding joints resistance to brittle failure. The technical diagnostic of welding joints, basis on specialized scientific and technical documents, involves investigation of metal composition, structure and properties.

* Corresponding author: TabakaevRB@tpu.ru

In this regard, a material structure investigation in the hazardous areas of welding joint is remained actual task.

2 Material and methods

Methods of x-ray spectral analysis with using of x-ray fluorescence analyzer EDX 2800 (SKYRAY, USA), x-ray structure analyzer DRON-3 (Russia) and microhardnessmetry with using of microhardness tester PMT-3 (Russia) were applied for the research.

Two zones of tube from steel 12H18N9T (fig. 1), exhibiting the structural instability at early stages of the decomposition of a supersaturated solid austenite solution, were subjected to investigation [7]. Zone No.1 is located at a distance 7.5 mm from axis of welding joint, zone No.2 is located at a distance 14.5 mm. Zone No.1 called heat-affected zone (HAZ), No.2 – a zone of thermal welding influence (in accordance with hot-tinting).

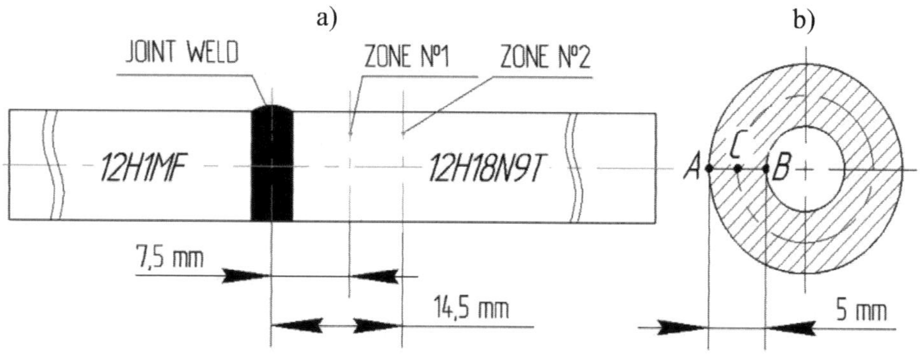

Fig. 1. The scheme of welding joint: a) zones No.1 and No.2 for investigation; b) AC, BC – the directions of microhardness measurements at tube wall thickness.

3 Results and discussion

The experimental results are shown in figures 2 and 3.

From figure 2 implies that along of thermal influence of welding take place a heterogeneous decomposition of the supersaturated solid austenite solution, which reveals itself by the phenomena of structural polymorphism, in result of both samples is acquired a mixed phase structure. One phase has a body-cented cubic (BCC) lattice, second phase is austenite (γ-Fe) with face-centered cubic (FCC) lattice. For zone No.1 the concentration of α-Fe and γ-Fe on outside surface are accordingly 81 and 19 mas%, for inside surface – 83 and 16%. For zone No.2 the concentration of α- and γ- phases on outside surface are accordingly 89 and 11%, inside surface is presented only α-phase of Fe.

The result of heterogeneous solid solution decomposition is the emergence of point defects in kind of exempt substitutional impurities and their inhomogeneous distribution (fig. 3a, b). The process of allocating is reversible and accompanied by dissolving of impurities at presence of temperature and strain gradients. This is significantly different processes, because dissolution is accompanied by the formation of vacancy porosity on the dissolution front, void coalescence and formation of microcracks. The final stage of this defect is macrocrack.

 DOI: 10.1051/matecconf/20167201059

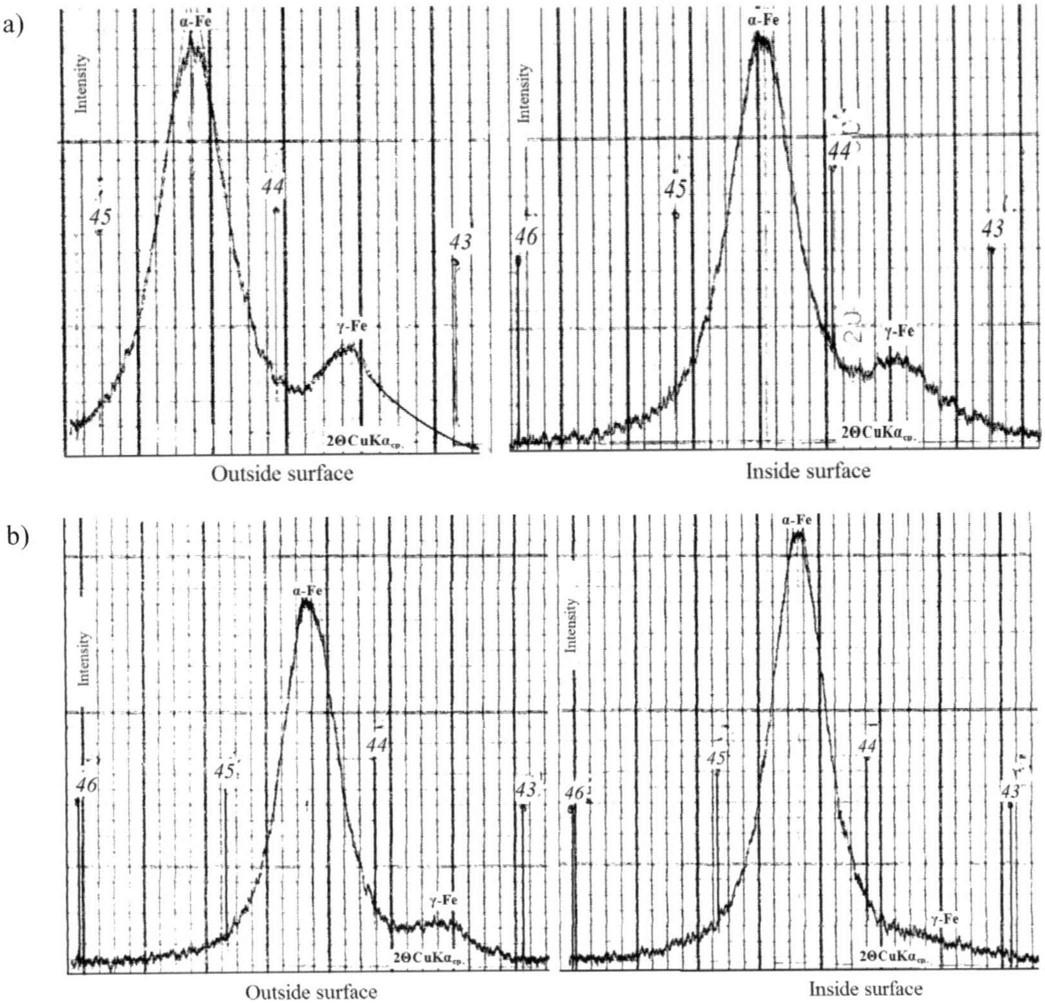

Fig. 2. Fragments of radiographs: a) sample from zone No.1; b) sample from zone No.2.

It is known [8 and others], that 80-85% forced stops of thermal power plants boilers is due to damage of tube heating surfaces. The tubes, from which produced heating surfaces, in supply condition due to difficult technological chain of their processing (drawing, rolling, annealing and etc.) is acquired the anisotropy of properties, caused by heterogeneity and discontinuity of structure that provides the spread in values of mechanical characteristics along the tube length and wall thickness. To confirm of this position in present work is measured the microhardness along wall thickness of superheater tube from heat-affected zone and zone of thermal welding influence of welding joint from heterogeneous steels 12H1MF and 12H18N9T (in accordance with fig. 1), results of which are shown in fig. 4.

Figure 4 is shown that the parameters of microhardness, and also its distribution, are different at measurement from outside and inside surfaces to centre of tube wall (in accordance with fig. 1).

The microhardness distributions along wall thickness are revealed its fibering by levels of microhardness. For example, in heat-affected zone microhardness at outside surface is 1700 MPa, at inside – 3400 MPa, at the middle of wall – 2500 MPa. The maximum 4000 MPa and the minimum of microhardness 1000 MPa are reached near the middle of wall. Inside of these ranges microhardness is strongly oscillated (fig. 4a).

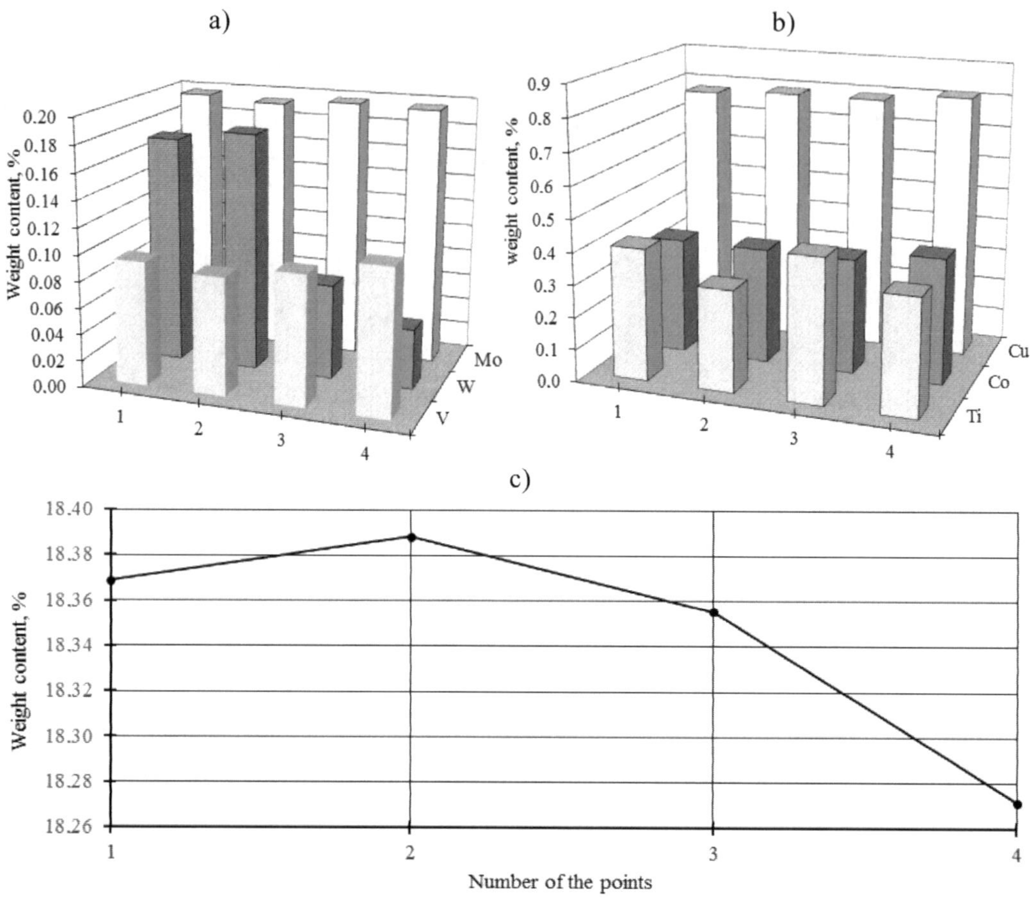

Fig. 3. The results of spectral analysis: a) the concentration of alloying elements *V*, *W*, *Mo*; b) the concentration of alloying elements *Ti*, *Co*, *Cu*; c) the concentration of chrome (points 1, 3 – inside surface of zones No.1 and No.2 accordingly; points 2, 4 - outside surface of zones No.1 and No.2).

In zone of thermal welding influence microhardness distribute thus: the maximum 3500 MPa is detected near the middle of tube wall, at which the microhardness decreases to outside and inside surfaces to a value about 1700 MPa. In this case microhardness also oscillated from 1400 to 3100 MPa (fig. 4b).

The microhardness is parameter of strength, which characterize a resistance of material to deformation and destruction [9]. The presence of significant spread in microhardness values provides spread of strength metal properties and, consequently, resource of heating spirals, working in the same conditions.

Presented in fig. 4 zonal microhardness distribution along wall thickness lead to decrease of hot-resistance in conditions of long-term operation due to emergence and development of ring cracks in the welding joint of pipeline.

This conclusion is confirmed by detected lamination of superheater tube wall (fig. 5) in result of deep oxidation at high temperature.

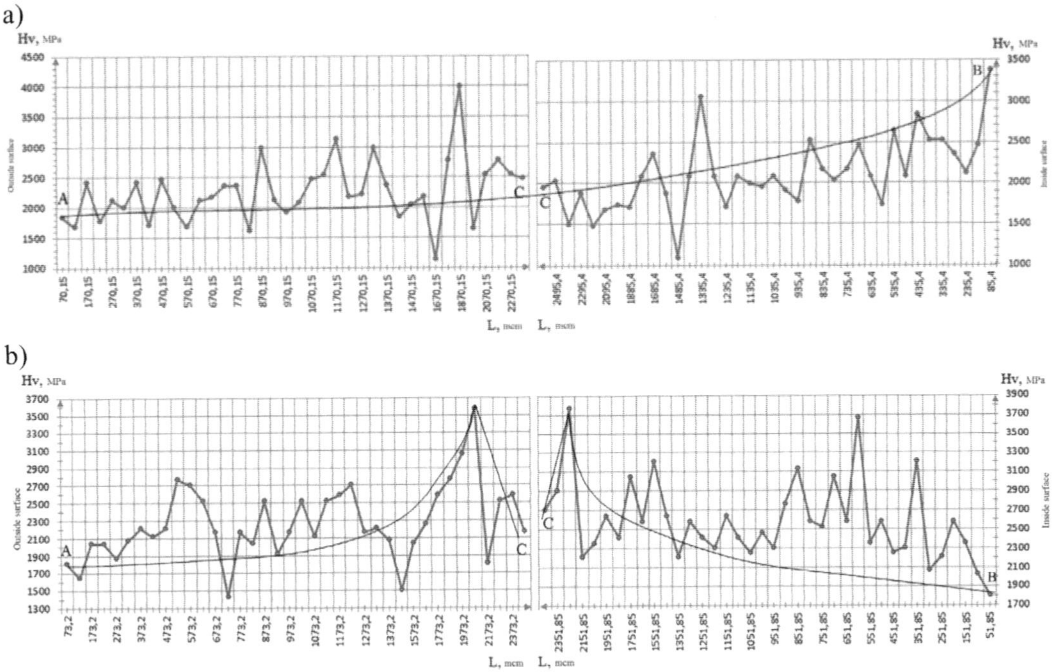

Fig. 4. The microhardness along wall thickness of welded tube: a) zone, located at a distance 7.5 mm from axis of the welding joint; b) zone, located at a distance 14.5 mm from axis of the welding joint; AC, BC – the directions of microhardness measurements at tube wall thickness (in accordance with fig. 1).

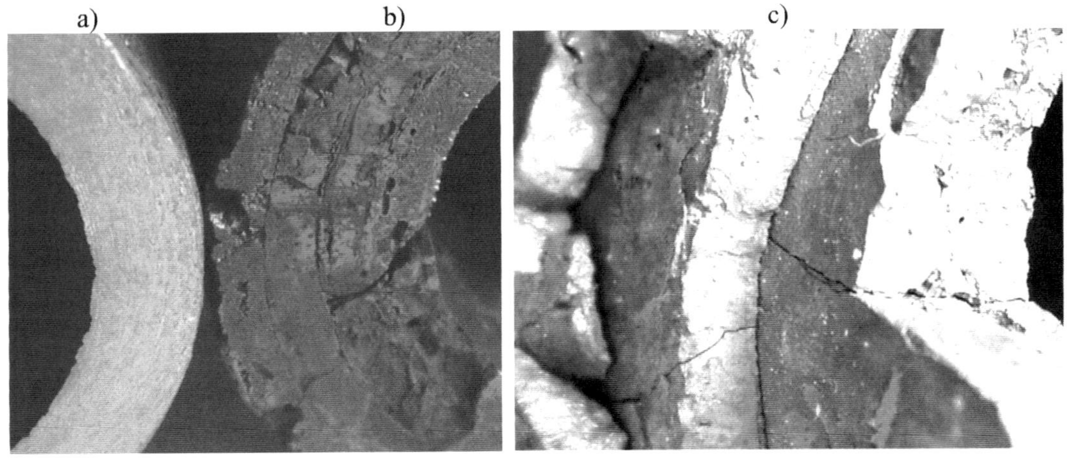

Fig. 5. The layered tube structure: a) the pipe wall in the initial state; b, c) the pipe wall after different stages of deep oxidation.

4 Conclusions

The results of completed investigation are follows:

1. Inside and outside tube wall surfaces of hazardous zones in welding joint have different technological and resource characteristics.
2. The microhardness very sensitive to changes of metal structure. During welding, heating, mechanical loads the microhardness changes in direction of yield strength change

and, as the hardness of separate micro-local areas of material microstructure, is a very difficult property, which can be regarded as integral characteristic of strength and ductility.

3. The welding processes are responsible for the further fibering of tube wall structure, creating zones with strongly different strength properties, reducing the resistance to destruction and impacting to the characteristics of hot-resistance and long-term strength due to development of ring cracks in the welding joint of pipeline.

4. The monitoring of microhardness and structural phase conversions can be used for control by changes of mechanical properties in result of post welding and reductive heat treatment of welding joints.

Acknowledgments

The reported study was funded by RFBR according to the research project No. 15-08-99544 A.

References

1. V.A. Bogachev, G.A. Urusova, M.A. Shumovskaya, Pow. Tech. Eng. **48**, 467 (2015)
2. V.A. Bogachev, D.S. Zmienko, A.E. Korneev, T.P. Pshechenkova, B.E. Shkol'nikova, Therm. Eng. **59**, 274 (2012)
3. J.S. Lee, K. Maruyama, Met. Mat. Int. **21**, 639 (2015)
4. L.L. Lyubimova, A.A. Makeev, A.A. Tashlykov, A.S. Zavorin, R.N. Fisenko, Steel in Transl. **44**, 117 (2014)
5. L.L. Lyubimova, A.A. Makeev, A.S. Zavorin, A.A. Tashlykov, A.I. Artamontsev, B.V. Lebedev, R.N. Fisenko, Therm. Eng. **61**, 600 (2014)
6. L. Lyubimova, R. Tabakaev, A. Tashlykov, A. Zavorin, V. Zyubanov, EPJ Web Conf. **110** (2016)
7. A.A. Makeev, L.L. Lyubimova, A.S. Zavorin, A.A. Tashlykov, Steel in Transl. **39** (2009)
8. E.Y. Priymak, V.I. Gryzunov, Met. Sc. Heat Treatm. **53**, 136 (2011)
9. Y. Jin, R. Li, Z. Yu, Y. Wang, J. Mat. Eng. Perf. **25**, 1327 (2016)

MATEC Web of Conferences 72, 01060 (2016) DOI: 10.1051/matecconf/20167201060
HMTTSC-2016

EXPERIMENTAL RESEARCH OF INFLUENCE OF A RELATIVE PARTICLES POSITIONING IN A GAS STREAM ON CHARACTERISTICS OF THEIR AERODYNAMIC TRACES

Roman S. Volkov[1,*], *Maria* M. Osmolovskaya[1], and *Irina* P. Ozerova[1]

[1]National Research Tomsk Polytechnic University, 634050 Tomsk, Russia

Abstract. The cycle of experimental studies on determination of length of aerodynamic traces of the particles which are flowed round by an air stream is executed. When carrying out researches, panoramic optical methods for diagnostics of multiphase flows of PIV and PTV were used. Velocities of an air flow were varied in the range of 1-3 m/s. The sizes of particles changed from 1mm to 5 mm. The defining influence of the sizes of particles and velocities of an air stream on length of aerodynamic traces is established. Influence of a relative positioning of particles on features of formation of an aerodynamic trace is shown.

1 Introduction

Nowadays huge number of technologies use single-phase and multiphase drop streams [1-3].The most widespread and at the same time complex challenge is the choice of initial parameters of such streams. It is especially difficult to predict such parameters when drops (or particles) move through a counter gas flow. In that case the drops in a stream going the first exert considerable impact on the subsequent drops. Such influence is shown, first of all, due to decrease in speeds of an air stream in a trace of a drop (particle).

For correct prediction of the initial parameters of a stream (speeds and the sizes of drops) it is necessary to know the features of aerodynamic traces of drops. Besides it is important to understand how the relative positioning of drops influences the characteristics of aerodynamic traces. Such researches can be carried out by using modern diagnostic PIV systems [2-5].

The purpose of this work - experimental research of influence of a relative particles positioning in a gas stream on characteristics of their aerodynamic traces.

2 Experimental methods and the experimental setup

For carrying out researches the experimental setup which was designed for these purposes was used. The scheme of the experimental setup is given in fig. 1. Experimental setup

* Corresponding author: romanvolkov@tpu.ru

contained in the basis a complex for diagnostics of parameters of multiphase streams. The stand (fig. 1) consisted of the following registration equipment: high-speed CCD video camera of type "IMPERX IGV B2020M" (image format is 2048×2048 pixels, frame rate is not less than 1.5 Hz, delay between two consecutive shots is no more than 5 μs); a solid double-pulse Nd:YAG laser of type Quantel "EverGreen 70" (wavelength is 532 nm, pulse energy is no more than 74 mJ, pulse duration is not more than 12 ns, repetition rate is less than 15 Hz); a synchronizing processor (sampling signals is no more than 10 ns, supports the modes of external and internal start-up).

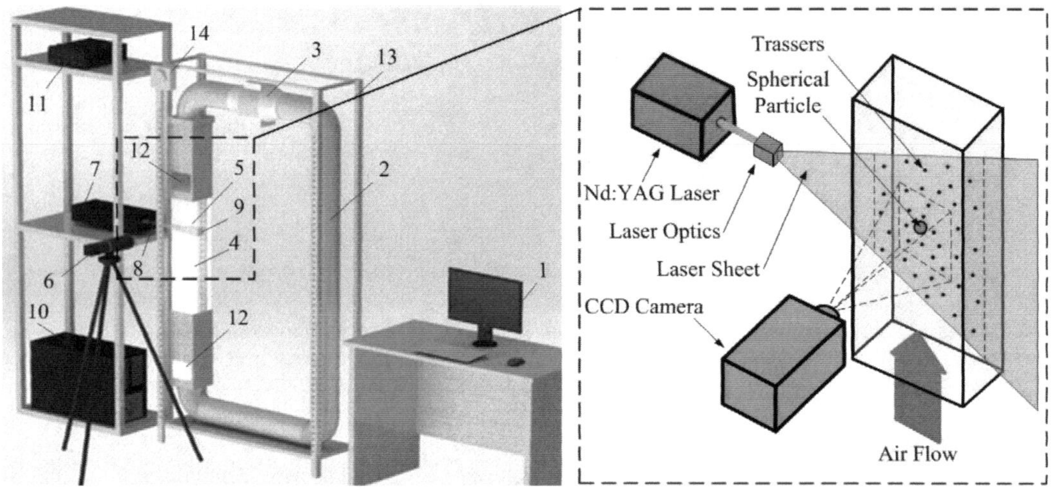

Fig. 1. Scheme of the experimental setup: 1 – work station; 2 – circulation loop; 3 – air fan; 4 – glass contour; 5 – spherical particle; 6 – CCD camera; 7 – double pulsed Nd:YAG laser; 8 – laser optics; 9 – laser sheet; 10 – power supply of Nd:YAG laser; 11 – synchronizer of signals; 12 – openings for installation in a contour of spherical particles and tracers; 13 – supporting structure of circulation loop; 14 – speed controller of air fan.

As registration area the glass part (4) of a circulating loop (3) was used. For these purposes opposite to glass part at an angle of 90 degrees, CCD camera (6) and the double pulsed Nd:YAG laser (7) were set. The researched spherical particle (5) was fixed in registration area of a CCD camera (6) in the glass channel (4) on its axis of symmetry.

As material of particles plasticine has been chosen. All particles were made spherical shape. The sizes (diameters – d_0) of particles changed from 1mm to 5 mm. In each experiment at the same time from 1 to 2 particles were used. For creation of an air stream in a circulating loop (2) the air fan (3) complete with the speed controller (14) has been used. Air speed (U_g) in the registration area changed in the range of 1-3 m/s. For each value of a particle diameter (d_0) and air velocity (U_g) not less than 10 experiments were made. Length of an aerodynamic trace (l_{trace}) of a particle, and also characteristics and features of air flow were defined in each experiment

For determination of an air stream velocities PIV [4] and PTV [5] methods have been used. PIV and PTV techniques are based on the photographing of flow images appeared in the plane intersected by laser light, given that the optical axis of the CCD camera is perpendicular to the plane of laser light. We introduced the tracers in the air flow. To do it, we used the specific device that is a metallic container. Air compressor was connected to this container. Using the metal hose, the compressed (about 200 kPa) mixture of TiO_2 (particle size is less than 1 μm, the maximum use temperature – 2300 K) and air was injected into loop (2) through an openings (12). After injecting, the tracers continued to move through the air flow produced by air fan (3) inside of the loop (2).

The error in determining of U_g using this approach did not exceed 2% [2, 3].

3 Results and discussion

As a result of carrying out experimental studies the two-dimensional velocity fields illustrating process of a spherical particle streamlining by air stream were received. During the analysis of videograms of process lengths of aerodynamic traces have been determined. After comparison of the received results dependences of the l_{trace} parameter on velocities of an air stream (U_g) and initial diameter of the used particles (d_0) have been constructed. The received results are presented at fig. 2.

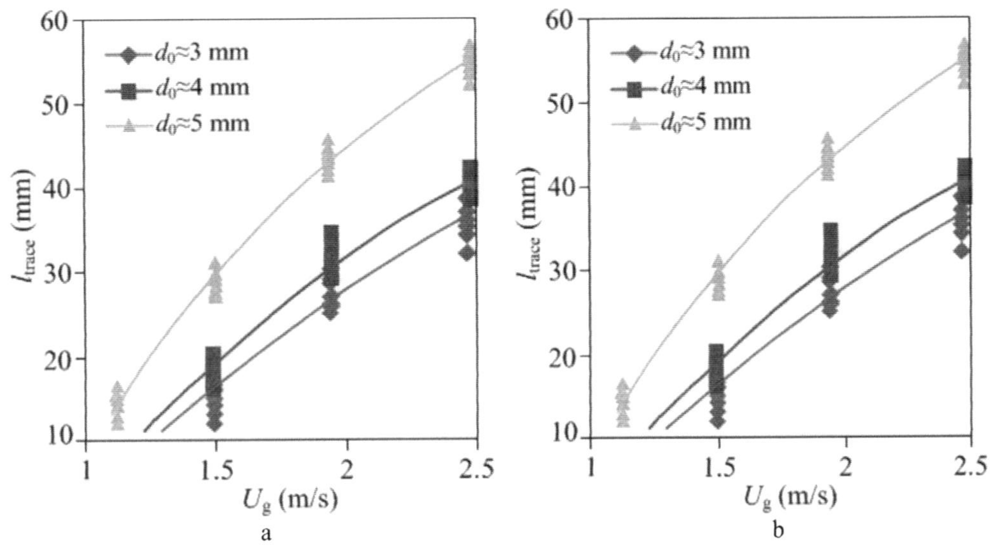

Fig. 2. Dependences of length of an aerodynamic trace of a particle on velosities of air flow (a) and sizes of a particle (b).

The analysis of fig. 2 allows to speak about the defining influence of speed of U_g on length of an aerodynamic trace (fig. 2a). So, growth of U_g by 2.5 times leads to increase in l_{trace} values more than by 3-5 times. And for rather large particles (d_0≈5 mm) influence more essential. Influence of initial diameter of a particle (d_0) on l_{trace} values has more moderate character (fig. 2b). Increase of d_0 in the range of 3-5.5 mm leads to increase of the l_{trace} parameter no more than by 1.5-2 times.

Such result (fig. 2) shows importance of accounting of such parameter as U_g when forecasting features of a streamlining of spherical particles. It is important to consider especially this feature in technologies which use the drop flows moving in opposite directed gas stream. For example, vaporizing cooling of water in coolers of ventilator type, thermal (fire) treatment of liquids and water fire extinguishing. Selecting the necessary parameters of a drop stream it is possible to provide conditions under which drops will pass almost freely a counter gas flow. Or opposite, it is possible to reach conditions of full braking of a stream.

Experimental researches of influence of a relative droplets positioning in a gas stream on characteristic of their aerodynamic traces have been conducted. The example of the obtained data illustrating features of a streamlining of two spherical particles (d_0≈5 mm) is given at fig. 3.

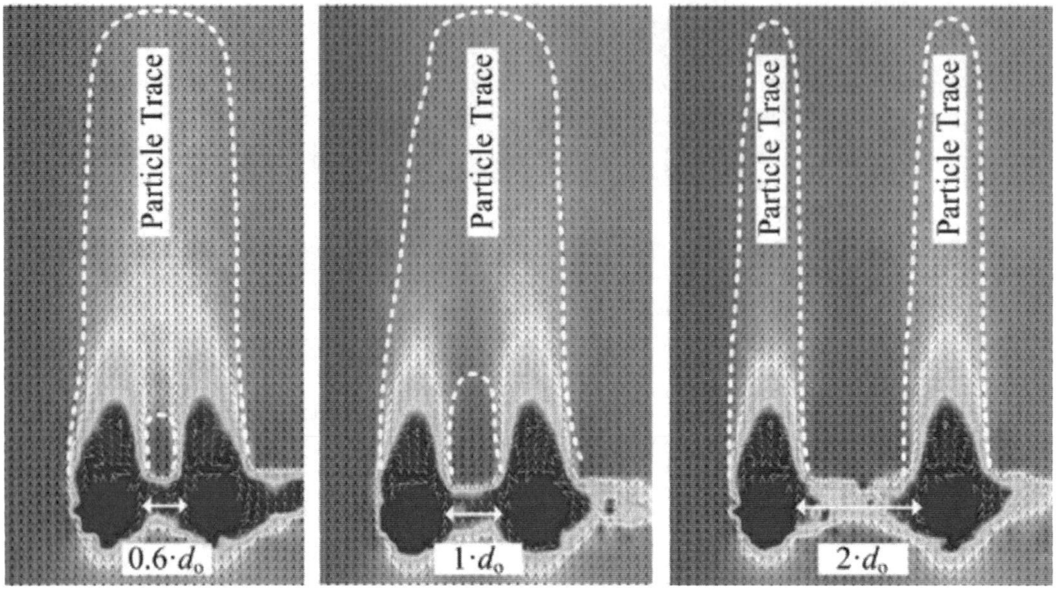

Fig. 3. 2D velocity fields of an air stream obtained with use of the PIV method.

During experiments it has been established that at rapprochement of particles on distance less than $1 \cdot d_0$ is observed one general aerodynamic trace. At the same time length of the integrated trace for 5-10% exceeds length of an aerodynamic trace of a single particle. During removal of particles on distance more $(1.5\text{-}2) \cdot d_0$ at each particle is observed the separate aerodynamic trace.

3 Conclusion

During experimental studies features of a streamlining by an air stream of spherical particles are established. The influence of velocity of an air stream on the length of an aerodynamic trace of a particle was shown. It is established that the size of a particle renders moderate value on the length of an aerodynamic trace. It is recorded that at rapprochement of two particles on distance less than $1 \cdot d_0$ the integrated aerodynamic trace which length for 5-10% exceeds length of a trace of a single particle arises.

Acknowledgments

The investigation was supported by Russian Science Foundation (project 14–39–00003).

References

1. A.Yu. Varaksin, High Temp. **51**, 3 (2013)
2. R.S. Volkov, G.V. Kuznetsov, P.A. Strizhak, Int. J. Heat Mass Transfer **79** (2014)
3. P.A. Strizhak, R.S. Volkov, Exp. Therm. Fluid Sci. **75** (2016)
4. J.V. Simo Tala, S. Russeil, D. Bougeard, J.-L. Harion, Exp. Therm. Fluid Sci. **50** (2013)
5. D. Damiani, E. Meillot, D. Tarlet, J. Therm. Spray Technol. **23**, 3 (2014)

 DOI: 10.1051/matecconf/20167201061

VERIFICATION OF CONJUGATE HEAT TRANSFER MODELS IN A CLOSED VOLUME WITH RADIATIVE HEAT SOURCE

Vyacheslav I. Maksimov[1,*], *Tatiana* A. Nagornova[1], and *Nikolai* I. Kurilenko[2]

[1]National research Tomsk Polytechnic University, 634050 Tomsk, Russia
[2]Tyumen State Architecture - and Construction University, 625001 Tyumen, Russia

Abstract. The results of verification of mathematical model of convective-conductive heat transfer in a closed volume with a thermally conductive enclosing structures are presented. Experiments were carried out to determine the temperature of floor premises in the working conditions of radiant heating systems. Comparison of mathematical modelling of temperature fields and experiments showed their good agreement. It is concluded that the mathematical model of conjugate heat transfers in the air cavity with a heat-conducting and heat-retaining walls correspond to the real process of formation of temperature fields in premises with gas infrared heaters system.

1 Introduction

The tasks of free convection in the closed volumes attract attention of many researchers in the recent decades. A large part of research is devoted to the processes of heat transfer without regard to the heat sink in the building envelope (eg [1-3]). To the tasks of Conjugate heat transfer is given much less attention (for example 4, 5) due to the problems of energy equations solution in the area with discontinuities transfer coefficient. The tasks of free convection in the regions with the heat-conducting walls and the heat-storing enclosing constructions with local radiant heating sources are even more complex [6-8]. The solutions [6-8] are obtained for the conditions of complex heat exchange – due to emitter energy, located in the upper part of solution region, when heated a thin near-surface layer of the lower part of the enclosing rectangular constructions with the subsequent warming up of air as a result of free convection under the conditions of heat sink in the vertical walls. But results [6-8] did not pass the verification stage according to the experimental data due to problems of the organization of experimental studies at the necessary for the comparison level of the metrological guarantee of temperature measurements.

The purpose of this research - the experimental and theoretical study of the processes conjugate heat transfer in a closed rectangular premises with a local radiant heat source and heat-conducting walls under identical conditions. The main objective is to verify the

* Corresponding author: elf@tpu.ru

mathematical formulation [6-8] on the results of experiments on large-scale models of typical objects with radiant heating systems.

2 The methodology of the research and experimental results

Research on measuring temperatures in typical sections held indoors (fig. 1) with a gas infrared emitter (Russian manufacture "Termoshvak") located in the upper part of the analysis region. Power of emitter – 30 kW. On the concrete base, starting from the axis of symmetry of the emitter in the directions *X* and *Y* (fig. 1) were established thermocouples at a distance 300 mm from each other.

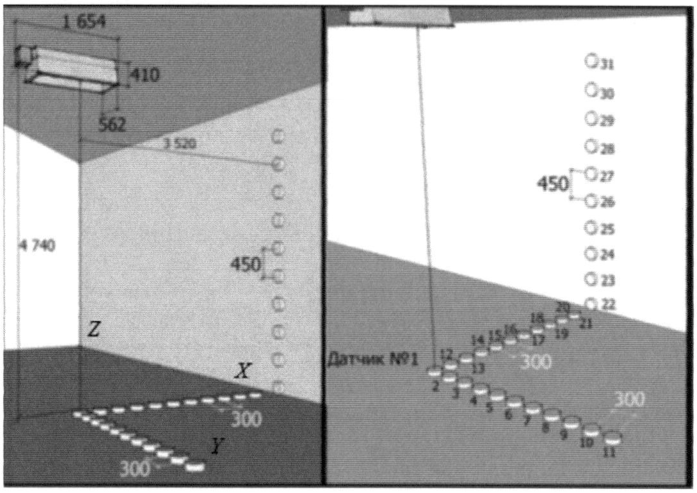

Fig. 1. The area of experimental studies and sensor layout: dimensions in mm.

The studies were conducted in winter period when the outside air temperature minus 20 °C. At the initial time the indoor temperature 18 °C. Dimensions of emitter 1654×410 mm. The vertical enclosing constructions are made from sandwich- panels with the known thermophysical characteristics. Temperature measurements were made in the floor surface during a time interval corresponding to the output of the thermal regime close to a stationary (1 hour) using of thermocouple and optical pyrometer with systematic errors ± 4%. As a results of experiment were obtained values of heat flux to the surface of concrete base (table).

Table. Heat flux distribution along the coordinate x of the section passing through the axis of GIE symmetry to the right side enclosing structure (fig. 1).

coordinate, m (fig. 1)	0	0.5	1	1.5	2	2.5	3	3.5	4
q, W/m^2	190.5	180.2	182.1	172.2	134.2	122.2	102.1	106	80.2

It was found that the proportion of thermal energy from the emitter to the floor surface, close to 57 % of the total allocated heat of GIE. It should be noted that the temperature of the concrete floor surface is selected as the object of measurements due to high heat-retaining capacity of concrete. It is substance with the thermophysical characteristics sufficiently uniform in the volume. Therefore the process of heating concrete floor is sufficiently inertia and improbable the random deviations of temperatures in individual sections of surface.

Fig. 2 shows the results of experiments in the form of a floor surface temperature distributions (T_f) in the direction X.

Distribution of T_f on X shows its nonmonotonic variation with distance from the point corresponding to the symmetry axis of the GIE.

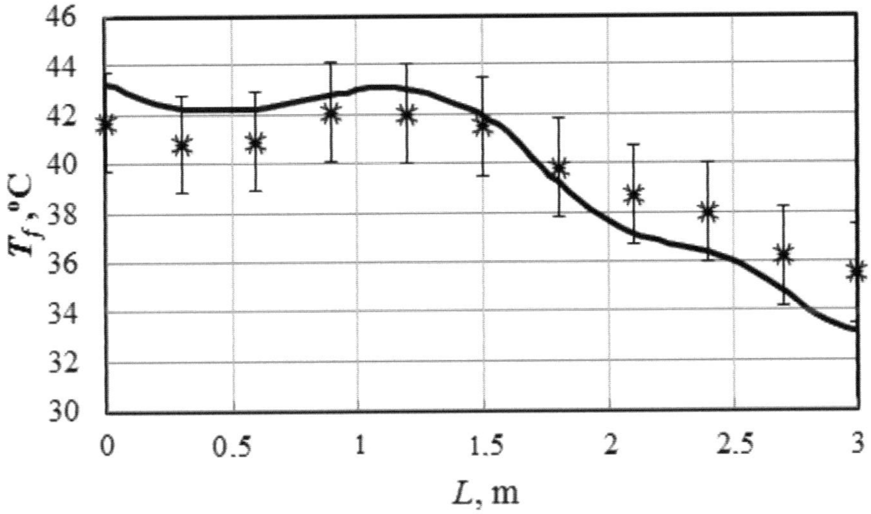

Fig. 2. The distribution of the floor surface temperature of the transmitter center to the side walling: * - experimental results with a confidence interval; solid line - the results of mathematical modeling.

It is clearly seen that at a distance of 3 m from the axis of symmetry the maximum difference T_f is 10 degrees. The character of changes T_f with increasing X complies with the conditions of the inhomogeneous heat transfer at the "concrete – air" section. Such distribution T_f (X) can only be the result of natural convection processes, due to the heating of the concrete to temperatures more 33 °C. One can reasonably conclude about shaping of vortex cells in influence zone of infrared radiation on the surface concrete floor.

3 Numerical investigations

Mathematical modelling of heat transfer processes carried out for the area shown in fig. 3. The dimensions of the solutions field 8×6 m.

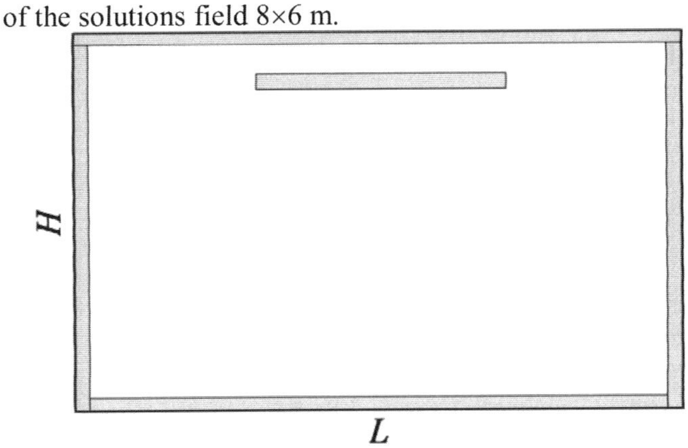

Fig. 3. Area of solving the problem: L, H - the height and length of solutions area respectively.

Statement of the Problem [6] accounted for convective heat transfer in the air and conductive in enclosing structures. For all the external borders of the area to account the convective and radiative heat exchange with the environment. On the left and right boundaries (fig. 3) is outside air with temperature minus 20 °C.

On the surface of the GII was set established in the experiment the surface temperature of the radiator "Tehnoshvak – 20". On three internal interfaces of media the boundary conditions of the fourth kind are assigned (equality of heat fluxes and temperatures). On the lower concrete base was assigned the value of heat flux, which varied depending on coordinate q (X), on the basis obtained experimental data (table). Initial conditions started as the analogous to initial data of experimental studies.

For describing the process of the combined heat transfer in the solution region (fig. 3) as the base was used mathematical model [6-8], developed for calculating the heat transfer in the locked air region with the heat-conducting enclosing walls and the local energy source.

System of equations, which describe the process of heat transfer in this conditions, includes the Navier-Stokes equations, recorded in the transformed variables (function of current – vorticity) [4-8], and the equations of energy for each of the region elements (fig. 3). For the variables the function of current and vorticity on the interfaces of media were written analogous [4-8] boundary conditions.

The system of partial differential equations is solved by the finite-difference method [4, 5] with the use of an algorithm [4, 5], that developed for the numerical solution of free convection in the locked rectangular regions with the local energy sources.

4 Results and discussion

As a result the solutions of the problem of convective- conductive heat transfer are obtained temperature fields and the contour lines of the current function, which describe the processes being investigated. In fig. 4 are represented the results of the numerical simulation, which illustrate basic laws governing the heat transfer in the considered region with the heating system on the base GIE for moment of time, which corresponds to experiment (1 hour).

It is established that near GIE are reached sufficiently high temperatures of air (to 200 °C). But at a distance the half meter from it of temperature considerably they are lowered. In the section Z = 4.5 m T it changes insignificantly: in the range from 32 to 35 °C. And only on the concrete base T = 40 (fig. 2 and 4 *a*). Above GIE also are formed the zones of high temperature (to 110 °C). With the minimally permissible distance of emitter from the surface of upper floors are formed the flow of heated air of the type of Benard cells (fig. 4 *b*).

In fig. 2 are represented the results of experimental and numerical studies in the form of the temperature distributions on the surface of concrete floor in the direction X from the center section of the region to right side vertical wall. In the figure the confidence interval of the measurements of temperatures, which composed ± 2 °C, is also noted. The results of mathematical simulation fall into this interval.

It is possible to make a conclusion about the good agreement of the obtained with the use of a mathematical model 6-8 of theoretical consequences and experimental values of the fundamental characteristic of the process –the temperature of floor surface. With the analysis of laws governing heating air under the conditions in question it is necessary to have in the form that the energy, which comes from emitter, is not absorbed by air, and accumulates in the thin near-surface layer of concrete floor.

Subsequently the concrete heats the air (arriving flying to the surface of floor), which rises upward, freeing the zone of intensive heating for cold air, which is descended. Thus, is formed the circulation flow, whose typical example is given in fig. 4.

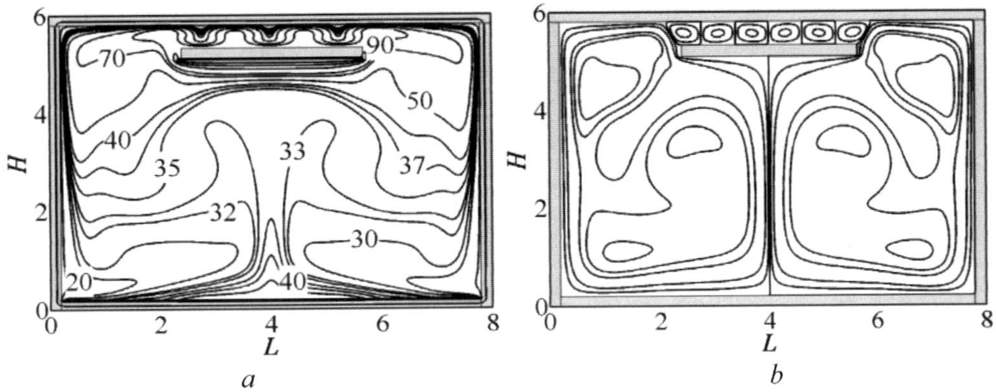

Fig. 4. The temperature distribution (*a*) and current functions (*b*) at time 1 hour. Values the temperature is listed in °C.

In conclusion, it should be noted that the model [6-8], as demonstrated by its verification by comparison with the results of the temperature measurements in an environment adequate accepted in mathematical modeling, has a good potential for development in terms of a more detailed description of the temperature field of emitter. This level of detail can create the preconditions for a more accurate calculation of air temperature and the corresponding specification of the parameters of the microclimate of industrial premises with heating systems based on gas infrared emitters.

5 Conclusion

Verification of the mathematical model of heat transfer process of closed volume with thermally conductive in enclosing structures showed good agreement between the experimental results and the previously developed [6-8] mathematical model of convective-conductive heat transfer in a closed system under consideration with radiant heating.

Acknowledgments

The work was supported by the Russian President's grant (Scientific School project - 7538.2016.8).

References

1. P. Kumar, V. Eswaran, Int. J. Therm. Sc. **67** (2013)
2. K. Lari , M. Baneshi , S.A. Gandjalikhan Nassab, A.S. Komiya, Int. J. Heat Mass Trasf. **54** (2011)
3. R. Roslan, H. Saleh, I. Hashim , A.S. Bataineh, Natural convection in an enclosure containing a sinusoidally heated cylindrical source, Int. J. Heat Mass Trasf. **70** (2014)
4. G.V. Kuznetsov, M.A. Sheremet, Fl. Dyn. **41** (2006)
5. G.V. Kuznetsov, M.A. Sheremet, Int. J. Heat Mass Transf. **54** (2011)
6. G.V. Kuznetsov, N.I. Kurilenko, V.I. Maksimov, G.Ya. Mamontov, T.A. Nagornova, JEPT **86** (2013)
7. T.A. Nagornova, Chem. Petr. Eng. **51** (July 2015)
8. V.I. Maksimov, T.A. Nagornova, EPJ Web Conf. **76** (2014)

MATEC Web of Conferences 72, 01062 (2016) DOI: 10.1051/matecconf/20167201062
HMTTSC-2016

DYNAMIC WETTING AND DEWETTING: COMPARISON OF EXPERIMENT WITH THEORIES

Evgeniya.G. Orlova[1,*], *Ekaterina R.* Malyhina[1], *Dmitriy.V.* Feoktistov[1], and *Larisa.S.* Zhidkova[1]

[1]National Research Tomsk Polytechnic University, 634050 Tomsk, Russia

Abstract. The dynamics wetting/dewetting of a metal surface by distilled water drop was studied experimentally. The advancing and receding dynamic contact angles were obtained as a function of a contact line speed. The hydrodynamic and molecular-kinetic models have been applied to the experimental data to interpret the obtained results. The independent variables of the molecular-kinetic and hydrodynamic models, and the determination coefficient were determined by fitting procedure. The receding contact angles are found to be fitted better to the wetting models in comparison with the advancing dynamic contact angles.

1 Introduction

Technological processes of spray cooling, cultivated crop dusting, spin soldering, spray coating, DNA synthesis are based on the drop spreading and evaporation phenomena [1-7].

When the liquid is moving over the solid substrate, a dynamic contact angle (DCA) is established. This angle reflects the balance between driving and drag forces and is a function of a three-phase contact line (CL) speed.

To describe the physical process of spreading, several theories predicting the relationship between the DCA and the CL speed were developed. These theories can be classified according to the mechanisms of energy dissipation in the vicinity of the contact line: molecular-kinetic (MKT) [8, 9] and hydrodynamic [10, 11] theories.

The molecular kinetic theory considers the molecular jumps at the CL as a rate-activated, with an equilibrium frequency k_o and a displacement distance λ . So, the relationship between the DCA θ_D and the CL speed U is described by [8]:

$$\theta_D = \arccos\left[\cos\theta_S - \left(\frac{2k_B T}{\sigma_{LV}\lambda^2}\right)\arcsin\left(\frac{U}{2k_0\lambda}\right)\right] \qquad (1)$$

where k_B is the Boltzmann constant, T the liquid temperature, θ_S the static contact angle, and λ and k_0 are parameters corresponding to the length of displacement and frequency constant for molecular displacements of the contact line.

* Corresponding author: lafleur@tpu.ru

Nowadays, there is no definitive way of predicting the constants λ and k_0 apriori for a given solid/liquid system. For this reason, these parameters are usially obtained from experiment by curve fitting procedures [12]. The experimentally determined parameters obtaind by different scientific groups are listed in Table 1.

Table 1. Fitting parameters of molecular-kinetic model obtained in some works.

Reference	Liquid	θ_S, °	λ, nm	k^0, Hz
[13]	water	-	0.3	$2\cdot10^{10}$
[14]	water	82	0.36	$8.6\cdot10^{9}$
[15]	water	54.1	1.02	$1.19\cdot10^{3}$
[16]	water	55.5	1.79	$6.67\cdot10^{5}$
[17]	glycerine solution	-	7.3	$3.4\cdot10^{7}$

According to the hydrodynamic model the moving process of the contact line is dominated by the viscous dissipation of the liquid; and the bulk viscous friction is the main resistance force for the contact line motion. The shear flow in the vicinity of the contact line leads to a viscous bending of the liquid/vapor interface. For this reason, the dynamic contact angle deviates from its equilibrium value, and can be defined as [10]:

$$\theta_D^3 = \theta_S^3 \pm \left(\frac{9\mu U}{\sigma_{LV}}\right)\ln\left(\frac{L}{L_m}\right) \tag{2}$$

where L is a characteristic capillary length, and L_m the slip length. In Eq. (2) the plus sign applies to the advancing liquid movement and the minus sign applies to the receding liquid movement.

The capillary length is given by [18]:

$$L = \sqrt{\frac{2\sigma_{LV}}{\rho g}} \tag{3}$$

where σ_{LV} is surface tension at the liquid/vapour interface, and ρ the liquid density.

The slip length L_m is the fitting parameter. It represents the length of the region where the no-slip boundary condition of classical continuum theory does not hold, so this value should be in the order of molecular dimension [15].

The aim of this work is to obtain the dependences of the advancing and receding dynamic contact angles from the contact line speed of a drop when it spreads over the metal surface. Also the reconciliation of experimental data with the hydrodynamic and molecular-kinetic models of wetting dynamics was undertaken.

2 Materials and methods

Experimental setup (fig.1) was used to study the wetting and spreading over the solid substrate [19, 20].

Experimental setup consists of the electronic syringe pump *1* to supply the distilled water through the opening 2 mm in diameter from the lower side of the substrate *2*. Thus, the bottom-up method [21] for drop formation (0.3 ml) with a liquid flow rate of 0.04 ml/s was implemented. The substrate was set on the special platform, which can be heated from the bottom side of the thermoelectric module *3* attached to the aluminium plate of the goniometer *4*. To record the temperature, the chromel-alumel thermocouples *5* were used.

The light beam from the source *8* passed through a fiber-optic illuminator, and converted to the plane-parallel light. The drop spreading over the surface was lighted, and the shadow images of this process were obtained by the high-speed camera Fastvideo-500M (with a maximum resolution of 1280×1024 pixels and at 10 frames per second) equipped with a macro lens.

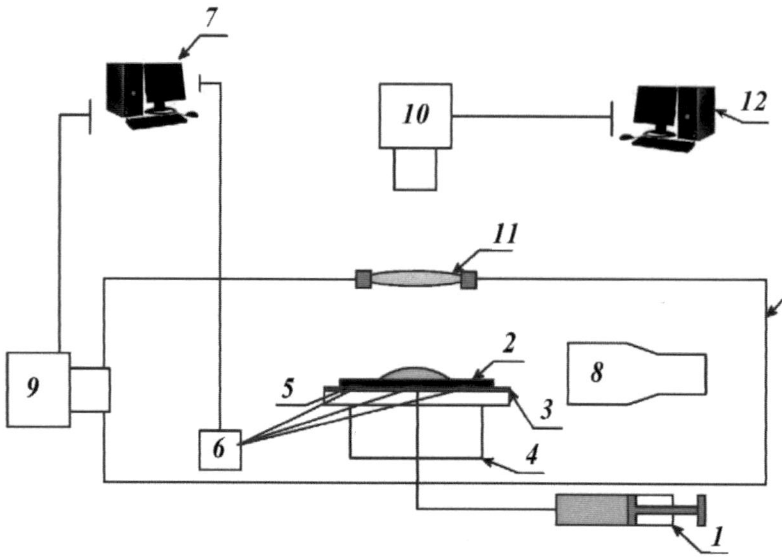

Fig. 1. The scheme of the experimental setup: 1 – syringe pump; 2 – substrate; 3 – thermoelectric module; 4 – goniometer; 5 – thermocouples; 6 – analog-to-digital converter; 7, 12 – personal computers; 8 – light source; 9, 10 – high-speed video cameras; 11 – lens of the upper optical system; 13 – transparent box.

Control of the geometrical shape of the drop was carried out by the high-speed camera *10*. To increase the size of the video frames, the lens *11* was used.

The experimental setup is isolated from the external factors changing the temperature, velocity of air flows and environmental humidity by the transparent box *13* made of polymer glass with 3 mm thick.

The substrate made of brass was used in the experiment. Its surface was studied by profilometer "Micro Measure 3D station". The arithmetic average surface roughness for the brass substrate is equal to 0.016 μm.

The error of dosing by the syringe pump does not exceed 2%. Random error is calculated from the results of repeated experiments. To do this, from two to six experiments were performed at fixed values of the studied factors. Thereafter, the mean values of the contact angle and standard deviations were defined. The variation coefficients were equal to from 10% to 14%. Random errors of measurements were mainly due to the small limits when dispensing liquid through a silicone tube.

3 Results and discussion

According to results of the experiments the advancing and receding DCA were defined during wetting and dewetting of the substrate. The spreading of the drop starts at contact angle of 83.6°±1.5° that corresponds to the maximal CL speed (1.76 mm/s). The spreading ends at the angle of 72.1° ± 1.5°. This value corresponds to the the pre-measured static angle of water/brass/air system $\theta_s = 70.8 \pm 1.5°$. The CL speed in this case is equal to zero.

Then the drop relaxes [13], and the dewetting starts. The liquid is being withdrawn by the syringe pump, and the receding DCA is being formed.

The hysteresis of the contact angle is defined as the difference between the advancing and receding DCA, when the CL speed is equal to zero (fig. 2).

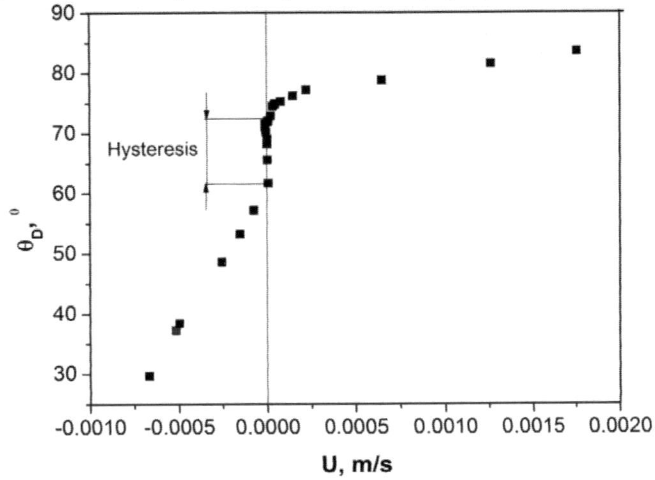

Fig. 2. The advancing and receding dynamic contact angles versus contact line speed. The contact angle hysteresis.

The equations of MKT (1) and hydrodynamic models were used for interpretation of the experimental data and obtaining of functional dependences of the DCA from the CL speed.

Figure 3 shows the dependences of the advancing and receding DCA from the CL speed.

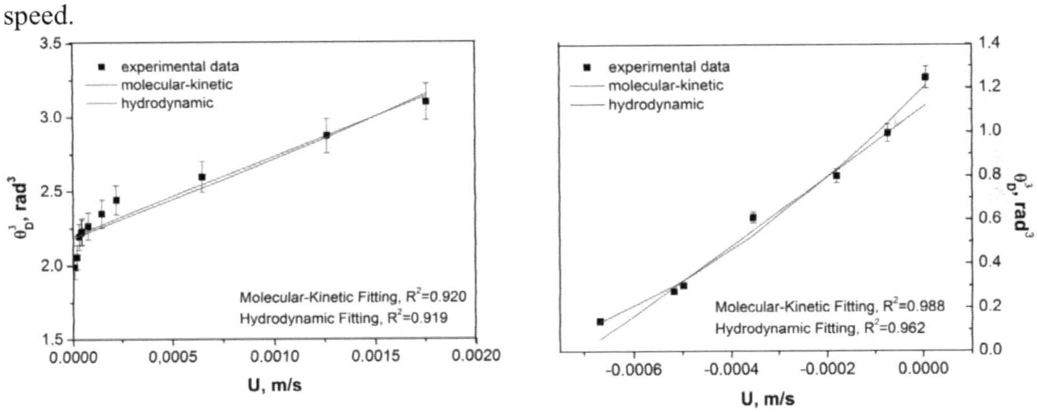

(a) (b)

Fig. 3. Fitting of the MKT and hydrodynamic models to (a) the advancing DCA and (b) the receding DCA.

The treatment of the experimental data was performed by the least squares method. After that, the obtained experimental data of advancing and receding DCA have been separately fitted to the MKT and hydrodynamic wetting models. To estimate the reconciliation of the experiment and models, determination coefficient R^2 has been obtained (fig. 3).

According to the results of the reconciliation with the molecular-kinetic model two independent variables were defined: the length of displacement and frequency constant for molecular displacements of the contact line. An initial guess input of k^0 и λ were taken from Table 1.

The value of the $\ln\left(\frac{L}{L_m}\right)$ parameter was determined at fitting the experimental data with the hydrodynamic model. The value of the L_m parameter are not presented separately, because it is not physically meaningful. Probably, this is due to the fact that the value of the capillary number $Ca = \frac{\mu U}{\sigma_{LV}}$ in Eq. (2) is smaller, in order of 10^{-6} for distilled water, in comparison with the value of θ_S^3.

The values of the molecular-kinetic and hydrodynamic parameters and R^2 values from the fitting for the advancing and receding contact line are listed in Table 2.

Table 2. Fitting parameters of molecular-kinetic and hydrodynamic models obtained in work.

	θ_S,° (measured)	Molecular-kinetic model				Hydrodynamic model		
		θ_S,°	k^0, 10^9 Hz	λ, nm	R^2	θ_S,°	$\ln\left(\frac{L}{L_m}\right)$	R^2
Advancing CL	70.8±1.5°	74.4	8.13	0.042	0.920	62.6	44400	0.919
Receding CL		60.9	6.51	0.025	0.988	58	12700	0.962

It was found that Eq. (1) and (2) fit well with the experimental data. However the receding contact angles $\theta_R = f(U)$ were fitted better in comparison with the advancing DCA $\theta_A = f(U)$. It is confirmed by the values of the determination coefficients for the receding contact line R^2=0.988 and R^2=0.962 for the molecula-kinetic and hydrodynamic models, respectively.

Acknowledgments

The reported study was supported by the grant of President of Russian Federation for the government support of young Russian scientists (MK-6810.2016.8).

References

1. W.M. Grissom, F.A. Wierum, Int J Heat Mass Transf. **24**, 261 (1981)
2. D.V. Zaitsev, O.A. Kabov, Microgravity Sci Technol. **19**, 174 (2007)
3. D.O. Glushkov, G.V. Kuznetsov, P.A. Strizhak, R.S. Volkov, Thermal Science **20**, 131 (2016)
4. D.V. Zaitsev, E.A. Chinnov, O.A. Kabov, I.V. Marchuk, Technical Physics Letters **30**, 231 (2004)
5. V.P. Lebedev, V.V. Lemanov, S.Ya. Misyura, V.I. Terekhov, Fluid Dynamics **28**, 624 (1993)
6. S.Ya. Misyura, International Journal of Heat and Mass Transfer **71**, 197 (2014)
7. K. Batischeva, E. Orlova, D. Feoktistov, EPJ Web Conf. **19**, 01001 (2014)
8. T. D. Blake, J. M. Haynes, J. Colloid Interface Sci. **30**, 421 (1969)
9. E. Ruckenstein, C. S. Dunn, J. Colloid Interface Sci. **59**, 135 (1977)
10. R. G. Cox, J. Fluid Mech. **168**, 169 (1986)
11. O. V. Voinov, J Fluid Dyn **11**, 714 (1976)

12. T.D. Blake, J Colloid Interf Sci **299**, 1 (2006)
13. R. Sedev, Adv Colloid Interface Sci. **222**, 661 (2015)
14. T.D. Blake, M. Bracke, Y.D. Shikhmurzaev, Phys. Fluid **11**, 1995 (1999)
15. S.R. Ranabothu, C. Karnezis, L.L. Dai, J. Colloid Interf. Sci. **288**, 213 (2005)
16. R. Fetzer, J. Ralston, J. Phys. Chem. C. **113**, 8888 (2009)
17. J.-H. Kim, H.P. Kavehpour, J.P. Rothstein, Phys. Fluids **27**, 032107 (2015)
18. M. Schneemilch, R. A. Hayes, J. G. Petrov, J. Ralston, Langmuir **14**, 7047 (1998)
19. E.G. Orlova, G. V. Kuznetsov, D. V. Feoktistov, EPJ Web Conf. **82**, 01053 (2015)
20. D. Feoktistov, E. Orlova, A. Islamova, EPJ Web Conf. **110**, 01018 (2016)
21. K. Gleason, S.A. Putnam, Langmuir **30**, 10548 (2014)

DROPLET SPREADING PROCESS IMPACT ON IGNITION CHARACTERISTICS OF CONDENSED MATERIALS

Gennady Ya. Mamontov[1, *], *Galina* V. Taratushkina[2], and *Elena* Yu. Kurilenko[3]

[1]National Research Tomsk Polytechnic University, 634050 Tomsk, Russia
[2]Seversk Institute of Technology of MIFI National Nuclear Research University, 636036 Seversk, Russia
[3]Civil engineering Institute of Tyumen Industrial University, 625000 Tyumen, Russia

Abstract. Mathematical simulation of condensed material solid-phase ignition in the context of the in-situ heating by the melted or heated to high temperature metal droplet was carried out. The authors developed the mathematical model that describes the heat transfer process in the "droplet – condensed material" system by the system of heat transfer equations with boundary and initial conditions. The problem is solved by the finite difference method. Four modes of condensed material ignition that are distinguished by the temperature range of every mode were identified for standard conditions of the in-situ heat effect.

1 Introduction

Ignition processes of condensed materials in the course of the in-situ heating by a single particle have been intensively studied for the last few years. The problems of solid [1, 2] and liquid [3, 4] fuels are solved. However, the results [1–4] are obtained for conditions when the solid particles had an effect on condensed materials. In the real world, there are some types of liquid-alloy droplet impact on condensed materials. These droplets spread over the cold surface in the course of solidification. Up to the present moment, the problem of fluid drop spreading in the course of its solidification on the cold surface is not solved. Therefore, it is worth analyzing conditions and characteristics of condensed material ignition by a droplet heated to high temperatures while describing the spreading process based on relatively simple models.

2 Problem statement and solution procedure

The problem of condensed material ignition by a single hemispherical droplet of a small radius is studied (fig. 1). When the process is simulated, the hemisphere is replaced by a segment with an initial size h_o, one edge of which corresponds to the "droplet – condensed material" interface and another one to the droplet upper bound. The type 3 boundary

* Corresponding author: gmamontov@tpu.ru

conditions that took into consideration the heat transfer with an outdoor environment by means of convection and emission were set at the upper bound. In such a case, the particle spreading was simulated by the hemisphere height change (segment length h) at a target speed v, the value of which was the problem parameter and varied at the rage from 0 to 1.67 m/s. This range was selected in terms of experimental data on fluid drop spreading speed over the solid surface [5, 6].

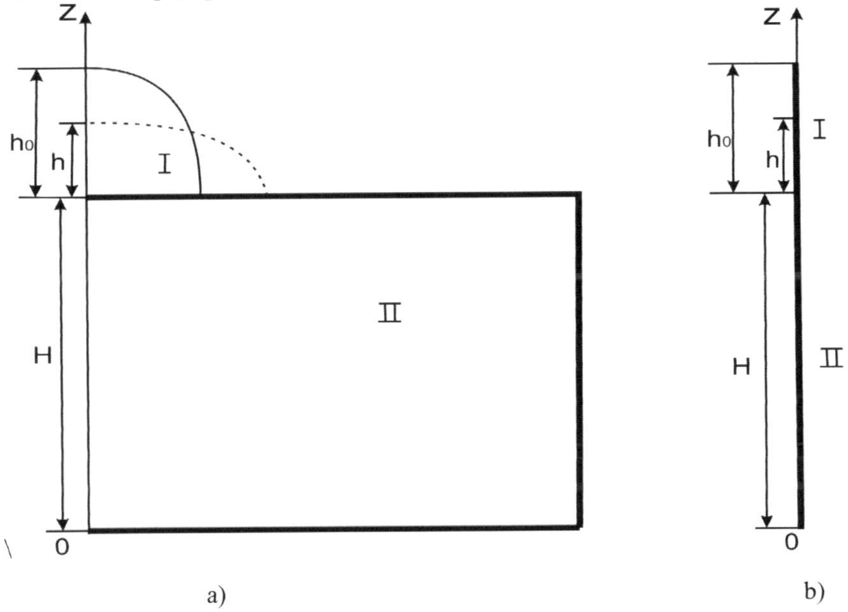

Fig. 1. Problem solution domain: a) geometric interpretation of the drop spreading process; b) stated geometric layout of the process.

The problem was solved in a fixed system of reference. The origin of the system was connected with a nonheating surface of condensed materials. The height of segment I was depressed in the course of spreading. The bed depth of condensed materials was selected in such a way that its initial temperature was preserved on the nonheating surface until the ignition moment. It is determined [1, 2] that when the condensed materials are ignited by the single particle heated to high temperatures, the condensed material ignition conditions are first reached on the particle symmetry axis (in axisymmetric problem solving) at the "particle – condensed material" interface point. Hence, the ignition delay times $t_з$ obtained in one-dimensional problem solving, at a first approximation can be considered as $t_з$ proper values for particles (or droplets) of large diameter (more than 100 µm).

When the problem was set up, it was supposed that the droplet has a temperature ($T_к$) that is much higher than an initial temperature (T_0) of condensed material and an exterior temperature T_v (the process of the melted metal droplet fallback from a "cold" air on a "cold" surface of condensed material was simulated). This set up corresponds to conditions of the hazardous material ignition by melted metal droplets that are formed while cutting off or welding metal structures or in an emergency in the chemical industry when as a result of an explosion the droplets of polymers softened to high temperatures are splashed around the site of explosion.

The problem is set under conditions that when particles do not spread over the condensed material surface, the phase transition heat of metal or polymer solidification is not generated and the steplike thermal and physical characteristics of the droplet material do not change when transferring from one phase state to another. The solid-state reaction of condensed material ignition was studied.

Under these conditions, the mathematical model of the process under study is a system of equations of the particle energy and the condensed material, and an equation that describes the "response depth" η changes under appropriate initial and boundary conditions:

$$\frac{\partial T_1}{\partial t} = \frac{\lambda_1}{c_1\rho_1}\frac{\partial^2 T_1}{\partial z^2} + \frac{Q}{c_1}\frac{d\eta}{dt}, \qquad 0 \le z \le H \tag{1}$$

$$c_2\rho_2\frac{\partial T_2}{\partial t} = \lambda_2\frac{\partial^2 T_2}{\partial z^2}, \qquad H \le z \le H + h \tag{2}$$

$$\frac{dy}{dx} = k_0(1-\eta)exp\left(-\frac{E}{RT_1}\right)$$

The boundary and initial conditions for the defined problem are:

$$-\lambda_1\frac{\partial T_1(0,t)}{\partial z} = 0, \qquad 0 < t < t_3 \tag{3}$$

$$-\lambda_1\frac{\partial T_1(H,t)}{\partial z} = -\lambda_2\frac{\partial T_2(H,t)}{\partial z}, \quad T_1 = T_2, \ 0 < t < t_3 \tag{4}$$

$$-\lambda_2\frac{\partial T_2(h+H,t)}{\partial z} = \alpha\left(T_g - T_2\right) + \varepsilon\sigma\left(T_g^4 - T_2^4\right), \ 0 < t < t_3 \tag{5}$$

$$T_1(z,0) = T_0, \qquad 0 < z < H \tag{6}$$

$$T_2(z,0) = T_{ch}, \qquad H < z < h + H \tag{7}$$

$$h = h_0, \ t = 0. \tag{8}$$

$$h = h_0 - vt, \qquad 0 < t < t_3 \tag{9}$$

$$v = 0, \ T_2 < T_{\text{pl}} \tag{10}$$

$$\eta(z,0) = 0, \quad 0 < z < H \tag{11}$$

The numerical analysis was carried out under the following values of the major material and problem parameter characteristics: λ_1=0.12 W/m·K; c_1=1670 J/kg·K; ρ_1=500 kg/m^3; λ_2=41.0 W/m·K; c_2=574 J/kg·K; ρ_2=7870 kg/m^3; E=111000 J/mmol; Q=2.9·10^6 J/kg; k_0=7·10^6 1/s; ε=0.5; α=300 W/m^{2}K; T_{pl}=930 K; T_0=T_g=300 K, where indexes 1correspond to the condensed material and 2 – to the droplet.

The stated boundary value problem is solved in a similar way [1, 2] by the finite difference method using an algorithm developed to solve non-linear problems of the heat transfer with movable boundaries of the simulation domain [7] and in-situ heat release sources [8]. An unequally-spaced and irregular grid [6] was used. The time step was chosen according to conditions of the iteration process convergence [7].

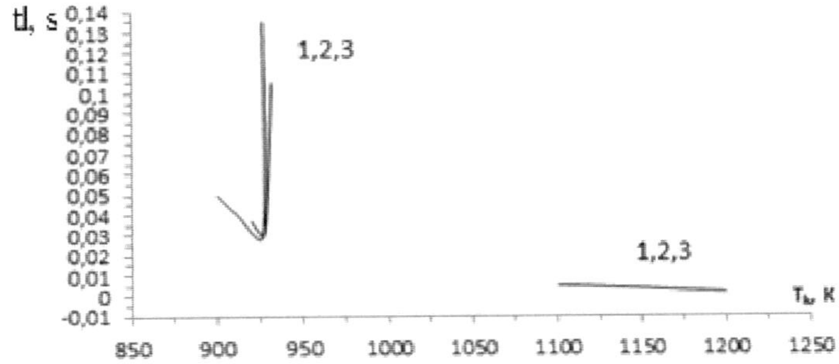

Fig. 2. Dependence of t_l on T_k at the "particle lifetime" of $3\cdot10^{-3}$s: 1 – h_0 = 0.001 m, v=0.333 m/s; 2 – h_0 = 0.003 m, v=1.0 m/s; 3 – h_0 = 0.005 m, v=1.667 m/s.

Figure 2 shows ignition delay time dependences of the studied pattern condensed material on the droplet initial temperature under various spreading rates υ and the initial droplet height h_0. υ and h_0 values were chosen to provide the constant droplet "lifetime" (i.e. time when the droplet height significantly differs from 0). Figure 2 displays that it is possible to distinguish four characteristic variation ranges of the droplet initial temperature that correspond to various types of condensed material ignition mechanisms.

The first is a lower temperature range (lower than the droplet metal melting point) that is characterized by $t_з$ value reduction when $T_к$ grows and it corresponds to the general requirements of condensed material ignition theory under the in-situ heating [9, 10]. At this range, the droplet is solid and when $T_к$ grows, the ignition delay time is reduced.

The second is a narrow range where $T_к$>T_{pl} and the condensed material ignition process is carried out under conditions of droplet spreading over the condensed material surface. In such a case, due to the high spreading rates, $T_к$ value increase (at the narrow range) causes a paradoxical result - $t_з$ increase due to the droplet initial temperature growth. In comparison with the first range, $t_з$ change is caused by the fact that under T_{ch}>T_{pl} the droplet of a certain height is solidified, and its bigger part spreads over the condensed material surface. Hence, the per-unit-volume heat content (per unit surface area) of the droplet is essential in comparison with the case when the droplet is solid. Thus, the higher the droplet initial temperature is, the bigger the area where it spreads is. In this case, after a complete solidification, the droplet height is lower and more time is necessary to heat the condensed material near-surface layer to meet the ignition requirements.

The reasonable progress of the second range is the third one when the droplet temperature is high enough and decreases quickly when the droplet spreads over the condensed material surface up to the complete solidification. In this case, the spreading time is much lower than the ignition delay time (the solidified layer height is very small, and the droplet is not able to heat the condensed material at a necessary depth to ignite it). If $T_к$ increases, the smaller part of the droplet is left in each section under consideration after the solidification and the heat current at the reaction zone is decreased.

When $T_к$ ($T_к$<*1100K*) values are high enough, the droplet heat content is sufficient to meet the ignition requirements with very low $t_з$ values (fig.2). At this range (that can be called the fourth one), the high temperatures of the low height droplet are sufficient and the droplet is able to serve as an in-situ condensed material ignition source.

Based on the obtained results it is possible to make a conclusion that if the droplet "lifetimes" are equal on the condensed material surface, the dependence $t_з(T_к)$ is almost identical (fig.2). In other words, when the droplet spreads over the condensed material surface, υ is also a key parameter as well as $T_к$.

The specified conformities become apparent when the spreading rates are completely different, υ change from 0.333 m/s to 1.667 m/s does not provoke any remarkable changes in t_3. The ignition condition analysis at high rates υ is inappropriate because at high impact rates, even the small droplets will splash and the probability of the condensed material ignition will be close to 0. Furthermore, it is possible to mention that under the real fire-hazardous conditions of metal welding or cutting off, the melted metal droplets splash from the ignition zone with the initial rate that is not higher than several dozens of meters per second. Then, the droplet rate dramatically falls.

In conclusion, it is worth mentioning that all obtained results refer to the nondegenerated droplet existence modes on the condensed material surface (for modes when at least 10% of droplet metal is free of distortion). If the impact velocity is that that the particle spreads immediately up to the thickness of micron units, the condensed material ignition is not possible. When the droplet material melting point rises, the first, the second and the third ignition ranges (fig. 2) are degenerated as at high $T_к$ the ignition delay time of the condensed material will not depend on $T_к$. However, when $T_к$ decreases, the ignition mechanism becomes more complex (fig. 2).

3 Conclusion

The results of the carried theoretical studies show the condensed material ignition process in the context of the in-situ melted metal droplet heating and show the ability to realize four mechanisms to initiate the combustion reaction at different droplet temperatures. The received theoretical results and the stated conformities are the basis to develop the condensed material ignition models.

Acknowledgments

The work was supported by the Russian President's grant (Scientific School project - 7538.2016.8).

References

1. Yu.A. Chumakov, A.G. Knyazeva, Combust. Explos. Shock Waves **48** (2012)
2. G.V. Kuznetsov, G.Ya. Mamontov, G.V. Taratushkina. Khim. Fiz. **23** (2004)
3. G.V. Kuznetsov, P.A. Strizhak., J. Eng. Thermophys. **17** (2008)
4. O.V. Vysokomornaya, G.V. Kuznetsov, P.A. Strizhak., Russ. J. Phys. Chem. B **5** (2011)
5. V.E. Nakoryakov, S.Y. Misyura, S.L. Elistratov. Boiling crisis in droplets of ethanol water solution on the heating surface, J. Eng. Thermophys. **22**, 1-7 (2013)
6. V.E. Nakoryakov, S.Y. Misyura, S.L. Elistratov., Int. J. En. Tech. **4** (2012)
7. V.L. Strakhov, A.N. Garashchenko, G.V. Kuznetsov, V.P. Rudzinskii., Combust. Explos. Shock Waves **37** (2001)
8. G.V. Kuznetsov, M.A. Sheremet., Fl. Dyn. **41** (2006)
9. R.S. Burkina, A.G. Knyazeva., Combust. Explos. Shock Waves **28** (1992)
10. G.V. Kuznetsov, G.Ya Mamontov, G.V. Taratushkina., Combust. Explos. Shock Waves. **40** (2004)

MATEC Web of Conferences **72**, 01064 (2016) DOI: 10.1051/matecconf/20167201064
HMTTSC-2016

MODELLING COAL COMBUSTION CONDITIONS AT ITS STORAGE IN A STACK

A.S. Masanik[1], and *A*.V. Vasilevskiy[1,*]

[1]National Research Tomsk Polytechnic University, 634050 Tomsk, Russia

Abstract. A mathematical model describing the heat and mass transfer processes during storage and ignition of coal stored in the stack, has been proposed. Numerical studies have allowed to determine the composition of gases and temperature profiles in the depth of the stack, depending on the air temperature and cloud cover. It was found that the cloudless hot weather (daytime temperatures of 303 K) within three days the top of the stack is heated to a temperature is its fire.

1 Introduction

Russian power industry is the largest consumer of primary fuel resources in the world. On a share of Thermal power stations account for about 30% of coal and more than 25% of natural gas produced in the country, and 10% of the produced oil. Thus the share of electricity in the internal coal consumption is 50%, and natural gas - 40%. These data indicate that the coal reserves are large in Russia.

It is known that coal can ignite spontaneously, especially when improper storage or in hot sunny weather. In many cases, the cause of fires in coal bunkers, and the open spaces is spontaneous combustion. Therefore, a central place in this area takes the problems of increasing the reliability of solid fuel storage. This problem can be solved by creating a scientific basis predictive modeling processes that occur during storage of solid fuels. Relevance topics of the research defined by its focus on creating the scientific foundations of security during storage and processing of natural solid fuels.

2 Formulation of the problem

Considered coal stack onto the upper surface of which can affect the convective and (or) the radiative heat flow. The aim is to determine the conditions of the surface layer stack heat and mass transfer with the environment (the conditions of the thermal effects on the stack), in which coal ignition in a stack may occur.

A mathematical model of heat and mass transfer, and coal ignition which is stored in the stack consists of a system of equations [1]:

mass conservation of the original condensed matter (coal)

[*]Corresponding author: subbot@tpu.ru

$$\rho_1 \frac{\partial \varphi_1}{\partial t} = -R_p, \qquad R_p = k_p \rho_1 \varphi_1 \exp\left(-\frac{E_p}{RT}\right); \tag{1}$$

mass conservation of water in a bound state

$$\rho_2 \frac{\partial \varphi_2}{\partial t} = -R_L, \quad R_L = k_L \rho_2 \varphi_2 \exp\left(-\frac{E_L}{RT}\right); \tag{2}$$

mass conservation of coke

$$\rho_3 \frac{\partial \varphi_3}{\partial t} = \alpha_C R_P - sR_C, \quad R_C = \frac{M_C}{M_{O_2}} k_C \rho_5 \varphi_5 \varphi_3 \exp\left(-\frac{E_C}{RT}\right); \tag{3}$$

mass conservation of ash

$$\rho_3 \frac{\partial \varphi_3}{\partial t} = \alpha_S R_P, \text{ where } \alpha_C = \frac{\nu'_3 M_3}{\nu_1 M_1}, \quad \alpha_S = \frac{\nu'_4 M_4}{\nu_1 M_1}, \quad \sum_{j=1}^{5} \varphi_j = 1; \tag{4}$$

the continuity of the gas phase

$$\frac{\partial}{\partial t}(\rho_5 \varphi_5) + \frac{\partial}{\partial z}(\rho_5 \varphi_5 v) = \alpha_G R_P + R_L + sR_C, \qquad \alpha_G = \sum_{\alpha=2}^{5} \frac{\nu'_{5\alpha} M_{5\alpha}}{\nu_1 M_1}; \tag{5}$$

mass conservation α-gas phase constituents

$$\frac{\partial}{\partial t}(\rho_5 \varphi_5 c_\alpha) + \frac{\partial}{\partial z}(\rho_5 \varphi_5 v c_\alpha) = \frac{\partial}{\partial z}\left(\rho_5 \varphi_5 D_\alpha \frac{\partial c_\alpha}{\partial z}\right) + R_{5\alpha} + R_{5S\alpha}, \quad \alpha = 1, \ldots, 6; \tag{6}$$

conservation of energy

$$\sum_{j=1}^{5} \rho_j \varphi_j c_{pj} \frac{\partial T}{\partial t} + \rho_5 \varphi_5 c_{p5} v \frac{\partial T}{\partial z} = \frac{\partial}{\partial z}\left(\lambda \frac{\partial T}{\partial z}\right) + q_P R_P - q_L R_L + q_C s R_C; \tag{7}$$

gas movement in the pores and state equation

$$v = -\frac{k}{\mu}\left(\frac{\partial p}{\partial z} - \rho_5 g\right), \quad p = \frac{\rho_5 RT}{M}, \quad M = 1 \Big/ \sum_{\alpha=1}^{7} \frac{c_\alpha}{M_{5\alpha}}. \tag{8}$$

The system of differential equations (1) – (8) was solved with the boundary conditions

$$t = 0: \quad \varphi_j = \varphi_{jH}, \ j = \overline{1,4}, \ c_\alpha = c_{\alpha H}, \ \alpha = \overline{1,6}, \ p_5 = p_H, \ T = T_H; \tag{9}$$

$$z = 0: \quad p = p_e, \ c_\alpha = c_{\alpha e}, \ \lambda \frac{\partial T}{\partial z} = \alpha_1 (T - T_e) + q_R; \tag{10}$$

$$z = H: \quad p = p_H, \quad \frac{\partial c_\alpha}{\partial z} = 0, \quad \lambda \frac{\partial T}{\partial z} = \alpha_2 (T - T_w). \tag{11}$$

Here v – filtration rate; α_G, α_C, α_S – proportion of gaseous products, coke and ash formed during the pyrolysis of coal; M – molecular weight gaseous products; quantities characterizing the condensed phase, attributed to index s; k – coal seam permeability coefficient; T – the temperature in the coal stack; E_p, k_p, E_L, k_L, E_C, k_C – activation energy and pre-exponential factor of coal pyrolysis reaction, evaporation of water and the heterogeneous reaction of coke oxidation, respectively; R – universal gas constant; $D_\alpha, \lambda_{5\alpha}, c_\alpha, M_{5\alpha}$ – coefficients of molecular diffusion, thermal conductivity, mass concentration and the molecular weight of α-components , parameter with index α =1 corresponds to the component O_2, α=2 – CO, α=3 – CO_2, α=4 – H_2, α=5 – CH_4, α=6 – H_2O, α=7 – N_2; p – pressure gaseous products in the pores; p_H – external atmospheric pressure; s – pore specific surface; q_R – radiative flux from the external environment on the surface of the coal; t – time; z – spatial variable, directed from the top surface stack vertically downwards; q_R и T_E – function of time (depending on time of day), α_1 и α_2 – heat transfer coefficients characterizing the convective heat transfer on the top and bottom surfaces of stack; ρ_i, c_{pi}, λ_i, φ_i – true density, heat capacity, thermal conductivity and the volume fraction of coal – i=1, water – i=2, coke – i=3, ash – i=4 and gas phase – i=5; λ_S, λ_5, λ_R – coefficient of thermal conductivity of the condensed phase (coal) and the coefficients of molecular and radiant thermal conductivity the gaseous mixture; σ – Stefan-Boltzmann constant; μ – dynamic viscosity factor; β – constant in the quadratic Darcy law; g – acceleration of gravity; v – filtration rate.

3 The method of solving and analysis of numerical results

The boundary value problem (1) – (11) was solved numerically, use iterative-interpolation method [2]. Based on the analysis of the results, it was found that in dry hot weather, coal stack will not light up if it does not get sunlight, that is when only the convective heat transfer is carried out. There were also calculations when on the surface of the coal stacks impacted convective and radiative heat flow. Before surface of the earth in western Siberia in cloudless weather, depending on the degree of atmospheric dustiness, reaches the maximum radiative heat flow equal to 800 ÷ 1200 W/m^2. Therefore, for the heat equation we considered the boundary conditions that take into account the air temperature at night dropped to 293 K, and in the daytime rose to 303 K. Radiative flux was zero at night, and during the day rose to 800 ÷ 1200 W/m^2.

Figure 1 shows a heating of coal at the upper surface of the stack, when the radiation stream by day increased to 1000 W/m^2. As can be seen from the graph, in such effect heat flow to stack have happened fire in the coal layer depth a distance of approx 0.05 m from the surface.

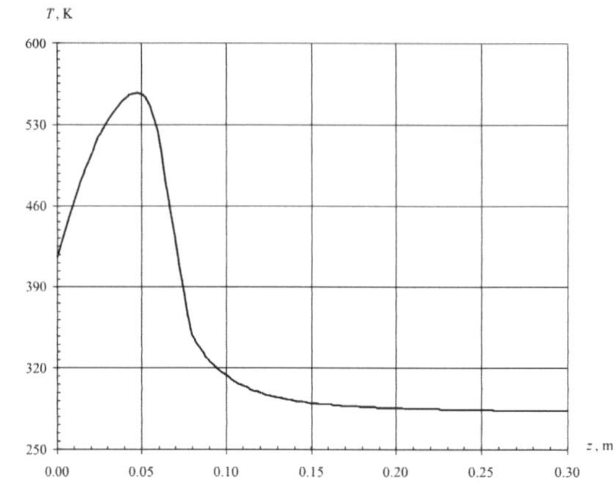

Fig. 1. Distribution of temperature at the upper surface of the stacks.

Figure 2 shows the temperature change with time at interior points the coal stacks: the upper curve 1 at a distance of 0.005 m from the surface, average 2 – 0.05 m and lower 3 – 0.1 m.

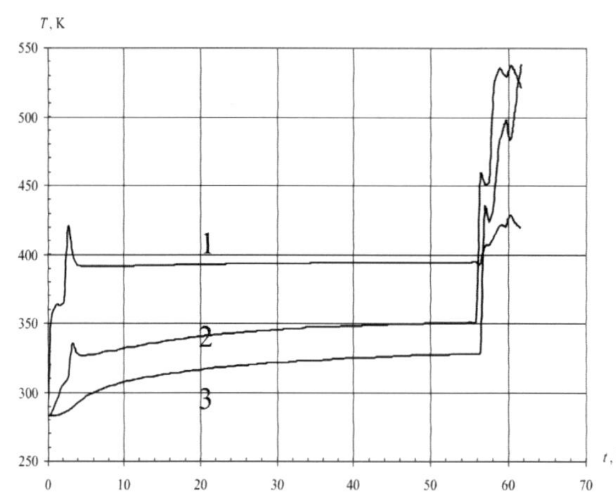

Fig. 2. The temperature change over time in some points of the coal seam.

Analyzing Figure 2 it can be stated, that at simultaneous effects of on the coal stack convective and radiative flux on the third day there was an ignition of coal stored in the stack.

4 Conclusions

Thus, we can conclude the following: during storage of coal in the stack are not protected from the sun in hot cloudless weather may cause it to catch fire. In the considered case, ignition has occurred in about three days.

References

1. R.N. Kulesh, A.N. Subbotin, Nat. Eng. Sc. **4** (2014)
2. A. M. Grishin, V. I. Zinchenko, K. N. Efimov, A. N. Subbotin, A. S. Yakimov. *Iteration-interpolation method and its application* (Publ. house TGU, Tomsk, 2004)

PHYSICAL MODELING OF SUPERSONIC FLOW OVER BODIES WITH PLANE AND AXISYMMETRIC SHAPE

E.A. Maslov[1, 2, *], *V*.V. Faraponov[1], *A*.A. Chupashev[2], and *V*.V. Matskevich[2]

[1]National Research Tomsk State University, 364050 Tomsk, Russia
[2]National Research Tomsk Polytechnic University, 364050 Tomsk, Russia

1 Introduction

One of the major stages in methods development of internal ballistic performance calculation for propulsion (thrust, density impulse, the fuel burning rate, pressure-chamber, etc.) is a simulation of the structure and definition of the gas encircling flow parameters. During the operation of the engine by supersonic flowing, the verification of the developed mathematical models on the basis of a comparison with experimental data is required due to the complexity of the intra-chamber processes and changes in performance duct tract geometry. Field tests of the engines are conducted to obtain the necessary data on the main parameters of the process, but it is due to the high material costs. A comparative analysis of numerical and experimental data obtained by laboratory model tests is the most efficient way to obtain objective experimental data.

This paper presents the results of an experimental study of one of the main parameters of supersonic air stream over bodies with plane and axisymmetric shape in the range of Mach number M=(2÷5).

2 Description of the experimental setup

The experiments were performed on a model aerodynamic unit (MAU) [1] in the open airwith ambient atmospheric pressure measuring of the equalization chamber pressure and of the installation air flow duct pressure and visualizing of the flow structure.

The experiments were conducted on two types of models: cone-shaped andwedge shaped. The semi-vertex angle was 15 °. Models were mounted on a standwith the help of a holder, allowing the pitch angle to vary. Fig. 1 shows photographs of models. Fig. 2 shows a photograph of the wedge-shaped model mounted on the MAU.

* Corresponding author: maslov_eugene@mail.ru

 DOI: 10.1051/matecconf/20167201065

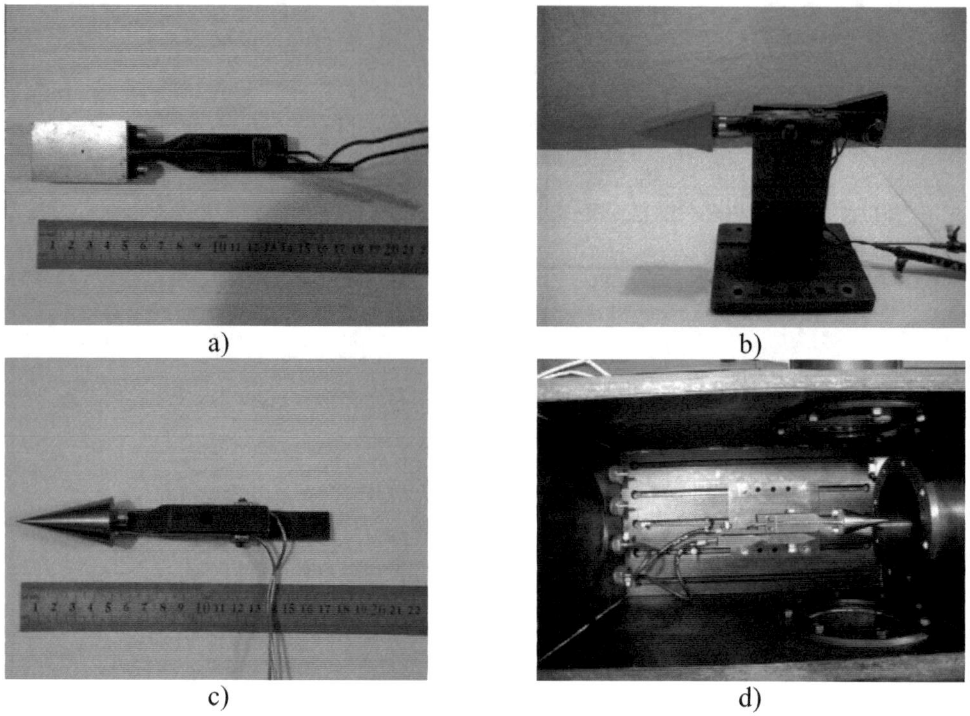

Fig. 1. Models: a) – general view of the wedge model; b) – wedge on a holder; c) – general view of the cone model; d) –working part MAU with the established cone model.

Fig. 2. The cone-shaped model installed on the MAU(top view).

The pressure in the prechamber and in the MAU air flow duct was measured in the course of the experiments. The pressure of gas cylinders with compressed air was controlled as well. The prechamber pressure was measured by a sensor type "DM 5007A – DIU2". The pressure was measured near the working chamber rear wall by two pressure sensor units, each of which contains five pressure sensors "TDM2-A" for MAU control mode during the test. All pressure sensors are connected to the digital recording equipment. The signals are received from the pressure sensors on ADC. Signals were recorded by the computer and saved in a file. The program developed by the authors was used for data processing. This program allows counting the received voltage values into values of pressure (in atmospheres). Combined pressure sensors developed by Central Aerohydrodynamic Institute, which make it possible to measure both pressures simultaneously (Pitot-Prandtltube) were used for measuring the total and static pressure in the experiments of supersonic air flow in the MAU working chamber. The measured values of the pressure allowed us to calculatethe air flow Mach number in the MAU working chamber.

Static pressure was measured by means of selective pipe connecting a hole on the model surface with pressure sensor. The holes were made on the surface of the model (fig. 1, a) for the gas selection. The holes on the surface were made in specific fixed positions in the vertical plane. In the experiments, pressure selection in two opposite points is required for monitoring the accuracy of determining the tangage angle.

3 Results

Fig. 3 shows the results of the series of experiments for the plane (wedge-shaped) and the axisymmetric (cone-shaped) models of air flow in the range of Mach numbers M=(2÷5) at an angle of tangage 0 °.

Analysis of the results showed a difference in pressure values on the surface of the wedge and the cone in this flow velocity range. This effect is explained by the fact that the flow past the cone has a spatial nature. This contributes to a smoother change in the gas flow direction as compared to flow past the wedge [2]. It should be mentioned that the pressure values difference for flow past the plane and axisymmetric models decreases with increasing circumfluent flow rate.

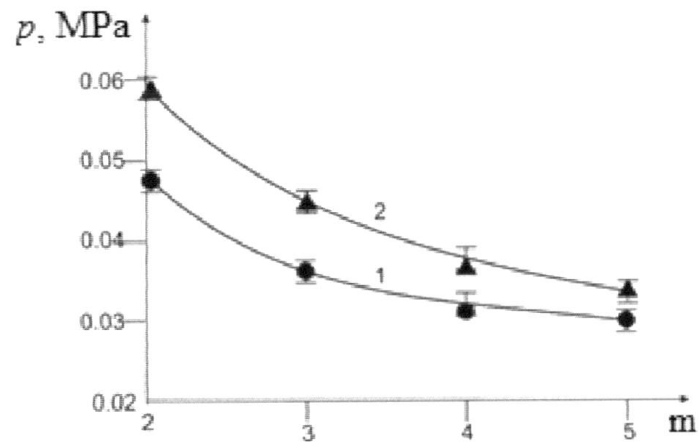

Fig. 3. The pressure at a fixed point on the side surface models: 1 – the cone; 2 – the wedge.

Based on the experimental results of the study it can be concluded that it is necessary to take into account spatial nature of the flow while conducting the mathematical modeling of dynamic processes in propulsion systems in the range of M=(2÷5).

Acknowledgments

The article was written as part of a research project No. 8.2.46.2015, carried out with the support of the Program "Tomsk State University Research Foundation them. D.I. Mendeleev" in 2015.

References

1. V. Zvegintsev *Gas-dynamic installation short-acting. Part 1: Settings for scientific research* (Parallel, Novosibirsk, 2014)
2. N. Krasnov, V. Mishka, A. Danilov, *The aerodynamics of questions and problems* (Higher. School, Moscow, 1985)

RADIANT HEATING OF LIQUID IN THERMAL ENTRY LENGHT

G. Medvedev[1,*], and *A*. Martynova[1]

[1]National Research Tomsk Polytechnic University, 634050 Tomsk, Russia

Abstract The given paper presents the calculation method of liquid flow and radiation heating in thermal entry length. For the solution of a problem with nonlinear boundary conditions the method of linear function is used [2] which allows to putting boundary conditions in a linear form.

1 Problem statement

Inside the thin-walled channel (fig.1) liquid fluid is moved and heated axially owing to outer high-temperature gas filling. The heat transfer on the outer surface keeps in both with the Stefan-Boltzmann's law. In the tube entry velocity traverse is parabolic, and the liquid temperature is constant and equal to θ_0. Assume the viscosity factor is dependent on temperature and all other physical properties are permanent. It is reasonable that density is constant, weakly dependent on temperature for most of liquid fluids excluding from consideration influence of natural convection. The friction heat and axial heat transfer are neglected.

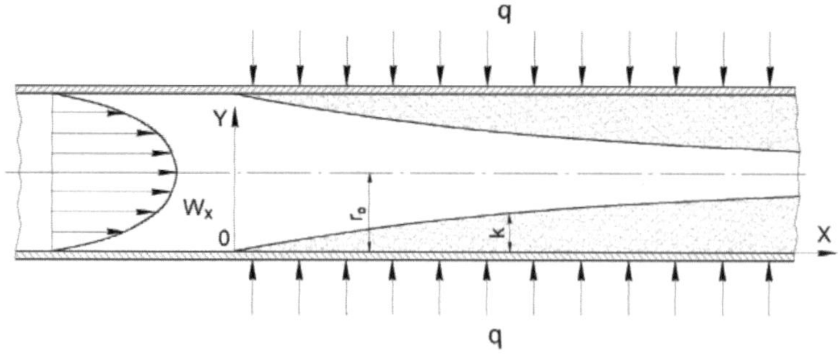

Fig.1. The physical model and coordinate system for thermal entry length.

Within specified conditions the heat transfer process is described by energy equation

* Corresponding author: ggm@tpu.ru

DOI: 10.1051/matecconf/20167201066

$$a\frac{1}{r}\cdot\frac{\partial}{\partial r}\left(r\frac{\partial\theta}{\partial r}\right)=W_x\frac{\partial\theta}{\partial x}+W_r\frac{\partial\theta}{\partial r}, \quad (1)$$

equation of through flow

$$\frac{\partial W_x}{\partial x}+\frac{1}{r}\cdot\frac{\partial}{\partial r}\left(r\cdot W_r\right)=0, \quad (2)$$

and motion

$$\frac{1}{r}\cdot\frac{\partial}{\partial r}\left(r\mu\frac{\partial W_x}{\partial r}\right)=\frac{\partial P}{\partial x}+\rho\left(W_x\frac{\partial W_x}{\partial x}+W_r\frac{\partial W_x}{\partial r}\right)-\frac{\partial}{\partial x}\left(\mu\frac{\partial W_x}{\partial x}\right)-\frac{\partial\mu}{\partial r}\cdot\frac{\partial W_r}{\partial x}. \quad (3)$$

Here $\theta_0=\frac{T_0}{T_c}\leq\theta=\frac{T}{T_c}<1$; T_c is temperature of heating medium, and T_0 is temperature of inlet flow.

The boundary conditions for the energy equation (1) reduce to inlet temperature

$$\theta=\theta_0 \text{ with } x=0 \quad \text{and } r=0, \quad (4)$$

to symmetry condition on pipe axis

$$\frac{\partial\theta}{\partial r}=0, \text{ with } \quad x\geq 0 \quad \text{and} \quad r=0 \quad (5)$$

and to the law of radiant heating on surface

$$\frac{\partial\theta}{\partial r}=\frac{\sigma T_c^3}{\lambda}\left(1-\theta^4\right)\frac{d}{d_0}, \text{ with } x\geq 0 \text{ and } r=r_0. \quad (6)$$

To linearize the boundary conditions (6) we introduce nonlinear integral transformation [2]

$$Z=\int_0^\theta\frac{d\theta}{1-\theta^4}=\frac{1}{2}\left(arth\theta+arctg\theta\right) \quad (7)$$

The transformation (7) linearizing the boundary condition (6)

$$\frac{\partial Z}{\partial r}=\frac{\sigma\cdot T_c^3\cdot r_0}{\lambda}\cdot\frac{d}{d_0}, \text{ with } x\geq 0 \text{ and } r=r_0,$$

reduces the energy equation (1) to form:

$$a\frac{1}{r}\cdot\frac{\partial}{\partial r}\left(r\frac{\partial Z}{\partial r}\right)-aF=W_x\frac{\partial Z}{\partial x}+W_r\frac{\partial Z}{\partial r} \tag{8}$$

where

$$F=\frac{4\theta^3}{\left(1-\theta^4\right)^2}\cdot\left(\frac{\partial \theta}{\partial r}\right)^2 \tag{9}$$

In this connection the symmetry condition (5) isn't changed

$$\frac{\partial Z}{\partial r}=0,\ \text{with}\ x\ge 0\ \text{and}\ r=0,$$

and the transformed inlet temperature may be written as

$$Z=\frac{1}{2}\left(arth\theta_0+arctg\theta_0\right)=Z_0\ \text{with}\ x\ge 0\ \text{and}\ 0\le r\le r_0.$$

As is shown in [2-3] with not-too-large amounts of temperature gradients $\partial\theta/\partial Z$ the nonlinear complex (9) in the energy equation (8) may be neglected.

With further analysis we distinguish two zones in thermal entry length: thermal boundary layer with thickness Δ and liquid flow core, where temperature is constant and equal to inlet heated length temperature. It is considered that the amount Δ is substantially smaller than the pipe radius r_0. The last condition means that the considered problem is constrained by a set of small calculated lengths.

The using nondimensional variables

$$t=\frac{Z}{Sk},\quad SK=\frac{\sigma\cdot T_c^3\cdot r_0}{\lambda}\cdot\frac{d}{d_0},\quad W_x=\frac{w_x}{\bar{w}},\quad W_y=\frac{w_y}{\bar{w}},$$

$$M=\frac{\mu}{\mu_0},\quad Pe=\frac{\bar{w}\cdot 2r_0}{a},\quad X=\frac{x}{r_0},\quad Y=1-\frac{r}{r_0},\quad \mathrm{K}=\frac{\Delta}{r_0}$$

and averaging right-hand sides of energy and motion equations relatively on thermal boundary layer thickness and pipe radius (like in [1]), we shall obtain

$$\frac{1}{1-Y}\cdot\frac{\partial}{\partial Y}\left[\left(1-Y\right)\frac{\partial t}{\partial Y}\right]=B\left(X\right), \tag{10}$$

$$\frac{\partial W_x}{\partial X}+\frac{1}{1-Y}\cdot\frac{\partial}{\partial Y}\left[\left(1-Y\right)W_y\right]=0, \tag{11}$$

$$\frac{1}{1-Y}\cdot\frac{\partial}{\partial Y}\left[\left(1-Y\right)M\frac{\partial W_x}{\partial Y}\right]=A\left(X\right) \tag{12}$$

$B(X)=\frac{Pe}{2k}\int\limits_0^k\left(W_x\frac{\partial t}{\partial X}+W_y\frac{\partial t}{\partial Y}\right)dY$, and $A(X)$ is defined in [1].

The solution of (10) and (12) must satisfy the boundary conditions

$$\frac{\partial t}{\partial Y}=-1,\quad W_x=W_y=0 \text{ with } X\geq 0 \text{ and } Y=0, \tag{13}$$

$$t_0=\frac{Z_0}{Sk},\quad \frac{\partial t}{\partial Y}=0 \qquad \text{with } X\geq 0 \text{ and } 1\geq Y\geq k, \tag{14}$$

$$\frac{\partial W_x}{\partial Y}=0 \qquad \text{with } X\geq 0 \text{ and } Y=1. \tag{15}$$

The integral transformed by means of the system (10) – (15) was outlined in the paper [1] and in general terms is expressed by dependence

$$t=\frac{Z_0}{Sk}+\frac{1}{k\left(2-k\right)}\left[-\left(1-k\right)^2 Ln\frac{1-Y}{1-k}-Y\left(1-\frac{Y}{2}\right)+k\left(1-\frac{k}{2}\right)\right] \tag{16}$$

Inserting then $Z=Sk\cdot t$ in relation (7) we define the target relative temperature θ.

The expression for nondimensional temperature on the wall surface we can obtain from (16), assuming $Y=0$

$$t_{suf}=\frac{Z_0}{Sk}+\frac{1}{k\left(2-k\right)}\left[\left(1-k\right)^2 Ln\left(1-k\right)+k\left(1-\frac{k}{2}\right)\right] \tag{17}$$

The most simple dependencies for temperature profile are obtained in the problems when liquid viscosity is constant. With $\mu=const$ the boundary layer thickness is defined from comparatively simple equation [17]

$$k^3\left(1-\frac{k}{5}\right)=24\frac{x}{Pe\cdot 2r_0} \qquad (18)$$

2 Results

The fig. (2) shows temperature profiles in thermal entry length which were obtained for that conditions. The calculations are made on the basis of (18), (16) and (7). The inlet temperature θ_0 is assumed equal to 0.5. Nondimensional boundary layer thickness k is chosen within ranges 0.10 and 0.20. The calculated lengths $x / Pe \cdot 2r_0$: $0.41 \cdot 10^{-4}$; $3.2 \cdot 10^{-4}$ match the chosen values of k.

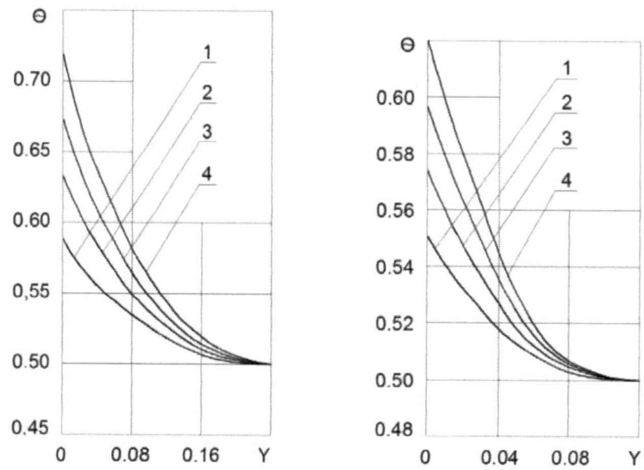

$k = 0.2;\quad x / Pe \cdot 2r_0 = 3.2 \cdot 10^{-4}$ $\qquad$ $k = 0.1;\quad x / Pe \cdot 2r_0 = 0.4 \cdot 10^{-4}$

Fig. 2. Temperature profiles in thermal entry length, 1–4: $Sk = 1.0;\ \ 1.5;\ \ 2.0;\ \ 2.5$.

Here it may be noted that in conditions of constant viscosity the velocity profile is parabolic and it is possible to use exact solution of energy equation without averaging its right-hand side.

References

1. B.S. Petukhov, *Heat transfer and resistance within laminar flow in tubes* (Energiya, Moscow, 1967) [in Russian]
2. V.V. Ivanov, G.G. Medvedev, Energy and transport **3** (1970) [in Russian]
3. V.V. Ivanov, The physics and chemistry of materials processing **5** (1972) [in Russian]

MATEC Web of Conferences **72**, 01067 (2016) DOI: 10.1051/matecconf/20167201067
HMTTSC-2016

EXPERIMENTAL INVESTIGATION OF GAS HYDRATE PRODUCTION AT INJECTION OF LIQUID NITROGEN INTO WATER WITH BUBBLES OF FREON 134A

Anton V. Meleshkin[1], *Ivan V.* Mezentsev[1, *], and *Dmitriy S.* Elistratov[1]

[1]Kutateladze Institute of Thermophysics SB RAS, Novosibirsk, Russia

Abstract. The hydrodynamic processes during the injection of the cryogenic liquid into the volume of water with bubbles of gas freon 134a are studding experimentally. A processes during the explosive boiling of liquid nitrogen in the volume of water are registered. Video recording of identified gas hydrate flakes formed during this process is carried out by high speed camera. These results may be useful for the study of the new method of producing gas hydrates, based on the shock-wave method.

1 Introduction

There are many methods for intensifying the process of hydrate formation: intensive mixing of water saturated with dissolved gas; finely dispersed atomization of a water jet; vibration and ultrasonic impact on the two-phase systems, etc. The main disadvantage of the proposed methods is the low rate of gas hydrate formation and, therefore, low performance of the systems based on these methods. The Institute of Thermophysics SB RAS is involved into artificial gas hydrate production by the shock-wave method since 2000. In 2003, the method for producing gas hydrates has been proposed (RF patent No. 2270053). The essence of this invention is as follows. The compressed and cooled gas is supplied into the reaction vessel filled with water, which is under the static pressure and the temperature below the equilibrium temperature of hydrate formation. The gas pressure is slightly higher than the pressure in the reaction vessel, and the gas temperature equals the temperature of water in the vessel. Gas is mixed with water. The pressure shock waves with the amplitude of up to the hundreds of atmospheres are generated in the gas-liquid medium. The shock waves can be generated by the electromagnetic pulse emitters, pneumostrikers and other devices [1, 2]. When the shock wave propagates in the gas-liquid medium, the pressure increases and the gas phase crushes throughout the reaction vessel due to weak dissipation of the wave. Therefore, the degree of medium metastability and the number of nucleation sites of gas hydrate increase, the size of gas inclusions decreases, interfacial surface increases, the relative velocity of gas inclusions in liquid increases, and liquid motion

* Corresponding author: mezn@itp.nsc.ru

becomes more turbulent. All these phenomena accelerate the mass transfer process at the interface and, hence, intensify the process of hydrate formation.

This paper presents the results of studying the shock-wave method of gas hydrate production by injecting liquid nitrogen. The authors of [3] studied the process of nitrogen supply into a vessel of a small volume (140 ml). The volume of injected liquid was about 2 ml. For example, at introduction of liquid nitrogen into water (7 bars), the amplitude of the pressure jump reaches 14 bars in 5 seconds. Previously, we have carried out the experiments on entering a jet of liquid nitrogen in the water volume [4]. The maximal amplitude of the pressure jump in these experiments was 53 bars, and the rate of pressure rise was 567 bar/s. These results were a part of the works required for the development of a new method of gas hydrate production, based on the shock-wave method.

2 Experimental setup and methods

Gas hydrate production was studied at the experimental setup with the viewing windows along the entire height of the working section. The working section is a parallelepiped with the height of 740 mm (cross-section of 150 x 150 mm) made of stainless steel with the wall thickness of 15 mm. Two viewing windows in the working section, allowed us to fix the processes occurring at applying a jet of liquid nitrogen into water, and namely to carry out the high-speed optical measurement. Gas was supplied through the bubble generator from the bottom of the setup.

Liquid nitrogen was fed into the working section as a jet through the injector (fig. 1). The volume of injected liquid nitrogen was 28 ml.

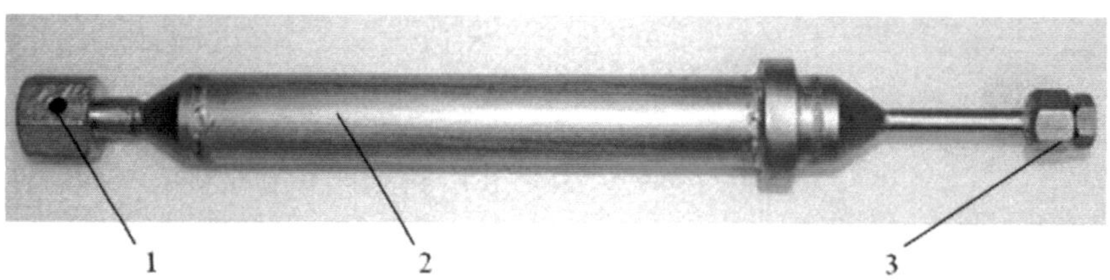

Fig. 1. General view of the injector: 1 – ring slot, 2 – injector chamber, 3 – input unit.

The storyboard of resulting high-speed shooting of liquid nitrogen injection into water saturated with bubbles of Freon gas is shown in fig. 2. In the first frame, the working section saturated with the bubbles of Freon R134a can be clearly seen; in the second frame, liquid nitrogen is injected. In the third and fourth frames, we can see rapid mixing of the working section due to liquid nitrogen boiling in water. In the fifth and sixth frames, the working section after liquid nitrogen boiling is presented. The gel-like gas hydrate flakes, whose generation rate is high, can be clearly seen in the frames; this gives an order advantage of the shock-wave method over all known analogical methods.

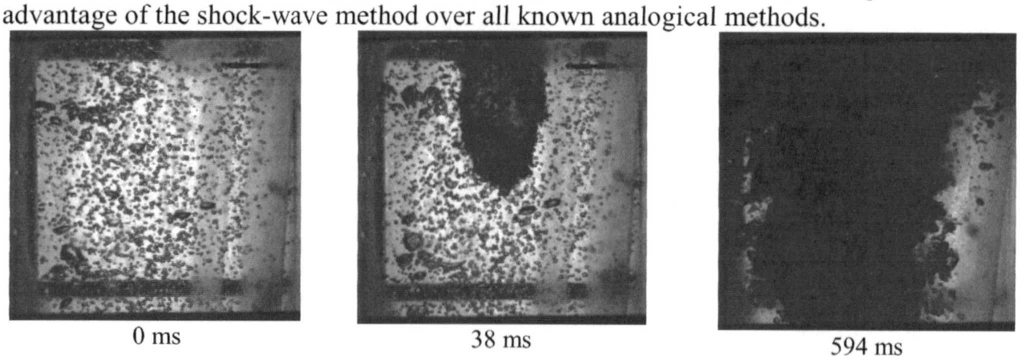

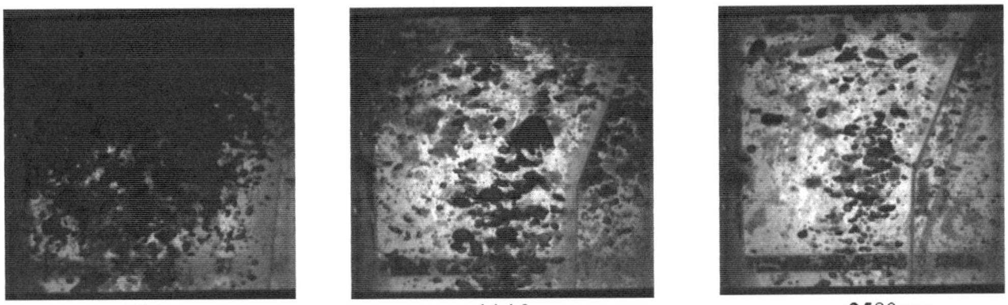

734 ms 1146 ms 2580 ms

Fig. 2. Frames of high-speed recording of the process of liquid nitrogen injection (10 000 fr/s).

This method of gas hydrate production does not require the additional external sources of the pulse pressure. Here, pulsating (oscillating) gas and steam bubbles are the powerful tool for pressure pulse generation; they can create powerful pressure pulses, which destroy even steel. To initiate bubble pulsation, it is necessary to give a significant initial rate to the interface, for example, by micro bursting. For this purposes, a jet of liquid nitrogen, which can be overheated to the spinodal temperatures, as a pure cryogenic liquid. Subsequent heating causes explosive boiling of a part of liquid nitrogen and disperses the boundary of a gas cavity.

During the experimental studies, the pressure amplitude at liquid nitrogen boiling in water with Freon bubbles, whose maximal value was 4.2 bars, was measured. This pressure jump allowed us to reach the metastable equilibrium zone of Freon R134a gas hydrate.

3 Conclusion

This method of gas hydrate production does not require the additional external sources of the pulse pressure. Here, pulsating (oscillating) gas and steam bubbles are the powerful tool for pressure pulse generation; they can create powerful pressure pulses, which destroy even steel. To initiate bubble pulsation, it is necessary to give a significant initial rate to the interface, for example, by micro bursting. For this purposes, a jet of liquid nitrogen, which can be overheated to the spinodal temperatures, as a pure cryogenic liquid. Subsequent heating causes explosive boiling of a part of liquid nitrogen and disperses the boundary of a gas cavity.

During the experimental studies, the pressure amplitude at liquid nitrogen boiling in water with Freon bubbles, whose maximal value was 4.2 bars, was measured. This pressure jump allowed us to reach the metastable equilibrium zone of Freon R134a gas hydrate.

Acknowledgments

The work was financially supported by the Russian Science Foundation 15-19-10025.

References

1. V.E. Dontsov, A.A. Chernov Doklady Physics **54**, 4 (2009)
2. V.E. Nakoryakov, V.E. Dontsov, A.A. Chernov, Doklady Physics. **51**, 11 (2006)
3. H. Clarke, A. Martinez-Herasme, R. Crookes, D.S. Wen, Int. J. Multiphase Flow **36**, 4 (2010)
4. V.E. Nakoryakov, A.N. Tsoi, I.V. Mezentsev, A.V. Meleshkin J. Eng. Thermophys. **23**, 1 (2014)

BOILING OF BINARY ZEOTROPIC BLENDS IN THE PLATE HEAT EXCHANGER OF THE HEAT PUMP

Nadezhda N. Mezentseva[1,*], *Ivan V.* Mezentsev[1], and *Anton V.* Meleshkin[1]

[1]Kutateladze Institute of Thermophysics, Novosibirsk, Russia

Abstract. In this paper, we consider the process of boiling in the evaporator of the heat pump. Zeotropic binary refrigerants R32/R152a (30/70%) and R32/R134a (30/70%) are used as working medium. Calculations are made for brazed plate heat exchanger during boiling of zeotropic blend refrigerants with account of peculiarities of this process. Results of calculation of the heat transfer coefficient for zeotropic blends are given.

1 Introduction

The evaporator is an integral part of the heat pump. The evaporator is a heat exchanger that takes heat from the low-potential source by evaporation of the working medium (refrigerant). To perform technical and economic analysis and select the heat exchanger, knowledge about the heat transfer coefficients at a change in the aggregate state during boiling is required.

The most common heat exchangers, used as evaporators of the heat pumps are the shell-and-tube ones. The processes of intra tube boiling of R32/R134a and R32/R152a refrigerants in the shell-and-tube heat exchanger were considered in [1, 2].

The shell-and-tube heat exchangers consist of tube bundles, mounted in tube plates, housings, covers, chambers, pipes and supports. The tube and inter tube spaces in these devices are separated, and each of them can be divided by partitions into several passages. The shell-and-tube heat exchangers have reliable construction, their sizes vary from small to large ones, they have broad limits of the pressure drops on both sides, i.e., their version are suitable for different operating conditions. However, such a variety of application conditions for the shell-and-tube heat exchangers and their constructions should not exclude the search for other, alternative solutions such as the use of the plate, helical, or compact heat exchanger in the cases, when their characteristics are acceptable, and their use can lead to solutions that are more economical.

The current study is aimed at determination of the heat transfer coefficient at refrigerant boiling in the evaporator of vapor-compression heat pump at a change in the qualitative index of blend composition.

* Corresponding author: mnn@itp.nsc.ru

2 Results and discussion

The plate heat exchanger is chosen as the evaporator. Currently, the plate and plate-finned heat exchangers are the most advanced. The plate heat exchangers present a type of the surface recuperative heat exchangers with heat transfer surface, made of a thin sheet. The defining feature of the plate heat exchangers is design and shape of the heat exchange surface and channels for the working medium. The heat exchange surface is formed by individual plates and the channels for the working medium have the slot-like shape.

Zeotropic mixed refrigerants R32/R152a (30/70) и R32/R134a (30/70) are considered as the working body. The features of heat pump operation on zeotropic blends are presented in [3, 4]. The thermodynamic cycle of the heat pump is shown in fig. 1. Boiling process (5-1) occurs there at variable temperature. Then, dry saturated vapor enters the super heater (1-1'). Further, superheated vapor is supplied to the compressor (1'-2'). Enthalpy at point 2' is obtained with consideration of the influence of indicated efficiency of compressor. Condensation of refrigerant vapors in process (2-3) takes place at variable temperature. In this case, we consider the heat pump cycle for the heating system. In this cycle, the heat of super cooling (3-4) is spent for overheating refrigerant vapors before entering the compressor (1-1'). A change in the position of point 5 at throttling characterizes the relative quality index of the blend (kg_{vapor} / kg_{total}). In the case, when throttling occurs along line 4'-5' or 4"- 5", the effect of vapor component of the blend on the boiling process reduces and the heat transfer coefficient increases.

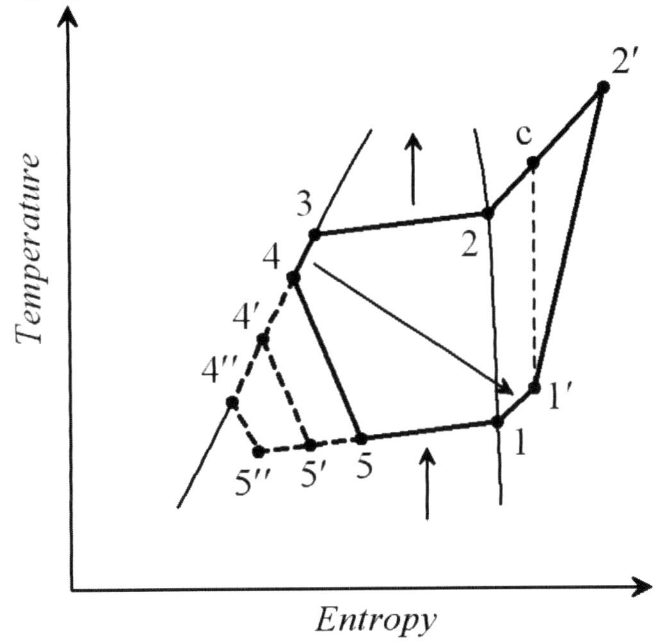

Fig. 1. Thermodynamic cycle of vapor-compression heat pump on zeotropic blends of refrigerants.

Relationship for determination of the Nusselt number in application to the plate heat exchangers is suggested in [5]:

$$Nu_R = a \cdot Nu_L^b \cdot Bo^c \cdot Fr^d \cdot \omega^e \cdot Co^f \cdot X_{tt}^g \cdot M^h \tag{1}$$

where *a, b, c, d, e, f, g, h* are the constant for the boiling process.

The Nusselt number for the liquid phase is determined by formula:

$$Nu_L = 0.16 \cdot \mathrm{Re}_L^{0.89} \cdot \mathrm{Pr}_L^{0.3(0.4)} \quad (2)$$

The Reynolds number is determined by formula:

$$\mathrm{Re}_L = \frac{w_R \cdot d_E \cdot (1-\bar{x})}{\nu_L} \quad (3)$$

$\bar{x}$ – average qualitative index of composition;
Pr_L – Prandtl criterion (at average refrigerant temperature);
w_R – velocity of refrigerant flow in a tube, m/s;
ν_L – kinematic viscosity of refrigerant, m^2/s;
d_E – equivalent hydraulic diameter, m.

The Froude number is determined by formula:

$$Fr = \frac{\omega^2}{g \cdot d_E} \quad (4)$$

g – acceleration of gravity, m/s^2;
ω – acentricity factor:

$$\omega = -\log_{10}(P/P_C) \quad (5)$$

P – blend pressure, MPa;
P_C – critical pressure of blend, MPa.

The convective number is determined by formula:

$$Co = \left(\frac{\rho_V}{\rho_L}\right)^{0.5} \cdot \left\{\frac{(1-x)}{x}\right\}^{0.8} \quad (6)$$

ρ_V, ρ_L – refrigerant densities in vapor and liquid phases, respectively, kg/m^3;
x – relative qualitative index of composition, kg_{vapor}/kg_{total}.

Results of calculation of the heat transfer coefficient for zeotropic blends R32/R152a (30/70) and R32/R134a (30/70), carried out by formulas (1-6), are shown in fig. 2.

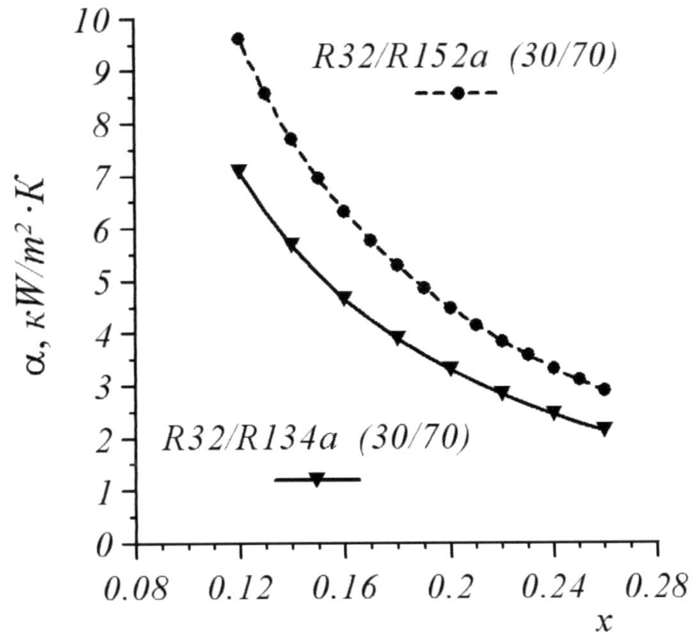

Fig. 2. A change in the heat transfer coefficient vs. relative qualitative index of zeotropic blend composition.

3 Conclusion

According to analysis of results, at a change in the relative qualitative index (vapor content) from 0.12 to 0.26 (an increase in the vapor component), the heat transfer coefficient increases by 3.3 times.

References

1. L.A. Ogurechnikov, Refrigeration Equipment **8** (2010) [in Russian]
2. L.A. Ogurechnikov, J. Altern. Energ. Ekol. **2** (2010)
3. N.N. Mezentseva, L.A. Ogurechnikov, Energy and Heat **12** (2008) [in Russian]
4. L.A. Ogurechnikov, N.N. Mezentseva, J. Altern. Energ. Ekol. 7 (2008)
5. S.C. Palmer, W.V. Payne, P. Domanski, *Evaporation and Condensation Heat Transfer Performance of Flammable Refrigerants in a Heat Exchanger* (NISTIR 6541, Departament of Commerce, Gaithersburg, 2000)

MATEC Web of Conferences **72**, 01069 (2016) DOI: 10.1051/matecconf/20167201069
HMTTSC-2016

THE RESEARCH OF THE BOILING CRISIS OF HYDROUS SOLUTION LITHIUM BROMIDE

Nina V. Mironova[1, *], *Vladimir* S. Morozov[2], *Sergey* L. Elistratov[1, 2], and *Roman* S. Ratkov[1]

[1]Novosibirsk State Technical University, 630073 Novosibirsk, Russia
[2]Institute of Thermophysics SB RAS, 630090 Novosibirsk, Russia

Abstract. Experimental and theoretical investigate the process of non-isothermal desorption of water solutions of lithium bromide in the spheroidal state on the horizontal heating surface. It was established experimentally that on reaching average weight of the salt concentration of 15 ÷ 23% and 32 ÷ 34%, respectively, at a temperature of the heating surface 400 and 600 ° C, the solution direct contact with the heating surface and the boiling crisis is replaced by a crisis mode of "explosive" boiling up. Some possible causes of the phenomenon were investigated: decrease in the thickness of the vapor layer with increasing concentration, the effect of temperature of the heating surface, radiation heating and capillary instability of interphase.

1 Problem Statement

For the development of absorption thermotransformers with the multi-stage regeneration is necessary to know the physical dependency of the desorption of water solutions of lithium bromide at high temperatures of the heating surface.

Known [1, 2] that evaporation of fluids may occur in the boiling crisis conditions when they stably separated from the heating surface of thin steam film (fig. 1).

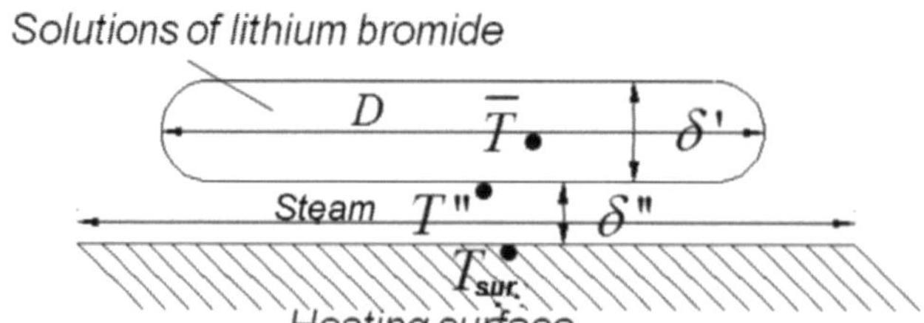

Fig. 1. Diagram evaporation solution of LiBr in the spheroidal state (bubble free spheroid).

In this case, the heat transfer from the heating surface to the fluid takes place by heat conduction and radiation through separating their thin resilient layer of steam. The expression the determination of his thickness in relation to the pure liquids, such as water, of an evaporating into spheroidal state on the horizontal heating surface, obtained in [2]:

* Corresponding author: mironovaNina@inbox.ru

$$\delta'' = \frac{1}{\sqrt[4]{32 \cdot \xi^2}} \cdot \sqrt{\frac{\lambda''(T_{sur} - T'')D}{r\left[1 + c\left(T'' - \overline{T}\right)/r\right]}} \sqrt[4]{g\delta'\rho'\rho''}, \quad (1)$$

where: ξ – coefficient of discharge, which takes into account the shape of the periphery of the flat spheroid and the nature of its steam flow; λ'', ρ'' – thermal conductivity and density of saturated steam, respectively; T_{sur}, T'' и $\overline{T}$ – heating wall temperature, temperature of the saturated vapor in the lower part spheroid and the average temperature of the spheroid, respectively; D, c, ρ', δ' и r – respectively diameter, specific heat capacitance, density, thickness and specific heat of evaporation of liquid. In the derivation of (1) does not take into account radiation heating and evaporation from the top and side surfaces of the flat spheroid, with a decrease in the values of D, its thickness $\delta' = const$ in the process of evaporation. Continuous steam film under evaporating in the spheroidal state clean liquid is stored until their complete evaporation. For water is possible take $\xi = 0.143$.

The solution (1) has a physical limitations. According to (1) the thickness of the layer of steam δ'' depends on D, and can take any value. Experimentally established [1] that the output of steam is carried out not only at the peripheral part of a spheroid, but also through the layer of liquid in the form of vapor bubbles. The average distance between periodically pop bubbles at the boil in a large volume [3] and on evaporation of in the spheroidal state of clean liquids [1,4] can be defined as:

$$L = 2\pi\sqrt{\frac{\sigma}{g(\rho' - \rho'')}}, \quad (2)$$

where: σ - capillary tension, ρ' и ρ'' - density of the liquid and of steam, respectively.

2 Results

Provided that $D \approx L$ expression (1) allows us to estimate δ'' for real processes of evaporation of liquids in the spheroidal state.

It was established experimentally [4], in contrast to the previously investigated the processes of evaporation in the spheroidal state clean liquids, the desorption process (evaporation) bubble free spheroids aqueous solutions of lithium bromide on the horizontal heating surface is completed "explosive" boiling up (fig.2) with the destruction of the continuity of the vapor film. At the time of "explosive" boiling up critical value weight concentration salt of $\varphi_{KP} = m_{LiBr}/(m_{H_2O} + m_{LiBr}) \approx 15...23\%$ for T_{ST}=400°C.

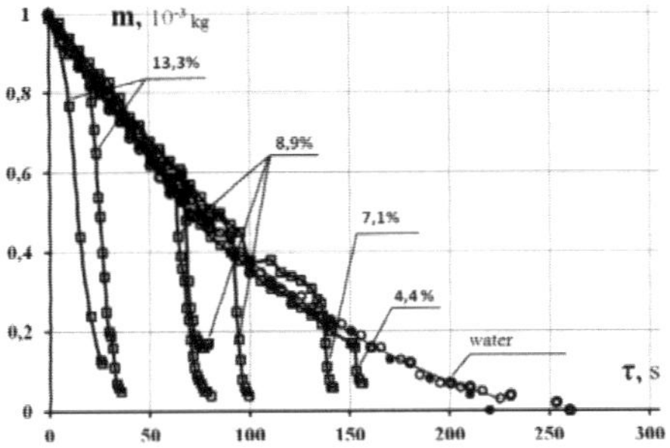

Fig. 2. The time variation of the mass of bubble free spheroids water and aqueous solutions of LiBr for various initial weight concentrations salt φ_0=13.3÷4.4% for T_{ST}= 400°C.

Obtained in this paper the experimental data (fig.3, No. 8-11 video footage) for bubbly spheroids aqueous solution of lithium bromide also confirm the “explosive” nature of the desorption process at achievement values φ_{KP}=32…34% and its further development (No. 12-14 video footage) on the heating surface for T_{ST}=600°C. Desorption process was completed by evaporation of the solution rests on the heating surface (video frame No. 15). Part of the solution in the form of a splashes at “explosive” boiling up flew in an experimental plot area inaccessible for weighing. In the process experiments performed synchronous in time τ mass determination solution m with an accuracy of 0.01 g and a temperature of the heating surface T_{ST} by a thermocouple, located in the heater at a distance of 1 mm from the heating surface. Given the known initial the mass of salt in the solution m_0=2.14 g is determined by its current values.

It is noteworthy that obtained in our experiments for bubbly and bubble free spheroids values φ_{KP} is 2-3 times fewer than concentrations of crystallization of aqueous solutions of lithium bromide [5].

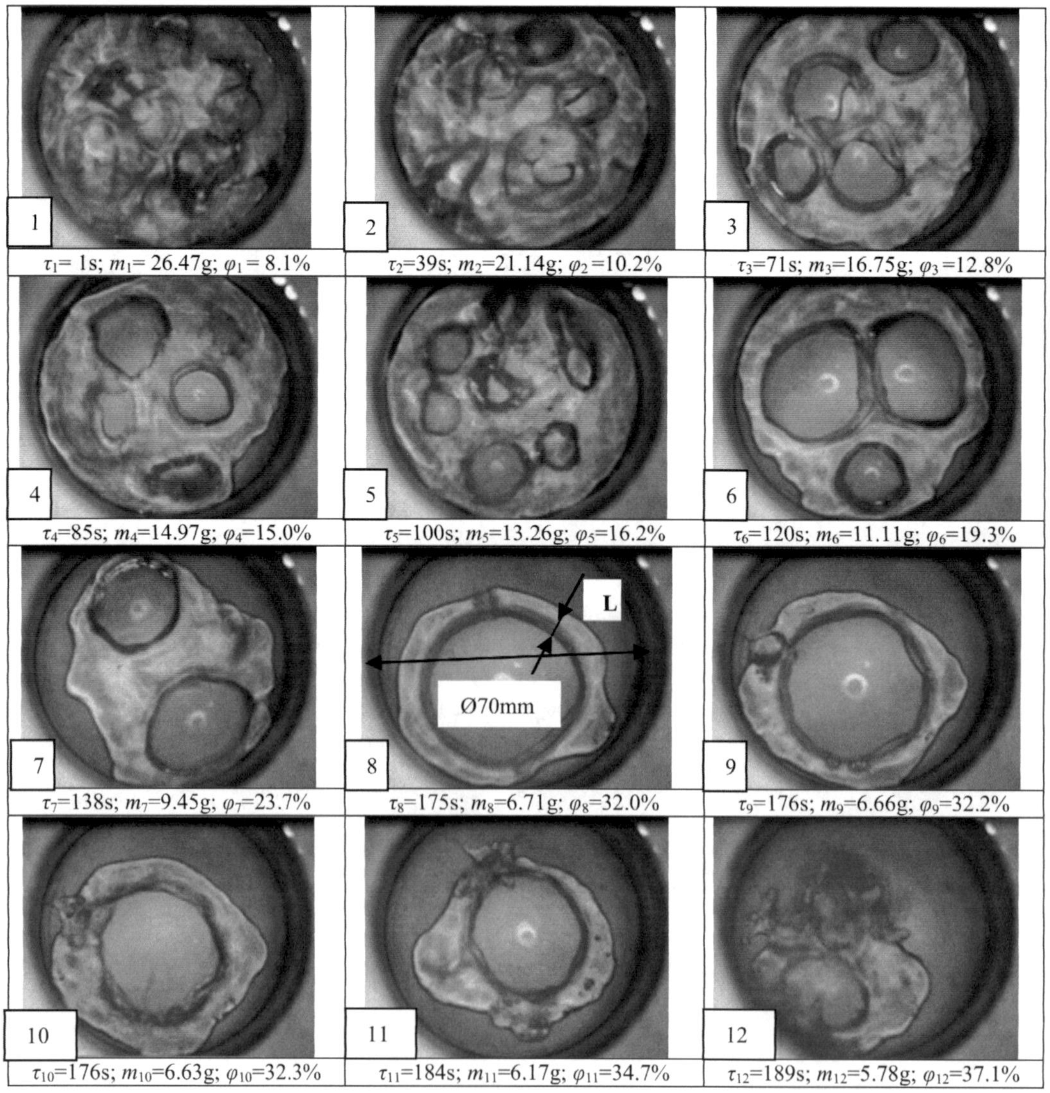

τ_1= 1s; m_1= 26.47g; φ_1 = 8.1%

τ_2=39s; m_2=21.14g; φ_2 =10.2%

τ_3=71s; m_3=16.75g; φ_3 =12.8%

τ_4=85s; m_4=14.97g; φ_4=15.0%

τ_5=100s; m_5=13.26g; φ_5=16.2%

τ_6=120s; m_6=11.11g; φ_6=19.3%

τ_7=138s; m_7=9.45g; φ_7=23.7%

τ_8=175s; m_8=6.71g; φ_8=32.0%

τ_9=176s; m_9=6.66g; φ_9=32.2%

τ_{10}=176s; m_{10}=6.63g; φ_{10}=32.3%

τ_{11}=184s; m_{11}=6.17g; φ_{11}=34.7%

τ_{12}=189s; m_{12}=5.78g; φ_{12}=37.1%

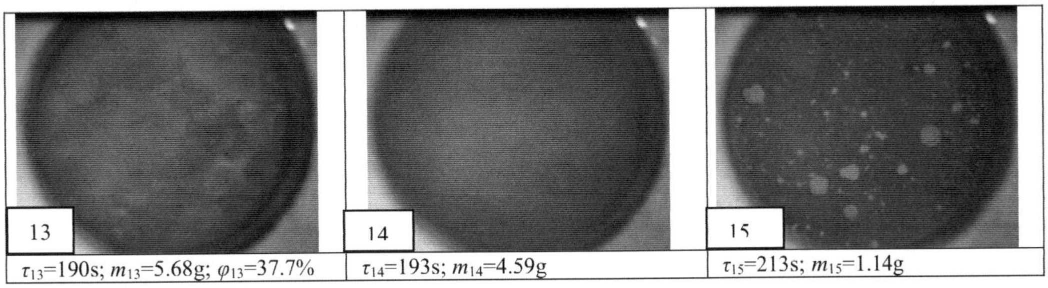

Fig. 3. The picture of desorption aqueous solution of LiBr on the heated surface with a temperature of 600 °C (current time value τ, the mass m and the weight concentration of salt in the solution φ are indicated with the beginning of filming).

Theory and experimental data for the evaporation of clean liquids in the spheroidal state can not explain the phenomenon of the "explosive" boiling up.

Presented our vision of the phenomenon. With increasing φ changes the values c, r, $\overline{T}$ and other parameters of the process in the expression (1). Based on the data about the characteristics of water and salt solutions [4] provided that $T''=\overline{T}$, $\delta'=4\cdot10^{-3}$mm и ξ= 0.143 were estimated values δ''. The average value of $D \approx L = 4\cdot10^{-3}$m is determined based on measurement at the initial moment of contact solution with the heating surface (fig. 3, video frame No. 8). From the presented in fig. 4 and 5 curves shows that the δ'' tends to increase with an increasing T_{ST} and decrease with increasing φ. The difference values of φ_{KP} at a T_{ST} =400 ^{0}C (fig. 2) and T_{ST} =600 ^{0}C (fig. 3) is explained by the increase in δ'' with increasing T_{ST}.

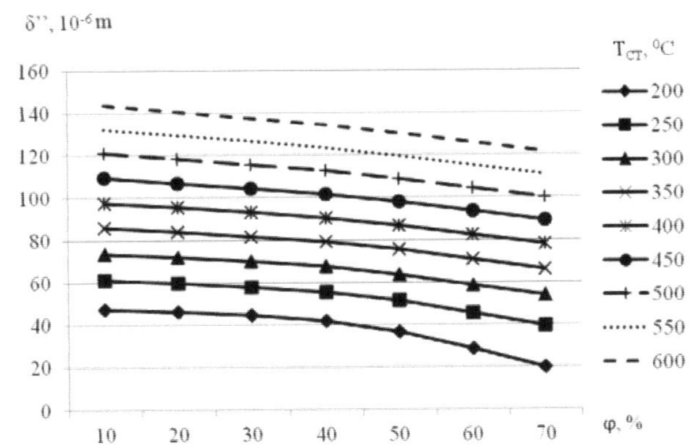

Fig. 4. Dependence of the thickness vapor layer on the temperature of the heating surface.

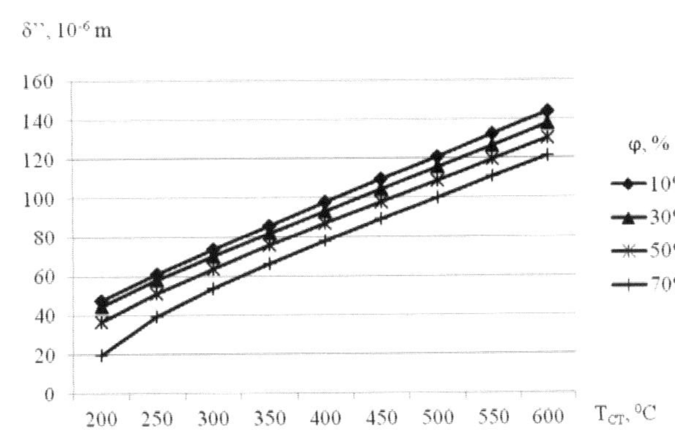

Fig. 5. Dependence of the thickness vapor layer on the concentration of the aqueous solution of lithium bromide.

In defining the degree of stability of the boiling crisis (evaporation of the liquid in the spheroidal state) is determined by the vapor layer's thickness δ'' under the evaporating liquid spheroid. When decreased as a result of the wave of instability interfacial surface of the liquid will probably solution direct contact with the heating surface. Thus, the contact mechanism "explosive" boiling up is triggered.

In the (1) unaccounted factor is the heating of the solution at the expense of radiant heat transfer. Its maximum value at the initial time "explosive" boiling up (fig.3, No. 8 and 11 video footage) can be determined by the expression:

$$q_1^{\mathrm{MAX}} = \varepsilon \cdot \sigma \cdot \left[T_{\mathrm{ST}}{}^4 - (T'')^4 \right] \cdot F_{\mathrm{SF}}, \quad (3)$$

and a minimum value corresponding to the flow of radiant from the heating surface to the bottom of the spheroid, the expression:

$$q_1^{MIN} = \frac{\sigma \cdot \left[T_{\mathrm{ST}}^4 - (T'')^4 \right]}{\dfrac{1}{\varepsilon_{\mathrm{ST}}} + \dfrac{1}{\varepsilon} - 1} \cdot S. \quad (4)$$

Here, ε and $\varepsilon_{\mathrm{ST}}$ - emissivity factor surface of an aqueous solution of lithium bromide and heating wall material (titanium), respectively; $\sigma = 5.67 \cdot 10^{-8} \mathrm{W/m^2 \cdot K^4}$; F_{SF} and S – respectively, the total area of the outer surface area of the solution and its projection on the heating surface, m^2.

The total heat exchange capacity q, determined based on mass changes spheroid to 2.74 g in 37 s (fig.3, video footage No.7 and 8), is 163 W, and according to (3) and (4) the contribution of radiant heat transfer amounted to 181.4 ... 42 W. Consequently, the radiant heat transfer is able to provide additional heating of the spheroid and increase the thermodynamic instability of the solution, as a factor of "explosive" boiling up.

Thus, as a possible reason for the "explosive" fracture weighed portions of an aqueous solution of lithium bromide, which are in spheroidal state on the heating surface are general and local increase in concentration of the solution, existence of its own capillary instability of phase contacting area of the solution with the formation of surface waves, leading to reduction in the size of a thin layer of steam to dimensions irregularities the heating surface. An additional factor reinforcing this effect is overheating of the solution due to the radiant heat transfer.

References

1. V.M. Borishansky, K.M. Arefiev, I.I. Paleev, Tekhtilov R.T. Trudy, TSKTI [Proceedings of CKTI] **78-83,** 62 (1962) [in Russian]
2. S.S. Kutayeladze, Atomizdat Publ. **416** (1979) [in Russian]
3. E.I. Nesis, Moscow Nauka Publ. **280,** (1973) [in Russian]
4. S.L. Elistratov, V.E. Nakoryakov, N.I. Grigoryeva *Research and development of the SB RAS in the field of energy-efficient technologies* (Nauka Publ., Novosibirsk 2009) [in Russian]
5. A.V. Baranenko, S.V. Karavan, O.A. Pinchuk, D.V. Karavan, Moscow, Pero Publ. **393** 2014) [in Russian]

CALCULATION OF INDUSTRIAL ENTERPRISE VENTILATION SYSTEM BY NETWORK INTEGRAL METHOD

Evgeniya I. Mihienkova[1, *], *Sergey* A. Filimonov[2], *Pavel* A. Neobyavlyayuschiy[2], and *Dmitry* V. Boykov[2]

[1]FGAOU VO "Siberian Federal University", Krasnoyarsk, Russia
[2]OOO "TORINS", Krasnoyarsk, Russia

Abstract. This paper describe a ventilation system calculation of the technology building industrial enterprise. On the basis of the calculation model for the enterprise offered technical decision of ventilation systems, subject to a compliance exchange multiplicity, purification efficiency, decontamination from the work area; provided the required volume of gas extraction from process equipment according to the sanitary standards and environmental requirements. Produced selection of ventilation equipment parameters, solved the problem of the air exchange balancing between ventilation systems to prevent the emergence of parasitic flows between the rooms building. SigmaNet software package was used for the implement the calculation.

1 Introduction

Accordance to current sanitary regulations and rules of work site arrangement, both domestic and industrial purposes, in the territory of the Russian Federation must be provided with the optimal parameters of the microclimate. Ventilation is one of the means to ensure the optimum performance of the microclimate. Governing rates for ventilation are air temperature, relative humidity, air velocity in the room and the intensity of the heat generation. At ventilation setting, the calculation appears guarantee of correct functioning system.

Calculation of ventilation systems was carry out by a network integral Theory of hydraulic circuits [1, 2], which gives an idea of the balance characteristics studied system (flow, pressure drop, and so on). The hydraulic circuit includes nodes, which provide the mass and energy conservation laws, expressed in the form balance equation and heat energy (Kirchhoff law), and connecting branches, which simulate various designs flue path. Each branch of the network reflects dependence linking the pressure drop at this branch of the gas flow through its cross section.

SigmaNet software package, used for the calculation, designed to simulate the stationary flux-distribution of non-isothermal multicomponent media in the branched pipe installation (ventilation, gas removal). Complex takes into account the chemical reactions in

* Corresponding author: mihienkova_evgeniya@mail.ru

the system of pipe installation and heat and mass transfer with the environment. The apparent advantage of this method is a significant reduction in calculation time compared with a computational fluid dynamics calculation, while accuracy preservation.

Initial data for calculation are: project statement of the ventilation system of industrial enterprise building; applicable rules, regulations and requirements.

The ventilation system consists of three independent systems: plenum, general ventilation; process ventilation. Process ventilation ensures the removal, transport and treatment of exhaust gases from the equipment (hoods and ovens), and general ventilation supplies conditioned air to each room and remove the exhaust air from the upper part of the work area to eliminate the local excess concentrations of pollutants.

For correct construction of a network model of ventilation systems built on a three-dimensional model of the building scale models of flues in the software package SolidWorks, and was made in the integration of SigmaNet.

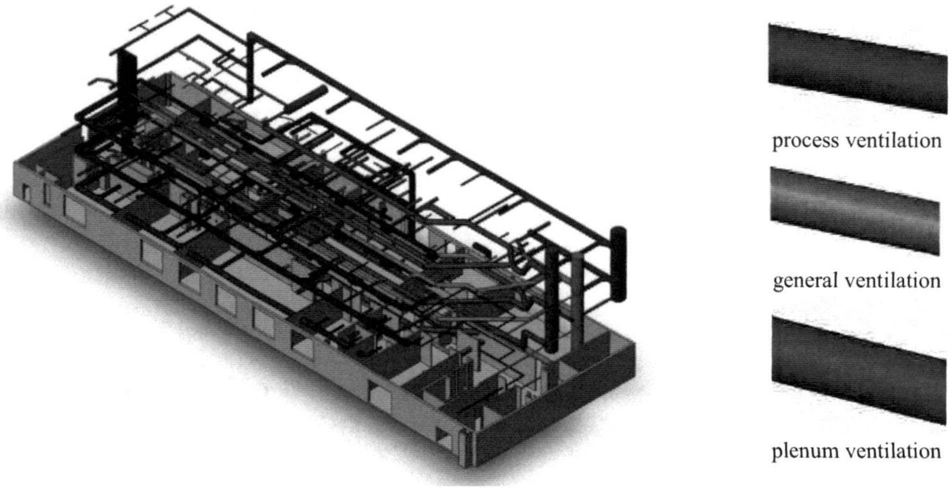

Fig. 1. Geometrical model of the flues plant.

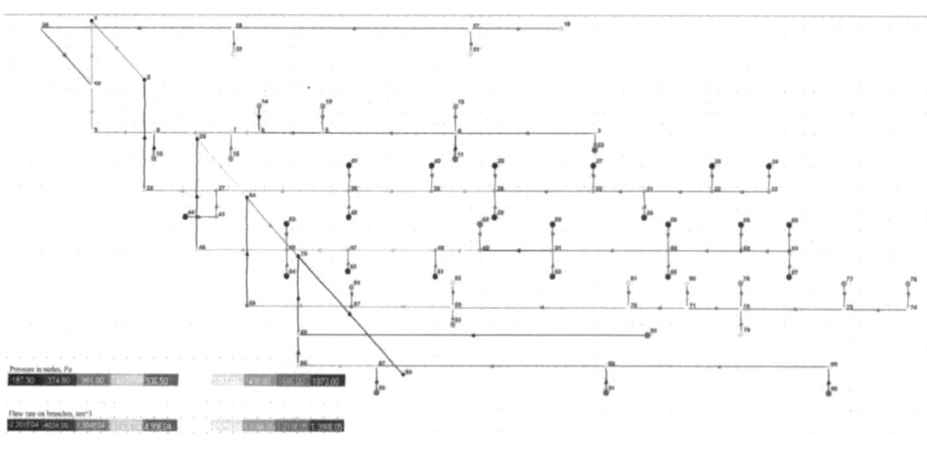

Fig. 2. Calculation of the manufacturing ventilation.

Technical solutions include the use of the general, plenum and process ventilations with dynamic management of operating modes on the technological cycle of the equipment, shift, presence of staff. It is possible to significantly reduce the load on the gas purification and ventilation equipment. Process ventilation collects used air only from working equipment with harmful substances.

DOI: 10.1051/matecconf/20167201070

2 Mathematical model

Implementation of network methods based on the mathematical model described in paper [1] and thesis [3].

2.1 Discretization of the continuum motion equations in the network model

Simulation of non-isothermal multicomponent media reduces to combine problems of flow distribution carrier flow carrier flow, the component and heat flux. At the same time for the relations between the properties of the carrier flow and characteristics of the medium used the equations of state [4].

To represent the network uses a directed graph [5], consisting of two sets: a nodes set N and a branches set (pipes) U. Each branch is put in relation pair of nodes, one of which is origin and the other end of this branch (fig. 3-4). *Oi* is a subset branches starting in *i*-node and I_i is subset branches ending at the *i*-node. Also introduced a set internal nodes network *N-inter*, in which the law of conservation.

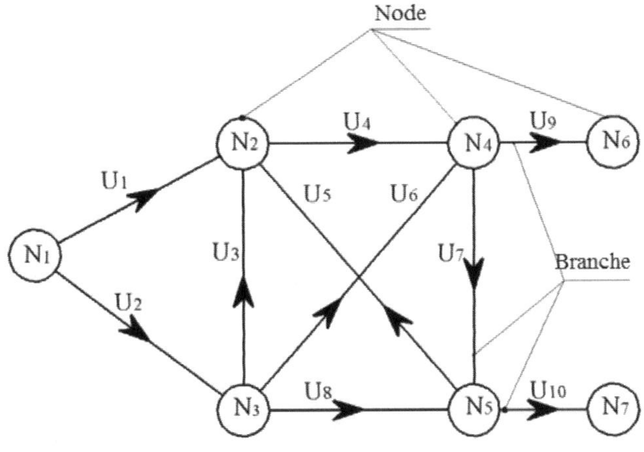

Fig. 3 Graph example.

Set U_i for any $i \in N$, is the set branches incident to that node.

The natural orientation of branches $O_i(u_l)$, $I_i(u_l)$ inserted on the graph to an arbitrarily predetermined flow u_l, $l \in U$.

Using expression (1), the flow distribution problem in the network can be reduced to a combination of the law of conservation of mass at the node (2) and in resistance law pipe (3).

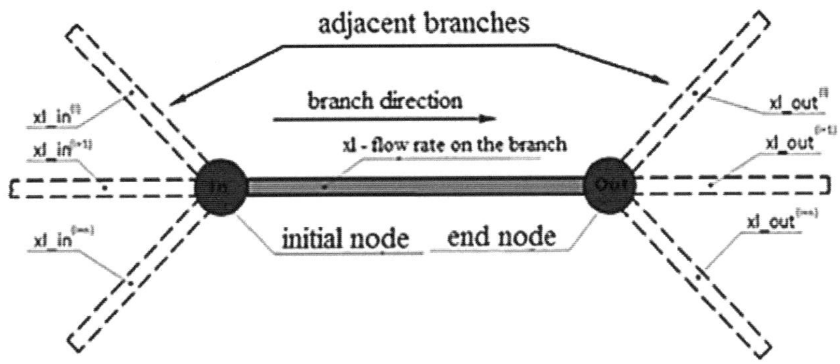

Fig. 4 Sampling network element.

$$D_{il} = \begin{cases} 1, if\ l \in O_i \\ -1, if\ l \in I_i \\ 0, else \end{cases} \quad (1)$$

$$\sum_{l \in U_i} D_{il} x_l = q_i, i \in N \quad (2)$$

$$s_l |x_l| x_l = \sum_{i \in N} D_{il} \cdot p_{Di} + h_l, l \in U \quad (3)$$

x_l – mass bearing flow on the branch (*kg / s*), q_i - source bearing flow in an existing node (*kg / s*), p_{Di}- node pressure (*Pa*), s_l - coefficient of resistance, h_l – branch head (*Pa*)

2.2 Hydraulic resistance network site

Hydraulic resistance is a fundamental value in modeling of the hydraulic circuit. Pressure distribution depends on the correct resistance assignment and flow distribution - in the case of a branched network. Therefore, we consider this value in more detail. According to the manual [6] under the hydraulic resistance (4) refers to total energy (power) loss on a branch to the kinetic energy (power) of the cross-section (for example, at the origin branch), or the ratio of total pressure drop at the branch to dynamic pressure in the cross-section.

$$\xi_{tot} = \frac{\Delta P_{tot}}{2 \cdot \rho \cdot v_l^2} \quad (4)$$

ΔP_{tot} - total pressure drop the length of branches, ρ - density, v_l- average flow rate in the branch section.

There are two types of hydraulic resistance (total pressure drop):

Friction resistance - ξ_{tr}. This resistance is due to a hydraulic friction which is caused by viscosity (both molecular and turbulent) real fluids arising in fluid flow.

Local resistance - ξ_{m}. The local total pressure loss occur at the local violation of the normal flow, flow separation from the walls, formation of vortex and intensity turbulent mixing in the field of pipeline configuration changes or at a meeting and flow around the obstacles.

Friction resistance (5) at a set subcircuit expresses in terms of the coefficient of hydraulic friction:

$$\xi_{\mathrm{fr}} = \lambda_{\mathrm{fr}} \frac{l}{d_{\mathrm{g}}} \quad (5)$$

λ_{fr} - coefficient of flow friction, d_{g} - hydraulic diameter of branch (*m*), *l* - branch length (*m*).

The coefficient of linear friction depends on the fluid flow conditions and the form of section pipe.

Using the superposition principle of loss in which take the arithmetic sum of friction loss and local loss (6) and formula (4) and (5) we obtain the expression for determining the total pressure drop on the circuit better known as the law of Darcy-Weisbach (7- 8)

$$\Delta P_{\mathrm{tot}} = \Delta P_{\mathrm{fr}} + \Delta P_{\mathrm{m}} \quad (6)$$

$$\Delta P_{\text{tot}} = \left(\frac{\lambda_{\text{fr}} \cdot l}{d_{\text{g}}} + \xi_{\text{m}}\right)\frac{\rho \cdot v_l^2}{2} \tag{7}$$

Or going to the symbol in the formula (3):

$$s_l = \left(\frac{\lambda_{\text{fr}} \cdot l}{d_{\text{g}}} + \xi_{\text{m}}\right)\frac{1}{2 \cdot \rho \cdot f^2} \tag{8}$$

f – pipe sectional area (m).

2.3 Boundary Conditions

For the numerical scheme is necessary to set the boundary conditions (9), which are set for network with the help of specific types of nodes, namely: "atmospheric" and "source" nodes. The first-order condition for the pressure and concentration of the components set in atmospheric node:

$$P_{\text{i}} = const;\ \eta_{\text{i}} = const \tag{9}$$

Subject to the pressure pattern is relative (system of pressure linear equations is solved up to a constant) in the constructed network model of the hydraulic circuit must be at least one "atmospheric" node.

As for the "source" node, it's characterized by non-zero source / mass flow.

Also worth noting is the following fact: if the node is a border, it does not mean that it is incident with only one branch. However, in the case of the base node number incident branches are 2 or more.

3 Problem equipment selection

For the calculation the models of all major elements of ventilation systems create: vertical collectors, gas-main pipeline, lateral pipeline in the room. For correct selection of pumping equipment the necessary flow rate set in each ventilation system for each room. Indoor ventilation network is expressed additional hydraulic resistance on each elements [5].

The following provision was made to ensure the efficiency of hoods: for each lateral pipeline process ventilation, which comes into the room with hoods, set the minimum required pressure.

As a result the calculation of the nodes, simulating rooms, were obtained values of excess pressure / vacuum necessary to provide the required volume of air. At the maximum value of the pressure / vacuum and total flow rate pumping equipment was chosen for each system.

4 Calculation of ventilation systems

To align the ventilation systems was developed dynamically balanced model that takes into account the air exchange between the ventilation systems in the rooms and the corridor (fig. 5). In this model, we built all three ventilation system, constraint in between performed using branches, simulating the room (low flow resistance) and branches, simulating the gap between the rooms and the corridor.

For each system boundary condition is pressure from the previous problem.

Air volumes supplied to the processing space are dynamically balanced with the volume of general and process hoods. On the basis of this task settings controlled damper to regulate the air flow in the building rooms was solved.

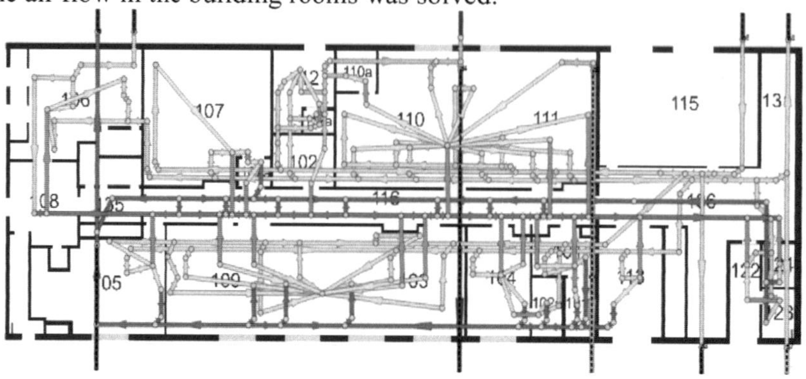

Fig. 5. Model flues 1st floor displaying the network of air exchange between rooms.

The calculations determine the coefficients of local resistance (10) lateral pipeline in the room providing the required flow rate of inflow and hoods. According to the obtained values of local resistance coefficient is determined by the angles of rotation damper. Local resistance coefficient to the rotary shutter was calculated according to the formula [6]:

$$\xi = (\frac{1{,}56}{1 - D^2 sin\delta} - 1)^2 \quad (10)$$

D - passing-pipe size (m), δ – rotation angle of the damper.

5 Conclusion

Based on simulation data, we propose technical solutions for ventilation systems in the building enterprise, taking into account the requirements of multiplicity of ventilation, cleaning efficiency, removing hazards from the work area.

A dynamic balanced system of process and general ventilation is developed, which prevents the occurrence of parasitic flows between different enterprise redistributions building.

The flow resistance for each element of the ventilation system calculated, parameters of ventilation equipment chosen.

Rated conditions damper control has been implemented, which provides the necessary air volume for each room.

References

1. D.V. Boykov, FESTU **38-49**, (2006)
2. A.P. Merenkov, V.Ya. Khasilev, *Theory of Hydraulic Circuits* (Nauka, Moscow 1985) [in Russian]
3. B.A. Elgin, The thesis for the degree of PhD of Physical and Mathematical (1999)
4. S.V. Patankar, *Numerical Heat Transfer and Fluid Flow* (Hemisphere, 1984)
5. R.Wilson, *Introduction to the theory of graphs* (Russian transl.: Mir, Moscow, 1977)
6. I.E. *Idelchik Manual of hydraulic resistance* (Mashinostroenie, Moscow, 1992) [in Russian]

MATEC Web of Conferences **72**, 01071 (2016) DOI: 10.1051/matecconf/20167201071
HMTTSC-2016

DYNAMICS OF EVAPORATION FRONT PROPAGATION IN FREON R21 WITH ADDITION OF NANOPARTICLES

Mikhail Moiseev[1, *], *Denis* Kuznetsov[1], and *Vladimir* Zhukov[1]

[1]Kutateladze Institute of Thermophysics SB RAS, Novosibirsk

Abstract. The paper presents the results of an experimental study on propagation dynamics of the self-sustained evaporation front in a large volume of Freon-R21 with the addition of SiO_2 nanoparticles. The experimental data on propagation velocity and structure of evaporation fronts were obtained; the spectral analysis of fluctuations of the evaporation front interface was carried out. The characteristic frequencies and amplitudes of interface fluctuations were determined depending on the velocity of evaporation front propagation. It was shown that the addition of nano-sized particles significantly affects the front velocity and character of interface fluctuations.

1 Introduction

The task of intensification of heat removal from the heated wall by means of phase transition has more than a century history. The steam boilers and reactors at nuclear power plants, heat pipes and film distillation apparatuses of cryogenic and chemical enterprises [1], spray heat exchanging systems [2], cooling systems for superconductors and processors of supercomputers: all of them use the heat of phase transition for heat transportation from the hot source. Usual water and organic, cryogenic liquids, and liquid mixtures and solutions are usually used as the working liquids [3, 4]. Limiting the intensity of heat removal is caused by formation of a vapor film near the heat transfer surface and, in particular, it is determined by the dynamics of this film propagation along the heating surface [5, 6].

In the last decade there has been a tendency to use nanofluids for heat transfer intensification both in the systems without phase transition, due to improvement of thermal physical properties of liquid [7], and in systems with phase transition, due to expansion of the area of nucleate boiling because of an increase in the value of the first critical heat flux in nanofluids [8]. Nanofluids may be a promising substitute for the classical liquids used as coolants in different applications. In most studies dealt with nanofluids, heat transfer is studied under the steady-state conditions. There are few studies of critical phenomena at non-stationary heat release in nanofluids. As an example, we can present [9], where the authors conducted the experiments on cooling the heated rods in nanofluid and observed a

* Corresponding author: moiseevmikhail@gmail.com

significant improvement in the cooling rate as compared to the pure liquid. The practical importance of these studies is caused by the necessity to develop the reliable protection systems for emergency decompression of devices containing nanofluids at high pressures and temperatures (e.g., future nuclear reactors, using nanofluids as a coolant).

The aim of this work is an experimental study of dynamics of propagation of the self-sustained evaporation front in nanofluids, including the study of the effect of nanoparticles on dynamic and structural characteristics of fluctuations of the evaporation front interface.

2 Research setup and experimental methods

Experiments were carried out in Freon R21 at reduced pressures $P/P_c = 0.037$ (0.193 MPa). For experimental studies of heat transfer and transient dynamics at non-stationary heat release under free convection the setup, shown schematically in fig. 1, was developed.

On bed 1, the frame 2 with refill tank 3 and working chamber 4, housing the working section 10, is hinged. The working chamber is a sealed cylindrical stainless steel vessel. The inner diameter of the chamber is 250 mm, and the height of the workspace is 250 mm. The working chamber is equipped with four windows 5. The bellows 6 with an adjusting screw 7 allows creating the required pressure (up to 0.4 MPa) in the working chamber at the closed valve 8. To set the desired temperature of liquid in the bottom of the chamber, heat exchanger 9 was located. Thus, using this working chamber, it is possible to perform experiments both with the fluid in equilibrium with vapor, and under the conditions of substantial subcooling (to saturation temperature). Frame hinging on the bed allows turning the chamber in relation to the vector of gravity.

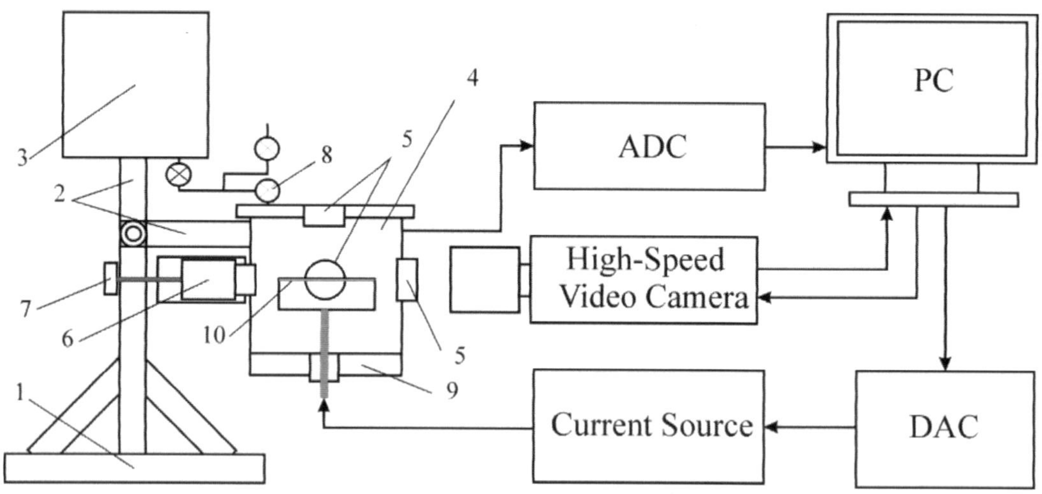

Fig. 1. Scheme of experimental setup.

The working section was made of the stainless steel tube. A thin-film platinum thermometer was installed inside the tube to measure the temperature in the quasi-stationary regimes. The tube ends were hermetically sealed. The tube was heated by electric current passing through the tube. Two conductors of 0.05-mm diameter were welded in the middle of the tube to measure the voltage drop in the region of 31-mm length. In the experiments, two tubular working sections of stainless steel were used. Sections *No. 1* and *No. 2* had outer diameter of 3 mm, wall of 0.5 mm and length of 50 mm. Surface roughness of section *No. 1* was presented by transverse marks with the width of 5-25 μm, and surface roughness of section *No. 2* was presented by large cavities with 100-150 μm radius and depth of 20 μm. In addition, the surface of section *No. 2* had a granular structure with the grains of 5 - 20 μm.

The step-wise current pulse was supplied to the working section. Since inertia in the section was rather high, wall heating occurred at the times of the evaporation front development (about 50 ms) with almost a constant rate (the initial section of the exponent). The heating rate of the heat transfer surface was 4500 K/s. The heat-transfer surface was heated to the desired temperature by pulse time variation. Stepped current interruption stopped heat generation in the working area, and the wall temperature remained virtually constant at the times of the evaporation front propagation ($dT/dt = 0$ K/s). Numerical calculation using thermal conductivity equations has shown that for the time of front passage (much lesser than the time of convection development) the wall temperature before the front varied not more than by 0.2 K. Thus in each experiment, the propagation of self-sustained evaporation front was investigated at the pre-set constant wall temperature.

The working section temperature, average over the length, was measured using temperature dependence of the section resistance. For this purpose, in the middle part of each working section, the voltage drop was measured in the segment of 31 mm. In each measurement, calibration was performed by the temperature of undisturbed fluid. This technique allowed measuring the average temperature of the working section with an error less than 3 K.

Dynamics of the measured wall temperature change was compared with the calculation for non-stationary thermal conductivity regime. Comparison of temperature difference, measured in transient process, at the times lesser than the time of convection development, showed excellent agreement of the experiment and numerical calculation. Until the moment of liquid boiling up, time dependence of the instantaneous wall temperature was determined on the volt-ampere characteristic of the working section. Since the vapor phase formation on the measured part of the working section, the wall temperature was determined based by the equations of non-stationary thermal conductivity. This approach is valid at rather large rates of the working area heating, when the convective mechanism of heat removal from the wall has no time to develop.

Visual observations of the dynamics of vapor phase formation and propagation on the heat-releasing surface were registered by a high-speed digital video camera Phantom v7.0. The shooting rate was usually 25 000 frames per second with the exposure of 26 microseconds. To measure the velocity of vapor front propagation, the camera software was used. When analyzing the results of high-speed digital video-shooting, using specially developed software, the line of the interfacial boundary in the propagating front was determined on each frame. The obtained data served to determine the average velocity of the front V_{fr}. Measurement error of the average velocity in the section of 6-12 mm along the heater did not exceed 0.05 m/s.

The used additives were SiO_2 particles that had an average size of 20 nm and hydrophilic properties. The powder was loaded into the working volume, then vacuum exhaust, filling with freon and multiple mixing were performed. Particle concentration was 0.0077% vol.

3 Experimental results

Dependence of the average velocity of evaporation front V_{fr} on wall superheat is shown in fig. 2. As can be seen from the diagram, there are two areas with different rates of a change in the vapor front velocity, depending on the temperature rise. In the area of low overheating, the growth rate of the front velocity vs. wall superheat is satisfactorily described through the calculation by the model [5]. In the area of high overheating, the significantly higher growth rate of the front velocity vs. wall overheating is observed. In [5] it was suggested that this effect relates to intensification of heat transfer across the interface due to development of Landau instability [10]. In pure liquid, the data for different heaters coincide within the error. In nanofluids on both heaters, front initiating takes place at substantially higher overheating than in the pure liquid (approximately 55 K and 34 K, respectively). In the overheating range of 55-65 K the front velocities in nanofluids are equal or less than those in a pure liquid, but when the higher temperature pressures are achieved, the front velocity in nanofluids increases significantly. This effect is more pronounced for the heater No.1 with a lesser degree of surface roughness. In successive series of experiments, the front velocity in nanofluids increased (fig. 3).

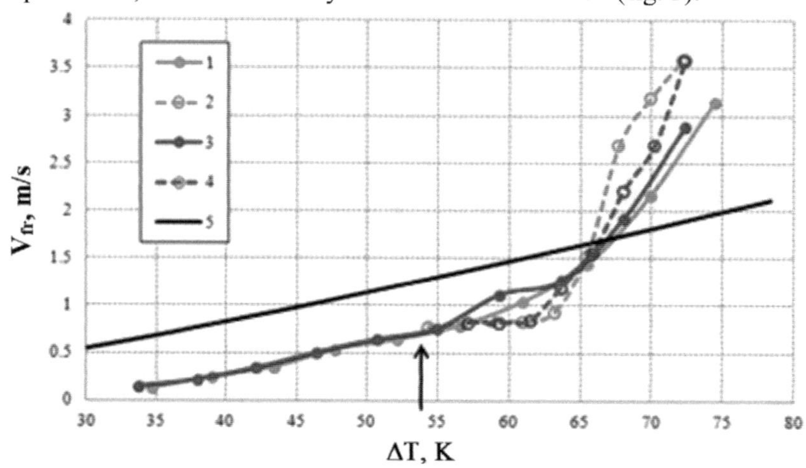

Fig. 2. Average front velocity vs. wall superheat. 1 – heater No.1, pure R21; 2 – heater No.1, R21 + 0.0077% SiO_2; 3 – heater No.2, pure R21; 4 – heater No. 2, R21 + 0.0077% SiO_2; 5 – calculation by model [5]. The arrow indicates the minimal value of overheating, when the evaporation front is initiated in nanofluid.

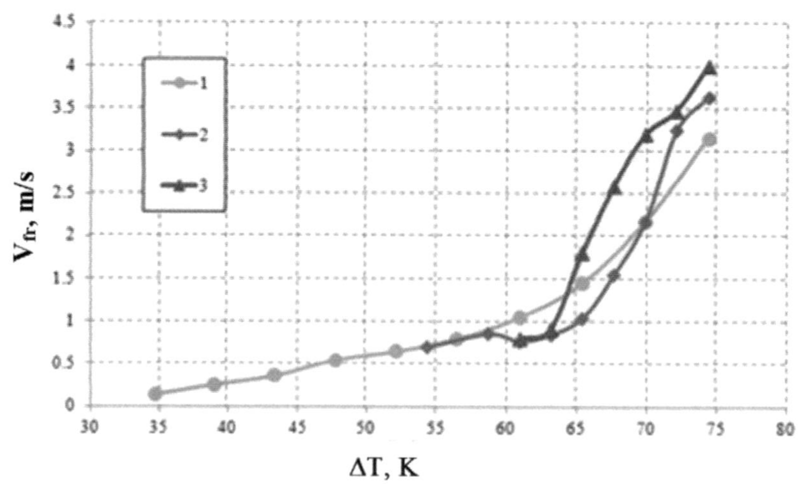

Fig. 3. Front velocity vs. wall superheat on heater No. 1 in different experimental series. 1 – pure R21; 2 – R21 + 0.0077% SiO_2 in two days nanoparticles addition; 3 - R21 + 0.0077% SiO_2 in nine days.

To investigate pulsations of the front interface, we used the dependence of x coordinate on time over one of the heater generatrices, i.e. for the fixed value of y-coordinate. The linear trend, complying with the distance run at constant speed V_{fr}, was deducted from the resulting dependence. The fast Fourier transformation was applied to coordinate deviations from the path at a constant velocity. In all the spectra, the pronounced peak of main harmonic was observed. Using the amplitude and frequency of the main harmonic, we determined the characteristic velocity pulsations and acceleration of the interface.

Dependences of frequency and amplitude of the main harmonic on average velocity for heater No. 1 are shown in fig. 4. In pure liquid, the frequency of the main harmonic increases with a rise of the average velocity, and the amplitude decreases first and then achieves a constant value. The similar tendency is observed, but fluctuation frequencies and amplitudes decrease in comparison with pure liquid.

The amplitude value of interface acceleration, defined as the product of main harmonic amplitude on the square of angular frequency, can be estimated by data obtained. In pure liquid, acceleration varied within 50-1500 m/s^2 depending on wall overheating. In nanofluids, interface acceleration was several times lower than in the pure liquid, and it decreased in the successive series of experiments (fig. 5).

It is shown in [11] that interface acceleration affects significantly the instability development and intensification of heat transfer through the interface. A decrease in acceleration at nanoparticles addition gives an opportunity for instability development at the lower velocities of evaporation front.

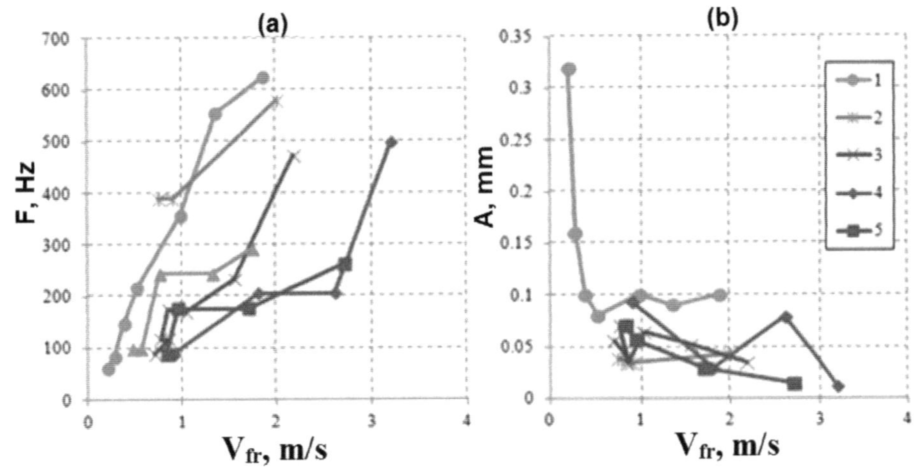

Fig. 4. (a) – frequencies and (b) – amplitudes of fluctuations of the front interface on heater No. 1 depending on average velocity. 1 – pure R21; 2 – R21 + 0.0077% SiO_2 in a day after nanoparticles addition; 3 – R21 + 0.0077% SiO_2 in two days; 4 - R21 + 0.0077% SiO_2 in nine days (series 1); R21 + 0.0077% SiO_2 in nine days (series 2).

The detected dependence of velocity and amplitude-frequency characteristics of fluctuations of evaporation front interface on the number of experimental series supports the effect of deposition of nanoparticles and their agglomerates on the heater wall at intense liquid evaporation at front propagation. However, a significant increase in the temperature difference, required for front initiation, was observed in the first experimental series with nanoparticles, and it is not explained by the effect of particles deposition.

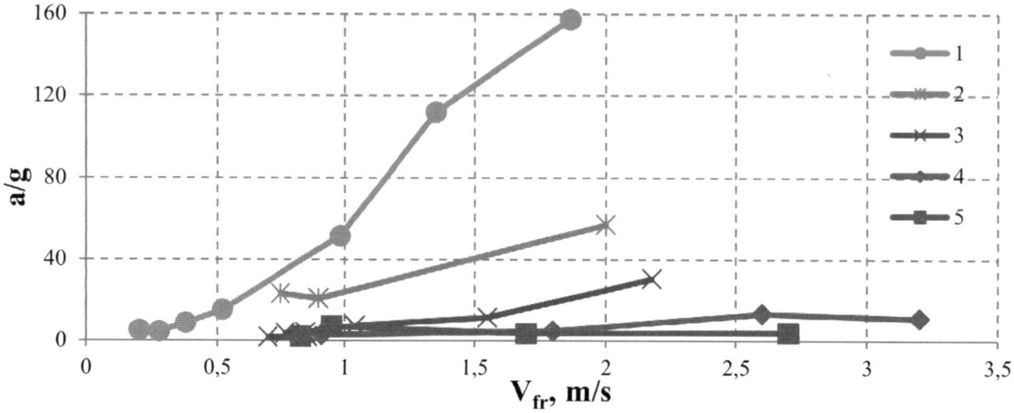

Fig. 5. Amplitude acceleration of the interf ace vs. average velocity. 1 – pure R21; 2 – R21 + 0.0077% SiO_2 in a day after nanoparticles addition; 3 – R21 + 0.0077% SiO_2 in two days; 4 - R21 + 0.0077% SiO2 in nine days (series 1); 5 - R21 + 0.0077% SiO_2 in nine days (series 2).

4 Conclusions

The evaporation front is initiated in the nanofluid at significantly higher wall overheating than in the pure liquid. Addition of nanoparticles into liquid significantly increases the velocity of evaporation front in the area of high overheating of the heating surface. Perhaps, this effect is associated with deposition of nanoparticles and their agglomerates on the heating surface. The effects caused by the addition of nanoparticles are more obvious for the heater with lower surface roughness. In pure liquid, we do not observe the effect of surface roughness on the characteristics of evaporation front. The effect of addition of nano-sized particles on the earlier development of instability has no an unambiguous explanation and requires following examination.

Acknowledgment

The work was financially supported by RFBR (project No. 13-08-00178-a).

References

1. A.N. Pavlenko, V.E. Zhukov, N.I. Pecherkin, V.Yu. Chekhovich, O.A. Volodin, A. Shilkin, and C. Grossmann, AIChE J **60(2)**, 690 (2014)
2. S.Y. Misyura, Int. J. Heat Mass Transfer **71**, 197 (2014)
3. V.E. Nakoryakov, S.Y. Misyura, S.L. Elistratov, R.A Dekhtyar, J. Eng. Thermophys. **23** (4), (2014)
4. V.E. Nakoryakov, S.Y. Misyura, S.L. Elistratov, JET **20(4)**, 1 (2011)
5. A.N. Pavlenko, V.V. Lel, Thermophysics and Aeromechanics **6(1)**, 105 (1999)
6. A.N. Pavlenko, A.S. Surtaev, Microgravity Sci. Tech. **22(2)**, 215 (2010)
7. K.V. Wong, O. De Leon, Adv. Mech. Eng. **2**, 519659 (2010)
8. S.J. Kim, I.C. Bang, J. Buongiorno, L.W. Hu. Bulletin of the Polish Academy of Sciences – Tech. Sciences **55**, 2 (2007)
9. H. Kim et al., ASME, ICNMM2009-82082 (2009)
10. L.D. Landau, JETP **14,** 6 (1944)
11. V.E. Zhukov, A.N. Pavlenko, M.I. Moiseev and D.V. Kuznetsov, Proc. of the 15th Int. Heat Transfer Conf. IHTC15-8766 (2014)

MATEC Web of Conferences **72**, 01072 (2016) DOI: 10.1051/matecconf/20167201072
HMTTSC-2016

SURFACE CRYSTALLIZATION OF AQUEOUS SALT SOLUTION UNDER OVERHEATING AND OVERCOOLING

Vladimir Nakoryakov[1], *Sergei* Misyura[1,*] and *Vladimir* Morozov[1]

[1]Institute of Thermophysics of SBRA, Novosibirsk, Russia

Abstract. The investigation of the surface crystallization with low negative and high positive temperatures were carried in the paper. Crystallization curves for distillate (T_{s0} = -9 °C) and different mass salt concentrations NaCl (T_{s0} = 80 °C) were obtained. Experimental data indicate that the crystallization centers influence each other and the number of centers does not change with time. The maximum speeds for the crystallization front reached 0.3-0.5 m/s. There are a significant anisotropy and a curvature of crystallization front. The surface kinetics should be considered to clarify the rate of freezing and melting of ice in modeling global warming.

1 Introduction

Typically, crystallization models examine large volumes, the independence of crystalline centers, three-dimensional crystallization and often do not take into account the surface phenomena in the description of global warming. A kind of kinetic curves depend on the dimensionality of the growth centers of the new phase [1]. A deviation from equilibrium depends on heat transfer and the high supercooling [2, 3]. Another important challenge is the crystallization of salt solutions at a high salt concentration in adsorption heat pumps and in apparatus for chemical technologies. Absorbers and desorbers operate with solutions having high salt concentration. There is a risk of local areas formation of partial surface crystallization during transient modes of pumps. The behavior of solutions and mixtures studied in [4-7]. Crisis phenomena at phase transitions are considered in [8]. Combustion of fuel in a gas-droplet mixture is considered in [9]. Thermal measurement technique discussed in [10].

2 Experimental data

The investigation of the surface crystallization with low negative and high positive temperatures were carried in the paper. The heating and the cooling rate was 1 °C/min. External pressure was 1 atm. Ambient temperature was 20 °C. The whole setup was in a closed housing with a relative humidity less than 10%. The diameter of the working

* Corresponding author: misura@itp.nsc.ru

cylinder with the solution was 15 mm, material was brass. The temperature of the liquid surface was measured by a high-speed thermal imager. The error of the surface temperature was within 1 °C. fig. 1 shows photographs of the thermal image for surface crystallization front. Red corresponds to 0 °C.

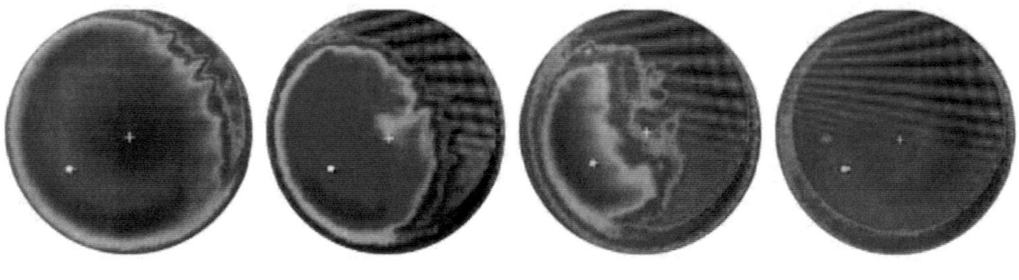

Fig. 1. Thermal pictures of the surface crystallization of distillate (T_{s0} = -9 °C).

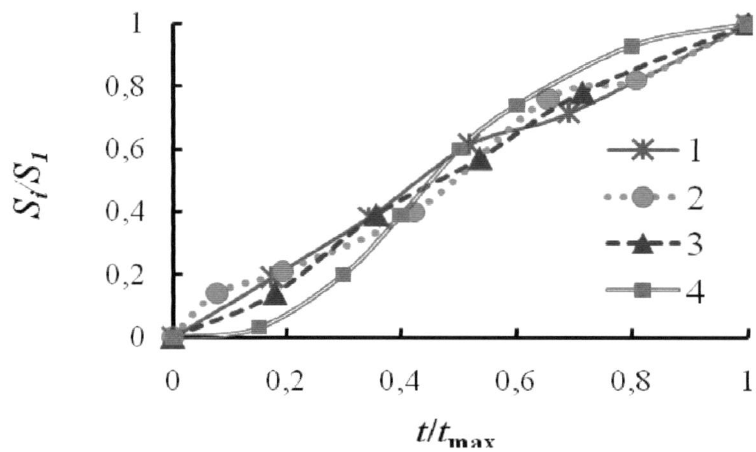

Fig. 2. Crystallization kinetics for the mass concentration of the salt solution NaCl: 1 – distillate, 2 - C_0 = 0.1%, 3 - C_0 = 1% (T_s = (263-264) K), 4 - C_0 = 26% (T_s = 353 K).

Fig. 2 shows crystallization curves for different mass salt concentrations NaCl. The growth rate of the curves (1-3) is quasi-linear and curve 4 corresponds to curve 1, 2 (fig. 3). It is obvious that the growth rate of the crystal film thickness decreases with increasing time. Then the experimental curve for the dimensionless volume $V_i/V_0 = f(t)$ will have the form 4 in fig. 3. The character of the crystallization curve for independent crystals centers depends on the dimensionality and has the general form V_i/V_0 = 1-exp($-kwc^{n1}t^{n2}$), where k is the coefficient, w is the speed of the new centers emergence, c is the crystal growth velocity, the coefficients n depend on the type of crystallization (dimensionality). Curve 1, 2, 3 correspond to crystal growth for three-dimensional, two-dimensional and one-dimensional growth, curve 4 for the two-dimensional growth when w = 0. Experimental data (Curve 4) indicate that the crystallization centers influence each other and the number of centers does not change with time (w = 0). The maximum speeds for the crystallization front reached 0.3-0.5 m/s.

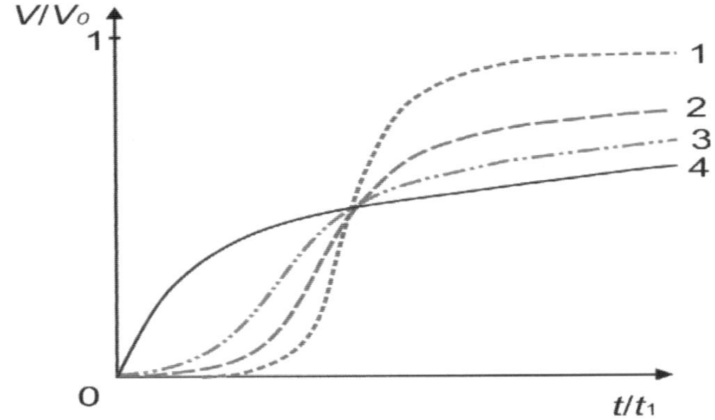

Fig. 3. Kinetic curves of crystallization.

3 Conclusion

The high rate of surface crystallization was observed at high overheating and high supercooling. At the same time there are a significant anisotropy and a curvature of crystallization front. The surface kinetics should be considered to clarify the rate of freezing and melting of ice.

Acknowledgments

This work was supported by the Russian Foundation for Basic Research, project No. 16-08-00488.

References

1. N.N. Sirota, *Crystallization and phase transitions* (Publisher BSSR, Minsk, 1962)
2. A.A. Chernov, J. Engineering Thermophys. **18**, 238 (2009)
3. A.A. Chernov, A.A. Pil'nik, JETP Letters **102**, 526 (8) (2015)
4. D. Feoktistov, E. Orlova, A. Islamova, EPJ Web Conf. **110**, 01018 (2016)
5. S.Ya. Misyura, Int. J. Therm. Sci. **92**, 34 (2015)
6. S.Y. Misyura, Exp. Thermal Fluid Sci. **75**, 43 (2016)
7. A.N. Pavlenko, V.E. Zhukov, N.I. Pecherkin, O.A. Volodin, A.S. Surtaev, X. Li, X. Gao, L. Zhang, H. Sui, H. Li H, J. Engineering Thermophys. **24**, 210 (2015)
8. S.Y. Misyura, Exp. Thermal Fluid Sci. **70,** 389 (2016)
9. S.Y. Misyura, Energy **103**, 430 (2016)
10. S.L. Elistratov, E.Yu. Slesareva, EPJ Web Conf. **82**, 01060 (2015)

NON-ISOTHERMAL DESORPTION OF THIN LAYERS OF AQUEOUS SALT SOLUTIONS

Sergei Misyura[1, *], and *Vladimir* Morozov[1]

[1]Institute of Thermophysics of SBRA, Novosibirsk, Russia

Abstract. Nonisothermal desorption of the aqueous salt solutions layers on the horizontal heating surface was experimentally investigated. Water solutions of NaCl, $CaCl_2$, LiBr, LiCl were used. Surface temperature was T_w=80 °C. After warming, the desorption rate for salt solutions continuously decreases with increasing time. Evaporation rate of distillate is constant for a long time period. There is a significant growth of the rate at the end of the evaporation, when the film thickness becomes substantially less than 1 mm. Desorber of absorption heat pump can operate without failure when there is a small local (partial) crystallization.

1 Introduction

The efficiency of absorption thermotransformers is determined by the processes of absorption and desorption. The theory of nonisothermal absorption is presented in monograph [1]. The processes of layers nonstationary desorption of aqueous salt solutions can be hardly described by the available theory of nonisothermal sorption, and they need experimental investigation in a wide range of operation parameters. To date, sorption kinetics in the vicinity of the melting point and kinetics at the start of crystallization is poorly studied. Evaporation of liquids was studied in [2-4] and desorption of solutions was considered in [5-7]. The crisis phenomena of liquids at high heating are considered in [8, 9]. Combustion of fuel in a gas-droplet mixture is considered in [10]. Measuring method for non-isothermal flows with thermal imager was presented in [11].

2 Experimental data

To register the current parameters of nonisothermal desorption of the aqueous salt solutions layers on the horizontal heating surface (see fig. 1), we used the experimental method for direct measurements of the weight of the layer. The working section was mounted on a laboratory balance. The values of mass solution concentrations C_0 were determined by the standard densimeters. The thermal field of layer surface (T_s) was measured by the thermal imager. The surface temperature of the heated wall was determined by thermocouples located near the wall with a relative error within 1%. The external pressure was 1 bar. Ambient temperature was 22 °C. Heat transfer law was T_w=const. We use titanium as the

* Corresponding author: misura@itp.nsc.ru

working material allowed experiments with corrosive water solutions of NaCl, $CaCl_2$, LiBr, LiCl at heating surface temperatures T_w=80 °C. In the process of desorption the layer thickness varied from 3 mm to 0.2 mm.

Experimental data on the desorption rate of layers for distillate and aqueous salt solutions are presented in fig. 1.

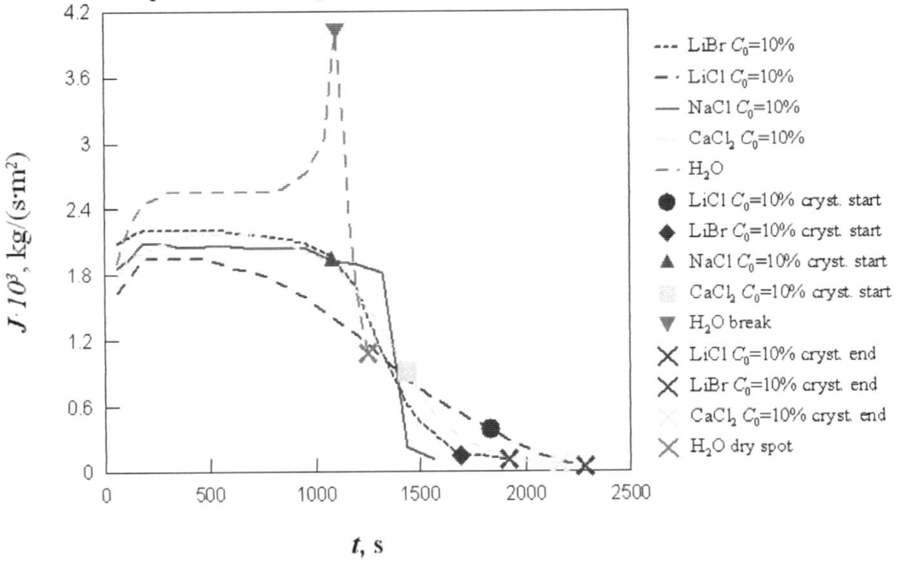

Fig. 1. The evaporation rate for distillate and the desorption rate for solutions (T_w=80 °C, the initial salt concentration is 10 %, the initial thickness of the liquid layer is 3.1 mm).

The bold dots (cryst. stat) indicate the beginning of the crystallization of solutions and the beginning of film disruption for distillate. The dots in the form of a cross indicate the end of the crystallization (crys. end). The temperature of heated wall (T_w) was 80 °C. At the initial moment the liquid is heated from the wall and the desorption rate increases. After this warming, the desorption rate for salt solutions continuously decreases with increasing time. Evaporation rate of distillate is constant for a long time period. There is a significant growth of the rate at the end of the evaporation, when the film thickness becomes substantially less than 1 mm.

3 Conclusion

Previously it was believed that in the vicinity of the crystallization the desorption rate is close to zero. However, because of low rate of crystallization front, the desorption rate remains essential for a long time after the onset of crystallization. In this regard, the desorber can operate without failure when there is a small local (partial) crystallization.

Acknowledgments

This work was supported by the Russian Science Foundation (grant No.15-19-10025).

References

1. V.E. Nakoryakov, N.I. Grigorieva, *Nonisothermal absorption in thermotransformers* (Nauka, Novosibirsk 2010)
2. M.V. Bartashevich, O.A. Kabov, V.V. Kuznetsov, Fluid Dyn. **45**, 924 (2010)

3. D.V. Feoktistov, I.A. Afanasyev, E.G. Orlova, EPJ Web Conf. **82**, 01054 (2015)
4. S.Y. Misyura, Exp. Thermal Fluid Sci. **70,** 389 (2016)
5. E.G. Orlova, G. V. Kuznetsov, D. V. Feoktistov, EPJ Web Conf. **82**, 01053 (2015)
6. S.Ya. Misyura, Int. J. Therm. Sci. **92**, 34 (2015)
7. S.Y. Misyura, Exp. Thermal Fluid Sci. **75**, 43 (2016)
8. A.N.Pavlenko, A.A. Tairov, V.E. Zhukov, A.A. Levin, A.N. Tsoi, J. Eng. Thermophys. **20** (2011)
9. A.N. Pavlenko, E.A. Tairov, V.E. Zhukov, A.A. Levin, M.I. Moiseev, J. Engineering Thermophys. **23**, 173 (2014)
10. S.Y. Misyura, Energy **103**, 430 (2016)
11. S.L. Elistratov, E.Yu. Slesareva, V.V. Ovchinnikov, J. Engineering Thermophys. **24**, 346 (2015)

MATEC Web of Conferences **72**, 01074 (2016) DOI: 10.1051/matecconf/20167201074
HMTTSC-2016

RESULTS OF STUDY OF SULFUR OXIDE REDUCTION DURING COMBUSTION OF COAL-WATER SLURRY FUEL THROUGH USE OF SULFUR CAPTURING AGENTS

Vasiliy I. Murko[1], *Viktor* I. Karpenok[1], *Yuliya* A. Senchurova[1], *Oleg* V. Tailakov[1], *Alexander* R. Bogomolov[2,*], and *Veniamin* A. Khyamyalyainen[1]

[1]Kuzbass State Technical University named after T.F. Gorbachev, 650000 Kemerovo, Russia
[2]Kutateladze Institute of Thermophysics SB RAS, 630090 Novosibirsk, Russia

Abstract. It is shown that an effective way of burning high sulfur coal is to burn coal-water slurry fuel (CWF) prepared on its basis containing a sulfur capture agent (SCA) entered in the slurry at the stage of preparation. The technique of thermodynamic analysis of chemical reactions during CWF burning has been developed including burning in the presence of SCA. Using the developed calculation program, the optimal temperature conditions have been determined as required for the effective reduction of sulfur oxides in flue gases when using different types of SCA. According to the results of calculating the composition of CWF combustion products when entering various substances in the burner space as SCA it has been determined that magnesite, calcite, and dolomite are the most effective natural minerals. The analysis of calculated and experimental data proves the efficiency of SCA addition as well as validity of the obtained results.

1 Introduction

It is known [1, 2] that addition of SCA in the fuel or feed of SCA directly in the combustion area is one of the effective methods of reducing the content of sulfur oxides in flue gases produced by combustion of various fuels, including coal, containing sulfur. When using the first option of sulfur capture it is preferable to use ecologically clean water-coal slurry fuel as coal fuel which already contains sulfur capture agent added at the stage of fuel preparation.

The main point of the second option of sulfur capture (addition of SCA in fuel combustion zone) is that certain chemical elements present in the SCA interacting with sulfur or sulfur-containing substances formed in the combustion process chemically interact to form new solid phase sulfur compounds. In this process a part of the resulting solid-phase sulfur compounds falls in the slag funnel of the boiler and is removed from the boiler with the boiler ash and slag, while the remaining part is carried out of the boiler with the flue gases in the form of fines and is captured by the gas treatment system. It is possible to

* Corresponding author: barom@mail.kuzstu.ru

evaluate the effectiveness of SCA elements impact on sulfur binding process during combustion of different types of coal fuel with the use of chemical thermodynamics methods. With regard to the chemical reactions occurring in the area of coal fuel combustion, it can be argued that subject to environmental conditions (pressure, temperature, composition of substance in the system) corresponding to the equilibrium state and sufficient time for such an interaction, the chemical response of substances located in the furnace space will provide the required (estimated) composition of combustion products. The required time of the reactive agents staying during combustion of various types of coal fuel is provided by rational choice of combustion technology and management of the combustion process.

Recently, swirl method of coal fuels combustion has become widely spread, including fuels significantly filled with water or mineral components [3, 4]. This method provides good mixing of combustion air and fuel, their rapid heating and thereby substantially uniform reactants distribution on the combustion chamber.

It is known [5] that in the chemical thermodynamics finding of conditions ensuring the reduction of the system to an equilibrium state is equivalent to the definition of the system parameters characterizing it when it reaches the state of equilibrium. It is necessary with the use of thermodynamic analysis to determine the chemical composition of this mixture and process conditions in the presence of SCA in which the equilibrium state of the system is achieved. Thus in the system in addition to conventional gaseous substances formed during combustion of wet coal fuel – carbon dioxide, water vapor, and solid ash and slag forming substances, there must be solids containing sulfur present and there should be no gaseous sulfur oxides. As a consequence, it is possible to obtain a complete binding of sulfur in the composition of solids, capture and removal of the latter from the process presents no particular technical difficulties. The results of the thermodynamic analysis need to be confirmed by the results of field experiments.

As van't Hoff has shown [5], as a measure of the chemical affinity, that is as a criterion indicating the direction of the chemical reaction, it should be Helmholtz potential for processes occurring at constant volume and Gibbs energy for processes occurring at constant pressure. We believe that the combustion process in the swirl furnace flows at constant pressure. In view of this in order to find the conditions sought by the system, it is necessary to define minimum isobaric potential (Gibbs free energy).

By definition Z Gibbs energy is a function of the following type:

$$Z = H - TS,$$

where H and S – enthalpy and entropy in the operating conditions respectively calculated according to the formulas:

$$H = H^0 + \int C_p dT; \; S = S^0 + \int (C_p / T) dT,$$

H^0, S^0 – enthalpy and entropy in the standard conditions respectively, C_p – specific heat at constant pressure; T – temperature.

Gibbs potential has a property of additivity, i.e.

$$Z = \sum Z_i,$$

where Z_i – Gibbs energy of the i-th substance (component) in the system.

It is obvious that due to the additivity we have:

$$Z_i = Z_{isv} a_i,$$

where Z_{isv} – specific value of the Gibbs energy of the i-th substance (based on 1 mol), a_i – quantity of the i -th substance, mol.

Thus, solution of the objective to study the furnace process in the physical sense comes to determination of the composition of the system (such a set of substances that makes up the system) and the temperature at which the Gibbs energy has a minimum value taking into account implementation of the laws of conservation of matter and energy. Therefore from the mathematical point of view it is necessary to find the minimum of a function of many variables:

$$Z = \sum a_i \left(H_i^0 + \int C_{pi} dT - T\left(S_i^0 + \int \left(C_{pi} / T \right) dT \right) \right) \quad (1)$$

with the additional conditions (restrictions) complying also with the corresponding laws of conservation of matter and energy.

2 Result and discussion

As can be seen from equation (1), Z function is linear relative to quantities of substances, but non-linear with respect to temperature. The temperature is as a member of the equation as are the amounts of combustion products, that is the temperature developed in the process of combustion depends on the conditions in the furnace, namely on the quality of the fuel, the balance of fuel and oxidant, heating (or cooling) of the fuel, and the blast air prior to feeding to the furnace. Thus, if we take the temperature in the equation (1) fixed, the problem is reduced to the problem of linear programming, the solution of which is the essence of the thermodynamic analysis. With this in mind, the programme was developed which allows to perform thermodynamic analysis of coal fuel combustion in the presence of SCA. The calculation algorithm is shown in fig. 1.

Combustion processes that take place in the furnace cause redistribution of the elements. For example, oxygen from air passes into carbon dioxide in flue gases. As a result of chemical transformation, the temperature of the system changes. However, the enthalpy of the system (adiabatic condition of the process) and the number of elements (the law of conservation of mass) remain unchanged; substances are transformed from ones into others, but there appear no additional atoms of any new elements. The values of amounts of each element are set in moles as initial data. This is necessary for drawing up of the equations of conservation of matter. The condition of conservation of the enthalpy of the system is taken into account as follows. Each cycle of calculation is carried out at a certain temperature. Throughout the cycle, the temperature is considered constant. Upon completion of the cycle enthalpy of the system is calculated. If the enthalpy is higher than the initial enthalpy of system, the temperature value is reduced, or increased, if otherwise. Next, the cycle of calculation is repeated with the changed temperature value. The calculation is carried out as long as the modulus of difference between the calculated and the initial values of the system enthalpy is less than the predetermined tolerance. In this case, the result of the last cycle is taken as the calculated composition of the combustion products, and the temperature corresponding to the given cycle as the temperature of combustion. In order to achieve a certain temperature in the calculation, the enthalpy of the system should be changed accordingly.

The calculation results using various SCAs showed that:

- in using slaked lime ($Ca(OH)_2$) as SCA, effective sulfur capturing is observed at temperature up to 900°C and at higher temperatures sulfur capturing efficiency decreases, and SO_2 appears in combustion products;

- in using limestone ($CaCO_3$), effective sulfur capturing is observed at temperature up to 1100°C, and at higher temperatures sulfur dioxide appears in gases;
- in using magnesite ($MgCO_3$), effective sulfur capturing is observed at temperature up to 1100°C, and at higher temperatures sulfur dioxide also appears in the combustion products. It must be noted that magnesium sulfate decomposes at a temperature above 1137°C, so calculations were performed at temperatures not exceeding 1100°C;
- in introduction of dolomite (which corresponds to simultaneous introduction of magnesite and calcite) effective sulfur capturing is observed at temperature up to 1100°C, and at the same time only magnesium dolomite component is "consumed" (which means higher efficiency of magnesite as compared with calcite) and there is an incomplete absorption of sulfur dioxide. Perhaps the presence of magnesium "suppresses" calcium activity, which will be explored experimentally;
- the use of the resulting slag phase (Ca_2SiO_4 и $CaSiO_3$) is completely ineffective for the purposes of binding sulfur in solid ash and slag compounds. This is likely due to the fact that the slag phase is not a combination of two oxides (CaO and SiO_2), but separate chemical compounds. For example, in [6] it is shown that this phase has a higher (by 72 kJ/mol) value of the enthalpy of the formation than the sum of enthalpies of the formation of its constituent oxides, as well as the higher value of the corresponding entropy. Calcium and silicon oxides stirred in the composition of ash at the molecular level form upon combustion conditions the slag phase which is unsuitable as an SCA, and the calculation has shown that.

Thus, introduction of technogenic product (slaked lime) which requires certain costs for its application and is not a natural mineral, was less effective than the use of natural mineral materials - calcite, magnesite, and dolomite.

With regard to magnesite it should be noted that its effective work up to 1100°C temperature is attended by some risk since in implementation of the firing process fluctuations (changes) of temperatures are observed quite often, and it is likely that conducting the process "at the upper limit" of SCA ability may from time to time result in appearing of sulfur dioxide in the products.

To assess compliance of the results of calculations and experimental data, the calculation was made corresponding to burning of fuel made from coal of the "Black Sea" mine (Bulgaria) (Table 1).

Table 1. Properties of CWF.

Parameter, symbol	Unit of measure	Value
Solid phase content	%	50.0
Low calorific value	kcal/kg	2950
Carbon, C^{daf}	%	72.0
Hydrogen, H_2^{daf}	%	6.0
Nitrogen, N_2^{daf}	%	1.5
Oxygen, O_2^{daf}	%	16.7
Sulfur, S^{daf}	%	3.8 Including combustible sulfur (organic and pyritic) - 2.4; incombustible (sulphate) - 1.4
Ash on dry basis, A^d	%	22.2

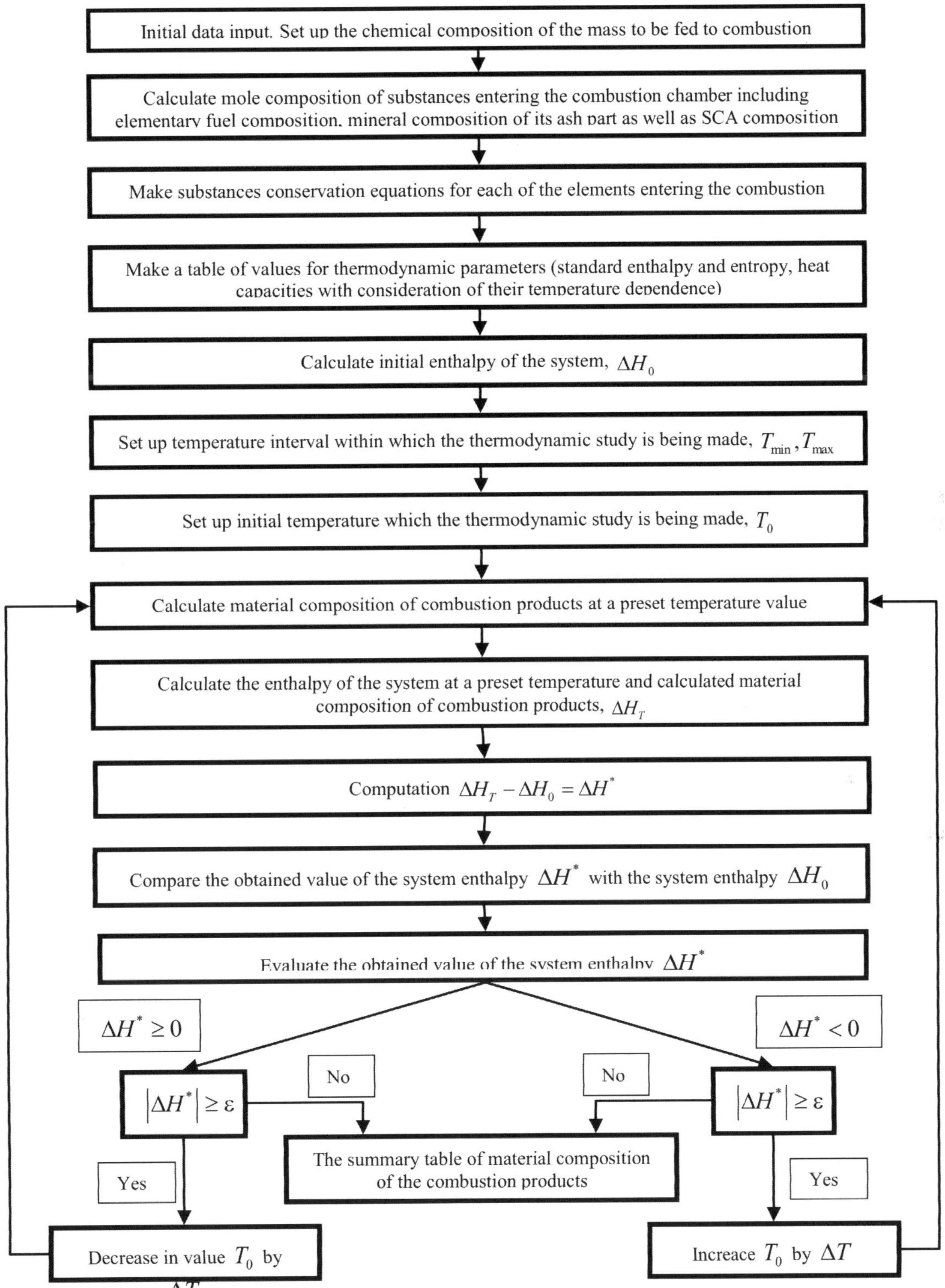

Fig. 1. Block diagram for execution of the program of thermodynamic analysis of coal fuel combustion in the presence of SCA.

Sulfur capture agent was prepared of limestone (46.3 CaO; 6.1 MgO; 98.1 ($CaCO_3$ +$MgCO_3$) – GOST 17498-72). Its properties were considered in the initial data for calculation.

The calculations were made for different values of the coefficient of excess air and combustion temperature (Table 2).

Table 2. Calculated composition of CWF combustion products.

	α = 0.9			α = 1.0			α = 1.3		
	1150K	**1320K**	**1550K**	**1150K**	**1320K**	**1550K**	**1150K**	**1320K**	**1550K**
C	0	0	0	0	0	0	0	0	0
O_2	0	0	0	0	0	0	3.87	3.87	3.87
H_2	5.63	0	0	0	0	0	0	0	0
S	0.19	0.19	0	0.03	0.03	0	0	0	0
Mg	0	0	0	0	0	0	0	0	0
Ca	0	0	0	0	0	0	0	0	0
N_2	55.75	55.75	55.84	60.39	60.39	60.40	63.89	63.89	63.89
CO_2	15.56	9.92	9.39	13.98	13.98	13.90	11.39	11.39	11.39
CO	0	5.63	6.19	0	0	0.07	0	0	0
H_2O	22.67	28.30	28.35	25.43	25.43	25.43	20.71	20.71	20.71
SO_2	0	0	0.03	0	0	0.027	0	0	0
SO_3	0	0	0	0	0	0	0	0	0
MgO	0.03	0.03	0.03	0.027	0.027	0.027	0	0	0
$MgSO_4$	0	0	0	0	0	0	0.022	0.022	0.022
CaO	0.16	0.16	0	0.006	0.006	0	0	0	0
$CaSO_4$	0	0	0.16	0.14	0.14	0.15	0.119	0.119	0.119

As shown in Table 2, a significant impact on the effectiveness of sulfur capture has atmosphere composition in the furnace. In the absence of an excess of oxygen ($\alpha = 0.9$ and $\alpha = 1.0$) elemental sulfur S is present in the combustion products (at a temperature of up to 1320 K) or sulfur dioxide SO_2 (at higher temperature values). The magnesium-containing part of SCA is "quiescent". In case of air excess ($\alpha = 1.3$) there was no sulfur or sulfur dioxide over the entire studied temperature range. Sulfur capture agent works fully - in its calcium containing and its magnesium containing part.

For comparison fig. 2 shows the results of calculations and experiments in similar conditions ($\alpha = 1.3$; T = 1750K; 1320K).

As is shown in fig. 2, use of SCA leads to almost two-fold reduction of the content of sulfur dioxide in the combustion products. Good agreement between the calculated and experimental values was obtained in burning of CWF without SCA. The calculation with SCA shows complete absence of SO_2 in the combustion products while in the experiments the amount of SO_2 was respectively 480 and 630 mg/m^3 at the combustion temperatures 1340 and 1150 K. The difference between calculated and experimental results is due to the fact that in the experiments it is not possible to sustain the full conditions of physical and mathematical model (in the furnace combustion products stay for a limited time, while the calculation shows the state sought to by the system). Thus, for practical effective removal of sulfur compounds from the flue gas stream it is necessary not only to correctly select an SCA, but also to arrange the combustion process accordingly. Perhaps finalization of design parameters of a furnace unit or working out modes of conducting the combustion process will allow to improve efficiency of SCA application.

 DOI: 10.1051/matecconf/20167201074

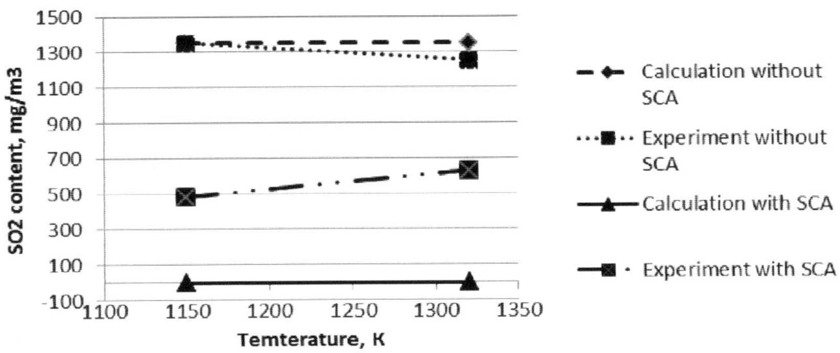

Fig. 2. Comparison of calculated and experimental results of SCA impact on the content of SO2 in the flue gases from combustion of CWF.

3 Conclusion

1. It is shown that the use of technology for preparation and vortex combustion of CWF produced on the basis of high-sulfur coal ensures high efficiency in reduction of sulfur oxides emissions in the flue gases.

2. The method has been developed for thermodynamic analysis of fuel combustion processes including sulfur fuel, in order to determine the effectiveness of performance of various SCAs fed into the combustion area as part of the fuel.

3. Based on the study results the most efficient SCAs were identified as natural minerals - magnesite, calcite, dolomite;

4. The developed method has provided good correlation of the results of calculations and field experiments on the demonstration stand during combustion of high-sulfur Bulgarian coals.

Acknowledgments

The study has been financially supported by the state in the face of the Minobrnauki of Russia (the Ministry of Education and Science) within the framework of the Federal Targeted Programme of Research and Development in Priority Areas of Russian Scientific and Technological Complex for 2014 to 2020, Agreement No.14.583.21.0004 for a grant of 16 July 2014. The unique identifier of research (project) is RFMEFI58314X0004.

References

1. Z.R. Ismaguilov, O.V. Tailakov, T.N. Teryaeva et al., Coal **9** (2015)
2. I.Ya. Sigal, *Protection of air pool at fuel combustion* (Nedra, Leningrad, 1988)
3. Technology of vortex combustion "TORNADO", http://sibpromenergo.ru/boiler/vih/tornado.html
4. V.I. Murko, A.K. Dzhundubaev, M.P. Baranova, A.I. Biybosunov, V.A. Kulagin, *Slurry fuel transporting and energy complexes* (Russian-Kyrgyz scientific and technical cooperation in the field of power engineering, Krasnoyarsk: Siberian Federal University, 2015)
5. A.N. Krestovnikov, V.N. Vigdorovitch, *Chemical thermodynamics* (Metallurgy, Moscow, 1973)
6. U.D. Veryatin, V.P. Mashyrev, N.G. Ryabtsev et al., *Thermodynamic properties of inorganic substances* (Directory. Moscow: Atomizdat, 1965)

STUDYING PERIODIC ROLLING WAVES ON THE SURFACE OF A FALLING LIQUID FILM

Vladimir E. Nakoryakov[1], *Vladimir V.* Ostapenko[1], and *Maria V.* Bartashevich[1,*]

[1]Kutateladze Institute of Thermophysics, Novosibirsk, Russia

Abstract. In this paper the mathematical model describing the falling thin liquid film on a vertical wall is considered taking into account heat and mass transfer at the interface in the regime of periodic rolling waves. The families of exact and numerical generalized solutions, where periodic traveling waves conjugate through the strong and weak discontinuities with each other or with the residual thickness, are constructed. Evolution of exact periodic generalized solutions is studied in time.

1 Problem statement

Heat transfer at condensation of immobile saturated vapor on a vertical surface was firstly considered in [1] for the case of the laminar flow of a condensate film. Later, theoretical and experimental studies of the film flows, including those with consideration of heat and mass transfer on the free surface, were carried out in many papers. It is shown [2] that the Kapitza waves are not the capillary, but the rolling ones. A possibility of traveling wave existence on the surface of a vertically falling liquid film without consideration of surface tension is shown in [3], and the equation, describing propagation of these waves, so-called kinematic equation, is also derived there (exact discontinuous solutions to this equation correspond qualitatively to the experimental results). In [4], the flow of a thin liquid film on a vertical wall was studied theoretically with consideration of condensation at the interface in the regime of the rolling waves. The families of discontinuous solutions were derived, where the traveling waves conjugate with each other or with a residual thickness through strong and weak discontinuities. In [5] the model that takes into account heat and mass transfer at the interface of a thin film of liquid flowing down a vertical wall in the regime of rolling waves for both cases – evaporation and condensation was studied. The full families of exact generalized solutions, which model the increasing and decreasing waves as well as the rolling waves, where the traveling waves are interfaced through strong or weak discontinuities with each other or with the "residual" thickness were constructed. Time evolution of these families of exact generalized solutions was studied. The maps of flow regimes of the liquid film on a vertical heat transfer surface were plotted. Based on the model developed in [3-5], the traveling waves on the surface of a falling liquid film were studied in [6], taking into account heat and mass transfer and surface tension.

* Corresponding author: bartashevichmv@gmail.com

In this paper, we consider the model that takes into account heat and mass transfer at the interface of a thin film of liquid flowing down a vertical wall in the regime of periodic rolling waves for both cases – evaporation and condensation. We have constructed the exact generalized solutions, which model the increasing and decreasing waves as well as the periodic rolling waves, where the traveling waves are conjugated through strong or weak discontinuities with each other or with the "residual" thickness.

The rolling wave equation was obtained in [5]

$$\frac{\partial h}{\partial t}+\frac{\partial}{\partial x}\left[\frac{h^3}{3}\left(1+\alpha\frac{\partial^3 h}{\partial x^3}\right)\right]=\frac{\beta}{h}, \tag{1}$$

where h is the liquid film thickness, and the coefficients are determined in the following manner:

$$\alpha=\frac{\sigma h_0}{g\rho l^3}\quad \beta=\frac{(T_s-T_w)\nu\lambda l}{\rho g r h_0^4}.$$

This is the rolling wave equation, which considers the capillary forces ($\alpha > 0$) and condensation at $\beta>0$, when $T_s>T_w$, or evaporation at $\beta<0$, when $T_s<T_w$. Let us estimate these parameters for water at T_s=310 K, T_w=300 K, l=2·10^{-2} m, $h_0=(\nu^2/g)^{1/3}$=0.5·10^{-4} m, g=9.8m/s^2, ν=10^{-6} m^2/s, ρ=10^3 kg/m^3, σ=72·10^{-3} N/m, λ=0.6 W/(mK), r=2.4·10^6 J/kg. Calculation gives the values α=4.6·10^{-5}, $\beta=1$. The discontinuous solutions to equation (1), limit at $\alpha\to 0$, are the stable generalized solutions to hyperbolic equation

$$\frac{\partial h}{\partial t}+\frac{1}{3}\frac{\partial h^3}{\partial x}=\frac{\beta}{h}. \tag{2}$$

The similar equation was derived in [5].

2 Periodic solution of equation of the rolling wave at phase transition

The hyperbolic equation (2) admits exact solution $h=H(t,\eta,\beta)=\sqrt{2\beta t+\eta^2}$, which describes a change in initial film thickness $H(0,\eta)=\eta$ in time. At $\beta>0$, i.e., at condensation, the film thickness increases, and at $\beta<0$, i.e., at evaporation, the film thickness decreases. We will derive the exact solutions, which model evolution of a single rolling wave at $\beta<0$, i.e., at evaporation. These solutions are set by formula

$$h=\begin{cases}\tilde{H}(t,\eta_l,\beta), & -\infty<x\le a(t),\\ h_-(x,t,x_0^-,V_-,\beta), & a(t)\le x<b(t),\\ h_+(x,t,x_0^+,V_+,\beta), & b(t)\le x\le c(t),\\ \tilde{H}(t,\eta_r,\beta), & c(t)\le x<+\infty\end{cases}. \tag{3}$$

and they are presented [5] by increasing h_- and decreasing h_+ waves, traveling with velocities V_- and V_+, conjugated with depths $\tilde{H}(t,\eta_l,\beta)$, $\tilde{H}(t,\eta_r,\beta)$ through weak discontinuities on lines $x=a(t)$ and $x=c(t)$, where

$$\tilde{H}(t,\eta,\beta)=\begin{cases} H(t,\eta,\beta), & 0\le t\le T(\eta,\beta), \\ 0, & t\ge T(\eta,\beta). \end{cases} \tag{4}$$

Here, $T(\eta,\beta)=\eta^2/(2|\beta|)$ is time, required for complete evaporation of the residual layer with initial depth η. It follows from formulas (3)–(4) that complete evaporation of the residual layer from the left of the rolling wave, i.e., at $x\le a(t)$, occurs at $T(\eta_l,\beta)$, and from the right of this wave, i.e., at $x\ge c(t)$, it occurs at $T(\eta_r,\beta)$.

We will study time evolution of infinite succession of waves of type (3) rolling in the residual layer $H(t,\eta,\beta)$, where $\eta_l=\eta_r=\eta$, in the regime of evaporation at $\beta<0$. We assume that the initial amplitude of these waves is relatively small. To set the initial position of rolling wave succession, we consider section $[x_*,x_*+X]$, corresponding to $[a(0),c(0)]\subset[x_*,x_*+X]$, and expand solution (3), limited in this section, to periodic solution $h=\bar{h}(x,t)$, meeting conditions

$$\bar{h}(x,0)=\tilde{h}(x,0)\ \ \forall x\in[x_*,x_*+X],\ \ \bar{h}(x+X,0)=\bar{h}(x,0)\ \ \forall x\in\mathbf{R}. \tag{5}$$

In the case of condensation, i.e., at $\beta>0$, time evolution of infinite succession of waves of type (3) rolling in the residual layer $H(t,\eta,\beta)$ at $\eta_l=\eta_r=\eta$, was considered in [4].

3 Results of the calculations

The solution to Cauchy problem is presented in the figure for eight successive time points for equation (2) with periodic initial conditions (5). This solution is continuous (lines 1-3 in fig.1.) during period $(0,\bar{T}_1)$, where $\bar{T}_1=\left(V_+^2-\eta^2\right)/\left(2\beta\right)$ is time point, determined from equation $H(\bar{T}_1,\eta,\beta)=\sqrt{V_+}$, where residual depth $H(t,\eta,\beta)$ achieves the lower existence limit of decreasing wave h_+. At $t>\bar{T}_1$, strong discontinuities (lines 4-8 in fig.1.) are formed on the leading edges of rolling waves. During period $(\bar{T}_1,T(\eta,\beta))$ succession of rolling waves with strong discontinuities on the leading edges falls down in the residual layer, whose depth decreases (lines 4-6 in fig.1.). At $t=T(\eta,\beta)$, the residual layer evaporates completely (line 7 in fig.1.), and at $t>T(\eta,\beta)$, these waves continue spreading over the dry wall (line 8 in fig.).

4 Conclusion

Here, we examine the model problem concerning the flow of a thin liquid layer on a vertical wall with consideration of heat and mass transfer at the interface in the regime of periodic rolling waves. The families of generalized periodic solutions to the rolling wave equation are presented, the evolution of exact solutions is studied.

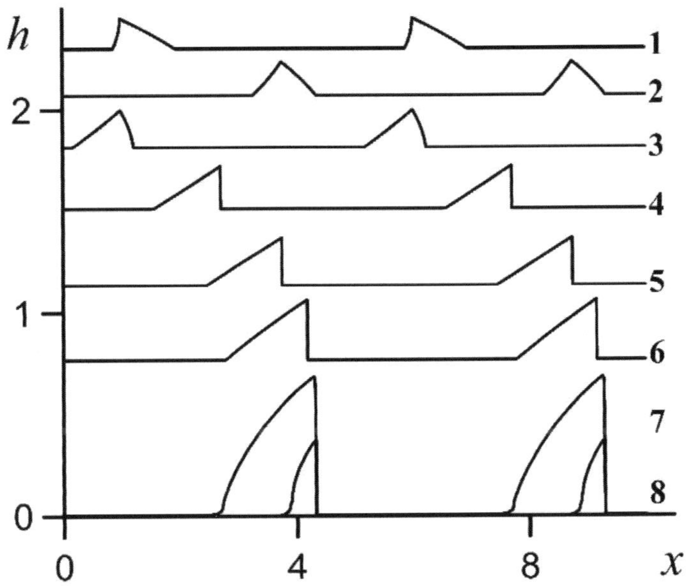

Fig. 1. Solution to the Cauchy problem for equation (2) with periodic initial conditions (5) for eight successive time points $t = 0,\ 0.5,\ 1.0,\ 1.5,\ 2.0,\ 2.35,\ 2.645,\ 2.8$ and following values of initial parameters $\beta = -1$, $V_+ = 3.0$, $V_- = 6.0$, $\eta_l = \eta_r = 2.3$, $x_0^+ = 4.25$, $x_0^- = 2.0$, $x_* = 0.875$, $X = 5.0$.

Acknowledgments

The work was financially supported by the grant MK-6384.2016.8.

References

1. W. Nusselt, Zeitschrift des VDI **60** (1916)
2. S.V. Alekseenko, V.E. Nakoryakov, B.G. Pokusaev, *Wave Flow of Liquid Films*. (Begell House, New York, 1994)
3. V.E. Nakoryakov, I.R. Shreiber, J. Appl. Mech. Tech. Phys. **14** (2) (1973)
4. V.E. Nakoryakov, V.V. Ostapenko, M.V. Bartashevich, Int. J. Heat Mass Transfer **55**, 23-24 (2012)
5. V.E. Nakoryakov, V.V. Ostapenko, M.V. Bartashevich, Int. J. Heat Mass Transfer **89** (2015)
6. O.A. Frolovskaya, V.V. Pukhnachev, Abstracts of International Symposium and School for Young Scientists Interfacial phenomena and heat transfer, Novosibirsk, Russia, 2-4 March 2016

MATEC Web of Conferences **72**, 01076 (2016) DOI: 10.1051/matecconf/20167201076
HMTTSC-2016

SIMULATION OF TURBULENT NATURAL CONVECTION IN A LOCAL ZONE OF LARGE-SCALE AREA UNDER CONDITIONS OF THE RADIANT ENERGY SUPPLY

Alexander Nee[1,*], and *Liliya* Valieva[1]

[1]National Research Tomsk Polytechnic University, Tomsk, Russia

Abstract. This paper presents the results of mathematical modelling of turbulent natural convection under conditions of the conjugate heat transfer in an open cavity heated by infrared emitter. Two-dimensional problem of heat transfer was formulated in the vorticity - stream function - temperature dimensionless variables and solved by means of the finite difference method on a uniform grid. According to the results of numerical analysis, the fields of differential and integral heat transfer characteristics, illustrating the unsteady nature of the thermophysical process under study, were obtained. The nonlinear dependence of the average Nusselt number versus dimensionless time (τ) was established. It was shown that the average Nusselt number monotonically increased in a range of $800 < \tau < 2000$

1 Introduction

Prediction of thermal regimes of large-scale heat supply objects with radiant heating system is a complex problem. The basic laws of jointly proceeding processes of conduction, convection and radiation under operating conditions of infrared emitters (IE) have not been studied at the level providing energy-efficient use of these systems. At the same time, it is known [1, 2] that investigation of heat transfer process in areas with radiant heating sources is advantageously conducted with applying of mathematical modelling methods. However, the development of adequate models of thermophysical processes occurring during IE operating is a challenge. Firstly, the turbulent flow regime should be considered. Secondly, the conjugate heat exchange and radiant energy transfer significantly influence on the microclimate formation in the solution domain. Well-known approaches [3, 4] to analyze the thermal regimes of systems with IE performed at a number of assumptions. Thus, the heat transfer model [3] based on the solution of the heat balance equation of the domain does not take into account convective heating of the air and heat-retaining properties of enclosure structures. Investigation [4] conducted under conditions of the laminar flow regime. Moreover, the radiant heating systems in many cases are applied for the local

* Corresponding author: nee_alexander@mail.ru

heating of specific working areas. Investigation of the heat transfer process in a local domain of large-scale area is of great interest.

The aim of this study is mathematical modelling of turbulent natural convection in a local domain of the large-scale area under conditions of the radiant energy supply.

2 Problem formulation and solution method

Local zone (fig. 1) of rectangular cross-section of large-scale area was considered. Solution domain was presented as a cavity filled by air with open boundaries (fig. 2 B1, B2, B3). The radiant flux coming from the IE was uniformly distributed along the air – wall interface (fig. 2 B4). It was assumed that the gas was viscous incompressible and absolutely transparent medium. "Mild" condition [5] was set for the open boundaries. Equalities of temperature and heat fluxes were accepted at the air – wall interface. Investigations were conducted under conditions of the turbulent flow regime.

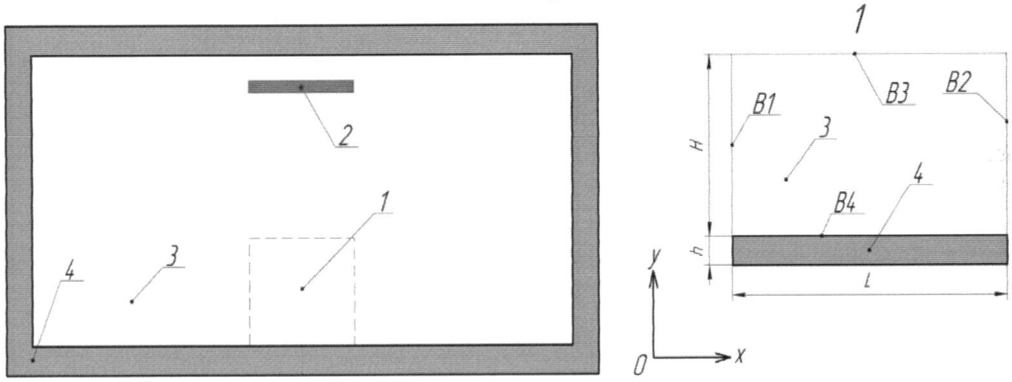

Fig. 1. Solution domain: 1 – the local working zone; 2 – infrared emitter; 3 – the air; 4 – thermally conductive wall; B1, B2, B3 – open boundaries; B4 – the air – wall interface.

Jointly proceeding processes of conduction and turbulent natural convection is described by the Navier - Stokes and energy equations. These equations in the vorticity – stream function – temperature dimensionless variables in the Boussinesq approximation are as follows[7]:

$$\frac{\partial \Omega}{\partial \tau}+U\frac{\partial \Omega}{\partial X}+V\frac{\partial \Omega}{\partial Y}=\frac{\partial^2}{\partial X^2}\left[\left(\sqrt{\frac{\Pr}{Ra}}+\nu_t\right)\Omega\right]+\frac{\partial^2}{\partial Y^2}\left[\left(\sqrt{\frac{\Pr}{Ra}}+\nu_t\right)\Omega\right]+\frac{\partial \Theta_1}{\partial X}, \quad (1)$$

$$\frac{\partial^2 \Psi}{\partial X^2}+\frac{\partial^2 \Psi}{\partial Y^2}=-\Omega, \quad (2)$$

$$\frac{\partial \Theta_1}{\partial \tau}+U\frac{\partial \Theta_1}{\partial X}+V\frac{\partial \Theta_1}{\partial Y}=\frac{\partial^2}{\partial X^2}\left[\left(\frac{1}{\sqrt{\mathrm{Ra}\cdot\Pr}}+\frac{\nu_t}{\Pr_t}\right)\Theta_1\right]+\frac{\partial^2}{\partial Y^2}\left[\left(\frac{1}{\sqrt{\mathrm{Ra}\cdot\Pr}}+\frac{\nu_t}{\Pr_t}\right)\Theta_1\right], \quad (3)$$

$$\frac{\partial \Theta_2}{\partial \tau}=Fo\left(\frac{\partial^2 \Theta_2}{\partial X^2}+\frac{\partial^2 \Theta_2}{\partial Y^2}\right). \quad (4)$$

Subgrid scale eddy viscosity was calculated as in [6]:

$$\nu_t = \left(C_S \cdot \Delta\right)^2 \cdot \sqrt{2 \cdot S_{ij} \cdot S_{ij}},\ S_{ij} = \frac{1}{2}\left(\frac{\partial U}{\partial Y} + \frac{\partial V}{\partial X}\right).$$

The initial conditions for the equations (1) - (4) are as follows:

$$\Psi(X,Y,0) = 0;\ \Omega(X,Y,0) = 0;$$

$$\Theta_1(X,Y,0) = \Theta_2(X,Y,0) = 0.$$

Boundary conditions for equations (1) - (4) are as follows:
at the external boundary of the solid wall:

$$\frac{\partial \Theta_2(X,Y,\tau)}{\partial n} = 0.$$

at the air – wall interface:

$$\Psi = 0,\ \frac{\partial \Psi}{\partial Y} = 0,\ \begin{cases} \Theta_2 = \Theta_1, \\ \dfrac{\partial \Theta_2}{\partial Y} = \dfrac{\lambda_1}{\lambda_2} \cdot \dfrac{\partial \Theta_1}{\partial Y} + Ki_1. \end{cases}$$

at the open boundaries:

$$\frac{\partial \Theta}{\partial X} = 0,\ \frac{\partial \Omega}{\partial X} = 0,\ \frac{\partial \Psi}{\partial X} = 0,$$

where Fo – Fourier number; Ki – Kirpichev number; Pr – Prandtl number; Pr_t – turbulent Prandtl number; Ra – Rayleigh number; U, V – dimensionless velocities corresponding u, v; X, Y – dimensionless coordinates corresponding x, y; τ – dimensionless time; Θ – dimensionless temperature; Ψ - dimensionless analogue of stream function; Ω – dimensionless analogue of vorticity; C_S – empirical constant (Smagorinsky constant); Δ – filter width. Indices: 1, 2 – elements of the system.

The problem (1) - (4) with the corresponding initial and boundary conditions was solved by means of the finite difference method analogically to [8, 9]. The implicit four-point pattern [10] was applied to approximate equations (1) - (4). Woods boundary condition [11] was used for the vorticity on a solid no-flip wall. One-dimensional finite-difference analogues of differential equations were solved by means of the sweep method [10].

The solution method and algorithm used were tested on a model problem of turbulent natural convection in a closed square cavity [12, 13]. Vertical walls were isothermal and horizontal boundaries were assumed adiabatic. Comparison of the fields of temperature and average Nusselt numbers with [12, 13] showed their satisfactory agreement.

2 Results and discussion

Investigation of differential and integral heat transfer characteristics conducted for the following values of similarity criteria: $Ra = 10^8$, $Pr_t = 1$, Pr=0.71, $Ki_1 = 42.86$, $Ki_2 = 21.43$. Fig. 2 presents the fields of temperature and flow pattern.

Figure 2 shows the fields of differential heat transfer characteristics illustrating the unsteady nature of the thermophysical process under study. It was clearly seen that an increase in time led to a rise in temperature at the B4 boundary (fig. 1). When $\tau = 300$, two convective plumes (fig. 2 a, c, e) were formed at the air – wall interface. The heated air ascended as the result of natural convection. An increase in time to $\tau = 600$ led to merge of convective plumes into the single torch in the section of X = 0.5. Air flowed into the cavity through the vertical boundaries (fig. 1 B1, B2) and removed through the horizontal boundary (fig. 1 B3). Convective torch reached the section of Y = 0.63. Further increase in the time to $\tau = 1200$ led to a rise in the average temperature in the analysis domain. The vertical size of the convective torch increased and reached the top horizontal boundary. The thermophysical process under study entered to the quasi-stationary mode. Further increase in time led to a rise in temperature in the section of Y = 0.05 while the air flow pattern and temperature field did not modified. It should be noted that the long-sustained establishment of the quasi-stationary fields of differential characteristics was due to the size of the solution domain corresponding to the turbulent regime of air flow.

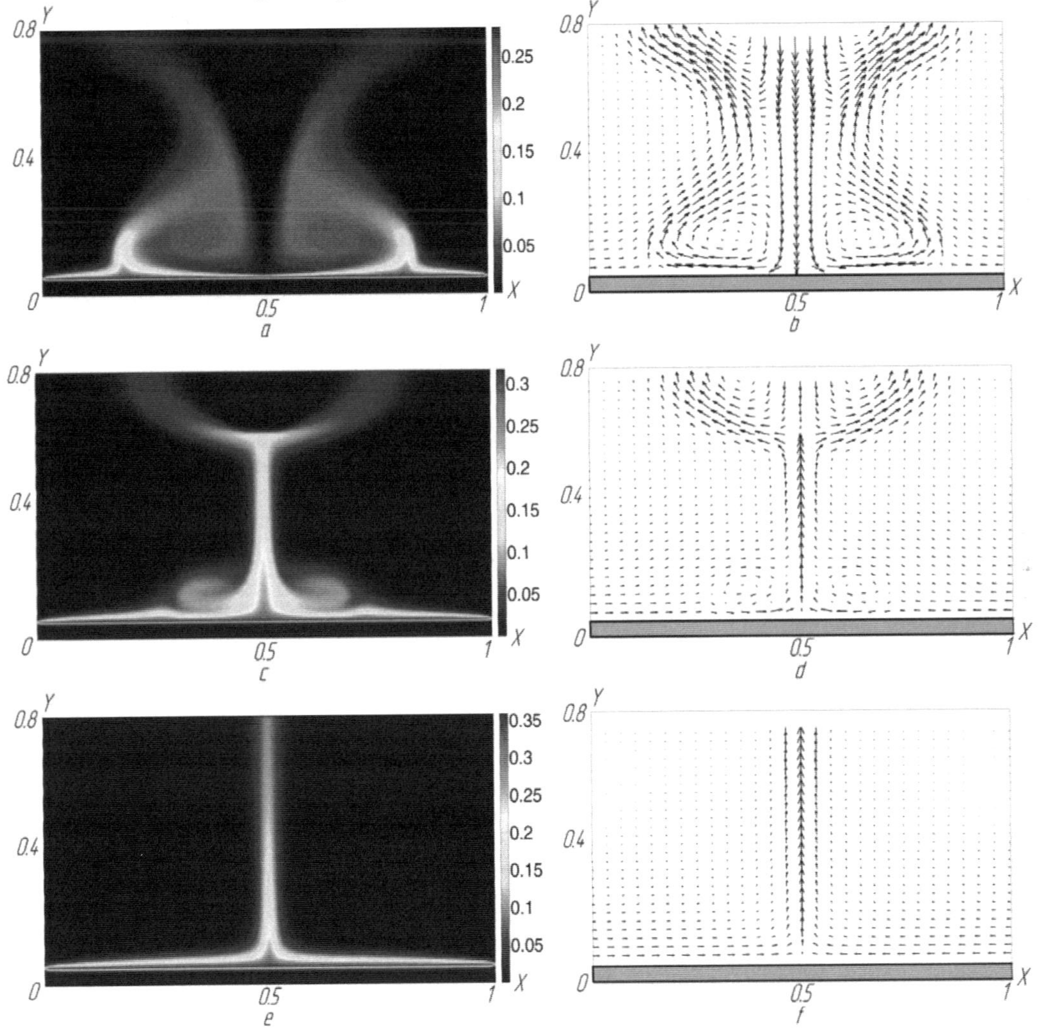

Fig. 2. Fields of temperature (a, c, e) and flow pattern (b, d, f) at $Ra = 10^8$: a, b) $\tau = 300$; c, d) $\tau = 600$; e, f) $\tau = 1200$.

Fig. 3 shows the temperature distribution in the air cavity in the Y = 0.5 section.

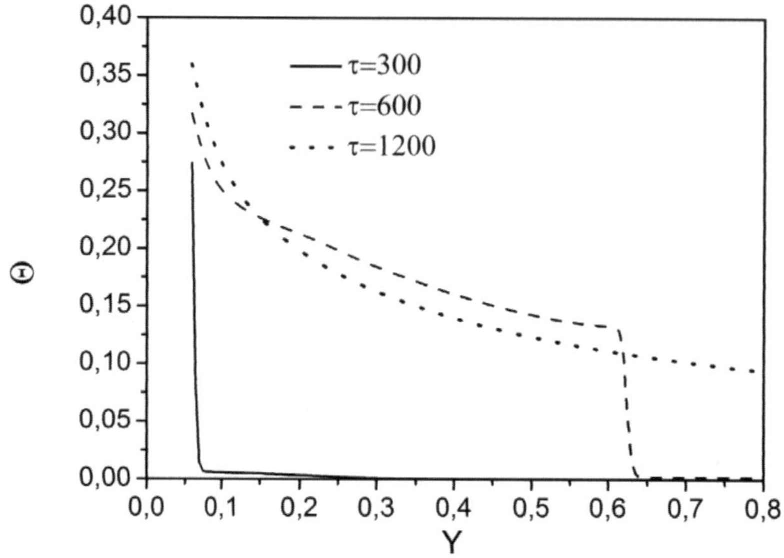

Fig. 3. The temperature distribution in the air cavity in the cross section X = 0.5.

It was clearly seen that an increase in time led to a rise in temperature at the air – wall interface. The heated air ascended in the direction of the Y-axis. The temperature gradient between the B3 boundary (fig. 1) and air – wall interface was reduced.

Figure 4 presents the dependence of the average Nusselt number at the air – wall interface versus time. The dimensionless heat exchange coefficient was calculated as in [5, 9]:

$$Nu_{av} = \int_0^1 \left| \frac{\partial \Theta}{\partial Y} \right|_{Y=0,05} dX.$$

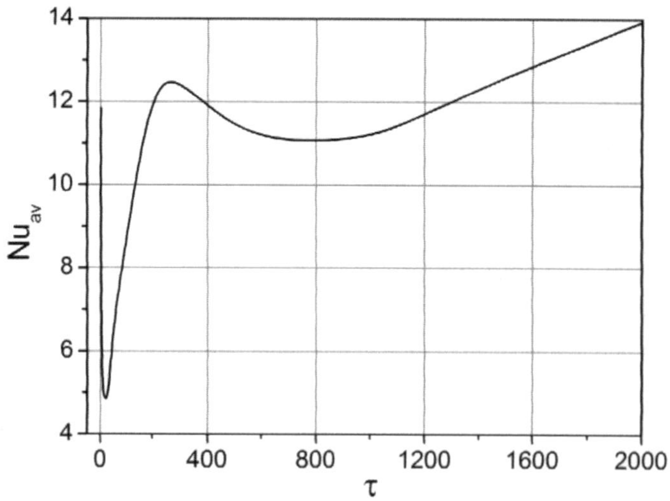

Fig. 4. Dependence of the average Nusselt number versus the dimensionless time.

Based on the analysis of the presented dependence (fig. 4), it was found that the average integral criterion (Nu_{av}) was reduced before the time $\tau = 30$, which was obviously due to the predominance of conductive heat transfer mechanism over convection. The average Nusselt number increased in a range of $30 < \tau < 300$, which was connected with the formation of convective plumes at the air – wall interface. A decrease in the average dimensionless heat exchange coefficient in a range of $300 < \tau < 800$ was due to the formation of the convective torch in the section of X = 0.5. Hereafter, the integral criterion (Nu_{av}) increased monotonically.

3 Conclusion

Two- dimensional problem of conjugate turbulent natural convection in the open cavity was formulated and solved. According to the numerical simulation results, it was established that the heat transfer process in a local domain of the large-scale area heated by infrared emitters had significant unsteady nature over a long period of time.

Acknowledgments

The reported research was supported by Russian Federation President Grant for state support of the Russian Federation leading scientific schools SS-7538.2016.8.

References

1. G.V. Kuznetsov, V.I. Maksimov, T.A. Nagornova, N.I. Kurilenko, G. Ya. Mamontov, J. of Engin. Ph. and Thermoph. **86**, 3 (2013)
2. G.V. Kuznetsov, T.A. Nagornova, A.E. Ni, JEPT. **88**, 1 (2015)
3. V. V. Buhmirov, S. A. Krupennikov, Yu. S. Solnyshkova, Her. Ivan. St. Pow. Un., **4** (2010)
4. A. Nee IOP CONF. SER. MATER. SCI. ENG **93** (2015)
5. G.V. Kuznetsov and M.A. Sheremet, Int. J. Heat Mass Transf. **52**, 1–2 (2009)
6. H. Sajjadi, R. Kefayatib, Therm. Sci. **19**, 1 (2015)
7. Sh. Chen, X.-h. Huang, Ch. Yang, G.-f. Liu, C.-j. Ding, B. Han, Appl. Math. Mech. - Engl. Ed. **34**, 11 (2013)
8. G.V. Kuznetsov, P.A. Strizhak J. Eng. Thermophys-Rus. **17**, 3 (2008)
9. G.V. Kuznetsov and M.A. Sheremet, Int J. Heat Mass Transf. **54**, 1 – 3 (2011)
10. A.A. Samarskii, *The Theory of Difference Schemes* (Nauka, Moscow, 1977) [in Russian]
11. P. J. Roache, *Computational Fluid Dynamics (*Hermosa Publishers, Albuquerque, NM, USA, 1976)
12. H.N. Dixit, V. Babu Int J Heat Mass Transf. **49** (2006)
13. N.C. Markatos, K.A. Pericleous, Int J Heat Mass Transf. **27** (1984)

MATEC Web of Conferences 72, 01077 (2016) DOI: 10.1051/matecconf/20167201077
HMTTSC-2016

THE OPPORTUNITY ANALYSES OF USING THERMOSYPHONS IN COOLING SYSTEMS OF POWER TRANSFORMERS ON THERMAL STATIONS

Atlant Nurpeiis[1, *], and *Tatiana Nemova*[2]

[1]National Research Tomsk Polytechnic University, 634050 Tomsk, Russia
[2]Tomsk State Architecture and Civil Engineering, 634003 Tomsk, Russia

Abstract. The opportunity analyses of using the thermosyphons as the main elements in the systems of thermal regime supplying has been conducted under the conditions of their usage in power transformers on thermal stations. Mathematical modeling of jointly proceeding processes of conduction, forced convection and phase transitions (evaporation and condensation) of coolant in the thermosyphon of rectangular cross section has been carried out. The problem of conjugated conductive-convective heat transfer was formulated in dimensionless variables "vorticity/stream function/temperature" and solved by finite difference method. The effect of the heat flux density supplied to the bottom cover of the thermosyphon from a transformer tank on the temperature drop in the steam channel was shown based on the analysis of numerical simulation results (temperature fields and velocities of steam). The parameters of energy-saturated equipment of thermal stations were found to be controlled by an intensification of heat removal from the top cover surface of the thermosyphon.

1 Introduction

It is known [1] that the most important equipment of thermal stations are the power transformers (PT). And the quality of generated electricity depends on the performance reliability of the power transformers [2]. Generally, the breakdown of the PT leads to large accidents and economical costs to repair [1–3]. During PT working, there are the short-term increases of voltage arising from the normal switching processes in the network or the emergency processes, as well as the pulse wave of the overvoltage arising from lightning atmospheric discharges [4]. One of the main factors affecting the performance reliability of such equipment is their efficient cooling [5–7]. The heat released by the active part is absorbed by the oil and transferred to the heat exchanger for further cooling [5, 8]. However, the use of additional equipment (pumps) for the oil circulation into the external radiators leads to electricity energy costs. For this reason, it is advisable to analyze the

* Corresponding author: nurpeiis_atlant@mail.ru

DOI: 10.1051/matecconf/20167201077

possibility of using self-contained heat transfer devices for cooling of the PT. Closed diphasic thermosyphons (CDTs) [10-11] have proven to be highly reliable, efficient heat transfer heat exchangers due to simplicity of manufacture, the absence of moving parts, regardless of the sources of electricity, etc. For this reason, the study of the application possibility of the CDT as a main element of the PT cooling system is an urgent problem.

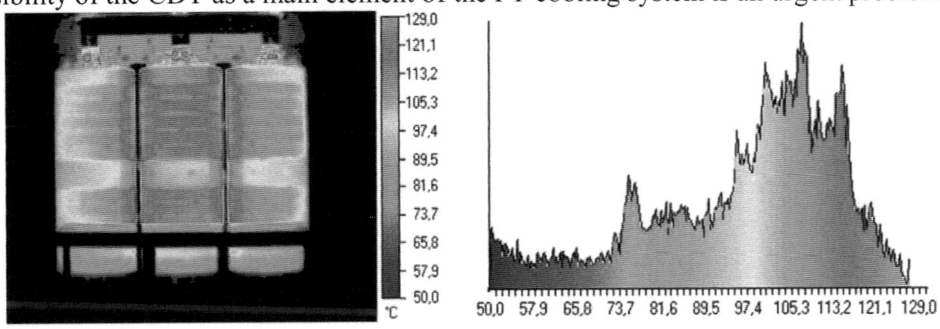

Fig. 1. A reference thermogram obtained for the oil-cooled hermetic transformer TMG "Transformer" with power of 1250 kV·A and voltage of 10/0.4 kV at 100% capacity [9].

To test the technical condition of PTs, thermal monitoring was used [9]. A processing method that allows to record small changes in temperature on the surface of the tank, caused by small defects in PT, was developed. A reference thermogram (fig.1.) for the oil-cooled transformer with power of 1250 kV·A and voltage of 10/0.4 kV at full capacity was obtained [9]. It allows to make a conclusion that the area of a high temperature is located at the top and on the sides of the PT tank. For this reason, the method of transformer tank cooling by a closed diphase thermosyphon was developed. It is proposed to install a vertical thermosyphon on the upper outer surface of the tank.

The aim of the present work is numerical simulation of the thermal regime of the two-phase closed thermosyphon under the condition of energy extraction from the power transformer.

2 Setting of the problem

The problem of conductive-convective heat transfer was solved in the area presented in fig.2. The closed thermosyphon of rectangular cross section made of copper and partially filled with liquid was under consideration. The energy was supplied to the CDT from the transformer through the surface $y=0$, $0<x<(l_1+l_2+L+l_3+l_4)$. The outer boundaries of the vertical walls were considered to be insulated. Convective heat transfer with the environment was taken into account on the top horizontal border. Vapors generated at the evaporation of the coolant were considered as viscous incompressible heat-conducting medium that satisfies the Boussinesq approximation [12–13]. The conditions of the equality of temperatures and heat flows were accepted on the boundaries $y=h_1$, $y=(h_1+h_2)$, $y=(h_1+h_2+H)$, $y=(h_1+h_2+H+h_3)$, $x=l_1$, $x=(l_1+l_2)$, $x=(l_1+l_2+L)$, $x=(l_1+l_2+L+l_3)$. The liquid layer forming due to the condensation of the vapor on the top cover flows into the evaporation zone under the conditions of gravity force. It was suggested that thermal-physical properties of the elements in this system do not depend on the temperature; the thickness of condensate film on the vertical and horizontal walls does not change in time.

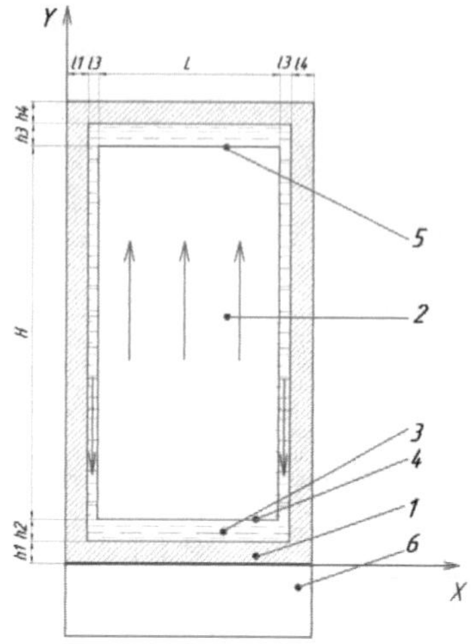

Fig. 2. The area of solution: 1 – metal enclosure; 2 – steam channel; 3 – liquid film; 4 – evaporation surface; 5 – condensation surface; 6 – top wall of the transformer tank. The vertical arrows indicate he direction of the liquid and vapor motion.

Heat transfer in CDT (fig. 2.) within the accepted physical model is described by the Navier-Stokes and energy equations in the converted dimensionless variables, as in [14-16] and have the form:

$$\frac{\partial \Omega}{\partial \tau}+\frac{\partial \Psi}{\partial Y}\frac{\partial \Omega}{\partial X}-\frac{\partial \Psi}{\partial X}\frac{\partial \Omega}{\partial Y}=\frac{1}{\mathrm{Re}}\cdot\left(\frac{\partial^2 \Omega}{\partial X^2}+\frac{\partial^2 \Omega}{\partial Y^2}\right), \tag{1}$$

$$\frac{\partial^2 \Psi}{\partial X^2}+\frac{\partial^2 \Psi}{\partial Y^2}=-\Omega, \tag{2}$$

$$\frac{\partial \Theta_2}{\partial \tau}+\frac{\partial \Psi}{\partial Y}\frac{\partial \Theta_2}{\partial X}-\frac{\partial \Psi}{\partial X}\frac{\partial \Theta_2}{\partial Y}=\frac{1}{\mathrm{Re}\cdot \mathrm{Pr}}\cdot\left(\frac{\partial^2 \Theta_2}{\partial X^2}+\frac{\partial^2 \Theta_2}{\partial Y^2}\right), \tag{3}$$

$$\frac{\partial \Theta_3}{\partial Fo_3}=\frac{1}{\mathrm{Re}\cdot \mathrm{Pr}}\cdot\left(\frac{\partial^2 \Theta_3}{\partial X^2}+\frac{\partial^2 \Theta_3}{\partial Y^2}\right), \tag{4}$$

$$\frac{\partial \Theta_1}{\partial Fo_1}=\frac{\partial^2 \Theta_1}{\partial X^2}+\frac{\partial^2 \Theta_1}{\partial Y^2}. \tag{5}$$

Initial conditions for equations (1) – (5):

$$\Psi(X,Y,0)=\Omega(X,Y,0)=0, \tag{6}$$

$$\Theta_1(X,Y,0)=\Theta_2(X,Y,0)=\Theta_3(X,Y,0)=0. \tag{7}$$

Boundary conditions of equations (1) – (5):

On the outer vertical boundaries: $\frac{\partial \Theta}{\partial X}=0$ (8)

On the boundary: $Y=0,\ 0<X<\frac{l_1+l_2+L+l_3+l_4}{H}:\ \frac{\partial \Theta_1}{\partial Y}=Ki$ (9)

On the boundary:

$$Y=\frac{h_1+h_2+H+h_3+h_4}{H},\ 0<X<\frac{l_1+l_2+L+l_3+l_4}{H}:\ \frac{\partial \Theta_1}{\partial Y}=Bi\cdot(\Theta_1-\Theta_e) \tag{10}$$

On the boundary of the coolant evaporation:

$$Y=\frac{h_2}{H},\ \frac{l_2}{H}<X<\frac{L}{H}:\ \begin{cases}\Theta_2=\Theta_3\\ \dfrac{\partial\Theta_2}{\partial Y}=\dfrac{\lambda_3}{\lambda_2}\cdot\dfrac{\partial\Theta_3}{\partial Y}-Q_e\cdot W_e\end{cases}\tag{11}$$

On the boundary of the coolant condensation:

$$Y=\frac{h_1+h_2+H}{H},\ \frac{l_2}{H}<X<\frac{L}{H}:\ \begin{cases}\Theta_3=\Theta_2\\ \dfrac{\partial\Theta_3}{\partial Y}=\dfrac{\lambda_2}{\lambda_3}\cdot\dfrac{\partial\Theta_2}{\partial Y}+Q_c\cdot W_c\end{cases}\tag{12}$$

On the interfaces: $$\begin{cases}\Theta_i=\Theta_j,\\ \dfrac{\partial\Theta_i}{\partial n}=\dfrac{\lambda_j}{\lambda_i}\cdot\dfrac{\partial\Theta_j}{\partial n},\end{cases}\ \text{где}\ \begin{vmatrix}i=\overline{1,3}\\ j=\overline{1,3}\end{vmatrix}\tag{13}$$

The following notations are taken: $Fo=\frac{a\cdot t_0}{H^2}$ – Fourier number; Re – Reynolds number; $\Pr=\frac{\nu}{a}$ – Prandtl number; $Bi=\frac{\alpha\cdot H}{\lambda}$ – Biot number; α – heat exchange coefficient, $\frac{\text{W}}{\text{m}^2\cdot\text{K}}$; L – crosswise size of the steam channel, m; H – height of the steam channel, m; Q_e – dimensionless analogue of phase-transition heat at evaporation; Q_c – dimensionless analogue of phase-transition heat at condensation; W_e – dimensionless analogue of the evaporation rate; W_c – dimensionless analogue of the condensation rate; *U, V*– dimensionless rates, corresponding to *u, v; x, y* – dimensional coordinates, m; *X,Y* – dimensionless coordinates, corresponding to *x, y;* λ – heat conduction coefficient, $\frac{\text{W}}{\text{m}\cdot\text{K}}$; τ – dimensionless time; Θ – dimensionless temperature; Ψ – dimensionless analogue of ψ; Ω – dimensionless analogue of ω.

The formulated system of equations with nonlinear boundary conditions was solved by finite difference method [17-18]. The flow of condensate film along the vertical walls was modeled as in [19]. The mass rates of evaporation and condensation were calculated according to the formula of Hertz-Knudsen [20]. An iterative algorithm [21-22] developed to solve the problems of conjugated heat transfer in a multiply-connected regions with intensive local heat release was used in this work.

The algorithms and methods of solution used were verified on the model problems of free and forced convection, published in the top-rated journals. In addition, a comparison of the numerical results with the experimental data was obtained [23].

3 Results and discussion

Numerical studies of fluid mechanics and heat transfer in the diphase closed thermosyphon under the conditions of energy removal from the top cover of the power transformer was carried out in a typical range of the heat flux variations, corresponding to the operating modes of the oil-cooled PTs [1-3]. The maximum permissible temperatures of the insulation surface of the PTs are limited by 333 ° K. At these temperatures, the heat flows to the bottom cover of the CDT equal from 10^3 W/m^2 to $2{\cdot}10^3$ W/m^2 depending on the conditions of heat transfer. The geometrical characteristics of the CDT were set in accordance with [24]. The dimensionless complexes were adopted in accordance with the test mode of the vapor flow and the specified dimensions of the heat exchanger under consideration. The numerical simulation results (temperature field and velocities of vapor) for clarity are presented in dimensional variables (fig. 3). The coordinate axes are set in meters; temperature is in degrees Kelvin.

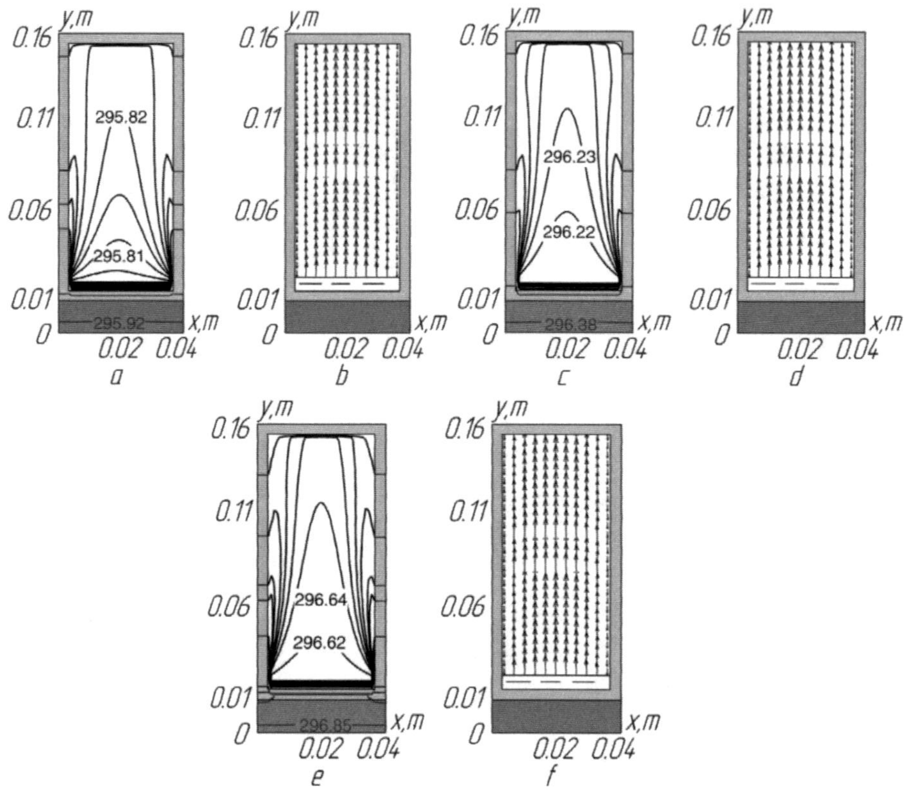

Fig. 3. Isotherms (a, c, e) and fields of vapor velocities (b, d, f) at Re=200 and α=25 W/(m[2]K): a, b – q=1000 W/m[2]; c, d – q=1500 W/m[2]; e, f – q=2000 W/m[2].

Based on the results of numerical simulation analysis, it was found that the increase in the supplied heat flow density q to the bottom cover leads to modification of the temperature fields (fig. 2 a, c, e) in the steam channel of the thermosyphon. At $q=1000$ W/m^2 the bottom cover and the coolant warm in the mode of conduction. It leads to starting of the vaporization process from the lower horizontal "liquid/vapor" interface. Vapor under the pressure gradient starts to move (fig.3 b, d, f) from the evaporation zone to the condensation zone of CDT. The vapor condensates on the boundary of $y=0.147$, $0.003<x<0.037$ (fig. 3). The generated layer of condensate flows along the vertical wall, coming back to the evaporation zone and creating the isolating film between the side wall of the CDT and rising vapor flow. Increase of the heat flow (fig.3 c, e) supplied to the bottom cover of the CDT leads to the increase of the temperature in the steam channel. As a result, the evaporation process becomes more intensive. Isotherms are drawn in the direction of the vapor movement. Extremum of T is observed in the in the section plane of symmetry (fig.3 a, c, e), which is associated with the choice of the heat removing boundary from the thermosyphon (top cover). The shape of the isotherms in the field of $0<y<0.15$, $0<x<0,01$ is due to the release of heat during the vapor coolant condensation on the boundary of $y=0.147$, $0.003<x<0.037$.

Figure 4 presents the temperature distribution in y direction in the central section with respect to x.

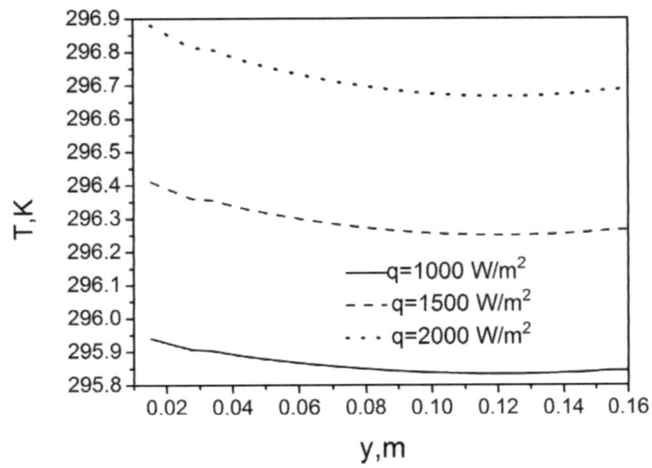

Fig. 4. Temperature distribution along the y axis in the section $x=0.02$.

Based on the analysis of the presented dependences $T=f(y)$ (fig. 4), it can be concluded that the increase of the heat flux density supplied to the bottom horizontal boundary leads to increase of the temperature gradient in the steam channel. The vapor velocity increases due to overheating of the coolant. A small temperature drop can be observed on a horizontal surface $y=0.027$, $0.003<x<0.037$. It is due to the release of the latent heat of vaporization. Temperature in the vicinity of the horizontal surface $y=0.147$, $0.003<x<0.037$ increases because of the release of heat at the vapor coolant condensation. Also it should be noted that the increase in heat flux q twice leads to a change of the temperature on the PT tank surface no more than by 1°K. It illustrates the efficiency of thermosyphon in load conditions corresponding to the thermal regimes of the PTs on the thermal stations.

It is known that the efficiency of thermosyphon cooling system is determined by the heat transfer characteristics of the process in the CDT, namely, by the temperature drop between the upper and lower covers. Analysis of the results allows to conclude that the proposed approach to the study of thermal modes in the CDT permits to predict the variations of the basic characteristics of heat transfer at the heat load changes.

It should also be noted that the results of the performed theoretical studies provide a basis for concluding that it is possible to use the thermosyphon as a regulating system of thermal modes in PTs of the thermal stations in the ranges of load changes from $q=1000$ W/m^2 to $q=2000$ W/m^2. Using the presented mathematical model and the method of solving the problem can allow to reduce significantly the number of the experiments on the stage of system designing.

4 Conclusions

A new approach to modeling of thermal modes in the CDT, connecting the main thermal and hydrodynamic processes that occur during the operation of such heat exchangers has been proposed. The possibility of PTs coil cooling in thermal stations due to the thermal mode control system working based on the thermosyphons was submitted. The Developed original code can be applied on the stages of designing of power transformers cooling systems on the thermal stations, which is based on the thermosyphons.

Acknowledgments

The work was supported by the Russian Federation President's grant (NSH project - 7538.2016.8).

References

1. V.A. Starshinov, *Electrical part of electrical power stations and substations.* (Publishing house MEI, Moscow, 2015) [in Russian]
2. V.V. Zhukov, *Electrical part of electrical power stations with the gas turbine and combined cycle power plant: a manual for high schools* (Publishing house MEI, Moscow, 2015) [in Russian]
3. S.D. Lizunov, *Power transformers:* (Hand book, ed. by S.D. Lizunova, A.K. Lohanina. Energoizdat, Moscow, 2004) [in Russian]
4. A.M. Golunov. *Cooling units of oil power transformers* (Energiya, Moscow, 1964) [in Russian]
5. K. Kish, *Heating and cooling of transformers* (Energiya, Moscow, 1980) [in Russian]
6. K.A., Batischeva, E.G., Orlova, D.V., Feoktistov, EPJ Web Conf. **82** (2015)
7. E.G., Orlova, D.V, Feoktistov, A.G., Islamova, MATEC Web Conf. **23** (2015)
8. E.G., Orlova, G.V., Kuznetsov, MATEC Web Conf. **19** (2014)
9. V.I. Zavidey, V.I. Pechenkin, S.V. Kalanchin, Energoekspert. **6** (2011)
10. M.K., Bezrodnyi, I.L., Pioro, N.O., Kostyuk. *Transfer processes in the biphasic thermal siphon systems* (Fact, Kiev, 2005) [in Russian]
11. A.E. Nurpeiis, EPJ Web Conf. **76** (2014)
12. Y. Dzhaluriya *Natural Convection: Heat and Mass Transfer* (Mir, Moscow, 1983 [in Russian])
13. Yu. A., Sokovishyn, O.G., Martynenko, *Introduction to the theory of free-convection heat transfer*.(Izd-vo Leningrad University, Leningrad, 1982) [in Russian]
14. G.V., Kuznetsov, M.A., Sheremet, Thermophys. Aeromech. **16**, 1 (2009)
15. G.V., Kuznetsov, M.A., Sheremet, Russ. Microelectron **37**, 2 (2008)
16. A.A. Samarsky. *The theory of difference schemes* (Nauka, Moscow, 1977) [in Russian]
17. G.V., Kuznetsov, A.E., Sitnikov. High Temp. (2002)
18. G.V., Kuznetsov, A.E., Sitnikov. Therm. Aeromech. **10**, 1 (2003) [in Russian]
19. H., Shabgard, B., Xiao, A., Faghri, R, Gupta, W., Weissman. Int. J. Heat Mass Transf. **70** (2014)
20. A.O. Zhdanova, G.V. Kuznetsov, P.A. Strizhak. JEPT. **87** (2014)
21. G.V., Kuznetsov, P.A., Strizhak. J. Eng. Thermosyph. **18**, 2 (2009)
22. G.V., Kuznetsov, P.A., Strizhak. Int. J. Heat Mass Transf. **53**, 5–6 (2010)
23. B. Fadhl, L.C. Wrobel, H. Jouhara. Appl. Therm. Eng. **78** (2015)
24. G.V. Kuznetsov, M.A. Al-Ani, M.A. Sheremet. J. Eng. Thermosyph. **20**, 2 (2011)

MATEC Web of Conferences **72**, 01078 (2016) DOI: 10.1051/matecconf/20167201078
HMTTSC-2016

INFLUENCE OF THE TILT ANGLE OF LANCES ON THE FLUID FLOW INSIDE THE GASIFIER SHAFT

P.V. Openyshev[1, *], and *M*.A. Sheremet[1, 2]

[1]Tomsk Polytechnic University, 634050 Tomsk, Russia
[2]Tomsk State University, 634050 Tomsk, Russia

Abstract. Fluid flow inside the vertical gasifier shaft having the granular porous phase has been numerically analyzed. The effect of the tilt angle of lances for air injection on the fluid flow has been studied. The distributions of the temperature, particles volume fraction and trajectory of particles motion inside the gasifier shaft have been obtained. The results clearly show an essential effect of the tilt angle of lances on fluid flow and heat transfer.

1 Introduction

There is strong interest to electric power installation for producing synthetic gas from solid fuels at the present time. Topicality of development of gas-generator installations consists in more effective utilization of waste of the industry and fuels with low fuel heat capacity. Moreover a transition to gas-generator technology promotes conservation of relatively small petroleum and natural gas reserves [1–3].

The production problem of gasifiers is the lack of reliable engineering methods for an analysis of such equipment. Development of mathematical models based on the analysis of physical-chemical characteristics of the gasification process is the actual scientific task due to its solution can significantly accelerate the design stages of gas-generator plants [4, 5].

The aim of this work is the numerical simulation of unsteady two-phase flow and heat and mass transfer in the gasifier shaft.

2 Physical model

We consider a two-dimensional channel model of the gas generator shown in fig. 1. Coal bed of the height of 525 mm is considered as the granular porous phase. Air is supplied to the bottom of the channel through the lances symmetrically inclined at an angle α and moves upward with velocity in the inlet section of 40 m/s and a temperature of 1000°C. It should be noted that the model does not take into account the phase transformations. In this case, we have analyzed two-phase flow and heat transfer without gasification process inside the considered channel.

* Corresponding author: o_o_v@mail.ru

DOI: 10.1051/matecconf/20167201078

Numerical simulation of two-phase flow and heat transfer in the channel of the gasifier has been conducted using the computational fluid dynamics package "Fluent" [6]. Geometric model with finite element mesh consisting of rectangular and triangular elements with a minimum size of 5 mm, has been created in the preprocessing program "Gambit" [7]. Further the mesh has been exported to the software package "Fluent". Further on using this software we have defined completely all phases properties, boundary conditions and parameters of the numerical method. It should be noted that the difference scheme "QUICK" has been used for an approximation of convective terms. Difference scheme "PRESTO" has been used for the pressure definition in the Poisson equation. Simultaneous determination of the velocity and pressure fields has been realized on the basis of the "SIMPLE" algorithm.

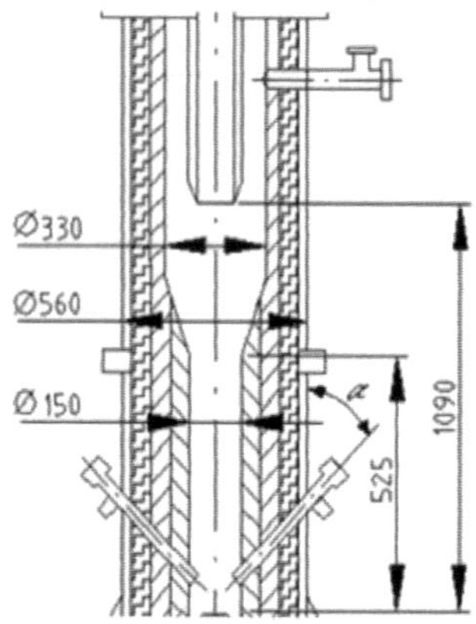

Fig. 1. The schematic of the gasifier shaft.

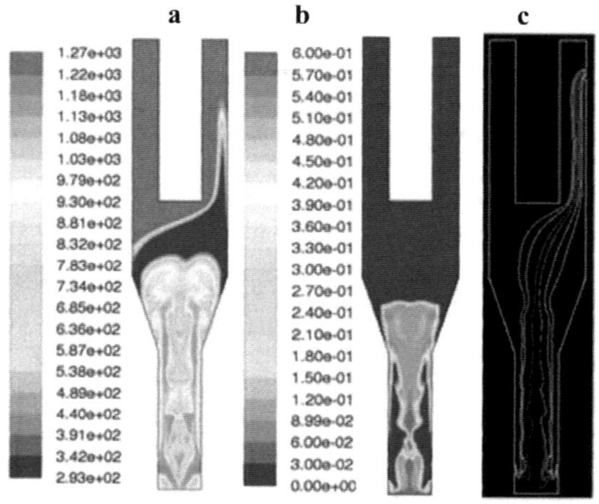

Fig. 2. Distributions of the temperature (a), particles volume fraction (b) and trajectory of particles motion (c) for $t = 0.35$s and $\alpha = 15°$.

We have analyzed an influence of the lances mounted in the lower part of the gasifier shaft, on the flow structure and heat transfer for different time moments 0.35, 1 and 2 seconds and various inclination angles of the lances $\alpha = 15°$, 45° and 75°.

It should be noted that the simulation of fluid flow and heat transfer in the case of two-phase flows has been conducted using the Euler approach (continuum model). Continuum model allows to analyze the two-phase flow as a motion of two interacting continuums associated with the carrying agent and particles [8]. It is assumed that the dispersed phase is represented as a continuum with the density that is continuously distributed in space. Parameters of this continuum are considered as local average values of particle parameters. The behavior of the two-speed and two-temperature continuum is described by the equations of continuum mechanics in Euler variables. Interaction of phases is achieved by interfacial exchange coefficients. Such interaction depends on the phase type. The properties of granular flow are determined based on the kinetic theory.

3 Results and discussion

At the initial time ($t = 0.35$s) distributions of thermal hydrodynamic characteristics (fig. 2) for the inclination angle of lances ($\alpha = 15°$) represent non-significant mixing at the bottom of the gas-generator shaft, that illustrates symmetry fields of temperature and particles volume fraction in a straight-flow zone. The complicated structure of trajectories of the dispersed and continuous phases is observed for the lances inclination angles of 45° and 75° for $t = 0.35$s, in contrast to the results presented in fig. 2. However, the high speed of injecting air ($\upsilon = 40$ m/s) for the inclination angle $\alpha = 15^\circ$ causes the severe non-uniformity in the distribution of the dispersed phase volume fraction inside the channel with time (fig. 3). Such heterogeneity is not observed for injecting angles of 75° and 45°, namely, the dispersed phase motion occurs along the channel surfaces.

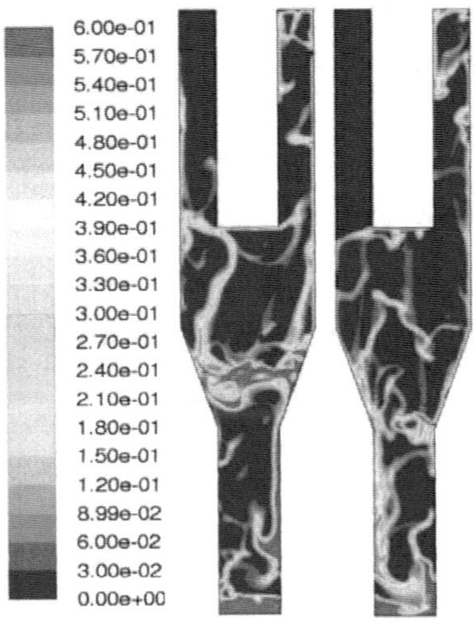

Fig. 3. Distributions of particles volume fraction for $\alpha = 15^\circ$, $t = 1$s (left contours) and $t = 2$s (right contours).

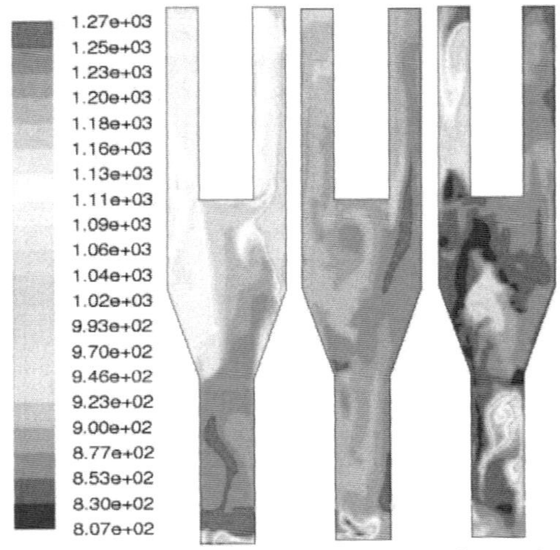

Fig. 4. Distributions of the temperature for $t = 2$s, $\alpha = 75^\circ, 45^\circ, 15^\circ$ (from left to right).

The typical temperature distributions for the gasification process can be found for $t = 2$s and $\alpha = 75^\circ$, where the zones of dry distillation and drying the fuel are allocated in the upper part of the shaft (fig. 4). It should be noted that the resulting high values of temperature in these areas are due to the absence of endothermic reactions occurring in the gasifier. Distributions of thermal hydrodynamic characteristics for $\alpha = 45^\circ$ differ non-significantly from the results obtained for the angle of 75° at different times.

This research is not sufficient and further work is required to identify the optimal range of inclination angles for the lances.

The research was realized with financial support of Minobrnauki of Russia in framework of FTP "Research and development in prior direction of scientific-technological complex of Russia in 2014-2020 years", unique R&D identifier RFMEFI58114X0001.

References

1. A.F. Ryzhkov, A.V. Popov, Energetic **10** (2012)
2. A.V. Remezov, V.V. Chumak, J. Vestnik KuzGTU **2** (2005)
3. S.A. Tsybulsky, *Modern coal gasification technology in industrial power* (Tomsk Publ., Tomsk, 2014)
4. I.G. Donskoy, *Modelling and optimization of operating modes of the dense layer of the gas generator for the combined-cycle thermal power plant* (Diss., Irkutsk, 2014)
5. P.V. Openyshev, M.A. Sheremet, Key Eng. Materials **685** (2016)
6. O.V. Baturin, N.V. Baturin, V.N. Matveev, *Computation of the fluid flow using the software package "Fluent"* (Samara Publ., Samara, 2009)
7. O.V. Baturin, N.V. Baturin, V.N. Matveev, *Construction of computational models in the preprocessor "Gambit" of the software package "Fluent"* (Samara Publ., Samara, 2009)
8. K.N. Volkov, V.N. Emelyanov. *Flows of gas with particles* (Moscow Publ., Moscow, 2008)

DYNAMICS OF UF_6 DESUBLIMATION WITH THE INFLUENCE OF TANK GEOMETRY FOR VARIOUS COOLANT TEMPERATURE

Aleksey A. Orlov[1], *Alexandr F.* Tsimbalyuk[1], *Roman V.* Malyugin[1,*], and *Anatolij A.* Glazunov[2]

[1]National Research Tomsk Polytechnic University, 634050 Tomsk, Russia
[2]Scientific Research Institute of Applied Mathematics and Mechanics at Tomsk State University, 634050 Tomsk, Russia

Abstract. Mathematical model of UF_6 desublimation in a vertical immersion tank is presented in the article. Results of calculations of the filling dynamics of the tanks with $1 m^3$ volume at various coolant temperatures, with and without ellipticity of the end walls are given. It is shown that allowance for the ellipticity of the end walls of the tanks leads to a significant increase in the time of desublimation of UF_6.

1 Introduction

Desublimation process is widely used in various sectors of the economy, including in the nuclear fuel cycle. Gaseous UF_6 is used as the working medium at the plants of uranium enrichment. Uranium hexafluoride desublimate into special containers at the output of the separation cascade. Tanks with depleted UF_6 are sent to the warehouse for long-term storage. Tanks with uranium enriched in ^{235}U transported to plants for transfer UF_6 to UO_2 and production fuel rods.

Several stationary or quasi-stationary mathematical models describing the UF_6 desublimation process in a single tank developed to date. Detailed descriptions of these mathematical models presented in the review [1]. Their common disadvantages are:

- The use of empirical data;
- Lack of consideration of the motion of gaseous UF_6;
- Lack of consideration of the ellipticity of the tank end walls (tank is represented in the cylinder form) and the availability of desublimation at its bottom section.

Actually UF_6 desublimation process is nonstationary and essentially depends on the coolant temperature and the gas pressure inside the tank. Desublimation layer thickness on the inner surface of the container increases during filling. The result is that there is deterioration in conditions of heat exchange between the UF_6 phase transition surface and coolant and the desublimation speed varies nonlinearly.

Two-dimensional non-stationary mathematical model of UF_6 desublimation in a vertical immersion tank was developed by us [2], in there are no disadvantages of known stationary

* Corresponding author: malyugin@tpu.ru

or quasi-stationary models. Verification of the mathematical model [2] is shown that it adequately describes occurring at UF_6 desublimation process heat and mass transfer.

The impact of geometry of the vertical submersible tank on the dynamics of filling the solid UF_6 at different coolant temperatures is investigated, using created mathematical model.

2 Object of study

UF_6 desublimation to a vertical cylindrical tank (fig. 1) was considered.

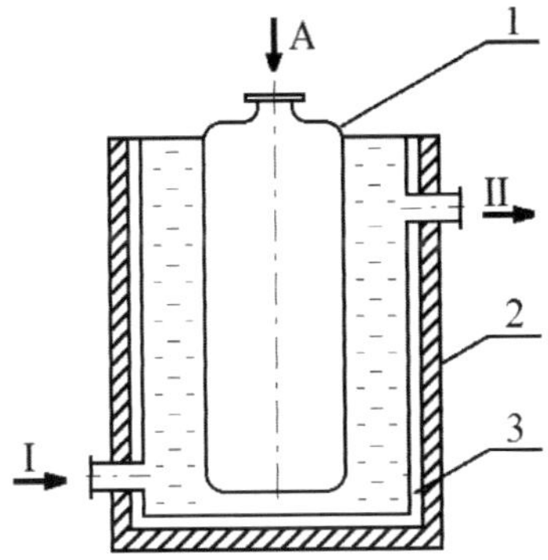

Fig. 1. Immersed vertical tank.

Tank (1) is in thermostat (2). The thermostat is filled with saturated $CaCl_2$ solution (streams I, II), which is cooled by liquid nitrogen (3). Gaseous UF_6 is fed through a valve located at the top of the tank (stream A). The upper part of the tank with the branch pipe protrudes from the thermostat filled with coolant, and is not involved in heat exchange.

3 Description of mathematical model

In the construction of the mathematical model was made the following assumptions:

1) UF_6 gas does not contain impurities;

2) desublimation process occurs only on the side and bottom parts of the tank (top part of the tank with the branch pipe protrudes from the thermostat with the coolant);

3) the outer surface temperature of the tank wall is considered to be constant;

4) the desublimation process of gas phase is determined by heat removal through the tank walls and the UF_6 solid layer;

5) UF_6 desublimation speed throughout the heat exchange surface is averaged at each time step;

6) the phase transition temperature is determined by the equilibrium temperature and pressure above the desublimate layer.

The computational domain was divided by uniform grid, which solved the integral equations of gas dynamics:

$$\begin{cases} \frac{d}{d\tau}\iint_S r\rho dxdr + \oint_G (\rho udr - \rho vdx) = 0, \\ \frac{d}{d\tau}\iint_S r\rho udxdr + \oint_G r\left[\left(\rho u^2 + p\right)dr - \rho uvdx\right] = 0, \\ \frac{d}{d\tau}\iint_S \rho vdxdr + \oint_G \left[\rho uvdr - \left(\rho v^2 + p\right)dx\right] = \iint_S pdxdr, \\ \frac{d}{d\tau}\iint_S Edxdr + \oint_G r\left[(E+p)udr - (E+p)vdx\right] = 0, \end{cases} \tag{1}$$

$$E = \rho\left(e + \frac{u^2 + v^2}{2}\right), \qquad e = \frac{p}{\rho(\gamma - 1)},$$

where G is arbitrary closed contour, which limits the area S in the plane of variables x, r of cylindrical coordinate system, the beginning of which is placed on the axis of symmetry in the inlet section; τ is time; p is pressure; ρ is density of gas; e is UF_6 specific internal energy; E is UF_6 apparent energy; γ is adiabatic exponent of gas; u, v are projections of the velocity vector on the axis x, r respectively.

For the tank wall heat conduction equation was solved:

$$\frac{\partial rT}{\partial \tau} = \frac{\partial}{\partial n}\left(\kappa\frac{\partial rT}{\partial n}\right) \tag{2}$$

where κ is thermal diffusivity of the wall material, n is normal coordinate measured from the outer surface inside of the tank, T is temperature.

Calculation of the temperature in the UF_6 desublimate layer was carried out using a mobile system coordinates of the form $\xi=\xi(x,r,\tau)$, $\eta=\eta(x,r,\tau)$, related to the interface of "tank wall-desublimate". Has been solved the Stefan problem with moving boundary of phase transition:

$$\begin{aligned} \frac{\partial(\psi T)}{\partial \tau} = &-\frac{\partial}{\partial \xi}\left[\left(-\lambda\frac{\partial T}{\partial x} + T\frac{\partial x}{\partial \tau}\right)\frac{\partial y}{\partial \eta} - \left(\lambda\frac{\partial T}{\partial y} + T\frac{\partial y}{\partial \tau}\right)\frac{\partial x}{\partial \xi}\right] - \\ &-\frac{\partial}{\partial \eta}\left[-\left(-\lambda\frac{\partial T}{\partial x} + T\frac{\partial x}{\partial \tau}\right)\frac{\partial y}{\partial \xi} + \left(\lambda\frac{\partial T}{\partial y} + T\frac{\partial y}{\partial \tau}\right)\frac{\partial x}{\partial \eta}\right] + \psi g. \end{aligned} \tag{3}$$

where ψ, $\partial x/\partial\xi$, $\partial y/\partial\xi$, $\partial x/\partial\eta$, $\partial y/\partial\eta$ is jacobian and the metric coefficients.

Equation (3) was used for the determination of temperature fields and positions of grid points within desublimate layer.

To determine the UF_6 desublimation speed was used Stefan condition:

$$\rho_U v_w(\tau) L = \frac{\lambda_U r}{l_w(\tau)}\frac{\partial T}{\partial \eta}, \tag{4}$$

where c is specific heat capacity, $l_w(\tau)$ is desublimate thickness, $v_w(\tau)$ is desublimation speed, η is dimensionless coordinate associated with the surface of "tank wall-desublimate", L is heat of desublimation, λ is thermal conduction coefficient. Index U refers to a solid UF_6, S to a tank wall, E to refrigerant.

Consideration of the UF_6 phase transition boundary was produced using a dynamic adaptation grid of the computational domain [3]. Method of V.I. Mazhukin [4] was used for the numerical solution of differential equations of heat conduction. System of the gas dynamics equations was changed to a dimensionless form and was solved numerically using the algorithm SIMPLE [5].

Temperature on the outer surface of the tank was set by boundary condition of the first kind (surface temperature is equal the coolant temperature); on the border "tank wall – desublimate" by wondered conjugation conditions; on the surface of the phase transition – by the boundary condition of the first kind (temperature of the phase boundary is equal to equilibrium temperature of the phase transition).

The condition of sticking and impermeability to UF_6 gas flow rate was set at the upper end of the tank wall. The flow rate of gaseous UF_6 to the heat transfer surfaces is equal to the speed of the desublimation.

Tank cooled to a coolant temperature before the start of desublimation. The desublimate thickness at the initial time is zero.

4 The calculation results

The calculation of the dynamics of filling tank of 1 m^3 (fig. 2a) up to 70% without the ellipticity of the ends tank walls (fig. 2b), with the ellipticity of the bottom of the tank wall only (fig. 2c), with the ellipticity of the two end tank walls (fig. 2d).

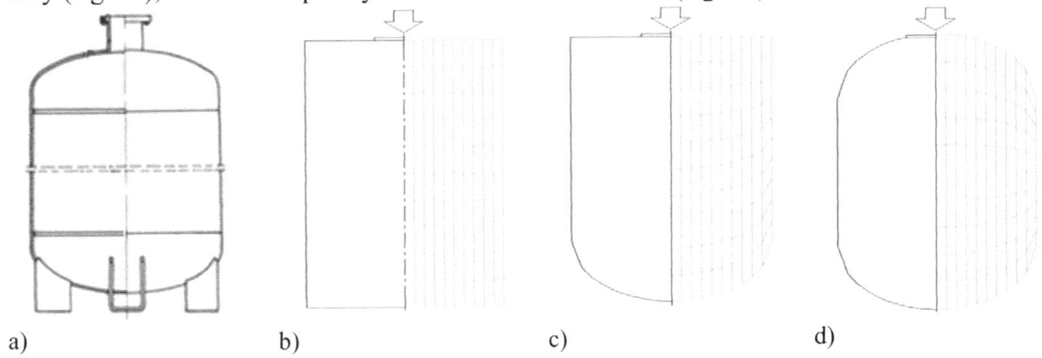

Fig. 2. Transport tank (a) and the corresponding computational domains (b, c, d).

The results of calculation of the dynamics of UF_6 desublimation in a tank volume of 1 m^3 with coolant temperatures –10°C and –20°C are shown in fig. 3 and fig. 4, respectively.

Results of the calculations in fig. 3 is show that the estimated filling time with considering the ellipticity both end tank walls is 405 hours, which 11.1% and 23.5% higher than the estimated filling time of considering only the ellipticity of the tank bottom and without ellipticity end tank walls respectively.

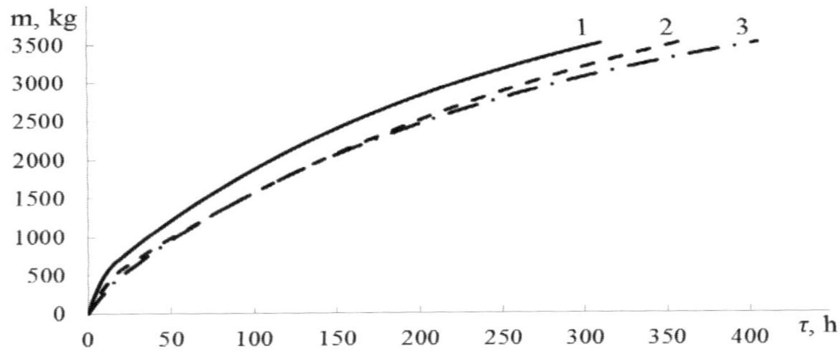

Fig. 3. Dynamics of 1 m^3 tank filling (t_{CaCl} = – 10°C):1 – flat ends of the tank walls, 2 – ellipticity of the bottom wall, 3 – ellipticity of the two end tank walls.

Results of the calculations in fig. 4 is show that the estimated filling time with considering the ellipticity both ends tank walls is 320 hours, which 10.9% and 23.4% higher than the estimated filling time of considering only the ellipticity of the tank bottom and without ellipticity end tank walls respectively.

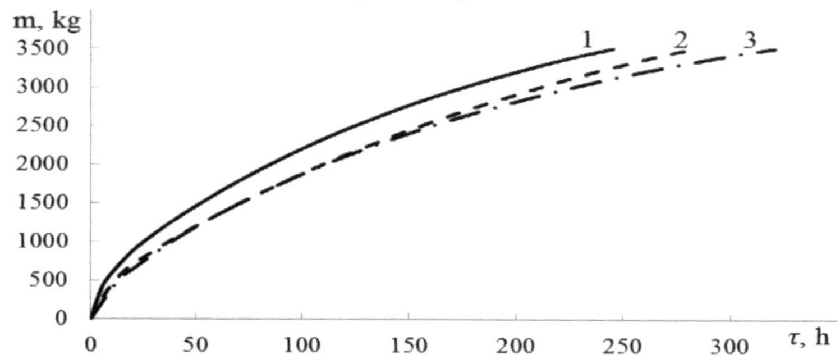

Fig. 4. Dynamics of 1 m^3 tank filling (t_{CaCl} = – 20°C): 1 – flat ends of the tank walls, 2 – ellipticity of the bottom wall, 3 – ellipticity of the two end tank walls.

As the visible pattern dynamics of tank filling from its geometry is maintained at different coolant temperatures. The time of tank filling changing almost in the same proportion for the different geometry.

5 Conclusion

1. Calculations of dynamics of the filling vertical immersion tank 1 m^3 to 70% without ellipticity of the ends tank walls, considering only the ellipticity of the tank bottom, with considering the ellipticity both ends tank walls is performed using the developed UF_6 desublimation nonstationary mathematical model.
2. The calculation results showed that the inclusion of the ellipticity of the ends tank walls substantially increases the value of the estimated time. The pattern of change in the estimated time of filling of the container from its geometry is maintained at different coolant temperatures (–10°C and –20°C).
3. The necessity of taking into account the tank geometry in calculating the dynamics of filling the solid UF_6 is shown.

References

1. A.A. Orlov, R.V. Malyugin, Adv. Mater. Res. **1084**, 620 (2015)
2. A.A. Orlov, A.F. Tsimbalyuk, R.V. Malyugin, Key Eng, Mater. **683**, 533 (2016)
3. S. Godunov, A. Zabrodin, N. Ivanov, A. Kraiko, G. Prokopov, *Numerical solution of multidimensional problems of gas dynamics* (Nauka, Moscow, 1976) [in Russian]
4. N.A. Dar'in, V.I. Mazhukin, Math. Modeling. **3**, 29 (1989)
5. S.V. Patankar *Numerical heat transfer and fluid flow* (Hemisphere, Washington D.C., 1980)

MATEC Web of Conferences **72**, 01080 (2016) DOI: 10.1051/matecconf/20167201080
HMTTSC-2016

RESEARCH OF ENERGY SAVING MODE OF BUILDING DISTRIBUTED CONTROL SYSTEM UNDER INFLUENCE OF MANY DISTURBING FACTORS

Maxim N. Morozov[1, *], and *Roman* B. Ushakov[1]

[1]National Research Tomsk Polytechnic University, 634050 Tomsk, Russia

Abstract. A model of the building heating system, consisting of energy source, a distributed automatic control system, elements of individual heating unit and heating system is designed. Application Simulink of mathematical package Matlab is selected as a platform for the model. There are the specialized application Simscape libraries in aggregate with a wide range of Matlab mathematical tools allow to apply the "acausal" modeling concept. Implementation the "physical" representation of the object model gave improving the accuracy of the models. Advanced weather block allowed to investigate the efficiency of the control system with dynamically changing the numerous factors affecting the building energy consumption for maintaining optimal climate parameters. The conducted research of heating system and automatic control system allowed estimate efficiency of control system regulating process. The results of research of thermal condition premises, situated in the different sides of the building facade are given. The conclusions about the possibility of increasing the operational efficiency of heating system in the implementation of distributed control systems are presented.

1 Introduction

In recent years issues of energy conservation, improving energy and resource efficiency are considered decisive in many countries [1, 2].

Particular attention, as a rule, is given to activities aimed at improving the energy efficiency of administrative and industrial purposes buildings and facilities [3-9].

There are several active requirements for a modern heating system. On the one hand it is necessary to maintain the microclimate parameters at a comfortable level and on the other – it is required increasing energy efficiency of engineering building life support systems.

The aim of the present work is to investigate influence of the different type disturbing factors on the climate control system functioning and on the thermal regime and heat consumption parameters.

[*]Corresponding author: benm6@yandex.ru

2 Simulation object

The object of research is an administration building which is located in West Siberia (Tomsk city). Heat supply of the building is carried out by a central heat supply station.

The object is the block of weather conditions, heat supply system and envelopes of the room. The heating system consists of the individual heating plant (IHP), main pipes (inlet and outlet pipes), water risers, heating appliances and control valves. The structure of the object is described in [10]. The possibility of the weather conditions variation in a wide range is the distinctive feature of the model under investigation in this paper.

Four rooms were selected in building for achieving goals of the article. All the external envelopes of one room are characterized by the same spatial orientation. Identical conditions of heat supplying are provided for heating system with parallel connection of heating appliances. For approximation to the truth operating conditions the spatial orientation of the rooms exterior wall envelopes was selected with following manner:

- room No.1 – West;
- room No.2 – South;
- room No.3 – North;
- room No.4 – East.

Rooms are characterized by identical space planning with following materials: square is 41.52 m^2, volume is 112.1 m^3 (figure 1).

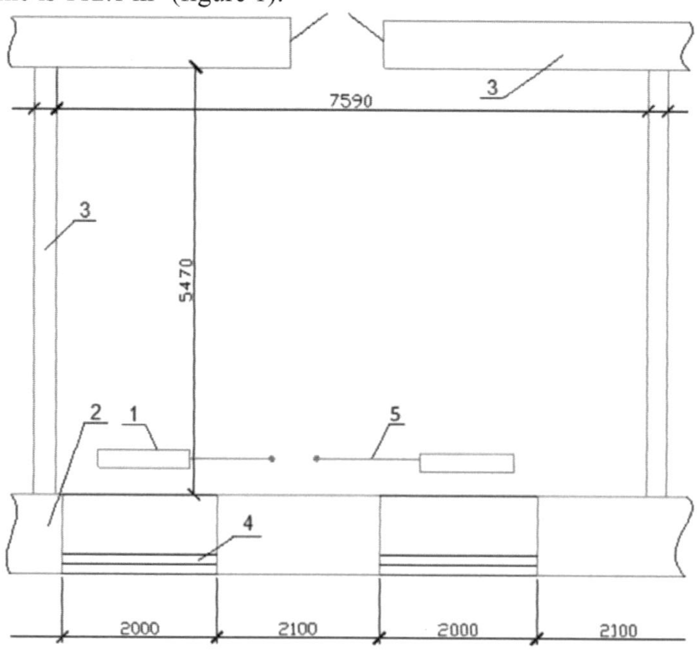

Fig. 1. A schematic illustration of simulation object (a) plan of a room, (b) axonometric scheme of the one-pipe heating system: 1 - heating appliance; 2, 3 – interior and exterior envelopes; 4 – multiple glass pane; 5 - water pipe.

Glazing ratio K_{gl} is set to 0.23, which corresponds to the surface area of opaque constructions – 17.85 m^2 and translucent structures – 5.38 m^2 (net of area of opaque multiple glass pane profile)

There is intelligent heat consumption system with distributed structure [11] were assumed: the main regulator adjusts the coolant temperature (supply and return) in IHP, local regulators adjust the individual room climate. Unique features are provided by multi-loop control system of heat consumption. In this case, local control loop is limited to the physical boundaries of controlled space. The main regulator spreads to the whole building and interfaces technological parameters of urban heating network and heating system of building.

However, when choosing of distributed control system (DCS) [11-14], automation specialists are faced with the complicated problem of design, software synthesis and adjustment of such systems. As a rule, to solve these problems, engineers apply computer simulation models [15, 16]. The use of computer simulation provides a small labor cost and reduces time of experimentation.

3 Thermal model of building

The thermal model of the building was developed to simulate operation of the heating system of the described above object. Simulink by Matlab was selected as the platform of a model. This choice is motivated by wide capabilities of software: tools to analyse system dynamics in time, connection with Workspace and other Matlab applications (for instance, Control System Toolbox), simulation of various disturbances (determined or random with different distribution laws) and situations (regular or non-regular, including emergency) which occur when in operation of the technical system. The block-oriented approach was applied to simulation and visual programming of heat supply system of building in combination with the wide range of Matlab mathematical tools. Simscape extension package which implements the concept of a "physical" simulation was also applied to develop the model [17].

Realization of the thermal model in Matlab is based on representing of each individual functional part in the form of s-functions and subsystems blocks. During simulation it is observed redistribution of heat flows between sources, medium of heat energy transfer and consumers.

4 Modelling results

To achieve the purpose of the present work series of numerical experiments were performed to analyze influence of selected disturbing factor on building heat consumption and thermal regimes of controlled rooms. Variations of the initial modelling data specified in Table 1.

Table 1. Initial data for modelling.

Variable parameters	**Number of experiment**					
	No.1	No.2	No.3	No.4	No.5	No.6
Average outdoor temperature T_{out0}, °C	-20	**-30**	-20	-20	-20	-20
Average flow velocity V_w, m/s	3	3	**4.5**	3	3	3
Insolation heat gains Q_{insol}	no	no	no	**yes**	no	no
Internal heat sources Q_{int}	no	no	no	no	**yes**	no
Low heat consumption mode	no	no	no	no	no	**yes**

Values of parameters, obtained in experimental No.1, are assumed as the basic in the following experiments and changes only one by one.

Researches of influence of different types disturbing factor on heat consumption characteristics and estimation of energy saving at specialize room thermal regimes application are of particular interest for HVAC-engineers and energy managers. An estimation of heat consumption values of controlled rooms $Q_{rm\,i}$ (i = 1…4) was given at present work.

Table 2 shows heat consumption data of research rooms.

Table 2. Daily heat consumption by control rooms.

Number of experiment	**Control rooms**			
	Room No.1	Room No.2	Room No.3	Room No.4
No.1 (basic)	21.49	20.37	18.89	21.49
No.2 (T_{out0} = -30 °C)	37.47	36.28	34.42	37.74
No.3 (V_w = 4.5 m/s)	23.93	22.67	21.12	23.93
No.4 (insolation heat gains)	20.16	14.44	18.17	19.65
No.5 (internal heat gains)	16.76	15.56	14.01	16.76
No.6 (energy saving thermal regime)	13.13	12.7	12.11	13.13

Data submitted in Table 2 allow carrying out a comparative analysis. The lowest consumption of thermal energy was obtained for room No.3 (North) in the 1-st experiment. This is explained by several factors: absence of solar irradiation, wind direction (to the North). Wind load creates aerodynamic shadow on the exterior of room No.3. This reduces the value of air environment velocity along the surface of the outdoor envelopes and thus the thermal transmittance of external surface.

Comparison of experiments No.2 and No.3 provides the following results: at 50% increase of the values of T_{out} и V_w relative to the basic values (experiment No.1), the highest growth of heat consumption was obtained by carrying out an experiment No.2 (increasing of parameter $Q_{rm\,2}$ was 82.2 %), whereas in experiment No.3 heat consumption increased by 11.8 %.

Comparison of heat gains from solar irradiation and internal heat sources (experiments No.5 and No.6 respectively), which was conducted for research object, shows a great importance of insolation influence. It should be noted a considerable spread of values of $Q_{rm\,i}$ for rooms with different spatial orientation: DCS provided savings 5.93 kW·h (29.11 %) in room No.2, whereas in room No.3 energy saving effect is only 0.72 kW·h (3.8 %). In case of availability of internal heat sources energy saving attains a maximum value of 4.88 kW·h (25.83 %) and average value of 4.79 kW·h (23.36 %).

Experiment No.6, characterized by applying specialized thermal regimes should be noted particularly. Analysis of obtained data allows to conclude about energy conservation of 8.36 kW·h or 38.9 % (in rooms No.1 and No.4). Actual value ΔT_{in} obtained for research object at T_{out0} = -20 °C is 2.2 °C (rooms No.1 and No.4), that can be explained by considerable value of specific thermal capacity. The same experiments at T_{out0} = -30 °C shows that temperature T_{in} was reduced to the value of 21 °C (ΔT_{in} = 3 °C) under otherwise equal conditions.

Conducted researches allows estimating individual influence of each factor on a heat consumption and thermal regimes of controlled rooms. Multifactorial analysis play a key role in investigations of buildings thermal regimes and functioning of control systems and technological equipment. In this case simultaneous influence of many factors is taken into account and simulation conditions as realistic as possible. Figure 2 shows changing of indoor temperature T_{in} in controlled rooms in case of applying the low heat consumption mode under simultaneous influence of following factors: outdoor temperature T_{out0} = -30 °C, wind velocity V_w = 3 m/s, availability of solar irradiation and internal heat gains.

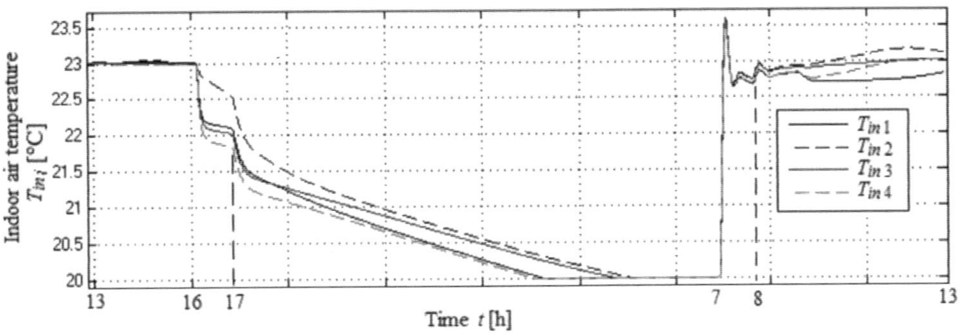

Fig. 2. Changing of indoor temperature T_{in} in controlled rooms in case of assumed factors influence.

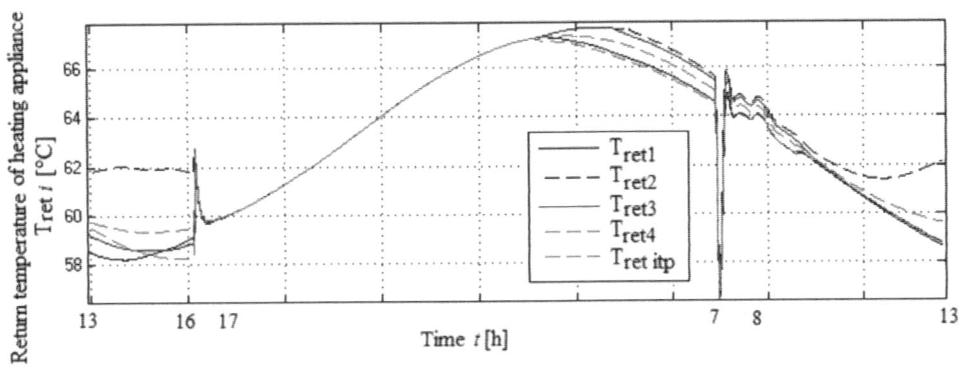

Fig. 3. Changing of return temperature of heating appliance $T_{ret\ i}$ and individual heating plant $T_{ret\ itp}$.

Figure 2 shows possibility of realization of thermal regime (low heat consumption mode) at ΔT_{in} = 3 °C. The maximum deviation of indoor air temperature does not exceed 0.59 °C (maximum dynamic control error) from the set value of $T_{in\ set}$= 23 °C. Activation of low heat consumption mode start at 16 o'clock (1 hour before finish a work day). It is interesting to note that specific shape of the graph caused by nulling the internal heat sources (end of work day). Further special energy saving mode is deactivated and parameter $T_{in\ set}$ assume a value 23 °C at 7 o'clock (1 hour before start a work day). Control error (deviation of T_{in} from set value) does not exceed 0.05 °C in steady state operating modes.

5 Conclusions

Modelling for the winter climate conditions (February) of Western Siberia was conducted. Researches of heat consumption of rooms, which are characterized by different spatial orientation were conducted.

Dynamics analysis of heat consumption parameters of controlled rooms allows to conclude about the potential of energy savings that can be achieved on real object in the implementation of intelligent distributed control system of building heat consumption, that can effectively compensate all considered disturbing influences. Applying the specialize low heat consumption mode allows the energy conservation of 8.36 kW·h or 38.9 % at the average outdoor temperature -20 °C.

Distributed control system effectively compensates the deviation of indoor air temperature in case of multifactorial influence: maximum dynamic control error does not exceed 0.59 °C after deactivation of low heat consumption mode.

Obtained modelling results (heat consumption, dynamic of indoor air temperature and the temperature distribution in the building envelopes) can be used to develop energy conservation measures for existing or planned buildings.

Acknowledgments

The reported study is supported by the Ministry of Education and Science of the Russian Federation Grant (project 16-38-00628).

References

1. V. Roshchanka, M. Evans, *Playing hot and cold: how can Russian heat policy find its way toward energy efficiency* (Pacific Northwest National Laboratory, Washington, 2012)
2. G. Keikkala, A. Kask, J. Dahl, V. Malyshev, V. Kotomkin, Energ. Policy **35**, 1452 (2007)
3. A. Korppo, N. Korobova, Energ. Policy **42**, 213 (2012)
4. A. Boute, Pace Envtl. L. Rev. **29**, 746 (2012)
5. Z. Wang, Y. Ding, G. Geng, N. Zhu, Energy Convers. Manage. **77**, 233 (2014)
6. G. Dall'O, L. Sarto, Energy Build. **67**, 298 (2013)
7. G. Mustafaraj, D. Marini, A. Costa, M. Keane, Appl. Energy **130**, 72 (2014)
8. T. Berger, C. Amann, H. Formayer, A. Korjenic, B. Pospischal, C. Neururer, R. Smutny, Energy and Build. **80**, 517 (2014)
9. M.D. Moldovan, I. Visa, M. Neagoe, B.G. Burduhos, Energy Procedia **48**, 924 (2014)
10. P. Strizhak, M. Morozov, Glavnyy energetic **7**, 39 (2014)
11. R. Scattolini, J. Process Contr. **19**, 723 (2009)
12. Q. Tian-Wei, H. Xiang, W. Jun-mei, Proc. Env. Sci. **11**, 643 (2011)
13. P.-D.Morosan, R. Bourdais, D. Dumur, J. Buisson, Energy Build. **42**, 1445 (2010)
14. F. Barata, N. Felix, R. Neves-Silva, Proc. Technology **17**, 772 (2014)
15. M. Wallace, R. McBride, S. Aumi, P. Mhaskar, J. House, T. Salsbury, Chem. Eng. Sci. **69**, 45 (2012)
16. S. Privara, Z. Vana, E. Zacekova, Energy and Build. **55**, 341 (2012)
17. D. Broman, P. Fritzson, *Proceedings of the 2nd International workshop on equation-based object-oriented languages and tools*, 59 (Paphos, Cyprus, 2008)

MATEC Web of Conferences **72**, 01081 (2016) DOI: 10.1051/matecconf/20167201081
HMTTSC-2016

EXPERIMENTAL RESEARCH OF THERMOPHYSICAL PROCESSES IN A CLOSED TWO-PHASE THERMOSYPHON

D.V. Feoktistov [1,*], *E*.A. Vympin [1], and *A*.E. Nurpeiis[1]

[1]National Research Tomsk Polytechnic University, 634050 Tomsk, Russia

Abstract. The temperature distribution in a thermosyphon was studied experimentally. To conduct the research, a closed two-phase thermosyphon was developed, which differs from the known by simple construction. The method of studying the rapid processes of conduction, convection and phase transitions was also developed. It will allow to highlight the operational modes of the thermosyphon, considering the load, cooling conditions of the condensation section, value of the heat supply. According to obtained results the instabilities of the temperature fields over the cross-section of the two-phase closed thermosyphon were observed by means of using the modern measuring equipment. It has been suggested that the instabilities can be caused by different modes of thermosyphon operation.

1 Introduction

Two-phase thermosyphons are simple in design, effective as thermal engineering devices that can remove the high local heat releases. Unfortunately, such cooling systems are not widely used in industry due to the lack of scientific basis for their design. However, their application has shown high effectiveness. For example, to regulate the thermal regime of heat engineering constructions in Yamal, to compensate the linear expansions of Trans-Siberian railways and backup abutment of oil pipelines in Alaska.

The known experimental studies of such heat removal systems are focused on the finding the optimal values of the amount, chemical composition and pressure of the coolant, which would ensure the most effective operation of the particular design of the thermosyphon [1-2].

At the same time the fundamental bases of jointly proceeding processes of conduction [3], convection [4-6] and phase transition [7-16] are not studied at level to provide the design of energy efficient heat removal equipment.

This paper aims to study experimentally the temperature distribution in a two-phase closed thermosyphon. The second and major task undertaken was to develop the methods of conducting experiments to study jointly proceeding processes of conduction, convection and phase transition.

* Corresponding author: fdv@tpu.ru

2 Materials and methods

Experimental studies were carried out on the experimental setup. Its schematic diagram is shown in figure 1.

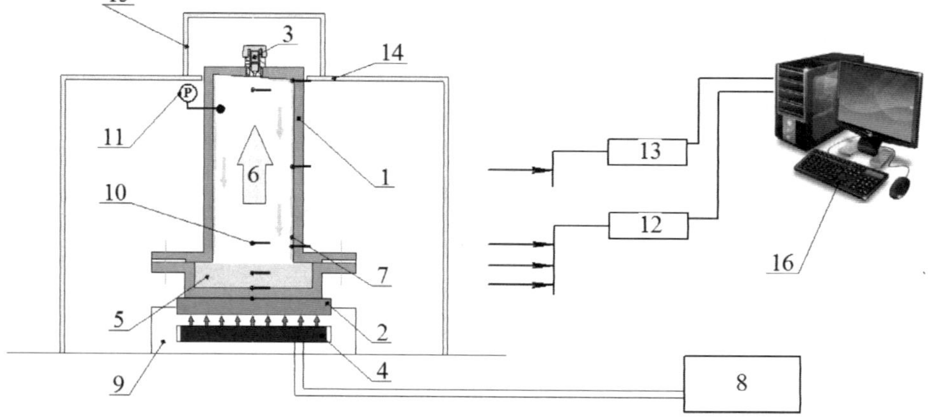

Fig. 1. Schematic diagram of the experimental setup: 1 – thermosyphon; 2 – metal substrate; 3 – valve; 4 – heating spiral; 5 – working liquid; 6 – liquid vapour; 7 – condensate; 8 – power supply, 9 – working table; 10 – thermocouples; 11 – pressure sensor; 12 – analog-to-digital converter NI 9214; 13 – analog-to-digital converter NI USB-6000; 14 – glass box; 15 – cover for an upper opening; 16 – personal computer.

The layout of the thermocouples and pressure gauge is presented in figure 2.

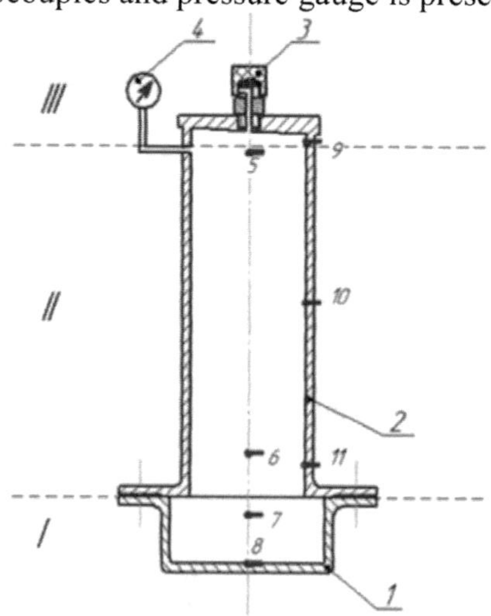

Fig. 2. Layout of the thermocouples and pressure gauge: I - evaporative part; II – vapour channel; III – condensation part; 1 – bottom cover; 2 – top cover; 3 – valve for pressure control; 4 – pressure sensor; 5 –11 – thermocouples.

Thermosyphon is made of copper. It consists of two sections: the lower *1* and upper *2* covers, fastened by eight screws with a diameter of 6 mm. The thickness of the side walls is 1.5 mm; the bottom wall is 2 mm. The upper wall is inclined at an angle of 3° to the horizon. It provides the movement of the condensate along the wall. A threaded opening is made for the valve *3* in the upper wall. The valve is designed to control the pressure in the thermosyphon.

The construction of thermosyphon can be divided into three working sections: evaporative I, vapour II and condensation III (fig. 2). The bottom cover of the thermosyphon with an inner diameter of 54 mm, a volume of 50 ml operates as an evaporative part. Two thermocouples recording the temperature of the liquid and the bottom of the thermosyphon are installed in this section.

The vapour channel is made of copper pipe with an internal diameter of 39.2 mm. Here, the temperatures of vapours are measured by thermocouples mounted in the lower, upper and central parts. Their position is shown in figure 2.

The upper wall of the thermosyphon operates as a condensation part III. The temperature of the condensate, moving on the side wall, is measured by the thermocouples *9, 10, 11*. Vapour pressure is registered by sensor *4*.

According to the preliminary series of experiments it was found that the operation of the thermosyphon significantly affected by the external factors, such as the presence of staff, changes in the flow of air masses in the laboratory because of the doors, windows opening, etc.). To reduce the impact of these factors on the research, thermosyphon was covered by glass box.

Box was made of plexiglass of 3 mm thick (fig. 3). The construction has the openings for mounting of the thermocouples *3* and cable from the heater *2*, and the opening on the upper wall.

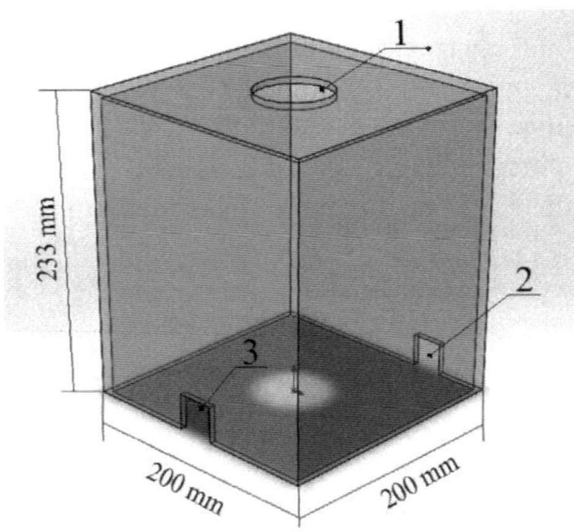

Fig. 3. Plexiglass box: 1 – opening for access to the condensation section of the thermosyphon, 2, 3 –openings for mounting thermocouples and heater cable.

3 Results and discussion

According to results of the preliminary experiments the possible ranges of the significant factors were determined (Table 1).

The maximum volume of filling (50 ml) of the thermosyphon was chosen based on the size of the evaporative section. Operating conditions of the thermosyphon, when it is filled less than on 20% (from the evaporative section), are considered as emergency due to the formation of dry spots on the walls. The variation of the heating power was made in accordance with the conditions of the unit. Distilled water and 3M™ Novec™ 7100 engineered fluid were chosen as the working liquids because of their physical properties. Water is cheap and readily available coolant. 3M™ Novec™ 7100 is low-boiling liquid with a boiling point of 61°C and a density of 1.52 g/ml, has a low surface tension, strength to thermal and chemical impact, non-flammability and low toxicity.

Table 1. The main significant factors.

Factors	Values
Volume of filling of thermosyphon, ml (% from evaporative section)	10 – 50 (20-100)
Power of heating, W	3 – 40
Working liquid	distilled water, 3M^{TM} Novec^{TM} 7100 engineered fluid

The temperature was measured using calibrated thermocouples K type (OMEGA 5TC-TT-K-40-36) with a thermojunction of 80 microns. The responses were tested with standard resistance thermometers ETS-100 with the error in the test temperature ranges of 0.05 °C. The measured values of the temperatures were saved on the personal computer with the help of analog converter National Instruments 9214 (NI 9214) and developed software LABVIEW Signal Express 2012. The usage of modern equipment allowed to record fast processes of conduction, convection and phase transitions. The estimation of measurement errors was carried out as described in [15]. According to results of calculation the error does not exceed 2%.

Figure 4 shows a typical thermogram obtained during the preliminary experiments. Distilled water with volume of 30 ml was used in these experiments; the heating value was 28 in watts. The experiment was conducted without measures for intensifying heat transfer (forced cooling of condensation section of thermosyphon by moistening or air blowing). Also the recorded values from the pressure sensor located in the condensation section are presented in fig. 4.

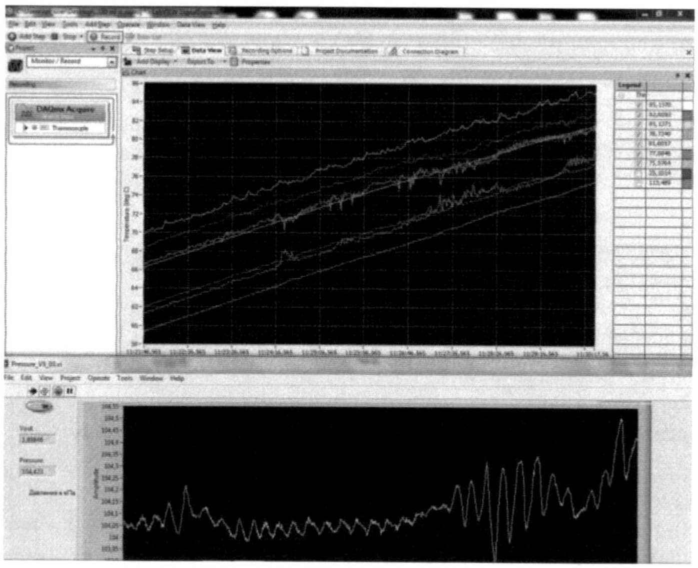

Fig. 4. Screenshots of software for recording the temperature and pressure.

According to analysis of the thermograms and corresponding pressure values it was found that there are significant instabilities of the temperature fields over the cross-section of the two-phase closed thermosyphon. Registered temperature fluctuations in the vapour section correspond to the temperature "jumps" of liquid at the bottom of the thermosyphon and pressure in the condensation section of the unit. Probably, these instabilities are caused by different operational modes of the thermosyphon, which depend on the boiling regimes. It is also found that the specific heat flux during operation of the thermosyphon is changed regardless the amount of heat supplied to the lower part of the thermosyphon and volume of filling.

Before reaching the boiling point of the liquid in the thermosyphon the instabilities of the temperature fields, as well as the change in specific heat flux are strongly marked. Presumably, this fact can be explained by the change in the boiling mode occurring in thermosyphon during its operation.

4 Conclusions

In conclusion it should be noted, that one of the main achievements of this work is to develop the two-phase closed thermosyphon, which differs from the known by simple construction. The instabilities of the temperature fields over the cross-section of the two-phase closed thermosyphon were observed by means of using the modern measuring equipment for registration of the temperature and pressure. To the best of our knowledge, these instabilities can be caused by different modes of thermosyphon operation.

The developed method of studying the rapid processes of conduction, convection and phase transitions will allow to highlight the operational modes of the thermosyphon, considering the load, cooling conditions of the condensation section, value of the heat supply. Also in this work the backgrounds were developed for an explanation of the physical mechanism of processes in the two-phase thermosyphon, the determination of transition conditions from one boiling mode to another.

Acknowledgments

The reported study was supported by the grant of President of Russian Federation for the government support of young Russian scientists (MK-6810.2016.8).

References

1. Marchuk I., Lyulin Y., Kabov O., Interfacial Phenomena and Heat Transfer **1** (2), 153 (2013)
2. D. Attinger, Z. Zhao, D. Poulikakos, Heat Mass Transfer **122**, 544 (2000)
3. A.E. Nurpeiis, EPJ Web Conf. **76**, 01016 (2014)
4. S.Ya. Misyura, Int. J. Heat Mass Transfer **71**, 197 (2014)
5. V. E. Nakoryakov, S. Y. Misyura, S. L. Elistratov, J. Eng. Thermophys. **22**, 1 (2013)
6. O.A. Kabov, D.V. Zaitsev, Multiphase Sci. Technol. **21**, 249 (2009)
7. D.O. Glushkov, G.V. Kuznetsov, P.A. Strizhak, Therm. Sci. **19**, 1541 (2015)
8. G.V. Kuznetsov, A.V. Zakharevich, N.S. Bel'kov, Chem. Pet. Eng. **50**, 424 (2014)
9. D.O. Glushkov, J.-C. Legros, P.A. Strizhak, A.V. Zakharevich, Fuel **175**, 105 (2016)
10. D.V. Zaitsev, O.A. Kabov, Microgravity Sci. Technol. **19**, 174 (2007)
11. Y.V. Lyulin, D.V. Feoktistov, I.A. Afanas'ev, E.S. Chachilo, O.A. Kabov, G.V. Kuznetsov, Tech. Phys. Lett. **41**, 665 (2015)
12. V.I. Maksimov, T.A. Nagornova, V.S. Chernyshev Conditions and characteristics of water crystallization on the working surface of evaporator heat pumps in reservoirs with low temperatures, MATEC Web Conf. **23** (2015)
13. K. Ponomarev, E. Orlova, D. Feoktistov, EPJ Web Conf. **110** (2016)
14. G.V. Kuznetsov, D.V. Feoktistov, E.G. Orlova, Thermophysics and Aeromechanics **23**, 17 (2016)
15. N.A. Yaryshev, *Theoretical Fundamentals of Non-Stationary Temperature Measurements* (Energoatomizdat, Leningrad, 1990)
16. A.S. Krasnoshlykov, V.I. Maksimov, EPJ Web Conf. **82** (2015)

EVALUATION OF EFFECTIVENESS OF LIGNOSULFONATE APPLICATION FOR ORGANIC COAL-WATER FUEL RHEOLOGICAL PROPERTIES IMPROVEMENT

Vitaliy Osipov[1, *], *Kirill* Larionov[1], *Alexander* Matveev[1], and *Alexander* Gromov[1, 2]

[1]Tomsk polytechnic university, DNTPP, 634050 Tomsk, Russia
[2]Nürnberg University of Technology Georg-Simon-Ohm

Abstract. The effect of lignosulfonate on viscosity of organic coal-water-fuel (OCWF) was investigated. Brown coal from "Borodinskoe" deposit, waste oil Motul 8100 X-Clean 5W-30-C3, distilled water and powdered lignosulfonate was used as raw materials for slurry preparation. OCWF viscosity were measured using a rotational viscometer BROOKFIELD DV-II + Pro EXTRA. Optimum lignosulfonate concentration was obtained (1.3-1.4%).

1 Introduction

Various technologies of solid fuel combustion are currently actively developing [1]. One of the promising technologies is preparation of composite fuels using coal. Preparation and combustion of such composite as coal-water fuel (CWF) allows to decrease the harm caused to the environment compared to conventional coal combustion in a flow [2].

Despite the environmental advantages of CWF, this technology did not receive widespread introduction at energy facilities. Primarily, this is due to the fact that the concerned type of fuel has a number of not completely solved problems associated with low calorific value and high viscosity characteristics.

The problem of low calorific value of CWF may be solved by adding to its composition organic liquid fuels, which have higher energy characteristics compared to coal. This kind of fuel is called organic coal-water fuel (OCWF). The most successful kind of liquid fuel is waste oil, because it solves two problems at the same time: improves properties of composite fuel and provides a way for recycling waste oil, which would improve the environmental aspects of energy production [3].

In turn, the addition of oils in coal-water fuel sharply worsens viscosity characteristics, resulting in decreased reliability of the fuel and transport systems [2].

* Corresponding author: vitosipov93@mail.ru

To solve this problem, the use of various surface-active agents (SAA) that can improve rheological properties of OCWF is proposed. In this paper, powdered lignosulfonate LST was used as a surfactant.

2 Experimental setup

Experimental studies of OCWF rheological properties using a powdered lignosulfonate LST were conducted using a rotational viscometer BROOKFIELD DV-II + Pro EXTRA at different spindle speeds and ambient temperature of 25°C.

2B brand brown coal from "Borodinskoe" deposit was used as initial solid fuel sample. After initial processing in a ball-mill pulverizer, it was sieved through a sieve with a mesh size of 80 microns. Further, received material was stirred in a galvanized drum with a volume of 0.5 liter with spherical grinding bodies with a diameter of 5 mm and a weight ratio of 1:1. In addition to pulverized coal, distilled water, waste oil Motul 8100 X-Clean 5W-30-C3 and powdered lignosulfonate LST with different concentrations presented in Table 1 were added in the drum.

After mixing of the materials, the drum with OCWF was placed in a Fritsch Pulverisette 7 Micro Mill planetary mill, in which wet milling was carried out for 5 minutes at a speed of 500 r/min.

Upon reaching the predetermined time of grinding, the resulting suspension was poured into a measuring cylinder with a volume of 0.25 liter and dynamic viscosity was measured.

Table 1. Mass concentrations of studied OCWF samples.

Sample number	Mass concentration, %			
	Coal	**Oil**	**Water**	**Lignosulfonate**
1	50	-	50	-
2	44	4	52	-
3	46	2	52	-
4	45	5	50	-
5	45	5	49	1
6	45	5	48.5	1.5
7	45	5	48	2
8	45	10	44	1
9	45	10	43.5	1.5
10	45	10	43	2

3 Results

The obtained results of prepared OCWF samples dynamic viscosity dependence from the spindle rotation speed and concentration of lignosulfonate are shown in figures 1 and 2.

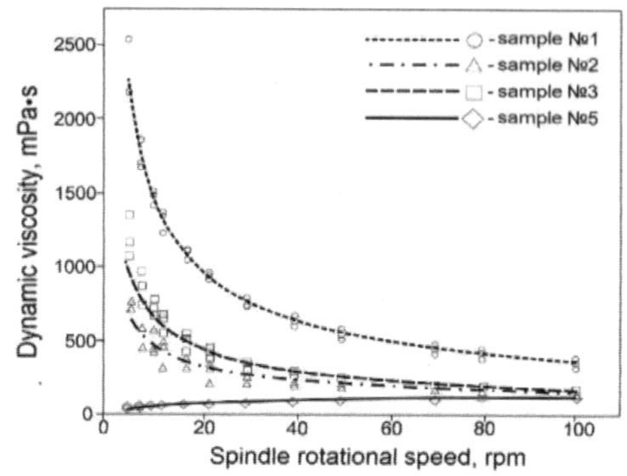

Fig. 1. The dependence of dynamic viscosity from spindle rotation speed.

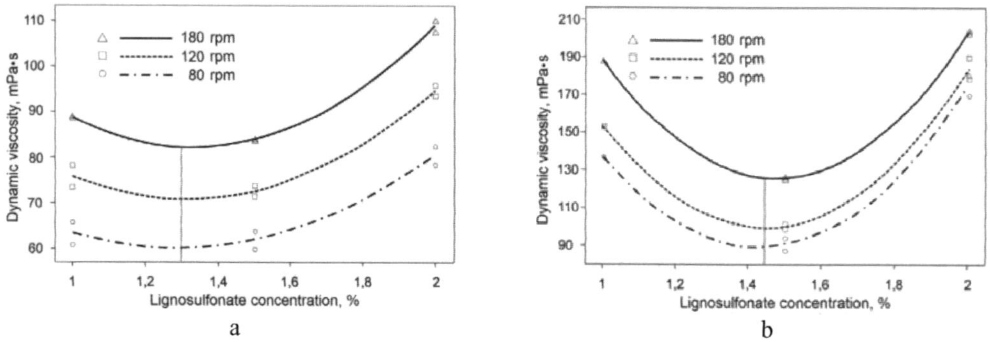

Fig. 2. The dependence of dynamic viscosity coefficient from lignosulfonate concentration (a – samples #5, 6, 7; b – samples #8, 9, 10).

Conducted experimental studies showed that a small amount of added oil can lead to a strong mixture viscosity increase. During the experiment, concentration of coal and oil 44% and 4%, respectively, with no added surfactant was achieved (sample No. 2). A further increase in the solid part (sample No. 4) led to a sharp increase in viscosity, making the fuel non-flowing.

Figure 1 shows that viscosity of the sample dramatically decreases with an increase of spindle speed, indicating pseudoplastic nature of the slurry.

The results presented in figure 2 represent the effect of lignosulfonate, which can significantly reduce fuel viscosity at low shear rates, thus changing the nature of the dependence - dynamic viscosity increases with an increase of spindle rotation speed, while its range of variation becomes narrower. Thus, the injection of this additive allows further increasing of the combustible portion of the fuel.

Figure 2 illustrates that the addition of lignosulfonate lowers OCWF viscosity with increasing its concentration up to only a certain limit. Further increase in the amount of lignosulfonate leads, on the contrary, to an increase in fuel viscosity.

Figure 2 shows that the optimum value of the lignosulfonate amount in OCWF is about 1.3-1.4% at concentration of 5% oil and 45% coal and 1.4-1.5% at concentration of 10% oil and 45% coal.

4 Conclusion

Experimental study of OCWF rheological properties showed that the presence of waste oil in the composite fuel greatly increases viscosity. Solution to this problem is achieved by adding a lignosulfonate surfactant, thereby reducing viscosity of the slurry and increasing content of the combustible portion in OCWF. Despite this, an excessive increase of lignosulfonate concentration in the composite fuel gives a negative effect.

Thus, it was determined that all samples of OCWF have optimum lignosulfonate concentration 1.3-1.5%.

Acknowledgments

The research was realized with financial support of Minobrnauki of Russia in framework of FTP "Research and development in prior direction of scientific-technological complex of Russia in 2014-2020 years", unique R&D identifier RFMEFI58114X0001.

References

1. K.V. Slyusarskiy, M.K. Bejsekov, J.V. Marysheva, Y.Ja. Rakov MATEC Web Conf. **37** (2015)
2. Osintsev K.V. Energy and resource saving in power engineering and the social sphere **1**, 242 (2013)
3. D.V. Gvozdyakov, V.E. Gubin, Chem. and Petrol. Eng. **6** (2015)
4. D.O. Glushkov, D.P. Shabardin, P.A. Strizhak, K.Y. Vershinina, Fuel Process. Technol. **143** (2015)
5. G. Atesok, F. Boylu, A.A. Sirkeci, H. Dincer, Fuel **81** (2002)
6. G.S. Khodakov, E.G. Gorlov, G.S. Golovin, Solid Fuel Chem. **40** (2006)
7. R. Chen, M. Wilson, Y.K. Leong, P. Bryant, H. Yang, D.K. Zhang, Fuel **90** (2011)

MATEC Web of Conferences **72**, 01083 (2016) DOI: 10.1051/matecconf/20167201083
HMTTSC-2016

MATHEMATICAL MODELING OF THE HEAT TRANSFER PROCESS IN POROUS COOLING SYSTEMS UNDER INFLUENCE LOW ENERGY DISTURBANCES

Vyacheslav Ovchinnikov [1,*], *Eugenia* Ruleva[1], and *Anatoliy* Yakimov[2]

[1]National Research Tomsk State University, Scientific Institute of Applied Mathematics and Mechanics, 634050 Tomsk, Russia
[2]National Research Tomsk State University, Department of Physical and Computational Mechanics, 634050 Tomsk, Russia

Abstract. The paper considers the interaction of a high-temperature flow with a transpiration cooling system exposed to low-energy perturbations of a coolant. It studies the possibility of controlling the heat exchange process in a composite material. The numerical calculations are compared with the experimental data.

1 Introduction

Vibration and oscillation processes accompany some natural phenomena and may be both harmful and beneficial. An oscillatory motion of gases and liquids intensifies a wide range of heat and mass transfer processes and stimulates the heat transfer of bodies, drying, diffusion, electrodeposition, etc. [1–3].

Transpiration cooling systems are widely used in engineering applications: evaporation systems, filters, thermal protection elements of flight vehicles [4–7]. The operation of such systems may be accompanied by some perturbations, such as acoustic oscillations, wall vibrations, pressure fluctuations and turbulent noises. Depending on the type of perturbations, oscillation amplitude and frequency, thermal and filtration parameters of porous materials may be distorted [7]. For example, in study [6], the thermal conductivity coefficient of a porous body increased in the presence of pressure fluctuations on its boundary.

The purpose of this paper is to carry out a theoretical study of transpiration cooling systems in the presence of low-energy perturbations and to compare the obtained findings with the known data.

* Corresponding author: ovc@niipmm.tsu.ru

2 Problem Statement

Based on the approximate calculated relations obtained in [9], study [8] proposes a formula for finding a heat flow on the permeable plate in a turbulent gas flow:

$$\frac{\mathrm{St}}{\mathrm{St}_0}=\frac{\gamma\exp\left(\frac{1-\mathrm{Pr}}{1+\mathrm{Pr}}\gamma\right)}{1+\frac{2\gamma}{\mathrm{Pr}}\exp\left(\frac{\gamma}{1+\mathrm{Pr}}\right)-\exp\left(\frac{1-\mathrm{Pr}}{1+\mathrm{Pr}}\gamma\right)},\tag{1}$$

where $\mathrm{St}=\frac{q_w}{(\rho v)_e(h_e-h_w)}$, $\gamma=\frac{(\rho v)_w}{(\rho v)_e\mathrm{St}_0}$, $\mathrm{St}_0=\frac{(\alpha/c_p)_0}{(\rho v)_e}$, $\mathrm{Pr}=\left(\frac{c_p\mu}{\lambda}\right)_e$, St is the Stanton number, Pr is the Prandtl number, $(\rho v)_w$ is the flow rate of the coolant gas in pores of the thermal protective material, h is the enthalpy, λ is the thermal conductivity coefficient, c_p is the specific heat capacity, μ is the dynamic viscosity, α/c_p is the heat transfer coefficient. Superscripts and subscripts: e is the outer edge of the boundary layer, w is the external heated surface of the body in flow, 0 means the absence of injection from the surface. Here, we consider the case when an injected substance and a substance of the primary flow are identical.

The varying (pulsating) convective heat flow acting on the thermal protective material for a definite time is assumed to occur in region *1* (fig. 1). Then, according to (1), we have:

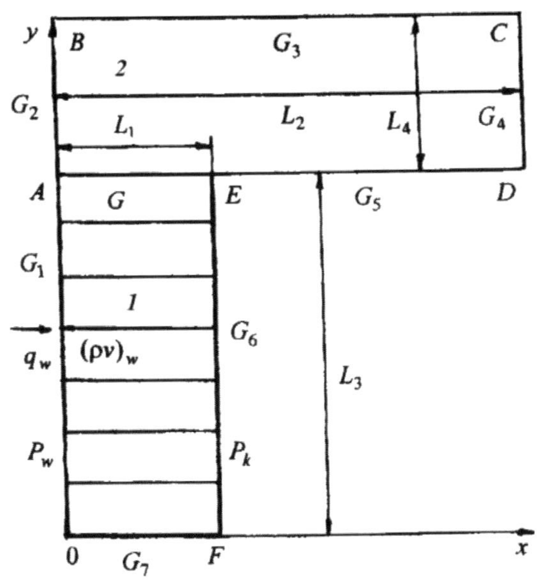

Fig. 1. Diagram of the flow around a body.

$$q_w^{(1)}=\frac{(\rho v)_w(h_e-h_w^{(1)})\exp\left(\frac{1-\mathrm{Pr}}{1+\mathrm{Pr}}b\right)}{1+\frac{2b}{\mathrm{Pr}}\exp\left(\frac{b}{1+\mathrm{Pr}}\right)-\exp\left(\frac{1-\mathrm{Pr}}{1+\mathrm{Pr}}b\right)},\tag{2}$$

where $b = \dfrac{(\rho v)_w}{(\alpha / c_p)_0}$, $\left(\dfrac{\alpha}{c_p}\right)_0 = \left(\dfrac{\alpha}{c_p}\right)_{in}\left[1 + \dfrac{A\cos(ft)}{(\alpha / c_p)_{in}}\right]$, $f = 2\pi\nu$, $h_w^{(1)} = c_{p2}T_{1w}$, A is the pulsation amplitude of the gas flow, ν is the frequency of periodic perturbations, c_{p2} is the specific heat capacity of the coolant gas at a constant pressure, t is the time, T is the temperature, a_1, a_2 are the constants. Superscripts and subscripts: *in* is the initial value (no pulsation), 1 and 2 below are, respectively, the frame and the gas in region *1*; (1) and (2) above are, respectively, region *1* and *2* in fig. 1.

In curtain zone *2*, we use the formula from [10] in the form of the Mugalev's modification, which takes into account the effect on injection on the heat flow [4]

$$q_w^{(2)} = \left(\frac{\alpha}{c_p}\right)_{in} (1 - k_1 b^{k_2})(h_e - h_w^{(2)}),\ h_w^{(2)} = c_{p2}T_w,$$

where $b = \dfrac{(\rho v)_w (y = 0)}{(\alpha / c_p)_{in}(\pi / 2 - z)}$, $1 \le z \le 1 + L_4 / L_3$, k_1, k_2 are the constants.

The thermal conductivity coefficient (λ_1) for the frame of the thermal protective material is known from [7, 11]

$$\lambda' = \frac{\pi c_{p2}\rho_2 \varphi B^2 \nu}{\sqrt{2}}, \tag{3}$$

where ρ_2 is the density of the coolant gas, φ is the porosity, B is the pulsation amplitude for the frame wall of the thermal protective material.

The expression for effective viscosity μ_{ef} in the Darcy law (8) can be taken as Einstein's modification [12]:

$$\mu_{ef} = \mu[1 + C\cos(tf)], \tag{4}$$

where C is the dimensionless coefficient ($0 < C \le 0.2$).

The oscillation frequency *f* can be found by the formula from [13]

$$f = \frac{1}{B}\left(\frac{2I}{\rho_2 a}\right)^{0.5},$$

where I is the intensity of perturbations, a is the speed of sound under normal conditions.

For the sake of simplicity, we make the following assumptions:

1) The Reynolds number in the incoming hypersonic air flow is sufficiently high ($\mathrm{Re}_\infty >> 1$) and the boundary layer has formed in the vicinity of the body surface;

2) The thermal protective material is a two-temperature medium, i.e. gas and condensed phases have different temperatures;

3) The mass flow rate at a normal to the plate in flow is significantly greater than the mass flow rate along this plate (fig. 1, $L_1 << L_3$);

4) A body interacting with a gas flow is not destroyed; no heterogeneous chemical reaction and phase transition occur inside or on its surface;

5) The density of the gas phase is determined from the ideal gas equation;

6) The filtered gas is assumed to be uniform, with the molecular mass being close to that of the air mixture;

7) The contents of the injected gas and the gas in the incoming flow are identical.

8) The boundary problem in spatial variables is assumed to be two-dimensional in the case of heat propagation by conduction.

The flow rate of the coolant gas will be determined from the combined solution of the steady-state equation of continuity and the nonlinear Darcy law [4, 14]. The fact is that high mass rates $(\rho v)_w$ (Re > 10) are associated with turbulent filtration behavior [14], which is characterized by the quadratic dependence of the pressure differential on the rate.

The problem of calculating heat and mass transfer parameters can be reduced to a system of conservation equations [7]:

in region *1* (see fig. 1)

$$\frac{\partial(\rho_2 v\varphi)}{\partial x} = 0, \tag{5}$$

$$\rho_1 c_{p1}(1-\varphi)\frac{\partial T_1}{\partial t} = \frac{\partial}{\partial x}\left[\lambda_1(1-\varphi)\frac{\partial T_1}{\partial x}\right] + \frac{\partial}{\partial y}\left[\lambda_1(1-\varphi)\frac{\partial T_1}{\partial y}\right] - A_v(T_1 - T_2), \tag{6}$$

$$c_{p2}\left(\rho_2\varphi\frac{\partial T_2}{\partial t} + \rho_2 v\varphi\frac{\partial T_2}{\partial x}\right) = \frac{\partial}{\partial x}\left(\lambda_2\varphi\frac{\partial T_2}{\partial x}\right) + \frac{\partial}{\partial y}\left(\lambda_2\varphi\frac{\partial T_2}{\partial y}\right) + A_v(T_1 - T_2), \tag{7}$$

$$\alpha_n \mu_{ef} v + \beta_n \rho_2 \varphi v|v| = -\frac{\partial P}{\partial x}, \tag{8}$$

$$P = \frac{\rho_2 R T_2}{M},\ \lambda_2 = \lambda_{2in}\sqrt{\frac{T_2}{T_{in}}},\ \mu = \mu_{in}\sqrt{\frac{T_2}{T_{in}}}, \tag{9}$$

in region *2* (solid steel edging in fig. 1)

$$\rho c_p \frac{\partial T}{\partial t} = \frac{\partial}{\partial x}\left(\lambda\frac{\partial T}{\partial x}\right) + \frac{\partial}{\partial y}\left(\lambda\frac{\partial T}{\partial y}\right). \tag{10}$$

The system of equations (5)–(8), (10) must be solved by taking into account the following initial and boundary conditions:

$$T_i\big|_{t=0} = T\big|_{t=0} = T_{in},\ i = 1,\ 2. \tag{11}$$

The following balance boundary conditions [15] are observed on the external heated surface of the porous plate $0A$ (region *1* in fig. 1) [15].

$$(q_w^{(1)} - \varepsilon^{(1)}\sigma T_{1w}^4)(1-\varphi) = -\lambda_1(1-\varphi)\left(\frac{\partial T_1}{\partial x}\right)\bigg|_{G_1}, \tag{12}$$

$$q_w^{(1)}\varphi = -\lambda_2\varphi\left(\frac{\partial T_2}{\partial x}\right)\bigg|_{G_1}; \tag{13}$$

on the symmetry axis $0F$

$$\left.\left(\frac{\partial T_i}{\partial y}\right)\right|_{G_7} = 0, \quad i = 1, 2. \tag{14}$$

The thermal insulation conditions are specified on the surface of the solid steel shell BCD:

$$\left.\left(\frac{\partial T}{\partial y}\right)\right|_{G_3} = 0, \quad \left.\left(\frac{\partial T}{\partial x}\right)\right|_{G_4} = 0; \tag{15}$$

on the internal surface DEF [15]:

$$\left.\left(-\lambda\frac{\partial T}{\partial y}\right)\right|_{G_5} = \delta(T|_{G_5} - T_{in}), \quad \left.\left(-\lambda_1(1-\varphi)\frac{\partial T_1}{\partial x}\right)\right|_{G_6} = \delta(T_1|_{G_6} - T_{in}), \tag{16}$$

$$T_2|_{G_6} = \frac{\delta}{c_{p2}(\rho v)_w}(T_1|_{G_6} - T_{in}) + T_{in}. \tag{17}$$

The ideal contact condition is specified on conjugation line AE of regions *1* and *2*:

$$\lambda_1(1-\varphi)\left.\left(\frac{\partial T_1}{\partial y}\right)\right|_{G_-} = \lambda\left.\left(\frac{\partial T}{\partial y}\right)\right|_{G_+}, \quad T_1|_{G_-} = T|_{G_+}, \quad T_1|_G = T_2|_G; \tag{18}$$

on the external heated surface of edging *2*:

$$q_w^{(2)} - \varepsilon^{(2)}\sigma T_w^4 = -\lambda\left.\left(\frac{\partial T}{\partial x}\right)\right|_{G_2}. \tag{19}$$

Pressure in the pores and the environment is equal both on the external and internal surfaces of region *1*.

$$P_w|_{G_1} = P_e, \quad P|_{G_6} = P_{L_1}. \tag{20}$$

Hereinafter, x and y are the transversal and longitudinal spatial coordinates, P is the pressure, v is the gas filtration rate in region *1*, ρ is the density, δ is the heat transfer coefficient on the internal surface of the plate, A_v is the volumetric heat transfer coefficient between the gas and the frame, R is the universal gas constant, α_n and β_n are respectively the viscosity and inertial coefficients in Darcy's law, σ is the Stefan–Boltzmann constant, $\varepsilon^{(i)}$, $i = 1, 2$ is the emissivity of the surface of the thermal protective material and the edging, respectively, M is the molecular mass of air, L_i, $i = 1, 2, 3, 4$ are respectively the thicknesses and lengths of shells *1* and *2* in fig. 1, Pe is the Peclet number, Nu is the Nusselt number. Superscripts and subscripts: ∞ is the value of the incoming gas flow at infinity, k is the end time of thermal exposure, the prime above means the pulsation of heat transfer parameters, *ef* is the effective value, L_1 is the inner side of the body in flow, n is nonlinearity.

3 Numerical Method and Input Data

The pressure on the external heated surface of the permeable body was assumed to be known from the experiment as a function of the longitudinal coordinate y [11]. The pressure on the inside "cold" surface of the plate (L_1) was taken as:

$$P_{L_1} = kP_{e0} . \tag{21}$$

The constant k was specified to ensure the necessary flow rate of the coolant (in particular, not to reach the melting temperature of the steel shell, i.e. 1600 K [4]) in the thermal exposure area from $t = 0$ to $t = t_k$.

The quasi-stationary equation of continuity $\rho_2\varphi v = -(\rho v)_w$ (the negative sign is explained by the fact that the normal coordinate x is directed deep into the body (see fig. 1), and the coolant flows in the opposite direction) can be integrated along with the first equation (9), the nonlinear Darcy law (8) and the boundary conditions (20) to find the gas flow rate and the pressure in region *1* [7, 15]:

$$(\rho v)_w(y) = \frac{\left[2\beta_n(P_{L_1}^2 - P_w^2)\varphi MD_{L_1} / R + E_{L_1}^2\right]^{0.5} - E_{L_1}}{2\beta_n D_{L_1}} ,$$

$$P(x,y) = \{P_w^2 + 2R(\rho v)_w[\beta_n(\rho v)_w D + E] / M\varphi\}^{0.5} ,$$

where $D(x,y) = \int_0^x T_2(s,y)ds$, $E(x,y) = \alpha_n \int_0^x \mu_{ef} T_2(s,y)ds$.

The volumetric heat transfer coefficient A_v between the gas and the frame was determined from the formulas [16]

$$\mathrm{Nu}_v = \nu_1 \mathrm{Pe}^{\nu_2}, \quad 0.5 < \mathrm{Pe} < 80, \quad \nu_2 = 1 - 1.3,$$

where $\mathrm{Nu}_v = A_v l^2 / \lambda_2$, $\mathrm{Pe} = (\rho v)_w l c_{p2} / \lambda_2$, $l = \beta_n / \alpha_n$.

The boundary problem (6), (7), (10)–(19) was solved numerically using the locally one-dimensional split-plot method [17]. The implicit, absolutely stable, monotone difference scheme was used with the cumulative approximation error $O(\tau + H_x^2 + H_y^2)$, where H_x is the spatial step along the coordinate x, H_y is the spatial step along the coordinate y, τ is the time step. The numerical method was tested for the basic option.

With all initial parameters being equal, the calculation was made for different spatial steps $H_x = 0.5 \cdot 10^{-4}$ m, $H_y = 2.4 \cdot 10^{-4}$ m, $h_{x1} = 2 \cdot H_x$, $h_{x2} = H_x$, $h_{x3} = H_x / 2$, $h_{x4} = H_x / 4$, $h_{y1} = 2 \cdot H_y$, $h_{y2} = H_y$, $h_{y3} = H_y / 2$, $h_{y4} = H_y / 4$. The temperature of the frame and the gas was recorded across the depth of the body at different times. In all cases, the problem was solved with the variable time step chosen on the assumption that the prescribed accuracy was equal for all spatial steps. The difference $\Delta = \max[\Delta_{T_1}, \Delta_{T_2}]$ in the relative temperature error was decreasing to reach $\Delta_1 = 10.3$ %, $\Delta_2 = 5.4$ %, $\Delta_3 = 2.6$ % at $t = t_k$. The calculation results presented below were obtained for spatial steps $h_{x3} = H_x / 2$, $h_{y3} = H_y / 2$.

The thermophysical and structural properties of the porous material were taken from for the sample made of sintered stainless steel powder [18, 19] and from [20] for air. The thermo-physical coefficients of solid steel (edging) $\lambda(T)$, $c_p(T)$ as a function of temperature were taken from [21], while the values α_n and β_n from [11, 18]. The results presented below were obtained at T_{in} = 293 K, $(\alpha / c_p)_{in}$ = 0.2 kg/(s·m^2), A = 0.03 kg/(s·m^2), μ_{in} = $1.81 \cdot 10^{-5}$ kg/(m·s), T_e = 3600 K, λ_e = 0.782 W/(m·K), c_{pe} = 4024 J/(kg·K), h_e = $1.449 \cdot 10^7$ J/kg, ρ_e = 0.088 kg/m^3, μ_e = $1.747 \cdot 10^{-4}$ kg/(m·s), R = 8.314 J/(mole·K), B = 10^{-3} m, M = 29 kg/kmol, σ = $5.67 \cdot 10^{-8}$ W/(m^2·K^4), $L_1 = 2 \cdot 10^{-3}$ m, $L_2 = 2 \cdot 10^{-2}$ m, $L_3 = 9.5 \cdot 10^{-3}$ m, $L_4 = 2 \cdot 10^{-3}$ m, $\lambda_1 = 2.92 + 4.5 \cdot 10^{-3} \cdot T_1$ W/(m·K), $\rho_1 c_{p1} = (1252 + 0.544 \cdot T_1) \cdot 10^3$ J/(K·m^3), δ = 100 W/(K·m^2), λ_{2in} = 0.0257 W/(m·K), ρ = 7800 kg/m^3, I = 0–0.5 kg/s^3, t_k = 10 s, a = 340 m/s, φ = 0.36–0.67, a_1 = 965.5, a_2 = 0.0735, $\varepsilon^{(1)} = 0.85$, $\varepsilon^{(2)} = 0.9$, $C = 0.2$, $\nu_1 = 0.015$, $\nu_2 = 1$.

4 Discussion of Numerical Solution Results

Table 1 shows the experimental $T_w(0)$ [11] and calculated $T_{1w}(0)$ temperatures of the external surface of permeable steel as a function of the coolant flow rate, parameter k in the formula (21) with the initial data from Section 2 and the porosity φ = 0.36, without periodic perturbations of the coolant gas (I = 0). As shown in Table 1, the difference in relative errors (Δ) between the calculated and experimental temperatures does not exceed 11.9 %.

Table 1. Relative error of the surface temperature as a function of the coolant flow rate and parameter k.

k	$(\rho v)_w(0)$, kg/(s·m^2)	$T_w(0)$, K	$T_{1w}(0)$, K	Δ, %
1.072	0.2	1224	1370	11.9
1.092	0.4	1044	968	7.2
1.105	0.6	792	763	3.7
1.122	0.8	648	589	9.1

Fig. 2 and 3 present the external temperature of the frame surface T_{1w} and the coolant flow rate $(\rho v)_w$ as a function of the longitudinal coordinate y at φ = 0.67 at the end time of thermal exposure $t = t_k$. Lines *1–4* in fig. 2 and 3 correspond to the intensity of pulsations: I = 0, 0.1, 0.3, 0.5, i.e. for cases without or with pulsations for heat and mass transfer parameters in the formulas (2)–(4).

Analysis of lines *1–4* (fig. 2 and 3) shows that periodic perturbations taken into account may both increase or weaken the intensity of heat and mass transfer. This suggests that transpiration cooling systems are sensitive to pulsations of the coolant gas flow and that the heat transfer process in the thermal protective material can be controlled.

Table 2 presents the temperature of the external surface of the porous steel frame T_{1w} as a function of the coolant flow rate at $\varphi = 0.4$, $\nu = 0$ and $\nu = 5.2\ s^{-1}$. The values $T_{1w}(0)$ at $x = y = 0$ correspond to the calculation results $T_w(0)$, while the values taken from [22] to the experimental results. The difference in the relative error between the numerical solution and the experiment does not exceed 18.2 % at $\nu = 0$ and 17.1% at $\nu = 5.2\ s^{-1}$.

Table 2. Relative error of the calculated temperature of the external frame surface as a function of the flow rate of the coolant gas without and with pulsations.

ν, s^{-1}	**0**				**5.2**			
$(\rho v)_w(0)$, kg/ $(s\cdot m^2)$	0.4	0.8	1.2	1.4	0.4	0.8	1.2	1.4
$T_w(0)$, K	1080	666	504	450	900	594	450	360
$T_{1w}(0)$, K	933	578	412	373	752	492	375	339
Δ, %	13.6	13.2	18.2	17.1	16.4	17.1	16.6	5.8

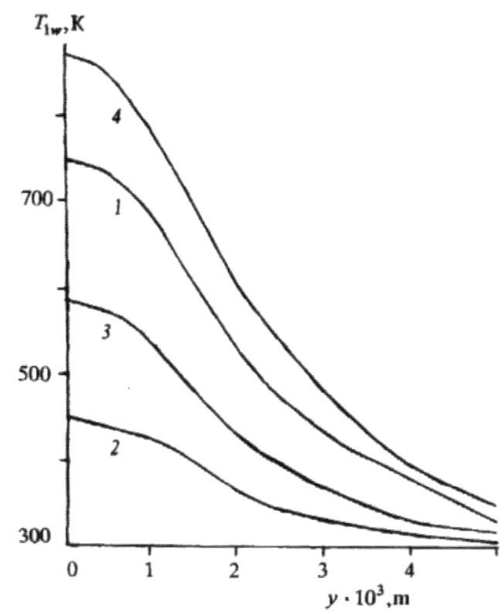

Fig. 2. The temperature of the external frame surface as a function of the longitudinal coordinate y at time $t = t_k$: *1* – *I* = 0, *2* – 0.1, *3* – 0.3, *4* – 0.5.

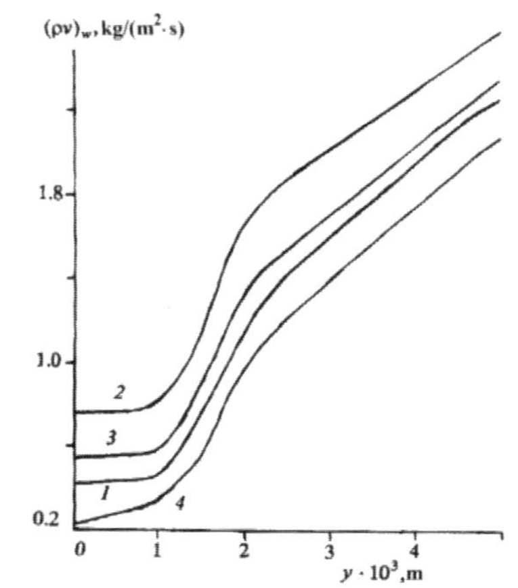

Fig. 3. Flow rate of the coolant gas as a function of the longitudinal coordinate y at time $t = t_k$. The symbols are the same as in fig. 2.

On fig. 4 shows the temperature distribution of the porous steel frame T_1 (solid-curves) and gas T_2 (dashed curves) in depth y layer at $x = 0$ and $\varphi = 0.4$ at different moments of time. Fig. 4 shows that for the balance of the boundary conditions (12) and (13) [15], have meaning for conservation laws, $T_{2w} > T_{1w}$. However, as we move deeper into the material quantity T_1 is greater T_2 . It related by the sharp increase in the temperature of air at the surface and in a certain neighborhood because of the heat exchange with the external environment and the framework and greater thermal conductivity of the air than the skeleton of the porous steel. Then, as the heat transfer process of establishing the condensed phase is warming up faster, since the thermal conductivity of permeable steel two orders of magnitude higher than the thermal conductivity of the gas.

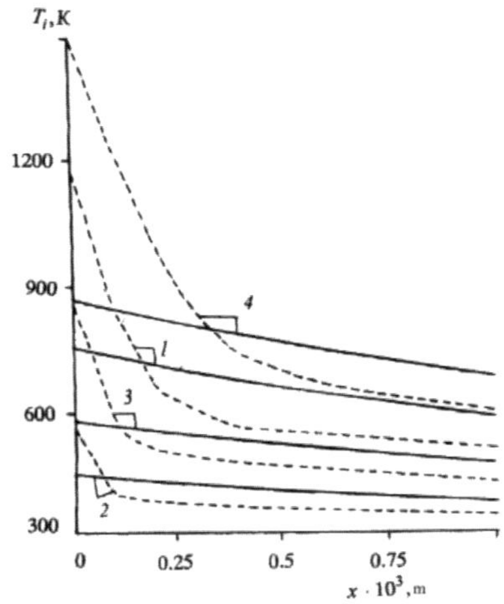

Fig. 4. Temperature distribution carcass gas layer depth with x ($y = 0$) at time $t = t_k$. The symbols are the same as in fig. 2.

5 Conclusion

The mathematical model was developed to calculate heat and mass transfer parameters in transpiration cooling systems with periodic perturbations of the coolant gas. Pulsations of the coolant gas flow have an ambiguous effect on the intensity of the heat mass transfer in transpiration cooling systems. The calculation results were proved to be consistent with the experimental data [11, 22] both in quantitative and qualitative terms.

Acknowledgments

The study was sponsored by the Ministry of Education and Science, Russian Federation, state assignment No. 9.1024.2014/k.

References

1. Yu.S. Kachanov, V.V. Kozlov, V.Ya. Levchenko, *Origin of the Turbulence in the Boundary Layer* (Novosibirsk, Science, 1982)

2. V.E. Nakoryakov, A.P. Burdukov, A.M. Boldarev, P.N. Terleev, *Heat and Mass Transfer in a Sound Field* (Siberian Branch of the USSR Academy of Sciences, Novosibirsk, 1970)
3. O.V. Abramov, *Crystallization of Metals in an Ultrasonic Field* (Science, Moscow, 1972)
4. Yu.V. Polezhaev, F.B. Yurevich, *Thermal Protection* (Energy, Moscow, 1976).
5. V.M. Polyaev, V.A. Mayorov, L.A. Vasilyev, *Hydrodynamics and Heat Transfer in Porous Elements of Flight Vehicle Structures* (Mechanical Engineering, Moscow, 1988)
6. N. L. Azhishchev, V.I. Bykov, Proceedings of the Siberian Branch of the USSR Academy of Sciences. Engineering Science Series **21**, 27 (1987)
7. A.M. Grishin, A.N. Golovanov, V.I. Zinchenko, E.K. Efimov, A.S. Yakimov, *Mathematical and Physical Modeling of Thermal Protection* (Tomsk University Publishing House, Tomsk, 2011)
8. V.D. Sovershenny, J. Eng. Phys. **12**, 538 (1967)
9. V.D. Sovershenny, Izv. Akad. Nauk SSSR. Mech. Fluids Gases **3**, 45 (1966)
10. A.V. Bureev, V.I. Zinchenko, Applied Mech. and Eng. Phys. **6,** 72 (1991)
11. A.N. Golovanov, E.V. Ruleva, A.S. Yakimov, High-Temperature Thermophys. **49**, 914 (2011)
12. A. Einstein, M. Smoluchowski, *Brownian Motion* (Main Publishing House of General Engineering Literature, Moscow-Leningrad, 1936)
13. A.N. Golovanov, Physics of Combustion and Explosion **35**, 67 (1999)
14. R.A. Andrievsky, *Porous Metal-Ceramic Materials* (Metallurgy, Moscow, 1964)
15. A.M. Grishin, A.N. Golovanov, A.S. Yakimov, Appl. Mech. Eng. Phys. **4,** 141 (1991)
16. V.F. Zanemonets, V.I. Rodionov, Heat and Mass Transfer. Minsk International Forum, **7**, 42 (1988)
17. A.A. Samarsky, *Introduction to the Theory of Difference Schemes* (Science, Moscow, 1971)
18. A.N. Golovanov, J. Eng. Phys. **66**, 695 (1994)
19. O.M. Alifanov, A.P. Tryanin, A.L. Lozhkin, J. Eng. Phys. **52**, 461 (1987)
20. N.B. Vargaftik, *Guide to Thermophysical Properties of Gases and Fluids* (Physical and Mathematical State Edition, Moscow, 1963)
21. V.F. Zinovyev, V.I. Rodionov, *Thermophysical Properties of Metals at High Temperatures. Handbook* (Metallurgy, Moscow, 1989)
22. A.N. Golovanov, E.V. Ruleva, Herald of the Tomsk State University. Mathematics and Mechanics **14**, 85 (2011)

MATEC Web of Conferences **72**, 01084 (2016) DOI: 10.1051/matecconf/20167201084
HMTTSC-2016

SUPERHEATER TUBE FLAT WALL STATIONARY TEMPERATURE FIELD

A.T. Parpiev[1], and *V*.S. Loginov[1, *]

[1]National Research Tomsk Polytechnic University, 634050 Tomsk, Russia

Abstract. The BKZ-220-100-9 steam generator platen superheater tube flat wall stationary temperature fields analysis have been made. The six steel grades, using in boiler fabrication, namely, St. 10, St. 20, 12H1MF, 15HM, 1H18N9T and 12H18N12T, have been used. The temperature curves calculation has been made by using outer and inner surface heat-transfer coefficients nine different combinations.

1 Problem physical model

The present paper takes the BKZ-220-100-9 steam generator platen superheater tube element as an investigation object. The investigation object wall is flushed with the following operation mediums such as flue gases with the temperature of 1374.15 K on the outer surface and steam with the temperature of 619.15 K on the inner surface [1, 2]. It follows that the steam heating process occurs in tubes using the heat released by flue gases by means of a heat-transfer through the cylindrical wall. The heat is transferred through the wall by means of the heat conductivity process and from the inner surface to the steam by convection. The steels heat conductivity coefficients are known [3, 4] and have constant values (Table 1). The object calculation geometrical characteristics and heat-transfer coefficients are known also [1, 2] and are the constant values (Table 1). The investigation object stationary temperature distributions analysis for a six steel grades at the various outer and the inner surfaces heat-transfer coefficients combinations are the present paper problem. The one-dimensional stationary heat conductivity equation with the third type boundary conditions is used for a task set solving.

2 Problem mathematical model

The flat wall stationary temperature field is determined by a mathematical model follows as:

$$\frac{d^2T(x)}{dx^2} = 0, \;\; L_1 < x < L_2,$$

* Corresponding author: loginovvs@tpu.ru

$$\lambda \frac{dT(x=L_1)}{dx} = \alpha_{m1}\left(T(x=L_1) - T_{m1}\right),$$
$$\lambda \frac{dT(x=L_2)}{dx} = \alpha_{m2}\left(T_{m2} - T(x=L_2)\right). \quad (1)$$

λ is the steel heat conductivity coefficient, measured by Wt per m per K. α_{m1}, α_{m2} are heat-transfer coefficients on the inner and outer surface, respectively, measured by Wt per square meter per K. T_{m1}, T_{m2} are medium temperatures on the outer and the inner surface, respectively, measured in K. $T(x=L_1)$, $T(x=L_2)$ are temperatures on the left and the right boundaries, measured in K.

The (1) equation analytical solution is obtained in compliance with [5, 6] and looks like so:

$$T(x) = T_{m1} + \frac{\left(T_{m2} - T_{m1}\right)}{\delta + \frac{\lambda}{\alpha_1} + \frac{\lambda}{\alpha_2}} \cdot \left(x + \frac{\lambda}{\alpha_1}\right). \quad (2)$$

δ is the wall thickness, measured in m.

This expression is as follows in a dimensionless form:

$$\Theta_{station}(\mathrm{X}) = \frac{\mathrm{Bi}_2\left(\mathrm{Bi}_1 \cdot \mathrm{X} + 1\right)}{\mathrm{Bi}_1 + \mathrm{Bi}_2 + \mathrm{Bi}_1 \cdot \mathrm{Bi}_2}, \quad (3)$$

$\Theta_{station}$ (X) = $(T(x)-T_{m1})/(T_{m2}-T_{m1})$ is the dimensionless temperature. X = x/δ is a dimensionless thickness. $\mathrm{Bi}_1 = \alpha_1 \cdot \delta / \lambda$; $\mathrm{Bi}_2 = \alpha_2 \cdot \delta / \lambda$ are the Biot numbers.

The obtained solution check are made by its substitution from an initial equation and corresponding boundary conditions (1). This equation dimensions check are made also. The made check results have confirmed the computed solution accuracy.

3 Results and discussion

The nine heat-transfer coefficients variations on the outer and the inner surfaces are used in the paper. α_{zh1} and α_{zh2} are reduced by 100, 10 4 and 2 times, respectively, from the first alternative to the forth alternative comparing with the seventh alternative. A_{zh2} is reduced by 10 times in the fifth alternative, but is increased by 10, 20 and 30 times, respectively, from the seventh to the ninth alternatives comparing with the calculation value [1, 2]. The made calculations results are showed below.

Table 1. Heat-transfer coefficients, Wt/(m^2·K) and heat conductivity coefficients, Wt/(m·K).

Temperature curves	1	2	3	4	5	6	7	8	9
α_{m1},	34.84	348.4	871	1742	3484	3484	3484	3484	3484
α_{m2}	11.03	110.3	275.8	551.5	11.03	110.3	1103	2206	3309
λ(St.10)	39.53	39.41	39.20	38.89	47.04	45.93	38.31	33.96	31.51
λ(St.20)	37.60	37.51	37.36	37.13	43.13	42.27	36.72	33.78	32.17
λ(12X1MF)	36.74	36.66	36.54	36.36	40.45	39.93	36.02	33.49	31.90
λ(St.15XM)	35.44	35.36	35.23	35.03	39.43	38.86	34.66	32.00	30.40
λ(St.12X18H12T)	23.52	23.59	23.71	23.87	20.64	21.20	24.14	25.44	26.12
λ(St.1X18H9T)	23.41	23.49	23.62	23.81	20.46	21.01	24.11	25.69	26.59

Table 2. Temperature differences and temperature drops, K.

steel grade		1	2	3	4	5	**6**	7	8	9
St.10	*А*	181.2	178.5	174.1	167.1	2.4	22.7	154.5	222.1	257.0
	Б	1.3	12.6	30.9	59.9	1.4	13.8	112.4	182.2	830.4
	В	572.5	563.9	549.9	528.0	751.2	718.5	488.1	350.7	270.6
St.20	*А*	181.2	178.4	173.8	166.5	2.4	22.7	153.5	221.8	258.6
	Б	1.3	13.3	32.4	62.5	1.5	15.0	116.5	183.0	830.4
	В	572.4	563.4	548.8	526.0	751.1	717.3	484.9	350.2	272.3
12X1MF	*А*	181.2	178.3	173.6	166.2	2.4	22.7	153.1	221.3	258.0
	Б	1.4	13.6	33.1	63.7	1.6	15.8	118.4	184.3	830.4
	В	572.4	563.2	548.3	525.1	751.0	716.5	483.5	349.5	271.6
15XM	*А*	181.2	178.2	173.3	165.7	2.4	22.7	152.1	218.8	254.2
	Б	1.4	14.0	34.3	65.9	1.7	16.3	122.3	190.6	830.4
	В	572.4	562.8	547.4	523.4	750.9	716.1	480.5	345.6	267.7
12X18H12T	*А*	181.0	176.5	169.6	159.2	2.4	22.3	142.1	205.4	242.0
	Б	2.1	20.9	49.8	92.9	3.2	29.3	164.1	225.1	830.4
	В	571.8	557.6	535.6	502.9	749.4	703.5	448.8	324.5	254.8
1X18H9T	*А*	181.0	176.5	169.5	159.1	2.4	22.3	142.1	206.0	243.4
	Б	2.2	20.9	50.0	93.2	3.2	29.5	164.2	223.6	830.4
	В	571.8	557.5	535.5	502.7	749.4	703.2	448.7	325.4	256.3

а is the temperature difference on the inner surface ($\Delta T_{wm1} = T(x = 0) - T_{wm1}$), K; *б* is temperature drop in the cylindrical wall ($\delta T = T(x = L) - T(x = 0)$), K; *в* is temperature difference on the outer surface ($\Delta T_{wm2} = T_{wm2} - T(x = L)$), K.

Analyzing the flat and cylindrical walls temperature curves showed in figure 1, as you can see the obtained results correspond to the physical meaning. The high temperatures are on the outer surface, but on the inner surface temperatures are below. The temperature curves from the first to the ninth alternatives are the same qualitatively, but there is their quantitative distinction.

The 1, 2, 3, 4 and 7 temperature curves vertical intercept is particularly useful. This flat wall intercept temperature is the same for all five temperature curves, in spite of the thermal physical properties distinction. At the same time this law may be observed for all six steel grades. This vertical intercept coordinates may be determined by using any two temperature curves, forming this vertical intercept, for various Biot numbers, found from expression (2).

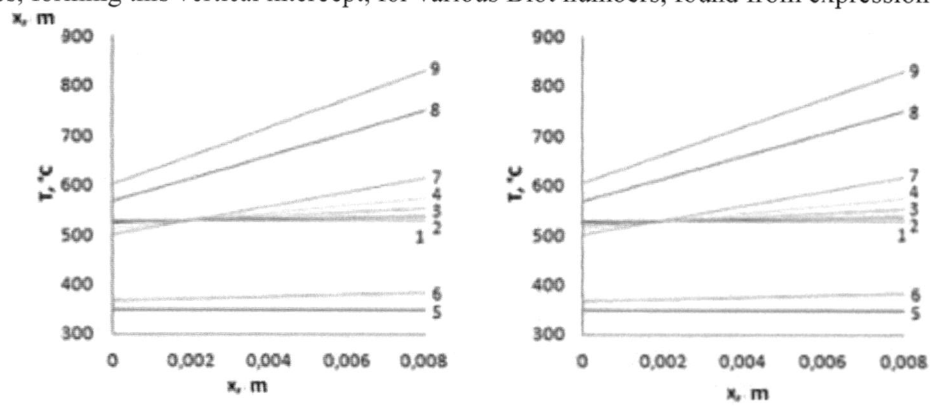

a) b)

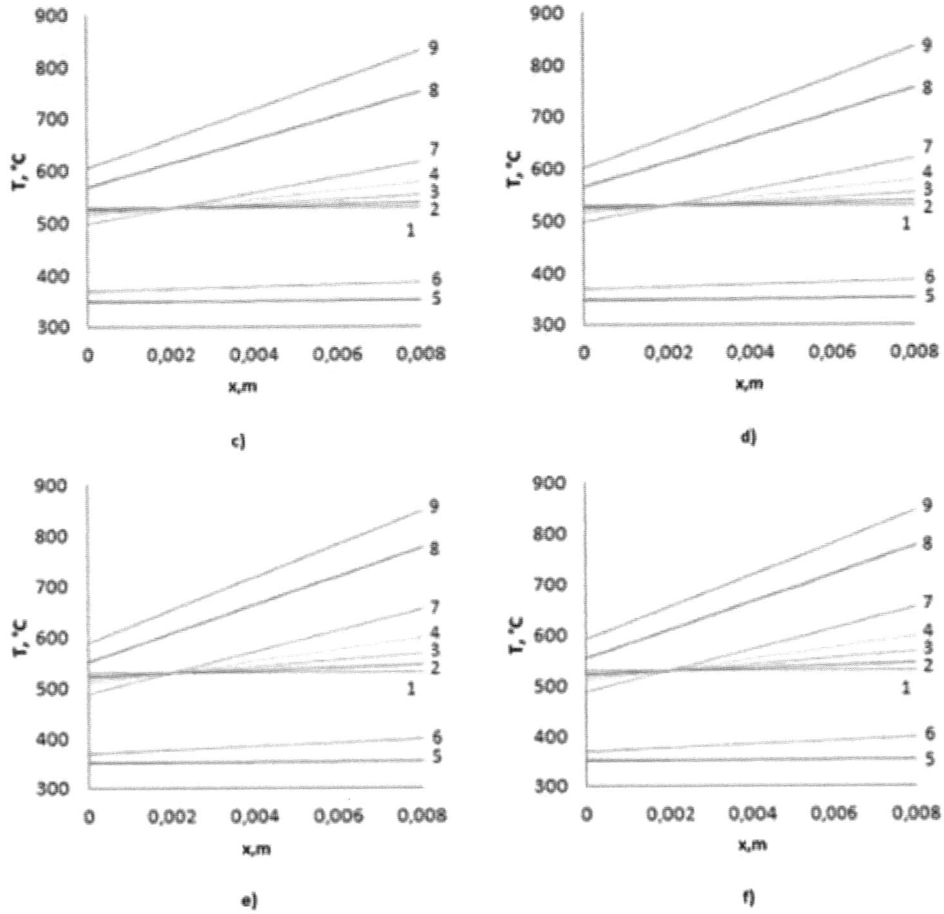

Fig. 1. Plane wall temperature distributions diagrams for steel, K: *a* – St.10; *b* – St. 20; *c* – 12Kh1MF; *d* – 15KhM; *e* – 12Kh18N12T; *f* – 1Kh18N9T.

These mathematical expressions are shown below:

$$x_{v.i.} = \frac{\left[\dfrac{1}{Bi_3\cdot\left(1+\dfrac{1}{Bi_4}\right)+1}\right]-\left[\dfrac{1}{Bi_1\cdot\left(1+\dfrac{1}{Bi_2}\right)+1}\right]}{\left[\dfrac{1}{1+\dfrac{1}{Bi_1}+\dfrac{1}{Bi_2}}\right]-\left[\dfrac{1}{1+\dfrac{1}{Bi_3}+\dfrac{1}{Bi_4}}\right]}\cdot\delta;$$

$$T_1(x_{v.i.}) = T_{m1}+\left(T_{m2}-T_{m1}\right)\cdot\frac{\dfrac{x_{v.i.}}{\delta}+\dfrac{1}{\mathrm{Bi}_1}}{1+\dfrac{1}{\mathrm{Bi}_1}+\dfrac{1}{\mathrm{Bi}_2}};$$

$$T_2(x_{v.i.}) = T_{m1} + \left(T_{m2} - T_{m1}\right) \cdot \frac{\dfrac{x_{v.i.}}{\delta} + \dfrac{1}{\text{Bi}_3}}{1 + \dfrac{1}{\text{Bi}_3} + \dfrac{1}{\text{Bi}_4}}.$$

$Bi_3 = \left(k\alpha_{m1}\delta\right)/\lambda_2$, $Bi_4 = \left(k\alpha_{m2}\delta\right)/\lambda_2$ are the Biot numbers. λ_1, λ_2 are the heat conductivity coefficients for 1 and 2 temperature curves, respectively, measured in Watt per meter per K. k is the proportional coefficient.

The cylindrical and flat walls vertical intercepts coordinates, determined according (3), are the following: $x_{v.i.} = 0{,}0019$ m, $T(x_{v.i.}) = 802{,}70$ K. For the vertical intercept forming laws determination make calculation, followed below.

Example

BKZ-220-100 steam generator platen superheater.

Steam temperature: $T_{m1} = 619.15$ K.

Flue gases temperature: $T_{m2} = 1374.15$ K.

Material: 1X18H9T steel.

Geometrical characteristics: $x = L_1 = 0$ m, $x = L_2 = 0.008$ m.

Heat-transfer coefficients: 1) $\alpha_{m11} = 348.4$ Wt/(m^2·K), $\alpha_{m21} = 110.3$ Wt/(m^2·K); 2) $\alpha_{m17} = 871$ Wt/(m^2·K), $\alpha_{m27} = 275.75$ Wt/(m^2·K); 3) $\alpha_{m18} = 3484$ Wt/(m^2·K), $\alpha_{m28} = 3309$ Wt/(m^2·K) (Table 1).

Heat conductivity coefficients: 1) $\lambda_1 = 23.49$ Wt/(m·K); 2) $\lambda_7 = 23.62$ Wt/(m·K); 3) $\lambda_8 = 26.59$ Wt/(m·K) [15, 16].

Now figure out the thermal flows ratios supplied to the outer surface and transferred from the inner surface of the cylindrical wall and heat-transfer coefficients and temperature differences ratios on the outer and the inner surfaces, respectively.

Solution

1) Now figure out the thermal flows densities supplied and transferred from the cylindrical wall (Table 2):

a) curve 1

$q_{s1} = \alpha_{m21} \cdot \Delta T_{m21} = 110.3 \cdot 557.5 = 61492.25$ Wt/m^2;

$q_{t1} = \alpha_{m11} \cdot \Delta T_{m11} = 348.4 \cdot 176.5 = 61492.60$ Wt/m^2;

b) curve 7

$q_{s7} = \alpha_{m27} \cdot \Delta T_{m27} = 275.75 \cdot 535.5 = 147664.13$ Wt/m^2;

$q_{t7} = \alpha_{m17} \cdot \Delta T_{m17} = 871 \cdot 169.5 = 147634.50$ Wt/m^2;

c) curve 8

$q_{s8} = \alpha_{m28} \cdot \Delta T_{m28} = 3309 \cdot 256.3 = 848096.70$ Wt/m^2;

$q_{t8} = \alpha_{m18} \cdot \Delta T_{m18} = 3484 \cdot 243.4 = 848005.60$ Wt/m^2.

2) Now figure out the thermal flows densities ratios supplied and transferred from the cylindrical wall:

$Q_{s7}/ q_{s1} = 2.40$; $q_{t7}/ q_{t1} = 2.40$; $q_{s8}/ q_{s1} = 13.79$; $q_{t8}/ q_{t1} = 13.79$.

3) Now figure out the heat-transfer coefficients ratios on the outer and the inner wall:

$\alpha_{m27}/ \alpha_{m21} = 2.5$; $\alpha_{m17}/ \alpha_{m11} = 2.5$; $\alpha_{m28}/ \alpha_{m21} = 30$, $\alpha_{m18}/ \alpha_{m11} = 10$.

4) Now figure out the temperature differences on the outer and the inner wall:

$\Delta T_{m27}/ \Delta T_{m21} = 0.960$; $\Delta T_{m17}/\Delta T_{m11} = 0.960$; $\Delta T_{m28}/ \Delta T_{m21} = 0.46$; $\Delta T_{m18}/\Delta T_{m11} = 1.38$.

The example demonstrates the temperature curves with the outer surface supplied heats and inner surface transferred heats equal ratios and platen superheater outer and inner surfaces temperature differences and heat-transfer coefficients equal ratios have vertical intercept.

Conclusions

There are temperature distributions vertical intercepts for steels discussed in the paper. It is possible if there are the outer surface supplied heats and inner surface transferred heats ratios equality and the superheater outer and inner surfaces temperature differences and heat-transfer coefficients ratios equality.

References

1. *Boilers thermal design. Normative method* (Prod. 3. St. NPO TsKTI Publ., Petersburg, 1998)
2. Yu.M. Lipov, Yu.F. Samoylov, T.V. Vilenskiy, *Construction and thermal design of steam boiler. Manual for graduate students* (Energoatomizdat Publ., Moscow, 1988)
3. B.E. Naymark, *Physical properties of steels and alloys, using in power engineering* (Energiya Publ., Moscow, 1967)
4. Yu.G. Dragunov, A.S. Zubchenko, Yu.V. Kashirskiy, *Steel and alloy guide* (Prod. 4., Moscow, 2015)
5. E.M. Kartashov, V.A. Kudinov, *Heat conductivity and applied thermoelasticity analytical theory* (LIBROKOM Publ., Moscow, 2012)
6. N.S. Piskunov, *Differential and integral calculus for technical colleges. vol. 2. Manual for technical colleges* (Prod. 13. Nauka Publ., Moscow, 1985)
7. E.A. Krasnoshchekov, A.S. Sukomel, *Heat transfer problem book Manual for graduate students. Prod. 4.* (Energiya Publ., Moscow, 1980)
8. *RD 10-249-98. Strength calculation norms of steam and hot water boilers and steam and hot water pipelines* (Gostekhnadzor Rossii Publ., Moscow, 1999)

MATEC Web of Conferences **72**, 01085 (2016) DOI: 10.1051/matecconf/20167201085
HMTTSC-2016

PERFORMANCE OF THE NUCLEAR POWER PLANT CONDENSING UNIT

Egor Pavelyev[1], and *Sergey* Lavrinenko[1,*]

[1]National Research Tomsk Polytechnic University, 634050 Tomsk, Russia

Abstract. In the article the analysis of the modernization of the condensing unit. Thermal calculations and design of the condensing unit of the Kalininskaya nuclear power plant was conducted. Extension of service life of the operating NPP power units is one of the most important tendencies of the modern stage of nuclear power engineering development and the most efficient funds investment for preserving generating capacities. The data indicate when modernizing the condensing unit, its cooling area was increased for about 4,000 m^2 due to increasing number of tubes. In addition, tube geometrical dimensions were changed; the wall thickness was reduced from 1 and 2 mm to 0.7 and 1.25 mm respectively.

1 Introduction

Service life of first generations of the nuclear power plants (NPP) of the Soviet Union and Russia was determined by absence of actual operational data on the NPP equipment wear and tear, and economic aspects providing necessity of compensating expenses on construction and operation of NPPs [1, 2], obtaining normative amount of profit and forming financial resources for subsequent decommissioning of NPPs.

Further operation experience of the Russian and foreign nuclear power plants provided basis for justifying a technical possibility of revising the power unit service life periods previously set. Comparison of expenses for service life extension of the operating NPPs with investments in construction of new generating capacities demonstrated obvious economic efficiency and attractiveness of investments in service life extension. As a result, long-term growth plans of the nuclear industry of Russia [3, 4] included extension of the operating NPP service life as one of the most important tendencies of the modern stage of nuclear power engineering development.

Processes of carbonate scale deposition on condenser heat-exchange surfaces (CHS) from oversaturated solutions of recirculating water of the closed-circuit cooling system (CCS) can be theoretically divided into several stages, separation of which is found quite impossible [5-9]. Increase of deposit amount in local overheating areas causes corrosion damages of the heat-exchange surfaces and can result in necessity to perform expensive works on repair and even replacement of a steam generator.

* Corresponding author: serg86@tpu.ru

2 Numerical analysis

For providing reliability and increasing steam generator lifetime, and improving reliability and efficiency of the secondary system power units of NPP with a pressurized water reactor PWR-1000 was made of the design and the thermal design of the condenser of the steam turbine K-1000-60/1500 of unit 1 of Kalinin NPP, using the following basic equations (1-6):

The equation of heat balance:

$$Q = G \cdot \Delta h_s \cdot \eta = D \cdot \Delta h_w \quad (1)$$

G – consumption of steam, kg/s;
$\Delta h_s, \Delta h_w$ – enthalpy of steam, water, kj/kg;
η – efficiency factor.

Number of tubes in the one water course:

$$N = \frac{4 \cdot D \cdot v_w}{\pi \cdot d_{in}^2 \cdot w_w} \quad (2)$$

v_w – volume of water, m^3/kg;
d_{in} – inner diameter of the tubes, m;
w_w – velocity of the water, m/s.

Heat transfer coefficient:

$$K = \frac{1}{d_a \cdot \left(\frac{1}{\alpha_1 \cdot d_{ou}} + \frac{1}{2 \cdot \lambda} \cdot ln\left(\frac{d_{ou}}{d_{in}}\right) + \frac{1}{\alpha_2 \cdot d_{in}}\right) + R} \quad (3)$$

d_a – average diameter of the tubes, m;
d_{ou} – outer diameter of the tubes, m;
α_1 – heat transfer coefficient during condensation of steam, W/(m^2K);
α_2 – heat transfer coefficient from tubes to water, W/(m^2K);
λ – thermal conductivity, W/(m∗K);
R – thermal resistance of a contaminated tube, (m^2K)/ W.

The surface area of the heat exchange:

$$F = \frac{Q}{K \cdot \Delta t_a} \quad (4)$$

Δt_a – average temperature difference, °C.

The length of the tubes:

$$L = \frac{F}{\pi \cdot d_a \cdot N} \quad (5)$$

The diameter of the tube bundle:

$$D_t = 1.05 \cdot d_{ou} \cdot b\sqrt{\frac{N}{\Psi}} \quad (6)$$

b – relative pitch;
Ψ – fill factor of the tube plate.

3 Practical implementation

After that JSC "Energospetsmontazh" performed works on replacement of the tube systems. These works were performed on the existing condenser of the steam turbine K-1000-60/1500 of power unit No. 1 of the Kalininsky NPP.

The works were performed during maintenance outage. The condensers with heat exchange tubes from copper-nickel alloy were replaced with condensers with corrosion-resistant steel tubes with weight 33 tons and over 8 m long each. In addition, metal structures for demounting condensers, trestle bridge for condenser replacement, and support assemblies with total weight 19 tons were mounted.

The developed tube systems and assemblies of the modular design condenser of turbine plant K-1000-60/1500 of power unit No. 1 of the Kalininsky NPP shall provide for:

- excluding copper-containing alloys from the NPP secondary system equipment;
- reliability, safety and performance of the condenser tube systems and assem-blies, with extended service life of the NPP equipment being taken into account;
- hydraulic closeness of the condenser tube systems due to fastening tubes in outer tube sheets by expanding and welding; elimination of thermal stresses in places of "cooling tube – tube sheet" attachment with relative motion of the shell and tubes;
- growth of the design vapor pressure in the condenser (of vacuum) with the respective gain of the turbine plant generator output on terminals of the generator with steam turbine K-1000-60/1500 manufactured by JSC "Turboatom", being confirmed by testing conducted by an independent specialized body.

The condenser modernization included replacement of the old tube systems (tubes from copper-nickel alloy grade MNZH5-1 and outer and intermediate sheets from carbon steel grade Steel 20) with new tube systems – of module No. 1…8 of every shell, with side walls and bottoms, with collected, expanded and welded tubes (tubes from corrosion-resistant stainless steel grade TP316L).

4 Comparative analyses

Among all the parameters determining the condensing steam turbine operation mode, vapor pressure at the end of expansion – spent vapor pressure in the condenser (vacuum level) – has the greatest influence on the turbine economical operation. Providing and maintaining the required vacuum in the condenser that ensures most economical operation of the turbine plant as a whole, is a very important task of the operating personnel. Therefore, increase of the vapor pressure in the condenser for 0.01 kg/cm^2 causes increase of a gross average heat rate of a unit for NPP turbine plant approximately for 1% for wet steam turbine plants - approximately for 2% of the steam turbine nominal capacity.

The main criteria of the condensing unit operation, characterizing the equipment performance under given conditions (rate of condensation per sq. m. of condenser surface, cooling water flow rate and temperature, etc.) are spent vapor pressure in the condenser and temperature drop at the condenser output δt (difference between spent vapor saturation temperature and cooling water temperature at the condenser output).

Table 1. Comparative analysis of condensers К-45600 and К-49210.

Characteristic	**Unit of measurement**	**Condenser type**	
		К-45610 before modernization	**К-49210 after modernization**
Type		Surface, single-pass, two-flow	Surface, single-pass, two-flow
Design cooling area in the condenser shells			
F1 F2 F3	m^2	12877 16366 16366	16403 16403 16403
Cooling tube length	mm	8970	8970
Cooling tube effective length	mm	8910	8910
Cooling tube type		"one-piece"	"welded"
Cooling tube assortment	mm	28x1.0x9000 28x2.0x9000	28x0.7x9000 28x1.25x9000
Number of cooling tubes in the condenser	pc.	57360/960, incl.:	61800/984, incl.:
Shell No. 1:			
28x1.0x9000/28x0.7x9000		16160	20600
28x2.0x9000/28x1.25x9000		304	328
Shell No. 2:			
28x1.0x9000/ 28x0.7x9000		20600	20600
28x2.0x9000/28x1.25x9000		328	328
Shell No. 3:			
28x1.0x9000/28x0.7x9000		20600	20600
28x2.0x9000/28x1.25x9000		328	328
Fastening of cooling tubes in the end (outer) sheets		expansion	expansion and welding
Number of passes / flows		1/2	1/2
Overall dimensions			Installation in the existing foundation
Number of outer sheets	pc.	2	2

4 Conclusion

Extension of service life of the operating NPP power units is one of the most important tendencies of the modern stage of nuclear power engineering development and the most efficient funds investment for preserving generating capacities.

The data given in table No. 1 allow making a conclusion that when modernizing the condensing unit, its cooling area was increased for about 4,000 m^2 due to increasing number of tubes in shell No. 1. In addition, tube geometrical dimensions were changed; the wall thickness was reduced from 1 and 2 mm to 0.7 and 1.25 mm respectively. It was necessary due to the lower thermal conductivity coefficient of stainless steel as compared to the copper-containing alloy previously used. Moreover, it should be noted that the condenser overall dimensions remained unchanged that is a very important parameter of modernization.

Acknowledgments

The research was realized with financial support of Minobrnauki of Russia in framework of FTP "Research and development in prior direction of scientific-technological complex of Russia in 2014-2020 years", unique R&D identifier RFMEFI58114X0001.

References

1. N.N. Galashov, S.A. Tsibulskiy, Power Technology and Engineering: Sci. J. **48**, (2015)
2. A.M. Zavalishen, S.L. Kim, V.V. Maltsev. The Higher Education Establishment Bulletin. Nuclear Industry **1** (2011)
3. Development Program for the nuclear power engineering of RF for 1998 – 2005 and for the period until 2010 approved by Decree No. 815 of the Government of RF on 21.07.1998
4. Development strategy for the nuclear power engineering in the first half of XXI century approved by the Government of RF on 25.02.2000
5. D. Hasson, M. Avriel, W. Resnick, Ind. Eng. Chem. Fundam. **7**, (1968)
6. J. Taborek, T. Aoki, R.B. Ritter. Chem. Eng. Prog. **68** (1972)
7. J. Taborek, T. Aoki, R.B. Ritter, et al. Chem. Eng. Prog. **68** (1972)
8. V.A. Ruppel, Yu.S. Tretyakova, S.V. Lavrinenko, et al. MATEC Web Conf. **37** (2015)
9. K. Larionov, S. Lavrinenko, V. Gubin, et al. XVth International Conference Linguistic and Cultural Studies: Traditions and Innovations (LKTI 2015), Procedia Social and Behavioral Sciences **206** (2015)

MATEC Web of Conferences **72**, 01086 (2016) DOI: 10.1051/matecconf/20167201086
HMTTSC-2016

SOLVING PROBLEM OF THERMAL CONDUCTION FOR PROVIDING STRENGTH OF ELECTRONIC UNITS ON THERMAL IMPACTS

Svitlana Petrashchuk[1,*], *Igor* Kovtun[1], and *Andriy* Voznyak[2]

[1]Khmelnitskiy National University, Drawing and Graphic Design Department, 29016 11 Institutska Str. Khmelnitskiy, Ukraine
[2]Khmelnitskiy National University, Radio Engineering and Communication Department, 29016 11 Institutska Str. Khmelnitskiy, Ukraine

Abstract. Paper represents new approach to physical and mathematical modeling of thermal fields distribution in sealed electronic units on the base of introducing "effective parameters" of the system. Effectiveness of developed mathematical model and its computer implementing are demonstrated on the example of standard unit sealed by compound.

1 Introduction

Modern development of cutting-edge electronic components and units by applying new composite materials and aimed at providing reliability and resource is impossible without applying mathematical modeling of thermal and mechanical processes, which run inside these components during their operation.

Developing technologies for manufacturing sealed electronic systems demands considering impacts which are produced on their strength by as physical and mechanical so thermal factors.

2 Solution

The paper [1] represents physical and mathematical models, which were obtained and researched, for radial and tangential stress calculation in material of electronic component

$$\sigma_{r_1} = \frac{E_1}{1-\mu_1}\left[-\frac{1}{r^2}\int_{R_1}^{r}\alpha_1\Delta t_1 r dr + \frac{r^2-R_1^2}{r^2\left(R_2^2-R_1^2\right)}\int_{R_1}^{R_2}\alpha_1\Delta t_1 r dr\right] + \frac{P_1R_1^2-PR_2^2}{R_2^2-R_1^2} - \frac{\left(P_1-P\right)R_1^2R_2^2}{r^2\left(R_2^2-R_1^2\right)}, \quad (1)$$

$$\sigma_{t_1} = \frac{E_1}{1-\mu_1}\left[\frac{1}{r^2}\int_{R_1}^{r}\alpha_1\Delta t_1 r dr + \frac{r^2-R_1^2}{r^2\left(R_2^2-R_1^2\right)}\int_{R_1}^{R_2}\alpha_1\Delta t_1 r dr - \alpha_1\Delta t_1\right] + \frac{P_1R_1^2-PR_2^2}{R_2^2-R_1^2} + \frac{\left(P_1-P\right)R_1^2R_2^2}{r^2\left(R_2^2-R_1^2\right)}, \quad (2)$$

and in material of the selected compound cylinder

* Corresponding author: svitlana.petrashchuk@gmail.com

$$\sigma_{r_2}=\frac{E_2}{1-\mu_2}\left[-\frac{1}{r^2}\int_{R_2}^{r}\alpha_2\Delta t_2 rdr+\frac{R^2-R_2^2}{r^2(R_3^2-R_2^2)}\int_{R_2}^{R_3}\alpha_2\Delta t_2 rdr\right]+\frac{PR_2^2-P_2R_3^2}{R_3^2-R_2^2}-\frac{(P-P_2)R_2^2R_3^2}{r^2(R_3^2-R_2^2)}, \quad (3)$$

$$\sigma_{t_2}=\frac{E_2}{1-\mu_2}\left[\frac{1}{r^2}\int_{R_2}^{r}\alpha_2\Delta t_2 rdr+\frac{R^2-R_2^2}{r^2\left(R_3^2-R_2^2\right)}\int_{R_2}^{R_3}\alpha_2\Delta t_2 rdr-\alpha_2\Delta t_2\right]+\frac{PR_2^2-P_2R_3^2}{R_3^2-R_2^2}+\frac{(P-P_1)R_2^2R_3^2}{r^2\left(R_3^2-R_2^2\right)}, \quad (4)$$

where μ_1, μ_2 – Poisson's ratios of component and compound materials correspondingly; E_1, E_2 – elasticity modulus of component and compound materials; α_1, α_2 – coefficient of linear expansion of component and compound materials; R_1, R_2 – internal and external radius of electronic component; R_2, R_3 – internal and external radius of compound; r – variable radiuses: $R_1 \le r \le R_2$, $R_2 \le R \le R_3$;

$\int_{R_1}^{r}\alpha_1\cdot\Delta t_1\cdot rdr$, $\int_{R_1}^{R_2}\alpha_1\cdot\Delta t_1\cdot rdr$, $\int_{R_2}^{r}\alpha_2\cdot\Delta t_2\cdot rdr$, $\int_{R_2}^{R_3}\alpha_2\cdot\Delta t_2\cdot rdr$ – temperature integrals; P – contact pressure:

$$P=\frac{2E_1E_2\left[\left(1+\mu_1\right)\left(R_3^2-R_2^2\right)\int_{R_1}^{R_2}\alpha_1\Delta t_1 rdr-\left(1+\mu_2\right)\left(R_2^2-R_1^2\right)\int_{R_2}^{R_3}\alpha_2\Delta t_2 rdr\right]}{E_2\left(R_3^2-R_2^2\right)\left[\left(1+\mu_1\right)R_1^2+\left(1-\mu_1\right)R_2^2\right]+E_1\left(R_2^2-R_1^2\right)\left[\left(1+\mu_2\right)R_3^2+\left(1-\mu_2\right)R_2^2\right]}, \quad (5)$$

Temperature drops Δt_1, Δt_2, present in formulas of thermal integrals (and stresses σ_{t_1}, σ_{t_2}) are defined as: $\Delta t_1 = t_{elcm}(r,\tau)-t_0$, $\Delta t_2 = t_{str}(r,\tau)-t_0$, where $t(r,\tau)$ – temperature of cylindrical surface of radius r in the moment of time τ, which is counted from the moment to transfer item from the constant temperature t_0 into the temperature t_1. For calculations t_0 is assumed to be the initial temperature of the body.

Using obtained formulas 1-5 for calculations required solving non-stationary problem of thermal conduction in compounded sealed unit, i.e. finding temperature distribution law over the volume of sealed unit from the moment as it is transferred from its highest temperature t_0 and to the lowest one t_1 (in our case from +70 deg C to -60 deg C) and until the whole volume receives the environmental (lowest) temperature. Solving this problem was offered by developing mathematical model for thermal fields calculation over the sealed unit volume on the base of introducing "effective parameters" [2], which consider inhomogeneity of the unit's material, since sealed unit has a complex structure and consists of many heterogeneous components: various electronic components and compound.

Effective parameters represent average physical and mechanical characteristics of all materials compiling the sealed unit. So temperature conductivity of the whole unit is defined as effective parameter by the formula:

$$\chi=\sum_{i=1}^{n}\upsilon_i\chi_i,$$

where $\upsilon_i=\frac{V_i}{V}$, (i = 1,2, ..., n), V_i – volume, occupied by i-component in sealed unit (m^3), V – total volume of the sealed unit, χ_i – temperature conductivity of the i-component (m^2ph).

Using effective parameter χ allows assuming sealed unit as homogeneous body made from mixture of compound and electronic components materials.

Compounded sealed unit having box shape with coordinate origin in the cross point of its diagonals and axes along its principal axes of inertia and sizes $-a \le x \le a$, $-b \le y \le b$, $-c \le z \le c$ (fig. 1) has the following solution for the problem of thermal conduction:

$$t = (t_0 - t_1) \cdot \Phi_1\left(\frac{a-|x|}{2\sqrt{\chi \cdot \tau}}\right) \cdot \Phi_2\left(\frac{b-|y|}{2\sqrt{\chi \cdot \tau}}\right) \cdot \Phi_3\left(\frac{c-|z|}{2\sqrt{\chi \cdot \tau}}\right) + t_1 \quad (6)$$

where t_0 – initial body temperature, t_1 – environmental temperature, Φ_1, Φ_2, Φ_3 – Laplace functions [3], τ – time.

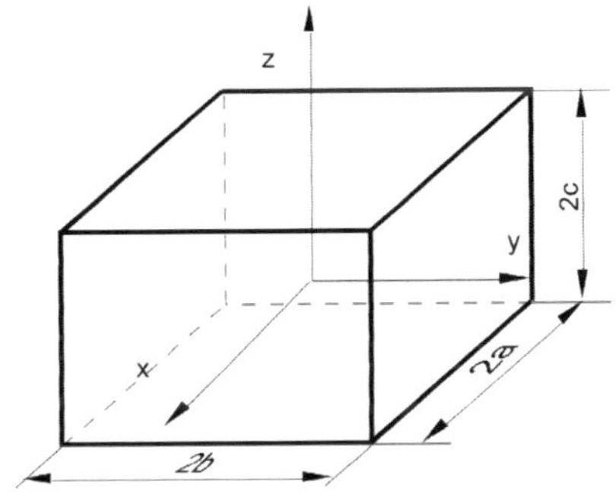

Fig. 1. Sealed unit represented as homogeneous rectangular box.

Obtained law of temperature distribution allows estimating temperature in any moment of time and in any point of the solid body. Moreover, this law makes possible to estimate the time of the complete cooling or heating a sealed unit.

The applied calculations of thermal fields in sealed units were to perform by using originally developed software on the base of MATHCAD. The developed software outputs on a display the two- and three-dimensional graphs of isotherms, which demonstrate thermal field in the selected section of a sealed unit in the specified moment of time, and also estimates time of complete cooling or heating sealed unit.

The thermal field for the sealed unit ZU 5.760.001 with sizes 21x14x8 mm was calculated for the thermal impact in temperature diapason from t_0 = +70 deg C до t_1 = -60 deg C.

The calculation was conducted at partial volume of compound (type TSK-25) V_1=1576 mm^3 and volume of electronic components equal V_2=776 mm^3. Physical and mechanical characteristics of resistor's and compound's materials are: temperature conductivity χ_1 = 0.18 mm^2/s, χ_2 = 0.414 mm^2/s; thermal conductivity K_1 = 0.837 W/(m·deg), K_2 = 0.33 W/(m·deg); thermal capacity C_1 = 0.253 W·h/kg·deg, C_2 = 0.395 W·h/kg·deg; density ρ_1 = 2000 kg/m^3, ρ_2 = 1600 kg/m^3.

Fig. 2 shows temperature distribution in horizontal section by the plane z = 0 (fig. 1). Isotherms are built in various moments of time up to the complete cooling.

The temperature calculation conducted in the set time interval resulted that actual cooling a sealed unit occurs in 1366 sec (22.7 min).

For experimental estimation of the complete cooling time the sealed unit ZU5.760.001 had been prepared by attaching microthermocouple XK_{68} before it was sealed by compound ESK-25. After compound had polymerized the unit was subjected to the range of thermal impacts (thermal cycling) in the temperature range from +70 deg C to -60 deg C. Indications of microthermocouple were recorded on the tape of light-beam oscillograph

H071.4M. As result of conducting 10 thermal cycles the average time of cooling sealed unit – 1160 seconds (19.33 min) has been measured, which deviation from estimated one is negligible. Noteworthy is that in manufacturing process when such units are subjected to technological testing, which is conducted in order to improve reliability of the product, the cooling (heating) time makes 1 hour: product is first placed in a heat chamber; then it is transferred to a cooling chamber (the transfer time is limited to few seconds); exposure period in chambers lasts until the thermal balance (1 hour). The total amount of thermal cycles equals three. Such testing is supposed to detect hidden design defects and reject potentially unreliable products.

Thus, modeling and estimating thermal field in sealed units and its time rate allows optimizing technological time for conducting thermal impacts in the process of their manufacturing. So for the case given above the exposure time for thermal impacts in temperature range from +70 deg C to -60 deg C is recommended to optimize and shorten down to 23-24 min as longest.

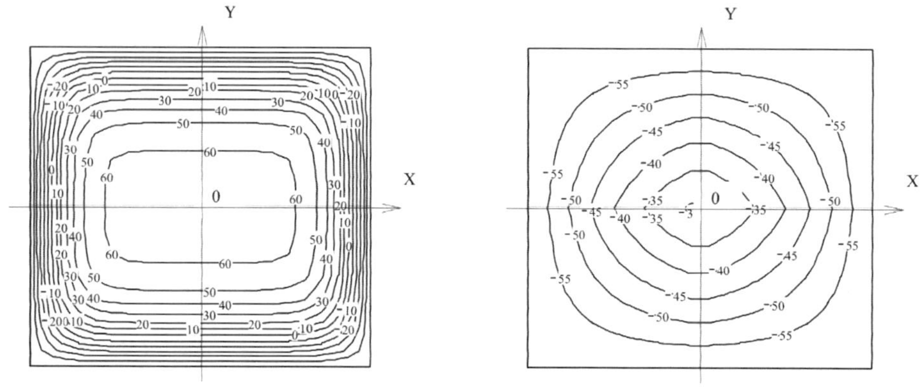

a) b)

Fig. 2. Isotherms for horizontal section of the sealed unit in time moments τ=8s (a), τ=200s (b).

After the problem of thermal conduction has been solved for thermal impacts in the volume of the whole sealed unit whose material was considered as a composite one and the shape of its individual components was not considered at all the further research of temperature distribution near the estimated electronic component and surrounding layer of compound appears to be reasonable. Therefore the problem of non-stationary radial distribution of temperature in electronic component and compound has been posed and solved; these results will be represented in the further papers.

3 Conclusion

The paper represents mathematical models for thermal fields calculation over the volume of the sealed unit on the base of introducing "effective parameters" which represent average physical and mechanical characteristics of all materials compiling the sealed unit. The developed model allows estimating thermal field distribution in time and over the volume of the solid body and also optimizing technological time for conducting thermal impacts in the process of manufacturing sealed units.

References

1. V. Royzman, S. Petrashchuk, I. Kovtun, A. Lokoshchenko, Journal of Vibroengineering **15**, 1 (2013)

2. Y. B. Zeldovich, J. Phys. Chem. **13** (1939)
3. G. Carslow, D. Eger *Thermal conduction of solid bodies* (Nauka, Moscow, 1964)

AN EXPERIMENTAL PROCEDURE TO ESTIMATE LIFETIME OF WATER DROP WITH GRAPHITE INCLUSION UNDER INTENSIVE VAPORIZATION WITH EXPLOSIVE BREAKUP

Jean C. Legros[1], *Vladislav* M. Gumerov[1], and *Maxim* V. Piskunov[1,*]

[1]National Research Tomsk Polytechnic University, Tomsk, Russia

Abstract. We developed an experimental setup equipped with tube furnace for continuous heating of heterogeneous drops at a constant temperature and high-speed camera to study characteristics of phase transitions at interfaces of these drops. Also, an experimental procedure was proposed to estimate time characteristics of processes occurring when heated heterogeneous drops in high-temperature environment. As an example, at temperature of heating of 1373 K, lifetime of 15 μl water drop with 1 mm solid inclusion made of natural graphite in high-temperature environment equals almost to 1 s. Experimental data also enabled to reveal minimum temperature at which intensive vaporization of 5 μl, 10 μl and 15 μl drops with inclusions in size of 2×2×2 mm proceeds with explosive breakup. This temperature equals to 803 ± 10 K depending on initial water volume in heterogeneous drops.

1 Introduction

In study [1], the authors discovered the phenomenon of intensive vaporization of heterogeneous (with single graphite inclusion) water drop with the explosive breakup of liquid layer around an inclusion. Moreover, the high temperature of combustion products was defined as a one of the primary condition for implementing the mentioned phenomenon. Using an air flow system enabled to vary temperature conditions in a tube with combustion products. However, the experimental procedure of research [1] includes the difficulty in temperature monitoring. This was quite complicated to measure the temperature at which intensive vaporization of heterogeneous drop with explosive breakup occurs. To solve this problem, the authors determined the temperature ranges conforming to the lifetimes of heterogeneous drops in high-temperature gas environment. Then, average temperatures of ranges were computed. Taking into account the key role of heating

[*] Corresponding author: piskunovmv@tpu.ru

temperature when reaching explosive breakup, determination of this temperature as an average value requires to be clarified.

In this paper, heating the water drop with single graphite inclusion proceeds at constant temperature of gas environment. Definitely, this will allow to determine more precisely temperature which conforms to intensive vaporization of drop with its explosive breakup. Thus, the purpose of this research is to develop and to test an experimental setup providing for continuous heating at constant temperature, as well as to determine temperature conditions for implementing intensive vaporization of heterogeneous drop with explosive breakup of water layer around single graphite inclusion. Moreover, finding the time characteristics of the processes under study is necessary.

Accurate measuring the parameters above mentioned is most critical in developing such technologies as firefighting by using flows with different additives [2], thermal treatment of waste water or industrial water [3], cooling heated surfaces [4–6], etc.

Importantly, the complex mechanism of phase transitions of homogeneous and heterogeneous liquids at temperatures more 1000 K remains poorly studied. Nevertheless, the operational conditions of the modern power engineering equipment, as well as other applications, in which working cycle is based on the processes of evaporation (condensation) and boiling, are characterized by such high temperatures.

2 Experimental setup and procedure

The main distinctive feature of the given research is an application of experimental setup equipped with tube furnace. fig. 1 presents the schematic illustration of such setup. The used furnace is a one by Nabertherm (maximum temperature of heating is 1573 K; length of ceramic tube is 250 mm). Built-in type S thermocouple enables to monitor temperature of heating with measuring error ±1 K and response time not more than 5 s. The furnace is also equipped with regulator *9* which allows keeping a constant temperature of heating.

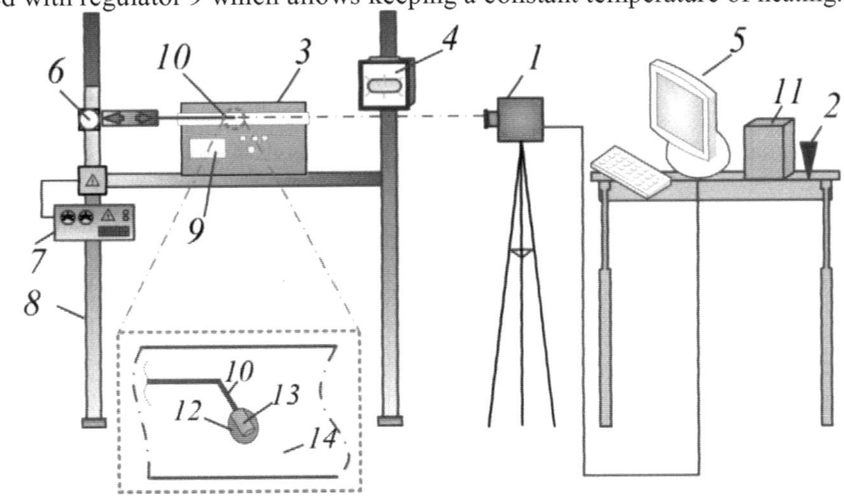

Fig. 1. Schemetic illustration of the experimental setup: *1* – high-speed camera, *2* – dosing device, *3* – tube furnace, *4* – spotlight, *5* – personal computer (PC), *6* – moving mechanism, *7* – electric power supply of moving mechanism, *8* – mount, *9* – heating temperature regulator, *10* – ceramic rod, *11* – weighing system, *12* – water drop, *13* – solid inclusion, *14* – tube body.

Owing to the design of furnace *3*, we can apply high-speed video camera *1*. It is arranged to the right of furnace *3* in such a way that axis of symmetry of furnace tube is in line with axis of lens (see fig. 1). To the left of furnace *3*, moving mechanism *6* transfers water drop *12* with graphite inclusion *13* to tube body *14*. Importantly, motion of heterogeneous drop inside furnace continues till center of tube. This distance equals to approx. 125 mm from the entrance of tube. Such approach provides for reaching temperature, which is displayed on screen of regulator *9*, with required accuracy. During preliminary tests, we measured temperature at the entrance of tube, it was almost two times less than the preset one. Thus, we became to perform experiments at the center of tube.

High-speed video shooting is possible by Nikon macro lens with focal length 200 mm and color CMOS camera *1* (Phantom V411; maximum resolution is 1280×800 px; maximum sample rate is $6\cdot10^5$ fps). Nevertheless, in the given research, sample rate is comparatively not high (up to 10^3 fps at resolution 512×512 px or 512×256 px). However, this speed is quite enough to study features and behavior of heterogeneous drops during evaporation and boiling in detail, as well as to determine time characteristics of the investigated processes. Maximum errors in determining time characteristics are equal to from 0.01 s to 0.001 s depending on the preset sample rate in Phantom Camera Control software. In particular, this software provides for time measurement. fig. 2 presents a scheme with times which are determined in the study.

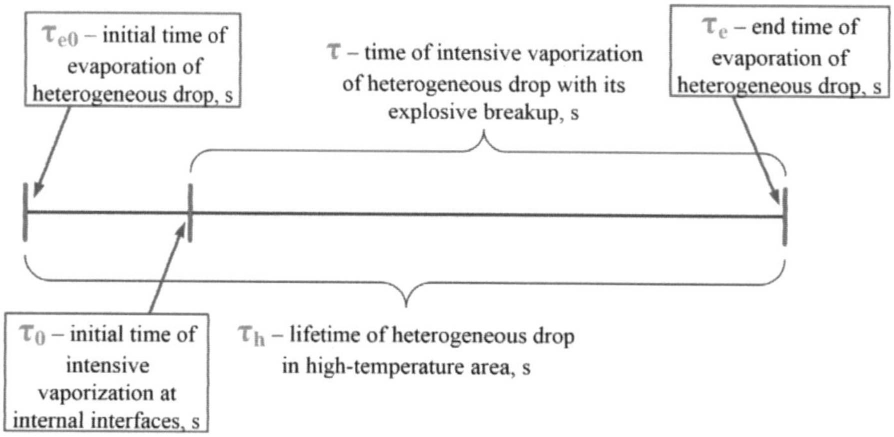

Fig. 2. Schematic illustration of times which are characteristic of the processes observed in the study.

There are several particularities concerning the scheme in fig. 2. The initial time of evaporation of a heterogeneous drop conforms to the time moment when it stops at the center of tube. The end time of evaporation of a heterogeneous drop is determined when analyzing video data during their processing. This time characterizes the moment when water layer of heterogeneous drop is completely evaporated.

Let us briefly consider additional equipment of the experimental setup. Spotlight *4* provides suitable amount of light during high-speed shooting. Dosing device *2* (minimum volume – 5 μl, maximum volume – 50 μl, error ± 3 %) enables to take water of required quantity out of the vessel. In all the experiments, the volumes of water drops are 5 μl, 10 μl or 15 μl. Importantly, the critically important conditions for performing experiments are full covering of solid inclusion by water droplet and using of new inclusion for each test. In the research, material of solids is a natural graphite [7–9]. We manufacture large inclusions in size: 2×2×1 mm, 2×2×2 mm and 2×2×3 mm; shape: cube and parallelogram. These inclusions are fixed on ceramic rod *10*. Then dosing device *2* lowers water drops on them. Also, we weigh solids by analytical balance *11* (Vibra AF 225DRCE; maximum and minimum masses – 220 g and 0.001 g, respectively; error – not more than 0.5 %) and monitor their masses during tests.

The experimental procedure of the given study includes the following main steps:
(1) Tube furnace 3 is heated up to preset temperature.
(2) Mechanism *6* moves water drop *12* with inclusion *13* fixed on rod *10* in tube body *14*.
(3) During motion of heterogeneous drop, high-speed recording is activated.

(4) After completion of water evaporation, ceramic rod *10* with inclusion *13* moves back.
(5) After preparation of the next heterogeneous drop, the steps (2)–(4) repeat.
(6) Analysis of time characteristics and temperature measurement occurs after performing necessary number of series of experiments.

3 Results and discussion

For better understanding processes which occur when heated heterogeneous water drop, fig. 3 demonstrates snapshots of heterogeneous drop according with time characteristics which are shown in fig. 2.

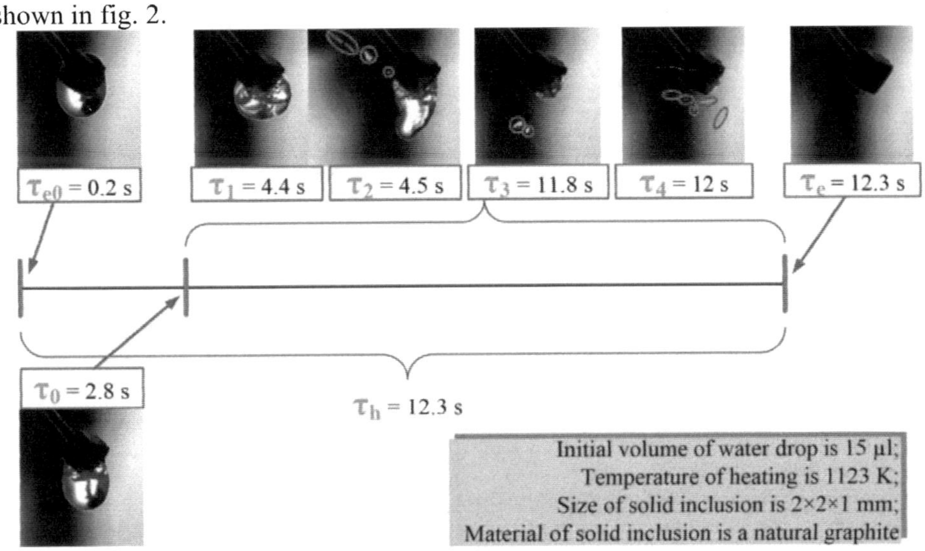

Fig. 3. Snapshots of heterogeneous drop according with time characteristics presented in fig. 2.

In this particular case, lifetime τ_h is about 12 s. Actually, time characteristics strongly depend on temperature of heating. In the experiments, they vary in the wide range for different configurations of heterogeneous drops (size of solids and water volume). As an example, fig. 4 illustrates lifetimes of 10 μl and 15 μl drops with solids in size of 2×2×1 mm when heated in the range from 1073 K to 1373 K.

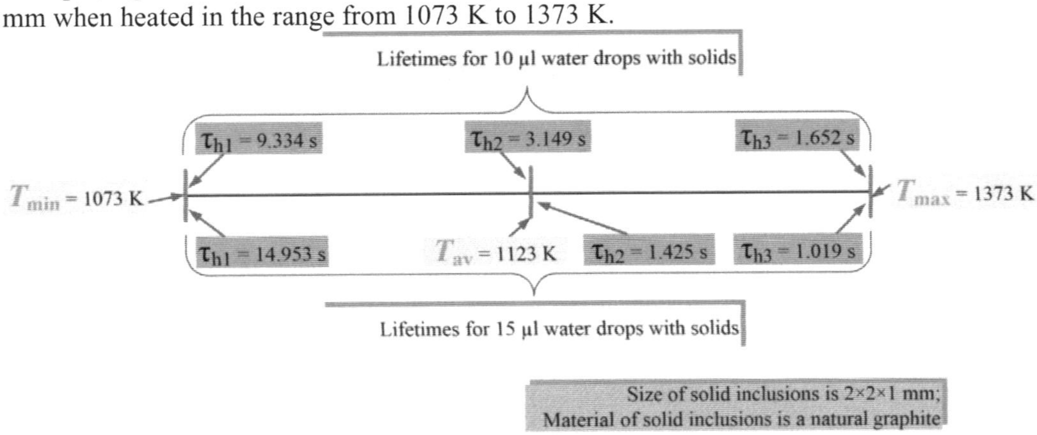

Fig. 4. Lifetimes of 10 μl and 15 μl water drops with graphite solids in size of 2×2×1 mm when heated in the range from 1073 K to 1373 K: τ_{h1}, τ_{h2}, τ_{h3} are lifetimes of heterogeneous drops at 1073 K, 1123 K, 1373 K, respectively.

For water drops with solids of other size, lifetimes at the same temperatures (see fig. 4) differ insignificantly.

In this research, we determined minimum temperatures of heating at which intensive vaporization of 5 μl, 10 μl and 15 μl water drops with solids in size of 2×2×2 mm proceeds with explosive breakup how it is shown in snapshots of fig. 3. Fig. 5 presents schematic illustration of operating temperature range with an indication of the onset temperature of intensive vaporization of heterogeneous drops with their explosive breakup. T_{ex5}, T_{ex10} and T_{ex15} are temperatures of the onset of intensive vaporization for heterogeneous drops with initial water volume of 5 μl, 10 μl and 15 μl, respectively. Note that for mentioned volumes of water, these temperatures are quite close.

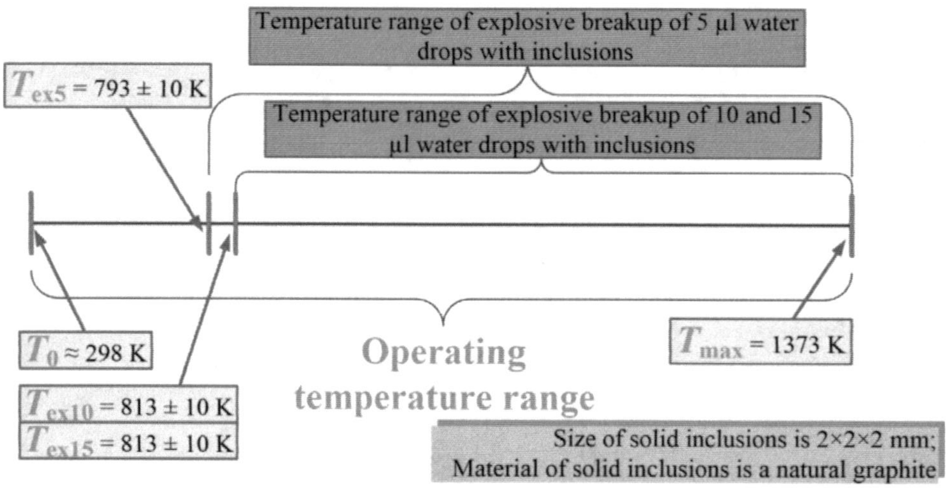

Fig. 5. Onset temperatures of intensive vaporization of heterogeneous drops with their explosive breakup.

The experimental results can be relevant for practical applications where phase transitions occur during operation at high temperatures of environment. In particular, we believe that time characteristics of the observed processes and temperature measurements performed during experiments will contribute to develop energy-efficient solutions in such high-temperature technologies as treatment of industrial and waste water, firefighting by flows of liquids with solid additives, etc. For these technologies, the heat transfer enhancement in the area of interaction between liquids with solids and high-temperature environment is a critically important problem. Thus, investigations on mechanisms, conditions and characteristics of phase transitions of heterogeneous drops at high temperatures are extremely necessary.

4 Conclusions

The experimental setup providing for continuous heating of heterogeneous drops at constant temperature was developed and tested. For studying phase transitions at interfaces, these experimental setup was equipped with high-speed camera.

- Intensive vaporization of 5 μl, 10 μl and 15 μl drops with inclusions in size of 2×2×2 mm occurs at temperatures of heating not less than 793 K, 813 K and 813 K, respectively.
- In the paper, lifetimes in high-temperature environment were analyzed for 10 μl and 15 μl drops with inclusions in size of 2×2×1 mm. At maximum temperature of 1373 K, lifetime of heterogeneous drop with initial water volume of 15 μl equals almost to 1 s. At 1073 K, lifetimes are much more both for heterogeneous drops with water volume of 10 μl, and for drops with volume of 15 μl. They equal to about 10 s and more.

Acknowledgments

This work was supported by the Scholarship of the President of the Russian Federation (SP-1049.2016.1).

References

1. G.V. Kuznetsov, M.V. Piskunov, P.A. Strizhak, Int. J. Heat Mass Transf. **92**, 360 (2016)
2. O.P. Korobeinichev, A.G. Shmakov, V.M. Shvartsberg, A.A. Chernov, S.A. Yakimov, K.P. Koutsenogii, V.I. Makarov, Fire Saf. J. **51**, 102 (2012)
3. A.Yu. Varaksin, High Temp. **51**, 377 (2013)
4. S.-H. Hsu, Int. J. Heat Mass Transfer. **86**, 65 (2015)
5. I. Sher, Appl. Therm. Eng. **36**, 219 (2012)
6. B.-R. Fu, Int. J. Heat Mass Transf. **95**, 206 (2016)
7. D.R. Lide, *CRC Handbook of Chemistry and Physics* (CRC Press, Boca Raton, FL, 1998)
8. A. Nayer, *The Metals Databook* (McGraw-Hill, New York, 1997)
9. *Metals Handbook, Volume 2 – Properties and Selection: Nonferrous Alloys and Special-Purpose Materials* (ASM International, 2001)

NUMERICAL MODELING OF THE EFFECT OF THAWING OF SOIL IN THE AREA OF PLACING TANKS FOR STORAGE FUEL OF THERMAL POWER PLANTS AND BOILER

V.Yu. Polovnikov[1,*], *V*.Yu. Ozhikenova[1], and *F*.T. Makhsutbek[1]

[1]National Research Tomsk Polytechnic University, 634050 Tomsk, Russia

Abstract. This paper describes the numerical modeling of heat transfer in the area placing of the tank for storage fuel of thermal power plant and boiler with considering the influence of thawing of the soil. We have established that the thawing of the soil in the area of placing of the tank for storage fuel of thermal power plant and boiler have little effect on the change of heat loss.

1 Introduction

Oil is used by more than 200 thermal power plants (TPP) in Russia as a primary or backup fuel. The main problem of its use - is ensuring uninterrupted supply of boilers by heated and filtered oil in the winter. Thus on heating oil during its preparation for burned (in order to reduce viscosity is exponentially dependent on the temperature) is expended considerable part of the generated electricity of TPP. Reducing these costs is one of the important tasks. Its solution is possible at optimal thermal conditions in fuel oil storage and systems supply that fuel [1].

The research of thermal modes of different objects (a residential building, a hot water pipes, a fuel storage tank et al.) is an important problem [2].

Currently, the method of calculation and design of a fuel storage tank does not take full account of all processes in a tank, that occur during the preparation of a fuel oil for combustion.

One of the most promising approaches to research of different equipment is the use of a computer programs. They take into account various effects and processes in the system of heat transfer.

The objective is the mathematical modeling of the thermal loss of fuel storage tank and numerical analysis of thermal conditions of the tank.

* Corresponding author: polovnikov@tpu.ru

2 PROBLEM STATEMENT

We consider a typical vertical cylindrical steel underground tank. figure 1 shows a schematic representation of the field solution of the problem. For the domain under consideration (fig. 1) we solve a 2D linear and stationary problem of heat transfer in conditions of thawing of the soil .

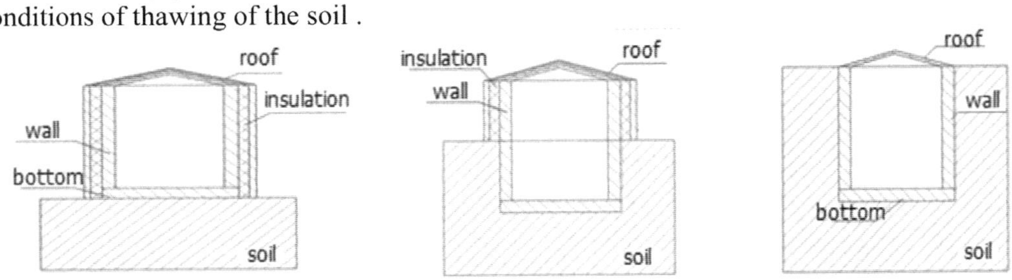

Fig. 1. Schematic representation of the field solution of the problem.

Formulating the problem, we used the following assumptions:

1. The heat transfer processes in the internal and the external environment are disregarded.
2. Thermophysical characteristics of materials used in the analysis are constant and known values.
3. There is an ideal thermal contact conditions at the boundaries.
4. The heat in the decision domain (fig. 1) is transferred only by conduction;
5. The interface between white and frozen ground has a constant temperature of 273 K.

The listed assumptions, on the one hand, do not impose constrains of principle on the physical model of the system (fig. 1), but, on the other hand, allow one to simplify in a certain manner the algorithm and method for solving the posed problem.

3 Mathematical model

The processes of heat transfer to the object under consideration (fig. 1) will be described by the following equation:

$$\nabla^2 T_{\mathrm{ins}} = 0; \tag{1}$$

$$\nabla^2 T_{tg} = 0; \tag{2}$$

$$\nabla^2 T_{\mathrm{fg}} = 0. \tag{3}$$

$$T_{\mathrm{ins},1} = T_{\mathrm{oil}} = \mathrm{const.} \tag{4}$$

$$\lambda_{\mathrm{ins}}\,\mathrm{grad}(T_{\mathrm{ins},2}) = \lambda_{tg}\,\mathrm{grad}(T_{tg,2});\ T_{\mathrm{ins},2} = T_{tg,2}; \tag{5}$$

$$\lambda_{fg}\,\mathrm{grad}(T_{fg,3}) = \lambda_{tf}\,\mathrm{grad}(T_{tg,3});\ T_{tg,3} = T_{\mathrm{fg},3}. \tag{6}$$

$$-\lambda_{\mathrm{g}}\,\mathrm{grad}(T_{\mathrm{g},4}) = \alpha(T_{\mathrm{g},4} - T_{\mathrm{ex}}); \tag{7}$$

$$\mathrm{grad}\,(T_{\mathrm{g}})=0,\ x_{\rightarrow}\pm\infty,\ y_{\rightarrow}-\infty; \tag{8}$$

4 Method of solution and Initial data

The system of equations (1)–(12) was solved by the software package COMSOL Multiphysics using the module General Heat Transfer.

In carrying out a numerical simulation assumed that the surface temperature of the inner tank Tm = 60 ° C. Ambient temperature tex = -40 ° C.

Table 1 shows the thermal characteristics of materials and substances used in research [2,3].

Table 1. Thermal properties of materials.

Material	Foundation	Thermal insulation	Ground			
			Clayey		Sandy	
			Frozen	Thawed	Frozen	Thawed
λ, [W/(m·K)]	1.54	0.046	1.1	1.3	2.3	3.7
c, [J/(kg·K)]	887	840	1231	959	1486	1005
ρ, [kg/m^3]	2200	150	1700	1700	2000	2000

5 Results of numerical simulation

The main results of numerical modeling of thermal conditions of the system under consideration (fig. 1) are listed in Table 2 and in fig. 2.

Table 2 lists the results of numerical experiments of the heat loss of a vertical cylindrical steel tank (Q1 is the heat loss excluding thawing of soil and Q2 is the heat loss, taking into account the soil thawing).

Table 2. Results of the calculation of heat loss.

Tank	α, [W/(m^2·K)]	Clayey soil			Sandy soil		
		Q_1, [W]	Q_2, [W]	δ, %	Q_1, [W]	Q_2, [W]	δ, %
Underground	5	1674.78	1653.66	1.26	2032.78	1986.31	2.30
	10	1964.89	1944.54	1.03	2341.20	2298.87	1.81
	15	2097.16	2075.04	1.06	2481.31	2438.14	1.74
	25	2219.69	2198.03	0.98	2610.27	2568.69	1.59
Semi-underground	5	1978.62	1962.83	0.8	2261.58	2237.19	1.08
	10	2298.00	2282.66	0.23	2596.17	2573.79	0.86
	15	2440.07	2424.01	1.15	2743.09	2721.59	0.78
	25	2574.73	2556.47	0.59	2879.79	2856.26	0.82
Aboveground	5	2461.77	2455.46	0.26	2574.17	2571.52	0.1
	10	2832.79	2826.54	0.22	2948.62	2941.88	0.23
	15	2993.21	2987.41	0.19	3110.68	3110.40	0.01
	25	3140.79	3135.03	0.18	3259.91	3258.75	0.04

Numerical analysis results indicate that the presence of thawing of soil in the container placement area for the storage of fuel thermal power plants and boiler houses does not lead to significant changes in the thermal regime of the object.

Figures 2 show the typical temperature fields in the area of storage tanks for fuel thermal power plants and boiler plants , taking into account the freezing of sandy and clay soil.

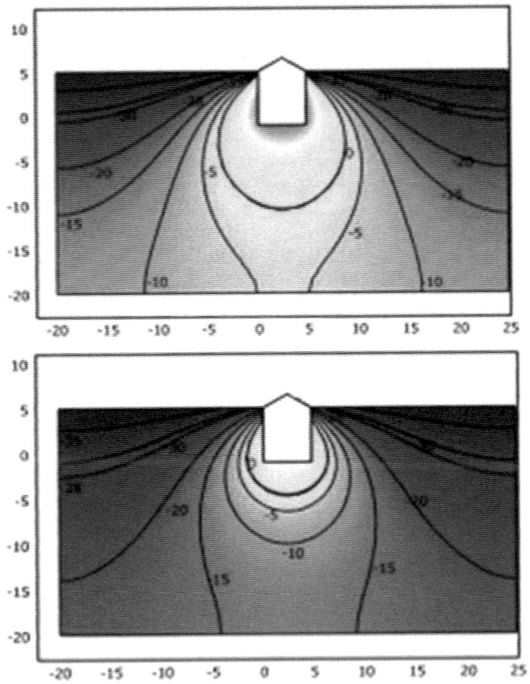

Fig. 2. Typical temperature field in the zone of the storage tanks for fuel thermal power plants and boiler.

6 Conclusion

The numerical analysis of thermal conditions and heat losses of tanks (tab.2 and fig.2) for storing fuel thermal power plants and boiler in the conditions of freezing and thawing of the soil, the area defined frozen and thawed areas. The relative change of the heat losses of tanks simulation results in terms of soil freezing in their distribution area is 0.98 - 2.30% for underground and 0.04 - 0.26% for the land, 0.23 - 1.15% for the semi-underground reservoirs. In addressing this problem it found that the inclusion of soil thawing in the area of the tank placement for storing fuel thermal power plants and boiler houses does not lead to significant changes in the thermal regime of the object.

Notations

T – temperature, K; λ – coefficient of thermal conductivity, W/(m·K); α – the heat transfer coefficient, W/(м^2·K).
Indexes: ins – a layer of heat insulation; g – ground; tg – thawed ground; fg – frozen ground; oil – oil; ex – external; 1 – the inner surface of the tank insulation; 2 – the boundary between "shell of tank – ground"; 3 – the boundary between "thawed ground – frozen ground"; 4 – the boundary between "soil – environment".

References

1. S.A. Kuznetsova EPJ Web Conf. **82** (2015)
2. V.Yu. Polovnikov, E.V. Gubina, J. Eng. Phys. Thermophys. **87** (2014)
3. V.Yu. Polovnikov, E.S. Glazyrin, EPJ Web Conf. **82** (2015)

DETECTION OF HEAT-STRESSED FURNACE WALL'S AREAS BASED ON RESULTS OF NUMERICAL SIMULATIONS OF THERMAL FLOWS

V.V Popov[1], *T*.S. Taylasheva[1,*], and *V*.M. Lebedev[2]

[1]Tomsk Polytechnic University, 634050 Tomsk, Russia
[2]Omsk State Transport University, 644046 Omsk, Russia

Abstract. In this paper, a numerical study of the formation of nitrogen oxides in the combustion chamber based on the model created by Mitchellom and Terbellom. The distribution of furnace temperature and the concentration of nitrogen oxides, as well as a comparison of numerical results with the data of field experiment.

1 Introduction

In the field of minor power engineering using steam for heating and industrial purposes the steam boilers of DKVR type manufactured by Biysk Boiler Plant became highly popular in the middle of last century. The line of these boilers includes several types with steam producing capacity from 2.5 to 20 tons of steam per hour and using various types of fuel with grate-fired furnaces of different types - for combustion of solid fuel; or with gas-and-oil burners for liquid and gaseous fuel combustion.

In the course of the long term operation the majority of boilers have been upgraded and modified. The main reasons for this upgrade and modification were as follows: the need to transfer the boiler to some other fuel beyond its design parameters; the attempt to make the boiler more cost-effective; the change in the steam parameters and changes in the boiler load (due to the changes in the technology of the industrial steam use); the need to fix the consequences of operation disorders.

2 Description of the model and the object of research

The DKVR-20 boiler has been investigated which has been designed for natural gas flaring. The boiler is equipped with upper and lower drums located along the axis of the boiler. The drums are connected by bent steam generating tubes spread inside the drums and forming advanced convection tube bank. The circulation system of the boiler is fairly complicated: the staged evaporation system is used, which from one hand expands the ranges of natural

* Corresponding author: taylasheva@tpu.ru

waters used for boilers, on the other hand reliability parameters of circulation system require special attention in case of long term operation of the boilers.

Due to this it is necessary to evaluate the changes in heat absorption by heating surfaces, reliability parameters of water-steam circuit hydrodynamics, temperature mode of pipes within evaporation elements. In order to evaluate the above the first stage includes variation calculations of conditions for fuel combustion, heat exchange and air mechanics in the furnace, results of these calculations are presented in this paper.

One of the techniques of variation evaluation of heat-mass-exchange conditions in the furnace is mathematic simulation of furnace processes, and the current level of this method development allows solving this task numerically. The FIRE 3D software [1, 2] is used in this paper after its adaptation to small furnaces and gaseous fuel.

Numerical simulation allows achieving the following for the DKVR boiler:

- Obtaining 3D visual models for each parameter being evaluated,
- Visualizing the measurement of each parameter in plane and shifting section planes in depth, width and height within following ranges: X – from 0 to 10 m, У – from 0 to 5.5 m, Z – from 0 to 2.7 m.
- Building up distribution dependence curves for temperature, radiation density, heat flow, flow rate of fuel-air mixture, pressure, oxygen concentration, density of gases,
- Defining extremes of all above parameters,
- Varying boiler load within ranges from 15 to 100 %.

3 Results and discussion

Fig. 1 shows temperature and aerodynamic fields in the furnace of the DKVR-20-23 boiler with 100% load in the vertical section through the axis of two burners (z=1.43 m). In this section one can observe the uneven temperature field, which is due to burners location on the front side of the boiler in two rows (y_1 = 1.22 m; y_2 =2.3 m). This distribution of temperatures is defined by two horizontal flows which fill the middle and lower part of the burner in the even manner, and it correlates with the distribution of fuel-air mixture.

In the lower part of the burner at the level of y=0.5 m the values of flow rate of combustion gases are within ranges 4-4.5 m/sec. The field of low flow rates is observed in the upper part of the furnace by the front screen and ceiling connection. The average flow rate in this field makes 1-2.5 m/sec. This is due to the fact that the whole flow is drawn to the horizontal gas duct through the furnace throat. The maximum gas flow rate achieves 15 m/sec near the burners.

Operation conditions of the boiler have been investigated within the ranges from 15 to 100 % of nominal steam producing capacity through measuring the gas flow rate in case of regular excess of air in the furnace. In case of the load at 50 % and 100 % the fuel supply was through two burners and if the load is 15 % the fuel is supplied only through the lower burner.

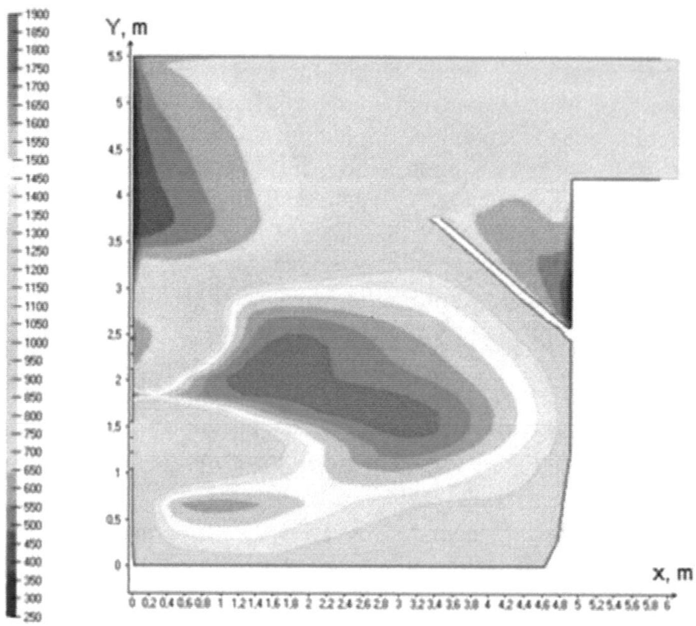

Fig. 1. Temperature field in the section as per burners' axis.

Figure 2 shows the measurement of the average value of the mean heat flow regarding the furnace height as per horizontal section. Changes in the average heat flow have certain typical features. For the minimum selected load of the boiler (15 % out of the nominal one) the maximum of the average heat flow is at the level of 0.6-0.8 m from the furnace floor (fig. 1, Curve 1). The heat flow has a similar distribution when the boiler operates at 50% of its nominal load with the lower burner being in operation (fig. 1, Curve 2). If the installed burners are equally loaded (fig.2, Curves 3 and 5) the field of maximum values shifts to the level between axes of burners and makes 653.4 and 795.7 kW/m^2 respectively. Should the upper burner be involved into operation (fig. 2, Curve 4) the maximum value of the average heat flow makes 562.6 kW/m^2.

The obtained result allows analyzing the fuel combustion process, intensity of heat flows in the furnace not only under various operation modes of the boiler but in the local fields of the furnace. The opportunity to evaluate the heat physical conditions in the near wall region for any part of the screen regarding the height and depth of the furnace is especially interesting.

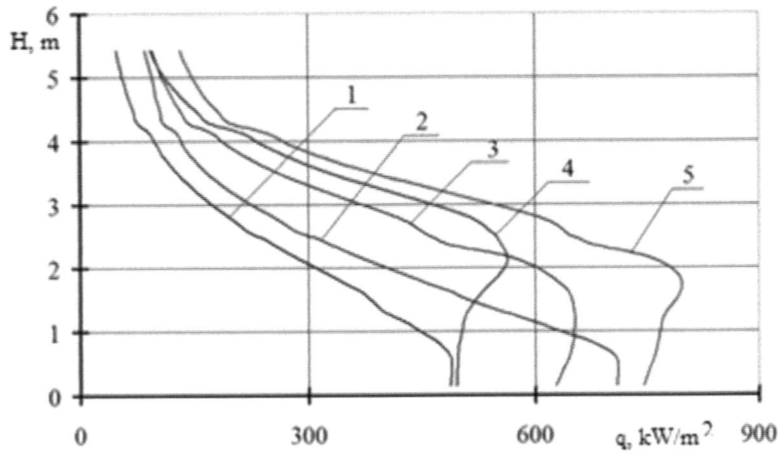

Fig. 2. The changes in the average heat flow regarding furnace height for various operation modes of the boiler: 1 – 15 % load out of the nominal one, the lower burner is operating; 2 – 50 % load out of the nominal one, the lower burner is operating; 3 – 50 % load out of the nominal one, two burners are operating; 4 – 50 % load out of the nominal one, the upper burner is operating; 5 – 100% load of the boiler.

4 Conclusions

This model can be used for defining temperature fields, fields of flow rates and heat flows when evaluating the reliability of operation modes of the DKVR boilers, including the head and hydraulic calculations and assessment of operation conditions for furnace screens.

References

1. N.V. Vizgavlyust, A.V. Starchenko, A.V. Gil, T.S. Taylasheva, EPJ Web Conf. **82** (2015)
2. T.S. Taylasheva, Bulletin of the TPU **314**, 4 (2009)

MATEC Web of Conferences 72, 01090 (2016) DOI: 10.1051/matecconf/20167201090
HMTTSC-2016

SIMULATION OF ROOM CLIMATE OF PUBLIC BUILDINGS

Maria Prorokova[1,*], and *Vyacheslav* Bukhmirov[1]

[1]State Educational Institution of Higher Professional Education Ivanovo State Power University named after V.I. Lenin, 1153003 Ivanovo, Russia

Abstract. The article presents the details of modeling of heat exchange and mass transfer in a room in the formation of a microclimate. The mathematical model is implemented in the program ANSYS Fluent and is used to predict the climate parameters after the implementation of energy saving measures in the building. Verification of the mathematical model by comparing the experimental data with the results of the measurement of microclimate parameters of the experiment.

1 Introduction

In [1, 2, 3] shows that in the assessment of the effectiveness of energy-saving measures for non-production buildings (residential, public and administrative), an important task is to predict climate parameters that determine the comfort of the premises after the implementation of energy saving measures. Currently, scientists are actively used by a number of software and computer systems, allowing defining the parameters of the environment in the room after making changes to the technical characteristics of the building or changing the operating mode of the power equipment. The most common software and computing systems in the field of modeling of heat and mass transfer processes in the formation of microclimate buildings include Ansys Fluent, STAR-CD, FlowVision, GasDynamicsTool (GDT), SimuLink.

In order to modeling of convective and radiative heat transfer in the formation of room climate using a distributed model (fully or partially) and lumped parameters. Models with distributed parameters include the continuity equation, motion, heat balance, the balance of impurities in the form of differential partial recorded for air as a viscous compressible fluid [4, 5].

2 Mathematical model

In the Ivanovo Power Engineering University (ISPU), has developed a mathematical model that describes the movement processes, heat and mass transfer on the assumption that the air in the rooms for a stay of people is a three-component mixture of gases (air directly, carbon dioxide and water vapor). In this case, the mathematical modeling of the formation

* Corresponding author: prorokova_mv@list.ru

of a microclimate is based on the continuity equation for a mixture of gases, transfer of momentum equation and the equation of conservation of energy [6, 7, 8].

In the equations of the mathematical model, the following assumptions:

1) in the equation of energy conservation compressibility accounted only the main component of the mixture - the air, since the concentration of carbon dioxide and water vapor in the gas mixture are small;

2) in the gas mixture is not diathermancy indoor environment;

3) the air flow is carried out uniformly over the entire surface of the air-permeable element enclosing structure room.

The mathematical model of indoor climate averaged parameters used gas mixture, in particular:

- the density of the gas mixture, kg/m^3:

$$\rho_m = \gamma_f \rho_f + \gamma_{CO_2} \rho_{CO_2} + \gamma_{H_2O} \rho_{H_2O}; \tag{1}$$

- dynamic viscosity mixture, Pa·s:

$$\mu_m = \gamma_f \mu_f + \gamma_{CO_2} \mu_{CO_2} + \gamma_{H_2O} \mu_{H_2O}. \tag{2}$$

where γ_i – bulk concentration of the i-th component of the mixture, m^3/m^3; ρ_i – density of i-th component of the mixture, kg/m^3; μ_i – viscosity component of the mixture, Pa·s. Indexes f, CO_2, H_2O, respectively designated air, carbon dioxide and water vapor.

The mathematical model is different, first of all, to use when setting the boundary conditions of actual air space, which is determined experimentally [9]. Secondly research shows [12], the presence of heated or cooled surfaces in the room can have a significant impact on the human feeling of comfort, in accordance with this radiation heat transfer is taken into account by introducing a special source term in the energy equation [4, 5].

For a description turbulent properties of the gas mixture used k-ε turbulence model [10, 11].

In [2] as an additional factor has been proposed level of climate comfort, which includes a coefficient comfort human thermal state (CCH), which was determined by solving the heat of human balance, as well as amendments to the radiative cooling, the asymmetry of the heat flow and air quality medium (the presence of impurities) in the room. Check the adequacy of the developed mathematical model is executed by comparing the CCH values obtained from the experimental results and the implementation of a mathematical model in the environment of ANSYS Fluent. To do this in the classroom ISPU an experiment was conducted on the Measurement of microclimate parameters (temperature, relative humidity, air mobility), as well as the temperature of the enclosing surfaces and air spaces.

Further based on the results of the experiment coefficient of thermal comfort the human condition was calculated. And then in the software and computer system created a geometric model of experimental facilities and to implement the proposed mathematical model of heat transfer in this area. According to the simulation results of the determined value of the coefficient of thermal comfort the human condition. In the experiment, and the mathematical modeling of controlled and the average air temperature in the room.

3 Results and discussion

As a result of experimental and theoretical study of the microclimate in the educational university auditorium was obtained value of the coefficient of thermal comfort condition of the person according to the experiment - 0.412, and the results of the simulation - 0,329. The average air temperature in the test room according to the experiment was equal to -

26.2 ^{0}C, and in the simulation results - 24.45 ^{0}C. Comparison of experimental results and mathematical modeling indicates the adequacy of the developed mathematical models of indoor climate in terms of determining the coefficient of thermal comfort the human condition.

4 Conclusions

1. Was developed a mathematical model of heat and mass transfer processes in the the premises of human presence, which allows on the basis of information on the actual air exchange, predict indoor climate parameters.

2. The adequacy of the proposed mathematical model was verified by comparing the experimental and calculated values of human comfort factor of the thermal state and the average air temperature in the room classroom.

References

1. Y.A. Tabunschikov, AVOK **5** (2008)
2. V.V. Bukhmirov, M.V. Prorokova, Bulletin ISPU **4** (2015)
3. P.O. Fanger, AVOK **4** (2003)
4. L.D. Landau, E.M. Lifshitz, *Continuum Mechanics: Hydrodynamics and elasticity theory* (Dep. publ tehnikoteoret. lit., OGIZ, Moscow, 1944)
5. S. Patankar, *Numerical methods for solving problems of heat transfer and fluid dynamics* (Energoatomisdat, Moscow, 1984)
6. ANSYS FLUENT User's Guide. Software Release Version 14 (2011)
7. Y.A. Tabunschikov, M.A. Brodach, *Mathematical modeling and optimization of thermal efficiency of buildings* (AVOC-PRESS, Moscow, 2002)
8. A.Y. Snegirev *High-performance computing in engineering physics. Numerical simulation of turbulent flows: Proc. allowance* (Publishing house of the Polytechnic. University Press, St. Petersburg, 2009)
9. V.V. Bukhmirov, M.V. Prorokova, Herald of the Russian Ministry of Energy **1,** 1 (2015)
10. I.A. Belov, S.A. Isaev, *Simulation of turbulent flows: Proc. Allowance* (Balt. state. tehn. University Press, St. Petersburg, 2001)
11. A.A. Yun, B.A. Krylov, *Calculation and simulation of turbulent flows with heat exchange, mixing, chemical reactions, and two-phase flows in the software package FASTEST-3D* (Publishing house of the Moscow Aviation Institute, Moscow, 2007)
12. A. Machkashi, L. Banhidi, *Radiant heating* (Stroyizdat, Moscow, 1985)

MATEC Web of Conferences **72**, 01091 (2016) DOI: 10.1051/matecconf/20167201091
HMTTSC-2016

MATHEMATICAL MODELING OF SALT LENS FORMATION PROCESS IN THE CATHODE ASSEMBLY OF ALUMINUM ELECTROLYSIS CELLS

Anton Orlov[1,*], and *Alexander* Proshkin[1]

[1]Siberian Federal University, Kirensky st. 26, Krasnoyarsk, Russia

Abstract. In this paper, with the help of mathematical modeling we investigate growth under saline lens bottom blocks of the cathode aluminum cell. The penetration of the melt in the hearth of the cathode of the device is one of the main causes of premature shutdown of cells, because it leads to excessive heat loss and destruction of the hearth. The study was built mathematical model of unsteady filtration and heat the porous body with an incompressible viscous fluid. With the help of the developed mathematical model we studied the dynamics of growth of the lens. Research has identified the rate of melt penetration into the hearth of the cathode unit in the course of its life. The calculation results were compared with the field data.

1 Introduction

Lifting bottom blocks in industrial aluminum cell caused by several mechanisms. One of the key mechanisms is the formation of the salt of the lens between the bottom blocks and refractory material. The growth of this lens leads to accelerate the erosion of the cathode blocks and consequently decrease the life of the cell. The mechanism of their formation is not fully understood, but the main reason is probably the reaction between the liquid electrolyte is filtered through pores with carbon and molten aluminum, have got through the cracks [1]. Mechanical disruption of the cathode blocks, leading to stop electrolysis will occur if overpressure exceeds the strength of the porous material. Along with this ascent of the cathode blocks may cause substantial deformation of blooms. The greatest deformation normally occur at the point where the blooms coming out of the bottom blocks and where there is also a high enough temperature to make easily deformable steel [2, 3].

Figure 1 clearly shows how the hosted lining materials impregnated with electrolyte. Impregnating the cathode electrolyte block rushes to the refractory materials where there are physical and chemical transformations of these materials. In particular, albite is formed which

* Corresponding author: an_riv@mail.ru

is highly viscous glassy substance during these reactions. Formation albite is the primary mechanism that prevents further impregnation materials. When albite forming reaction front reaches the solidus isotherm, the melt cannot penetrate into the bath hearth. He begins to accumulate between the cathode blocks and refractory material, forming a so-called lens. Growth leads to lens ejection cathode blocks leading to their destruction.

Fig. 1. Spent a cathode assembly (Part).

The aim of this work is the calculation of the growth rate of lens and identifying ways to reduce it by means of mathematical modeling. Mathematical modeling is a valuable tool that can help to optimize the design of the cathode liner device that will minimize lens growth during the entire period of operation of the cell, thus increasing its service life.

During continuous operation of the cell occurs in the electrolyte filtering hearth cathode. The hydrostatic pressure and capillary forces electrolyte displaces air in the pores of lining materials. The melt is introduced through the carbon block and responds within a certain range with a refractory material to form a dense vitreous and impervious barrier at 800-840 °C, which stops or significantly hinders the further introduction of the hearth. In the calculation of this process is taken into account an increase in the viscosity of the electrolyte as it passes through the refractory layer.

2 Statement of the problem

This task is considered a two-phase filtration of incompressible immiscible liquids in a heterogeneous non-deformable porous body. It is believed that the dynamic viscosity of the phases are constant, fluid flow is slow and there is no phase transition.

Table 1. Initial data.

Temperature of the electrolyte	°C	956
Temperature of the environment	°C	20
Temperature of the electrolyte solidus	°C	820
Heat transfer coefficient: melt / unit cover / air	 $W/(m^2 \cdot {}^oC)$ $W/(m^2 \cdot {}^oC)$	 830 25
The density of the electrolyte	kg/m^3	2100
The viscosity of the electrolyte	Pa·s	$2{,}79 \cdot 10^{-3}$
The permeability of the bottom blocks	m^2	10^{-13}

The computational domain is a half portion of the cathode unit of the cell geometry is shown in figure 2. The initial data are given in Table 1. The weakest point in the design of the cell are printed interconnects and peripheral joints, so the simulation was carried out taking into account these features. The calculation takes into account all the components of an aluminum electrolytic this design, including interconnect joints. electrolyte filtration occurs through the porous structure of the hearth block and interconnect joints.

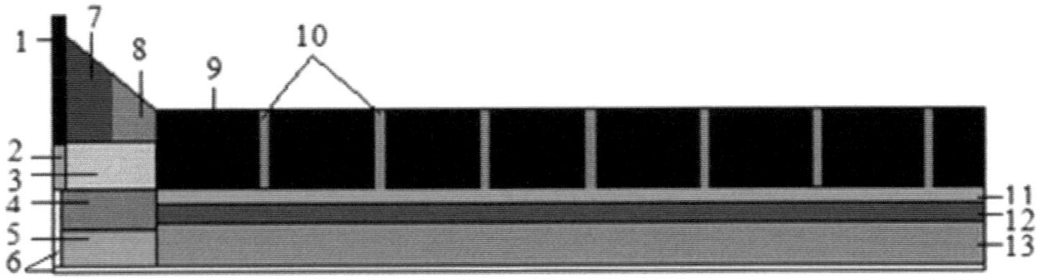

Fig 2. The geometry of the computational domain: 1 - SiC block, 2 - compensator, 3 - refractory concrete, 4 - refractory, 5 – heat insulation, 6 - leveling backfilling, 7 - inserter, 8 - peripheral seam, 9 - bottom block, 10 - interblock seams, 11 and 12 - refractory, 13 - heat insulation.

3 Mathematical model

Mathematical model of non-stationary filtration process and porous body heat transfer with incompressible viscous liquid consist of following equation system.

- liquid-flow equation in porous medium

$$\vec{w} = -[D] grad\ P \tag{1}$$

- liquid continuity equation

$$\frac{\partial \varepsilon \rho}{\partial \tau} + div(\rho\, \vec{w}) = 0 \tag{2}$$

- energy equation for liquid

$$(\rho\, c)_G \varepsilon \frac{\partial T_G}{\partial \tau} + div((\rho\, c)_G\, \vec{w} T_G - \lambda_G grad\ T_G) = \alpha_V (T_s - T_G) \tag{3}$$

Where the first term of equation on the left describes the increase of enthalpy, second term of equation heat convection, and on the right described heat transfer between liquid and solid body:

- heat-transfer equation in the porous medium

$$(\rho\, c)_s (1-\varepsilon) \frac{\partial T_s}{\partial \tau} + div(-[\lambda]_s grad\ T_s) = \alpha_V (T_G - T_s) + q_V \tag{4}$$

where:

DOI: 10.1051/matecconf/20167201091

$$[\lambda]_G = \begin{bmatrix} \lambda_1 & 0 & 0 \\ 0 & \lambda_2 & 0 \\ 0 & 0 & \lambda_3 \end{bmatrix}$$

- effective heat-conduction tensor, recorder in the coordinate system (X, Y, Z) c coincides with the main axes of the heat conduction, and q_V - volumetric heat generation in porous material;

- equation of condition for liquid

$$\rho_G = \rho(p, T_G, \tau) \tag{5}$$

In this study, to determine the boundaries of the free electrolyte fluid used in the method of cells (VOF), proposed by Hirt and Niholsom [4]. This method is considered to be one of the most effective methods for solving problems with the free surface of the liquid. Application of this method to solve similar problems can be found in [5, 6]. The idea of this method is that the liquid and gas are considered as a single two-component medium, and the spatial distribution of the phases within the computational domain is defined by a special marker functions *F (x, y, z, t),* which value is the volume fraction of liquid phase in the computational cell as follows:

$$f(x) = \begin{cases} 0, & if\ the\ cell\ is\ empty \\ 1, & if\ the\ cell\ is\ filled \end{cases}$$

and $0<F(x,y,z,t)<1$ – if the phase boundary passes through cell.

Since the free surface of the electrolyte moves with liquid, then tracking free boundary moving in space accomplished by solving transport equation of liquid phase volume ratio in cell.

$$\frac{\partial \varepsilon F}{\partial t} + \vec{w} \cdot \nabla \varepsilon F = 0 \tag{6}$$

Here $\vec{w}$ - velocity vector of two-phase medium, found by solving the system of equations of hydrodynamics.The density and molecular viscosity two-component medium are considered by the volume fraction of the liquid in the cell mixture according to the rule:

$$\rho = \rho_1 F + (1 - F)\rho_2,$$

$$\mu = \mu_1 F + (1 - F)\mu_2.$$

Here ρ_1, μ_1 – density and viscosity of electrolyte, ρ_2, μ_2 – respectively, density and viscosity of air. Thus obtained density values ρ and viscosity μ are included in the equation of motion and determine the physical properties of the two-phase medium.

4 Results and discussion

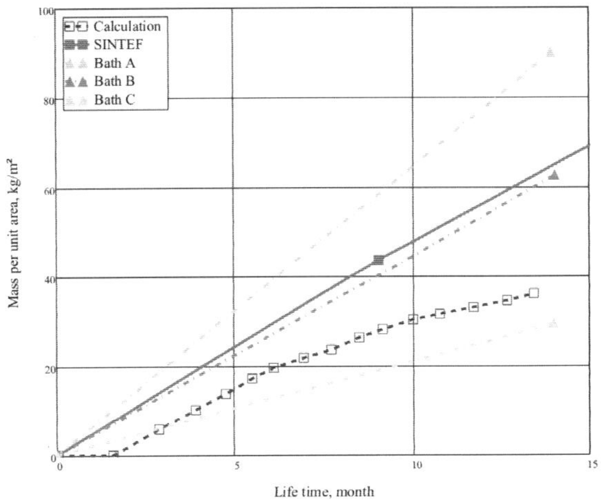

Fig. 3. Dynamics of the melt in the cathode assembly.

Figure 3 shows a comparison of the calculated results to the rate of arrival of the melt into the refractory material of the cathode device with the experimental data obtained at the Norwegian Institute of Technology (SINTEF) and field data on the melt flow of some industrial baths (A, B, C). Since the baths are different in size, comparison of data produced by the melt mass, coming through the unit area. It can be seen that the calculation agrees qualitatively with the field data. Also on the calculated line it is seen that the active electrolyte penetration into the refractory material begins after two months of operation of the cell. During this time, the electrolyte is impregnated only bottom blocks. Modeling in this way makes it possible to predict only 1,5 years of use of baths.

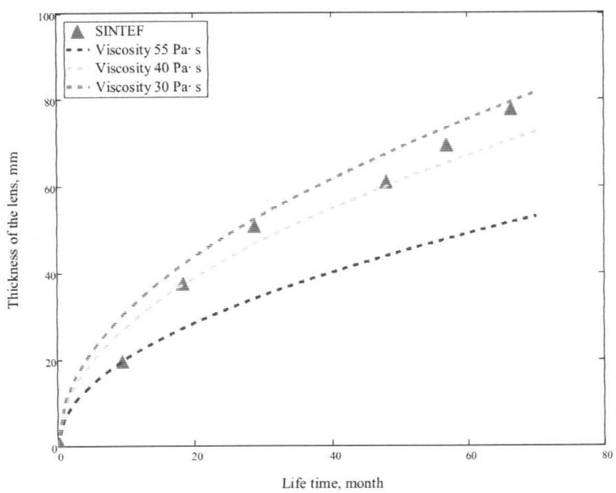

Fig 4. The dynamics of growth of the lens.

Figure 4 shows a comparison of the calculation results with the experimental data for SINTEF lens growth with electrolyte penetration speed in the hearth of the cell. The viscosity of the electrolyte was varied between 30 and 55 Pa·s. It is evident that the calculations carried out in this way allows a fairly high quality characterize the dynamics of the growth of the lens

5 Conclusions

A mathematical model of the lens of the growth process. The comparison of the calculation results with the field data. With the help of the developed mathematical models of the dynamics of the impregnation of the lining of the cathode electrolyte device. Studies have shown a dynamic melt infiltration hearth bath during operation of the cell. Further results will evaluate the thermal processes in the electrolyte penetration electrolyzer cathode lining materials, and enable the liner to optimize the design to reduce costs and increase the thermal lifetime of the cathode device.

The developed mathematical model allows to simulate thermal processes and electrolyte penetration education lens up to 1.5 years of operation of the cell. To predict the dynamics of the growth of the lens after one and a half years require adjustment model, which would allow to carry out changes in the computational domain in the simulation.

References

1. M. Sorlie, H.A. Oye, *Cathodes in Aluminium Electrolysis. 2nd edition* (Aluminium-Vertag, Düsseldorf, 1994)
2. S. Pietrzyk, P. Palimaka, W. Gebarowski, Arch. Metall. Mater. **59** (2014)
3. Femi Fakoya, Donald Picard. *Characterisation of the material behaviour of cathode steelcollector bar at high temperatures and low stress levels* (Light Metals, 2014)
4. C.W. Hirt, B.D. Nichols, J. Comput. Phys. **39** (1981)
5. A.A. Shebeleva, A.V. Minakov, A.A. Yagodnitsina, V.G. Andyuseva. MATEC Web Conf. **23** (2015)
6. A.S. Orlov, A.V. Minakov, A.V. Proshkin. MATEC Web Conf. **23** (2015)

MATEC Web of Conferences **72**, 01092 (2016) DOI: 10.1051/matecconf/20167201092
HMTTSC-2016

THEORETICAL AND NUMERICAL CHARACTERIZATION OF THE THERMAL PHYSICAL PROPERTIES OF CARBON CERAMIC MATERIAL

Pavel V. Prosuntsov[1], and *Nikita* Yu. Taraskin[1, *]

[1]Bauman MSTU, Moscow

Abstract. The article presents models representative volume element of the porous carbon-ceramic material. Simulating of heat transfer process in a representative volume element is carried out and thermal characteristics of material are determined by a settlement way. Influence of characteristics of material structure – its porosity and type of reinforcement – on values of heat-physical characteristics is investigated. It is shown that in the field of high temperatures (more than 1500 T) the defining role in the course of heat transfer in high-porous (P> 40%) materials belongs to radiation heat exchange in a pores.

1 Introduction

Russia, the USA, EU and China are currently developing manned spacecraft, such as Orion, Dragon, PTKS for the Earth orbit, Moon and Mars missions. One of the most important and complex scientific and technological challenges is to create a reliable and weight efficient thermal protection system.

Thermal protection coating (TPC) used for the Vostok, Soyuz, Mercury and Apollo reentry modules was based on ablative materials. In this case, a significant part of the energy from high enthalpy gas flow to the front surface is absorbed due to disintegration and ablation of specialized polymer composite materials, while heat transfer to the frame is blocked with a thermal protection layer. Ablative TPC have proved highly reliable, however, they are characterized by significant linear density and non-reusability [1].

New generation spacecraft industry requires multi-use reentry vehicles and TPC. Such TPC can be manufactured on the basis of carbon ceramic composite materials (CCCM) with the operating temperature of 2000°C, density of 2000 kg/m^3 [2]. The load-bearing capacity of CCCM is achieved by means of high strength carbon fiber and SiC ceramic matrix serving as antioxidant. The possibility of creating TPC based on CCCM and an extra layer of thermal protection material [3, 4] is currently under consideration.

* Corresponding author: trzzz@mail.ru

TPC linear density can be reduced if porous CCCM are used, where carbon fibers are coated against oxidation with a thin (1-2 μm) layer of SiC. It can be assumed that the CCCM heat transfer coefficient can be significantly decreased by regulating the material porosity and selecting the rational layout of the reinforcement fibers.

Theoretical rationale of the possible characteristics of porous CCCM is presented in [5], however, only a limited set of material structure variants was analysed, excluding the radiative heat transfer in the porous CCCM volume. Thus it remains relevant to research the possible thermal physical properties of porous CCCM with 1000 to 2000 kg/m^3 density.

2 Analysing the effect of the CCCM layers layout

The model of the porous CCCM structure was based on microimages from an optical microscope (fig. 1). It was assumed that the porous CCCM comprised separate orthogonal layers of carbon fibers 6 mm in diameter with 1 μm SiC ceramic coating.

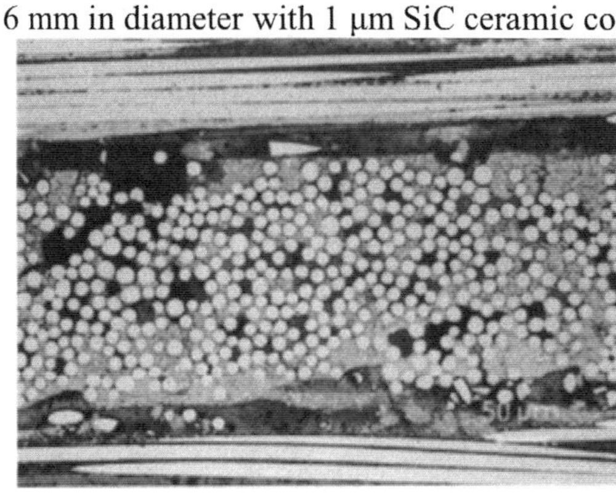

Fig. 1. Image of carbon ceramic material microstructure [2].

MSC.Digimat software package was used to create geometric models of the representative element for the 18% porous volume (fig. 2). The models comprised various combinations of the carbon layers thicknesses in orthogonal directions.

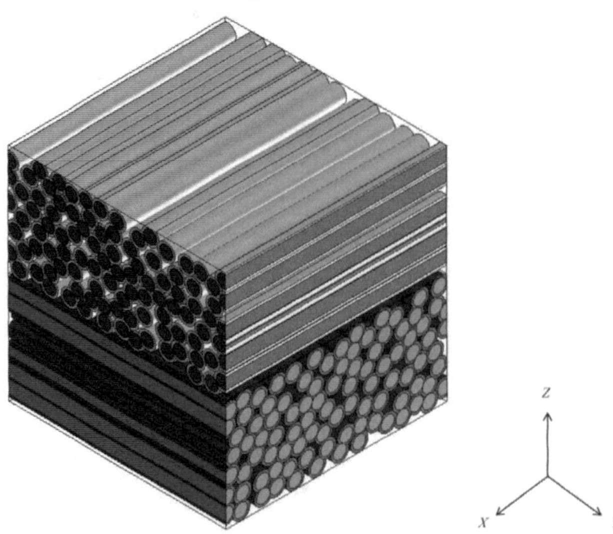

Fig. 2. Geometric model of a representative element of porous CCCM with orthogonal pattern reinforcement from SiC coated carbon fibres.

The models in STEP format were exported into Ansys Workbench 16 software package [6], where a finite element model of the representative element was created. The next stage was the simulation of heat transfer in the volume representative element. It was assumed that the temperatures in the two faces normal to one of the axes are different but constant over the surface. The remaining faces of the volume representative element were assumed to be thermally insulated. The simulation results were used to estimate the density of the heat flow in the direction of the selected axis, which enabled determing the coefficient of molecular heat transfer based on Fourier law. The initial data on the properties of the porous CCCM components were taken from [5].

The results in fig. 3 demonstrate that the combination of the carbon layer thicknesses in the orthogonal directions does not affect the heat transfer coefficient in the direction of the Z axis. However, the heat transfer is considerably affected by this ratio due to the high heat transfer along the fiber axis. Moreover, to enable XY-plane isotropy the carbon layer thickness ratios should be uniform in the orthogonal directions.

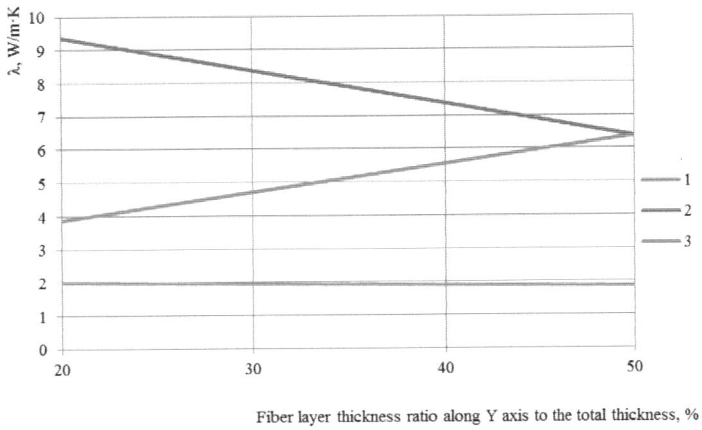

Fig. 3. Molecular heat transfer coefficient of CCCM with 18% density and orthogonal reinforcement pattern as a function of layer thickness ratio at 1700°C. 1 – molecular heat transfer coefficient along Z-axis; 2 – along X-axis; 3 – along Y-axis.

3 Analysis of the CCCM porosity effect

Two groups of models were used to estimate the porosity effect on the CCCM thermal physical properties. The first group included the models of the representative element with the orthogonal layers pattern and the equal layer thickness ratio in the orthogonal directions (fig. 4a). The second group employed the model of the representative element with the random fiber layout (fig. 4b). The both group models were based on homogeneous fibers 7 μm in diameter. These homogeneous fibers are assumed to possess thermal physical properties equivalent to those of SiC coated carbon fiber. The heat transfer process in a single SiC coated carbon fibre was simulated in order to determine the thermal physical properties of the homogeneous fibers, and the heat flux through the fiber section was determined. The condition of the thermal flux equality for the coated fiber and the homogeneous fiber was used to determine the equivalent heat transfer coefficient for the homogeneous fiber shown in fig. 5.

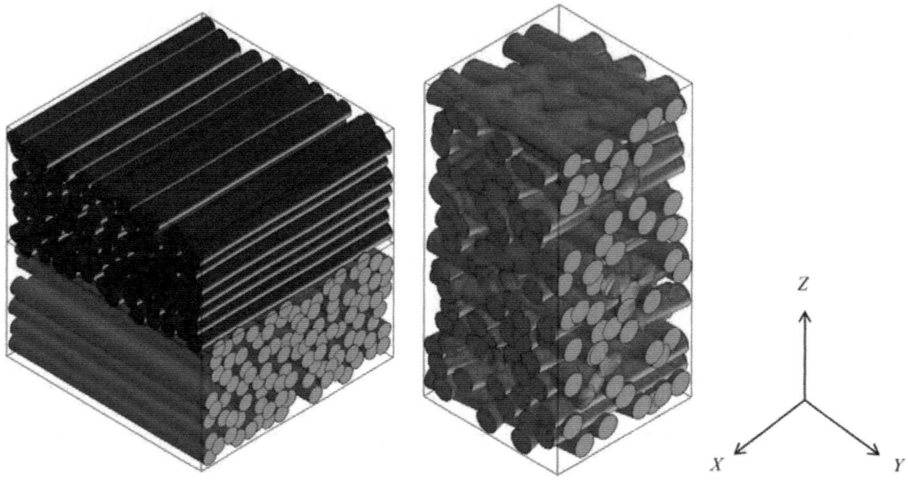

Fig. 4. Geometrical model of the representative element:a) with the orthogonal reinforcement pattern; b) with the random reinforcement pattern.

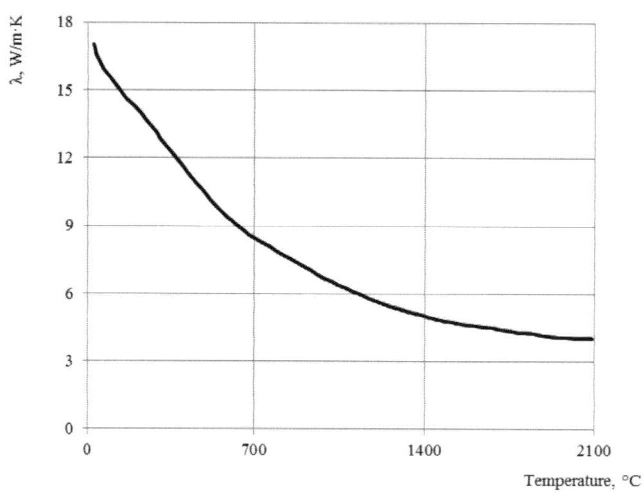

Fig. 5. Heat transfer coefficient for the homogeneous fiber.

CCCM porosity for the both models was in the range from 10 to 50%. The thermal conductivity coefficient of the porous CCCM was determined with the same method as presented above. The results (fig. 6) demonstrate a strong correlation between the heat transfer coefficient and the material porosity. The porous CCCM with the orthogonal layers has lower heat transfer than the random layers material. It is evident that increasing the CCCM porosity up to 50% reduces the heat transfer coefficient approximately 10 times, which makes porous CCCM desirable for thermal insulation purposes. The 50 % constraint with regard to the maximum porosity is due to the fact that the mechanical properties become unsatisfactory at greater porosity.

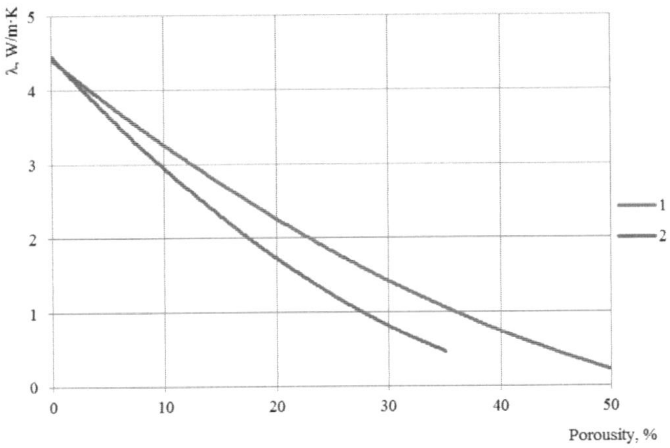

Fig. 6. Molecular heat transfer coefficient for CCCM with different reinforcement patterns as a function of residual porousity:
1 – model with the random reinforcement pattern; 2 – model with the orthogonal reinforcement pattern.

4 Effective heat transfer coefficient of the porous CCCM

The specific characteristic of the porous CCCM is that there are two simultaneous and interrelated heat transfer processes occurring within the material volume, i.e. conductive heat transfer along the frame and radiative heat transfer in the pore volume. The ratio of the conductive and radiative heat depends on the material porosity and temperature. The total heat exchange for porous media can be characterized by the effective heat transfer coefficient, which is a ratio of the total heat transfer through the porous medium to the local temperature gradient. Molecular heat transfer coefficient, unlike effective heat transfer coefficient, only accounts for the conductive heat transfer along the fibers, excluding the radiative process in the pore volume.

In order to estimate the effective heat transfer coefficient for the porous CCCM, a finite-element model of a volume representative element (fig. 7) was built using Siemens NX software package [7]. This model comprised the porous material zone (2, fig. 7) and two contacting plates (1 and 3, fig. 7).

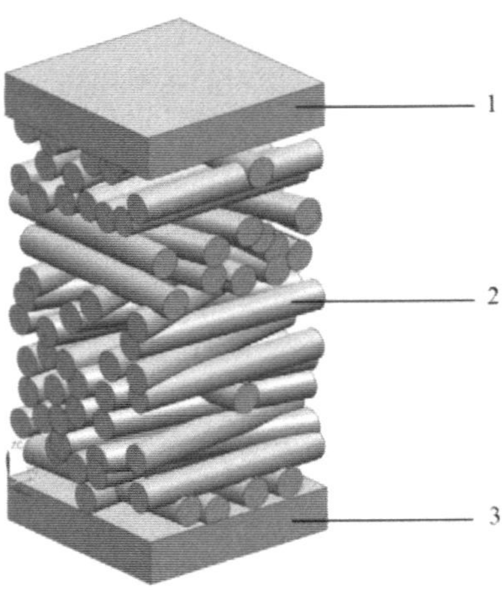

Fig. 7. Finite-element model of a representative element. CCCM with random reinforcement pattern.

The temperature for the outer surfaces was set as uniform across the surface but different in value. The plates side surfaces were assumed to be thermally insulated. The thermal contact between the plates and the porous material fibers were assumed to be ideal. Additional reflecting surfaces with a zero thickness and the refractive index of 1 were located on the side surfaces of the porous material. This ensured one-dimensional radiative heat transfer in the porous CCCM. The plates and fibers surfaces were assumed to be grey, diffusely reflecting, with the emissivity factor of 0.8. It was assumed that the combined radiative-conductive heat transfer is taking place in the porous CCCM volume.

Simulating the heat transfer process in such volume representative element enabled determining the magnitude of the total heat flux through the porous material and estimating its effective heat transfer coefficient. The calculation results demonstrate that in the middle and low temperature range (below 1600°C) the CCCM with 50% porosity possess the minimal effective thermal conductivity. However, as the temperature increases and, consequently, the radiative heat transfer intensifies in the inter-fiber space, their effective heat transfer coefficient grows rapidly. On the contrary, the effective heat transfer coefficient of low porosity (20-30%) CCCM in the low and middle temperature range is sufficiently higher than that of the high porosity materials, however its temperature-driven increase is much smaller due to a lesser role of radiative heat transfer. As a result, the effective heat transfer coefficient of the low porosity CCCM is smaller than that of the high-porosity CCCM in the high temperature range. Thus, a conclusion can be drawn that thermal protection systems require varied porosity CCCM layers: low porosity in the "hot" zone, with high porosity in the area adjacent to the load-bearing structure. The exact porosity distribution can be determined by solving a thermal design problem.

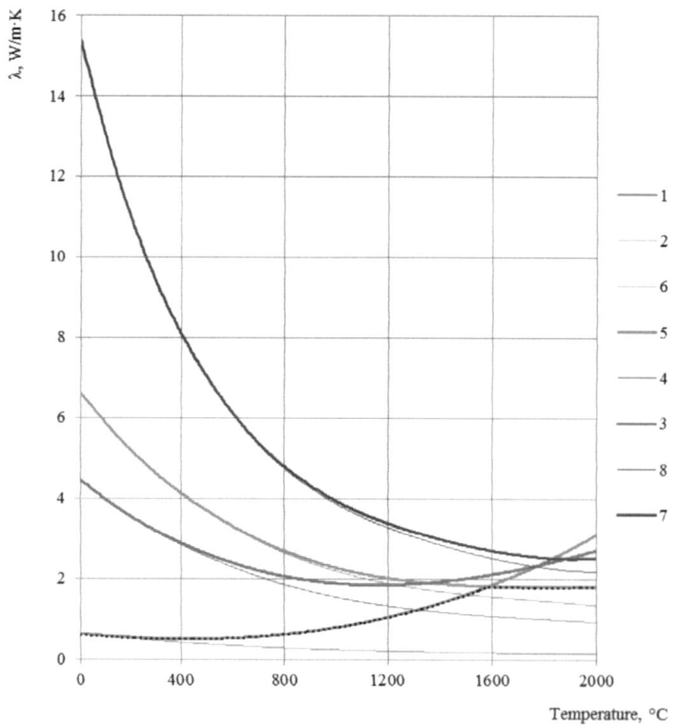

Fig. 8. Effective and molecular heat transfer coefficient of porous CCCM. 1,3,5,7 – effective heat transfer coefficient for 50, 40, 30, 20% porosity correspondingly; 2,4,6,8 – molecular heat transfer coefficient, 50, 40, 30, 20% porosity correspondingly.

5 Conclusion

The structure models were created for the porous carbon ceramic material with orthogonal and random carbon fibre reinforcement patterns. The thermal physical characteristics of the porous CCCM were determined based on the analysis of the heat transfer process in the representative volume element. It was demonstrated that increasing the CCCM porosity up to 50% would reduce the heat transfer coefficient approximately 10 times, making the porous CCCM a feasible choice as a thermal protection coating. The combined radiative-conductive heat transfer in the porous CCCM was simulated. It was demonstrated that for the high-porosity (over 40%) materials in the high temperature range (over 1400°C) the radiative heat transfer plays the key role in the heat transfer process. The conclusion was made that thermal protection systems require CCCM with variable porosity through the thickness.

Acknowledgments

Some results of this paper were obtained as part of the grant agreement No. 14.577.21.0099 by the Russian Federation Ministry of Education and Science. Unique identifier for Applied Scientific Research – RFMEFI57714X0008.

References

1. Yu.V. Polezhaev, F.B. Yurevich, *Thermal protection* (Moscow, 1976) [in Russian]
2. P. Bansal Narottam, *Handbook of Ceramic Composites* (Kluwer academic publishers, Boston, 2005)
3. Sandra P.Walker, Kamran Daryabeigi, Jamshid A. Samareh, Sasan C.Armand, Scott V. Perino, AIAA, NASA Langley Research Center; Hampton, VA, United States, Jan 13 (2014)
4. Kamran Daryabeigi, George R. Cunnington, Steve D. Miller, Jeffry R. Knutson, AIAA, NASA Langley Research Center; Hampton, VA, United States, Jun 28 (2010)
5. S.V. Reznik, P.V. Prosuntsov, K.V Mikhailovsky, J. Eng. phys. **88,** 3 (2015)
6. A.V. Bruyaka *Engineering analysis in Ansys Workbench* (Samara State Technological University, Samara, 2010)
7. P.S. Goncharov, I.A. Artamonov, T.F. Halitov, S.V. Denisihin, D.E. Sotnik *NX Advanced Simulation Engineering analysis* (DMK Press, Moscow, 2012)

RESEARCH OF FERROMAGNETIC NANOPARTICLES TRANSPORT UNDER MAGNETIC FIELD

Maxim Pryazhnikov[1,*], *Dmitriy* Guzey[1], *Andrey* Minakov[1,2], *Vladimir* Zhigarev[1], and *Anna* Shebeleva[1]

[1]Siberian Federal University
[2]Institute of Thermophysics SB RAS

Abstract. During the research we conducted an experimental study and numerical simulation of ferromagnetic nanoparticles transport behavior in a constant magnetic field. The growth dynamics of nanoparticle deposits on channel walls was studied depending on the Reynolds number and the intensity of the magnetic field. We obtained flow pattern, the concentration field and the trajectory of nanoparticles depending on the Reynolds number. The simulation results were compared with data from experiments. There was good qualitative and quantitative agreement between calculation and experiment.

1 Experimental study

During the research we conducted the experimental study of the behavior of ferromagnetic nanoparticles flowing in a liquid in a constant magnetic field. According to the study an experimental setup was a cylindrical tube, made of glass with an internal diameter of 6 mm and a length of 50 cm. The tube was connected to a pump and a container with liquid via hoses. Below a permanent neodymium magnet of 50x30x10 mm was brought to the glass tube. Distilled water with the nanoparticles was pumped through the tube. The particles interacted with the magnetic field, and deposited on the tube wall. For pumping liquid we used gear pump WT30001JA. Pump drives allowed to pump various fluids at a rate of 85.7 to 2571.4 ml / min.

Investigation of particle deposition process was conducted with a ferromagnetic fluid based on distilled water with the nanoparticles Fe_3O_4. Nanopowders had been acquired in JSC "Plazmoterm" (Moscow). Volumetric concentration of particles was very low, 10^{-3} %. The size of the nanoparticles was 50 nm. In order to prepare the ferrofluid we used a standard two-step method. The required amount of the powder was added to the liquid, after which the resulting suspension was thoroughly stirred. To destroy nanoparticle conglomerates the suspension was treated in the ultrasound bath "Sapphire" SC-10338 during 45 minutes.

* Corresponding author: arrivent@yandex.ru

We considered the permanent magnetic field produced by the neodymium magnet, the size of 50x30x10 mm. Near the surface of the magnet field reached a value of about 0.25 Tesla. The field was measured with the help of a milliteslameter TPU. A high-speed camera was used to fix the deposition of particles in the tube in a constant magnetic field.

The flow rate and the magnetic field strength varied during the experiments. Two series of measurements were conducted. The first series was held to study the effect of nanofluid flow on shape of particles deposition on the tube. The magnet was located under the tube, ym = 0 mm. Five mass flows were considered. Mass flow correspond to the following Reynolds numbers: 1200, 2810, 4220, 5760, 7145. The results of the study are presented in the form of sequential frames for the flow of liquid when the number of *Re* = 2810 (every 10 seconds) in figure 1.

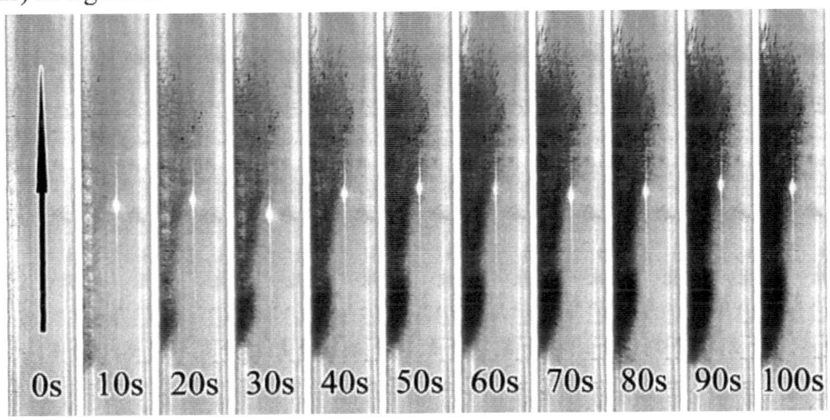

Fig. 1. The deposition of particles in a constant magnetic field at Re = 2810.

We obtained a dynamics of the formation of particles deposition at different Reynolds numbers. In addition, we made a comparison of deposition shapes of Reynolds numbers 60 seconds after the start of the nanoparticles deposition process in the tube. With the increase of the Reynolds number the growth rate of deposition and the height itself nanoparticle deposition height were being reduced. It was shown that with increasing Reynolds number, length of the deposition zone in the tube is reduced.

The second series was conducted in order to investigate the influence of the magnetic field intensity on particle formation deposition during the time. Liquid was pumped at a constant rate, the value of which corresponded to the Reynolds number *Re* = 1200. Three variants of magnet position were considered: below a tube onto $y_m = 0$, below tube onto y_m = 14 mm, and onto y_m = 28 mm.As a result of experimental studies it was found that with the weakening of the magnetic field the deposition height is reduced. The height of the particles increased during deposition time. About a minute after the start of the experiment the growth rate reached a maximum. Then a dynamic equilibrium between deposition and entrainment of the particles was achieved, and deposits form was stabilized. It is also obvious that the position of the magnet (magnetic field) influences the shape of the particles deposition: the smaller the magnetic field is, the greater the flow of entrained particles is and the less deposits are on the tube.

2 Numerical simulation

Numerical simulation of transport of nanoparticles in a constant magnetic field was conducted.To simulate the transport of nanoparticles in the magnetic field we used a novel Euler-Lagrange combined two-component approach. As part of the Lagrange approach for the base fluid the equations of continuity, momentum and energy were being solved, and the movement of the particles was modeled on the basis of Newton's second law by solving an ODE. Euler two-component model describes nanoparticles in a liquid as a binary mixture, one of which components is nanoparticles. Statement of the problem completely repeated the experimental study. The computational domain is a circular channel with an inner diameter of 6 mm and a length of 50 cm. A computational grid consisting of 500 thousand nodes, was used for the simulation.. fig. 2-3 shows the results of calculations (vector velocity field in the channel, the trajectory of the particles) at Re = 2810.

By increasing the Reynolds number the flow was being complicated and the shape of the vortex field in the channel wall was being changed. The deposits form was also changed. Nevertheless, the calculation perfectly reproduces the shape of deposition observed in the experiment (see fig. 4). With a flow rate of increase in the action of magnetic force compared to inertial forces are weakened.

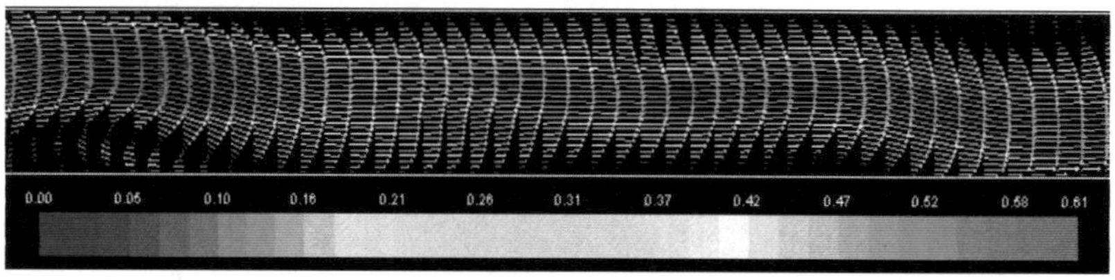

Fig. 2. The vector velocity field in the channel at Re = 2810.

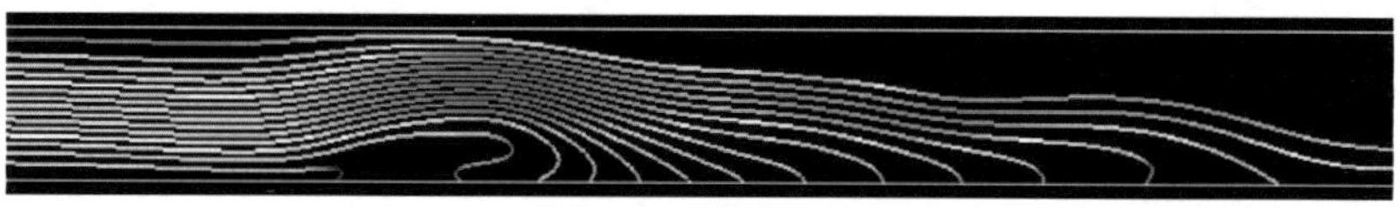

Fig. 3. Trajectories of nanoparticles at Re = 2810.

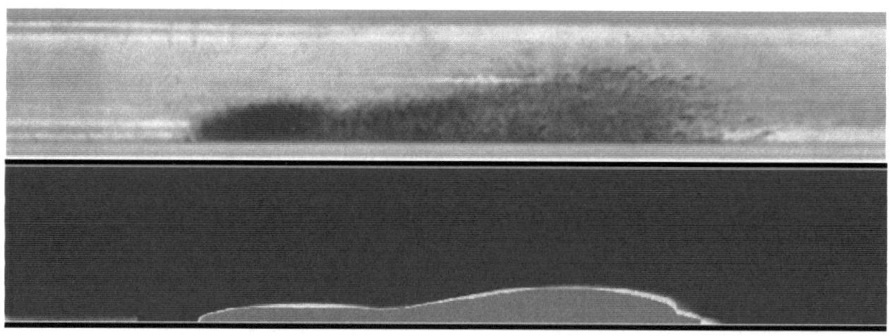

Fig. 4. Comparison of shape of particle deposition in the channel at Re = 2810 (top - experiment, bottom - numerical simulation).

3 Conclusions

1. In this research we conducted the experimental study of nanoparticle transport in a constant magnetic field. During our experiments we studied the dynamics of nanoparticles depositions growth on the channel walls depending on the Reynolds number and the magnetic field strength. For the first time we obtained characteristic profiles of nanoparticle deposition zones, determined the dimensions of the deposition rate of their formation depending on the Reynolds numbers and the magnetic field strength.

2. We conducted a numerical simulation of transport of nanoparticles in a constant magnetic field. To simulate the transport of nanoparticles in the magnetic field we used a novel Euler-Lagrange combined two-component approach.

3. We conducted the verification and adaptation of mathematical models of magnetic nanoparticles transport under the influence of external electromagnetic fields on the data of

the experiment. We obtained the flow pattern, the concentration field and the trajectory of nanoparticles as a function of the Reynolds number. The simulation results were compared with experimental data, there was good qualitative and quantitative agreement between calculation and experiment.

Acknowledgments

This work was financially supported by the Ministry of Education and Science of Russia (Contract No. 14.607.21.0104 (RFMEFI60714X0104)).

References

1. C. M. Oldenburg, S. E. Borglin, G. J. Moridis, Transport in Porous Media **38** (2000)

THE DISTRIBUTION OF PARTICLES IN THE WALL REGION OF THE CYCLONE CHAMBER

Aleksandr Razva[1,*], *Michail* Vasilevsky[1], *Alice* Pleshco[1], and *Aleksandr* Naumkin[1]

[1]National research Tomsk polytechnic university, 634050 Tomsk, Russia

Abstract. Held cyclone operation analysis at different flow rates. In the absence of dust deposits on the surface of the device in countercurrent conical cyclone particle concentration distribution in the circumferential direction in the wall region is uneven rope-formed particles to form clots. At low concentrations where clots stably transported stream at any position of the cyclone can be made based on evaluation of the speed distribution of gas and particles in the bunches movement. In work the analysis of work of a cyclone is carried out at various speeds of a stream.

1 Introduction

Processes of a current of the two-phase environment in the cyclone chamber are difficult and still there is no physically reasonable model of transformation of the disperse environment from an aerosol state in bulk in the receiver. At the boundary of the swirling flow near the boundary surface of gas flow tangential speed decreases and becomes zero at the wall of the flow itself. Thus, the particles are near the wall in dynamic equilibrium at the interface and the transfer of particles flow in a radial direction on average no (condition not pickup particles) [1].

In the absence of dust deposits on the device surface in a conical counter-cyclone in cinders-output hole concentration of particles increases hundreds of times, causing the flow of braking due to the increased value of the shear stress on the wall. Bundles already formed on ¼ traffic from the injection point of the particles at high concentrations [2]. For larger particle concentrations at which the weight of bundles becomes sufficient for the manifestation of the forces of gravity, calculation of flow characteristics are shown in [3]. At low concentrations, at which the bundles stably transported cyclone flow at any position can be made based on evaluation of the speed distribution of gas and particles in the harness movement.

* Corresponding author: razva@mail.ru

2 The distribution of particles in the wall region of the cyclone chamber

To determine the time of passage of the material through the cyclone and comparing it with a gas residence time were conducted measuring the air speed at the inlet of the entire apparatus during retraction cinders-output air through the hole, flow of particulate material. Conducted immediate disconnection from the cyclone collector vacuum, determined the amount of dust precipitated in the cyclone. [4] Visually it was evident (figure 1), that the dust in the cone moves to a bandpass harnesses.

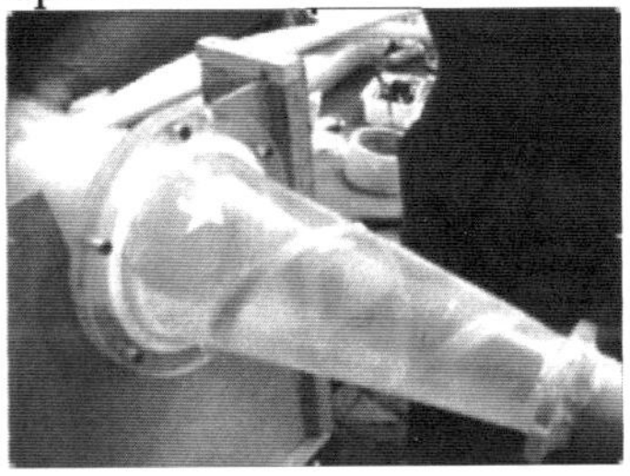

Fig. 1. The configuration of the dispersed phase in the cyclone cone.

The number of turns, their length was determined visually, as well as the calculation. The air speed in the turns was determined taking into account the acceleration of the flow in the cone, and averaged over the length of the coil. Determined during the passage of air through the cyclone cone as the cone volume divided by the flow rate, and while loop length divided by the tow speed. We determined the residence time, as the mass is left out in the cyclone with a cut-off the flow divided by the mass flow rate. Experiments were made on the powder M40 (corundum). Consumption was 1,1 g/s, flow rate 9–17 g/s, the speed of air in the inlet section 7 – 13 m/s. The ratio of the time of passage of the material to the air passage of time depending on the concentration and speed of air is in the range 2 ÷30 [4].

In order to determine the parameters of clotting (dust distribution M40 in the circumferential direction) in a countercurrent cyclone via the rotary section, air samples were taken at different concentrations of particles (or the number of allocated dust on the amount of the introduced dust) on the circumference of the cyclone at a fixed height equal to 1.5 the diameter of the input. It was found that concentration distribution is uneven in the circumferential (and concentrations are burst gaps) (figure 2).

We will enter coefficient of unevenness of distribution of particles in the district direction Ku=qmax/qmin. At concentration of dust 2–4 g/m^3 monofractional dust Ku ≈3, at concentration of the same dust 70–95 g/m^3 in an entrance stream Ku ≈5. For polyfractional dust with concentration 11–19 g/m^3 Ku ≈7. At concentrations of dust more 200 g/m^3 uneven particle distribution shown in circumferentially tows, which move with low speed in the circumferential direction and are detected visually. The computational model of concentration of particles in countercurrent cylindrical cyclone device disclosed in [1].

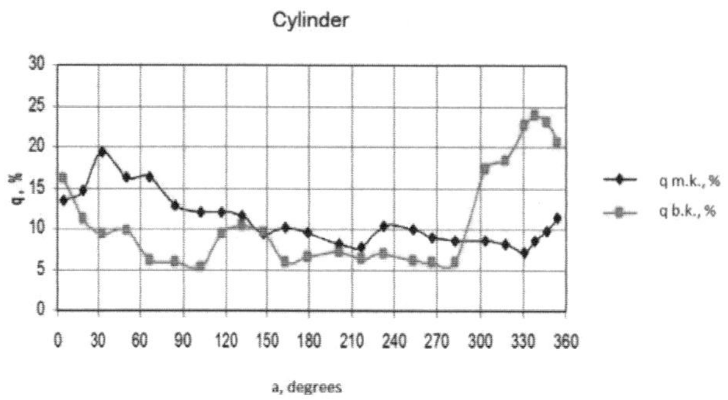

Fig. 2. The relative amount of monofractional dust output from the periphery of cylindrical part of a cyclone by 15% amount of air depending on angular coordinate (the direction of counting in the direction of rotation of a stream, a reference mark interface of input to the chamber cylinder):♦– concentration 2–4 g/m^3, ■– concentration 70–95 g/m^3.

The relative yield of cement dust is presented in figure 3.

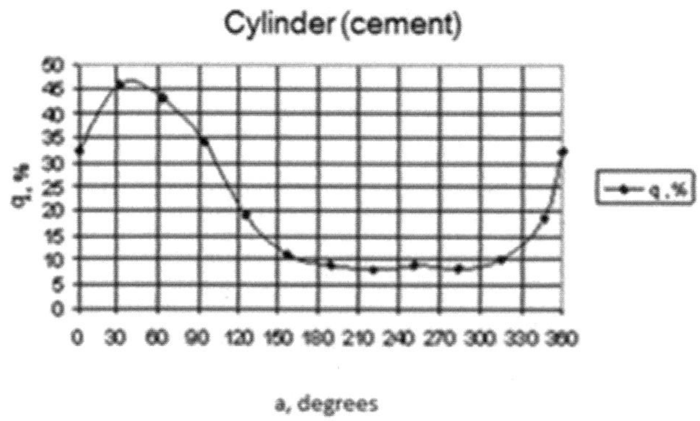

Fig. 3. The relative amount of polyfractional dust output from the periphery of cylindrical part of a cyclone by 15% amount of air depending on angular coordinate (the direction of counting in the direction of rotation of a stream, a reference mark interface of input to the chamber cylinder): ♦– concentration 11–19 g/m^3.

3 Estimates of parameters tows at high particle concentrations

Let C_* – the absolute speed harness (figure 4), then $C_* \cos\alpha$ – the circumferential speed of the harness, α– the angle between the C_* and district, ℓ – harness length, t – while its location on the surface, $t = \ell / C_*$, WΠ – dust flow, m – harness weight, $\eta_* = m/W_\Pi t$ – coefficient of dust into a bundle.

Length of a plait is equal to length of one round increased by number of rounds $\ell = \frac{\pi D}{\cos\alpha} \frac{H}{\pi D \mathrm{tg}\alpha} = \frac{H}{\sin\alpha}$, where H – height of separation part where the plait moves, D=2R – diameter of separation part, $m = A_S \ell (1-\varepsilon)\rho_\delta$; $A_S = \pi r_S^2$ – plait cross-sectional area, ε – the fractional void volume porosity of the dust in a plait, ρ_δ– dust density, $W_\Pi = \rho K_{in} Q = mC_* / \ell\eta_*$; $m = \rho K_{in} Q \ell \eta_* / C_*$. In the circumferential direction of the friction force on the tow surface is balanced by the wind carrying away the power flow:

$$\frac{mC_*^2\cos^2\alpha}{R}\varphi_S=\xi_S r_S H\frac{\rho}{2}(V_\varphi-C_*\cos\alpha)^2, \quad (1)$$

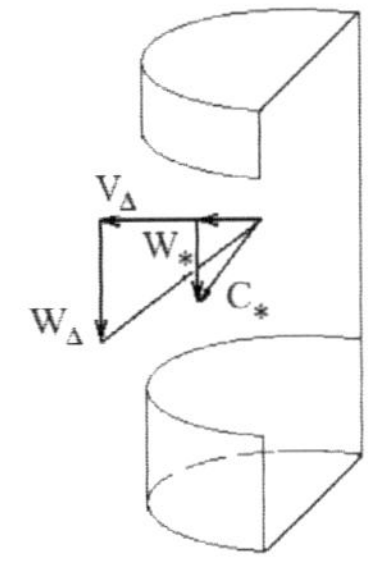

Fig. 4. The scheme of distribution of speed of a plait in the cyclonic device.

Here φ_S, ξ_S – coefficients of friction of particles about a surface and the aerodynamic resistance of a plait, V_φ – district speed of gas at distance from a wall $y=r_S$,

$V_\varphi(y)=\sqrt{\frac{\tau_w}{\rho}}(9.5+1.32\ln\eta)$; $\eta=\frac{y}{\nu}\sqrt{\frac{\tau_W}{\rho}}$ [8], $\tau_W=0.5C_f\rho V_\Delta^2/2$, $C_f\approx0.006$; V_Δ – value of district speed of gas on border of a wall zone; $V_\Delta=0.95V_{in}$.

Tow speed in the axial direction $W_*=W_\Delta(r_S/\Delta)^{1/14}$, here W_Δ– harness the maximum speeds at a distance from the wall $y=\Delta$; Δ – distance border flow core and the wall region [5]. Substituting $C_*=\frac{W_*}{\sin\alpha}=\frac{W_\Delta}{\sin\alpha}\left(\frac{r_S}{\Delta}\right)^{1/14}$ in (1) and considering that $r_S/R=(r_S/\Delta)(\Delta/R)$ and designating $(9.5+1.32\ln\eta)=\eta_y$, after transformations we obtain

$$\left[\frac{\varphi_S}{\xi_S}\frac{2\pi\Delta(1-\varepsilon)\rho_\delta}{R\quad\rho}\right]^{1/2}\frac{W_\Delta}{V_\Delta}\left(\frac{r_S}{\Delta}\right)^{8/14}=\sin^{1/2}\alpha\left[\eta_y tg\alpha-\frac{W_\Delta}{V_\Delta}\left(\frac{r_S}{\Delta}\right)^{1/14}\right] \quad (2)$$

When converting into account that $H/\ell=\sin\alpha$. Meaning $\Delta\approx0.03R$ и $\varepsilon=0.7...0.95$ correspond to the size of the wall region and the porosity of particles in a bundle.

From the relation $\pi r_S^2(1-\varepsilon)\rho_\delta\ell=\rho K_{in}Q\ell\eta_*/C_*$ find the value of r_S.

$$r_S=\sqrt{\frac{\rho}{\rho_\delta}\frac{K_{in}V_{in}f_{in}\eta_*}{(1-\varepsilon)\pi C_*}}, \quad (3)$$

here $Q=V_{in}f_{in}$; V_{in}, f_{in} – stream speed in the entering branch pipe, the section of a branch pipe through passage. The equations (2, 3) allow to define C_* и r_S. Estimates show that at low concentrations $K_{in}=(1...10)\cdot10^{-4}$ kg / kg and a flow of less than 5 micron particles, the quantity r_S it is 100 ... 300 mm, that particle size reaches effectively released from the stream at the receiver. Thus, the reason for trapping fine particles in the cyclone - binding bundles and their transportation in the receiver rope-formed state. In very small concentrations of fine particles occurs probably periodic accumulation of particles in the volume and precipitation cord factor receiver. In this case there are pressure pulsations in the receiver. In [2] provides information about the flow of brake particles in the harness.

4 Conclusion

The swirling flow efficiency of dust formed from bundles depends on the degree of dispersion in the dust receptacle harnesses gas flows. The flow of the cyclone effectively work towards opening dust-collecting bag wasting their energy on the generation of the second kind of vortices and the harness transport. For the dispersion of the bundles in the receiver must also be expended energy. This energy will be low if the receiver will continue cord factor process and attenuation flow twist. The cyclone receiver individually, depending on its design, the generation of turbulence is the nature and intensity gradient it is insufficient to break the harness. In receivers such as inertial relief in which turbulence damping occurs not only, but also increases the expense of additional vortex formation, separation of fine particles is not possible due to dispersion harnesses. The same situation is multicyclones receivers in which holes exiting dust-collecting bag torches interact and as a result, turbulence intensity in the common dust collecting chamber is comparable to the intensity of turbulence in the flame. This fact can be explained by the low efficiency of gas dedusting cyclone dust collector. [2] The most radical solution to the problem of increasing the stability and efficiency of cyclone systems is the use of transit "receivers", in which there is a flow attenuation and naturally formed layer of particles [2].

Acknowledgments

The work was held within the research state assignment "Science" No. 13.1339.2014/K (Code of Federal Target Scientific and Technical Program 2.1410.2014).

References

1. M.V. Vasilevsky, E.G. Zykov., and A.S. Razva. Theor. Found. Chem. Eng. **45**, 3 (2011)
2. M.V. Vasilevsky *Dedusting gas inertial devices* (Publishing house of Tomsk Polytechnic University, Tomsk, 2008)
3. V. Strauss. *Industrial cleaning gases*, (Chemistry, Moscow, 1981)
4. M. V. Vasilevsky, V.I. Romandin, E. G. Zykov, V.A. Polyushko., A. S. Razva. Bulletin of the Tomsk State University. Mathematics and Mechanics **23,** 3 (2013)
5. A.N. Shtym. *Aerodynamics of cyclonic swirl chambers* (Publishing house of Far East university, Vladivostok, 1985)

MATEC Web of Conferences **72**, 01095 (2016) DOI: 10.1051/matecconf/20167201095
HMTTSC-2016

THE LIQUID FILM FLOW WITH EVAPORATION: NUMERICAL MODELLING

Ekaterina Rezanova[1,2,*]

[1]Altai State University, 656049 Barnaul, Russia,
[2]Kutateladze Institute of Thermophysics SB RAS, 630090 Novosibirsk, Russia

Abstract. The flow of thin liquid layer on an inclined substrate is investigated numerically. The mathematical modelling is based on the Oberbeck-Boussinesq equations and the generalized conditions on the thermocapillary boundary simplified during the parametrical analysis. In the framework of the long-wave approximation the evolution equation which determines the thickness of the liquid layer in the case of moderate Reynolds numbers is derived. The results of numerical modelling of the liquid flow with evaporation at the interface are obtained.

1 Introduction

A large number of both theoretical [1-5] and experimental investigations [6] are devoted to the study of viscous incompressible fluids with evaporation. The interest to such currents is caused, in particular, by the wide possibilities of their practical application in modern industrial processes. Great attention is paid to the thin liquid films. The mathematical models of the flows in the thin layer approximation and their numerical study are presented in [1-5, 7-10]. Special attention while modelling of the flows with interfaces should be given to the formulation of the boundary conditions [11-13].

In this paper we study the flow of a thin layer of viscous incompressible liquid with evaporation on the thermocapillary boundary. The motion of the liquid is accompanied by a co-current gas flow. The mathematical modelling is based on the long-wave approximation of the Oberbeck-Boussinesq convection equations and the generalized kinematic, dynamic and energetic conditions at the interface [11-13]. The evolution equation, determining the liquid layer thickness, is obtained when the effects of gravity, evaporation, capillarity, thermocapillarity and additional tangential stresses from the gas medium are taken into account (see [7-9]).

[*] Corresponding author: katerezanova@mail.ru

2 The problem statement. The evolution equation of the liquid layer thickness

The thin liquid layer flow on an inclined, non-uniformly heated solid substrate in a two-dimensional case is investigated. Let the coordinate system be chosen so that the x-axis coincides with the direction of the substrate (see fig. 1). The position of the thermocapillary interface is determined by the equation $z = h(x, t)$, the solid plate is given by the equation $z = 0$.

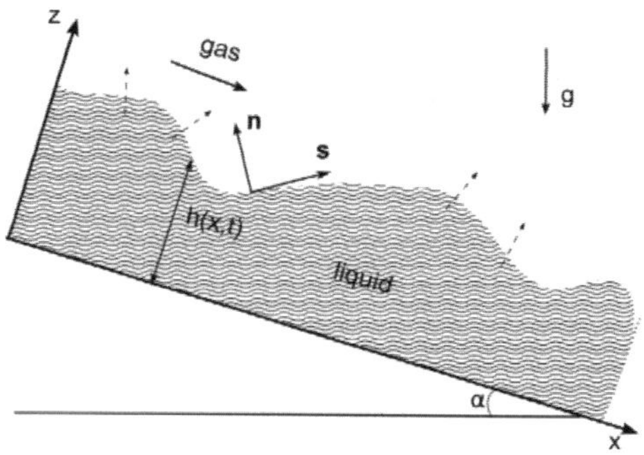

Fig. 1. Geometry of flow domain.

During the mathematical modeling of the investigated process two characteristic length scales are introduced: the transverse – d and the longitudinal – l, the ratio of the first to the second is the small parameter of the problem ε ($\varepsilon=d/l$). The one-sided model is based on the Oberbeck-Boussinesq equations [9]. The generalized kinematic, dynamic and energetic conditions at the thermocapillary interface after the parametric analysis in dimensionless form can be written as follows:

$$\begin{gathered} -\varepsilon\left(h_t + h_x - w\right) = J_{ev}E, \\ u_z = \alpha_\tau \tau(x,t) - \alpha_{Ma}(T_x + h_x T_z), \qquad p = p^g - \alpha_{Ca} h_{xx}\left(1-\alpha_\sigma T\right), \\ -T_z + \bar{\beta}_2\left\{T div_\Gamma \mathbf{v}\right\} = \bar{\beta}_3 J_{ev} + \bar{\beta}_6 h_{xx} J_{ev}. \end{gathered} \tag{1}$$

Here $\mathbf{v}=(u, w)$ is the velocity vector; p is the pressure; T is the temperature; J_{ev} is the local mass flux at the thermocapillary interface, $E=\kappa T_*/(\lambda_U \rho \nu)$ is the evaporation coefficient, κ is the heat conductivity coefficient, λ_U is the latent heat of vaporization, ρ is some relative value of liquid density, ν is the kinematic viscosity coefficient, T_* is the characteristic temperature drop; $\alpha_\sigma=MaCa/Pr$, $\alpha_{Ma}=Ma\varepsilon/Pr$, $\alpha_{Ca}=\varepsilon/Ca$, $\alpha_\tau = \bar{\rho}\bar{\nu}\bar{\upsilon}\varepsilon / \bar{h}$, where $Ma=\sigma_T T_* l/(\rho\nu\chi)$ is the Marangoni number, $Ca=u_*\rho\nu/\sigma_0$ is the capillary number, $Pr=\nu/\chi$ is the Prandtl number, χ is the thermal diffusivity coefficient, σ_T is the temperature coefficient of surface tension, σ_0 is the value of the coefficient of surface tension σ at the same temperature ($\sigma=\sigma_0-\sigma_T(T-T_0)$), $\tau(x,t) = \left(u_z^g + w_x^g\right)|_\Gamma$, $\bar{\rho}, \bar{\nu}$ are the ratio of the densities and coefficients of kinematic viscosity of the gas and liquid, $\bar{\upsilon} = u_*^g / u_*$, $\bar{h}$ is the ratio of the characteristic scale of gas layer to l, u_* is the characteristic longitudinal velocity of the liquid; $\bar{\beta}_2 = Ma\varepsilon / (RePrE\bar{U})$, $\bar{\beta}_3 = \varepsilon$, $\bar{\beta}_6 = \left(1-1/\bar{\rho}\right)\varepsilon / (\bar{U}Ca)$, $\bar{U} = \lambda_U / u_*^2$. The dependence of the local mass flux on the temperature is determined with the help of the Hertz-Knudsen equation [3, 7-10, 13]. The no-slip conditions on the solid substrate are fulfilled and the temperature distribution Θ_0 is set. The Reynolds number ($Re=u_* l/\nu$) is assumed to be unity.

The problem of calculation of the unknown functions (that are the components of the velocity vector u and w, temperature T and pressure p) is studied in the thin layer approximation. The thickness of the liquid layer should be found for the determination of these functions. Confining to the main members of the expansions of functions *u, w, T, p* in powers of the small parameter ε, and using the kinematic condition on the thermocapillary boundary, we get the evolution equation that specifies the position of the interface *h*:

$$h_t + h_x\left\{\frac{h^2}{2}((C_0)_x + \gamma_1 \sin\alpha\Theta_0) + hC_1 + \frac{h^4}{24}\gamma_2\cos\alpha A_x + \frac{h^3}{6}(\gamma_2\cos\alpha(\Theta_0)_x + \gamma_1\sin\alpha A)\right\} +$$
$$+\left\{\frac{h^3}{6}(C_0)_{xx} + \frac{h^2}{2}(C_1)_x + \frac{h^5}{120}\gamma_2\cos\alpha A_{xx} + \frac{h^4}{24}(\gamma_2\cos\alpha(\Theta_0)_{xx} + \gamma_1\sin\alpha A_x) + \right. \quad (2)$$
$$\left. + \frac{h^3}{6}(\Theta_0)_x\gamma_1\sin\alpha\right\} + \frac{E\alpha_J}{\varepsilon}\{Ah + \Theta_0\} = 0.$$

Here $\gamma_1 = Gr/\varepsilon$; $\gamma_2 = Gr$; $Gr = d^3 g\beta T_*/\nu^2$ is the Grashof number, β is the thermal expansion coefficient, $\alpha_J = \alpha_A \rho_s \lambda_U T_* / J_* \left(M/\left(2\pi R_g T_s^3\right)\right)^{1/2}$, α_A is the accommodation coefficient, ρ_s is the vapor density, J_* is the characteristic value of the vapor mass flux, M is the molecular weight, R_g is the universal gas constant, T_s is the saturated vapor temperature. Functions A, C_0 and C_1 are determined by the following relations:

$$A = \frac{\left(-\bar{\beta}_2(C_1)_x h + \bar{\beta}_3\alpha_J + \bar{\beta}_6 h_{xx}\alpha_J\right)\Theta_0}{1 + \bar{\beta}_2(C_1)_x - \bar{\beta}_3\alpha_J h - \bar{\beta}_6 h_{xx}\alpha_J h},$$

$$C_0(x,t) = p^g - \alpha_{Ca}h_{xx}\left(1 - \alpha_\sigma\Theta^0\right) - \gamma_2\cos\alpha\left(A\frac{h^2}{2} + \Theta_0 h\right) - \frac{\gamma_1}{Bu}x\sin\alpha + \gamma_2 h\cos\alpha,$$

$$C_1(x,t) = \alpha_\tau\tau(x,t) - \alpha_{Ma}\tilde{\Theta} - (C_0)_x h - \gamma_1 h\sin\alpha\left(A\frac{h^2}{2} + \Theta_0 h\right) - \gamma_2\cos\alpha\left(A_x\frac{h^3}{6} + (\Theta_0)_x\frac{h}{2}\right),$$

where $\tilde{\Theta} = A_x h + \left(\Theta_0\right)_x + h_x A$, $\Theta^0 = Ah + \Theta_0$.

The implicit finite-difference scheme, where the finite-difference analogs of the second-order approximation are used for all derivatives with respect to x, for the numerical solution of the equation (2) is constructed. The problem is supplemented by periodic conditions at the ends of the studied area $x = L$, $x = -L$ and the initial position of the interface [10].

3 The results of numerical investigation

The results of numerical modelling of the thin layer flow of the liquid type as ethanol with consideration of the evaporation process in the conditions of normal gravity (g=9.81·10^{-2} m/s^2) are presented. The initial position of the thermocapillary boundary is given by the equation $h_0 = 1 - \delta_1 \cos(kx)$ (δ_1=0.01, $k=\pi/2$), heating of the solid substrate is carried out according to the law $\Theta_0 = 1 + \delta_0 \cos k_1 x \cdot \cos k_2 t$ (δ_0=0.25, $k_1=\pi/2$, k_2=2).

Figure 2 shows the influence of the effects, taken into account in the study of the layer dynamics, on the boundary position. The angle of the heated substrate inclination relative to the horizon α is $\pi/6$ (see figure 1). The coefficient $\bar{\beta}_6$ is assumed to be zero in the studied case (lines 1 and 3 of the figure 2). In cases when $\bar{\beta}_3 \neq 0$ (lines 1 and 3 of the figure 2) the heat loss on the deformation of the interface as the result of the liquid evaporation are taken into account. If the term in the energy condition with a coefficient $\bar{\beta}_2$ is assumed unequal to zero than the consumption of energy required to overcome the surface deformation by the thermocapillary forces along the surface is taken into account (see lines 2 and 3 of the figure 2) [13]. Note that the most intense evaporation of the layer is observed when both the considered effects are assumed.

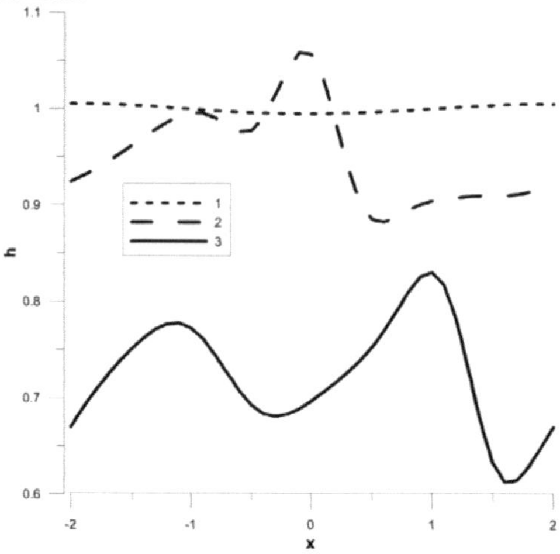

Fig. 2. The influence of various effects on the liquid layer thickness at the time t = 10^{-3}: 1. – the position of the boundary when $\bar{\beta}_3 = 0.1, \bar{\beta}_2 = 0$; 2. – the position of the boundary when $\bar{\beta}_2 = 0.3$, $\bar{\beta}_3 = 0$; 3. – the position of the boundary when $\bar{\beta}_2 = 0.3$ $\bar{\beta}_3 = 0.1$.

Acknowledgments

We gratefully acknowledge the support from the Ministry of Education and Science of Russia (Agreement No. 14.613.21.0011, project identifier RFMEFI61314X0011).

References

1. A. Oron, S.H. Davis, S.G. Bankoff, Rev. Mod. Phys. **69** (3) (1997)
2. J. Klentzman, V.S. Ajaev, Phys. Fluids **21**, (2009)
3. S. Miladinova, S. Slavtchev, G. Lebon, J.-C. Legros, Proc. 6th Workshop on Transport Phenomena in Two-Phase Flow, Bourgas (2001)
4. Yu.O. Kabova, V.V. Kuznetsov, O.A. Kabov, T. Gambaryan-Roisman, P. Stephan, Int. J. Heat Mass Transfer **68** (2014)
5. O. Shklyaev, E. Fried, J. Fluid Mech., **584** (2007)
6. Yu.V. Lyulin, O.A. Kabov, Int. J. Heat Mass Transfer **70** (2014)
7. O.N. Goncharova, E.V. Rezanova, Ya.A. Tarasov, Izvestia AltGU **81** (1/1), (2014)
8. O.N. Goncharova, E.V. Rezanova, Izvestia AltGU **81** (1/2), (2014)
9. O.N. Goncharova, E.V. Rezanova, Izvestia AltGU **85** (1/1), (2015)
10. E.V. Rezanova, Izvestia AltGU **89** (1), (2016)
11. C.S. Iorio, O.N. Goncharova, O.A. Kabov, Microgravity Sci. Technol **21** (1) (2009)
12. C.S. Iorio, O.N. Goncharova, O.A. Kabov, Computational Thermal Sci. **3** (4) (2011)
13. O.N. Goncharova, Izvestia AltGU **73** (1/2) (2012)

MATEC Web of Conferences **72**, 01096 (2016) DOI: 10.1051/matecconf/20167201096
HMTTSC-2016

ACHIEVEMENTS, PROBLEMS AND PROSPECTS OF THE SCIENTIFIC APPROACH AIMED FOR ENSURING RESOURCE-EFFICIENT TECHNOLOGIES IN THE CONTEXT OF PERSONNEL POLICY IN THE ENERGY SECTOR

Veronika M. Rostovtseva[1,*], and *Gavriil* A. Nizkodubov[1]

[1]National Research Tomsk Polytechnic University, 634050 Tomsk, Russia

Abstract. The article is devoted to the increasing role of human resources in the field of energy policy. Basic directions for the development of personnel policy, the ways of their implementation in the modern HR environment have been outlined. The results of the content analysis of the research on personnel policy and personnel training for the energy sector in the context of the whole spectrum of science have been introduced; certain specific features of the process have been studied.

1 Introduction

The energy sector has one of the strategic roles in the development of national industrial complexes, and the world economy as a whole. It is usually regarded as one of the factors of competitiveness of the national economy in the global market. Modern industry is based on the knowledge economy, so the success of this development is largely determined by the degree of the relationship between science and production, efficient use of resources. Energy is a knowledge-intensive industry and over the last twenty years about thousands of Ph.D. dissertations have been devoted to the energy sector. These researches have been carried out mainly in the framework of technical, physical, mathematical, and economic sciences. Resource-saving technologies are considered as a new direction in the development of modern energy sector.

People carry out promotion of scientific ideas, their practical implementation. The role of human resources is steadily increasing. It has been proved that the economic model based on knowledge is the most promising because it allows us to use existing intellectual and labor resources more efficiently [1]. Since the economic model constitutes the main part of the regular employees, professional and social organizations of the energy sector, subsequently, the personnel policy must also meet the changing situation, i.e. undergo certain changes. The problem of status and human resources development cannot be considered solely in economic terms, it is also relevant to the field of sociology, psychology

* Corresponding author: kachalov@tpu.ru

and pedagogy, which are associated with the development of an individual, its education and training, as well as issues of professional competence. However, similar studies have been carried out ten times less. This fact prompted us to attempt to identify the achievements and problems of an individual, as well as to offer our own vision of the prospects for the scientific approach aimed for ensuring resource-saving technologies in the context of personnel policy in the energy sector.

2 Personnel policy in the energy sector

Personnel work is the activity of individual enterprises or the whole branch aimed at the formation of their personnel. By personnel policy is usually meant a set of those principles, methods and forms of institutional mechanisms that allow developing goals and objectives in order to maintain, strengthen and develop human resources; creating a highly efficient and cohesive team capable of responding promptly to changing market requirements. In general, the range of issues is wide enough - staff, personnel policies, personnel strategy, personnel management, development of "labor potential". In order to solve the problems some domestic and international experience has been accumulated. However, the lack of uniform solutions in human resources management, and the specifics of the industry cause a need to specify issues in relation to the energy sector.

Personnel policy is implemented through multiple strategies. In practice, the development of human resources strategies is a concept design of the industry as a whole or of individual enterprises on the development of human resources and the implementation of this plan [2]. To solve the problems associated with human resources management a new human resources strategy is required. On the one hand, the strategy should match the challenges of the modern times, and, on the other hand, - the specifics of the current conditions bearing the imprint of the prolonged economic crisis. In respect of the matter, according [2] this resulted in the outflow of skilled labor, the reorientation of the educational priorities of the youth.

There are several areas of human resources policy. For example, introduction of new technologies in the prediction of new job places and opportunities. Currently, resource-efficient technologies have become popular where leading positions belong to resource conservation. The existing achievements will be briefly described below.

In scientific terms, the broad issues of energy saving of individual production processes, including fuel economy have been. Basically, energy saving solutions, issues of energy conservation and methods of their solutions on the example of a particular region are studied. The issue is being actively considered from the perspective of economic science. A large number of scientific studies is related to the development and maintenance of special technical equipment in various branches of modern industry, guest industry, agriculture, etc.

Some scientific research works are devoted to the search of methods and development of organizational and economic power management mechanisms. In our opinion, there is unreasonably little specialized research work on developing resource-saving technologies – only a small number of scientific studies can be mentioned [3-5].

Author [6] has introduced new and interesting ideas in terms of the factors that have a significant influence on modern human resources policy in the energy sector. The quoted author considers the process of finding market niches in the field of energy saving and energy efficiency management of science-based management solutions. The proposed conceptual characteristics of marketing environment reflect its condition and the basic tendencies of the development. In addition, author [7] reveals the role of the energy audit as an organizational-economic mechanism of energy companies. This, of course, must be taken into account for introduction of new technologies aimed at creation of new job places and opportunities.

Another area of personnel policy is closely related to the issues of human resources management. There are several ways to implement the policy. We are going to describe only three of them, which, in our opinion, are the most important. The first way is to develop special programs in order to address current and future challenges in energy industry or enterprises/companies. On the one hand, this is due to issues of **innovation** in the energy sector, the tools of their distribution [8]. On the other hand, - this is due to projects, which aim is to ensure resource-efficient, energy-saving technologies. For example, issues of methodology and practice of the development of comprehensive long-term energy conservation and energy efficiency programs are covered in the doctoral thesis [9]. In this respect, scientific works of S.M. Klyugin, S.M. Naumenko, I.V. Eliseeva must be noted as well. Author [10] has offered a scientific justification of technical solutions to organization of the personnel in the management system of energy enterprises; however, the author sees it through the prism of fiscal responsibility.

System management of personnel innovative activity should be considered as one of the resources of enterprises competitiveness in any industry. The essential difference between the present stage of the development of the energy sector from the previous one is due to the new look on the perception of the role and functions of the personnel resource in energy industry. Firstly, the human resource is capable of becoming human capital and can become profitable for the industry/enterprises. Secondly, the human resource management as a subject becomes a "domestic consumer", which allows the industry/company to develop successfully. According to [2] it is tthe point of view, which makes it possible to overview a modern personnel resource as a competitive advantage for industries/businesses.

The second way of implementation of the given direction of the personnel policy is to develop a special motivational tool for raising interest and satisfaction of personnel. Traditionally, it is associated with the introduction of certification procedures that can be implemented at different levels. For example, they can be implemented as a set of key performance indicators of industrial enterprises [11], or as a form of performance appraisal [12]. It is important to emphasize that the regulation of competitive relations and stimulation has been mainly undertaken in the search for economic mechanisms to address human resources management issues.

The third way of implementation of the personnel management is to create a modern staff selection system. The third way is usually seen in the wider context of management. Traditionally, there is a difference between open and close personnel policies. The open policy is characterized by willingness to employ any worker on condition that he/she has appropriate qualifications and skills. The close personnel policy is focused on hiring new staff only from a lower-level official job position. All subsequent replacements deal only with the number of employees at the enterprise. There are some social studies of the issue. However, the specificity of the energy sector is not taken into account. The selection of personnel depends on what knowledge, skills, and abilities, motivation of personnel resources are required for the industry/enterprise, and thus will be profitable. The current challenge is to determine the feasibility of training personnel on the ground (in terms of corporate training as well), encourage personnel to update existing competencies, or to dismiss personnel and attract staff from outside.

The main condition is the current level of competences that can be a result of personnel training and experience. Due to the competency component, view on human resources and its assessment proceeds from the category of economic issues into psychological and pedagogical categories. There are many issues in connection with the quality of training and assessment of intellectual capital in training of experts in technical universities. According to the just assertion of [13], current challenges have led to the need to form a human of innovative type. The author has also determined the status of twelve innovative human traits due to imperative challenges of modern innovation environment. Mobility and the qualification and professional potential are among them.

Mobility of staff resources may occur in three forms: territorial, that is associated with the change of residency and finding a new job in a city or in the country; vertical, i.e. climbing up the career ladder within the same type of activity; professional, that means a change in the type of activity.

Qualification and professional potential as a status characteristic of personnel resources can be represented by a number of factors: *education* – further training; *additional qualities and skills* that arise under the pressure of modern challenges and innovative environment, in particular, there is a need to know at least one foreign language, as well as to focus on the world, rather than a local practice; *competency moment*, the essence of which is seen as the ability of a human to independently obtain new information and to put acquired knowledge and skills into practice.

If we turn to the definition of the nature and content of the competences, it can be found out that the task cannot be completely resolved within the framework of economic sciences, some incentives will not be enough. These issues must be addressed mainly in the context of pedagogy and psychology.

3 Psycho-pedagogical perspective view on the development of human resources in the energy sector

As a whole, pedagogy does not ignore questions of training for various sectors of the domestic industry where much attention is paid to training of engineers. Interesting historical review of this issue, determination of the process characteristics, depending on the characteristics and needs of a particular region, has been made. As a part of a series of historical sciences, the research devoted to training of workers and engineering staff for some industries on the example of development in the second half of the twentieth century has been carried out. Despite that, the energy sector has not been researched yet.

With regard to vocational education and training of engineering students, the study mainly deals with interdisciplinary connections, formation of a mathematical thinking culture, technologies for formation of a system-evolutionary style of thinking, revealing of pedagogical conditions for formation of professionally significant personality traits, abilities of students et al.

However, the issues concerning training of future professionals in the field of energy and electrical and heating engineering have been rarely raised. Basically, much attention is paid to the technologies used in the educational process on the example of a single discipline. Often it is training in terms of a general education cycle in a technical college, for example, physics. Author [11] mentions application of remote technologies in the system of independent student work in computer science. T.N. Petikova characterizes the competence-oriented technology for training competitive engineering specialists in non-technical universities. Author [14] describes the technology of realization of continuity of professional training in colleges and universities. M.K. Medvedeva analyzes effectiveness of e-learning. There is also more common research on the formation of certain types of competencies in foreign languages: the reflective foreign language competence of students of power specialties [15].

The scientific research on the pedagogical issues concerning formation and development of professional competences is particularly scarce. In this regard, scientific study of E.V. Shischenko on the formation of professional competencies of technical students on the basis of integration of electrotechnical disciplines has become particularly interesting. We also consider it important to draw attention to scientific work [16], which deals with the formation of professional competences in the field of personnel management of university students. Despite the fact that the subject "Fundamentals of personnel management" is rarely included in training of future engineering specialists, the results obtained by the author are rather interesting and significant.

At present, almost all industries are experiencing an increased demand for qualified personnel in the production of different levels, from skilled workers to senior professionals. In that case, when the issue concerning formation of professional competences of the expert in the field of modern energy, it is important to have a complete idea of what knowledge, skills, abilities, should become target directions. They must meet modern requirements of the industry and help to overcome the deficit of technical specialists. A clear qualitative analysis and forecast of the staff situation at different levels substitutes the basis of personnel policy for the energy sector to ensure resource-efficient technologies.

4 Conclusions

Personnel policy is based on the appropriate strategies, and its implementation is carried out in several directions. The strategies have become of paramount importance since there is the need to introduce resource-efficient technologies and to manage staff in accordance with the requirements of the competitive environment and the specificity of the energy sector.

Historical retrospective of the training arranged for workers and engineers of the domestic power industry, as well as comparative benchmarking with similar foreign practice have become of great interest.

There is a growing popularity of the competence-based approach in training of competitive specialists for various industries. Despite the advances in the scientific study of vocational training for engineers in general, it is necessary to overcome the shortage of scientific educational research on the problems of formation and development of proper professional competences for specialists in the field of energy.

Modern personnel policy of the energy sector should be based on a clear qualitative analysis of human resources and the forecast of the staff situation at different levels.

References

1. A.A. Shiryaev *Knowledge Management in the reorientation of the Russian economy on an innovative path of development* (Ph.D., Diss., Moscow, 2012)
2. E.V. Mochalnikova *Formation of personnel strategy of manufacturing enterprises in the competitive environment* (Ph.D., Diss., Irkutsk, 2005)
3. I.F. Kurunov *Development of resource blast smelting technology, based on its research and mathematical modeling* (Dokt, Diss., Moscow, 2003)
4. S.N. Romanov *Organizational-economic mechanisms of development and industrial development of energy-saving technologies: the case of domestic gas savings* (Ph.D., Diss., Moscow, 2010)
5. A.S. Savenko *Resource saving management in the enterprise, based on the analysis of energy reserves* (Ph.D., Diss., Moscow, 2012)
6. V.P. Vershinin *Conceptual characteristics (state and basic trends) of the marketing environment in search for market niches in the field of energy saving and energy efficiency management of science-based management decisions* (Dokt, Diss., Sochi, 2011)

7. A.S. Mozgova *The organizational and economic mechanism of increase of energy efficiency of the oil and gas enterprises through energy audits* (Ph.D., Diss., Moscow, 2014)
8. T.A. Krolin *The mechanism and instruments of distribution of innovations in the field of energy saving* (Ph.D., Diss., Moscow, 2012)
9. I.A. Bashmakov *Development of integrated long-term programs in energy saving and energy efficiency: methodology and practice* (Dokt, Diss., Moscow, 2013)
10. T.A. Prokhorova T.A. *Organization of personnel management system of power plants as the center of financial responsibility* (Ph.D., Diss., Barnaul, 2007)
11. N.M. Kuznetsov N.M. *Development of motivation system based on key performance indicators of the enterprise efficiency: on the example of electric power* (Ph.D., Diss., Moscow, 2009)
12. E.V. Spiridonova E.V. *The development of the system of certification for electrical safety of electrical installation personnel* (Ph.D., Diss., Chelyabinsk, 2004)
13. R.G. Shangaraev *Modern challenges and the formation of the person of innovative type* (Ph.D., Diss., Ivanovo, 2006)
14. Z.M. Shakurova *Technology of realization of continuity of professional training in the system professional school - higher school. On the example of power specialties* (Ph.D., Diss., Kazan, 2002)
15. L.A. Sobinova *Students training in the professional focused foreign-language reading using an electronic manual* (Ph.D., Diss., Nizhny Novgorod, 2016)
16. L.L. Kifa *Formation of professional competence in the field of personnel management at students of higher education institution: on the example of the discipline "A human resource management basis"* (Ph.D., Diss., Tolyatti, 2012)

ABOUT THE RESULTS OF THE DESTRUCTION OF THE MOLECULES OF LIQUID HYDROCARBONS IN THE FIELD OF ACOUSTIC CAVITATION

Michael Rudenko[1,*], *Mark* Vasiltsov[1], *Stepan* Gorbasenko[1], and *Tatiana* Lavrenyuk[1]

[1]Angarsk State Technical University, Angarsk, Russia

Abstract. Analysis was conducted of fractional composition of hydrocarbon fuel. It is found that the excitation of cavitation in the fuel leads to a change of its fractional composition. This result can be explained by the destruction of liquid hydrocarbon molecules under high intensity of the unsteady pressure field

1 Introduction

To intensify the process, which is conducted in liquid, various methods are used, including cavitation. In [1], Professor Ivchenko V.M. highlighted the main hydrodynamic phenomena accompanying the collapse of the cavitation bubble. Firstly, it is the emergence of shock waves with an intensity of up to 100 MPa; forming streams 30-70 microns in diameter and up to 200 m / s; unsteady pressure fields of different frequency band from the collapsing bubbles.

More recent studies have shown the feasibility of processing the fuel-water mixture in the field hydrodynamic cavitation [2,3].

The data obtained is determine the interest to possible changes in the molecular composition of a liquid hydrocarbon fuel when it is processed in conditions the non-stationary pressure fields of high intensity.

In experiments [4] it was found that:

- when exposed to ultrasonic cavitation on hydrocarbon (C_9H_{20}) free radicals are formed;
- colliding with molecules of the original fuel, the free radicals is generated formation of new radicals;

* Corresponding author: mg-rud@mail.ru

– in products of reaction accumulate in hydrocarbons of lower molecular weight.

Decrease in molecular weight may be a special case, which is typical for the experimental conditions, or used hydrocarbon.

In this connection, was conducted an experiment to determine the fractional composition of the base diesel fuel (without additives) before and after exposure acoustic cavitation.

2 A brief description of the equipment and technology of the experiment

In experiments using a device of ultrasonic technology UZTA-0.8 / 22 with a frequency of ≈ 22 kHz.

Fuel test sample of 100 ml was placed in a sealed reactor made of stainless steel

When acoustic radiation imposed on the fuel, its temperature is increased, resulting in the need to use a cooling and controlling the temperature using a thermocouple.

Distillation of diesel fuel before and after cavitation ,was carried out in accordance with GOST 2177-82, using standard equipment.

3 Results of experiment

The results of measurements of fractional composition of diesel fuel before and after the cavitation effects are presented in figure 1, where the horizontal axis represents the value of the percentage composition of the fractionated fuel, and the vertical axis represents the temperature of boiling of the appropriate fractions. Markers "plus" correspond to the control sample. The samples treated in the field of acoustic cavitation, corresponds to a "X" mark.

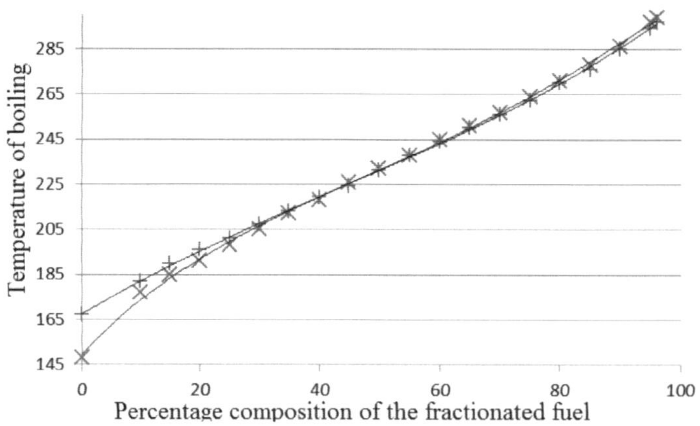

Fig. 1. The boiling points of the control sample and the treated diesel fuel.

The part of the experimental data, and the nature of the dependencies obtained prior to treatment and after fuel treatment in the field of acoustic cavitation is coincide, making it difficult to analyze the results.

Consider the difference between the boiling temperatures of the appropriate fractions (see. figure 2).

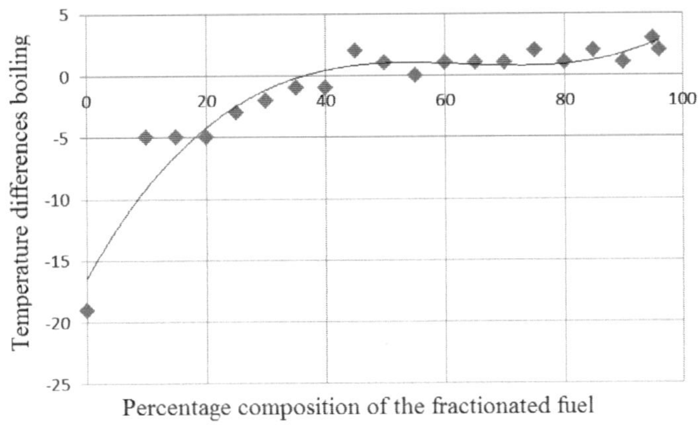

Fig. 1. Temperature differences boiling of experimental and control samples of the diesel fuel.

4 Discussion of the results

Note the following features:

- as a result of cavitation, distillation temperature of the light ends somewhat reduced. This may indicate the formation of hydrocarbons with lower molecular weights than the original fuel.
- Simultaneously with decreasing temperature of distillation light ends, there is a slight increase of temperature factions of high-boiling components of the diesel fuel. This may indicate the formation of hydrocarbons with a higher molecular weight than the original fuel.

We assume that the marked differences can be explained as follows:

- under acoustic cavitation occurs formation of radicals, which agrees with previous data and
- as a result of transformation, hydrocarbons accumulate with a higher and a lower molecular weight

References

1. V.M. Ivchenko, E.D. Malimon, Prilkladnaya mechanica I teplophisica [Applied Mechanics and Thermophysics] **5** (1975) [in Russian]
2. M.G. Radchenko, V.I. Grintsevich Hydrodynamica bol'shih scorostey: mejvusovskiy sbornik (Hydrodynamics of high speeds: Interuniversity collection, 1982) [in Russian]
3. V.M. Ivchenko, M.G. Rudenko, N.G. Rudenko, Sposop obrabotky topliva dvigateley vnutrennego sgoraniya (Internal combustion engines fuel processing method of the copyright certificate) USSR No.125419 [in Russian]
4. S.P. Zubrilov, V.I. Seliverstov, M.I. Braslavsky, LIVT Trudy **45** (1982) [in Russian]

MATEC Web of Conferences **72**, 01098 (2016) DOI: 10.1051/matecconf/20167201098
HMTTSC-2016

ANALYTICAL STUDY OF MICROWAVE HEATING OF THE COAL LAYER WITH MIXED BOUNDARY CONDITIONS OF I AND III TYPES

Vladimir Salomatov[1, *], *Vadim* Karelin[2], *Sergey* Paschenko[2], and *Vasiliy* Salomatov[1]

[1] Institute of Thermophysics SB RAS, Russia
[2] Novosibirsk State University, Russia

Abstract. This work is dedicated to the search for the exact analytical dependences of microwave heating due to absorption of a plane electromagnetic wave by coal layer with asymmetric and non-uniform heat dissipation conditions I and III kind. Some of simplifications have been made, such as one-dimensional problem, uniformity and isotropic coal material, and the constancy of the electrical properties of thermal coal during heating of microwave radiation. This has led to the fact that the Maxwell's task is solved separately from the Fourier's task, and a heat source generated in the carbon layer is subject to Bouguer law. For the system of equations of heat transfer has been found a new dependent variable, which is to simplify the search for a final solution. All this has given the possibility of finding rigorous analytical solution of the problem of microwave heating of the coal layer in the presence of asymmetric and inhomogeneous boundary conditions I and III kind.

1 Features of microwave heating of materials

In traditional conditions the heating of materials in the heat treatment process is performed by heat transfer from the surface to the inside by heat conduction. Absolutely other heating mechanism is connected with the microwave absorption in volume dielectric material, including coal. In this mode, the material forming the internal heat sources, which provide its heating. Energy conversion efficiency of the electric field to the heat increases in direct proportion to the oscillation frequency and square of the electric field. It should be noted the easy feeding of microwave energy for almost any of the heated area of the body. An important advantage of microwave heating is thermal absence of inertia i.e. the ability to almost instantly turn on and off the thermal influence on the material being processed. Hence, high precision of adjustment and reproducibility of the heating process are observed. The advantage of microwave heating is also fundamentally high conversion efficiency of the microwave to heat energies that is allocated in the amount of heated bodies. The theoretical value of this efficiency is close to 100%. An important advantage of

[*] Corresponding author: salomatov.vv@mail.ru

microwave heating is the possibility of the implementation and practical application of new and unusual his species, for example, selective, uniform, ultra-pure, self-regulating.

Selective heating is based on dielectric loss depending on the wavelength, i.e. depending of the dielectric loss tangent δ as a function of wavelength λ. In the multicomponent mixture of dielectrics it will be heated only those portions where high tg δ. So, in the coal array it will be harder to heat the organic part, rather than a mineral.

Uniform heating. Typically, the heat transfer is by convection and radiation. Hence it is inevitable the temperature gradient from the surface to the depth of the material, and the larger, the smaller the thermal conductivity. Reduce or eliminate the temperature gradient is almost possible by increasing the processing time. In many cases only by slow heating it is possible to avoid overheating of the surface layers. With microwave energy it is possible not only to uniformly heating over its volume, but also to receive any desired predetermined temperature distribution. Therefore, in the microwave heating there are a lot of opportunities for acceleration of a number of technological processes.

Ultrapure heating. In compare with heating by a gas flame and by arc torches which generally pollute materials, the microwave energy can be supplied to the material through the protective shell of solid dielectrics. The heat transfer medium may be absent at all. As a result the surface contamination from the product is almost completely eliminated. In addition, placing the heated material in a controlled volume environment or an inert gas, it is possible to eliminate surface oxidation. Pollution of the dielectric through which the supplied microwave energy is very small, because in the case of small losses, even with the passage of a large microwave power it remains practically cold.

Self-regulating heating. Upon heating for drying purposes the quality of the material is greatly improved due to the fact that the heating is stopped automatically as places dried. This is explained by the fact that the dielectric loss tangent of such materials, such as coal, is directly proportional to moisture. Therefore, with decreasing moisture the loss of the microwave energy during drying is reduced and heating is continued only on those portions of the processed material, which has remained high humidity.

The search for optimal microwave heating modes, designing microwave setups, the calculation of maximum temperature and heat flux and others parameters can be most effectively done using the theoretical approaches. Most interesting are primarily analytically rigorous solutions of microwave heating tasks. With their help it is readily accomplished to make a parametric analysis of microwave heating, to conduct a rapid calculation of thermal processes, identifies the fundamental laws. The purpose of this study is related to the search for such solutions, in particular, are often encountered in practice cases of microwave exposure to carbon layer with the mixed conditions of heat removal from its external surfaces of the first and the third kind.

2 The mathematical formulation of problems and build their solutions

We will solve the two tasks of this class.

Task 1

Diagram of the task is shown in figure 1.

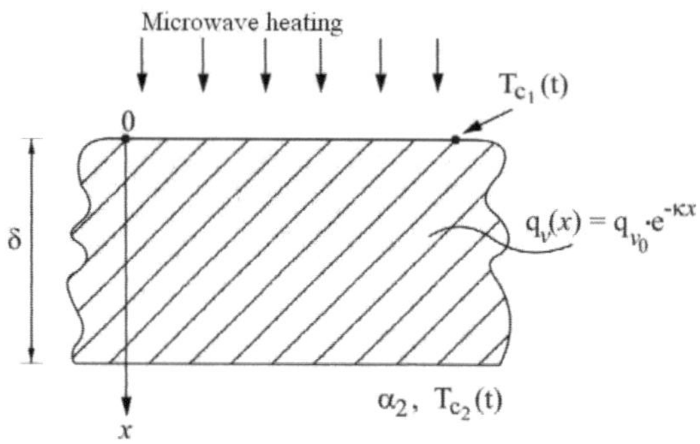

Fig. 1. Scheme of microwave heating task of the coal's flat layer with a mixed asymmetric boundary conditions I and III kind.

The microwave heating of the coal layer system of equations with the adopted simplifications is as follows:

$$\frac{\partial T(x,t)}{\partial t} = a\frac{\partial^2 T(x,t)}{\partial x^2} + \frac{q_{v_0}}{c\rho}e^{-kx} \tag{1.1}$$

$$T(x,0) = f(x) \tag{1.2}$$

$$T(0,t) = T_{c_1}(t) \tag{1.3}$$

$$\lambda\frac{\partial T(\delta,t)}{\partial x} = \alpha_2\left[T(\delta,t) - T_{c_2}(t)\right] \tag{1.4}$$

$T(x,t)-$ current temperature,K; $x-$ transverse coordinate, m; $a-$ thermal diffusivity, m^2/s; c – specific heat, J/kg·K; ρ – dencity, kg/m^3; t – time, s; q_{v_0} – the maximum value of the internal heat source, W/m^3; k - bsorption coefficient in the Bouguer law,1/m; λ-coal thermal conductivity, $W/(m\cdot K)$; $T_{c_1}(t)-$ time-variable temperature of the outer surface, K; α_2 – heat transfer coefficient on the lower surface, $W/(m^2\cdot K)$.

This system (1.1-1.4) by introduction of a new dependent variable:

$$u(x,t) = T(x,t) - T_{c_1}(t) - \frac{T_{c_1}(t)-T_{c_2}(t)}{\frac{\lambda}{\alpha_2}-\delta}x \tag{1.5}$$

It is reduced to a simpler system of equations with homogeneous boundary conditions:

$$\frac{\partial u(x,t)}{\partial t} = a\frac{\partial^2 u(x,t)}{\partial x^2} + q_v(x,t), \tag{2.1}$$

$$u(x,0) = F(x), \tag{2.2}$$

$$u(0,t) = 0, \tag{2.3}$$

$$\lambda\frac{\partial u(\delta,t)}{\partial x} = \alpha_2 u(\delta,t), \tag{2.4}$$

$q_v(x,t) = \frac{q_{v_0}}{c\rho}e^{-kx} \text{-} \frac{\partial T_{c_1}(t)}{\partial t} - \frac{\partial (T_{c_1}(t)-T_{c_2}(t))}{\partial t\,(\frac{\lambda}{\alpha_2}-\delta)}x$ – a new distribution of heat sources,

$F(x) = f(x) - T_{c_1}(0) - \frac{T_{c_1}(0)-T_{c_2}(0)}{\frac{\lambda}{\alpha_2}-\delta}x$ – new initial temperature distribution

The transformed system (2.1-2.4) with an inhomogeneous equation (2.1) and inhomogeneous initial condition (2.2) is more convenient to solve, divided into two [1]. We construct the general solution in the form of a superposition of two solutions $u(x,t)=u_1(x,t)+u_2(x,t)$, for each part of which should be solved the following boundary value problems:

for u_1

$$\frac{\partial u_1(x,t)}{\partial t} = a\frac{\partial^2 u_1(x,t)}{\partial x^2} + q_v(x,t), \tag{3.1}$$

$$u_1(x,0) = 0, \tag{3.2}$$

for u_2

$$\frac{\partial u_2(x,t)}{\partial t} = a\frac{\partial^2 u_2(x,t)}{\partial x^2}, \tag{4.1}$$

$$u_2(x,0) = F(x) , \tag{4.2}$$

under the previous boundary conditions (2.3), (2.4)

To solve each system of equations is necessary to find the eigenvalues and eigenfunctions of the homogeneous problem (system 4.1 and 4.2). The unknown function u_i - can be represented as the product of two independent functions:

$$u_i=\psi_i(x)\varphi_i(t)$$

Then, find the eigenvalues γ_k and eigenfunctions $\Psi_k(x)$ of homogeneous problem. This leads to two equations:

$$\varphi_i{}'(t) + (a\gamma)^2\varphi_i(t) = 0 \tag{5.1}$$

$$\psi_i{}''(x) + \gamma^2\psi_i(x) = 0 \tag{5.2}$$

Solutions for $\psi_i(x)$ generally following $\psi_i(x) = c_1\sin(\gamma x) + c_2\cos(\gamma x)$
Substituting in equations of the boundary conditions, we find:
1) $c_2=0$
2) $\mathrm{tg}(\gamma\delta) = \frac{\lambda}{\alpha_2}\gamma$. From this equation, we find the eigenvalues γ_k
Then for the functions $\Psi_n(x)$ and $\varphi_n(x)$ we have the following expression

$$\Psi_n(x) = \sin(\gamma_n x),$$

$$\varphi_n(x) = a_n e^{-(\sqrt{a}\gamma_n)^2 t}$$

a_n – integrating constant.
This is enough to find private solutions for the boundary value task u2:

$$u_{2n} = a_n e^{-(\sqrt{a}\gamma_n)^2 t}\sin(\gamma_n x)$$

Then we draw up a series of:

$$u_2 = \sum_{n=1}^{\infty} a_n e^{-(\sqrt{a}\gamma_n)^2 t}\sin(\gamma_n x)$$

Coefficients a_n can be found method of orthogonality,
Using the initial distribution $u_2(x,0)$:

$$F(x) = \sum_{n=1}^{\infty} a_n \sin(\gamma_n x)$$

This is a decomposition of a given function in a trigonometric Fourier series of sines. The expansion coefficients are found by the formula:

$$a_n = \frac{1}{\|\varphi_n\|^2} \int_a^b r(\xi) F(\xi) \varphi_n(\xi) d\xi \ ,$$

$\varphi_n(\xi)$ – eigenfunction ($\sin(\gamma_n x)$), r(ξ) – weighting function (in this case equal to 1)

a=0, b=δ, $\|\varphi_n\|^2 = \int_0^\delta \sin(\gamma_n x)^2\, dx = \left(\frac{\delta}{2} - \frac{\sin(2\gamma_n\delta)}{4\gamma_n}\right)$

Next, we find the desired function:

$$u_2(x,t) = \sum_{n=1}^{\infty} \frac{1}{\frac{\delta}{2} - \frac{\sin(2\gamma_n\delta)}{4\gamma_n}} e^{-(\sqrt{a}\gamma_n)^2 t} \sin(\gamma_n x) \int_0^\delta F(\xi) \sin(\gamma_n \xi) d\xi$$

To find the solution of (3.1-3.2) u_1 use applicable in this case, the theorem of Steklov VI [1], according to which $u_1(x,t)$ can be represented as a Fourier series in eigenfunctions of the homogeneous problem (4.1-4.2):

$$u_1(x,t) = \sum_{n=1}^{\infty} b_n(t) \Psi_n(x)$$

To find the coefficients $b_n(t)$ expand a given function q (x, t) in a Fourier series in eigenfunctions $\Psi_n(x)$:

$$q_v(x,t) = \sum_{n=1}^{\infty} q_n(t) \Psi_n(x),$$

$$q_n(t) = \frac{1}{||\Psi_n(x)||^2} \int_0^\delta q_v(x,t)\, \Psi_n(x) dx$$

Next, we substitute the expression for $q(x,t)$ и $u_1(x,t)$ в (3.1):

$$\sum_{n=1}^{\infty} b_n{}'(t) \Psi_n(x) = \sum_{n=1}^{\infty} a\, b_n(t) \Delta\Psi_n(x) + \sum_{n=1}^{\infty} q_n(t) \Psi_n(x)$$

Then, using (5.2) and comparing the coefficients of $\Psi_n(x)$, we find for $b_n(t)$:

$$b_n(t)' + \left(\sqrt{a}\gamma_n\right)^2 b_n(t) = q_n(t)$$

with the initial condition (3.2), which implies: $b_n(0) = 0$
Solution of linear inhomogeneous differential equation has the form:

$$b_n(t) = \int_0^t e^{-\left(\sqrt{a}\gamma_n\right)^2 (t-\tau)} q_n(\tau) d\tau$$

As a result for $u_1(x,t)$:

$$u_1(x,t) = \sum_{n=1}^{\infty} \frac{1}{\frac{\delta}{2} - \frac{\sin(2\gamma_n\delta)}{4\gamma_n}} \sin(\gamma_n x) \int_0^t \int_0^\delta e^{-(\sqrt{a}\gamma_n)^2(t-\tau)} q(\xi,\tau) \sin(\gamma_n \xi)\, d\tau d\xi, \quad (5.3)$$

The interim solution u(x,t)= $u_1(x,t) + u_2(x,t)$ of system (2.1 – 2.4):

$$u(x,t) = \sum_{n=1}^{\infty} \frac{1}{\frac{\delta}{2} - \frac{\sin(2\gamma_n\delta)}{4\gamma_n}} \sin(\gamma_n x) \int_0^t \int_0^\delta e^{-(\sqrt{a}\gamma_n)^2(t-\tau)} q(\xi,\tau) \sin(\gamma_n \xi) d\tau d\xi$$
$$+ \sum_{n=1}^{\infty} \frac{1}{\frac{\delta}{2} - \frac{\sin(2\gamma_n\delta)}{4\gamma_n}} e^{-(\sqrt{a}\gamma_n)^2 t} \sin(\gamma_n x) \int_0^\delta F(\xi) \sin(\gamma_n \xi) d\xi \quad (6.1)$$

After inverse substitutions we get the final solution of the system (1.1-1.4)

$$T(x,t) = T_{c_1}(t) + \frac{T_{c_1}(t) - T_{c_2}(t)}{\frac{\lambda}{\alpha_2} - \delta} x + \sum_{n=1}^{\infty} \frac{1}{\frac{\delta}{2} - \frac{\sin(2\gamma_n\delta)}{4\gamma_n}} \sin(\gamma_n x) \int_0^t \int_0^\delta e^{-(\sqrt{a}\gamma_n)^2(t-\tau)} \left\{ \frac{q_{v0}}{c\rho} e^{-k\xi} - \frac{\partial T_{c_1}(\tau)}{\partial \tau} - \frac{\partial (T_{c_1}(\tau) - T_{c_2}(\tau))}{\partial \tau \left(\frac{\lambda}{\alpha_2} - \delta \right)} \xi \right\} \sin(\gamma_n \xi) d\tau d\xi + \sum_{n=1}^{\infty} \frac{1}{\frac{\delta}{2} - \frac{\sin(2\gamma_n\delta)}{4\gamma_n}} e^{-(\sqrt{a}\gamma_n)^2 t} \sin(\gamma_n x) \int_0^\delta \left\{ f(\xi) - T_{c_1}(0) - \frac{T_{c_1}(0) - T_{c_2}(0)}{\frac{\lambda}{\alpha_2} - \delta} \xi \right\} \sin(\gamma_n \xi) d\xi \quad (6.2)$$

Task 2.

Now consider the case when the upper boundary of the layer satisfies the III kind of condition, the bottom - I kind.

The scheme of this problem is shown in figure 2.

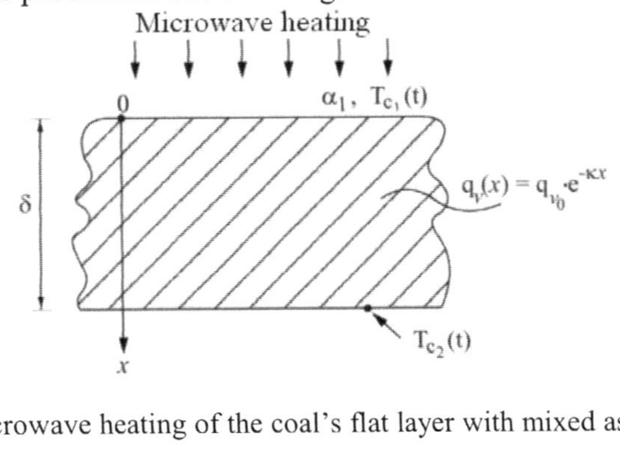

Fig. 2. Scheme of task of microwave heating of the coal's flat layer with mixed asymmetric boundary conditions I and III kind.

The microwave heating system's equations to the coal layer of this formulation of the problem is of the form:

$$\frac{\partial T(x,t)}{\partial t} = a \frac{\partial^2 T(x,t)}{\partial x^2} + \frac{q_{v0}}{c\rho} e^{-kx}, \quad (7.1)$$

$$T(x,0) = f(x), \quad (7.2)$$

$$-\lambda\frac{\partial T(0,t)}{\partial x} = \alpha_1\left[T(0,t) - T_{c_1}(t)\right], \quad (7.3)$$

$$T(\delta,t) = T_{c_2}(t), \quad (7.4)$$

In this case, a new variable, which simplifies the system (7.1-7.4), is as follows:

$$u(x,t) = T(x,t) - \frac{\delta^2-x^2}{\delta^2}[T_{c_1}(t) - T_{c_2}(t)] - T_{c_2}(t) \quad (7.5)$$

Now a simple system with homogeneous boundary conditions as follows:

$$\frac{\partial u(x,t)}{\partial t} = a\frac{\partial^2 u(x,t)}{\partial x^2} + q_v(x,t), \quad (8.1)$$

$$u(x,0) = F(x), \quad (8.2)$$

$$-\lambda\frac{\partial u(0,t)}{\partial x} = \alpha_1 u(0,t), \quad (8.3)$$

$$u(\delta,t) = 0, \quad (8.4)$$

$q_v(x,t) = \frac{q_{v0}}{c\rho}e^{-kx} - (\frac{\delta^2-x^2}{\delta^2})\frac{\partial(T_{c_1}(t)-T_{c_2}(t))}{\partial t} - \frac{\partial T_{c_2}(t)}{\partial t}$ -2 $a[T_{c_1}(t) - T_{c_2}(t)]$ – a new distribution of heat sources,

F(x) = f(x) $- \left(\frac{\delta^2-x^2}{\delta^2}\right)[T_{c_1}(0) - T_{c_2}(0)] - T_{c_2}(0)$ – new initial temperature distribution.

Further, by performing actions similar to the solution of problem 1, we divide the task into two subtasks with two solutions u_1 и u_2 . Then we find a new own functions:

$$\Psi_n(x) = \cos(\gamma_n x) - \frac{\alpha_1}{\lambda\gamma_n}\sin(\gamma_n x),$$

$$\varphi_n(x) = a_n e^{-(\sqrt{a}\gamma_n)^2 t},$$

The eigenvalues are found from a transcendental relations:

$$\text{tg}(\gamma_n\delta) = \frac{\lambda}{\alpha_1}\gamma_n,$$

a_n – integrating constant.
This is enough to find a private solution of the second sub-tasks (u2):

$$u_{2n} = a_n e^{-(\sqrt{a}\gamma_n)^2 t}(\cos(\gamma_n x) - \frac{\alpha_1}{\lambda\gamma_n}\sin(\gamma_n x))$$

Then we draw up a series of

$$u_2 = \sum_{n=1}^{\infty} a_n e^{-(\sqrt{a}\gamma_n)^2 t}(\cos(\gamma_n x) - \frac{\alpha_1}{\lambda\gamma_n}\sin(\gamma_n x))$$

Coefficients a_n can be found from the initial condition for the distribution of u_2 (8.2):

$$F(x) = \sum_{n=1}^{\infty} a_n(\cos(\gamma_n x) - \frac{\alpha_1}{\lambda\gamma_n}\sin(\gamma_n x))$$

This expansion is a given function by a trigonometric series of eigenfunctions.

The coefficients of decomposition are found a_n from:

$a_n = \frac{1}{\|\varphi_n\|^2}\int_a^b r(\xi)F(\xi)\varphi_n(\xi)d\xi,$

$\varphi_n(\xi)$ - eigenfunction - $(\cos(\gamma_n x) - \frac{\alpha_1}{\lambda\gamma_n}\sin(\gamma x))$, r($\xi$) – weighting function (in this case equal to 1)

We introduce a new value $c_n = -\frac{\alpha_1}{\lambda\gamma_n}$, simplifies subsequent calculations record:

a=0, b= δ, $\|\varphi_n\|^2 = \int_0^\delta (\cos(\gamma_n x) + c_n\sin(\gamma_n x))^2\,dx = \frac{\delta}{2}(1 + {c_n}^2) + \frac{1}{4\gamma_n}[(1 - {c_n}^2)\sin(2\gamma_n\delta) - 2c_n(\cos(2\gamma_n\delta) - 1)],$

Next, we find the desired function:

$$u_2(x,t) = \\ = \sum_{n=1}^{\infty}\frac{1}{\frac{\delta}{2}(1 + {c_n}^2) + \frac{1}{4\gamma_n}[(1 - {c_n}^2)\sin(2\gamma_n\delta) - 2c_n(\cos(2\gamma_n\delta) - 1)]}e^{-(\sqrt{a}\gamma_n)^2 t}\,.$$

$$[\cos(\gamma_n x) + c_n\sin(\gamma_n x)]$$

$$\int_0^\delta F(\xi)\,[\cos(\gamma_n\xi) + c_n\sin(\gamma_n\xi)]\,d\xi,$$

Searching solutions for another subtask u_1 performed similarly. So:

$$u_1(x,t) = \\ = \sum_{n=1}^{\infty}\frac{1}{\frac{\delta}{2}(1 + {c_n}^2) + \frac{1}{4\gamma_n}[(1 - {c_n}^2)\sin(2\gamma_n\delta) - 2c_n(\cos(2\gamma_n\delta) - 1)]}[\cos(\gamma_n x) + c_n\sin(\gamma_n x)]$$

$$\int_0^t\int_0^\delta e^{-(\sqrt{a}\gamma_n)^2(t-\tau)}q_v(\xi,\tau)\,[\cos(\gamma_n\xi) + c_n\sin(\gamma_n\xi)]\,d\tau d\xi$$

The total solution u(x,t)= $u_1(x,t) + u_2(x,t)$ of (8.1 – 8.4) is:

$$u(x,t) = \\ \sum_{n=1}^{\infty}\frac{1}{\|\varphi_n\|^2}[\cos(\gamma_n x) - \frac{\alpha_1}{\lambda\gamma_n}\sin(\gamma_n x)]\int_0^t\int_0^\delta e^{-(\sqrt{a}\gamma_n)^2(t-\tau)}q_v(\xi,\tau)\,[\cos(\gamma_n\xi) - \frac{\alpha_1}{\lambda\gamma_n}\sin(\gamma_n\xi)]\,d\tau d\xi + \\ \sum_{n=1}^{\infty}\frac{1}{\|\varphi_n\|^2}e^{-(\sqrt{a}\gamma_n)^2 t}\,[\cos(\gamma_n x) - \frac{\alpha_1}{\lambda\gamma_n}\sin(\gamma_n x)]\int_0^\delta F(\xi)\,[\cos(\gamma_n\xi) - \frac{\alpha_1}{\lambda\gamma_n}\sin(\gamma_n\xi)]\,d\xi, \quad (9.1)$$

After inverse substitutions final solution of the system (7.1-7.4) following:

$$T(x,t) = \\ \frac{\delta^2 - x^2}{\delta^2}\left[T_{c_1}(t) - T_{c_2}(t)\right] + T_{c_2}(t) + \\ \sum_{n=1}^{\infty}\frac{1}{\|\varphi_n\|^2}[\cos(\gamma_n x) - \frac{\alpha_1}{\lambda\gamma_n}\sin(\gamma_n x)]\int_0^t\int_0^\delta e^{-(\sqrt{a}\gamma_n)^2(t-\tau)}(\frac{q_{v0}}{c\rho}e^{-k\xi} - \\ (\frac{\delta^2 - \xi^2}{\delta^2})\frac{\partial(T_{c_1}(\tau) - T_{c_2}(\tau))}{\partial\tau} - \frac{\partial T_{c_2}(\tau)}{\partial\tau} - 2a[T_{c_1}(\tau) - T_{c_2}(\tau)])\,[\cos(\gamma_n\xi) - \frac{\alpha_1}{\lambda\gamma_n}\sin(\gamma_n\xi)]\,d\tau d\xi +$$

$\sum_{n=1}^{\infty} \frac{1}{\|\varphi_n\|^2} e^{-(\sqrt{a}\gamma_n)^2 t} [\cos(\gamma_n x) - \frac{\alpha_1}{\lambda\gamma_n} \sin(\gamma_n x)] \int_0^\delta (f(\xi) - \left(\frac{\delta^2-\xi^2}{\delta^2}\right) [T_{c_1}(0) - T_{c_2}(0)] - T_{c_2}(0)) [\cos(\gamma_n \xi) - \frac{\alpha_1}{\lambda\gamma_n} \sin(\gamma_n \xi)]\, d\xi,$ (9.2)

The derived analytically rigorous solution of the problem of microwave heating of the coal layer with mixed boundary conditions I and III type of (6.2), (9.2) in general allow for a wide range of private simplistic solutions with the possibility of a detailed parametric analysis. The information on the distribution of temperature fields is the base for the evaluation of microwave-ignition of coal fuel parameters determining the thermo-destructive forces, the search control of automated microwave heating effects, the implementation of optimal parameters of microwave technology coal arrays and other.

From these solutions can extract important information about the parameters of microwave heating of the coal massif:

1) To investigate the extremum of temperature function (6.2), (9.2), it is possible to find the position and the value of the maximum temperature;
2) In case of equality the maximum temperature of the coal layer and ignition temperature it is possible to find moment of ignition;
3) These solutions can also be used to find the maximum temperature difference between the coal layer, which directly affects the level of hazardous thermo-destructive forces.
4) Other process parameters for optimization, automation, etc.

3 Conclusion

Rigorous analytical solutions of two problems of microwave heating of the coal layer with mixed and fairly arbitrary boundary conditions I and III kind were built. Getting these solutions due to the adoption of the constancy of the electrical and thermal characteristics of the coal. With the help of two new substitutions the initial system of equations is reduced to a transformed with homogeneous boundary conditions that simplify the search for the final decision. The final solution was found by the method of making a superposition of two less complex subtasks. Internal heat source of the absorption of electromagnetic waves modeled by Bouguer law. With these solutions can be obtained vital information to a wide range of special cases: the temperature distribution on the layer with the passage of time, the maximum temperature, the ignition timing, the greatest temperature difference allotted heat and other important parameters of the coal heating layer of microwave radiation. The derived analytical solutions allow to obtain a plurality of individual results, which demand by engineering practice, which, of course, is the foundation for the creation of energy efficient and environmentally friendly thermal coal treatment microwave technology.

References

1. E.M. Kartashov, *Analytical methods in the theory of heat conduction of solid bodies* (Vysshaya Shkola, Moscow, 2001) [in Russian]
2. Vl.V. Salomatov, S.O. Sladkov, S.A. Pascenco, Vas.V. Salomatov, Combustion and plasma chemistry **12**, 2 (2014)

MATEC Web of Conferences **72**, 01099 (2016) DOI: 10.1051/matecconf/20167201099
HMTTSC-2016

INVESTIGATION OF BOILING IN HIGH-POWER LED COOLING SYSTEM

Evgeniy E. Shatskiy[1,2,*]

[1] Kutateladze Institute of Thermophysics SB RAS, 630090 Novosibirsk, Russia
[2] Novosibirsk state University, 630090 Novosibirsk, Russia

Abstract. The technology of solid-state light source is associated with the future of a number of sectors of the economy. The duration of failure-free operation, optical radiation power and other output characteristics of the LEDs are closely linked with p-n junction temperature, which makes the development of the cooling systems an important step in creating LED systems. In this work, we have created a new experimental setup for studying heat transfer from a local heat source, modelling the LED chip. The boiling heat transfer on the local heaters with a diameter of 5 and 10 mm has been investigated at the saturation temperature in bulk liquid. It is shown that on the finned surfaces the overheating relative to the saturation temperature in comparison with a smooth surface decreases up to one and a half times for the heater with a diameter of 10 mm, and up to three times for the heater with a diameter of 5 mm. There is up to two times increase in heat transfer coefficient on finned surfaces as compared to the smooth ones. For finned surfaces on the heater with a diameter of 1 mm the surface overheating relative to the saturation temperature decreases four times. More than three times increase is observed for the heat transfer coefficient on finned surfaces as compared to the smooth ones.

1 Introduction

Cooling of heat-emitting surfaces in different technological applications is an important problem, which can be solved by such processes as film flows [1, 2], etc. In our days the problem of creating an effective compact cooling system for powerful LEDs whose efficiency and durability are highly dependent on the efficiency of the power removal remains unsolved.

Study of evaporation mechanism on the surfaces of small size (SSS) was started in [3]. A review of works on heat transfer at evaporation on SSS may be found in [4]. In the literature, there is a dependencies for the critical heat flux density for each type of SSS, having the general form:

$$q_{cr}/q_{cr\infty} = k(d_s/l_\sigma)^n \quad (1)$$

* Corresponding author: shatskiy.itp@gmail.com

but differing in correlation multipliers; and d_s is the diameter of the heating surface. Here

$$q_{cr\infty} = 0.13 r_{lv} \rho_v \left(\frac{\sigma(\rho - \rho_v) g}{\rho_v^2} \right)^{0.25} \quad (2)$$

r_{lv} is the latent heat of vaporization, ρ_v is the vapor density, ρ is the fluid density, σ is the surface tension, and g is the gravity acceleration.

For locally heated horizontal plates k=3.4, n=-0.5 [4], for horizontal cylinders k=1.4 [5], and for spheres k=2.5 [5]. The reason for such significant effect of the size of SSS on q_{cr} is connected with the feature of hydrodynamic situation that occurs around the surface of small size (SSS).

2 Experimental setup

To study the heat transfer processes that occur during cooling of LED equipment in the boiling liquid, a new experimental setup was created. Schematic diagram of the experimental setup is shown in figure 1.

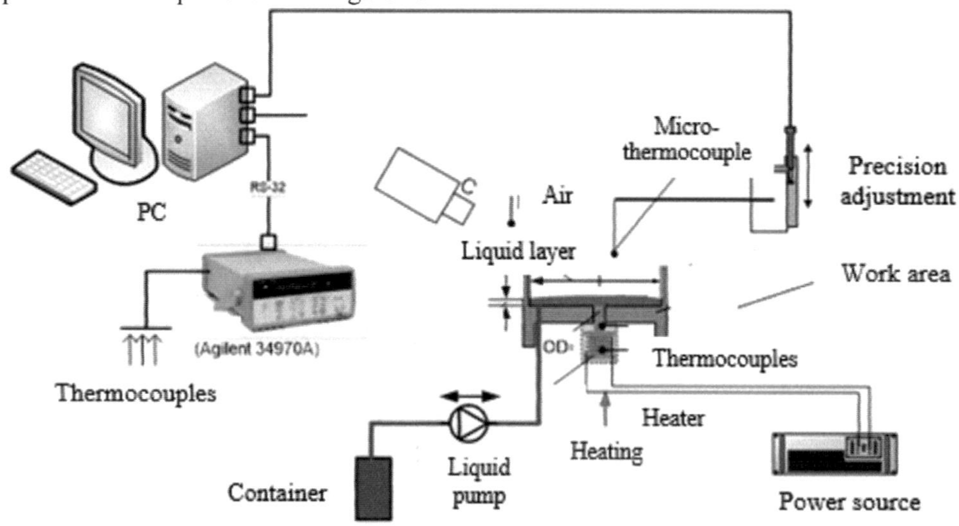

Fig. 1. Scheme of experimental setup.

The working fluid is supplied to the work area by means of a peristaltic pump. There the horizontal layer of liquid is formed. The thickness varies depending on the operating parameters of the experiment. In the center of the layer from the side of the substrate there is local heating performed by a copper heater. The power of the heating element is controlled by the power source. The temperature of the heater, the fluid and the medium is measured by a number of thermocouples. The data from thermocouples is collected using a control and measuring system, and the temperature measurement precision is 0.1 ^{0}C.

3 Results and discussion

Figure 2 presents a comparison of the obtained experimental data on the critical heat flux density with the calculated dependence (1) of [4] and results of work [6]. The figure shows that the density of the critical heat flux in a bulk liquid for heaters with D=1.6 and 5 mm is in good agreement with the dependence (1) and experimental data for vertical heaters. The heater size has a significant influence on the critical heat flux density (CHFD). The radial finning of the heater has no effect on CHFD, and the decrease of the layer height and the flow rate of the supplied liquid leads to a significant decrease in CHFD. The value of CHFD for D=1 mm is higher than the estimated one. Moreover, the highest increase was achieved by using finned surface.

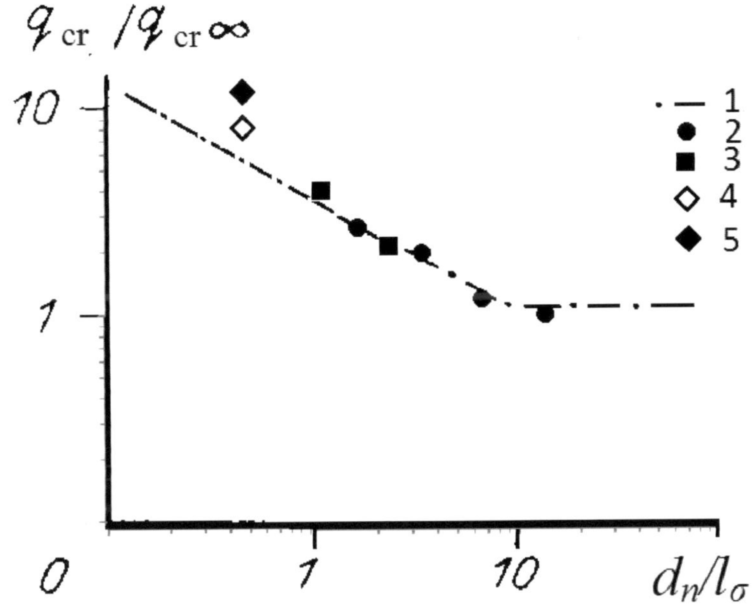

Fig. 2. – The dependence of CHFD on the size of the heater. 1 – calculation by formula (1) from [4], 2 – data from [6], 3 – D=1.6 and 5 mm, 4 – D=1 mm, 5 – D=1 mm with finning.

Experiments on the heaters with 5 mm diameter and smooth and finned surfaces have showed that for the finned surfaces the overheating relative to the saturation temperature decreases up to three times. Finning on the heater with 1 mm diameter decreases the overheating relative to the saturation temperature more than four times. Figure 3 shows the effect of the liquid temperature around the heat exchange surface on the dependence of the heat flux density for saturation temperature (points 1-3) and at a temperature of 25 – 30 ºC. Such regimes are possible in thermosiphons and heat pipes, if the regime of subcooling of condensed liquid is realized. It is seen that in this case, the temperature of the LED chip model is significantly reduced. It is shown that for the heat flux densities up to 1000 W/cm^2 the temperature of the thermal model of the LED chip does not exceed 120 ºC.

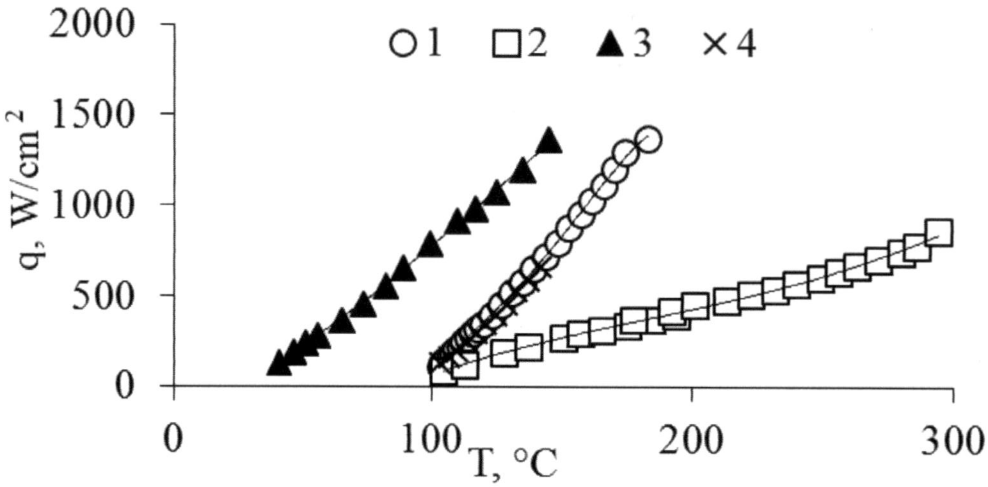

Fig. 3. – The dependence of the removed heat flux density on temperature, D=1 mm. 1 – large fins, 2 – smooth surface, 3 – nozzle with small fins, T= 25 – 30 ° C, 4 – small fins under saturated conditions.

Acknowledgments

The work was supported by the RF Ministry of Education and Science (Agreement No. 14.604.21.0053, project ID RFMEFI60414X0053).

References

1. E.A. Chinnov, J. Heat Transfer **138** (2016)
2. O.A. Kabov, D.V. Zaitsev, V.V. Cheverda and A. Bar-Cohen, Exp. Therm Fluid Sci. **35**, 5 (2011)
3. F. D. Moor, R. B. Mesler, AlChE J. **7**, 4 (1961)
4. V.I. Tolubinskiy, V.A. Antonenko, G.R. Kudritskiy, Yu.N. Ostrovskiy, Industrial Heat Engineering **9**, 4 (1987)
5. D. Linard, B. Dir, Teploperedacha **2** (1973)
6. E.A. Chinnov, A.V. Dyatlov, D.N. Kravchenko, J. Siberian Physico-Technical **5** (1992)

MATEC Web of Conferences **72**, 01100 (2016) DOI: 10.1051/matecconf/20167201100
HMTTSC-2016

NUMERICAL SIMULATION OF TURBULENT FLOW OVER OF SEMI-CYLINDRICAL PROFILE AT ZERO ANGLE OF ATTACK

Alexander Shebelev[1, *], and *Andrey* Gavrilov[1, 2]

[1] Siberian Federal University, Krasnoyarsk, Russia
[2] Institute of Thermophysics, Novosibirsk, Russia

Abstract. Results of unsteady two-dimensional numerical simulations for the turbulent flow over the semi-cylindrical profile at Reynolds number 45000 are presented. Three different turbulence RANS eddy viscosity models are used for the simulation. The calculations show the formation of alternating large-scale vortices shed in the wake of the body. The calculated averaged pressure distribution on the contour demonstrate a good agreement with the experimental data. A series of calculations have been performed for the estimation of influence of mesh resolution in the wake of streamlined body on the unsteady solution accuracy. It was found that a mesh size ~0.04 of chord length is sufficient to resolve the vortex street.

1 Introduction

A practical interest to the simulation of detached high Reynolds number flows is associated with numerical resolution of unsteady vortex structures, which are strongly influence on the integral characteristics of such flows. This class of flows is difficult to predict by methods of computational fluid dynamics because of unfixed separation point and development of non-stationary flow pattern.

2 Numerical algorithm

Flow around semi-cylindrical profile at zero angle of attack was investigated numerically. For this flow, there are the experimental data for drag and lift forces and for pressure distribution along the contour of body [1]. The Reynolds number defined by the length of chord L, velocity of undisturbed flow U and the fluid viscosity and density is $\mathrm{Re} = \rho UL/\mu = 45000$. Calculations have been performed in two-dimensional formulation. The height of computational domain is $40L$, the length in the streamline direction is $20L$.

[*] Corresponding author: aleksandr-shebelev@mail.ru

DOI: 10.1051/matecconf/20167201100

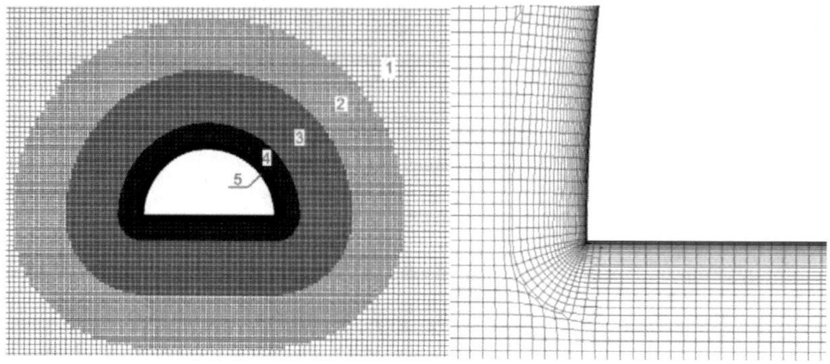

Fig. 1. Visual representation of the mesh.

The calculations were carried out on a hybrid hexahedral mesh composed by several uniform meshes and near wall body-fitted mesh (fig.1). To assess the influence of mesh resolution in the wake of body on the lift coefficient (*Cy*) and drag coefficient (*Cx*) four mesh with different mesh size in wake region were generated. Meshes parameters are specified in Table 1.

Table 1. Parameters computational meshes.

Mesh	mesh size in wake	number of cells, thousand
M1	0.02	250
M2	0.04	150
M3	0.08	126
M4	0.16	121

Non-reflecting outflow boundary conditions were set at the outlet. Symmetry conditions were set on the bottom and top faces of the computational domain. Non-slip boundary conditions were imposed at the wall. Mesh size near wall is enough for full resolution of near wall turbulent layer including viscous sublayer. The turbulence at the inlet boundary was defined by the length scale of turbulence $L_t=0.1\ L$ and turbulence intensity u/U=0.015.

Following Reynolds-averaged Navier-Stokes (RANS) models were used: the *k-ω SST* model with curvature correction [2], model with elliptic blending *k-ε-ζ-a* and model with elliptic relaxation *k-ε-ζ-f* [3].

The time step for unsteady simulation was *0.01 L/U*. The statistics was collected by averaging more than 100 of transit time. The calculations were performed with a finite volume in-house CFD code, σFlow [4].

3 Results and conclusions

The integral results of simulations performed with different turbulence models on mesh M1 and available experimental data are presented in the Table 2.

Table 2. Integral results.

Turbulent model	Period	*Cx*	*Cy*	*rms(Cx)*	*rms (Cy)*
k-ω SST cc	2.2	0.488	-0.89	0.046	0.314
k-ε-ζ-a	2.2	0.526	-1.014	0.064	0.252
k-ε-ζ-f	2.3	0.54	-0.944	0.064	0.208
Experiment	–	0.3	-1.1	–	–

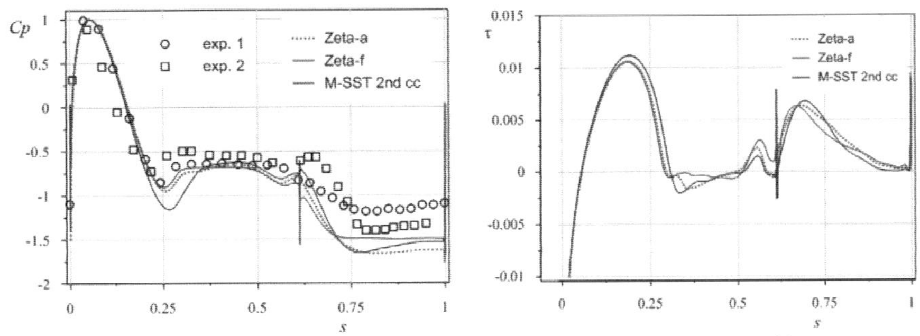

a) b)

Fig. 2. Distributions of average pressure (a) and wall shear stress (b) on surface. Symbols correspond to the experimental data [1].

Fig. 3. Streamlines of average flow for model *k-ε-ζ-f*.

Models with elliptic relaxation better predict the position of separation point and correctly reproduce average value of pressure near the separation point (fig. 2a). On the bottom of semi-cylinder all turbulence models underestimate the value of pressure. The low level of pressure can be explained by the high intensity of vortex flow generating under the body (fig. 3). All turbulence models overestimate the drag coefficient *Cx*. Models with elliptic relaxation slightly better predict the lift coefficient *Cy* with divergence from experimental data about 10% (Table 2).

In general, the unsteady RANS modeling of detached flow shows good agreement with experimental data. The unsteady formulation of RANS models makes it possible to resolve the development of vortex street behind the body.

Table 3. Influence of mesh size in the wake on the integral force (*k-ε-ζ-f* model).

Mesh	Cx	Cy	rms(Cx)	rms(Cy)
M1	0.5912	-0.96	0.0532	0.318
M2	0.5914	-0.96	0.0532	0.318
M3	0.5914	-0.96	0.0534	0.318
M4	0.5916	-0.958	0.0532	0.318
Experiment	0.3	-1.1	–	–

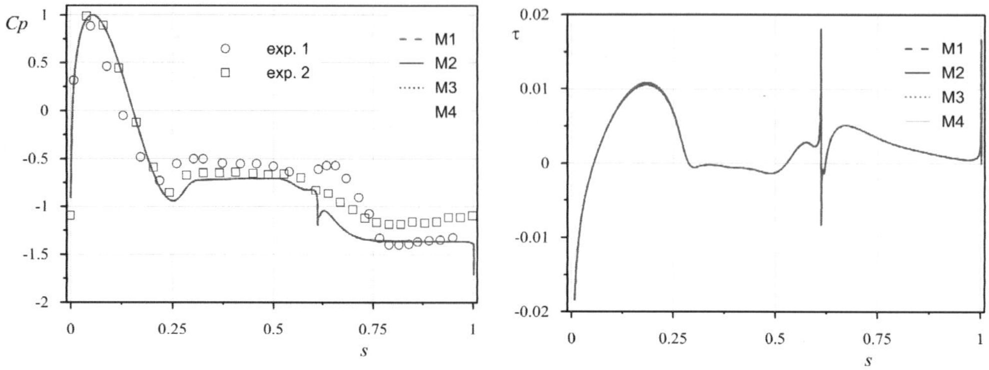

a) b)

Fig. 4. Influence of mesh resolution in the wake behind the body for the turbulence model *k-ε-ζ-f*. Averaged pressure distribution (a) and wall shear stress (b) on the body surface.

Figure 4 and Table 3 demonstrate that the mesh resolution in the of wake behind the streamlined body at a distance more than 2 *L* from the body does not affect on the integral forces acting on the body. In addition, it was found that a mesh size ~0.04 of chord length is sufficient to resolve the vortex street.

References

1. Z.P. Sluchanovskaya, Sci. works Instit. Mech. **24** (1973)
2. S.A. Isaev, P.A. Baranov, Yu.V Zhukova., A.E Usachov., V.B. Kharchenko, J. Eng. Phys. Thermophys. **87** (2014)
3. K. Hanjalić, M. Popovac, M. Hadžiabdić, Int. J. Heat Fluid Flow **25** (2004)
4. A.A. Gavrilov, A.V. Minakov, A.A. Dekterev, V.Ya. Rudyak , Siber. J. Industr. Math. **13** (2010)

NUMERICAL STUDY OF DEFORMATION DROPS OF WATER IN GAS STREAM

Anna Shebeleva[1,*], and *Mikhail* Chernetskiy[1]

[1] Siberian Federal University, Krasnoyarsk, Russia

Abstract. Mathematical model developed on basis of experimental data about processes of deformation organohydrocarbon drops of fuel in stream of oxidant. Numerical investigation of integral characteristics of deformation of droplet organohydrocarbon fuel with size of drops – 4.5 mm , speed of oxidant – 2 m / s, temperature of 298 K has been completed (in accordance with the conditions of the experiment). Specific change of surface longitudinal and cross section and of organohydrocarbon drops of fuel during its gravity precipitation in time has been shown. It showed satisfactory agreement theoretical and experimental data. The developed model is the basis for solving problem of collapse (crushing) of drops of coal-water and organohydrocarbon slurry during their motion in a gaseous medium at a subsonic speed.

1 Introduction

Behavior form droplets in the gas stream is important for practical applications is pleased, for example in the field of medical diagnostics, agriculture, improvement and optimization of burners nozzles, inkjet printing, etc.

Exists numerous publications which are devoted, for example, to the investigation of droplet formation, their deformation, breakage, collisions, and evaporation [1-4]. Continuous change of shape and area occurs at the droplet surface drops deformation in flight. Therefore, the integral characteristics of heat and mass transfer between the droplet surface and the surrounding air are changed.

The review [5], published on the use of numbers only Weber and Reynolds does not allow to fully describe the patterns of strain drops. Main characteristics of the "deformation cycles" drops depend on the initial droplet size, droplet velocity, gas temperature and fluid properties (density, viscosity and surface tension). This feature should be considered in simulating the deformation of droplets in the gas stream.

The aim of this study was to: testing of numerical methods and research of features of deformation processes of water droplets as they move into a gas stream (air), which takes into account the impact of the initial droplet size, velocity, liquid properties and gaseous medium.

* Corresponding author: an_riv@mail.ru

2 Mathematical model

Numerical methods are used in this paper to simulate two-phase flows in a T- microchannel, based on the volume of fluid method in the cells. This works well for the calculation of macroscopic flows with free surface [6-9]. The idea of the method is that the liquid and gas are treated as a single two-component medium and the spatial distribution of the phases within the computational domain is determined using a marker function *F(x,y,z,t,)*, which value specifies the volume fraction of the liquid phase in the computational cell as follows:

$$F(x,y,z) = \begin{cases} 0, \textit{if the cell is empty} \\ 1, \textit{if the cell is completely filled with the liquid} \end{cases}$$

$0<F(x,y,z,t)<1$ – if the phase boundary passes through the cell.

Since the free surface moves together with the liquid, tracking of the free boundary in space is performed by solving the transport equation for the volume fraction of the liquid phase in the cell:

$$\frac{\partial F}{\partial t} + V \cdot \nabla F = 0 \tag{1}$$

Here $\boldsymbol{V}$ is the velocity vector of the two-phase medium obtained by solving the hydrodynamic equations, consisting of the equations of conservation of mass, or the continuity equation:

$$\frac{\partial \rho}{\partial t} + \nabla(\rho \boldsymbol{V}) = 0 \tag{2}$$

and the equations of motion and the law of conservation of momentum:

$$\frac{\partial \rho V}{\partial t} + \nabla(\rho V \times V) = -\nabla p + \nabla(\tau) + F \tag{3}$$

Where τ - viscous stress tensor, $\boldsymbol{F}$ – vector of body forces, p - static pressure, ρ - the density of the two-phase medium.

The components of the viscous stress tensor τ_{ij} defined as:

$$\tau_{ij} = \mu\left(\frac{\partial U_i}{\partial x_j} + \frac{\partial U_j}{\partial x_i} - \frac{2}{3}\delta_{ij}\frac{\partial U_k}{\partial x_k}\right) \tag{4}$$

Where μ - viscosity coefficient of the liquid, U_{ij} – the components of the velocity vector.

The density and viscosity coefficient of the two-component medium considered obtained from the volume fraction of liquid in the cell:

$$\rho = \rho_1 F + (1-F)\mu_2 \tag{5}$$

$$\mu = \mu_1 F + (1-F)\mu_2 \tag{6}$$

Here, $\rho 1$, $\mu 1$ - density and viscosity coefficient of the liquid, $\rho 2$, $\mu 2$ - the density and viscosity coefficient of the second fluid.

Thus obtained density values ρ and viscosity μ are included in the equation of motion and determine the physical properties of the two-phase medium. The phenomenon of surface tension occurs when considering the fluid flows with the interface, which cannot be ignored in the case of flow in microchannels. Surface tension plays a key role in microchannels. The study of currents, controlled by surface tension, is a very difficult task. VOF method makes it relatively easy to take into account the effect of surface tension forces. Most often, the VOF method uses the so-called continuum surface force (CSF) algorithm to model surface tension [10,11]. The essence of this algorithm is that the equations of motion include an additional volume force F_s whose magnitude is given by the relation:

$$F_S = \sigma k \nabla F \tag{7}$$

Where σ is the surface tension coefficient and k is the curvature of the free surface determined by the divergence of the normal vector:

$$k = \nabla \left(\frac{n}{|n|} \right) \tag{8}$$

The normal to the free surface is in turn calculated as the gradient of the volume fraction of the liquid phase in the cell:

$$n = \nabla F \tag{9}$$

On the solid wall, the magnitude of the normal vector is determined from the contact angle θ:

$$n = n_W cos\theta + \tau_W sin\theta \tag{10}$$

Where n_w, τ_w – are the normal and tangential vectors to the wall.

3 Formulation of the problem. Results

In this study, a numerical method of calculation of two-phase flows with a free surface was tested. Experimental data were taken from a paper [5]. The geometry of the task is shown in figure 1. Geometry is a rectangle with a tube from which the water flows at a speed of 0.12 m / s. axisymmetric formulation used. The inner diameter of the nozzle of the dispenser is 2.5 mm.

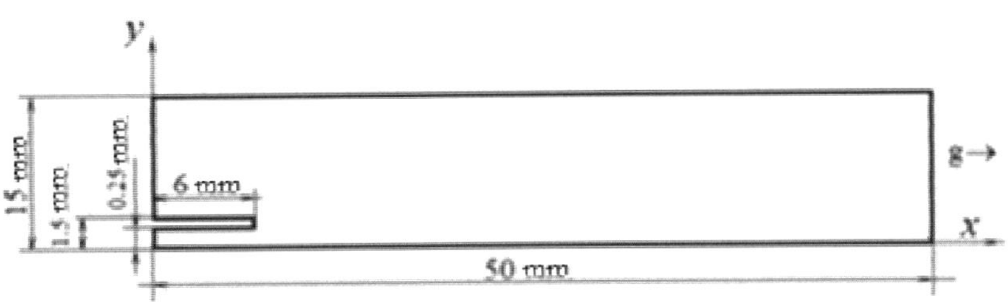

Fig. 1. Geometry of the case.

A drop of water moving with a certain “deformation cycles” sequence. The complexity is the simulation of the formation of the expiry of the jet of droplets from a nozzle at the initial stage. Nozzle geometry, fluid dispensing speed and the contact angle must be taken into account for an adequate description of this process.

Comparison of simulation results with experimental data [5] presented in figure 2 at the stage of the formation of drops. As can be seen, the agreement between calculation and experiment shows good dynamics for the formation of drops and form of the free surface.

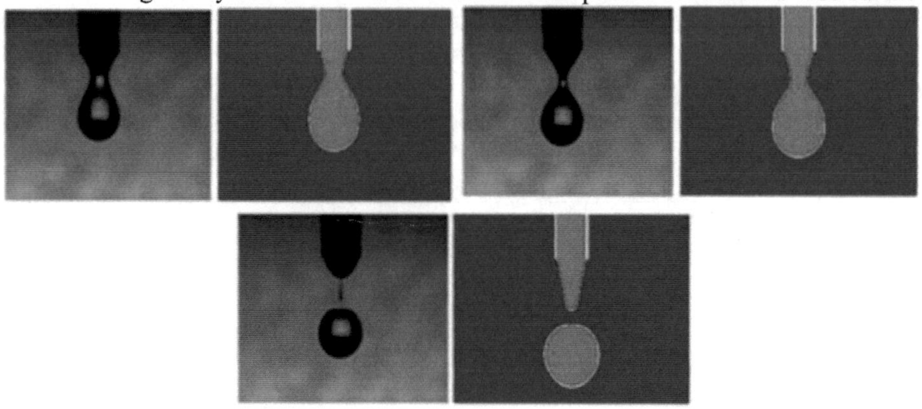

Fig. 2. Evolution of behavior of drop at initial stage. Black and white image - experiment, color images - the simulation results.

Movement of droplets in the stream is accompanied by periodic oscillations of its form, after the drop formed a comparison of the dynamics of the behavior of a drop in the calculation and the experiment shown in figure 3. It may be noted, the form of droplets in the experiment and the calculation agrees well.

Comparison of the calculated drop shape with experimental photographs shown in figure 4. at various times .we visualize droplet sectional while the experiment is presented in a three-dimensional droplet shape in perspective. However, in the main points, the calculated drop shape similar to experimental photos.

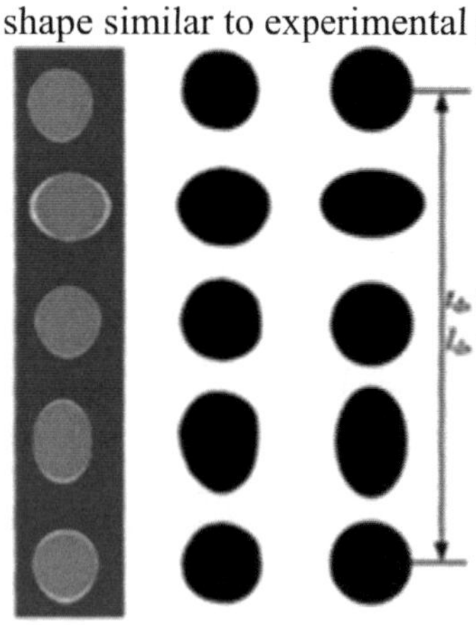

Fig. 3. Typical shapes of water drops a) The results of the calculation; b) The actual shape of the drop [5]; c) The ideal form of drops [5].

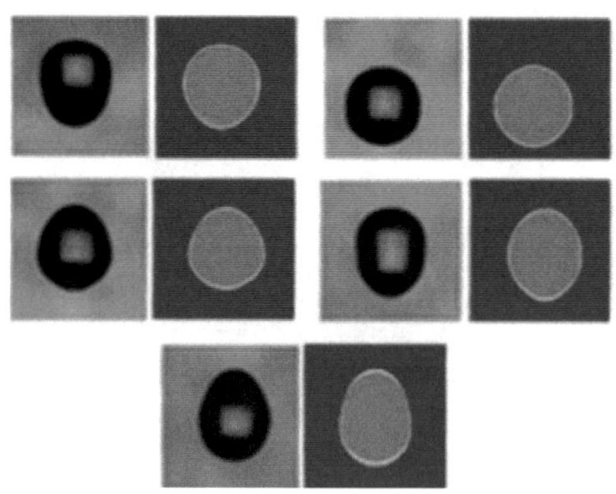

Fig. 4. Form of droplet surface in calculation and experiment at various time. Black and white image - experiment, color images - the simulation results.

Quantitative comparison of calculation and experiment is shown in figure 5. The graph shows the behavior of the relative amplitude of the vertical oscillations drops in time, related to the period of oscillations. As can be seen, the oscillation amplitude decays with time. The frequency and amplitude of the oscillations in the same experiment and calculation.

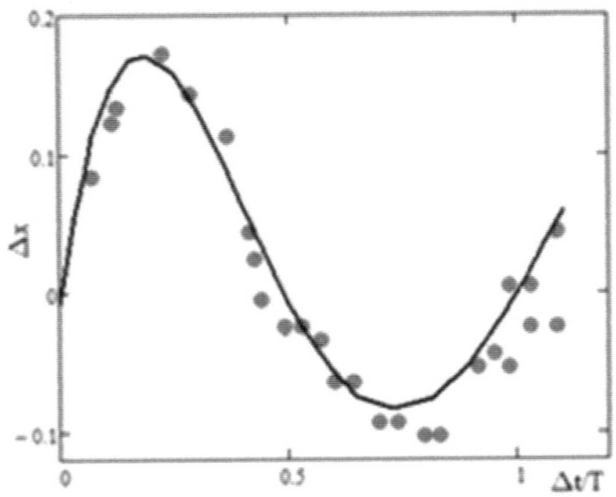

Fig. 5. Dependence of the relative vertical oscillation amplitude drops in the stream. Red facilities - calculation results.

4 Conclusion

Mathematical model and numerical method of calculating the deformation of droplets in the gas flow is developed. The model is based on the Volume of fluid method and algorithm CSF (continuum surface force) to calculate the surface tension. A series of calculations was performed fall and deformation of water droplets. The simulation results are compared with experimental data a satisfactory agreement. In the future, we plan to develop this model to the problem of deformation of drops coal-water slurry.

Acknowledgments

This research was done at the expense of the Russian Science Foundation (Project No. 15-19-10003).

References

1. S.K. Aggarwal, F. Peng, J. Eng. Gas Turbines Power **117** (1995)
2. R.S. Volkov, A.O. Zhdanova, M.V. Zabelin, G.V. Kuznetsov, P.A. Strizhak, Eur. Phys. J. Web Conf. **76** (2014)
3. G.V. Kuznetsov, P.A. Strizhak, High Temp. **52** (2014)
4. R.S. Volkov, G.V. Kuznetsov, P.A. Strizhak, Int. J. Heat Mass Transfer. **79** (2014)
5. R. S. Volkov, G. V. Kuznetsov, P.A. Strizhak. Int. J. Heat Mass Transfer **85** (2015)
6. A.V Minakov. Comp. Math. Math. Phys. **54** (2014)
7. D.V. Guzei, A.V Minakov, M.I Pryazhnikov, A.A Dekterev, T. A., Thermophysics and Aeromechanics **22** (2015)
8. R.I Iegorov, P.A.Strizhak, M.Y. Chernetskiy. EPJ Web of Conf. **110** (2016)
9. A.V. Minakov, M.V. Pervukhin, D.V. Platonov, M.Y. Khatsayuk. Comp. Math. Math. Phys. **55** (2015)
10. A.A. Shebeleva, A.V. Minakov, A.A. Yagodnitsina, V.G. Andyuseva, MATEC Web Conf. **23** (2015)
11. A. S. Orlov, A. V. Minakov, A. V. Proshkin, MATEC Web Conf. **23** (2015)

MATEC Web of Conferences **72**, 01102 (2016) DOI: 10.1051/matecconf/20167201102
HMTTSC-2016

MATHEMATICAL MODELLING OF THE PROCESS OF TUNGSTEN FLUORIDES REDUCTION BY HYDROGEN

Roman Brendakov[1,*], *Alexander* Shvab[1], and *Vladimir* Brendakov[2]

[1]National Research Tomsk State University, 634030 Tomsk, Russia
[2]Seversk Technological Institute NRNU MEPhl, 636070 Seversk, Russia

Abstract. The process of tungsten fluorides reduction by hydrogen is a component part of Fluoride technology of tungsten conversion. Nowadays the researchers are definitely interested in studying this process. It is connected with common use of metal tungsten products in different sectors of the economy, which is the result of unique qualities of this metal. With the help of physical and mathematical modelling of the process of tungsten hexafluoride reduction by hydrogen, it becomes possible to create an import substitution technology of metal tungsten conversion. Fluoride technology of tungsten conversion allows putting different coverings and make tungsten products of different shapes, which is impossible to get traditionally. The process of tungsten fluorides reduction by hydrogen can be referred to CVD processes (Chemical Vapor Deposition). Common use of CVD technologies for getting metal products and coverings is limited by definite problems, connected with access difficulties to initial components of research and the lack of information about their basic thermal characteristics. Therefore, mathematical description of the initial components mass-moving process, which provides with optimal value of their concentration in gas flow and in precipitation zone, is a question of current importance.

The process of tungsten fluorides reduction by hydrogen is realized according to chemical reaction, represented by a formula

$$WF_6 + 3H_2 = W + 6HF \tag{1}$$

Mathematical model of reduction process should describe non-stationary processes of heat and mass exchange when two-component mixture moves in the reactionary zone. The flat channel in the article is described and presented in the figure 1, in which the mixture of two gaseous substances, tungsten hexafluoride and hydrogen enters vertically through the bottom entry section at the T_0 temperature. The balance of these gases concentration is consistent with stoichiometric coefficient in equation (1). This mixture is considered to be the first component with mass concentration equal 1 at the entry section. The second component in this working zone is fluoric hydrogen, which shows up as a result of reaction

* Corresponding author: brend989@gmail.com

(1) near the heated bottom layer, situated on the left-side boundary line with the constant T_W temperature. It is possible to neglect the metal thickness, which shows up on the bottom layer, as it has a small size.

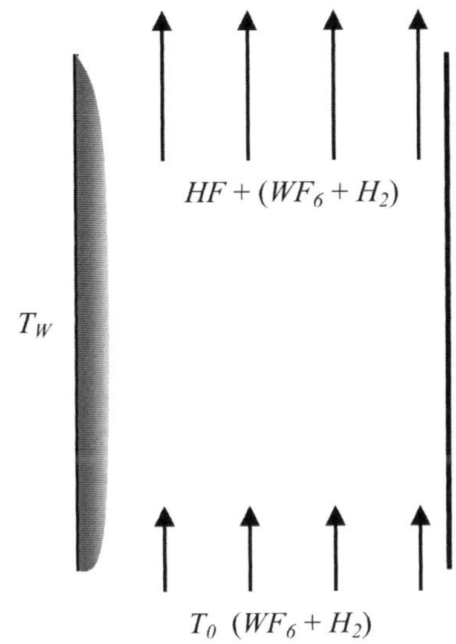

Fig. 1. Working zone scheme.

Taking into consideration all the allowances, the equations of convective heat- and mass-moving in the flat channel are:

$$c\rho\,(\partial\theta / \partial\tau + u\,\partial\theta / \partial z) = \partial(\lambda\,\partial\theta / \partial z) / \partial z + \partial(\lambda\,\partial\theta / \partial x) / \partial x;$$

$$\partial C / \partial\tau + u\,\partial C / \partial z = \partial(D\,\partial C / \partial z) / \partial z + \partial(D\,\partial C / \partial x) / \partial x; \qquad (2)$$

$$\theta = (T - T_0) / (T_w - T_0);$$

where c, ρ, λ are thermal capacity, density and thermal conductivity of the gas flow accordingly; C, D is mass concentration and diffusion coefficient of key flow component $(WF_6 + 3H_2)$.

For quasi-stationary speed type of stabilized gas flow in the flat channel, the expression is the following, taking into account formula (1):

$$u = 6\,\bar{u}[x / x_{max} - (x / x_{max})^2];$$

$$\bar{u} = (3 - C)\,\bar{u}_0 / 2;$$

$\bar{u}$ is a gas flow speed, averaged through a cross-section, x_{max} is a value of a crossed coordinate on the right border of the area, $\bar{u}_0$ is an averaged speed through a cross-section.

Therefore, the mathematical model of the process of tungsten fluorides reduction by hydrogen is a system of varied equations in partial derivatives (2), transport equation type, written for expansible temperature θ and concentration of the key component of mixture C can be solved computationally under certain limited circumstances known. Taking into consideration the equation of chemical reaction (1), it is enough to know the algorithm of calculation the speed of the reaction of reduction v_w to know the values of mixture components concentration on the surface of the bottom layer through the characteristics of the flow in the calculated area, completing the task this way.

In the work (1) the formula is offered to find the speed of tungsten concretion from the mixture $WF_6 + H_2 + HF$ at the atmosphere pressure of

$$v_w = \frac{3,6 \cdot 10^{19} N_{H_2} N_{WF_6} e^{-\frac{58500}{RT}}}{\left(\frac{1}{\rho} + \frac{21500}{\sqrt{T}} N_{H_2} e^{\frac{840}{RT}} + \frac{25600}{\sqrt{T}} N_{WF_6} e^{\frac{6700}{RT}} + \frac{20000}{\sqrt{T}} N_{HF} e^{\frac{4180}{RT}}\right)^2 T^3}$$

While deducing this formula there have been made some allowances.

First of all, it is considered, that the process of tungsten concretion is done at the constant temperature. Second, the adsorptive balances in the system are reached faster, than the reaction goes, which means, that at any moment there is a certain balance between partial pressure of the component of the reaction in the gas phase and the part of reactionary surface, occupied by its molecules. The comparison of big number of experimental data [2, 3, 4] with theoretical calculation concerning this equation revealed a satisfactory correspondence in a wide interval of process parameters varying, though it shouldn't be omitted, that it is only right in the kinetic part of the process. In the diffusive part, as expected, experimental data is rather lower, than that in the formula above.

Taking all the limitations and allowances into account, there has been made a test calculation of the changing of tungsten layer thickness through the length of bottom layer in comparison with the experiment ♦ [1]. The result is given in the figure 2.

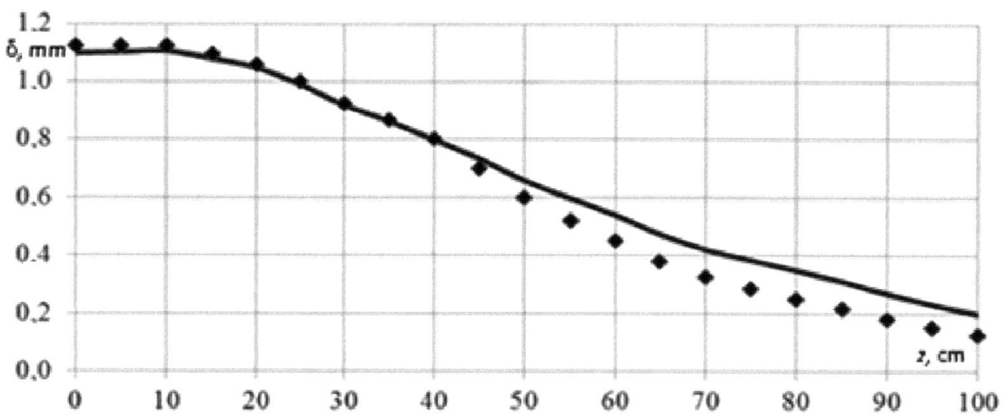

Fig. 2. The changing of tungsten layer through the bottom layer length.

A good correspondence of calculation and experimental data allows making a conclusion about the adequacy of created model. This model allows making numeral research of the optimization of the process of refractory metal fluorides reduction.

References

1. Yu. M. Korolev, V. I. Stolyarov, *Restoration of fluorides of refractory metals hydrogen* (Metallurgy, Moscow, 1981)
2. U. N. Golovanov., S. N. Kadimet., A. I. Krasovskii etc. The electronic technics. Technology, production process and equipment **43** (1971)
3. V. A. Neiberlein, J. Amer. Ceramic. Soc. Bulletin **48** (1965)
4. I. B. Averin, U. P. Plotnikov, Mech. Eng. **107** (1979)

MATEC Web of Conferences **72**, 01103 (2016) DOI: 10.1051/matecconf/20167201103
HMTTSC-2016

METHOD EXPERIMENTAL DEFINITION OF EFFICIENCY FUELBURN IN A GAS REACTORS WITH MINI CHANNELS

Ekaterina Slesareva[1,*], *Sergey* Elistratov[2], and *Valeriy* Ovchinnikov[1]

[1] SB RAS, Kutateladze Institute of Thermophysics, 630090 Novosibirsk, Russia
[2] Novosibirsk State Technical University, 630092 Novosibirsk, Russia

Abstract. Approbation of a method of experimental definition of efficiency of a fuelburn in minichanneling gas reactors has been carried out. The method of visualisation of a field of temperatures by the temperature transducer made of fine-meshed with low-inertia use for definition of thermal structure of gas streams. A field of temperatures on an outlet the simulator of a gas reactor with minichannels registered on thermogram by infrared imager. Experiments have shown that the thermal imaging method provides the good resolutions on spatial and on time for a field of temperatures both in stationary and in non-stationary regime of gas flow and of heat release. Qualitative and quantitative character a change a temperatures on thermograms correspond to classical physical representations about nature of a gas flow in lengthy channels. The study showed that the method of thermal imaging allows in a regime real time defined a parameters of gas flow on outlet a channels of reactor which are necessary for definition of fuelburn efficiency.

1 Introduction

The mini-microchannel gas reactors and the heat-exchange apparatus become an integral part of innovative directions a machine industry of heat power and energy-chemical. Data about composition and temperature the gaseous reaction products, gained immediately on an outlet from such apparatuses, allow to optimise their operation.

The thermographic method an experimental study a definition of thermal structure a gas flow [1] is applied. For visualisation a pattern of temperature a medium which transparent in an infrared light, possible used thin filaments or fine-meshed heated up in a stream as a indicators of temperature. The thermographic method possible to trace changes both common and an local pattern a changing a temperature in a gas flow. In the present paper results of approbation the thermographic method [2] for heat exchangers with minichannel are presented.

* Corresponding author: styuardessa@yandex.ru

2 A measurement procedure

Experiments on measuring non-stationary a field of temperature on an outlet from six longitudinal minichannels of quasitriangular cross-section have been carry out on the model framework organised by seven cylindrical heaters of equal diameter. On fig. 1 the cross-section and the passing channel in framework are schematically presented. In a quartz tube 1 outer diameter 23 mm and length 0.5 m disposes framework from seven electric heaters 2 diameter 5.9 mm and length 0.5 m which organising six passing minichannels 3 (№ 1 ÷ № 6) the quasitriangular shape (triangle ABC organised by archs AB, BC, CA). Kaolinite wool 4 in thickness 7 mm and landlocked an air spaces 5 reduced heat loss from heaters in a circumambient.

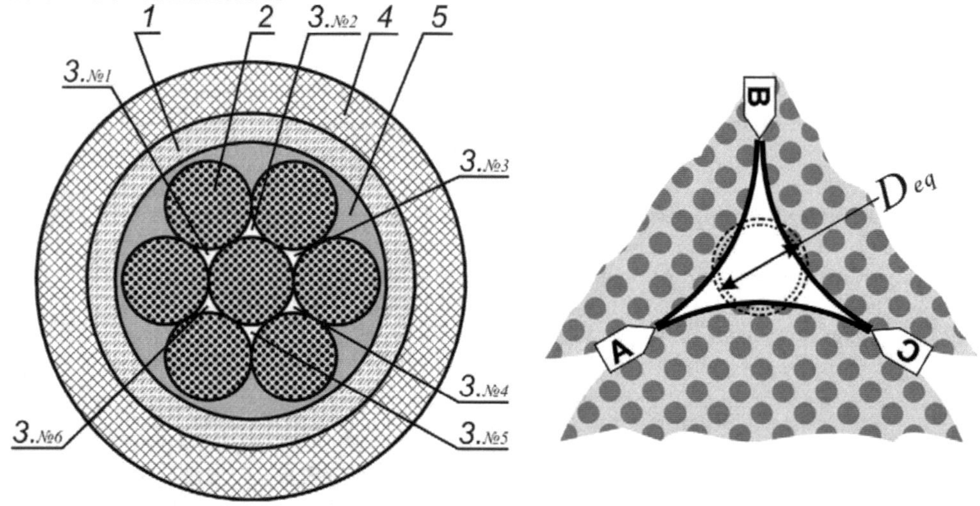

Fig. 1. A cross-section (a) and the shape (b) the passing quasitriangular channel of minichannel framework: 1 - a quartz tube; 2 - a heater; 3 - the passing channel (№ 1 ÷ № 6) for gas; 4 - thermo insulation; 5 - an air space; A, B and C - vertex of the quasitriangular channel.

The total heat exchange area passing quasitriangular channels is 0.0275 m^2. On fig. 1b the visual relationship a sizes the passing channel and its hydraulic diameter D_{eq} is shown. By shaped lines rounds for minimum (1.2 mm) and maximum (1.3 mm) meanings of hydraulic diameters in framework are shown. The odds in diameters is caused by inaccuracy at mount a electric heaters in a bundle and some ovality of cylindrical heaters. According to planimetric measuring of photographs the sectional area the passing channels varied from 2.7 to 3.0 mm^2. The brass mesh-thermode from a wire in the thickness about 80 microns with meshes 0.3 × 0.3 mm press oneself to surface of channels on outlet framework. Thermogram have been obtained by infrared imager NEC TH7100 with wavelength thermal radiation $\lambda = 8 \div 12$ microns.

3 Discussion of results

The thermographic method has been approved for various a flow rate of air and an intensity of its heating by cylindrical heaters in stationary and nonsteady conditions. Processing a thermograms has allowed determine a changes type a flow temperature in each a passing minichannels of framework. On fig. 2 four sample thermograms from the beginning (000) and after 99, 304, and 557 seconds after heat load 4.8 Wt surge on a central heater are shown, under the total flow rate a air 0.45 g/s. It is possible to see inhomogeneity a warm-up of framework because of a thermogravitational convection by volume assembly and

 DOI: 10.1051/matecconf/20167201103

variations a heating on the separate passing channels, caused by differences of their flow areas, noted previously.

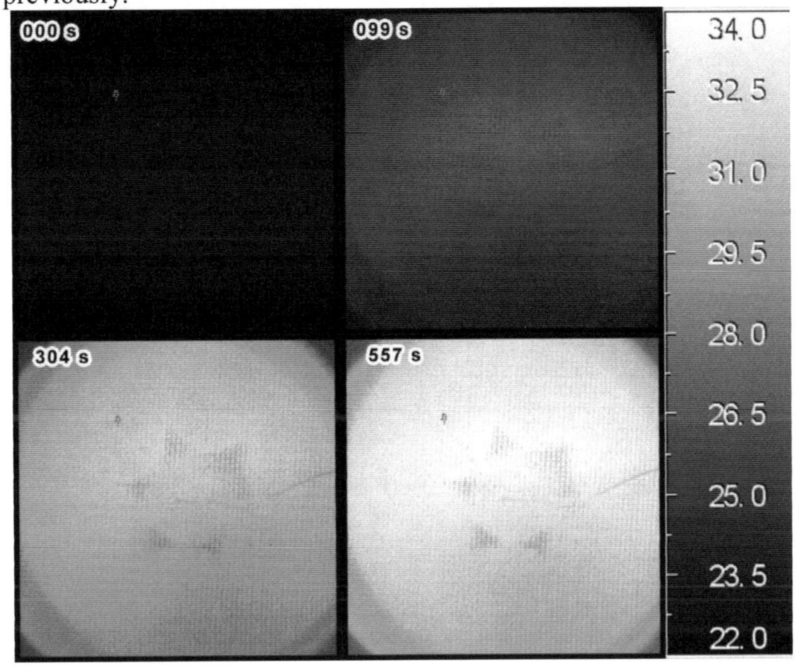

Fig. 2. The thermograms of mesh-thermode on outlet of a framework after heat load surge only on a central heater in a assembly.

Analogously on fig. 3 thermograms for case a decrease total flow rate air with 0.45 g/s to 0.26 g/s are shown, under a heat load 33.6 Wt proportioned uniformly on all a heaters. The thermograms allow to see as all panoramic pattern of nonsteady preheating a air in quasitriangular minichannels, and full details a dynamic preheating of model framework.

Processing a thermograms has shown, that after transition to a stationary conditions the temperature of air on outlet a framework correspond to a calculated values for a stationary flow, taking into account changing a conditions of heat exchange along length of the channel [3, 4]. Let's note, that for uprating accuracy of measuring by operation with channels of a smaller size it is necessary to use infrared imagers with smaller wavelength thermal radiation λ for traceable caloradiance.

4 Conclusion

Approbation the thermographic method flow with using fine-meshed grids-thermode for definition a thermal structure a gas flow on outlet the simulation framework with minichannels has shown its good a spatial-time resolution. Qualitative and quantitative character a change a temperatures on thermograms correspond to classical physical representations about nature of a gas flow in lengthy channels [3].

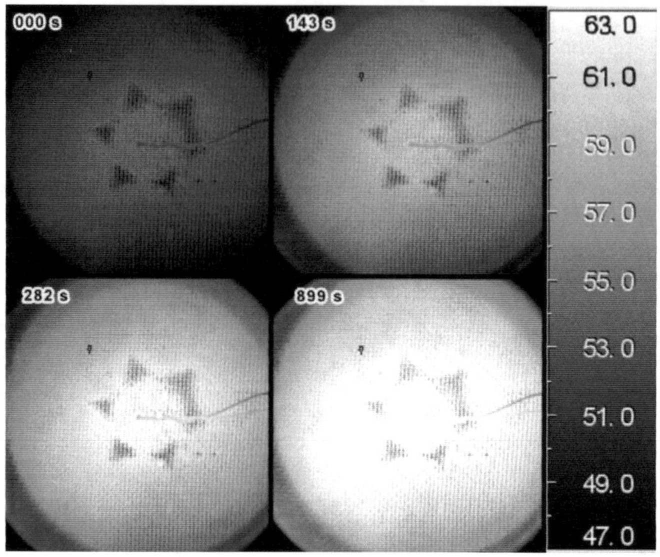

Fig. 3. The thermograms of mesh-thermode on outlet of a framework after a decrease total flow rate air.

The thermographic method is universal, its specific application for study a processes of stationary and nonsteady heat exchange, dependent from a specificity a subject of inquiry and engineering possibilities a equipment for thermal imaging.

Acknowledgments

This work was supported by the Russian Foundation for Basic Research, project No. 16-38-00502-mol-a.

References

1. A.V. Efimova, A.V. Zaitsev, B.P. Zhilkin, D.N. Tokarev, K.V. Zaitsev, Kh. Dashpuntsag, Bulletin of USTU: Therm. Eng. **33** (2004)
2. S.L. Elistratov, E.Yu. Slesareva, EPJ Web Conf. **82** (2015)
3. B.S. Petuhov, *Heat exchange and resistance under a laminar flow fluid in tubes.* (Energy, Moscow, 1967)
4. S.Y. Misyura, EPJ Web of Conf. **110** (2015)

MATEC Web of Conferences **72**, 01104 (2016) DOI: 10.1051/matecconf/20167201104
HMTTSC-2016

EFFECT OF OXIDIZING MEDIUM ON SYNTHESIS GAS CONTENT AT SOLID FUEL GASIFICATION

Alexander Korotkikh[1,2,*], *Konstantin* Slyusarskiy[1], and *Ivan* Sorokin[1]

[1]National Research Tomsk Polytechnic University, 634050 Tomsk, Russia
[2]National Research Tomsk State University, 634050 Tomsk, Russia

Abstract. Solid fuel gasification is promising technology in sphere of clean energy. The synthesis gas content for air-blown fixed bed gasification may be defined using Gibbs free energy minimization procedure. The minimization procedure was realized via steepest descent method. The feed consisted of steam, air and coal at standard conditions. The temperature and gas content were calculated at different ratios of coal/steam/air. It was found that optimal syngas content resulted at component ratio of 1.0/0.5/2.2 with the ambient temperature of 1300 K and syngas heating power of 7.7 kJ/m^3.

1 Introduction

The energy consumption drastically grows last decades in the world: the electricity and primary energy consumption increased by 2.5 times [1]. However, the greenhouse gas specific emission per person of generated electricity is on the level of 1970 [1], despite recent progress in energy-generating technology. Together with population growth it makes the greenhouse problem more and more relevant as well as development of energy-effective technologies.

Gasification (organic matter conversion into combustible gas by means of partial oxidation) is one of the most promising technology to improve efficiency of fossil fuel usage and decrease its negative impact on environment. In general, syngas consists of carbon mono- and dioxide, hydrogen and steam. The exact concentration depends on many factors such as the temperature, pressure and fuel/oxidizer weight ratio. The large number of simultaneous chemical reactions in system makes their description extremely complex. The equilibrium composition of syngas may be defined using Gibbs free energy minimization procedure [2].

The equilibrium composition of synthesis gas for steam-air blown coal gasification at different coal/steam/air ratios as well as the adiabatic temperature were calculated. The feed properties were taken for T-grade bituminous coal of Kuznetskiy deposit (Kapitalnaya-2 mine, Osinniki).

* Corresponding author: korotkikh@tpu.ru

2 Mathematical model

Gibbs free energy minimization methods are widely presented in scientific works. The method of steepest descent for such application was presented at the first time by White [3] and further modified by Koukkari [4] and is the most widespread in thermochemistry because of its simplicity and rapid convergence. Other methods (such as Newton method or method of internal points) are not so widespread and applied in some special spheres like geology or plasma physics.

Free Gibbs energy of enclosed single-phased multicomponent system may be calculated via equation:

$$G = RT\sum_{i=1}^{m} n_i\left(\left(\mu_i / RT\right)_i + \ln\left(n_i / \sum_{i=1} n_i\right)\right)$$

where R is the gas constant, J/(mole K); T is the temperature, K; n_i is the substance amount i, mole; μ_i is the Gibbs molar potential of substance, J/mole. It may be defined with using equation:

$$\mu_i = h_i - s_i T + RT\ln\left(p / p_0\right)$$

where h_i is substance molar enthalpy of formation at constant temperature, J/mole; s_0 is the substance molar entropy at constant temperature, J/(mole K); p_0 / p are standard/current pressures in system, respectively, Pa.

To meet material balance restrains for system which consisted of N molecules and M atoms on possible solution domain the following equation system was taken into account:

$$\sum_{i \in N} a_{ij} n_i = b_j, j \in M, n_i > 0$$

where a_{ij} is stoichiometric coefficient (number of *j* type atoms in molecule *i*); b_j is a total amount of *j* type atoms in system, mole.

Gibbs free energy minimum was found using the steepest descent method by Koukkari algorithm [4]. The large number of substances with close to zero concentration badly influences on algorithm result preciseness. That's why the decision to choose six substances with expected significant content was made. The steam, hydrogen, methane, oxygen, carbon mono- and dioxide were chosen because of high concentration in feed and low free enthalpy values at standard conditions. The nitrogen presence in air and ash presence in coal was taken into account only in heat balance equations.

The thermal effect of system transition from initial into equilibrium state was defined via law of constant heat summation (Hess's law) [4]:

$$\Delta H_r = \sum n_i^{pr} H_f^{pr} - \sum n_i^f H_f^f$$

where H_f is *i* substance molar enthalpy of formation, J/mole. Indexes *f* and *pr* refer to feed and products of reaction, respectively.

The process was considered as autothermal than the heat effect of reaction was equal to difference between enthalpies of products and feed at preset temperature. The syngas heating power was defined using equation [5]:

$$Q = \sum v_i Q_i$$

where v_i is molar concentration of *i* substance in mixture; Q_i is the heating power of *i* substance in mixture, J/m^3.

The following assumptions were made:

1. The interactions between nitrogen and ash with syngas was not included.
2. The carbon conversion was considered equal to unity.

3. The thermal losses during gasification were not taken into account.
4. The ash was considered to be solid at all temperatures.
5. The presence of condensed phase in system was neglected.

The thermochemical properties of substances were defined using open-source software CoolProp. The heat capacity value of ash was taken according to [6]. Algorithm was realized in MATLAB programming package.

The fuel properties were taken for T-grade coal of Kuznetskiy deposit. Coal chemical content was defined according to presented method [7]. The results of chemical content of studied solid fuel are presented in table 1.

Table 1. Solid fuel chemical content.

Elemental content, wt. %				Humidity, wt. %	Ash content, wt. %
C^d	H^d	N^d	S^d		
73.1	3.5	2.9	–	10.5	18.4

3 Results and discussion

Results of the adiabatic temperature and heating power of syngas at the different ratio of coal/steam/air at gasification of solid fuel are presented in figure 1.

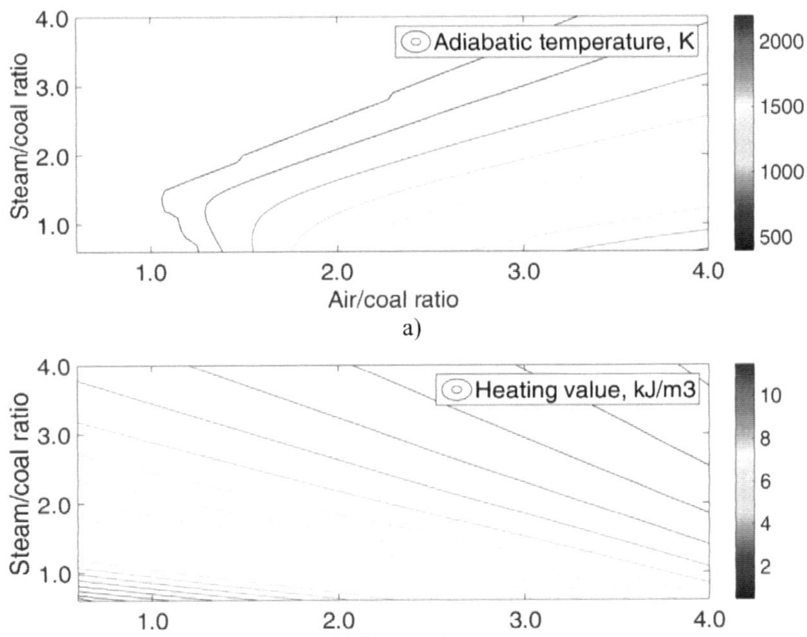

Fig. 1. The adiabatic temperature (a) and syngas heating power (b) at various coal/steam/air ratio.

The maximal possible value of syngas heating power 11.1 kJ/m^3 is obtained at coal/steam/air ratio of 1.0/0.5/0.5 and resulted at the adiabatic temperature of 300 K. This appear because of system stoichiometric properties – lack of oxygen promotes formation of combustible substances, even if they are thermochemically ineffective. In exploitation conditions it means that high heating power values as well as gasifier high efficiency values are correspond to allothermal processes. In real conditions, low temperature processes are rarely used because of process low intensity. The lowest temperature used are usually close to 1300 K. The syngas maximal heating power at 1300 K is 7.7 kJ/m^3 and obtained at coal/steam/air ratio with values of 1.0/0.5/2.2. The adiabatic temperature, in contrary, is increased with growth of coal/steam/air ratio, because of air content and reaches its maximum with coal/steam/air ratio of 1.0/0.5/4.0, the temperature value of 2100 K and the heating power of 4.5 kJ/m^3.

4 Conclusion

The mathematical model of single-phased multicomponent system equilibrium content was realized via Gibbs free energy minimization algorithm. The steepest descent method was used. The syngas heating power and adiabatic temperature were defined at different weight ratio of steam and air in the range of 0.5–4.0. The Kuznetskiy deposit bituminous coal was used as model fuel for calculation. The maximum value of syngas heating power corresponds to minimal oxidizer flow, while the adiabatic temperature increases with air expenditure and decreases with steam expenditure. The maximum heating power value is 11.1 kJ/m^3 at the temperature of 300 K and coal/steam/air ratio 1.0/0.5/0.5 and the maximum temperature is 2100 K at 1.0/0.5/4.0 coal/steam/air ratio with heating power of 4.5 kJ/m^3. The maximum heating power of 7.7 kJ/m^3 is obtained for coal/steam/air ratio 1.0/0.5/2.2 and at temperature of 1300 K.

Reference

1. BP Statistical Review of World Energy, June (2015)
2. V. Liso, G. Cinti, M.P. Nielsen, U. Desideri Appl. Therm. Eng. (to be published)
3. W.B. White, S.M. Johnson, G.B. Dantzig, J. Chem. Phys. **28** (1958)
4. P. Koukkari, *Introduction to constrained Gibbs energy methods in process and materials research* (VTT Technology, Espoo, 2014)
5. Z. Chen, Y. Shi, D. Lai, S. Gao, Z. Shi, Y. Tian, G. Xu, Fuel **176** (2016)
6. R.V. Krishnan, G. Jogeswararao, K. Ananthasivan, J. Therm. Anal. Calorim. **121** (2016)
7. V. Patil-Shinde, S. Saha, B.K. Sharma, S.S. Tambe, B.D. Kulkarni, Chem. Eng. Commun. **203** (2016)

MATHEMATICAL MODEL OF HEAT TRANSFER IN LAYERED STRUCTURE OF HUMAN SKIN INFLUENCED BY HEATED UP TO HIGH TEMPERATURES PARTICLE

Alexandr A. Stuparenko[1, *], and *Andrey* S. Solodkin[2]

[1] National Research Tomsk Polytechnic University, 634050 Tomsk, Russian Federation
[2] National Research University "MPEI", 111250 Moscow, Russian Federation

Abstract. Numerical research results of heat transfer in layered system "air-heated particle-skin layer" presented. Skin structure includes epidermis, derma and hypodermis layers. Skin was influenced by heated up to high temperatures particle. The problem is solved in simple one-dimensional statement in Cartesian system of coordinates. The typical range of influence parameters of heated particle considered. Temperature distributions in different moments of time obtained. Condition of burn occurrence by heated particle is under consideration in this work.

1 Introduction

Skin is a natural external cover of a body. The area of skin surface approximately is equal to 1-2 m^2. The skin can carry out various functions for the person: sensitive, protective, secretory and thermoregulatory [1]. The protective role of skin is multipurpose. It can protect internal parts of body from mechanical damages, detain water evaporation, interfere with penetration of microorganisms, protect from ultra-violet beams. Skin receptors make interaction of organism with the environment. About 15-20 % of blood lump can be collected in hypodermic network of blood vessels. The skin consists of three layers: epidermis, derma and hypodermic fatty cellular tissue (hypodermis) [2].

Epidermis is external layer which is formed by multilayered epithelium. Top epithelium layers are formed by the dead keratinous cells. This layer is most grown in place, where the skin is exposed to mechanical influence (palms and soles) [2].

Derma is layer of dense fibrous connecting tissues which is after epidermis layer. This layer consists from sweat glands, receptors of skin sensitivity, hair bulbs, sebaceous glands, blood and lymphatic vessels. Cells of derma layer are capable to division. Distinctive ability of these cells consists that they can secrete a pigment. Skin cells are bound by elastic fibers which give to skin elasticity [2].

Hypodermis is formed by fatty tissue which carries out role of thermoinsulator which protects organism from cooling. Fat in integument plays role of reserve of nutrients and can be spent in starvation [2]. Hypodermis comprises sweat glands and large blood vessels which allow to skin to adapt to local thermal influences in certain limits of temperatures [3].

* Corresponding author: firedanger@tpu.ru

Thermal burns occur as a result of high temperature source action on human skin [4]. Majority of such damages depend on source temperature and influence time. Heated up to high temperatures particles of metals and nonmetals are probable sources of high temperature at natural and technogenic fires [5-8].

The purpose of the present work is mathematical simulation of heat transfer in layered structure of human skin at influence of heated up to high temperatures particle.

2 Physical and Mathematical Statements

The human skin is rather difficult structure and one have a various texture in different parts of a body. It is possible to separate out three basic layers: epidermis, derma and hypodermic (fig. 1) [2].

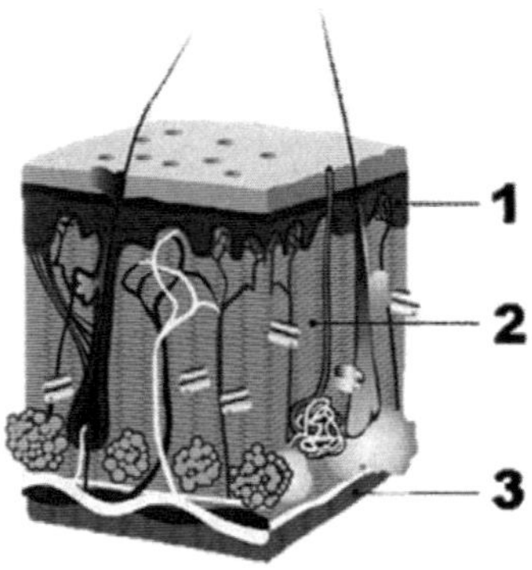

Fig. 1. Skin structure: 1 - epidermis, 2 - derma, 3 – hypodermis.

The basic assumptions [9, 10]:

1. Carbonaceous (natural scenario), steel and ceramic (technogenic scenario) particles are considered.
2. The three-layer system of human skin structure with effective thermophysical characteristics of each layer is considered.
3. It is supposed, that thermophysical characteristics of skin and particle do not depend on temperature.
4. Thermal destruction of skin tissue under the influence of the raised temperatures is neglected.
5. Cellular moisture evaporation from skin tissue also is neglected.
6. Conduction is basic mechanism of heat transfer in skin layers.
7. Monolithic structure of skin layers is considered.

Skin site which influenced by heated up to high temperatures particle of the definitive material with known thermophysical parameters is considered in present work. As a result of ideal contact there is an inert warming up of skin. The given physical phenomenon is considered as process of a heat transfer by heat conductivity. Borders of considered system are air and eternal tissues. The particle is considered in plate approach, that is one-dimensional statement of problem is used.

Heat transfer process in layered structure of skin is described by system of the non-stationary differential equations of heat conductivity with corresponding initial and boundary conditions:

$$\rho_1 c_1 \frac{\partial T_1}{\partial t} = \lambda_1 \frac{\partial^2 T_1}{\partial z^2},$$

$$\rho_2 c_2 \frac{\partial T_2}{\partial t} = \lambda_2 \frac{\partial^2 T_2}{\partial z^2},$$

$$\rho_3 c_3 \frac{\partial T_3}{\partial t} = \lambda_3 \frac{\partial^2 T_3}{\partial z^2},$$

$$\rho_4 c_4 \frac{\partial T_4}{\partial t} = \lambda_4 \frac{\partial^2 T_4}{\partial z^2}$$

$$T_i \big|_{t=0} = T_{i0},$$

$$z=0,\ -\lambda_1 \frac{\partial T_1}{\partial z} = \alpha_1 (T_1 - T_e),$$

$$z=H1,\ -\lambda_1 \frac{\partial T_1}{\partial z} = -\lambda_2 \frac{\partial T_2}{\partial z},$$

$$z=H2,\ -\lambda_2 \frac{\partial T_2}{\partial z} = -\lambda_3 \frac{\partial T_3}{\partial z},$$

$$z=H3,\ -\lambda_3 \frac{\partial T_3}{\partial z} = -\lambda_4 \frac{\partial T_4}{\partial z},$$

$$z=Lz,\ T_4 = T_{4t},$$

Where T_i, ρ_i, c_i, λ_i - temperature, density, thermal capacity and heat conductivity (1 - particle, 2 - epidermis, 3 - derma, 4 - hypodermis). z - spatial coordinate. t - time coordinate. The index 0, e, t corresponds to parameters during the initial moment, environment and internal tissues of organism. H_i – layer borders of system "particle-skin".

The problem is solved by a method of finite differences. Finite-difference analogues of the differential equations are solved by a marching method [11].

3 Results and Discussion

Temperature distributions in system "skin-particle" are presented in fig. 2-4. Numerical research of heat transfer processes in skin is carried out at influence of steel, ceramic and carbonaceous particles.

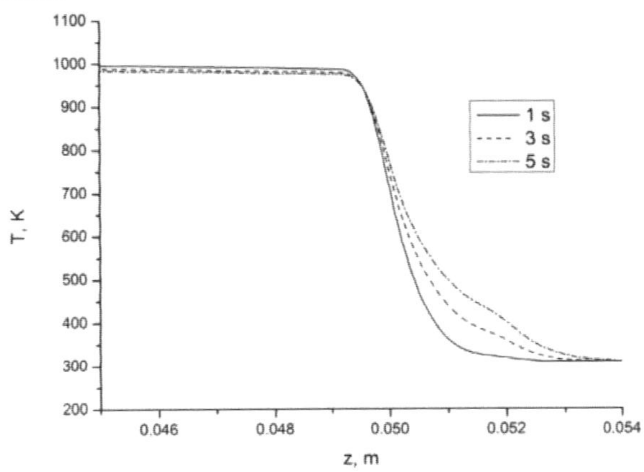

Fig. 2. Distribution of temperatures in system "skin-particle" during the various moments of time (steel particle, initial temperature T = 1000 K).

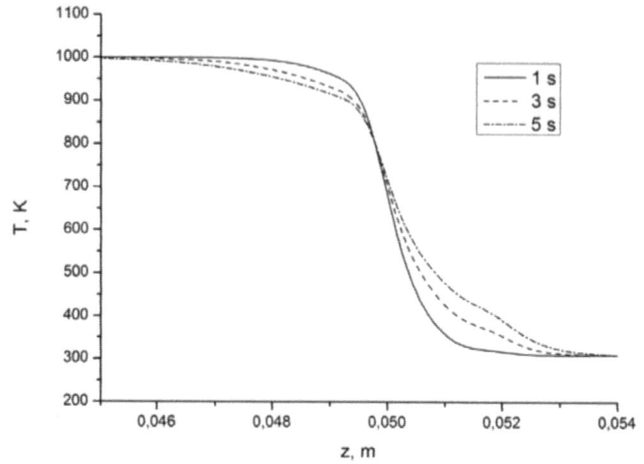

Fig. 3. Distribution of temperatures in system "skin-particle" during the various moments of time (ceramic particle, initial temperature T = 1000 K).

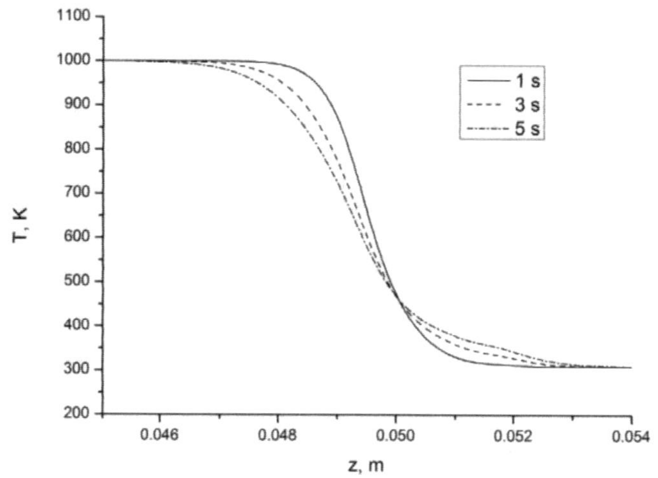

Fig. 4. Distribution of temperatures in system "skin-particle" during the various moments of time (carbonaceous particle, initial temperature T = 1000 K).

Short-term influence of heated up to high temperatures particle to skin was considered in the present work. This assumption corresponds to the scenario of human behavior which has no traumas received as a result of a fire and is capable to exclude long influence of "hot" particles on skin.

The analysis of the obtained distributions (fig. 2,3 and 4) shows., that appreciable cooling of particle occurs in the field of contact to skin and air. It is caused by the heat transfer at interaction with the environment and skin layer. Computing experiments have confirmed an obvious hypothesis that the particles having higher temperature are most dangerous to interaction with skin. So they have the greatest heat stock transferred to skin layer at contact. The top layer of investigated skin site on depth of the order of 0,5 - 2 mm is exposed to the most appreciable influence. Short-term influence of heated up to high temperatures particle mainly effect to epidermis and derma layers. Though, particle with initial temperature of the order of 1000 K cause appreciable heating of hypodermis layer.

The comparative analysis of numerical modeling results shows, that the greatest danger is represented by steel and ceramic particles which are characteristic for various technogenic fires and industrial failures. The least warming up of skin is observed at influence of a carbonaceous particle that corresponds to natural fires.

4 Conclusion

It is earlier shown, that the decision in one-dimensional single-layered statement allows to obtain enough exact results at research of heat transfer processes in human skin. Results accuracy increase slightly with extension of spatial dimensions [12]. It is established, that use of three-layer statement of problem allows to reveal spatial localization of zones with the raised temperature taking into account more exact structure of skin.

The obtained results, namely program realizations of mathematical models for heat transfer in system “particle-skin”, can be used for development of new generation of medical information systems [13-15]. Such systems, in turn, can find application in structures of the Ministry of Emergency Measures, medical institutions and departments of protection works or a civil defense at the industrial enterprises. Development of the complex deterministic-probabilistic approach on the basis of methods for forest fire danger forecast [16] and using information-computational technologies [17,18] taking into account the developed mathematical models of heat transfer in human skin is possible. Such system will allow prediction both forest fire danger and impact of fires on the persons.

Acknowledgments

The work was supported by the Russian President's grant (Scientific School project - 7538.2016.8).

References

1. J.T. Whitton, J.D. Everall, Br. J. Dermatol **89** (1973)
2. F. Xu, T.J. Lu, K.A. Seffen, J. Mech. Phys. Solids **56** (2007)
3. Standard D 4108-87, *Standard test method for thermal protective performance of material for clothing by open flame method, American society for testing and materials* (1994)
4. R.Sh. Enaleev, A.M. Zakirov, Yu.S. Chistov, E.Sh. Telyakov, Bulletin of the Kazan technological university **15** (2012) [In Russian]
5. A.V. Zakharevich, N.V. Baranovlskiy, Rus. J. Phys. Chem. B+ **9** (2015)
6. N.V. Baranovskiy, A.V. Zakharevich, D.S. Osotova, EPJ Web of Conf. **82** (2015)
7. S.L. Manzello, T.G. Cleary, J.R. Shields, J.C. Yang, Fire and Materials **30** (2006)
8. S.L. Manzello, T.G. Cleary, J.R. Shields, J.C. Yang, Int. J. Wildland Fire **15** (2006)
9. N.V. Baranovskiy, A.S. Solodkin, MATEC Web Conf. **23** (2015)
10. D.V. Korobkina, N.V. Baranovskiy, MATEC Web Conf. **19** (2014)
11. L. Vulkov, Lect. Notes in Comp. Sci. **9045** (2015)
12. N.V. Baranovskiy, A.S. Solodkin, A.A. Stuparenko, EPJ Web Conf. **110** (2016)
13. S. Petter, A. Fruhling, Int, J. Medical Inform. **80** (2011)
14. P.G. Goldschmidt, Communication ACM **48** (2005)
15. D.V. Korobkina, N.V. Baranovskiy, Cloud Sci. **1** (2014) [In Russian]
16. N.V. Baranovskiy, E.P. Yankovich, J. Autom. Inform. Sci. **47** (2015)
17. N.V. Baranovskiy, J. Autom. Inform. Sci. **47** (2015)
18. N. V. Baranovskiy, Cybernetics and Systems Analysis **51** (2015)

MATHEMATICAL MODELING OF NONSTATIONARY SEPARATION PROCESSES IN GAS CENTRIFUGE CASCADE FOR SEPARATION OF MULTICOMPONENT ISOTOPE MIXTURES

Alexey Orlov[1], *Anton* Ushakov[2,*], and *Victor* Sovach[2]

[1] National Research Tomsk Polytechnic University, 634050 Tomsk, Russia
[2] SC "PA Electrochemical Plant", 663690 Zelenogorsk, Russia

Abstract. This article presents results of development of the mathematical model of nonstationary separation processes occurring in gas centrifuge cascades for separation of multicomponent isotope mixtures. This model was used for the calculation parameters of gas centrifuge cascade for separation of germanium isotopes. Comparison of obtained values with results of other authors revealed that developed mathematical model is adequate to describe nonstationary separation processes in gas centrifuge cascades for separation of multicomponent isotope mixtures.

1 Introduction

Nonstationary processes can appear in the operation of gas centrifuge (GC) cascade for separation of multicomponent isotope mixtures (MCIM). During these processes isotope composition changes in cascade and it is necessary to minimize losses of cascade productivity. At present, there are mathematical models of nonstationary separation processes [1, 2], which have such disadvantage as a limited field of application: simulation is only possible in the case of stationary hydraulic parameters of GC cascade. Using of these models is impossible for the joint calculation of nonstationary hydraulic and separation processes. For elimination of this disadvantage we firstly developed the mathematical model of nonstationary hydraulic processes in GS cascade for MCIM separation [3]. This model was tested for separation Si, Xe, W, Ni isotopes. Further studies were aimed at creating the mathematical model of nonstationary separation processes in GC cascade for the MCIM separation. This article is devoted to its description.

* Corresponding author: ushakovaa2015@sibmail.com

2 The description of the mathematical model

MCIM separation takes place into a cascade (fig. 1). MCIM is composed of *n* components identified by *j* index ($j=\overline{1,N}$). A cascade consists of *S* separation stages numbered by *i* index ($i=\overline{1,S}$) and connected by counter-current type [4]. The cascade has three flows: input feed flow *F*, output light fraction flow *P* and heavy fraction flow *W*.

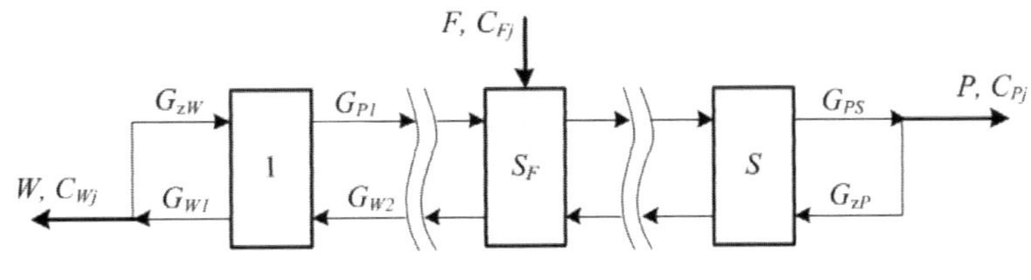

Fig. 1. Separation cascade.

During stationary hydraulic process the flows and concentrations of components in stage are interrelated by material balance equations and balance equations of each component:

$$F=P+W \tag{1}$$

$$FC_{Fj}=PC_{Pj}+WC_{Wj} \tag{2}$$

The stage has three flows: incoming feed flow G_{Fi}, output light fraction flow G_{Pi} and heavy fraction flow G_{Wi}:

$$G_{Fi}=G_{Pi}+G_{Wi} \tag{3}$$

$$G_{Fi}C_{Fij}=G_{Pi}C_{Pij}+G_{Wi}C_{Wij} \tag{4}$$

The separation of components in the *i*th stage is given by:

$$\chi_{ijl}=\frac{C_{Pij}C_{Wil}}{C_{Pil}C_{Wij}} \tag{5}$$

where χ_{ijl} is a separation factor of the *j*th and *l*th components in the *i*th stage.
For square cascade there are dimensionless parameters [1, 4]:

$$y_P=S\ln\chi_0 \tag{6}$$

$$\tau=\frac{G_F t\ln^2\chi_0}{H'} \tag{7}$$

where y_P is cascade length; χ_{0i} is separation factor per unit difference between the mass numbers; τ is time; *H'* is cascade holdup, kg; *t* is time, s.

The method of calculation of the nonstationary separation processes is based on the following points:
- Each stage is composed of finite number of capacities. MCIM is distributed in these capacities.
- At any time the value of process gas temperature, pressure and concentrations of each component are the same at any points of capacity.

- Basic equations of the mathematical model are the balance equations of process gas and components in each capacity and the equation of MCIM separation at the GC.

In the regards, the holdup of cascade is distributed in 4*S* capacities (fig. 2).

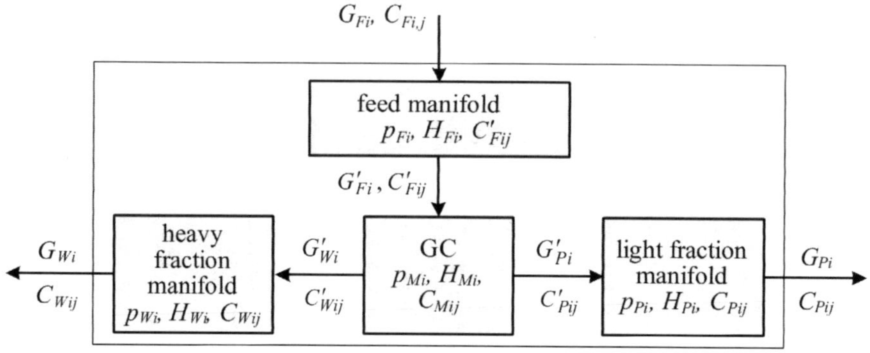

Fig. 2. Capacities of *i*th stage.

Differential equation describing nonstationary separation processes in each capacity is written in the general form as:

$$\frac{d\left(HC_j\right)}{dt} = G_{IN}C_{INj} - G_{OUT}C_j \tag{8}$$

Replacing derivatives to difference equations by backward Euler method [5], we have received:

$$\frac{H^{(k)}C_j^{(k)} - H^{(k-1)}C_j^{(k-1)}}{\Delta t} = G_{IN}^{(k)}C_{INj}^{(k)} - G_{OUT}^{(k)}C_j^{(k)} \tag{9}$$

$$\Delta t = t^{(k)} - t^{(k-1)} \tag{10}$$

where Δt is time step; k is the index of time layer.

The system of linear algebraic equations received from equations (10), is solved by the Gaussian elimination [5]. The result of calculation is values of component concentrations in each capacities and flows at every time layers. We have completed the algorithm described herein as a software product, developed on Delphi programming language using Embarcadero Delphi XE2.

3 The results of calculation cascade for germanium isotope separation

We have calculated parameters of nonstationary separation process in model GC cascade for germanium isotope separation for the purpose of verification of developed mathematical model and created software product. The two initial conditions are:

1. Hydraulic parameters of cascade correspond to the stationary values, isotope concentrations in all stage are equal to their concentrations in feed flow [1, 2] (initial conditions 1).
2. Hydraulic parameters of cascade and isotope concentrations in all stage are the result of calculation of the nonstationary process of a filling cascade with process gas (initial conditions 2) (fig. 3).

The calculation results of nonstationary separation process in the cascade for germanium isotope separation are given on fig. 4.

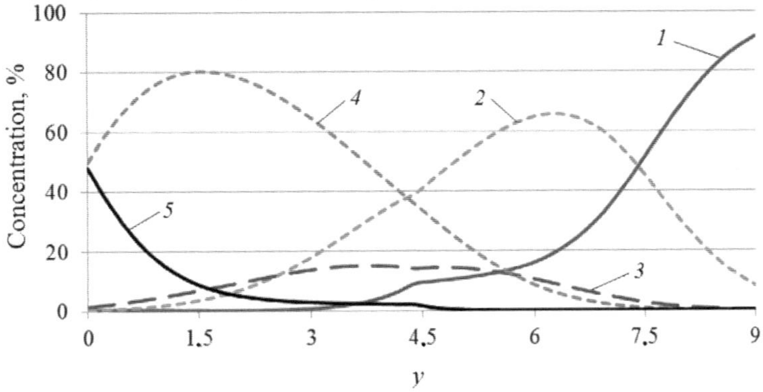

Fig. 3. Isotopes distribution along length of cascade after filling cascade by process gas: *1* – ^{70}Ge, *2* – ^{72}Ge, *3* – ^{73}Ge, *4* – ^{74}Ge, *5* – ^{76}Ge.

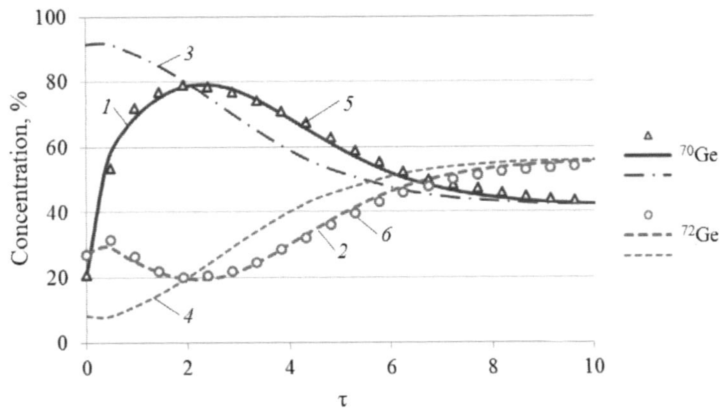

Fig. 4. Time variation of ^{70}Ge and ^{72}Ge concentrations in light fraction flow of cascade: *1*, *2* – developed model (initial conditions 1); *3*, *4* – developed model (initial conditions 2); *5*, *6* – models of other researchers [1, 2].

As it's seen on fig. 4, the calculation results obtained by developed model for initial conditions 1 (the curves 1 and 2) are well correlated with the results of other researchers [6, 7], which demonstrates the adequacy of developed mathematical model.

4 Conclusion

1. The result of research is creation of mathematical model of nonstationary separation processes in gas centrifuge cascade for separation of multicomponent isotope mixtures, representing the differential equation system of first order with given initial conditions.
2. The solution algorithm of equation system describing nonstationary separation processes is developed and realized as a software product.
3. We have made the calculation parameters of nonstationary separation process in model GC cascade for germanium isotope separation for the purpose of verification of developed mathematical model and created software product. Comparison of obtained values with results of other authors revealed that developed mathematical model is adequate to describe nonstationary separation processes in gas centrifuge cascades for separation of multicomponent isotope mixtures.

4. This mathematical model together with model of nonstationary hydraulic processes [9] makes it possible to perform calculations of isotope concentrations in cascade stages during its filling by process gas.

References

1. N.I. Laguntcov, E.V. Levin, B.I. Nikolaev, G.A. Sulaberidze, At. Energ. **6**, 62 (1987)
2. G.A. Sulaberidze, V.A. Palkin, V.D. Boricevich, V.D. Borman, A.V. Tichomirov *Theory of cascades for separation of binary and multicomponent isotope mixtures* (MEPhI, Moscow, 2011)
3. A.A. Orlov, V.P. Sovach, A.A. Ushakov Altern. Energ. Ekol. **23**, 187 (2015)
4. K. Cohen *The theory of isotope separation as applied to the large scale production of U-235* (McGraw Hill, New York, 1951)
5. U.G. Pirumov *Numerical methods* (Drofa, Moscow, 2003)

MATEC Web of Conferences **72**, 01107 (2016) DOI: 10.1051/matecconf/20167201107
HMTTSC-2016

STUDYING THE DYNAMICS OF BREAKDOWN OF THIN HORIZONTAL LIQUID LAYERS WITH LOCAL HEATING

Serafim Spesivtsev[1, 2, *], and *Yuriy* Lyulin[1,3]

[1] Kutateladze Institute of Thermophysics SB RAS, 630090 Novosibirsk, Russia
[2] Novosibirsk State University, 630090 Novosibirsk, Russia
[3] Tomsk Polytechnic University, 634050 Tomsk, Russia

Abstract. Experimental study of liquid layers breakdown when heated locally from the substrate side was made. Water and ethanol were used as working liquids with a layer thickness of 300 μm. Basic steps of the breakdown process were found and mean velocities of the dry spot formation were determined; the values are 0.06 mm/sec for ethanol and 5.15 mm/sec for water. The formation of residual layer over the hot-spot before the breakdown has been found for both liquids. The creation of a droplet cluster near the heating region is observed when using water as a working fluid. It was shown that evaporation is one of the general factors influencing the process of layer breakdown and dry spot formation as well as thermocapillary effect.

1 Introduction

Investigation of heat removal from the local heat source is currently one of the most complex problems in thermophysics due to its direct connection to the problem of microelectronic equipment cooling [1]. The average heat flux density on the surface of chips of commercially available computers and other electronic devices is currently known to reach 100 W/cm^2. Furthermore, there are limited areas of relatively small size where the heat flux density can exceed the average value by several times and reach 1 kW/cm^2 [2]. This effect takes place due to the design features of computer chips, where the processor cores cause the formation of "hot" spots. The highest heat release occurs in these local areas and the increased probability of heat transfer crisis takes place, leading to the loss in the chip efficiency and its destruction [3]. There are several approaches for solving the problem of heat removal from the "point" hot zones such as: the use of thermoelectric modules [4], boiling in microchannels [5] and spray cooling [6]. A similar problem also is linked with the LEDs cooling, where it is necessary to remove high heat flux from the local heat sources. One of the most promising methods for cooling of the electronic equipment is technology that uses processes with phase transformation. It is, for example, evaporation of a thin liquid layer, moving in a planar micro channel under the influence of the gas flow [7]. In this case the main mechanism of heat removal from the heat source is an intense

* Corresponding author: simafir94@gmail.com

evaporation due to latent heat of vaporization. Dynamics of evaporation and, thus, the heat removal from the heat source substantially depend on the conditions in a thin layer of liquid [8]. In particular, the liquid layer rupturing due to the action of thermocapillary forces and intense evaporation leads to an aggravation of heat removal from a local heat source [9]. Thus, it is necessary to define the conditions under which the liquid layer breakup occurs.

The aim of this work is to study the breakdown dynamics of the layers under heating from a point heat source using different liquids (ethanol, water).

2 Experimental rig

The research was conducted on experimental rig whose scheme is presented in fig. 1. Fluid from the syringe pump enters the work area, forming a horizontal liquid layer opened to the atmosphere. The working area consists of a caprolon base, a metal substrate and a heating element. There is a local heating from the side of the substrate in the center of the horizontal liquid layer. The heating element represents a brass core with round tip of 1.6 mm in diameter. Caprolon base has a special notch in the upper part to mount the metal substrate and the through hole in the center with a diameter of 1.6 mm. The metal substrate is made of stainless steel with a diameter of 50 mm and a thickness of 1 mm. At the center of the substrate there is a blind hole with a diameter of 1.6 mm and a height of 0.8 mm. The heating element was a brass core with round tip with a diameter of 1.6 mm and a height of 3 mm. The tip is tightly inserted into the blind hole of the substrate through the base. For a better thermal contact between the heater and the substrate the thermal paste was used. The distance between the tip and the upper part of the substrate was 0.2 mm. The heat source of the heater is a nichrome band wound on the core stem. Heating power is controlled by the power source. The heater core was placed in the notch of the base on its lower side, and between the heater and the base there was an air gap of 2 mm. From the bottom of the heater there was an insulating material to minimize heat loss. The temperature was measured by thermocouples using measuring system NI 9214 with an accuracy of ± 0.1°C. The relative humidity and the ambient air temperature are measured using the thermohygrometer Testo 645 with an accuracy of 2% and 0.1°C, respectively.

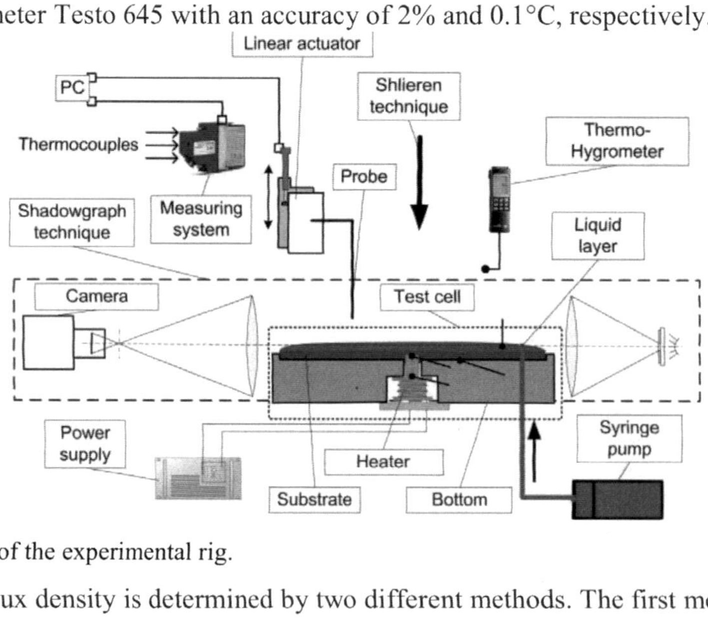

Fig. 1. Scheme of the experimental rig.

The heat flux density is determined by two different methods. The first method involves measuring the temperature difference in two different cross-sections along the length of the heater tip. For this purpose two thermocouples were mounted along the cylindrical part of the tip, and the heat flux density is calculated according to the formula:

$$q_1 = \lambda \, \Delta T / l, \qquad (1)$$

where λ is the thermal conductivity of the heater material, W/cm·K; l is the distance between the two sections of the heater tip, cm; and ΔT is the temperature difference in two cross sections along the length of the heater tip, K. The second method includes determining the power of Joule heat according to the formula:

$$q_2 = U \, I / S, \qquad (2)$$

where U is the voltage, V; I is the current intensity, A; and S is the section area of the heater tip, cm^2. While increasing the heat flux the difference between the methods increases up to 20%. This difference arises due to heat loss, the value of which is expected for such experiments. The main method for measuring the heat flux density is considered to be the first method as the most accurate.

To control the liquid layer thickness we use the shadowgraph method and the linear actuator with the probe. With the help of high-precision syringe pump we select the flow rate, which ensure a constant level of liquid, taking into account the evaporation. In the experiment, the liquid level was maintained 5-10 μm below the probe. The horizontal layer surface and the probe are observed using video camera Imaging Source DFK 23GP031 with a resolution of 2592x1944 pixels. To visualize surface deformations and register the breakdown we use an optical schlieren system with high-speed video camera Photron FASTCAM 675K-M3 (speed of 2500 fps at a resolution of 640x640 pixels and a scale of 25 μm/pix). The working section is mounted in horizontal position using a goniometer. The surface roughness of the substrate was determined by profilometer "Micro Measure 3D station" and the average roughness value amounted to R_a=0.327 μm. The advancing contact angle on the working surface in the heating area was determined by the sitting drop method (Young-Laplace) [10] at a room temperature of 25±2°C and amounted to θ_1=6±1° for ethanol and θ_2=76±1° for water.

3 Experimental results

The investigation was conducted at atmospheric pressure, temperature of 28±2°C and relative humidity of 25±3%. The used working liquids were ethanol (95% (mass.), GOST R 51723-2001) and ultrapure water. For water purification we used the system Merck Millipore Direct-Q 3 UV, which allows providing the water of type I (ultrapure water). Characteristics of the quality of the obtained ultrapure water: resistance of 18.2 MO/cm, TOC (total organic carbon) < 5 μg/l, microorganisms < 0.1 CFU/ml, pyrogens (bacterial endotoxins) < 0.001 units. end./ml; RNA < 0.01 ng/ml, DNA < 4 pg/ml; and particles > 0.22 μm < 1 particles/ml. The height of the liquid layer was 300 μm. For both working fluids we measured the critical heat flux density at which the breakdown of the liquid layer occurs. The value of the critical heat flux density for ethanol equals to 12.6 W/cm^2 at the substrate temperature in the heating area of 37.1°C, for water it is 117 W/cm^2 at the substrate temperature of 133°C, respectively.

It was found out that for both working liquids the layer breakdown occurs according to one scenario [11]. First, a thermocapillary deformation of the layer above the point heating area appears (fig. 2a, b). Further thinning leads to the formation of a residual liquid layer in the area of the point heating, fig. 2c [12]. Then, the residual liquid layer evaporates to a critical thickness, at which the layer breakdown takes place, fig. 2d. After the breakdown the entire heating area is intensely dried, and the round dry spot is formed, fig. 2e.

The average rate of dry spot formation was measured for two working fluids and was determined as the ratio of the characteristic radius of the resulting dry spot to the time of its formation in the heating area. The time of dry spot formation is counted from the moment of the residual layer breakdown to its complete evaporation. The average rate of dry spot formation for ethanol is 0.06 mm/s, and for water it is 5.15 mm/s, respectively. The time of dry spot formation for ethanol as a working fluid is 7.85 seconds, and for water it is 0.13 seconds. The difference in the breakdown rates is first of all connected with different rate of the residual layer evaporation, which directly depends on the heat flux density and the substrate temperature. In addition, it is influenced by the difference in properties and hydrodynamic parameters of working fluids: contact angles and surface tension.

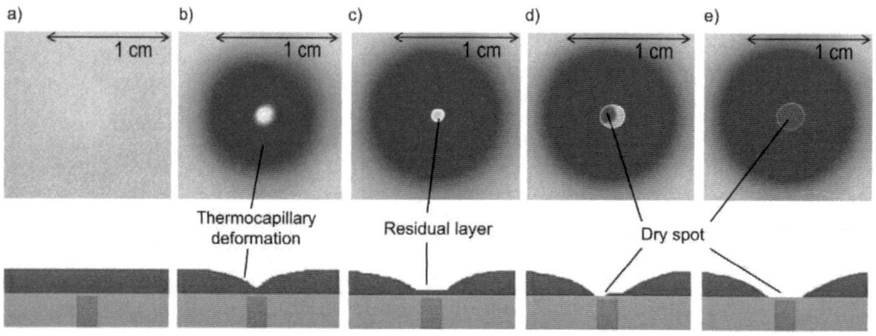

Fig. 2. The dynamics of the liquid layer breakdown in case of point heating. Liquid is ethanol and the layer height is 300 μm.

In the study of thermocapillary breakdown of the water layer with point heating the existence of the droplet cluster was found, fig. 3. The phenomenon of droplet cluster was investigated in detail in [13-15]. Experiments show that the droplet cluster is always formed near the heating area and practically does not influence the dynamics of the liquid layer breakdown.

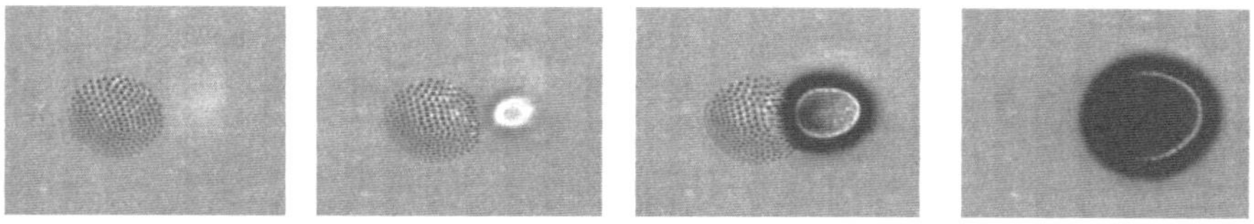

Fig. 3. Visualization of the droplet cluster and the breakdown dynamics. The liquid is water and the layer height is 300 μm.

4 Conclusions

Experiments have been performed to measure the critical heat flux and to visualize the breakdown dynamics of thin layers of ethanol and water using schlieren technique. The water layer breakdown was found to require a heat flux density by an order exceeding the critical heat flux density for a layer of ethanol of the same thickness. At that, times and the characteristic breakdown rates differ by two orders of magnitude, which is directly related to the difference in the critical heat fluxes and intensities of evaporation. It is shown that before the breakdown the residual layer appears in the area of local heating. When using water as the working liquid the formation of a droplet cluster may be observed in the vicinity of the heating area of the substrate. Along with the thermocapillary effect, evaporation is one of the main factors influencing the residual liquid layer breakdown and dry spots formation in the heating area.

Acknowledgments

The study was financially supported by the Russian Science Foundation (Project 14-19-01755).

References

1. A. Bar-Cohen, P. Wang, J. Heat Transf. **134**, 5 (2012)
2. R. Mahajan, C. Chiu, G. Chrysler, Proc. IEEE **94**, 8 (2006)
3. A. Watwe, R. Viswanath, Proc. Inter PACK'03 (2003)
4. A. Bar-Cohen, P. Wang, Microgr. Sc. Tech. **21** (2009)
5. P. Di Marco, W. Grassi, Int. J. Therm. Sci. **41** (2002)
6. E.A. Silk, E.L. Golliher, R.P. Selvam, Energy Convers. Manag. **49** (2008)
7. O.A. Kabov, D.V. Zaitsev, V.V. Cheverda, A. Bar-Cohen, Exp. Therm. Fl. Sc. **35**, 5 (2010)
8. A.A. Fedorets, I.V. Marchuk, O.A. Kabov, Int. Phen. Heat Transf. **1**, 1 (2013)
9. D.V. Zaitsev, O.A. Kabov, Microgr. Sc. Tech. **3/4** (2007)
10. I.V. Marchuk, V.V. Cheverda, P.A. Strizhak, and O.A. Kabov, T. A. **22**, 3 (2015)
11. Yu.V. Lyulin, S.E. Spesivtsev, I.V. Marchuk, O.A. Kabov, Tech. Ph. Let. **41**, 11 (2015)
12. D.V. Zaitsev, D.A. Rodionov, O.A. Kabov, Microgr. Sc. and Tech. **19** (2007)
13. A.A. Fedorets, I.V. Marchuk and O.A. Kabov, Tech. Ph. Let. **37** (2011)
14. A.A. Fedorets, I.V. Marchuk and O.A. Kabov, JETP Let. **99,** 5 (2014)
15. A.A. Fedorets, I.V. Marchuk, P.A. Strizhak and O.A. Kabov, T. A. **4** (2015)

FORMING THE COMPOSITION OF UNDERGROUND COAL GASIFICATION PRODUCTS IN THE SIMULATION OF VARIOUS HEAT AND MASS TRANSFER CONDITIONS IN THE COAL SEAM

A.S. Masanik[1], *A*.N. Subbotin[1,*], and *A*.S. Tarasanov[1]

[1]National research Tomsk Polytechnic University, 634050 Tomsk, Russia

Abstract. The mathematical model describing the heat and mass transfer processes in underground coal gasification is proposed. Numerical studies have allowed to determine the composition of gases depending on the temperature, pressure products of gasification, and the composition of the heated oxidant injected. Relations the composition of the concentration of combustible gas component of the oxidant injected: dry air, a mixture of oxygen, nitrogen and water vapor in different proportions were prepared. It is found that, depending on the oxygen content in the oxidizer low-temperature gasification mode is implemented (up to 15%). At higher values of the oxygen concentration in the oxidizer the high-temperature mode is realized, in which the fuel gas output increases significantly.

1 Introduction

Coal - a mineral resource, which is a unique material. A wide range of different products prepared on the basis thereof. Besides coal is one of the basic elements of modern fuel and energy balance in the world. Many countries are now actively developing modern coal deep processing technology to reduce the dependence of the production of chemical products from imports of oil and natural gas [1]. Look for new solutions to chemical and energy problems, natural resource extraction issues [2].

At present, almost all the major coal producing countries of the world are showing great interest in the underground coal gasification technology. It should be noted that the intensive research work and practical nature are underway in China, the US and many other countries [3]. Gasification is a low-temperature or high-temperature processes, which occur in the interaction of the organic mass of solid or liquid fuel or products of their thermal processing with oxygen, carbon dioxide, air, steam, or mixtures thereof. As a result of these processes is an organic part of the processed fuel turns into a combustible gas. [4]For gasification as feedstock, generally use peat, oil shale, lignite or stone. Typically, the oxidizing agent (gasifying agent) is air, oxygen or steam.

* Corresponding author: subbot@tpu.ru

The aim of this study is a numerical analysis of the conditions for the implementation of the underground coal gasification process, and analyzing the output composition of the synthesis gas depending on conditions, under which the gasification of the coal seam is carried out.

2 Physical formulation of the problem

Considered the most simple and quite typical diagram of underground coal gasification. The coal seam is drilled into which is inserted a pipe. Inside this pipe is placed a pipe of smaller diameter, with the extension and perforated holes at the bottom of the base. On the inside of the pipe in the coal seam is injected heated to a certain temperature, air or steam-air mixture. Over time the coal seam warms up and it will start the process of pyrolysis further can be an ignition of coal. Through the holes on the side surface of the outer tube, the products of pyrolysis and combustion penetrate into the outer pipe and moved from the coal seam to the surface.

Changing characteristics of the supplied vapor oxidant, will get different gas composition at the outlet. Will conduct the study, how it will change the composition of the synthesis gas when changing the conditions of flow of oxidant and thermophysical properties.

For typical coals when they are heating, depending on temperature, can occur quite a number of reactions [3]. In the present work, it was assumed that the analyzed process is accompanied by gasification reactions [5].

3 Numerical results of mathematical modeling

A mathematical model describing the processes that occur during the gasification of the coal seam is shown in [5]. For the numerical solution of the boundary value problem given in [5] used the method of making and difference scheme [6]. All gasified fuel characteristics were calculated: temperature distribution in the coal seam, field of velocity of filtration and pressure, and the volume fraction of condensed gaseous pyrolysis products, the moisture concentration of the gaseous pyrolysis products. [6]. For the numerical calculations, some positive pressure was set at the entrance $\Delta P = P_0 - P_H$, the moisture content of the coal seam W, the mass concentration of oxygen and water vapor are ranged. From these numerical results imply that warmed area formed in the coal seam, in which the coal gasification process takes place. figure 1 shows the isotherms in pumping the oxidant into the formation, the temperature is 800 K, and mass concentration of oxygen 0.1.

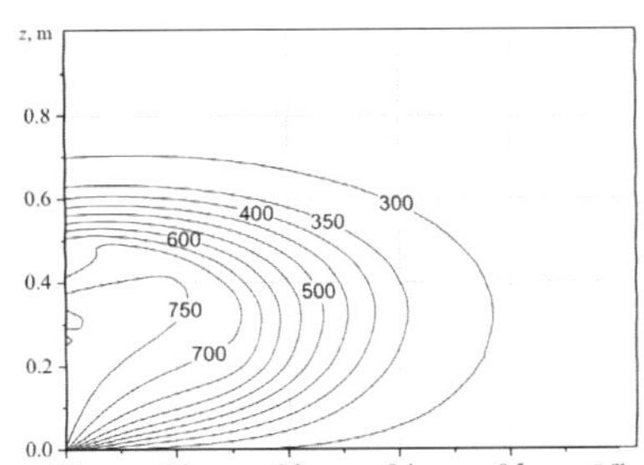

Fig. 1. The distribution of temperature in the coal when it overheats.

Increasing the mass concentration of the oxygen to 0.2 in vapor mixture injected into the coal seam, is leads to coal combustion. Even after reducing the injected oxidant temperature to 400 K it is combustion (fig. 2).

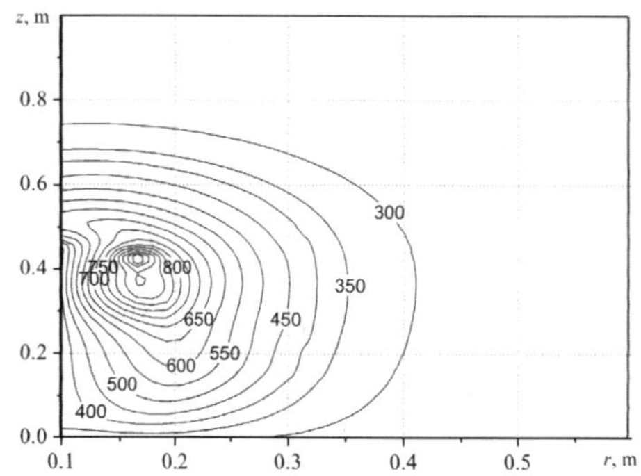

Fig. 2. The distribution of temperature in the coal seam after its ignition (Mass concentration of oxygen equal to 0.2).

Profiles of the volume fraction of coke for the same points in time are shown in figure 3.

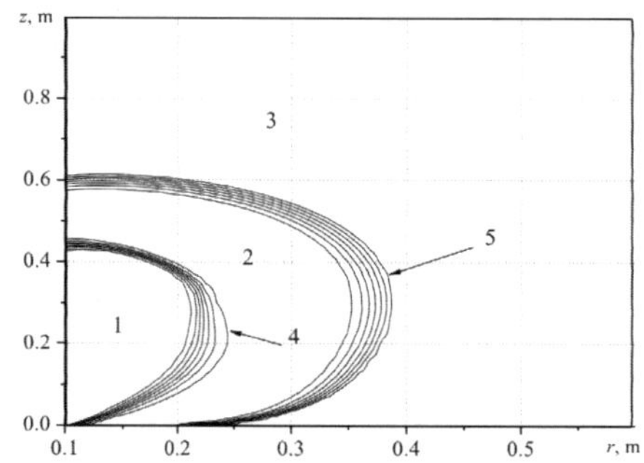

Fig. 3. Distribution of the volume fraction of coke upon heating the coal seam: 1 - an area in which the slag was; 2 - an area filled with char; 3 - unheated area filled with coal; 4 - boundary of coke oxidation; 5 - coal pyrolysis border.

Analyzing the graphs in figure 3, we conclude that formed three areas: 1 - where coal is burned down completely, it remained part of the non-combustible mineral; the area 2, in which the coke oxidation occurs, the area 3 - coal seam, which has not yet come any chemical processes, since it is not warmed up. As shown by numerical calculations, the bulk of the resulting products of pyrolysis and combustion from the bottom base of the coal seam moves along processualnoe region 2, moves along the boundary of pyrolysis, filtered in the outer tube and removed via the outer tube to the outside.

The bulk of the coal supplied to the layer of oxygen consumed in the areas 2 and 4 fig. 3 (oxidation products of pyrolysis), the rest of the volume of the test coal seam contains oxygen in small quantities. Formed in the oxidation of the pyrolysis coke and coal, carbon dioxide is moved together with the other gaseous components of the oxidation zone in the coke bed and enters the external tube wells. Therefore, in the coke focuses always the maximum concentration of all the resulting products of pyrolysis and combustion. In particular, in fig. 4 shows contours of carbon oxide.

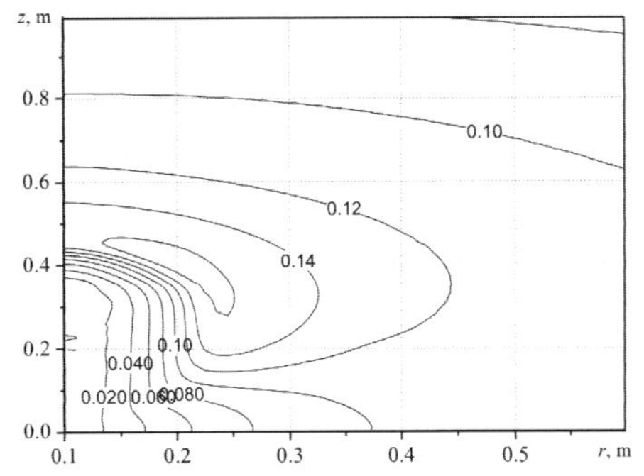

Fig. 4. Contours of the mass concentration of carbon monoxide.

Analyzing (fig. 4), the distribution of the mass concentration of carbon monoxide in gasification come to the conclusion that in the gas coming from the well, the mass concentration of CO is approximately equal to 0,12 – 0,14. If to carry out the pyrolysis of coal in an inert atmosphere, the mass concentration of carbon monoxide formed during the pyrolysis of coal will be three times less.

The distribution of the mass concentration of carbon dioxide in a coal seam is shown in fig. 5. From this distribution we can conclude that, for given parameters of coal gasification in a coal seam, the content of carbon dioxide decreases sharply compared with coal pyrolysis in inert atmosphere (pyrolysis of coal in an inert atmosphere concentration of CO_2 is three times more). The concentration of carbon dioxide can be further reduced by increasing the temperature of the gasification process.

From their analysis of the fields of mass concentrations of hydrogen and methane means that the concentration of these components is very small. During pyrolysis of coal in an inert environment mass concentration of hydrogen in twenty times, but methane is twenty five times higher. Such a strong decrease of the hydrogen concentration can be explained by its oxidation with the formation of water vapor, and methane – and carbon monoxide.

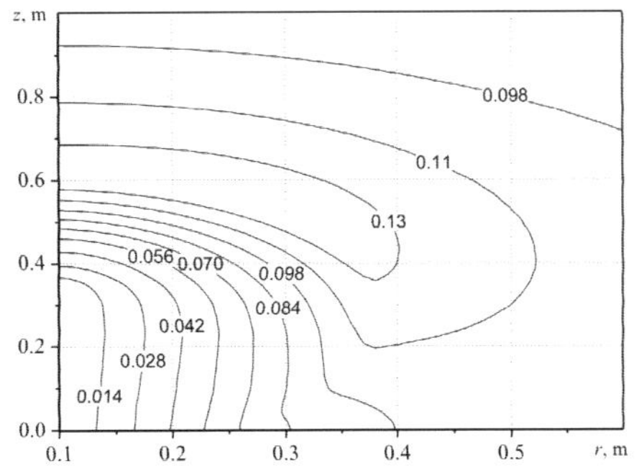

Fig. 5. Contours of the mass concentration of carbon dioxide.

This conclusion is confirmed by a significant increase in the mass concentration of water vapor in a coal seam (fig. 6).

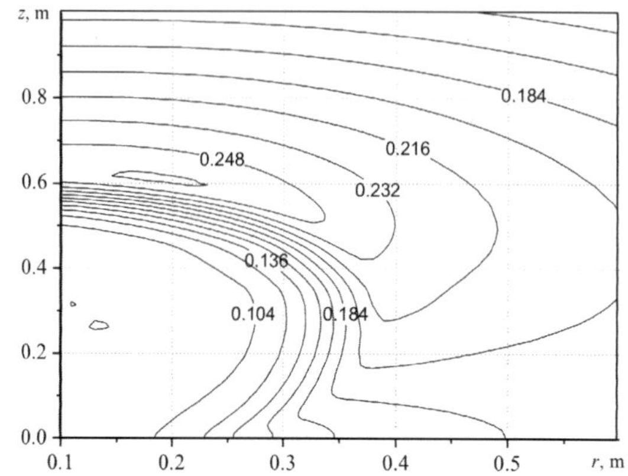

Fig. 6. The distribution of the mass concentration of water vapor.

In the vapor-air mixture supplied to the coal seam, the mass concentration of water vapor equal to 0.1. Inside the reservoir, it increases more than doubled, partly due to the evaporation of moisture of the coal seam (asked the initial moisture content of the coal seam is equal to 5%) and partly due to the fact that the reaction of hydrogen oxidation. It should be noted that the concentration of combustible gas in the gasification products is dependent on the concentration of water vapor, the gasification temperature and oxygen concentration in the vapor-air mixture supplied to the coal seam. In particular, if you increase the pressure drop, assuming it is equal to 2·103 PA, the velocity of the feed vapor-air mixtures in the coal seam increases, and this changes the ratio of the concentrations of gaseous components at the outlet. In particular, the oxygen concentration at a given differential pressure increased slightly compared with the previous calculation. Ultimately, this led to the increase in the concentration of carbon monoxide in the exhaust from gasification gas. Profiles of the mass concentration of carbon monoxide is shown in fig. 7, from which it follows that in the outer tube receives a combustible mixture with the mass concentration of carbon monoxide 0,14–0,18 (in the previous calculation it was equal to 0,12–0,14). The concentration of other components remained almost unchanged. When this differential pressure, the filtration velocity was doubled. Thus, the higher pressure drop in 2 times has led to the increase in the concentration of carbon monoxide and the volume of the resulting combustible mixture.

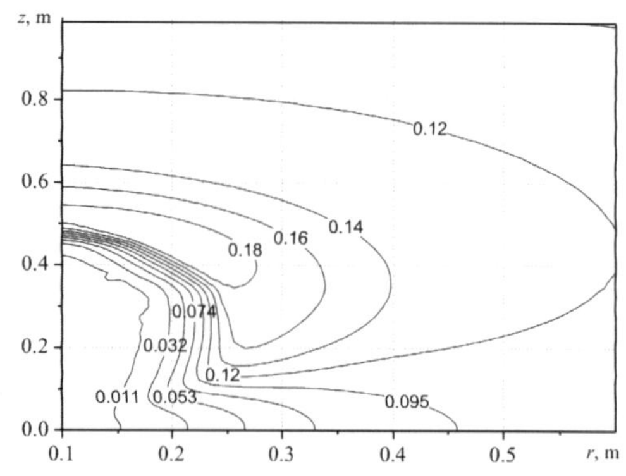

Fig. 7. The distribution of the mass concentration of carbon monoxide (C_{10}=0.1, $\Delta P=2\cdot10^3$ Pa).

In addition to the above results, calculations were carried out with an increased concentration of oxygen in air-steam mixture injected into the coal seam. Mass concentration of oxygen in the mixture was assumed equal to 0.15 and not 0.1 as in the previous calculations. The result is that the coal seam is increased, the oxygen concentration, which led to the increase in the concentration of carbon monoxide in the gasification products. In fig. 8 shows the volume distribution of the mass concentration of carbon monoxide while increasing by 50% the initial concentration of oxygen in air-steam mixture injected into the coal seam. Comparing fig. 8 and 4, we conclude that the increase in the initial oxygen concentration of 50% led to the increase in the concentration of carbon monoxide in the resulting gas mixture by about 50%. The mass concentration of other components remained almost unchanged. Further increase in the concentration of oxygen in the vapor-air mixture supplied to the coal seam, leading to regime change, coal seam is ignited (there is an area of burning).

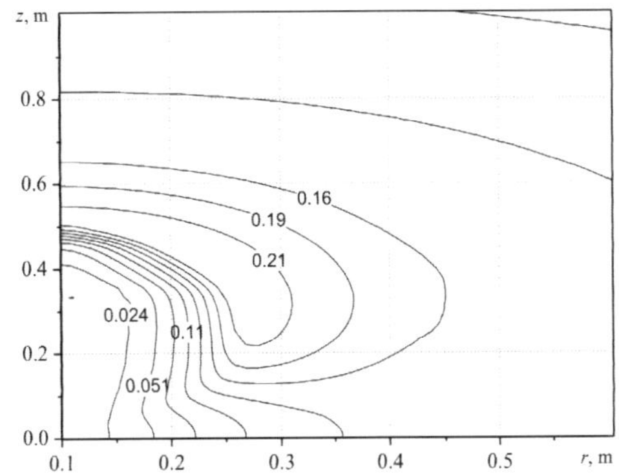

Fig. 8. The distribution of the mass concentration of carbon monoxide (C_{10}=0.15, $\Delta P=10^3$ Pa).

The temperature in this region rises to 1300 K, even when the flow in the coal seam, after ignition, the relatively cool air-steam mixture (T_0=400 K). The temperature rise in a coal seam led to the increase in the concentration of combustible gases. In fig. 9 shows the profiles of mass concentration of carbon monoxide. As follows from the analysis of this figure, the CO concentration at the outlet of the well was equal to 0.2÷0.3, and in the area of maximum temperature is equal 0.45.

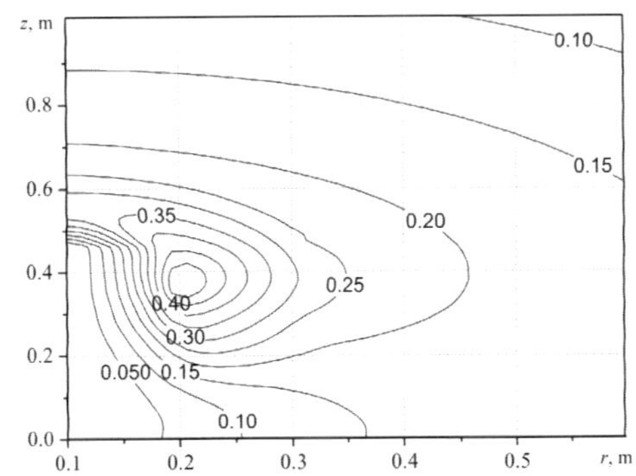

Fig. 9. The distribution of the mass concentration of carbon monoxide.

4 Conclusions

The results of numerical analysis give grounds to draw the following conclusions:

1. The mathematical model allows to predict the structure of products of gasification in a coal seam.
2. The supply and composition of the oxidant significantly affect the final yield of the main product of underground coal gasification.

3. Large gradients, all the major required functions, the spatial movement area of the main reactions and the size of the object of study greatly complicate the process of numerical solution of the problem.

References

1. V.J. Arens, A.A. Vertman, G.H. Hcheyan, Vestnik RAEN [RANS reporter] **3** (2007) [in Russian]
2. A.Y. Zorya, E.V. Kreynin, S.N. Lazarenko, J. Coal Kuzbass. Interreg. Res. Prac. **4** (2009)
3. R.N. Kulesh, A.S. Mazanik, A.N. Subbotin, Bul. TPU **4** (2014)
4. Y.V. Vasychkov, B.M. Vorobiev, *Invention patent No. 2100588* [in Russia]
5. G.V. Kuznetsov, A.N. Subbotin, Heat proc. in Tech. **9** (2010)
6. A. M. Grishin, V. I. Zinchenko, K. N. Efimov, A. N. Subbotin, A. S. Yakimov. *Iteration-interpolation method and its application,* (Publ. house TGU, Tomsk, 2004)

MATEC Web of Conferences **72**, 01109 (2016) DOI: 10.1051/matecconf/20167201109
HMTTSC-2016

INFLUENCE OF ABSORPTION OF THERMAL RADIATION IN THE SURFACE WATER FILM ON THE CHARACTERISTICS AND IGNITION CONDITIONS

Samen V. Syrodoy[1,*], *Nadegda* Yu. Gutareva[1], and *Roman* I. Taburchinov[1]

[1]National Research Tomsk Polytechnic University, 634050 Tomsk, Russia

Abstract. The results of the mathematical modeling of homogeneous particle ignition process of coal-water fuel covered with water film have been presented in article. The set co-occurring physical (inert heating, evaporation of water film) and thermochemical (thermal degradation, inflammation) process have been considered. Heat inside the film has been considered as the model of radiation-conductive heat transfer. Delay times have been determined according to the results of numerical modeling of the ignition. It has been shown that the water film can have a significant impact on performance and the ignition conditions. It has been found that heating main fuel layer occurs in the process of evaporation of water film. For this reason, the next (after the evaporation of the water film) thermal preparation (coal heating, thermal decomposition of the organic part of the fuel) and inflammation occur faster.

1 Introduction

Prospects for the use of coal-water fuel (WCF) as the main thermal in the power plants and local boilers have been sufficiently justified [1, 2]. To date, the investigations of water-coal technologies are being carried out intensively both in Russia and abroad [3, 4]. However, despite this, the widespread adoption of WCF in the fuel and energy complex (FEC) has not yet occurred. The latter may be due to a number of objective and subjective reasons. Subjective reasons include economic factors (substantial cost advantage of gas over coal [5]) and technological reasons. The objective reasons are insufficient knowledge of the main thermo-physical processes (slow heating, evaporation of water) and thermochemical processes (thermal degradation, inflammation) occurring together during the thermal preparation of the WCF particles to inflammation [6-9].

Experimental studies of coal-water fuel drops ignition is difficult, due to the need for placement of equipment of high-speed video fixation in high-temperature gaseous atmosphere (combustion camera)[10]. Additional difficulties with the registration of the moment of ignition (the appearance of open flames) occur on the background of the furnace environment. For this reason, the use of mathematical modeling is the most appropriate for studying the thermal preparation and ignition of coal-water fuel particles. However, it

*Corresponding author: ssyrodoy@yandex.ru

should be noted that the main problem of mathematical description of such inhomogeneous and high-wet fuels is complex (joint) occurring of basic physical and chemical (inert heating, water evaporation, thermal decomposition of the organic part of the fuel, ignition) processes under conditions of high-temperature heating.

It is known that the processes of phase conversions (in particular water evaporation) are one the most difficult in terms of the mathematical formulation. This is because that the phase transition (in particular evaporation) occurs in a very narrow (much less than the linear dimension of the particle) constantly moving zone (front of evaporation) [13, 14]. The models [15-20], describing the phase transitions (such as evaporation) have been developed at present, but most of them are based on the considerably simplifying assumptions of modeling procedure. For example, there are such means as the equilibrium parameters (temperature, evaporation rate) of the phase transition at the boundary front [15], the use of the heat balance method [16] or damp-potential model [17]. Previously, it has been found [18] that two drop systems can be formed during WCF spraying. The first system is consisted of carbon particles coated by water film. The second system has only coal-water droplets. The ignition of coal-water droplets has been considered in detail in [19]. At the same time, the question of ignition of homogeneous coal particles coated with a film of water remains little studied [20]. A film of water can influence significantly on ignition conditions of these particles. This influence is caused by the high endothermic effect of the water vaporization (up to 2.5 MJ / kg) and the ability of the water to absorb and dissipate the radiation (dependent on incident wave spectrum) [21]. It is safe to say that the surface layer of water may have a significant effect on the characteristics of the particles and WCF ignition conditions. This is due to several factors. The first factor is the high endothermic effect of water evaporation and heat capacity of water. The second important factor is that water is optically thin environment (depending on the incident radiation of wavelength) [21]. Accordingly, in a high-temperature heating is expedient to take into account not only conductive but also radiative heat transfer within the film. The problem about emission of electromagnetic radiation (in a range of) to a thin film of water up to the present days has been hardly studied.

The aim of this study is mathematical modeling of heat and mass transfer processes during the ignition of coal particles coated by water film, taking into account the absorption of thermal radiation by water.

2 Formulation of the problem

The following physical model of the ignition has been accepted. At the initial time ($t = 0$), a drop of WCF particles coated by water film enters the conditions corresponding to the combustion chamber of the boiler unit. Under the influence of high-intensive radiative-convective heat flow film of water begins to evaporate. As a result, the particle size decreases. As a result of intense radiation and convection heating is initiated by water evaporation. The mass velocity of the "dehydration" (and hence the endothermic effect) depends on the temperature of the surface water film. Radiation heat exchange between the coal particles and the external environment is carried out at high temperatures (T>1000K). The part of the thermal radiation is absorbed by the water film.

Boiling water occurs by heating the system boundary "coal – water" to vaporization temperature. The process of thermal decomposition and ignition is initiated after removing of the film particles WCF in a continuing heating process. It is assumed that particles on the surface of carbon oxidation reaction occur: It is assumed that the coal is composed predominantly of carbon, respectively, on the particle surface oxidation reaction proceeds as follows:

$$C + O_2 = CO_2 + 402{,}4kJ$$

The mathematical formulation of the problem, the corresponding stated higher of physical model is a non-stationary system of differential equations in partial derivatives:

- the energy equation for a homogeneous coal:

$$C_1\rho_1\frac{\partial T_1}{\partial t}=\lambda_1\cdot\frac{1}{r^2}\frac{\partial}{\partial r}\left[r^2\frac{\partial T_1}{\partial r}\right]\pm\sum Q_i\cdot W_i \tag{1}$$

$$t>0,\ 0<r<r_1.$$

- the energy equation for the film of water:

$$C_2\rho_2\frac{\partial T_2}{\partial t}=\lambda_2\cdot\frac{1}{r^2}\frac{\partial}{\partial r}\left[r^2\frac{\partial T_2}{\partial r}\right]-C_sW_v\frac{\partial T_2}{\partial r}+q_r\cdot\exp(-\chi_\lambda\cdot r) \tag{2}$$

$$t>0,\ r_1<r<r_0(\tau).$$

- the equation of chemical kinetics for homogeneous part of fuel:

$$\frac{\partial\eta}{\partial t}=\left[1-\eta(r,t)\right]\cdot k_0\cdot\rho\cdot\exp\left[-\frac{E}{R\cdot T(r,t)}\right] \tag{3}$$

$$t>0,\ 0<r<r_1$$

The system of equations (1-4) has been solved with the following boundary conditions and closing ratios:

$$0<r<r_0,\ T(r,0)=T_0,\ \eta(r,0)=\eta_0$$

$$-\lambda_2\frac{\partial T_2}{\partial r}\bigg|_{r=r_0}=\alpha\cdot[T_e-T_2(r_0,t)]+\varepsilon_2\cdot\sigma\cdot[T_e^4-T_2^4(r_0,t)]-Q_vW_v \tag{6}$$

$$-\lambda_1\frac{\partial T_1}{\partial r}\bigg|_{r=r_1}=\alpha\cdot[T_e-T_1(r_1,t)]+\varepsilon_1\cdot\sigma\cdot[T_e^4-T_1^4(r_1,t)] \tag{7}$$

$$\frac{\partial T_1(0,t)}{\partial r}=0 \tag{8}$$

There is (4) the boundary condition of the fourth kind at the interface of the “homogeneous coal — water film”, taking into account the radiation heat transfer in the system “homogeneous coal — environment” and the thermal effect of vaporization:

$$\lambda_1\frac{\partial T(r,t)}{\partial r}\bigg|_{r=r_1-0}=\lambda_2\frac{\partial T(r,t)}{\partial r}\bigg|_{r=r_1+0}+q_r-W_vQ_v \tag{4}$$

$$T_1(r_1;t)=T_2(r_1;t)=T_{w-c}(t)$$

The radiant heat flux on the boundary of the system interface “homogeneous coal — water film”, has been calculated from the mathematical expression of the law of Bouguer [22]:

$$q_r=\varepsilon_1\cdot\sigma\cdot\left[T^4(r_1,t)-T_e^4\right]\cdot\exp\left[-\chi_\lambda\cdot\left(r_0(\tau)-r_1\right)\right] \tag{5}$$

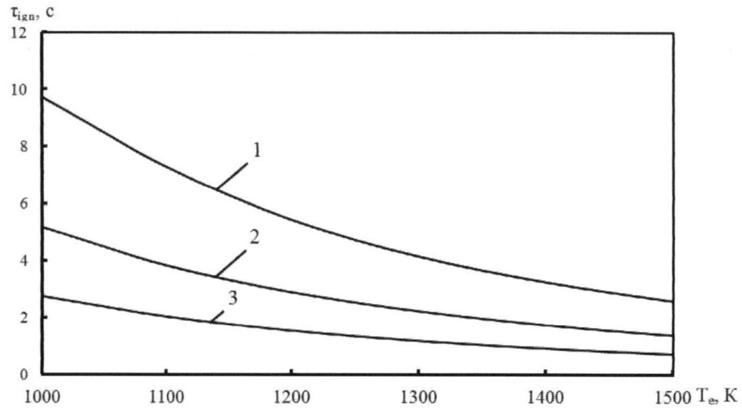

Fig. 1. The dependence of the ignition delay times of the ambient temperature at a rate (δ) particles: 1-δ=0, $5 \cdot 10^{-3}$m; 2-δ =0,$3 \cdot 10^{-3}$m; 3-δ =0,$2 \cdot 10^{-3}$m.

The rate of chemical reactions has been calculated from the mathematical expression of the Arrhenius law:

$$W_i = k_i \cdot \rho_i \cdot \exp\left(-\frac{E}{R \cdot T(r,t)}\right) \tag{9}$$

The rate of chemical reactions has been calculated from the mathematical expression of the Arrhenius law:

$$W_{td} = \left(1-\eta(r,t)\right) \cdot k_0 \cdot \rho \cdot \exp\left(-\frac{E}{R \cdot T(r,t)}\right) \tag{10}$$

The designations: T_0— is the initial temperature of the particles, K; T_e — is the temperature of the environment, K; $T_{w\text{-}c}$ — is the temperature at boundary section of the "coal — water", K; r_1 — radius of coal particle, m; r_0 — external WCF particle radius, m; σ — is a constant radiation of a black body; α — coefficient of convective heat transfer, $W/(\text{m}^2 \cdot \text{K})$; λ_1 — the coefficient of thermal conductivity of the homogeneous part of the WCF, $W/(m \cdot \text{K})$. λ_2 — the coefficient of thermal conductivity of water, $W/(m \cdot \text{K})$; C_1 — specific heat homogeneous part of WCF, $J/(kg \cdot \text{K})$; C_2 — the specific heat of water, $J/(kg \cdot \text{K})$; W_{eva} — mass the rate of water evaporation, $kg/(\text{m}^2 \cdot \text{s})$; Q_{eva} – is the thermal effect of the water evaporation , J/kg; Q_i — is the thermal effect of chemical reaction, J/kg; k_i — preexponent of chemical reaction $1/s$; E_i – the activation energy of a chemical reaction, $J/mole$; χ_λ — average-spectral coefficient of radiation absorption 1/m.

The mass evaporation rate has been calculated from the expression [23]:

$$W_v = W_0 \cdot \exp\left[\frac{Q_v \cdot \mu \cdot (T_s - T_f)}{R \cdot T_f \cdot T_s}\right] \tag{11}$$

Where: $\boldsymbol{T_1}$ — is the temperature at the boundary of evaporation, K; $\boldsymbol{T_0}$ — is the temperature of the liquid corresponding to the freezing point, K; $\boldsymbol{W_0}$ — the velocity of evaporation at temperature, $\boldsymbol{kg/(m^2 \cdot s)}$; $\boldsymbol{\mu}$ — molar mass of water vapor, $\boldsymbol{kg/mole}$; R — universal gas constant, $\boldsymbol{J/(mole \cdot K)}$.

3 The results

Figure 1 shows the dependence on the delay time of ignition homogeneous water coal particles coated with a water film, the temperature of the environment. The sizes of the particle have been varied in the range of $\delta = 0,2 \div 0,5 \cdot 10^{-3}$ m at a film thickness Δ=0,05$\cdot10^{-3}$m.

Dependency analysis (fig. 1) shows that the temperature of the external oxidant has a significant impact on the dynamics of the processes of thermal preparation and ignition of WCF particles. So we can say that for a particle with the diameter of $\delta = 0,5\cdot10^{-3}$ m temperature increase at 500K (from 1000 K to 1500 K), accelerates significantly ignition [approximately 3 times (delay time decreases from 10 seconds to 3)]. The latter can indicate the important role of the complex processes of evaporation of water film radiative-convective heat transfer, the heating up of base layer of fuel and the thermal decomposition of the organic carbon in the temporal characteristics of the thermal preparation steps. At the same time it should be noted that in the case of ignition of WCF drops with the diameter $\delta = 0,2 \cdot 10^{-3}$ m, the degree of this influence is negligible, which may also indicate the important role of the temperature distribution along the radius of the particle. Consequently, the application of balance-type models [16] during the predictive modeling of ignition of fuel can lead to large errors in determining the ignition delay times.

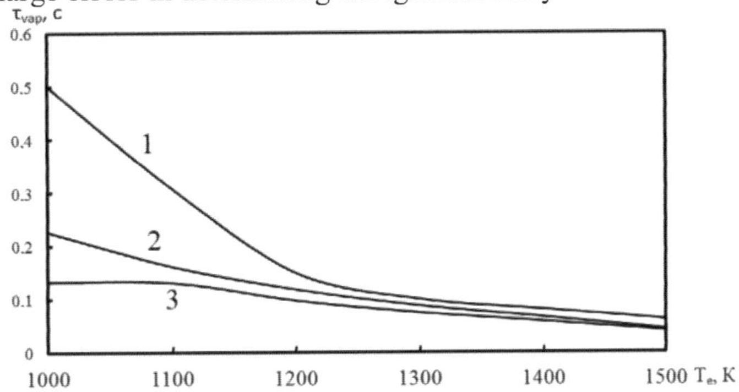

Fig. 2. The dependence of the evaporation times particle diameters rum (δ = 0,5 · 10-3m) on the ambient temperature at a film thickness (Δ): 1-0,15·10-3 m; 2- 0,1·10-3 m; 3-0,05·10-3 m.

Figure 2. shows the dependence of the evaporation times (particles of coal-water fuel with the diameter $\delta = 0,5 \cdot 10^{-3}$ m) on the ambient temperature by varying the thickness of the water film ($\Delta = 0,05 \div 0,15 \cdot 10^{-3}$ m). The analysis of the graph shows that the evaporation of water film can affect the dynamic of the thermal preparation. At this evaporation time of water film can take up to 10% of the total induction period. This is because in a small film thickness and penetration depth of the water heat radiation layer increases temperature of the latter which is much faster than excluding radiative intra-film heat transfer. Accordingly, it can be concluded that heating the base layer of fuel occurs in the water film evaporation process and the subsequent ignition takes place faster. The analysis of the curves (1) — (3) shows that at high temperatures (T_e>1200K) difference of evaporation time of a film is negligible, despite the substantial (more than 3 times) changes in Δ. At the same time difference in values τ_{vap} increases significantly at ambient temperature T_e<1200K. The latter indicates a significant effect of intra-film radiative heat transfer.

4 Conclusion

The basic temporal characteristics of processes of heat and ignition homogeneous preparation of coal-water fuel particles coated with a water film have been determined according to the results of numerical modeling. It has been found that the water film can affect significantly the dynamic of the thermal preparation. At this evaporation time of water film can take up to 10% of the total induction period. Radiative intra-film heat transfer plays a significant role this is due to the fact that it is under conditions of high heat. Accordingly, it can be concluded that heating of the base layer of fuel occurs in the water film evaporation process and steps of thermal preparation (inert heating, the thermal decomposition of organic fuel, ignition) proceed significantly faster than without the water film.

Acknowledgments

The work at the field of modeling of heat and mass transfer processes near boundary between two layers was supported by the scientific schools grant NSH-7538-2016.8.

Reference

1. E.S. Rubin, J. Wilson, Prog. Energy Combust. Sci. **21** (1995)
2. Ja´nos M. Beer, Prog. Energy Combust. Sci. **33** (2007)
3. N.I. Red'kina, G.S. Khodakov, E.G. Gorlov, Solid Fuel Chem. **47** (2013)
4. T.X. Phuoc, W. Ping, M. Dustin, S. Lawrence, Fuel Process. Technol. **127** (2014)
5. M.R. Betza, M.D. Partridged. M. Farren, L. Lobaoc, Energy Economics **50** (2015)
6. D.O. Glushkov, S.V. Syrodoy, A.V. Zhakharevich, P.A. Strizhak, Fuel Process. Technol. **148** (2016)
7. G.E. Liua, C.K. Law, Fuel **65** (1986)
8. P.L. Murdoch, M. Pourkashanian, A. Williams, Symp. (Int.) on Combust. **20** (1985)
9. A.P. Burdukov, V.I. Popov, V.G. Tomilov, V.D. Fedosenko, Fuel **81** (2002)
10. S.R. Correa da, H.J. Krautz, Appl. Thermal Eng. **82** (2015)
11. E. Hetmaniok, D. Słota, R. Wituła, A. Zielonka, Applied Mathematical Modelling **39** (2015)
12. R.M.S. Kushwaha, Applied Mathematical Modelling **37** (2013)
13. A.T. Il'icheva, G.G. Tsypkin, European Journal of Mechanics B/Fluids **27** (2008)
14. V.A. Shargatov, A.T. Il'ichev, G.G. Tsypkin, Int. J. Heat Mass Transfer. **83** (2015)
15. S.S. Sazhina, P.A. Krutitskiib, I.G. Guseva, M.R. Heikal, Int. J. Heat Mass Transfer **53** (2010)
16. Kijo–Kleczkowska Agnieszka, Fuel. **90** (2011)
17. R. Fernanda, G.B. Silvaa, G. Giane, M.K. Lenzic, E.K. Lenzi, Int. J. Heat Mass Transfer **63** (2013)
18. V.V. Salomatov, S.V. Syrodoy, N.Y. Gutareva, MATEC Web Conf. **23** (2015)
19. S.V. Syrodoy, V.V. Salomatov, N.Y. Gutareva, EPJ Web Conf. **76**, (2014)
20. S.V. Syrodoy, G.V. Kuznetsov, V.V. Salomatov, Therm. Eng. **62** (2015)
21. S.V. Syrodoy, N. Yu. Gutareva, K.A. Bugaeva, R. I. Taburchinov, MATEC Web Conf. **37** (2015)
22. S.A. Sulivan, J. Opt. Soc. Am. **53** (1963)
23. A.G. Pakes, W. Lambert's, J. Math. Anal. Appl. **378** (2011)
24. I.I. Markov, Collection of Scientific Works A series "Physic-chemical" Stavropol (2004)

MATEC Web of Conferences 72, 01110 (2016) DOI: 10.1051/matecconf/20167201110
HMTTSC-2016

EXPERIMENTAL SETUP FOR STUDYING THE CAPTURE OF SULFUR AND NITROGEN OXIDES FROM FLUE GASES

Andrey S. Sysolyatin[1,*], *Sergey* V. Kryukov [1], *Maxim* A. Chemakin [1], and *Igor* V. Dvorovenko[1]

[1]Institute of Power Engineering, Kuzbass State Technical University named after T.F. Gorbachev, 650000 Kemerovo, Russia

Abstract. The pilot unit is developed for adaptation of the absorptive mathematical model in a scrubber of the Venturi on a finding of factors of phase equilibrium by algorithm working off in the absorptive processes of extraction SO2 on water. The technique of work by test tests of trapping CO2 for water which have shown adequate work of a scrubber of the Venturi in the cascade with a spray separator-cyclone separator is completed.

1 Introduction

At the present stage of human development more than 850 mln t of nitrogen oxides, 150 mln t of sulfur dioxide, 310 mln t of carbon monoxide, 250 mln t of carbon dioxide, and 3 mln t of hydrocarbons are released into the atmosphere in the world annually [1].

Energy companies supply to the atmosphere more than 23% of emissions from all stationary sources in the Russian Federation. And among these emissions, nitrogen oxides accounts for about 97-98% (vol.), and others are sulfur oxides and various substances. Among sulfur oxides, SO_2 accounts for only 97-98%, and the rest is the share of SO_3 [2].

Nitrogen and sulfur oxides in the flue gases are the toxic substances, and in the case of their mixing with air moisture, they form acids, which, falling to the earth in the form of so-called acid rains, damage the farmlands, destroy engineering structures, and harm human health and all living things on Earth.

Therefore, a very important problem is development of the technological processes for cleaning the gas emissions of HPP from harmful gas components and fine solid suspensions.

The main methods used for cleaning gas emissions are absorption, adsorption and catalytic technique; each of them has its own advantages and disadvantages.

Industrial gases are emitted into the atmosphere in large quantities, exceeding in some cases, a million cubic meters per hour. The current cleaning devices cannot be used due to their low capacity by gas. The only types of devices meeting the given cleaning

* Corresponding author: teu.pmahp@rambler.ru

requirements are the hollow vortex apparatuses with high capacity, small size and low flow resistance, such as the Venturi scrubbers (tubes) and hollow spray-type absorbers [3-5].

Turbulent Venturi scrubbers are one of the most effective devices for wet gas cleaning. The main advantages of Venturi scrubbers are easy manufacturing, installation and maintenance, small sizes and high efficiency of fine particles catching [6]. In devices of this type, cooling, humidification and absorption of gases can be carried out simultaneously with purification.

Quite a long time, the following absorbents are commonly used in practice for cleaning from sulfur oxides: water and water solutions of: Na_2S0_3 (18-25%), NH_4OH (5-15%), $Ca(OH)_2$, Na_2C0_3, NaOH (15-20%), KOH, $(NH_4)_2S0_3$ (15-20%), $(NH_2)_2CO$, $ZnS0_4$, K_2C0_3, etc. [4, 7].

Recently, to improve the efficiency of flue gas cleaning, the methods of selective non-catalytic reduction (SNCR) [8] and SO_2 absorption by molten eutectic mixture of alkali metal carbonates (lithium, sodium and potassium) [9] are used in practice.

2 Experimental setup

The purpose of this work is creation of the experimental, rather uniform setup for studying the effectiveness of catching sulfur oxides, nitrogen oxides and mercury vapors using the most active absorbent-absorbed substance system, whose absorptive capacity is higher than that of the currently existing absorption pairs, and water consumption is reduced.

The schematic diagram of experimental setup for studying the efficiency of capturing sulfur and nitrogen oxides and mercury vapors by different absorbents, based on [10], is shown in fig. 1.

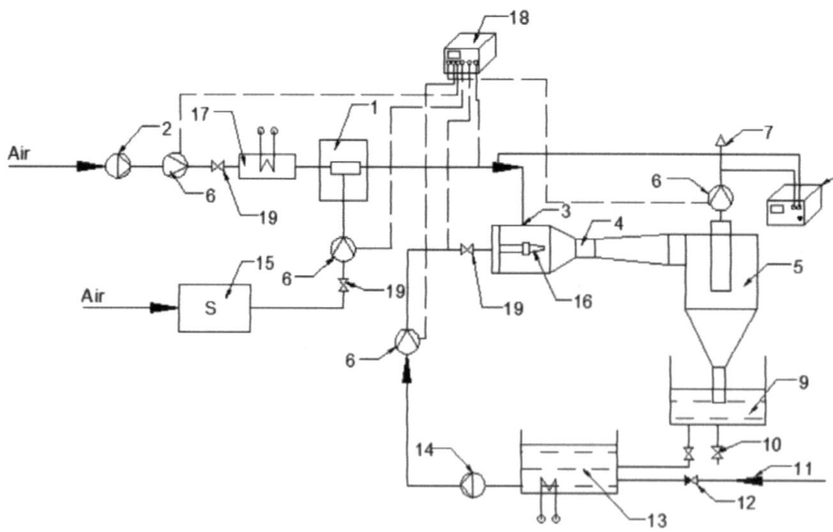

Fig. 1. Scheme of experimental setup: 1 – injector; 2 – air blower; 3 – introduction of gas-air mixture into scrubber; 4 – Venturi scrubber; 5 – cyclone; 6 – flow meter; 7 – output of cleaned gas mixture; 8 – gas analyzer; 9 – tank; 10 – valve; 11 – inlet of make-up absorbent; 12 – check valve; 13 – tank for absorbent; 14 – pump; 15 – furnace; 16 – nozzle; 17 – air heater; 18 – analogue-digital system for control and measuring; 19 – regulation valve.

An experimental setup consists of Venturi scrubber 4, cyclone 5, furnace 15, blower, pump, tanks, control devices, devices for measuring the flow rates and temperatures of gases and water, and concentrations of gases in air. Air is supplied by the blower into injector 1, where sulfur oxides or other gases from the furnace are injected; then, the formed gas mixture enters the Venturi scrubber, where oxides are absorbed by water with various additives. Then, gas-water mixture passes through the cyclone, where it is separated into the gaseous and water components, water is discharged from the cyclone from the bottom into the receiving tank, and gases are released from the top into the atmosphere. Gas and water flow rates are measured at the scrubber inlet, gas concentrations are measured in air at the scrubber inlet and gas outlet from the cyclone. The gas and water flow rates can be changed by the control valves, which allow research at various concentrations of impurities in air and different water flow rates in a wide range of determining parameters. The operation and equilibrium lines can be plotted using data obtained, which can be also used to evaluate the effectiveness of capturing and determine the most optimal operation regime of the setup.

3 Result and discussion

At the initial stage of research, there is the task to adapt the absorption mathematical model [11] for the Venturi scrubber at the sorption process in the framework of solving the inverse problem of finding the coefficients of phase equilibrium, fulfilling the algorithm for SO_2 recovery on water.

In this connection, the tests on performance of the experimental setup on CO_2 absorption by water were carried out. The tests were carried out on the setup for the following constant parameters: absorbent temperature (water) of 6°C; absorbed substance temperature of 22°C; inert gas flow of 0.01697 kg/s; gas velocity in Venturi scrubber throat of 44 m/s. The amount of absorbed component *M* vs. irrigation density *q* is shown in fig. 2.

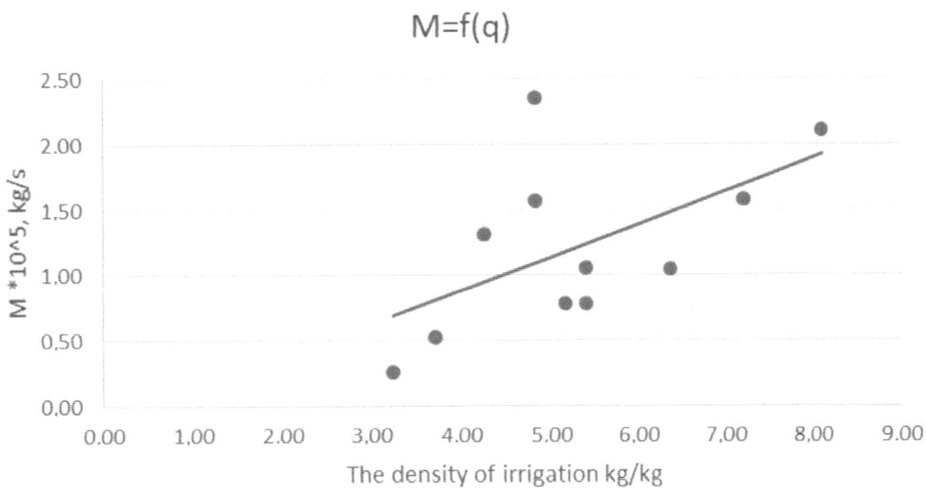

Fig. 2. The amount of absorbed component CO_2 from air-gaz mix vs. irrigation density.

4 Conclusion

The experimental setup for studying the capture of sulfur oxides, nitrogen oxides and mercury vapors was developed;

The primary experiments on capturing carbon dioxide were carried out; they showed the possibility of using the setup for adaptation of chemisorption processes.

Acknowledgments

The work was financially supported by the Russian Ministry of Education and Science n the framework of the Federal target program "Research and development in priority directions of development of scientific-technological complex of Russia for 2014-2020", according to Agreement No.14.583.21.0004 about the subsidies on July 16, 2014. The unique identifier of research (project) is RFMEFI58314X0004.

References

1. S.E. Belikov, *Integrated development of the methods for reducing nitrogen oxide emissions of thermal power plants by optimizing the combustion process and methods of fuel combustion* (Moscow, 2006)
2. E.R. Zvereva, T.M. Farakhov, A.R. Iskhakov, KSEU Bul. **8** (2011)
3. A. V. Dmitriev, *Cleaning gas emissions of HPP, working on liquid and solid fuel, in vortex apparatuses* (Kazan, 2006)
4. Yu. V. Zykin, *Development of resource-saving technology of wet scrubbing of HPP flue gases from ash* (Kazan, 2003)
5. A. S. Kazaryan, *Complex cleaning of flue gases of heat generating installations* (Rostov-on-Don, 2005)
6. V. M. Ramm, *Absorption of gases* (Khimiya, Moscow, 1976)
7. I.A. Volchin, S.V. Mezin, L.N. Rudenko, A.A. Yasinetskiy, Energ. **6** (2015)
8. O.N. Kulish, S.A. Kuzhevatov, I.Sh. Gleizer, M.N. Orlova, E.V. Ivanova, A.M. Zykov, O.N. Bragina, K.I. Kilchin, and S.N. Anichkov, Energ. **4** (2014)
9. N.K. Dosmukhamedov, V.A. Kaplan, E.E. Zholdasbay, D.N. Dosmukhamedov, I. Lyubomirskiy, Coal **8** (2015)
10. A.I. Kalachev, TEK. Str. Dev. **39** (2015)
11. M. I. Shilyaev, E. M. Khromova, A. R. Bogomolov, S. N. Shirokova, Izv. Vuz. Str. **3** (2015)

SMALL DISTRIBUTED ENERGY RUSSIA: COMBINED HEAT AND POWER GENERATION

Alexander Astafev[1,*], *Alexander* Kazakov[1], and *Alexander* Zavorin[1]

[1]National Research Tomsk Polytechnic University, 634050 Tomsk, Russia

Abstract. The issues and current trends of research in the field of decentralized energy supply for the period up to 2030 were considered. The analysis of the cogeneration market in Russia was done. The questions of gasification and hydrogen technologies as applied to the market of cogeneration plants were elucidated. Promising technologies for autonomous decentralized energy supply and the evaluation of their applicability to small energy were presented.

1 Introduction

The transition conception to innovate way of economic development, adopted in Russia, envisages a transition of country from export of raw materials to resource-innovative development with qualitative energy upgrade (both fuel and non-fuel) and related industries [1].

In energy strategy of Russia for the period to 2030 one of the tasks is modernization and creation of new energy infrastructure based on large-scale technological upgrade of energy sector in national economy. Is noted, [1] that energy safety is one of the most important components of national safety. Ensuring of energy safety is determined by the resource adequacy, economic availability, environmental and technological allowability. The resource adequacy determines the physical capabilities of sufficient energy supply for national economy and population, economic availability - the profitability of such supply at a suitable price conjuncture, environmental and technological allowability – the extraction, production and consumption possibility of energy sources under the frame of existing at each stage technologies and environmental constraints, which determines the functional safety of energy facilities [1].

The result of Russia energy strategy for the period to 2020 is existing problems, reflected in the strategy-2030:

–the weak development of the energy infrastructure in Eastern Siberia and the Far East;

–the underactive development of renewable energy and the use of local fuels in regional energy balances;

–the insufficient development of small energy and low involvement of local energy sources to energy balances of regional and local relevance.

* Corresponding author: TabakaevRB@tpu.ru

The priority directions of scientific and technical progress in energy sector of strategy-2030:

–the creation and deployment of technological equipment complex on a modular basis for new building and existing heat supply transfer to cogeneration basis;

–the development of use renewable energy technologies and multi-functional energy complexes for autonomous energy supply in areas without centralized energy supply;

–the development effective technologies of mains energy and heat supply based on renewable energy sources;

–the expanded production and using of new fuels from different biomass types;

–the development of small energy in decentralized energy supply zone by efficiency increasing of local energy sources use, consumption reducing of imported light oil products;

–the development of heat supply in Russia and its regions based on district heating with using of modern economically and environmentally efficient cogeneration units, which have a wide power range;

–the optimal combination of centralized and decentralized heat supply with allocation of relevant zones;

–the development of systems with centrally-distributed heat generation and different resources types, located in heat consumption areas;

–the modernization and development of decentralized heat supply systems with using of high-efficiency cogeneration and other units, as well as new automated individual heat generators for combustion of various fuels;

–the equipping of consumers by stationary and mobile heat supply units as a reserve and (or) emergency heat supply.

The strategy-2030 assumes that the regional structure of generating capacities will be formed as follows:

–the development of small energy on renewable energy sources, including be replacing of local diesel generation;

–the small energy and decentralized energy supply with active use of local and waste energy sources all types will be developed, especially in areas with low density of load; also units with coal gasification and energy technological units will be developed;

–the development and implementation of regional energy programs, energy saving problems, the economic-effective maximizing of local energy and fuel sources using, the economic-effective development of decentralized and individual heat supply systems.

2 The actual directions of researches in the decentralized energy supply field

In the interest of Russian energy development, also distributed on the small energy facilities, necessary to developed investigations in the field of heat and power generation for autonomous decentralized consumers. As practice shows, the problem of decentralized consumers is farness from transport schemes that increases the delivery cost of fuel sources in several times [2]. Often, the reserves of local low grade fuels, which can be used as an energy carrier, has in such remote areas, but characteristics of this fuels so far as complicate the process of direct combustion in boiler furnaces, because rarely used in practice.

The memorandum about creation and activity of technological platform “Small distributed energy” states that the main applications of new technological solutions in small distributed energy [3]:

–the cogeneration units for modernization of communal infrastructure in settlements;

–the typical sets of equipment and modular technological solutions for energy supply of remoted and isolated consumers;

–the complex local energy systems with maximum possible use of local fuel resources;

The main directions of technological development, supported under the frame of technological platform [3]:

–the technologies of heat and power generation, including fuel elements, hydrogen technology;

–the local energy sources using, including the modern technologies of peat using as a fuel resource, the gasification of local fuel resources, industrial and domestic wastes with using of synthesis gas for power generation;

Thus, the works on topics, related with investigations in the field of remote consumers energy supply, based on cogeneration units, which generates synthesis gas from local fuels with using of fuel elements, are actually.

3 The market of cogeneration units in Russian Federation

The development of cogeneration market in Russia is one of the strategic directions in the field of small energy. At the present time, 12 major national and 17 foreign producers of cogeneration equipment, presented at the Russian market, are existed. The most part of cogeneration stations are intended to work on natural gas and diesel fuel. Obviously, that the demand for such units is determined by relative service simplicity and its sufficient maneuverability. 6 foreign companies (GE Jenbacher, MTU, Motorgas, Tedom, Turbec, MWM) supplies equipment, worked on synthesis gas. 1 company from national (LLC ARZ "Sintur-NT") produces equipment for work on generating gas. The power range of cogeneration units is quite large: from 3.8 kW to 4.3 MW of electric and from 8.9 kW to 5.2 MW of heat. The main types of cogeneration units for heat and power generation are now gas reciprocating units. The efficiency of such foreign cogeneration systems generally is greater than 90%, while the efficiency of national systems is about 80%.

The conclusion: The modern market of Russia does not experienced deficit of cogeneration units based on combustion engines. The application area of supplied cogeneration units is limited by possibility of natural gas and diesel fuel delivery. The power of existing equipment is able to cover the needs of all consumers. The tying of power and heat generation is exists, it is not very convenient to the consumer. The foreign cogeneration stations are more efficient.

4 The gasification role

It is thought that the gasification of solid fuel is actual and perspective direction of research, given that this technology is allows to use low grade fuels with producing of high quality gas and solid products [4–6]. The developments, reduced heat consumption for own needs by autothermal mode of gasification, are exist. Is noted that problems of gas treatment and ensuring of work reliability for large gas generators are exist. Despite of the attractiveness of gasification as an energy technological recycling of solid fuels, is noted its low commercial attractiveness.

The conclusion: The current research level in the field of gasification is allows to create units, recycled a high quality coals and low grade fuels, but their large-scale deployment is limited by competitive advantages of existing major stations. The gasifier technologies are perhaps only option for recycling of low grade local fuels in remote decentralized areas.

5 The hydrogen technologies

The relevance of hydrogen energy development is proved by following criteria:

–the ability of use in autonomous energy units;
–the demand to hydrogen in industry;
–the high competitive of hydrogen technologies;
–the environmental friendliness during hydrogen combustion.

There is an opinion that a hydrogen production by conversion of natural gas is the most expedient in terms of hydrogen production on the industrial scales [7–9]. At the same time is believed that perspective direction is hydrogen generation from biomass and solid fuel with its subsequent use in fuel elements [9]. Today there is a problem, which limited the spread of hydrogen technologies:

–the low electrical efficiency of fuel elements;
–the problem of hydrogen storage and transportation.

The low electrical efficiency of fuel elements can be offset by using of generated on them heat that allows to achieve acceptable unit efficiency in generally. The hydrogen production at place of its consumption or in one energy unit is allows to minimize problems, associated with hydrogen storage and transportation.

Is noted that the peculiarity of hydrogen units in fuel elements is its cleanliness, however, there are developments of fuel elements, worked on synthesis gas. Thus, the development of solid oxide elements, which can be worked on gas with high CO2 content, is actually, the working temperature range (800-1000 °C) is allows to use such fuel elements in gas-turbine installations. For gases with high CO content are being developed a fuel elements with polymeric membranes (the working temperature range 120-300 °C) and fuel elements with molten carbonates (400 °C).

The conclusion: Of course, at the present time hydrogen technologies are actually and are demanded. Is traced that diversity of approaches to technologies of hydrogen production, which primarily determined by final purpose of use. The efficient increase of source fuel using is seen in the hydrogen and energy generation in one unit with simultaneous utilization of heat, generated by fuel element. The current state of researches in the field of fuel elements shows that the requirements to cleanliness of hydrogen are decreased.

6 The state of small and unconventional energy

The small energy as an integral component of energetic space is recognized around the world, including Russia. Moreover, the expediency to development of artificial small distributed energy is seen positive, economic and social perspective fact. Due to geographic and economic facts in Russia the large areas are not covered by centralized energy supply and problems with provision of such areas of fuel resources are exist. Currently, the power generation in decentralized areas is mainly produced by diesel stations that not economically efficient. Wherein such areas often have reserves of local low grade fuels, but their consumption is accompanied by appropriate technical devices [11–13].

The small energy as an integral component of energetic space is recognized around the world, including Russia. Moreover, the expediency to development of artificial small distributed energy is seen positive, economic and social perspective fact. Due to geographic and economic facts in Russia the large areas are not covered by centralized energy supply and problems with provision of such areas of fuel resources are exist. Currently, the power generation in decentralized areas is mainly produced by diesel stations that not economically efficient. Wherein such areas often have reserves of local low grade fuels, but their consumption is accompanied by appropriate technical devices.

The conclusion: Today is actually the issue of energy supply for decentralized consumers each of communal service and production. The use of cogeneration systems based on combustion engines, worked on imported diesel fuel, is a relic of the past and requires the replacement by deployment technologies of combined heat and power generation from local low grade fuel resources.

7 The perspective technologies for autonomous decentralized energy supply

The current researches in the field of autonomous heat and energy supply can be divided on the follows directions:

a) the co-combustion of several solid fuels (coal, biomass);
b) the production of synthesis gases;
c) the production and combustion of fuel briquettes;
d) the production of hydrogen and its use in fuel element;
e) the creation of hybrid units.

The autonomous units for power generation, combusted a coal and biomass in complex, for all that more oriented on the high quality coal, and biomass addition is allows to reduce coal consumption. This direction can't be oriented to remote areas, in which coalfields or existing transport infrastructure is absent. The all types of solid organic raw materials are suitable for synthesis gases production. The creation of such production near to existing fields of local low grade fuels is seen the perspective. Necessary to consider that the creation of power stations based on high temperature gas generators for small settlements is not expedient, because it pretty complicated technical device, specified in maintenance and required a certain qualification from personnel. The production and consumption of fuel briquettes is perspective direction in view of modern developments in the field of briquette production from low grade fuels [14, 15]. The briquette fuel can be combust in exist furnace units with grate firing without their reconstruction.

The heat and power production from hydrogen in fuel element can be regarded as breakthrough in the energy of future. However, the certain vigilance about deployment perspectives for the field of small generation is exist, because it pretty high-tech process, required the technologies of high clean hydrogen production, now realized by conversion of natural gas.

The creation of hybrid units is one of the diversification directions in small generation. Indeed, the presence of several devices for energy generation from various sources (renewable and non-renewable) simultaneously in such units is allows to increase reliability of energy supply to consumers. But, dependence of quality energy sources (coal, gas, diesel fuel) supply is remains.

The conclusion: The choice of autonomous energy supply to decentralized consumers should be viewed based on two conditions: 1 – the presence of own quality coal and (or) natural gas resources or well-established transport schemes with acceptable delivery cost; 2 – the presence of local low grade fuel resources. The directions *a* and *e* are expedient at carrying out of first condition. The directions *b* and *c* are expedient at carrying out of second condition. The technology d for conversion of natural gas is not expedient because natural gas in and of itself a high quality fuel.

Acknowledgments

The reported study was partially supported by the Ministry of education and science of The Russian Federation, state order No. 13.948.2014/K.

References

1. Energy strategy of Russia for the period to 2030 (*confirm by the government edict of the Russian Federation No. 1715-рб* 2009)
2. O. S. Popel, B. F. Reutov, A. P. Antropov, Therm. Eng. **57** (2010)
3. http://www.e-apbe.ru/distributed_energy/memo_TP_SDE.php
4. V. V. Lopatin, S. M. Martemyanov, Rus. Ph. J. **55** (2012)
5. I. Sircar, A. Sane, W. Wang, J.P. Gore, Fuel **119**, 38 (2014)
6. S. M. Martemyanov, A. L. Maslov, Adv. Mat. Res. **1040**, 620 (2014)
7. A. S. Zavorin, R. B. Tabakaev, P. Y. Novoseltsev, A. V. Astafev, MATEC Web Conf. **19** (2014)
8. A.V. Kazakov, R.B. Tabakaev, P.Y. Novoseltsev, A.V. Astafev, MATEC Web Conf. **19** (2014)
9. R.B. Tabakaev, A.V. Astafev, A.V. Kazakov, A.S. Zavorin, MATEC Web Conf., **23**, (2015)
10. P.S. Gergelizhiu, S.A. Khaustov, R.B. Tabakaev, P.Y. Novoseltsev, A.V. Kazakov, A.S. Zavorin, *Proc. of 2014 International Conference on Mechanical Engineering, Automation and Control Systems* (TPU, Tomsk, Russia)
11. S. A. Khaustov, A. S. Zavorin, K. V. Buvakov, N. A. Zakharushkin, MATEC Web Conf. **19** (2014)
12. R. B. Tabakaev, A. V. Kazakov, A. S. Zavorin, Sol. Fuel Ch. **49** (2015)
13. R. B. Tabakaev, A. V. Astafev, A. V. Kazakov, A. S. Zavorin, MATEC Web Conf. **23** (2015)
14. R. B. Tabakaev, A. V. Astafev, A. V. Kazakov, A. S. Zavorin, M. Polsongkram, IOP Conf. Ser.: Mat. Sc. Eng. **93** (2015)
15. R. B. Tabakaev, P. S. Gergelizhiu, A. V. Kazakov, A. S. Zavorin, IOP Conf. Ser.: Mat. Sc. Eng. **66** (2014)

MATEC Web of Conferences **72**, 01112 (2016)
HMTTSC-2016 DOI: 10.1051/matecconf/20167201112

EXPERIMENTAL STUDY OF TWO-PHASE THERMOSYPHON USING R-22 AS A WORKING FLUID

Maathe. A. Theeb[1,*]

[1]Department of Mechanical Engineering, Al-Musansiriyah University, Baghdad, Iraq.

Abstract. The two-phase closed thermosyphon (TPCT) is an effective heat transfer device decreasing the ground temperature around it in cold season. In this paper an experimental study has been done by using R-22 as a working fluids, the temperature and the pressure of the TPCT where measured with time (Transient study).

1 Introduction

Thermosyphon are enclose, passive two- phase heat transfer devices. They make use of the high efficient thermal transport process of evaporation and condensation to maximize the thermal conductance between a heat source and a heat sink. They are often referred to as thermal superconductors or thermal" short circuits" because they can transfer large amount of heat over relatively large distances with small temperature differences between the heat input and heat output zones. The amount of heat that can be transported by these devices is normally several orders of magnitude greater than pure conduction through a solid metal [1,2].They are proven to be very affective, low cost and reliable heat transfer devices for applications in many thermal management and heat recovery systems. They are used in many applications including but not limited to passive ground/road anti freezing, baking ovens, heat exchangers in waste heat recovery applications, water heaters and solar energy systems and are showing some promise in high-performance electrons thermal management for situations which are orientation specific.[3] studied Heat transfer and hydrodynamics of two phase closed thermosyphon (TPCT) , The mathematical model is formed for both vapor phase and liquid film in a non-dimensional form. The momentum and energy equations as parabolic equations have been solved by means of the locally one-dimensional scheme of A.A. Samarskii. The Poisson equation for the stream function has been approximated on the basis of the discrete scheme "cross". The obtained system of algebraic equations has been solved by the successive over relaxation method. The effect of the Rayleigh number on the velocity and temperature fields in the domain of interest and on the average Nusselt number at the solid-fluid interfaces is investigated. It was found that increase in the Rayleigh number leads to the increment of both the average Nusselt number and the stream function in the vapor zone.

* Corresponding author: bulba@tpu.ru

A cross section of a closed two-phase thermosyphon is illustrated in fig (1).The thermosyphon consist of an evacuated sealed tube that consists a small amount of liquid. The heat applied at evaporator section is conducted across the pipe wall causing the liquid in thermosyphon to boil in the liquid pool region and evaporate and/or boil in the film region [4]. In this way the working fluid absorbs the applied heat load converting it to latent heat. The vapor in the evaporator zone is at a higher pressure than in the condenser section causing the vapor to flow upward. In the cooler condenser region the vapor condenses thus releasing the latent heat that was absorbed in the evaporation section. The heat then conducts across the thin liquid film and exits the thermosyphon through the tube wall and into the external environment. Within the tube, the flow circuit is completed by the liquid being forced by gravity back to the evaporator section in the form of thin a thin liquid film. As a thermosyphon relies on gravity to pump the liquid back to the evaporator section, it cannot operate at inclination close to horizontal position.

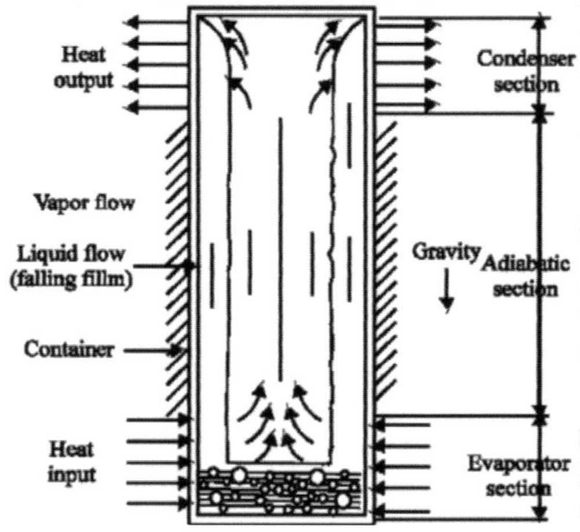

Fig. 1. Two-phase closed thermosiphon.

The studied thermosyphon consists of three classical sections including; an evaporator, an adiabatic section and a condenser. This thermosyphon has a length of 1 m with an internal and external diameter of 28.575 mm and 30.575 mm, respectively. The length of the evaporation, adiabatic and condensation sections were design to be 30, 30 and 30 cm, respectively. In addition, a 10 cm space was left above the tube to collect any non-condensable gas.

The fluids tested were R-22 and R-134a , so we can make a comparison between each other. We should considered that the experiment was made in winter season.

The evaporator section was heated using an electrical heater. The heat input to the evaporator was adjusted using an electrical energy regulator with a maximum heat input of 1.5 kW. We use the clamp meter to choose the proper heat value by choosing the proper current based on the assumption of the constant voltage.

$Q=VI$; $V=220$ volt

Hence: $Q \alpha I$

For steady state cases The regulator was connected to thermostat in order to stabilize the temperature in the lower tank (50,60,70,80)degree centigrade respectively in other words we stabilized the amount of the addition heat .The lower tank was mostly filled by water which transports heat by convection to the evaporator(absorbing latent heat)as shown in figure).It was closed by a plate and insulated by Glass wool in order to reduce the heat losses.The adiabatic zone was insulated by arm flex to minimize heat losses to the ambient. The condenser section of the thermosyphon was cooled by a flow of water, Condenser water was supplied via a constant-head water tank 30cm , This consisted of a tank and control valves as in figure , The temperature of the cold-water inlet was kept constant by those valves ,the tank was uncovered and insulated by glass wool insulation ,and the water temperature was about (20) degree centigrade. The temperature distribution along the thermosyphon was measured using four thermometers distributed respectively at lower water jacket (T1),adiabatic section (T2),condenser section (T3) and upper water jacket (T4)as shown in figure(3-1),the thermometers at adiabatic and condenser sections putted to be as much as possible directly contact within the pipe. The pressure along the thermosyphon was measured using a pressure gauge.

2 MEASURING SYSTEM

The measuring system includes the measurements of two parameters:
1. Temperature measurements system. Thermocouple probes of type-K with (1.4 m) long and (3.18mm) in diameter were fixed at each point of the test the test rig, the calibration cureve shown in fig(2-c).
2. Refrigerant Pressure gages
low pressure gauges was fixed to measure the pressure inside the thermosyphone as shown in fig(2). This gauges is calibrated, the calibration shown in fig(2-d)

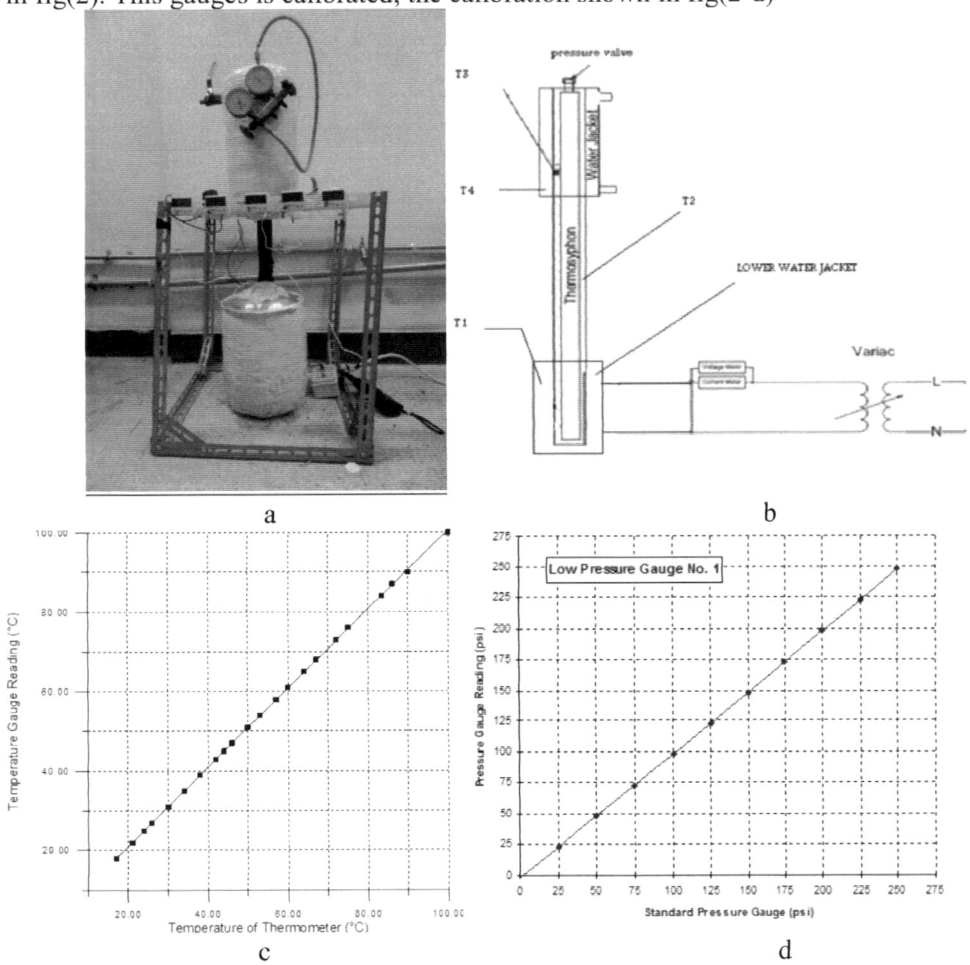

Fig. 2. (a,b)schematic diagram of thermosyphon test rig, (c) Calibration of temperature thermocouple type K, (d) Calibration of the low pressure gauge.

3 Results and discussion

At first we should know in what experimental circumstances the test was worked, the test is made in winter season at air temp 17 degree centigrade and that lead to increase the time for reaching the steady state .and the water which was used both in the test (the water jacket's) and in the calibration of thermometers was not distilled water so it effect at the result of this test .the other problem was in electricity, the voltage was not constant .The thermosyphon pipe was compressed with R-22 &R-134a gasses at 0.225g.

From the plots and tables below we can make a conclusion that the temperatures increased by time increasing and it's varied from temperature to another, we find the temperature of lower water jacket had large changes in temperature by time and this is true because is directly subjected to the heat source which is the heater in this test. The temperature of lower water jacket was increasing rapidly with reasonable changing in pressure, and the pressure is increasing by time increasing.

We can see in those tests that by current increasing the temperature increased better in other word the needed time for heating the same amount of water was less by current increases as we can see from tables that for the first 30 minutes the temperature

of the upper water jacket when current equal five AMPER is more than the other ones for the same current but different currents. We can find that by increasing the heat addition (q heater) the time needed for reaching the steady state case was less as we see it takes only 80 minutes otherwise it tokes 170 and 200 minutes for less heat subjected as it shown in R - 22 tables as an example.

The R-22 is better in transporting heat than R-134a but the last one takes more time than the first one to reach the study state condition. For steady state condition we can make a conclusion which is how long time expends there is no changing in temperature of all sections and this means the heat amount is constant.

Fig (3) shows the temperature for different positions on the thermosyphon for the case of 2 Amp. Power supplies the evaporation zone (440 Watt). It shows that the at 180 min the TPCT will reach to the steady state conditions, while fig(4) shows the pressure gradient inside the TPCT with time, here we can conclude that the pressure will increase so we must be sure that the heat is rejected in the upper zone(condensation zone) , fig(5) shows the effect of time with constant heat flux on the pressure of the TPCT , we can see that the pressure increasing with time while reaching the steady state.

Fig.(6,7,8) show the effect of time on the temperature, pressureinside thermosyphon, time on pressure respectively for power supplies 3.5 Amp.(constant heat flux equal to 770 Watt.and (9,10) for 1100 Watt. Fig. 2 shows the results of an experimental study of the heat transfer process in the steam thermal siphon tube channel (operating fluid - water):

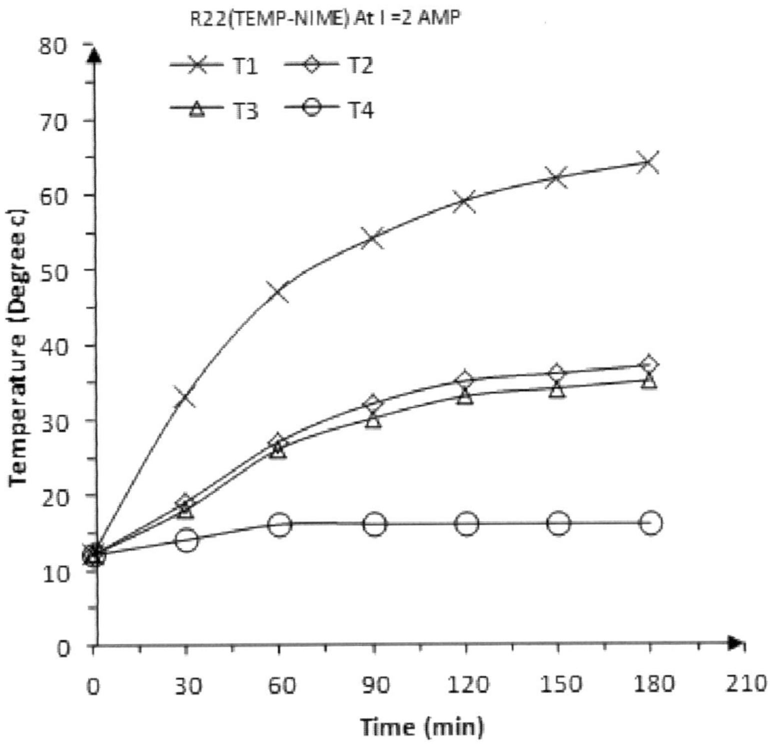

Fig. 3. Effect of time on the temperature.

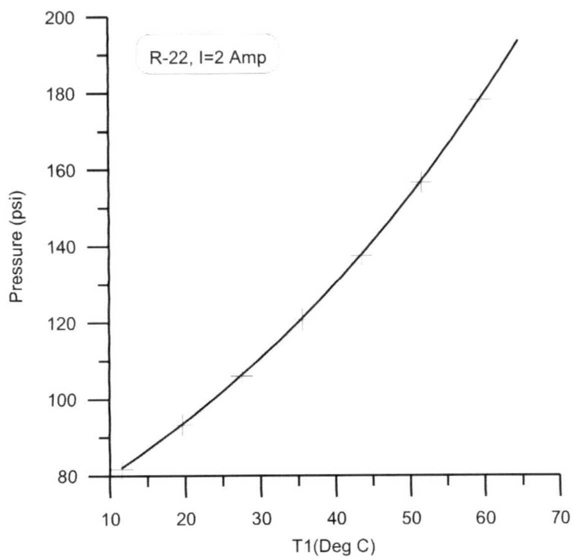

Fig. 4. Effect of temperature on the pressure.

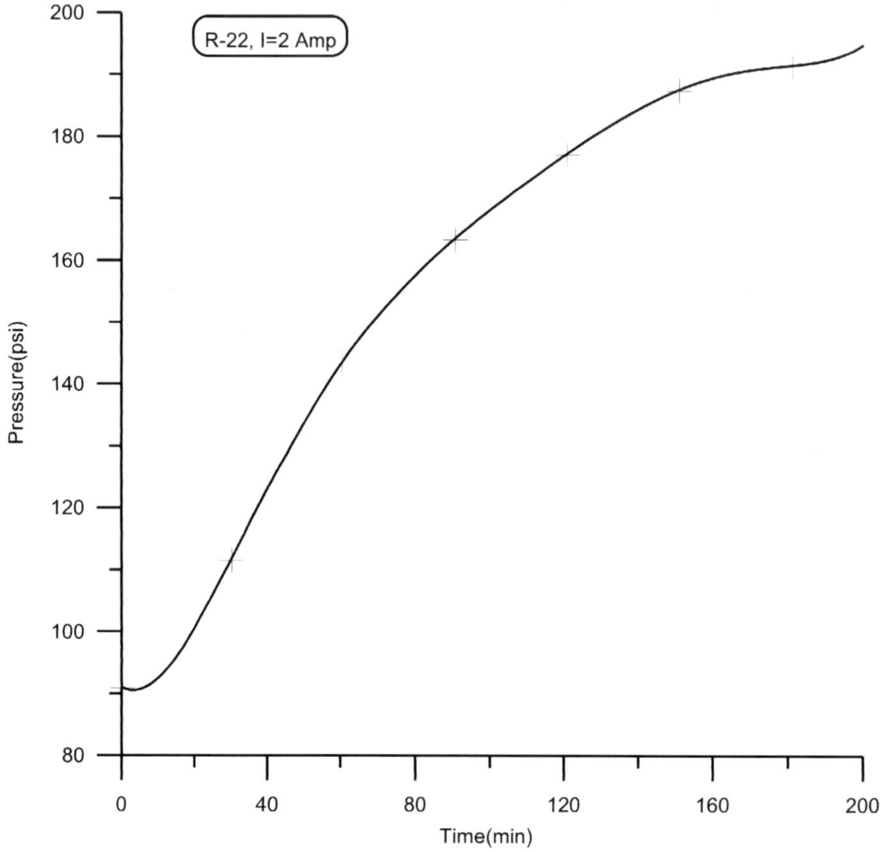

Fig. 5. Effect of time on the pressure.

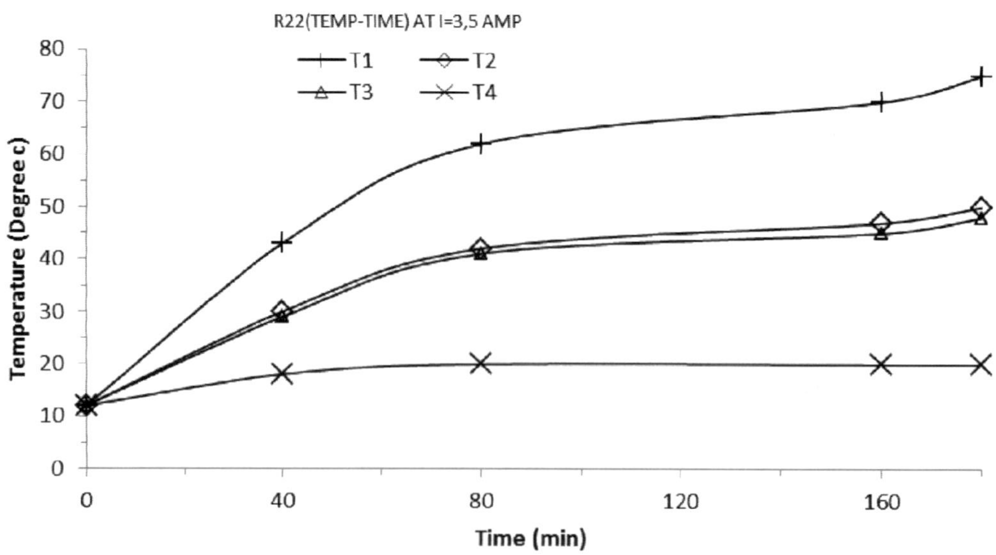

Fig. 6. Effect of time on the temperature.

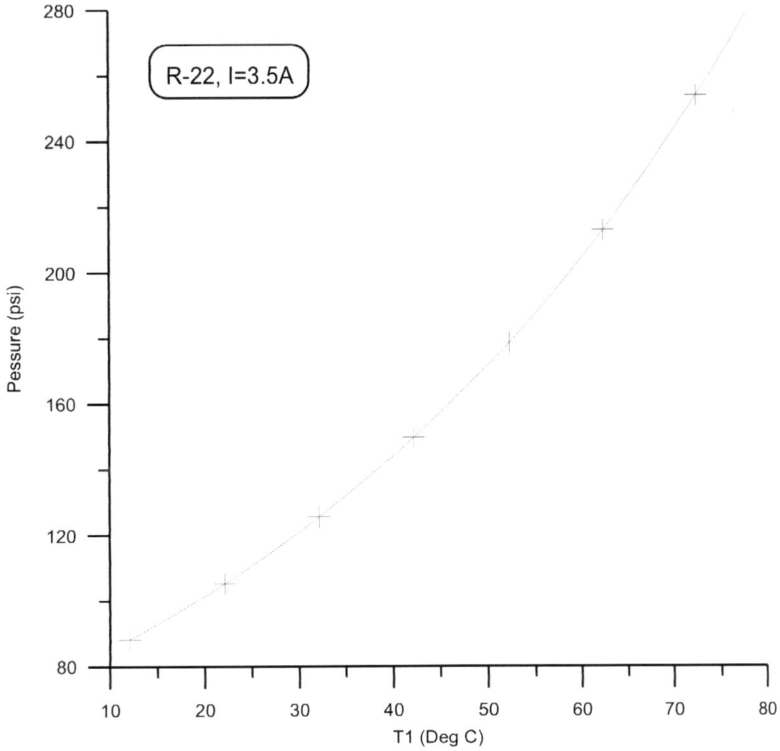

Fig. 7. Effect of temperature on the pressure.

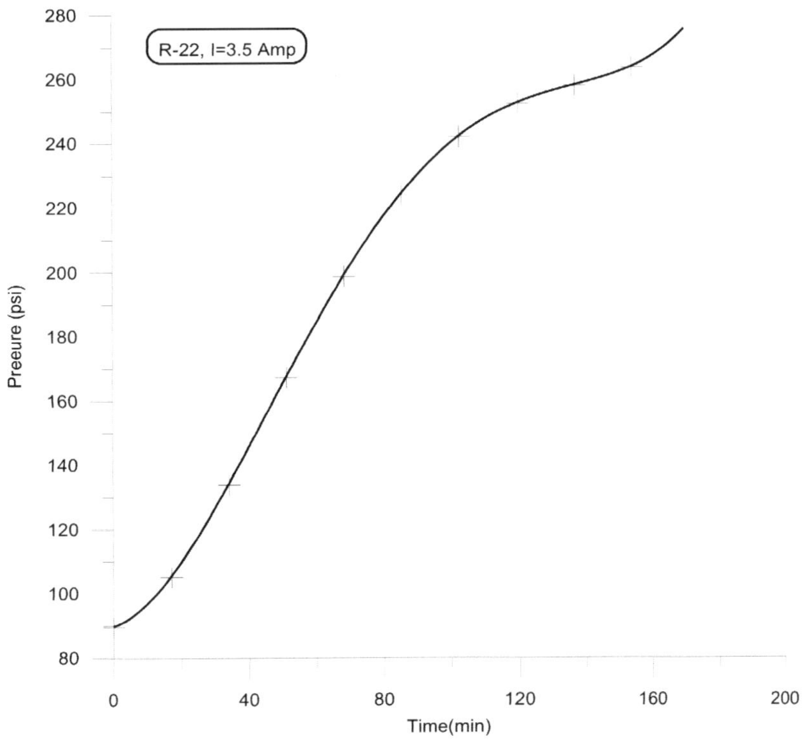

Fig. 8. Effect of time on the pressure.

 DOI: 10.1051/matecconf/20167201112

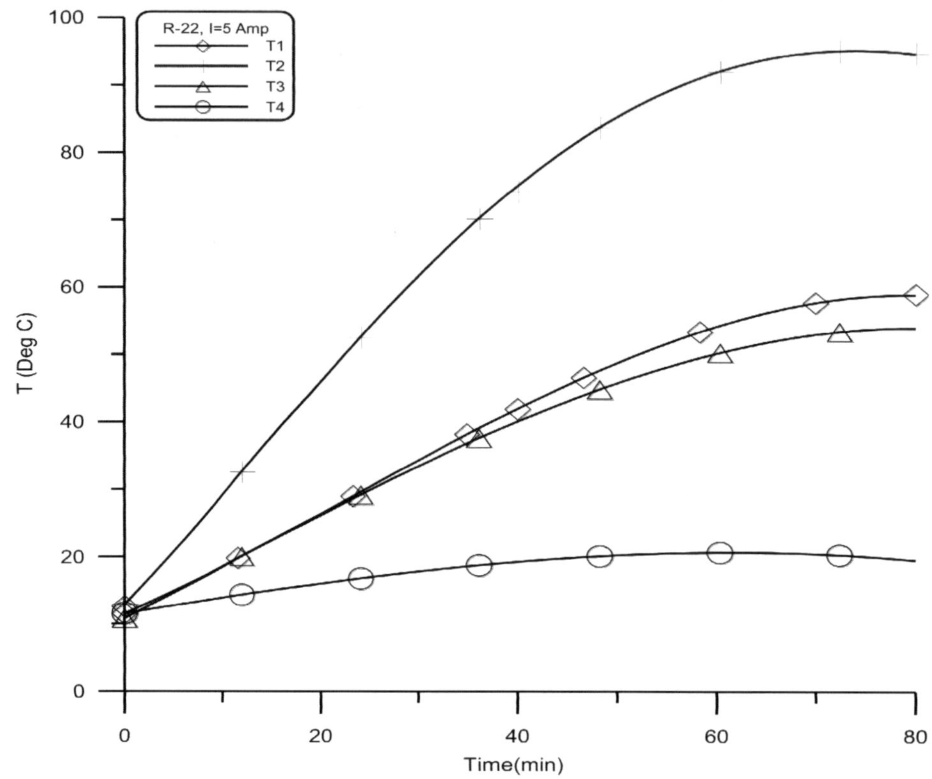

Fig. 9. Effect of time on the temperature.

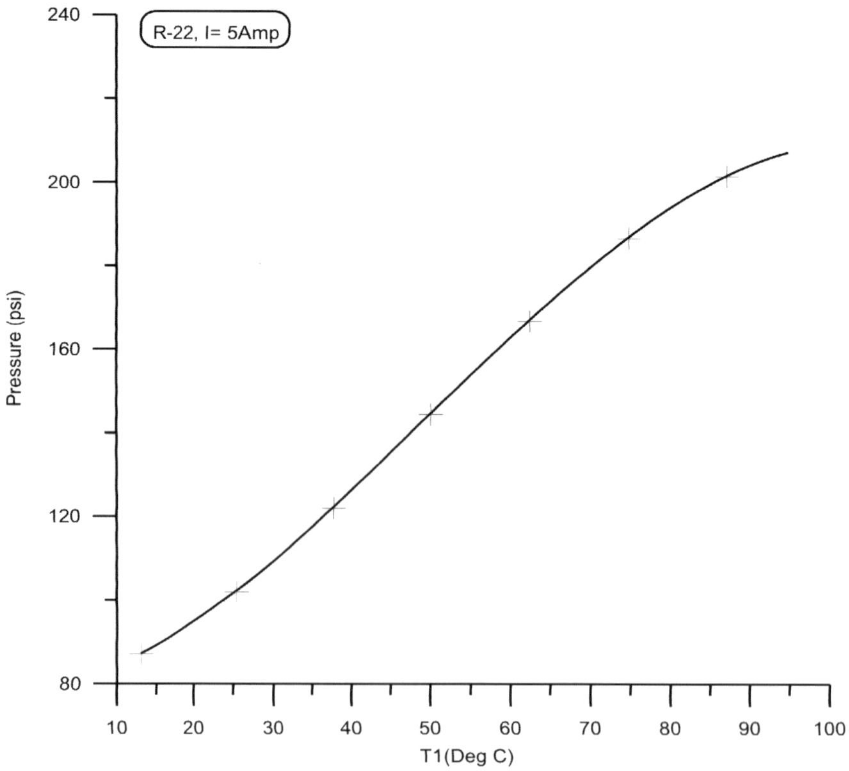

Fig. 10. Effect of temperature on the pressure.

4 Conclusion

A small diameter thermosyphon was tested experimentally with two different working fluids: R-22 and R-134a. We find that:

1-The R-22 is better in transporting heat than R-134a but the last one takes more time than the first one to reach the study state condition.

2-Pressure inside TPCT increase with increasing the supply temperature so it is better to use a high efficiency heat transfer method to avoid increasing the pressure.

3-TPCT pressure increasing with time while reaching the steady state.

4-Increasing the heat flux will increase bothe the temperature and pressure for the TPCT.

References

1. S.H. Noie, Appl. Therm. Eng. **25** (2005)
2. D.A. Reay, P.A. Kew, *Heat Pipes* (Butterworth –Heinemann, Oxford, 2006)
3. G.V. Kuznetsov, M.A. Al-Ani and M.A. Sheremet, J. of En. and Pow. Eng. **5** (2011)
4. M.S. El-Genk, H.H. Saber, J. Heat Transf. **120** (1998)

EXPERIMENTAL DETERMINATION OF HEAT TRANSFER COEFFICIENT UNDER FREE CONVECTION IN AN UNBOUNDED SPACE

Evgenya Timoshenko[1], *Viktoria* Afanaseva[1], and *Evgeny* Kravchenko[1,*]

[1]National Research Tomsk Polytechnic University, 634050 Tomsk, Russia

Abstract. This article deals with the Experimental determination of heat transfer coefficient under free convection in an unbounded space. Experiments were performed with different types of resistors when changing power, the heat transfer coefficient is calculated. On the basis of the work dependences of temperature of the resistor from power, the heat transfer coefficient from power are revealed, the interpolation polynomial equation is received.

1 Introduction

The electric scheme of almost any modern device has the heat allocating elements. Thermal control components of technical equipment such as power transformers [1], electro contact connections [2], semiconductor devices [3] is a critical task in maintenance of the set level of reliability [4]. As the object of research was chosen the typical element of electronic schemes - the resistor. The characteristics of the resistor are the value of nominal resistance and power, which shows how much energy it is able to dissipate without overheating [5,6]. In addition, it's necessary to provide that the thermal effects on the fuel element located near the device absent or minimized.

The purpose of the work is the experimental determination of the heat transfer coefficient of a typical fuel element in the conditions of free convection.

2 Scheme of installation and experiment

Researches of the thermal mode were conducted with use of the modern Testo-885 thermal imager having the following characteristics:

- Temperature sensitivity – less than 30 mK;
- Infrared resolution – 320 x 240 pixels.

The scheme of experiment is submitted in fig. 1. Determination of heat transfer coefficient was carried out with the presence of the radiator (fig. 1b) and without radiator (fig. 1a). To reduce the thermal resistance between the radiator and resistors was used heat-conducting paste ALSI-3.

* Corresponding author: kevatp@tpu.ru

DOI: 10.1051/matecconf/20167201113

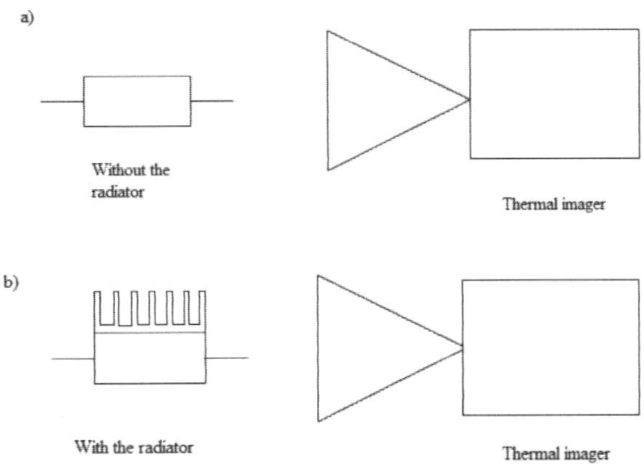

Fig. 1. Schematic image of experimental installation: a) the resistor without radiator; b) the resistor with a radiator.

Processing of thermograms it was carried out in the specialized software Testo. The essence of experiment is to determine the heat transfer coefficient of the resistor with a power limit of 10 watts. Heating of the resistor was due to increased resistance in the electrical circuit to which it was connected.

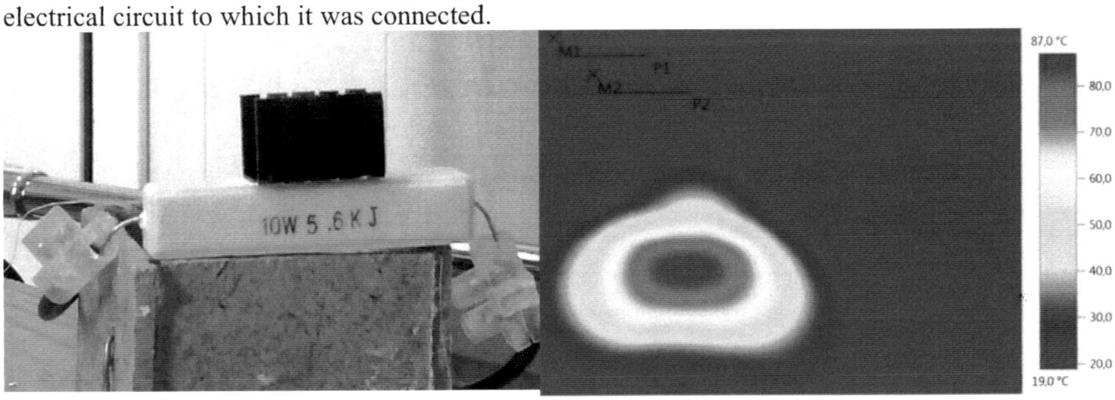

Fig. 2. The resistor with the radiator and temperature field at a power value of 8 watts.

For obtaining more exact values on the resistor by means of paste which reduces thermal resistance was placed the radiator prototype. The radiator was necessary for us for providing temperature condition.

As the main characteristic of the resistor is power, it was decided to check the values of temperature which will produce the resistor depending on the power change.

The mathematical formulation of the problem. Calculation of heat transfer coefficient was carried out according to the following expression [4]:

$$\alpha_{\kappa} = (1.42 - 1.4 \cdot 10^{-3} t_m) \cdot N \cdot \left(\frac{t - t_c}{L}\right)^{\frac{1}{4}} \quad (1)$$

where: N - coefficient considering orientation of a surface in space; L - defining size; tm - average temperature; t - case temperature; tc - ambient temperature;

The power in the electrical circuit was calculated as:

$$P = U \cdot I \quad (2)$$

where: I - amperage; U - voltage.

The power in the electrical circuit was measured by an ampermeter and the voltage was measured by a voltmeter connected in parallel to the resistor. Graphic temperature dependence of the resistor on the power is shown in figure 3.

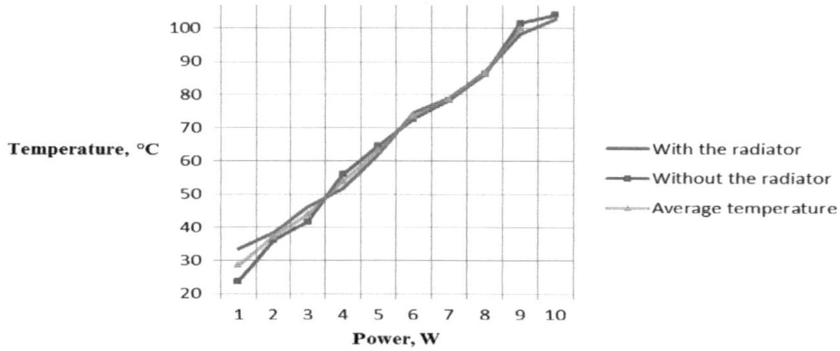

Fig. 3. Experimental dependence of temperature on power of the heat allocating element.

Value of coefficient of a thermolysis depends on the defining size, and the disseminating stream. For the average temperature values using Matlab program was obtained interpolating polynomial:

$$y = p_1 \cdot x + p_2 \tag{3}$$

Polynomial coefficients: p1=8.5439; p2=19.9443. The norm of discrepancy is 5.1644.

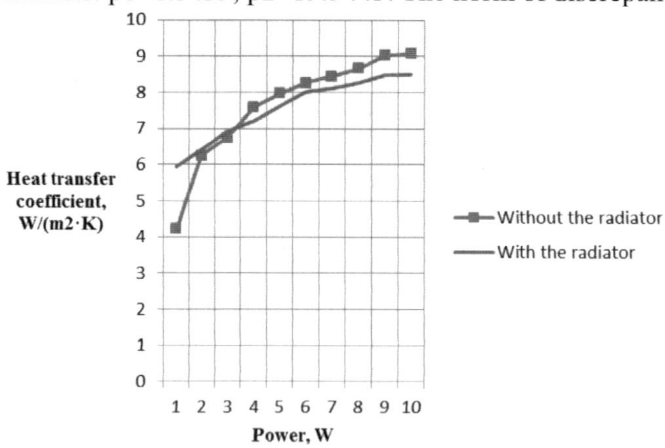

Figure 4. The dependence of the heat transfer coefficient of the resistor to the power side of the unit heatallocating element.

3 Conclusion

Experimental research has shown that the heat transfer coefficient for the resistor without the radiator is more than for the resistor with the radiator. These values can be explained by the fact that the thermal energy of the resistor without the radiator is dispersed directly in space and not wasted on heating of the radiator. On the other hand, in the conditions of free convection use of a radiator allows to lower thermal loading of an element.

Acknowledgments

This work was funded within the framework of realisation of Strategic Programme on National Research Tomsk Polytechnic University Competitiveness Enhancement in the Group of Top Level World Research and Academic Institutions (project VIU_ENIN_25_2016).

References

1. E.V. Kravchenko, D.Yu. Ivleva, MATEC Web Conf. **23** (2015)
2. E. Timoshenko, E. V. Kravchenko, MATEC Web Conf. **37** (2015)
3. V.G. Kuznetsov, E.V. Kravchenko, J. Eng. Phys. Thermophys. **88**, 6 (2015)
4. G. V. Kuznetsov, E. V. Kravchenko, EPJ Web Conf. **82** (2015)
5. R.S. Volkov, O.V. Vysokomornaya, G.V. Kuznestov, P.A. Strizhak, J. Eng. Phys. Thermophys. **86**, 6 (2013)
6. M.D. Kats, I.M. Kats, Meas. Tech. **58**, 2 (2015)

THE EFFECT OF INCLINATION ANGLE ON CRITICAL HEAT FLUX IN A LOCALLY HEATED LIQUID FILM MOVING UNDER THE ACTION OF GAS FLOW IN A MINI-CHANNEL

Egor M. Tkachenko[1,2,*], and *Dmitry* V. Zaitsev[1]

[1]Kutateladze Institute of Thermophysics SB RAS, 630090 Novosibirsk, Russia
[2]Novosibirsk State University, 630090 Novosibirsk, Russia

Abstract. Intensively evaporating liquid films moving under the action of the cocurrent gas flow in a microchannel are promising for the use in modern cooling systems of semiconductor devices with high local heat release. This work has studied the dependence of the critical heat flux on the inclination angle of the channel. It has been found that the inclination angle in the plane parallel to the flow has no significant effect on the critical heat flux. Whereas the inclination angle in the plane perpendicular to the flow, on the contrary, significantly changes the value of the critical heat flux. However, for a given flow rate of fluid there is a threshold gas velocity at which the critical heat flux does not differ from the case of zero inclination of the channel. Thus, it can be concluded that the cooling system based on shear-driven liquid films can be potentially used when direction of the gravity changes.

1 Introduction

The development of modern microelectronic equipment requires efficient cooling systems because of the need to remove high heat fluxes, up to 1 kW/cm^2 from local areas of the processor [1]. Thin and ultrathin (<10 μm) liquid films moving under the action of a forced gas or vapor flow in the channel are promising for the use in thermal control systems of modern semiconductor devices [2]. Works [3-5] are devoted to analytical and numerical study of hydrodynamics and heat and mass transfer during joint motion of intensively evaporating liquid film and gas flow in microchannels. Authors of experimental works [6-8] established the basic laws of the flow and crisis phenomena in the liquid film, moving under the action of the gas flow in the horizontal channel with 1x1 cm heater (simulating an electronic device). Thermal control of electronic devices in transportation, aircrafts and space vehicles requires the analysis of the influence of the gravity direction on the cooling system efficiency. This paper presents experimental data on the critical heat flux for different inclination angles of the working area to the horizon.

* Corresponding author: egor_tkachenko@mail.ru

2 Experimental setup

Figure 1 shows the scheme of the work area. Its main part is a thin plate of stainless steel with imbedded copper rod. The surface of the rod has a shape of a square head of 1 × 1 cm and imitates a computer chip. The rod is heated using a spiral wound around its lower part. This design of the heater provides a constant temperature on the rod surface, T = const (as confirmed by thermocouple measurements). The work area is covered with a transparent cover made of optical glass to form a flat channel. The height of the channel is 1.5 mm. Gas is fed into the work area with a compressor, passes through it and is vented to atmosphere at the end of the work area. The gas flow rate is measured with a digital flow meter Bronkhorst with an accuracy of about 1%.

The liquid is supplied from the thermostat, enters the channel through the liquid nozzle and moves under the action of friction from the gas over the stainless steel plate in the form of a film. Fluid then accumulates in the bottom of the work area and returns to the thermostat. The fluid flow rate is measured by the float-type flow meter with an accuracy of 1.5%. The distance from the gas inlet to the liquid nozzle is 57 mm, and from the liquid nozzle to the heater, is 32 mm. This ensures the stability of gas and liquid flows by the time they reach the heater. The channel width is 30 mm. Several thermocouples are embedded in the stainless steel plate and in the copper rod, as shown in fig. 1. This allows determining the working temperature of the surface. The surface temperature of the heater is calculated based on the depth at which the thermocouples are installed (2 mm). All thermocouples are calibrated individually with an accuracy of 0.1°C.

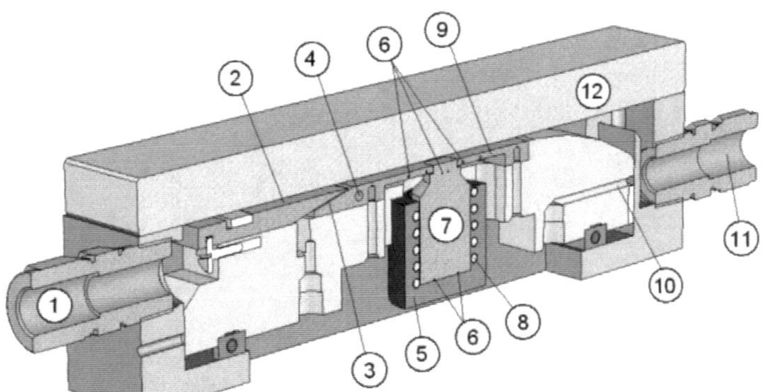

Fig. 1. The scheme of the work area: 1 — gas inlet; 2 — channel; 3 — liquid nozzle; 4 — thermal stabilizer; 5 — insulation; 6 — thermocouples; 7 — copper rod, 8 — nichrome spiral; 9 — stainless steel plate; 10 — liquid outlet; 11 — gas outlet; 12 — glass cover.

Heat flux is determined according to the electric power of the heating spiral. The thermal conductivity of copper is 400 W/mK, which is almost 30 times higher than the one of stainless steel (15 W/mK). This provides a reasonable heat spread from the heater to the stainless steel plate. According to evaluation based on the measurements of thermocouples embedded in the stainless steel substrate, the heat spreading is of 10-15% at q>250 W/cm^2, about 20% at q=100 W/cm^2 and up to 40% for small heat fluxes. To minimize the heat spreading from the heating spiral to the atmosphere, the spiral was wrapped with a layer of thermal insulation material (mineral wool, see 5 in fig.1).

Distilled water with initial temperature of 24°C is used as the working fluid. Air with a temperature of 24-27°C and a relative humidity of 15-30% is used as working gas. Atmospheric pressure during the experiments varies from 98500 to 101600 Pa. The experiments were carried out in stationary conditions. The gas velocity (volumetric gas

DOI: 10.1051/matecconf/20167201114

flow rate divided by the area of the channel cross section) U_{Sg} varies from 0.1 to 34 m/s. Maximum pressure drop along the channel is observed at the maximum gas flow rate and estimated to be less than 500 Pa.

3 Results

In this work we investigated how the inclination angle in the flow symmetry plane (the section plane in fig. 1), Θ, affects the flow, breakup and critical heat flux. The inclination angle changes with the step of 45° from 0 (horizontal channel) to 315° (-45°). The slope of the channel in this plane does not break the symmetry of the flow. Figure 2 shows the critical heat flux q_{cr} for different inclination angles, Θ, at Reynolds number of liquid Re_l=14. It is seen that the critical heat flux weakly depends on the inclination angle, except for the case Θ = -90°, where two points fall out of the generalization. This is probably due to the fact that these two points lie in the region of the channel flooding, where the heater is periodically washed with liquid, delaying the onset of the heat transfer crisis.

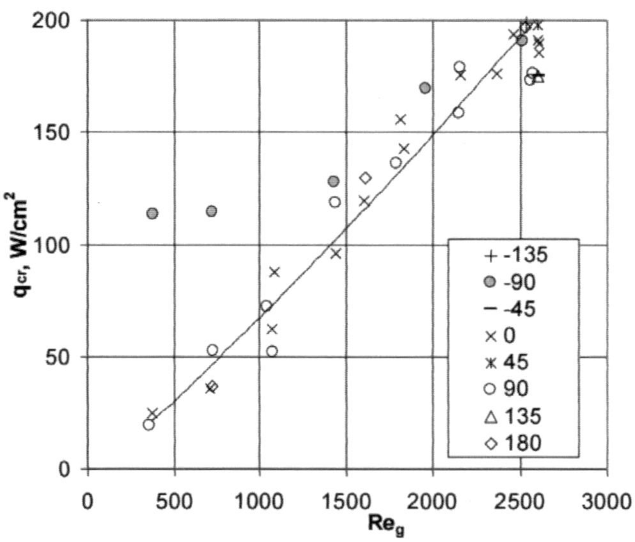

Fig. 2. The influence of the Reynolds number of the gas on critical heat flux for different Θ (shown in the legend), Re_l = 14.

We have studied only one case of channel inclination along the plane perpendicular to the flow: Ω = 90°. However, due to the symmetry of the channel, the case Ω = -90° does not differ from the case Ω = 90°. In addition, the case Ω = 180° is the same as the case Θ = 180°. The channel slope in this plane breaks the flow symmetry. Figure 3 shows the critical heat flux q_{cr} for Ω = 90°, at Re_l = 30 and Re_l = 46. Lines represent values q_{cr} corresponding to the inclination angles Ω = 0 and Θ = 0. It is seen that the inclination angle along the plane perpendicular to the flow has a significant influence on the critical heat flux (flow symmetry is broken). However, for each Re_l there exists a critical value of gas velocity, starting from which the critical heat flux almost coincides with the case Ω = 0 and Θ =0.

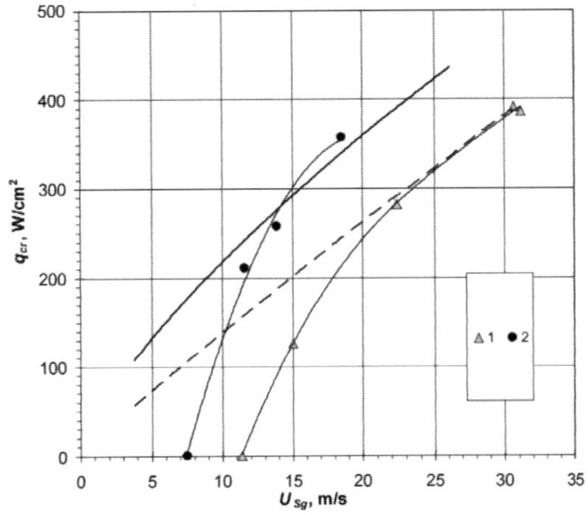

Fig. 3. Critical heat flux q_{cr} for $\Omega = 90°$, 1 - $Re_l = 30$, and 2 - $Re_l = 46$. Straight lines represent the values of q_{cr}, corresponding to the angles $\Omega = 0$ and $\Theta = 0$.

4 Conclusions

Thus, it can be concluded that the flow, the breakup and the critical heat flux of shear-driven liquid films weakly depend on the inclination angle in the channel symmetry plane (when the symmetry of the flow is preserved). And, on the contrary, the inclination angle in the plane perpendicular to the flow (when the flow symmetry is broken) has a significant impact on the patterns of the film flow and the crisis phenomena. However, for each value of the liquid flow rate there is a critical value of the gas velocity, starting from which the flow, the breakup and the critical heat flux have practically no dependence on the inclination angle in both planes. This shows that the cooling system, based on the evaporating liquid film moving in the channel under the action of gas flow, can be potentially used in transportation, aircraft and spacecraft, when the direction of gravity changes.

Acknowledgments

This work was supported by the Ministry of Education and Science of Russia (Agreement No. 14.604.21.0053, project identifier RFMEFI60414X0053).

References

1. A. Bar-Cohen, P. Wang, ASME. J. Heat Transf. **134** (2012)
2. O.A. Kabov, Yu. V. Lyulin, I.V. Marchuk, and D.V. Zaitsev, Int. J. Heat Fl. Flow **28** (2007)
3. E.Ya. Gatapova, O.A. Kabov, Int. J. Heat Mass Transf. **51** (2008)
4. Y. O. Kabova, Tech. Ph. Let. **39** (2013)
5. Y. Kabova, O. Kabov, T. Gambaryan-Roisman, P. Stephan, V.V. Kuznetsov, Int. J. Heat Mass Transf. **68** (2014)
6. D.V. Zaitsev, D.A. Rodionov, O.A. Kabov, Tech. Ph. Let. **35** (2009)
7. O.A. Kabov, and D.V. Zaitsev, Multiph. Sci. Tech. **21** (2009)
8. O.A. Kabov, D.V. Zaitsev, V.V. Cheverda, and A. Bar-Cohen, Exp. Therm Fluid Sci. **35** (2011)

MATEC Web of Conferences 72, 01115 (2016) DOI: 10.1051/matecconf/20167201115
HMTTSC-2016

SIMULATIONS OF METHANE MOTION IN A VERTICAL COAL BED WHEN FILLED WITH WATER

Aleksey Bubenchikov[1, 2], *Valentina* Tsyrenova[2], and *Sevan* Tsydypov[2,*]

[1]National Research Tomsk State University, 634050 Tomsk, Russia
[2]Buryat State University, 670000 Ulan-Ude, Russia

Abstract. The present paper reviews pressure-gravitational movement of a gas and liquid medium in a vertical isotropic bed of a canonical rectangular shape with two permeability windows on the side borders. The work proposes a simple computing technology based on symmetric approximation of differential members and the Gauss-Seidel linearization which allows to significantly expand the range of the model's applicability from the standpoint of moisture saturation compared to traditional numerical.

1 Introduction

The output of methane from coal beds to the mine area and the surface layer of the atmosphere [1-4] and its combustion [5-7] present a certain danger to human life and activity. Therefore, the work directed on defining levels of methane concentration in case of flooding or releasing water from coal mines is not only appropriate, but also essential both in the environmental aspect and in terms of health and safety. The importance of this issue is proved by the recent tragic events in the coal mine "Severnaya".

In modern conditions, a correct prediction of methane distribution over the volume of a deposit should be based on an adequate and verified mathematical model of two-phase filtration in a coal bed and extensive geological information in the vicinity of the deposit. Along with this, the mathematical model must contain data on complex spatial architecture of mine workings.

The same as in mechanics of turbulent flows, the predictive function of a filtration model is higher if positive experience of its application is wider. Thus, along with the already implemented applications of the proposed mathematical model [1-4], in this paper, we estimate the effect of gravity on the dynamics of a two-phase medium in a vertical bed. Leaving aside the problem of fluid motion out of the porous space, i.e. in the area directly related to mine development, we will focus on pressure-gravitational movement of a gas and liquid medium in a vertical isotropic bed of a canonical rectangular shape with two permeability windows on the side borders. It is assumed that the wetting and the non-wetting phase each move in their channel systems. The movement of each of the phases

* Corresponding author: hudobina@mail2000.ru

occurs under the influence of its own phase pressure and permeability of the bed with respect to each of the media is determined by its own permeability. In this case, the defining equations can be written in the form of the mass conservation laws for the gas and the liquid phases [8-10]:

$$\frac{\partial \rho^G m(1-s)}{\partial t} - div\left(\frac{\text{IK}k_r^G}{\mu^G}\rho^G\left(\vec{\nabla}p^G - \rho^G\vec{g}\right)\right) = q^G, \tag{1}$$

$$\frac{\partial \rho^F ms}{\partial t} - div\left(\frac{\text{IK}k_r^F}{\mu^F}\rho^F\left(\vec{\nabla}p^F - \rho^F\vec{g}\right)\right) = q^F, \tag{2}$$

here $\rho^G = \rho^G\left(p^G\right) = \rho^{G,0}\, p^G / p^0$ is the density of the gaseous phase, p_0 is the atmospheric pressure; $m = m\left(p^G\right)$ is the porosity (V_n/V, V_n are the pore volume); s is the effective moisture saturation (in principle, varies from 0 to 1), IK is the permeability tensor, generally, fully-filled; $k_r^G = \sqrt{1-s}\left(1-s^2\right)$ is the relative permeability of the gas phase, μ_G is the gas viscosity, q_G is the source member in the balance of the gas phase, q_F is the source member in the fluid balance; $\rho^F = \rho^{F,0}\, p^F / p^0$, $\rho^{F,0}$ is the density of the fluid at atmospheric pressure; $k_r^F = \sqrt{s}\left(1-\sqrt{1-s^2}\right)^2$ is the relative permeability for the liquid phase, μ_F is the viscosity of the fluid; $p^F = p^G - p^c(s)$; $p^c(s) = p^d\sqrt{1-s^2}/s$ is the capillary pressure, p^d is the threshold pressure, while $p^d = \rho^F g/\alpha$, where α is the parameter characterizing the porous medium.

2 Calculation results

The equations of mass balance of the liquid and the gaseous phases which include densities or pressures of the phases are non-linear equations. Furthermore, the coefficients at the derivatives in these equations at particular points of the area may become zero. This peculiarity creates obstacles to the use of specific numerical techniques. Typically, these obstacles are overcome at the cost of narrowing the range of moisture saturation changes. However, in this paper we propose a numerical algorithm which allows carrying out calculations in the whole range of moisture saturation changes.

The numerical algorithm is based on the use of the simplest method of calculating nonlinear coefficients by deflecting the included designed parameters to the previous iteration layer without using gradient methods for accelerating the convergence. The efficiency of the constructed algorithm is confirmed by good agreement of the calculation results with the analytical distribution of Buckley-Leverett for plane-radial filtration flow.

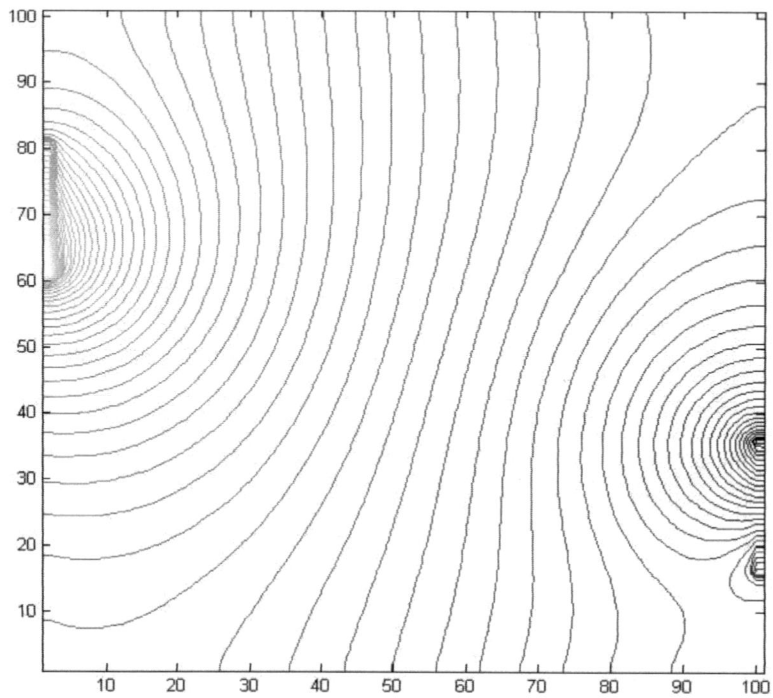

Fig. 1. Lines of equal pressure in gas phase.

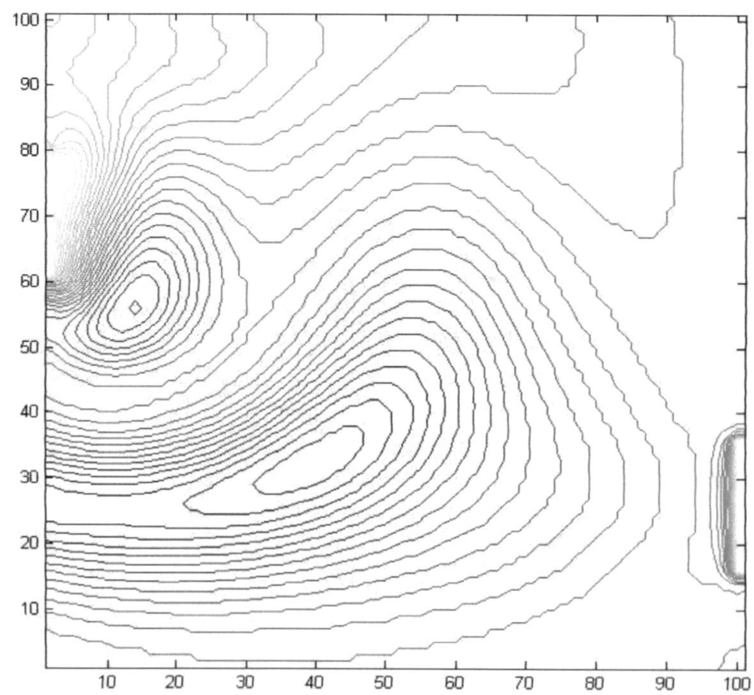

Fig. 2. Isolines of fluid pressure distribution.

The results shown in fig.1, 2 correspond to the case of steady flow which is obtained from the initial state of rest by the establishment method. At a certain point in time, taken as the start, the hydraulic fluid pressure increases in the permeability window on the left. Thereafter, movement of both phases begins. Calculations show that the gas phase appears to be more mobile at the stage of establishment, but, ultimately, the gas distribution pattern in the porous space becomes easier.

Acknowledgments

This paper was written within the frame of Competitiveness Improvement Program of Tomsk State University.

References

1. D.O.Dil, A.M.Bubenchikov, Vestnik TGU. Meh. Matem. **6,** 26 (2013) [in Russian]
2. D.O.Dil, A.M.Bubenchikov, Vestnik TGU. Meh. Matem. **5,** 25 (2013) [in Russian]
3. D. O. Dil, A. M. Bubenchikov, Key Eng. Mat. **685** (2016)
4. A.M. Bubenchikov, V.B. Tsyrenova, S.G. Tsydypov, Vestnik TGU. Meh. Matem. **1** (2014) [in Russian]
5. O.V. Matvienko, A.M. Bubenchikov, J. of Eng. Phys. and Therm. (to be published)
6. A.M. Bubenchikov, V.A. Korobitsyn, D.V. Korobitsyn, P.P. Kotov, Y.I. Shokin, J. Comp. Math. and Math. Phys. **54**, 7 (2014)
7. O.V. Matvienko, A.I. Baigulova, A.M. Bubenchikov, J. of Eng. Phys. and Therm. **87**, 6 (2014)
8. C. Shikuo, Y. Tianhong, W. Chenhui, Mater. of Int. Symp., China (2010)
9. M.Th. Van Genuchten, Soil Sci. Soc. Am. J. **44** (1980)
10. M.G. Schaap, M.Th. Van Genuchten, Vadose Zone J. **5** (2006)

MATEC Web of Conferences 72, 01116 (2016) DOI: 10.1051/matecconf/20167201116
HMTTSC-2016

COST PRICE OF PRODUCTS IN THE SYSTEM OF HEAT, REFRIGERATION AND ELECTRIC ENERGY PRODUCTION COMBINED AT THERMAL POWER PLANT

Alexander Tubolev[1,*], *Olga* Romashova[1], and *Leonid* Belyaev[1]

[1]National Research Tomsk Polytechnic University, 634050 Tomsk, Russia

Abstract. Nowadays combination of electric, heat and refrigerating energy production (trigeneration) is one of the modern technological solutions for energy efficiency increase and ecological problem solution [1]. Two types of refrigerating machines can be used for both energy and heat production combined: compression aggregates consuming electric energy and absorption aggregates using hot water heat, vapor or other heat conductors.

1 Goal setting

There is a goal of combined determination of tariffs for cold, heat and electric energy in trigeneration system and also purposeful area of different types of refrigerating machines use.

There are methods of expenses division for three types of products in the system of trigeneration developed on the basis of Ginter's triangle in the present work [2]. It allows calculating cost of refrigerating products developed in compression and absorption types of machines depending on coefficient of expenses distribution between heat and electric energy at thermal power plant and also determine economically efficient area of application of these types of machines.

2 Methods of expenses distribution for heat and electric energy production according to Ginter's triangle

Peculiarities of cost determination and formation of tariffs at thermal power plant producing two kinds of energy – heat and electric are presented in [2, 3]. As a basic method there is one offered by the authors according to which expenses for electric energy and heat production can be distributed between two kinds of produce in any way, but as the sum of expenses does not change because of it, the results of all possible variants of distribution are on one line (Picture 1, line ST) – 'Ginter's triangle'.

* Corresponding author: tubolev@tpu.ru

3 Calculation of production cost in the system of trigeneration

3.1 Basic data for calculation

As an example calculation of release cost price from thermal power plant with three types of turbine plant T-118/125-130-8 (electric energy, heat for hot water supply and refrigeration in the conditioning system) in the summer at turbine working according to heat graph on the basis of real energy characteristics has been made [4]. Climate region – Omsk for which design temperature for heating is $t_{DH.}$ = -37°C [5]; design temperature for conditioning is $t_{D.C}$ = +35°C [5].

The following refrigerating machines have been considered:

1) single-staged absorption bromide-lithium refrigerating machine with refrigeration capacity Q_0 = 1000 kW (absorption refrigerating machine -1000). Refrigeration coefficient ε_A depends on the temperature of direct network t_{DN} and is determined according to the catalogue of absorption refrigerating machine of the company Broad [6];

2) vapor compression refrigerating machine. Refrigeration coefficient of vapor compression refrigerating machine has been adopted according to the data of electronic resource of the company Termogaz and is equal to ε_K = 3 [7].

3.2 Basic calculation

Nominal load of selection of T-118/125-130 $Q_{\text{Sel}}^{\text{Nom}}$ = 219 MW. Load for hot water supply Q_{HWS} is equal to 20% from maximum heat load of thermal power plant Q_{TPP}, designed with account of district heating coefficient [8]. Cost of reference fuel $C_{R.F}$ = 3200 R/t.

$$Q_{\text{TPP}} = \frac{Q_{Sel}^{\text{Nom}}}{\alpha_{\text{TPP}}} = \frac{219}{0.6} = 365 \text{ MW}, \tag{1}$$

$$Q_{\text{HWS}} = 0.2 \cdot Q_{\text{TPP}} = 0.2 \cdot 365 = 73 \text{ MW}. \tag{2}$$

Maximum refrigerating load has been estimated for the region of joined heat consumers in accordance with relation:

$$\frac{Q_{\text{TPP}}}{Q_{\text{C}}^{\max}} \approx \frac{(t_{\text{Ins.r}} - t_{\text{DH}})}{(t_{\text{DC}} - t_{\text{Ins.r}})}, \tag{3}$$

where $t_{\text{Ins.r}} = 18\,^{\circ}\text{C}$ - average temperature inside rooms.

A number of hours of maximum load of refrigerating machines use $\tau_{\text{C}}^{\max}$ has been determined as:

$$\tau_{\text{C}}^{\max} = K \cdot \tau = 0.5 \cdot 2160 = 1080 \,\text{h}, \tag{4}$$

where K – coefficient of maximum load use for refrigeration production (equal to 0.5 on the basis of recommendations presented in [8]): τ – total number of hours for three summer months.

Maximum production of heat from turbine Q_{T}^{C} selection for absorption refrigerating machine for refrigeration production has been determined as $Q_{\text{T}}^{\text{C}} = Q_{\text{C}}^{\max} / \varepsilon_{\text{A}}$, where ε_{A} -

refrigerating coefficient of absorption refrigerating machine. Additional electric capacity of thermal power plant for the work of vapor compression refrigerating machine has been estimated in accordance with refrigeration coefficient of compressive machine ε_C.

Production of electric energy and heat for hot water supply from thermal power plant has been considered as a primary variant. Limitations on maximum consumption of water via network installation (for T-118/125-130-8: $G_{NW}^{max} = 1250$ kg/s [4]) have been considered in the new mode of combined release of heat from thermal power plant for hot water supply and conditioning. It has been accepted that working conditions of hot water supply consumers have not been changed in the new mode.

The sum of operating costs in the primary variant has been received from costs for electric energy and heat production:

$$C = C_E + C_H \tag{5}$$

Refrigeration is used as consumed product besides heat and electric energy in the system of trigeneration, and there are also operating costs for its production C_H^C at thermal power plant:

$$C' = C_E' + C_H' + C_H^C \tag{6}$$

4 Graph solution of the goal on the basis of Ginter's triangle

Heat and electric energy cost values were received at calculation of the basic variant according to physical method of expenses distribution (point R on the line ST, figure 1). Point R has been received at arbitrary expenses distribution. Thus, straight line PR, line segment TO and SO form Ginter's triangle.

Triangle NOM (fig. 1.) has been received for the system of combined release of heat, electric and refrigerating energy according to the results of design. Point 'C' corresponds to physical method of expenses distribution between electric energy and heat released from turbine differentiation by outer consumer. Line OE depicts dependence of refrigeration cost developed at absorption refrigerating machine on the method of expenses distribution for electric energy and heat in the system. There is a case depicted in the figure 1 (line OG) when electric energy at vapor compression refrigerating machine is received according to the cost price from the system of cogeneration (line ST) which allows comparing this variant with the variant of refrigeration production at absorption refrigerating machine (absorber works in the system of trigeneration) according to technical and economic data.

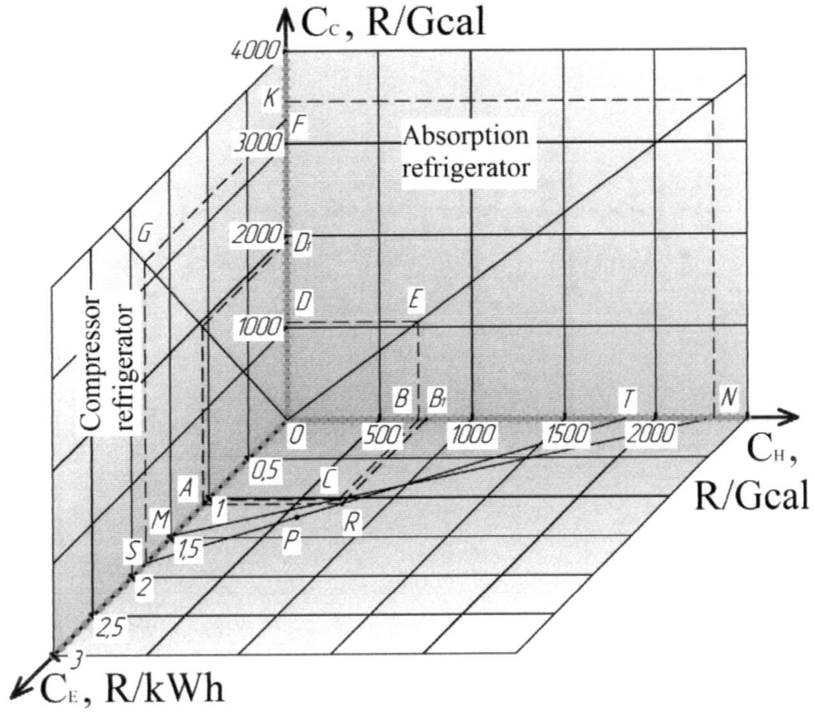

Fig. 1. Dependence of refrigeration cost price on heat and electric energy cost price in the system of trigeneration.

5 Conclusion

There are methods of graphic and analytical determination of production cost price in cogeneration and trigeneration systems for analysis of competitiveness of vapor compression refrigerating machine and absorption refrigerating machine using heat and electric energy of thermal power plant in the present work. Economically efficient areas of vapor compression refrigerating machine and absorption refrigerating machine application depend on the method of expenses distribution at thermal power plant between electric and heat energy. Trigineration decreases production cost price at expenses distribution between heat and electric energy according to physical method.

References

1. A. B. Bogdanov, Kot. Ros. b. nac. mas **7** (2012)
2. K. Arakeljan, N.N. Kozhevnikov, Teploj. **11** (2006)
3. V.A. Malafeev, I.A. Smirnov, Teploj. **4** (2003)
4. V.N. Ruzankov, B.V. Mihajlov, *Tipovaja jenergeticheskaja harakteristika turboagregata T-118/125-130*
5. E.Ja. Sokolov *Teplofikacija i teplovye seti: Uchebnik dlja vuzov* (Izdatel'stvo MJeI, Moscow, 2001)
6. D.N. Aref'ev, *Katalog BROAD,* (Olimp-biznes, Moscow, 2005)
7. http://www.termogazgroup.ru/
8. V.Ja. Ryzhkin, *Teplovye jelektricheskie stancii: Uchebnik dlja vuzov* (Jenergoatomizdat, Moscow, 1987)

MATEC Web of Conferences 72, 01117 (2016) DOI: 10.1051/matecconf/20167201117
HMTTSC-2016

EXPERIMENTAL STUDY OF HEAT TRANSFER IN THE STEAM THERMAL SIPHON TUBE CHANNEL

Mikhail A. Ulchiekov[1,*]

[1]National research Tomsk Polytechnic University, Tomsk 634050, Russia

Abstract. The results of experimental studies of heat transfer into the thermo-syphon tube with different thermal loads are represented. Are examined two types of liquid as the working fluid. The values of temperatures in the steam channel of thermo-syphon tube are registered and the dependences of the coefficients of heat exchange on the power of heating element are obtained. The influence of thermal load and properties of liquid on the intensity of the processes of evaporation and condensation of working fluid in the thermo-syphon installation are established.

1 Introduction

In connection with the development of electronic technology and the decrease of sizes of typical articles appears the need for devices for the intensive outlet of heat.

Due to the development of electronic technology and decrease of typical articles sizes appears the need for devices for intensive heat removal. These can be thermosiphon tube [1-3]. But the use of thermosiphons is not limited only to cooling devices of electronic engineering. They are effective with the industrial heat recovery, blade cooling of turbines, in the heating systems, etc [4]. A main difference in the thermo-syphon from the usual thermal tube in the fact that in the thermosiphon is used gravity, but not the force of surface tension for the recovery of condensate into the zone of the evaporation [5-6]. Studies in this area much [1-6], but little experimental data [1, 5], in connection with which this problem is urgent.

Purpose of the work – an experimental study of heat exchange in the thermo-syphon tube over a wide range of a change in the heat flux of applied into the boiling zone on lower boundary of steam channel.

The subject of a study is glass thermo-syphon tube lower surface of which it is located on the heating element. As the working fluids were investigated water and ethyl alcohol with the space factors of steam channel from 0.1 to 0.4.

* Corresponding author: ulchieckovmihail@mail.ru

2 Methods of experimental research

Experimental installation consisted of the glass tube (external de=0.04 m, internal di=0.038 m diameters, height l=0.28 m.), closed on top, heating element, autotransformer, multichannel device for measuring the temperature, connected to the computer through the adapter, heat insulation.

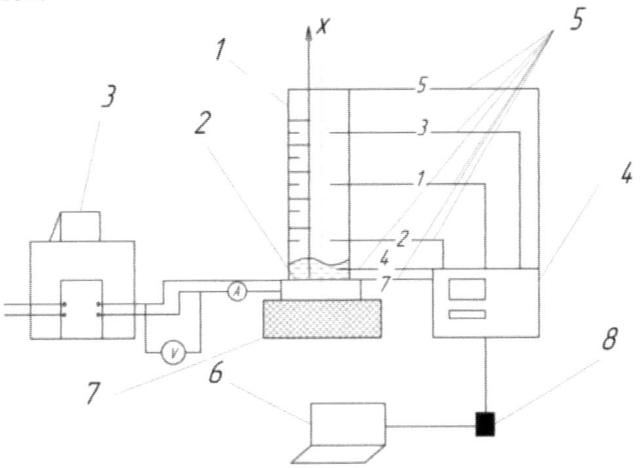

Fig. 1. Schematic diagram of installation: 1 - thermal siphon tube; 2 - a heating element; 3 - auto-transformer; 4 - Temperature-measuring device UKT38, 5 - thermocouple, 6 - a computer; 7 - thermal insulation; 8 – Adapter.

Temperature measurements were carried out using a chromel-alumel thermocouples are connected to the multi-channel measuring device OWEN UKT 38 as in [7, 8]. The values of the temperature every 2 seconds recorded on a personal computer software National Instruments LabView 8.1. Location of thermocouples in the thermosyphon are shown in table, where n-number entry into multi-channel measurement device:

Table 1. Location of thermocouples.

n	1	2	3	4	5	7
h, mm	140	70	210	1	the top cover	heater

The heating element is a tungsten wire, laid in a flexible hollow channel of the thermal insulation material. On the wire is put electrical insulator layer and a thin aluminum plate with a diameter 0.04 m. and a thickness 0.002 m.

Regulator power heater - autotransformer connected to the AC frequency of 50 Hz and a voltage of 220 V. In order to register the output voltage and current used by digital multimeter M266 series with contactless current sensor (accuracy class 0.1).

In the voltage regulator, which is connected to the heater, and the measured voltage exhibited a multimeter. Heater power ranged from 10 watts to 30 watts.

A metering device 3 ml were pumped liquid (alcohol or water) in the thermosyphon with thermocouples. The experiment is complete when the temperature of the tube height did not change over time (stationary mode).

3 Results and discussion

Fig. 2 shows the results of an experimental study of the heat transfer process in the steam thermal siphon tube channel (operating fluid - water):

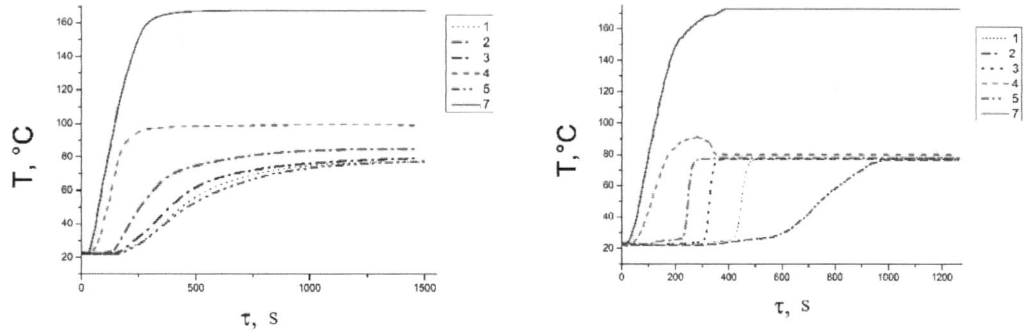

Fig. 2. The temperatures at different points of the thermosyphon, depending on the time when the power of the heating element q = 12 kW/m2: a) water; b) alcohol.

After the switching on heater the liquid temperature began gradually to grow to the boiling point (temperature of alcohol it rose above the boiling point, after which it occurred sharp effervescence and a decrease to the boiling point). The value of the temperature of heating element rose to the moment of intensive vaporization. An increase in temperatures on the height of thermosiphon began after effervescence and continued to the specific values (in this case in alcohol it was observed an uneven increase in the temperature on the height of tube).

Fig. 3 shows the dependence of the boiling and condensing the working fluid from the fuel cell power. The procedure was used [9]:

$$\alpha_{boil} = Nu_* \cdot \frac{\lambda}{l_*}, \quad \alpha_{cond} = \frac{G \cdot r}{\pi \cdot d_{in} \cdot l \cdot (t_{SK} - t_C)},$$

here α_{boil}, α_{cond} - heat transfer coefficients during condensation and boiling, respectively, W/m^2·K; Nu_* - Nusselt number, depends on boiling regime [9]; λ – the coefficient of thermal conductivity of the boiling liquid, W/mK ; l_* - reduced size, m [9]; *G* - steam flow, kg/s; *r* - heat of vaporization, J / kg; d_{in}, *l* - the inner diameter of the steam passage length, m; t_{sk}, t_c - temperature of steam and the condensation surface, K. The steam consumption amounted condensate flow, since the calculation was performed for the steady state.

Is established (fig. 3), that the heat load on the thermosyphon growth leads to a proportional increase (linearly) the heat transfers coefficient of condensing fluid. Due to differences between the thermal and hydrodynamic parameters of the heat transfer coefficient of condensing an alcohol vapor is higher than that of water at 150 W / (m^2·K) in the whole range of the thermal load on evaporation below 90 W / (m^2·K). Since ethanol is more volatile than water, the flow of alcohol vapor during the boiling is more. Since ethyl alcohol is more volatile component than water, then the expenditure of vapors of alcohol during the boiling is more.

Therefore, for the efficient operation of thermosiphon systems use low-boiling liquids as working medium is preferable from the standpoint of the efficiency of such heat exchangers.

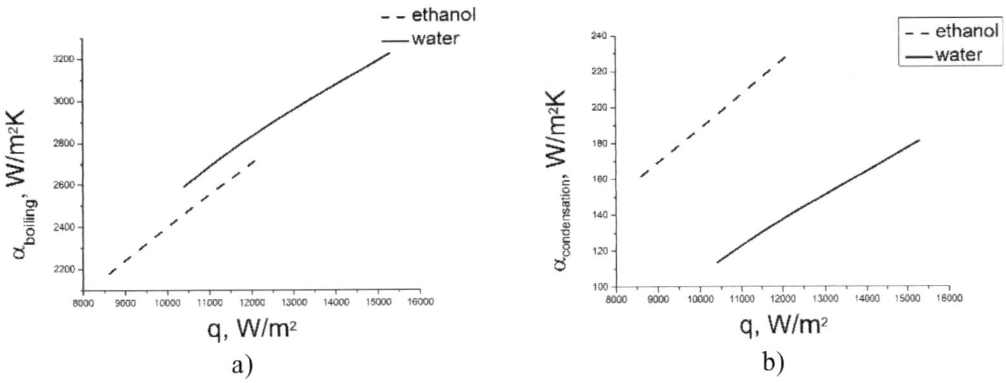

Fig. 3. Dependencies heat transfer coefficient with: a) boiling (1 - water 2 - alcohol); b) condensation (1 - water 2 - alcohol).

4 Conclusion

Are experimentally obtained the temperature fields in the thermo-syphon installation with different modes of its operation. Are obtained the dependences of the heat-transfer coefficients of working medium during its boiling and condensation in the thermosiphon with a change in the thermal load and the use of different working fluids. As a result, an experimental study it is established that with an increase in the thermal load the condensing coefficient and boiling of water and alcohol increases practically in the linear law. It is also established that the use of the low-boiling-point liquid in the thermo-syphon installation leads to the intensification of the process of heat transfer in the steam channel of thermo-syphon tube. Obtained in the work experimental data and dependences can be used both for the design of new thermo-syphon installations and modernization of the already working heat exchange equipment.

Acknowledgments

The work at the field of modeling of heat and mass transfer processes near boundary between two layers was supported by the scientific schools grant NSH-7538-2016.8.

References

1. Seok-Ho Rhi, *An Experimental and Analytical Study on Two-Phase Loop Thermosyphons* (Canada, 2000)
2. G.V. Kuznetsov, M.A. Al-Ani, M.A. Sheremet, J. Eng. Thermoph. **20** (2011)
3. J.G. Reed, C.L. Tien, ASME. J. Heat Trans. **109** (1987)
4. Kh. Kafeel, A. Turan., Heat Mass Trans. **49** (2013)
5. H. Farsi, J.-L. Joly, M. Miscevic, V. Platel, N. Mazet, Appl. Th. Eng. **23** (2003)
6. G.V. Kuznetsov, A.E. Sitnikov, High Temp. **40** (2002)
7. A.S., Krasnoshlykov, V.I. Maksimov, MATEC Web Conf. **19** (2014)
8. G.V. Kuznetsov, V.I. Maksimov, M.A. Sheremet, J. of Appl. Mech. Tech. Ph. **54** (2013)
9. S.S. Kutateladze, *Heat Transfer in Condensation and Boiling* (Mashgiz, Moscow, 1952)

MATEC Web of Conferences 72, 01118 (2016) DOI: 10.1051/matecconf/20167201118
HMTTSC-2016

MATHEMATICAL MODELING OF HEAT TRANSFER IN THE SYSTEM "THE OBJECT OF HEATING – THERMAL INSULATION LAYER – THE EXTERNAL ENVIRONMENT" AND MICROCLIMATE PARAMETERS OF INDUSTRIAL PREMISES

M. A. Ulchiekov[1,*]

[1]National research Tomsk Polytechnic University, Tomsk 634050, Russia

Abstract. This paper been solved the task of thermal conductivity for the system "the object of heating – thermal insulation layer – the external environment." The impact of the main groups of factors on the temperature of heating the object surface has been shown.

1 Introduction

Development of heat supply facilities industrial objects is carried out using standard parameters [1,2]. The choice of heating system parameters, usually is intended to achieve the temperature in the room T_A, which provide thermal comfort [3] conditions of work – regulations of the microclimate. Selection T_A same indoor air temperature is carried out mainly ''heuristic'' (based on operating experience of heating systems). Thus, as a rule, merely postulated that is comfortable for the operating temperature range of variation from *18* °C to *24* °*C*. This approach provides acceptable for traditional heating systems technology of work, without taking into account the possible individual characteristics of the main object of heat - working and production conditions (intensity, the presence of breaks, clothing type, mode of operation of the ventilation system, and others.) In this context, it seems appropriate to analyze the processes of heat removal (cooling) from the working and theoretical foundation of the currently accepted on the basis of empirical data on how working senses temperature industrial premises.

The purpose of the study is the solution of heat conduction problem for a system of ''the object of heating – thermal insulation layer – the external environment'' and the establishment of the numerical values of the heat flows from the surface of the body working, which correspond to the currently accepted conditions of the ''thermal comfort''.

* Corresponding author: ulchieckovmihail@mail.ru

2 Formulation of the problem

At the problem formulation it was assumed that the temperature of the main body of the human T_I is constant and may vary within certain limits, from $35.5\ ^\circ C$ to $36.8\ ^\circ C$. Out of this range is possible, but this is the exception rather than the rule. More intense heat dissipation into the environment may lead to a decrease in T_I, but in real conditions such a deviation can not be more than $0.8\ ^\circ C$, even in the lowest permissible ambient temperature.

System, which is regarded (fig. 1), consists of seven heating elements: the main part of the human body with a constant temperature for a long time; adjacent to the surface layer of the skin, consisting of three elements (subcutaneous fat, dermis and epidermis), which in real conditions there is a temperature difference, the air gap of small thickness between the human skin and clothing; clothes, performing the function of heat insulation; the external environment, the temperature of which may be taken as a first approximation constant.

In this formulation, the problem is reduced to the determination of the working body surface temperature T_S in certain thicknesses of the layers (fig. 1), thermal insulation properties of materials and substances, as well as the heat sink conditions on the outer boundary layer of insulation. The T_S temperature will determine the ''comfort'' heat mode working at a certain ambient temperature and cooling rate. Formulated heat conduction problem can be solved in one-dimensional formulation (cross-flow heat on the other two coordinates are negligible in most practically important variants) for the five-layer plate.

It should be noted that in general it is possible to vary parameters of the system - as the thicknesses of the individual layers and the heat insulating characteristics of the materials of these layers. Therefore, one characteristic of the process which determines at the simulation can be a surface body temperature of working T_S assuming the temperature T_I at a certain distance from the surface is constant (which is justified as a rule).

Solutions field geometry is shown in fig. 1.

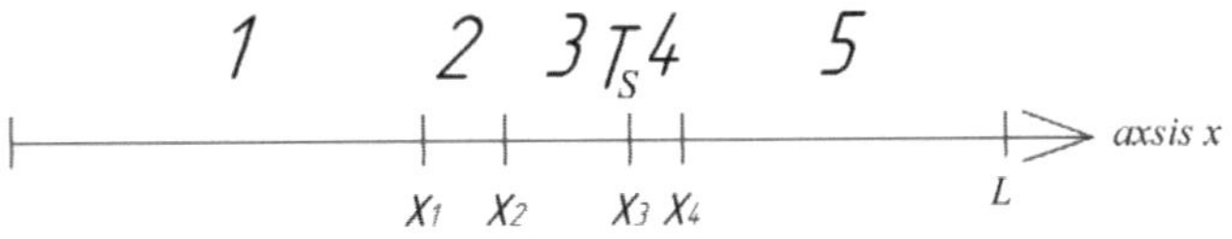

Fig.1. Field of solving the problem: 1) the subcutaneous fat; 2) the dermis; 3) the epidermis; 4) the air layer; 5) the thermal insulation (clothing).

Mathematical modeling of the process under study is reduced to solving a system of five equations of heat conduction for the system ''the object of heating - air gap - a layer of thermal insulation - the external environment'' with the appropriate boundary conditions.

In this case, we consider a linear one-dimensional heat conduction problem.

The mathematical formulation of the problem:

$$\begin{cases} \rho_1 c_1 \dfrac{\partial T_1}{\partial t} = \lambda_1 \dfrac{\partial^2 T_1}{\partial x^2}, \quad 0 < x < x_1; \\ \rho_2 c_2 \dfrac{\partial T_2}{\partial t} = \lambda_2 \dfrac{\partial^2 T_2}{\partial x^2}, \quad x_1 < x < x_2; \\ \rho_3 c_3 \dfrac{\partial T_3}{\partial t} = \lambda_3 \dfrac{\partial^2 T_3}{\partial x^2}, \quad x_2 < x < x_3; \\ \rho_4 c_4 \dfrac{\partial T_4}{\partial t} = \lambda_4 \dfrac{\partial^2 T_4}{\partial x^2}, \quad x_3 < x < x_4; \\ \rho_5 c_5 \dfrac{\partial T_5}{\partial t} = \lambda_5 \dfrac{\partial^2 T_5}{\partial x^2}, \quad x_4 < x < L; \end{cases}$$

Here ρ_i, λ_i, c_i - density, coefficient of thermal conductivity and specific heat of the material of *i*-layer system, T_i – current temperature value ($i = 1$ to 5).

On the outside ($x = L$) surface of the insulation layer the heat exchange takes place with the environment, which is described by the boundary conditions of the third kind, and thermal interaction between the layers of the system is described by the the fourth kind boundary conditions.

The initial and boundary conditions:

$$x = 0: \ T = T_1, \ T_1 = const;$$
$$x = x_1: \ \lambda_1 \frac{\partial T_1}{\partial x} = \lambda_2 \frac{\partial T_2}{\partial x}, \ T_1 = T_2;$$
$$x = x_2: \ \lambda_2 \frac{\partial T_2}{\partial x} = \lambda_3 \frac{\partial T_3}{\partial x}, \ T_2 = T_3;$$
$$x = x_3: \ \lambda_3 \frac{\partial T_3}{\partial x} = \lambda_4 \frac{\partial T_4}{\partial x}, \ T_3 = T_4;$$
$$x = x_4: \ \lambda_4 \frac{\partial T_4}{\partial x} = \lambda_5 \frac{\partial T_5}{\partial x}, \ T_4 = T_5;$$
$$x = L: \ \lambda_5 \frac{\partial T_5}{\partial x} = \alpha (T_L - T_A);$$
$$t = 0: \ T_1 = T_1^0, \ T_2 = T_2^0, ..., \ T_{x1} = T_{x1}^0, \ 0 < x < x_3;$$
$$T = T_0^a, x_3 < x < x_4;$$
$$T = T_0^{cl}, x_4 < x < L;$$

Here T_A - ambient temperature, α - heat transfer coefficient.

The system of differential equations with appropriate initial and boundary conditions solved by finite difference method, using an algorithm [4,5] developed for solving problems of heat transfer in the fields of multi-layer with discontinuities coefficients of thermal conductivity.

3 Numerical investigations

Investigations were carried out for different temperature of the environment T_A (from *291 K* to *297 K*), thicknesses of the air layer *Δ4* (*0.0005 m* to *0.002 m*), thicknesses of the clothes *Δ5* (from *0.002 m* to *0.007 m*).

Thermophysical characteristics of the layers the system under consideration [6,7]:

$\lambda_1 = 0.2 \ W/m \cdot K, \ \lambda_2 = 0.45 \ W/m \cdot K, \ \lambda_3 = 0.21 \ W/m \cdot K, \ \lambda_4 = 0.024 \ W/m \cdot K,$

$\lambda_5 = 0.03 \ W/m \cdot K$

$\rho_1 = 850 \ kg/m^3, \ \rho_2 = 1000 \ kg/m^3, \ \rho_3 = 1200 \ kg/m^3, \ \rho_4 = 1.2 \ kg/m^3, \ \rho_5 = 120 \ kg/m^3$

$c_1 = 2250 \ J/kg \cdot K, \ c_2 = 3200 \ J/kg \cdot K, \ c_3 = 3600 \ J/kg \cdot K,$

$c_4 = 1005 \ J/kg \cdot K, \ c_5 = 2260 \ J/kg \cdot K,$

The geometrical characteristics of the layers skin [8]:

$$\Delta_1 = 0.015 \ m, \ \Delta_2 = 0.0005 \ m, \ \Delta_3 = 0.001 \ m$$

Conditions of the heat exchange with the environment:

$$T_0^{cl} = 301 \ K, \ T_0^a = 303 \ K, \ T_0^{x1} = 307, \ \alpha = 5 \ W/m^2 \cdot K.$$

The duration of the modeled process: *t=7000 s.*

Fig. 2 shows the temperature field in the investigated systems at various ambient temperatures.

 DOI: 10.1051/matecconf/20167201118

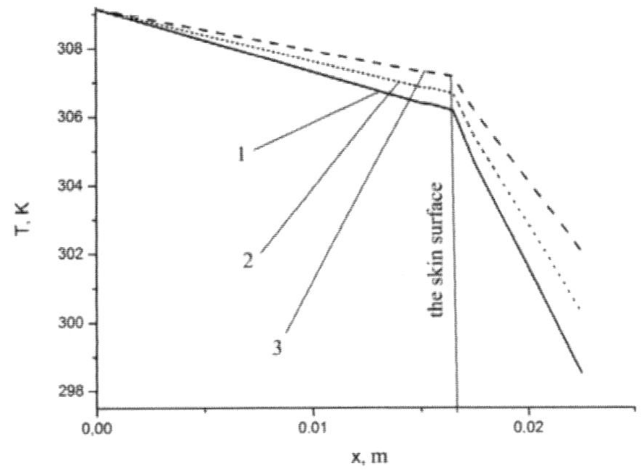

Fig. 2. The temperature distribution along the *x* coordinate at: 1 – T_A=291 *K*; 2 – T_A=294 *K*; 3 – T_A=297 *K*.

Fig. 3 shows the temperature of the skin surface dependence from the time at various ambient temperatures. It has been established that certain T_S falls to values corresponding to the conditions of heat transfer at the boundary $x = L$, and further does not change.

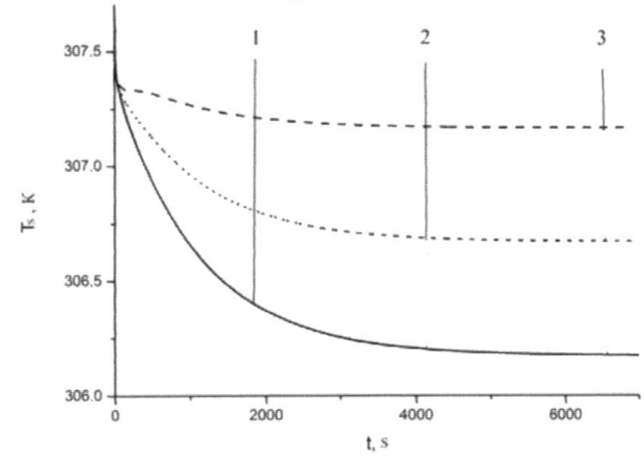

Fig. 3. T_S dependence on time at: 1 – T_A=291 *K*; 2 – T_A=294 *K*; 3 – T_A=297 *K*.

Fig. 4 shows the dependence temperature of heating surface of the object on the value of the T_A. This dependence is linear and the change of the T_S is 0.9 K at the ambient temperature change of 6 K.

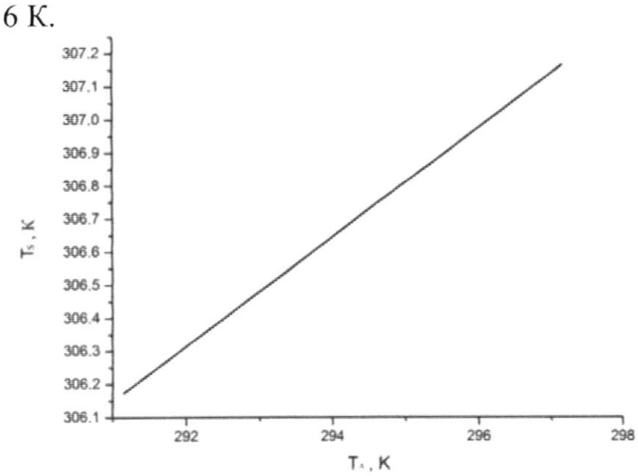

Fig. 4. The dependence of the T_S on the ambient temperature.

Fig. 5 shows the dependence T_S of the thickness air gap at different ambient temperatures. The change is from *0.24 K* to *0.3 K* with change *Δ4* to 0.0015 m.

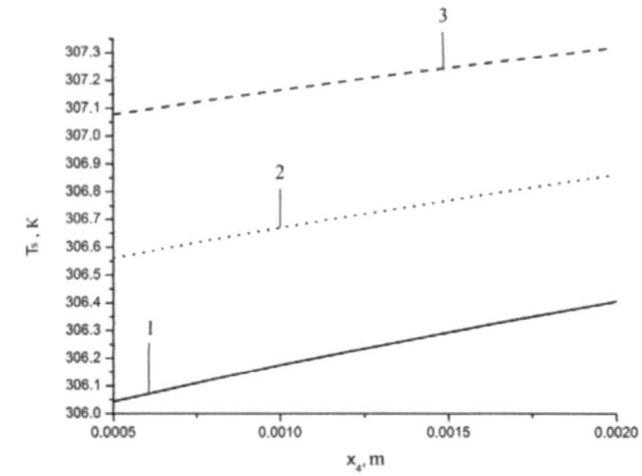

Fig. 5. The dependence of the T_S on the thicknesses of the air layer : 1 – T_A=291 K; 2 – T_A=294 K; 3 – T_A=297 K.

Fig. 6 shows the dependence T_S of the thickness of the outer thermal insulation at various ambient temperatures. The change is from *0.748 K* to *1.124 K* at changing Δ5 to *0.005* m.

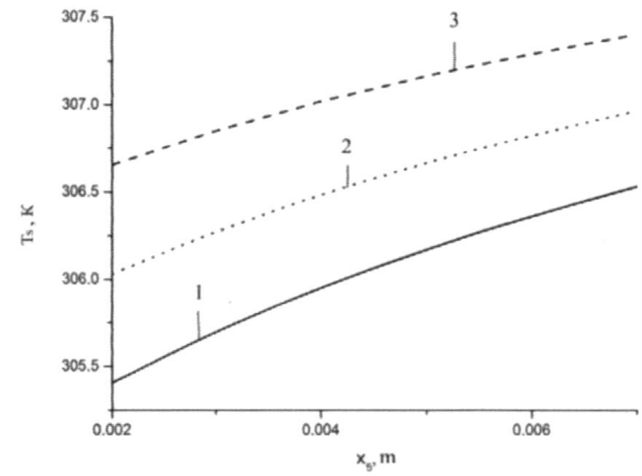

Fig. 6. The dependence of the T_S on the thicknesses of the thermal insulation: 1 – T_A=291 K; 2 – T_A=294 K; 3 – T_A=297 K.

4 Conclusions

1. The mathematical model heat transfer in the system ''heating facility - thermal insulation layer - the external environment'', by heat exchange with the environment in a natural convection mode is formulated, which differs from the known a high-level of field sampling solutions in the description of the heat transfer processes in the considered system and a possible forecast of microclimate parameters .
2. The influence the main groups of factors on the surface temperature of the heating object.

Acknowledgments

The work was supported by the Russian Federation President's grant (NSH project - 7538.2016.8).

References

1. P. Domingues, P. Carreira, R. Vieira, W. Kastner, Comp. St. Int. **45** (2016)
2. M Wallance, R. McBride., P. Mhaskar, , J. House, T. Salsbury, Ch. Eng. Sc. Energy **69** (2012)
3. M. Taleghani, M. Tenpierik, A. Dobbelsteen, Ren. Sust. En. R. **26** (2013)
4. O.V. Vysokomornaya, G.V. Kuznetsov, P.A. Strizhak, Rus. J. Ph. Ch. B **5** (2011)
5. V.L. Strakhov, A.N. Garashchenko, G.V. Kuznetsov, V.P. Rudzinskii, Comb., Exp. Sh. W. **37** (2001)
6. I.K. Kikoin *Tables of physical quantities* (Atomizdat, Moscow, 1976)
7. A.P. Babichev, A.P. Babushkin, A.M. Bratkovsky *Physical values,* (Eneroatomizdat, Moscow, 1991)
8. F. Xu, T. J Lu, K. A., Sef. Acta Mech. S. **24** (2008)

EVALUATION OF THERMOSYPHON APPLICATION FOR COOLING THE MODULAR AUTOMATED CONTROL SYSTEMS

Konstantin Yu. Ushakov[1], *Andrey* N. Peters[1], *Alexander* R. Bogomolov[2], and *Elena* Yu. Temnikova[1,*]

[1]Kuzbass State Technical University named after T.F. Gorbachev, 650000 Kemerovo, Russia
[2]Kutateladze Institute of Thermophysics SB RAS, 630090 Novosibirsk, Russia

Abstract. The experimental setup for studying operation effectiveness of the closed two-phase thermosyphon s aimed at the development of heat exchangers without disturbing the process of air circulation inside the module of control system, which can maintain the required temperature that provides effective heat removal from the zone of module equipment are elaborated. The primary results of experimental researches are presented.

1 Introduction

The autonomous modules of control systems have high heat release. To avoid equipment overheating and ensure its trouble-free operation, it is necessary to remove released heat. At that, the certain quality requirements should be obeyed, namely the purity of the working medium used in the cooling system. For example, to cool the modular automated control systems of electric calorific installations for air supply to the mines, the refrigerating units that match the requirements are used. However, the cost of one unit in average is about 2 mln. rubles; moreover they are quite cumbersome.

To date, there are the cooling systems used in various heat-exchanging devices based on the evaporation-condensation cycle: the closed two-phase thermosyphon s of different designs [1-2]. They are considered to be effective in comparison with other elements of cooling systems because they have high heat transfer characteristics and easy to be manufactured [1, 3-4]. Until now, thermosyphon introduction is not large-scaled, because, despite the relatively simple mechanism of thermosyphon s operation, the general theory of heat transfer processes in thermosyphon s, which takes into account a large set of physical processes in the zones of evaporation and condensation, in the vapor channel, and moving film of condensate, has not been developed [15].

The objective of this work is the creation of experimental setup for studying operation effectiveness of the closed two-phase thermosyphon s aimed at the development of heat exchangers without disturbing the process of air circulation inside the module of control

* Corresponding author: teu.pmahp@rambler.ru

system, which can maintain the required temperature (about 60°C) that provides effective heat removal from the zone of module equipment.

2 Experimental setup

The scheme of experimental setup for studying the process of heat removal by means of two-phase closed thermosyphon s is shown in fig. 1.

It is box 1 with confusers 2 attached from different sides to feed cold air to the upper part and hot air to the lower part (see fig. 1, arrows). To prevent hot and cold air mixing, there is sealed partition 3. In box 1, there are thermosyphons, consisting of a series of vertical copper tubes with the 8-mm diameter, 400-mm height, connected by a single horizontal tube: manifold. The working substance is freon R141b. It is a colorless transparent liquid, whose physical and chemical properties are close to the properties of R11 and R113. The chemical formula is CH_3CFCl_2. The boiling point is 32.1°C. The degree of tube filling is 20%, excluding the manifold.

The closed two-phase thermosyphon is a vertical wick-free heat pipe, where coolant transfer from the zone of evaporation to the zone of condensation occurs due to the lifting force. The working substance (freon) allows transfer of a large amount of energy due to evaporation in the bottom part of the unit through the heat supply from hot air. Vapor produced in the evaporation zone rises in the central part to the condensation region due to the buoyancy effects, and latent heat of phase transition is released there due to cold air. Condensed vapor goes back to the evaporation zone over the inner side surface of thermosyphon under the action of gravity [3].

The temperatures of hot and cold air and their flow rates were measured during the experiment at setup inlet and outlet.

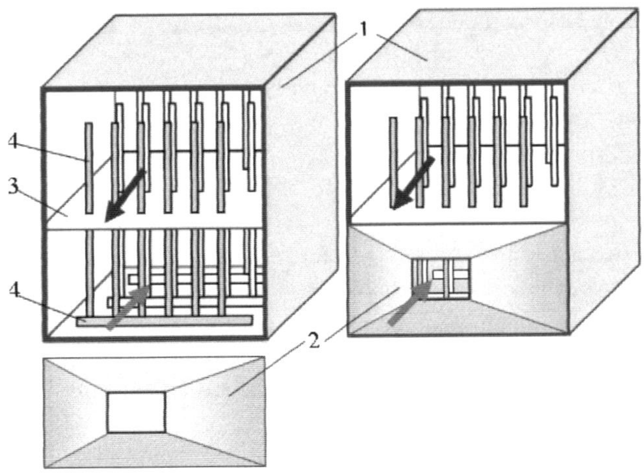

Fig. 1. Scheme of experimental setup: 1 – box; 2 – confusor; 3 – partition; 4 – thermosyphon.

3 Results

The diagram of changes in the hot and cold air temperatures along the setup length was plotted on the basis of temperature measurements; it is shown in fig. 2. It can be said that hot air cooling by 21.6°C leads to cold air heating at the top by 4.2°C.

The average temperature difference of 70°C was determined. The heat balance was calculated, and it was found out that heat removed from hot air through the thermosyphon (evaporation area) was 0.889 kW, and heat supplied to cold air (condensation zone) was 0.747 kW, and the difference (losses) was 19%. The heat transfer surfaces were calculated from the geometrical dimensions of thermosyphons in the zones of condensation and evaporation; they were 0.104 m^2 for the evaporation zone and 0.219 m^2 for the condensation zone. Then, the heat flux densities for these zones were 8.58 kW/m^2 and 3.41 kW/m^2, respectively.

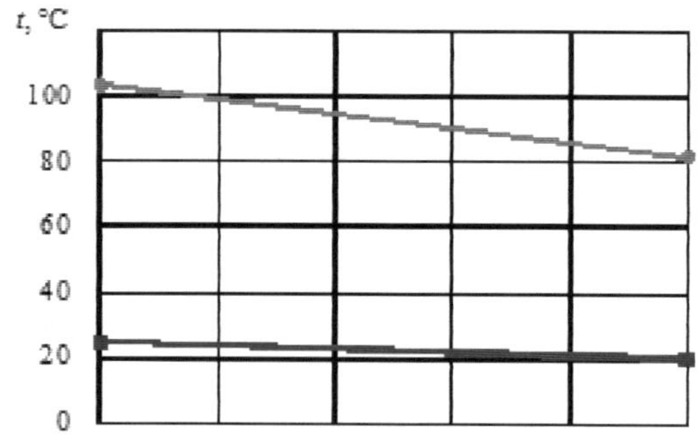

Fig. 2. Temperature change along the length.

4 Conclusion

It can be concluded that thermosyphon operation should be studied more accurately and comprehensively, the optimal degree of working medium filling in the tubes should be determined, and the lengths of the zones of freon boiling and condensation should be found.

Further, it is expected to carry out temperature measurements on the thermosyphon walls in the lower and upper parts aimed at determination of air heat transfer coefficient, temperatures of freon boiling and condensation by different methods using the data of numerical and experimental studies of other authors [1-6]. Thermal resistance, which is the main criterion for effective work of thermosyphons, will be determined, and the main contribution is made by the evaporation zone [6]. It is known that the heat transfer characteristics of thermosyphons are affected by many factors: the amount of charged coolant, its thermal-physical properties, lengths of the zones of heating (evaporation) and condensation and cooling conditions in the condensation zone. The large number of influencing factors poses the problem of finding the optimal design for the cooling system on the basis of the two-phase closed thermosyphons, which is important for the specific conditions of their application [4].

The developed model of experimental setup will be the prototype of a heat exchanger, which can be used to cool various equipments. In particular, to cool the modular automated control systems for electric calorific installations for supplying air to the mines, where the refrigeration units with the cost of 2 mln. rubles each are used today.

References

1. M.K. Bexrodnyi, I.L. Pioro, T.O. Kostyuk, *Transfer processes in two-phase thermosyphon systems. Theory and practices*, (Fakt, Kiev, 2005)
2. Al-Ani Maaz Abdulwahed Zib, *The features of hydrodynamics and heat and mass transfer in thermosyphons for application in thermal-power equipment (*Tomsk, 2011)
3. G.V. Kuznetsov, M.A. Al-Ani, M.A. Sheremet, Bul. TPU **318**, 4 (2011)
4. V.Yu. Kravets, V.I. Konshin, N.S. Vaneeva, , East. Eur. J. Ent. Tech. **2**, 5 (2014)
5. A.S. Krasnoshlykov, G.V. Kuznetsov, Int. En. Syst. **1** (2015)
6. V.Yu. Kravets, V.I. Konshin, E.N. Pismennyi, Mod. Inf. El. Tech. (2014)

TWO-DIMENSIONAL PROBLEM OF TURBULENT NATURAL CONVECTION IN A SEMI-OPEN CAVITY WITH RADIANT HEATING OF INTERNAL BOUNDARIES

Alexander Nee[1,*], and *Liliya* Valieva[1]

[1]National Research Tomsk Polytechnic University, Tomsk, Russia

Abstract. Mathematical modelling of turbulent natural convection in a semi-open cavity with a heat-conducting walls of finite thickness with radiant heating of internal boundaries was performed. Two-dimensional problem of the conjugate heat transfer was solved by means of the finite difference method. Scale influence of open boundaries and radiant heating of the gas – wall interfaces on the formation of differential and integral heat transfer characteristics was established. An increase in the dimensionless time (τ) led to displacement of extremum temperatures in the typical cross section (Y = 0.5) to an open vertical boundary. The average Nusselt number monotonically increased at the gas - a wall interfaces in a range of $400< \tau <1200$.

1 Introduction

It is known [1] that the convective heating of air masses is appropriate only for heating of small premises. In the case of large industrial buildings, where only a small part of the useful area (local working areas) is used prospectively to apply radiant heating systems based on gas infrared emitters (GIE) [2,3]. However, the use of radiant heating systems is currently not widespread due to the lack of a basic theory of the processes occurring when operating the GIE. For this reason, it is advisable to investigate the basic laws of heat transfer under conditions of radiant energy supply to the heating facilities. To solve these complex problems of heat transfer is prospectively to use the methods of mathematical modelling based on the solution of mass, momentum and energy transfer equations in the conjugate formulation.

It should be noted that the main specificity of the heat transfer process in the large-scale areas is turbulent flow regime. At the same time, the working zones of industrial facilities are locally distributed along the useful building area. In these cases, thermal regime modelling of locally based working area is of interest.

The purpose of this study is mathematical modelling of turbulent natural convection in a semi-open cavity under conditions of conjugate heat exchange and radiant heating of the internal boundaries.

* Corresponding author: nee_alexander@mail.ru

2 Problem formulation and solution method

Two-dimensional problem of turbulent natural convection was considered. Solution domain was presented as a cavity of rectangular cross section with an open vertical boundary. It was assumed that the radiant flux was uniform distributed along the bottom horizontal and vertical gas – wall interfaces. Turbulent flow was calculated by means of the large eddy simulation method [4]. In order to calculate the subgrid scale viscosity, I used a classical Smagorinsky model [5]. It was assumed that the thermal characteristics of the air and the walls were temperature-independent. Conditions of temperature and heat flux equalities were set at the gas - wall interfaces. Thermal insulation conditions were accepted at the external boundaries. I considered the model of a viscous gas in the Boussinesq approximation. "Mild" condition was set for the open boundary.

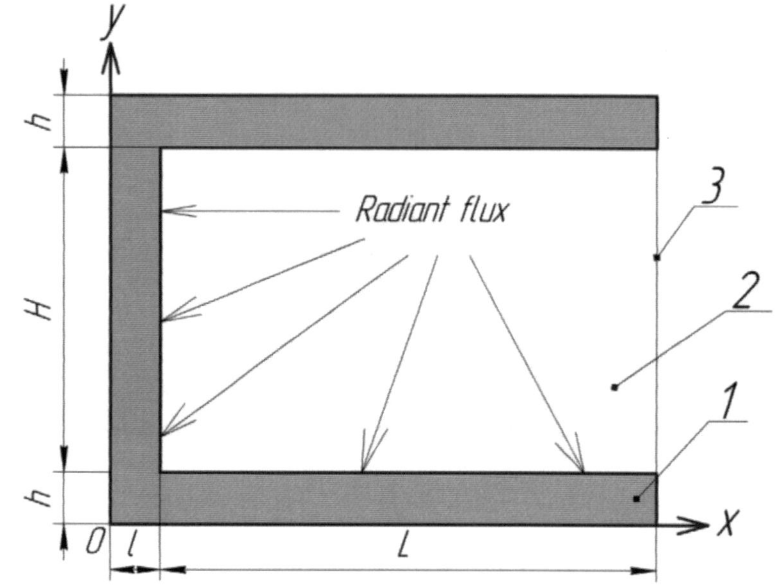

Fig. 1. Solution domain: 1 – thermally conductive wall; 2 - air; 3 - open boundary.

The turbulent natural convection and conjugate heat transfer process under study is described by unsteady two-dimensional Navier - Stokes and energy equations. These equations in vorticity - stream function - temperature dimensionless variables are as follows [4]:

$$\frac{\partial\Omega}{\partial\tau}+U\frac{\partial\Omega}{\partial X}+V\frac{\partial\Omega}{\partial Y}=\frac{\partial^2}{\partial X^2}\left[\left(\sqrt{\frac{\mathrm{Pr}}{Ra}}+\nu_t\right)\Omega\right]+\frac{\partial^2}{\partial Y^2}\left[\left(\sqrt{\frac{\mathrm{Pr}}{Ra}}+\nu_t\right)\Omega\right]+\frac{\partial\Theta_1}{\partial X}, \quad (1)$$

$$\frac{\partial^2\Psi}{\partial X^2}+\frac{\partial^2\Psi}{\partial Y^2}=-\Omega, \quad (2)$$

$$\frac{\partial\Theta_1}{\partial\tau}+U\frac{\partial\Theta_1}{\partial X}+V\frac{\partial\Theta_1}{\partial Y}=\frac{\partial^2}{\partial X^2}\left[\left(\frac{1}{\sqrt{\mathrm{Ra}\cdot\mathrm{Pr}}}+\frac{\nu_t}{\mathrm{Pr}_t}\right)\Theta_1\right]+\frac{\partial^2}{\partial Y^2}\left[\left(\frac{1}{\sqrt{\mathrm{Ra}\cdot\mathrm{Pr}}}+\frac{\nu_t}{\mathrm{Pr}_t}\right)\Theta_1\right], \quad (3)$$

$$\frac{\partial\Theta_2}{\partial\tau}=Fo\left(\frac{\partial^2\Theta_2}{\partial X^2}+\frac{\partial^2\Theta_2}{\partial Y^2}\right). \quad (4)$$

Subgrid scale eddy viscosity was calculated as in [5]:

$$\nu_t = (C_S \cdot \Delta)^2 \cdot \sqrt{2 \cdot S_{ij} \cdot S_{ij}},\ S_{ij} = \frac{1}{2}\left(\frac{\partial U}{\partial Y} + \frac{\partial V}{\partial X}\right).$$

The initial conditions for the equations (1) - (4) are as follows:

$$\Psi(X,Y,0) = 0;\ \Omega(X,Y,0) = 0;$$
$$\Theta_1(X,Y,0) = \Theta_2(X,Y,0) = 0.$$

Boundary conditions for equations (1) - (4) are as follows:
at the external boundary of the solution domain:

$$\frac{\partial \Theta_2(X,Y,\tau)}{\partial n} = 0.$$

at the gas – wall interfaces parallel to the X – axis:

$$\Psi = 0,\ \frac{\partial \Psi}{\partial Y} = 0,\ \begin{cases} \Theta_2 = \Theta_1, \\ \dfrac{\partial \Theta_2}{\partial Y} = \dfrac{\lambda_1}{\lambda_2} \cdot \dfrac{\partial \Theta_1}{\partial Y} + Ki_1. \end{cases}$$

at the gas – wall interfaces parallel to the Y – axis:

$$\Psi = 0,\ \frac{\partial \Psi}{\partial X} = 0,\ \begin{cases} \Theta_1 = \Theta_2, \\ \dfrac{\partial \Theta_1}{\partial X} = \dfrac{\lambda_2}{\lambda_1} \cdot \dfrac{\partial \Theta_2}{\partial X} + Ki_2. \end{cases}$$

at the open boundary:

$$\frac{\partial \Theta}{\partial X} = 0,\ \frac{\partial \Omega}{\partial X} = 0,\ \frac{\partial \Psi}{\partial X} = 0,$$

where Fo – Fourier number; Ki – Kirpichev number; Pr– Prandtl number; $\Pr_t$ – turbulent Prandtl number; Ra – Rayleigh number; U, V– dimensionless velocities corresponding u,v; X,Y – dimensionless coordinates corresponding x,y; τ – dimensionless time; Θ – dimensionless temperature; Ψ - dimensionless analogue of stream function; Ω– dimensionless analogue of vorticity; C_S – empirical constant (Smagorinsky constant); Δ – filter width. Indices: 1,2 – elements of the system.

Formulated boundary value problem was solved by means of the finite difference method analogically [6, 7]. Implicit two-layer scheme [8] was applied for equations (1) – (4) approximation. One-dimensional finite-difference analogues of differential equations were solved by the sweep method [9].

In order to validate mathematical model, solution method and algorithm used were tested on a model problem of turbulent natural convection [10, 11] in closed square cavity. Comparison of obtained temperature fields and average Nusselt numbers with [10, 11] showed their satisfactory agreement.

3 Results and discussion

Investigations of conjugate natural convection regimes were conducted for the following dimensionless criteria corresponding to the turbulent flow: $Ra = 10^8$, $\Pr_t = 1$, Pr=0.71, $Ki_1 = 42.86$, $Ki_2 = 21.43$. fig. 2 presents typical fields of temperature and stream function illustrating the influence of the non-stationarity factor and presence of the open boundary on differential heat transfer characteristics formation in the solution domain under consideration.

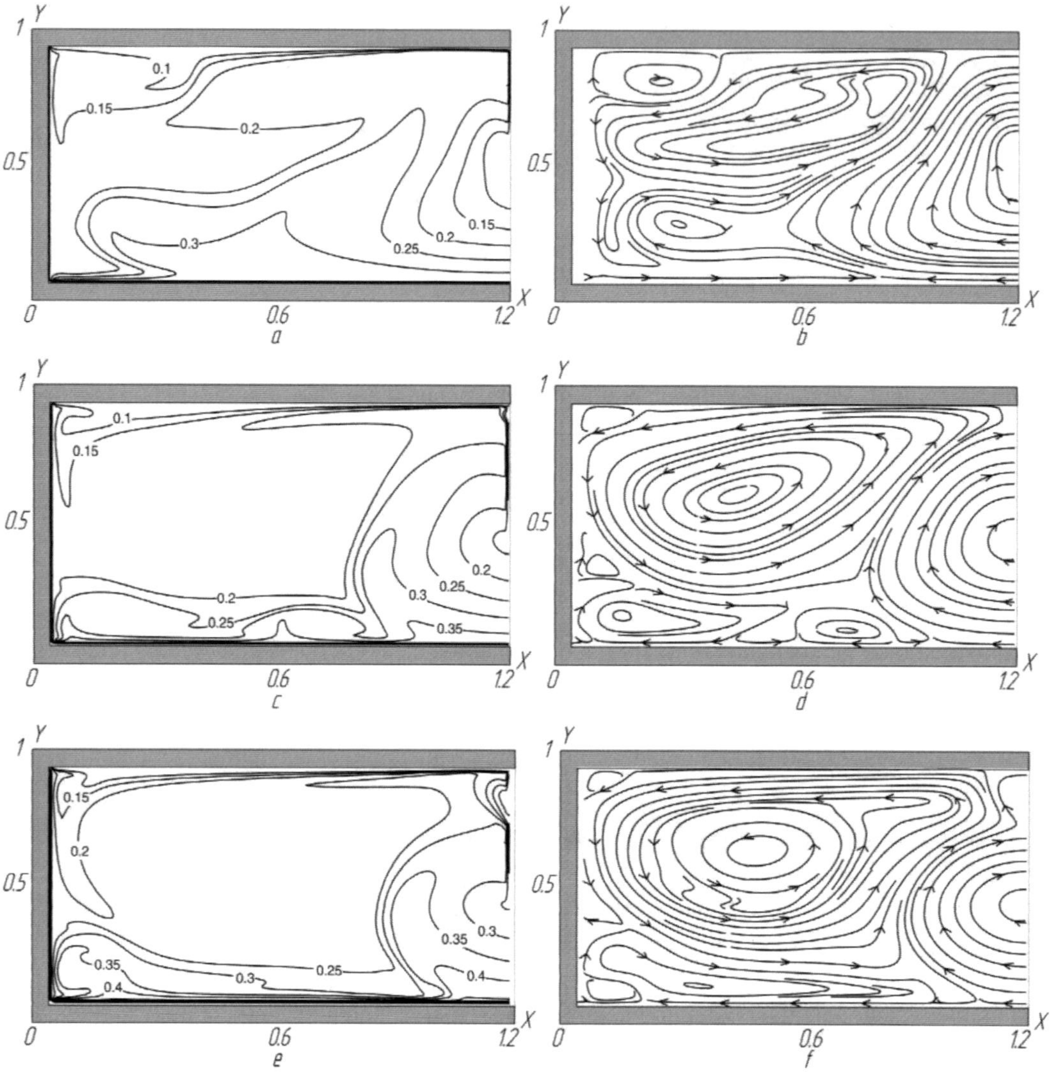

Fig.2. Fields of temperature (a, c, e) and stream function (b, d, f) at $Ra = 10^8$:
a, b) $\tau = 300$; c, d) $\tau = 600$; e, f) $\tau = 1200$.

Based on the analysis of the presented fields of differential heat transfer characteristics (fig. 2), it can be concluded that the thermophysical process under study had significant unsteady nature. When $\tau = 300$, a large-scale convective cell in the center of the air cavity and a small circulation flow were formed in the solution domain, which was obviously due to the radiant heat supply to the air – wall interfaces and the presence of the open boundary. An increase in the dimensionless time to $\tau = 600$ led to a rise in the temperature of the bottom horizontal interface. Large-scale convective cell in the center of the domain displaced small circulation flow in the left top and bottom corners of the cavity. The isotherms were shifted to the open boundary and followed the shape of the streamlines in the area of 0.9 <X <1.2, 0.1 <Y <0.75. Further increase in the dimensionless time to $\tau = 1200$ did not lead to significant modification of the differential heat transfer characteristics in geometrical and physical conditions under consideration.

Fig. 3 shows the temperature distribution in the gas cavity in the Y = 0.5 section.

 DOI: 10.1051/matecconf/20167201120

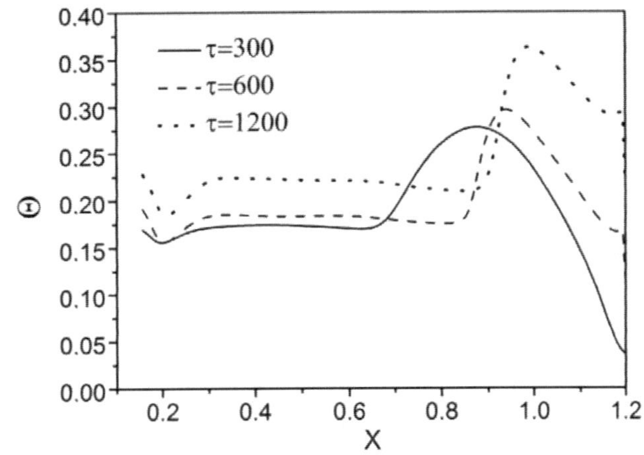

Fig. 3. The temperature distribution in the gas cavity in the Y = 0.5 section at $Ra = 10^8$.

It was clearly seen that an increase in dimensionless time led to a rise in the temperature in the typical cross-section of Y=0.5. Extremum of temperature was shifted to the open boundary, which was obviously due to the air flow pattern in this area (fig. 2 b, d, f).

In order to estimate the heat transfer rate, integral heat exchange characteristics were calculated. Dimensionless heat exchange coefficients at the top (Nu_{av1}), bottom (Nu_{av2}) and vertical (Nu_{av3}) gas – wall interfaces are defined as in [8]:

$$Nu_{av1} = \int_{0,2}^{1,2} \left|\frac{\partial\Theta}{\partial Y}\right|_{Y=0,1} dX; \quad Nu_{av2} = \int_{0,2}^{1,2} \left|\frac{\partial\Theta}{\partial Y}\right|_{Y=0,9} dX; \quad Nu_{av3} = \frac{1}{0,8}\int_{0,2}^{1,2} \left|\frac{\partial\Theta}{\partial X}\right|_{X=0,1} dY.$$

Figure 4 shows the dependence of the average integral Nusselt numbers versus dimensionless time.

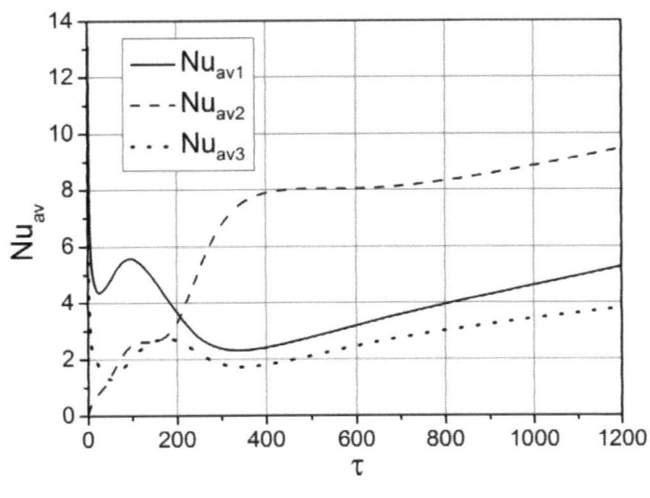

Fig. 4. Dependence Nu_{av} on τ when $Ra = 10^8$.

The obtained dependences (fig. 4) of the integral heat transfer characteristics illustrated the unsteady nature of the conjugate heat transfer process under study. The average Nusselt number at the bottom horizontal (Nu_{av1}) and vertical (Nu_{av3}) gas – wall interfaces had a nonlinear dependence before $\tau = 300$, which was connected with the radiant energy supply to these interfaces. The air layer was formed with an increase in temperature in the sections of X = 0.1 and Y = 0.1 (fig. 2 a, c. e) in a range of $1 < \tau < 40$. As the result, Nu_{av1} and Nu_{av3} reduced. Hereafter (before $\tau = 100$), the average dimensionless heat exchange coefficients at the bottom horizontal and vertical gas - a wall interfaces increased, which was connected with the formation of convective plumes as the result of natural convection. The average integral criteria Nu_{av} monotonically increased in a range of $400 < \tau < 1200$, which was obviously due to a rise in the circulation gas flow rate. At the same time, the thermophysical process under study gradually entered the quasi-stationary mode. The average temperature increased in the solution domain, but the shape of the isotherms did not substantially modify.

4 Conclusion

An approach for the modelling of turbulent natural convection in the semi - open cavity under conditions of the radiant heating of gas – wall interfaces was suggested. According to the numerical simulation results, scale influence of open vertical boundary and radiant heating of the gas – wall interfaces on the formation of differential and integral heat transfer characteristics was established. The formulated boundary value problem allows to decrease the spent of computing resources by reducing the modelling area size when the analyzing the thermal regimes of systems heated by infrared emitters.

Acknowledgments

The reported research was supported by Russian Federation President Grant for state support of the Russian Federation leading scientific schools SS-7538.2016.8.

References

1. K.-N. Rhee, K. W. Kim, Build Environ **91** (2015)
2. Buhmirov, S. A. Krupennikov, Yu. S. Solnyshkova, Her. Ivan. St. Pow. Un. **4** (2010)
3. G.V. Kuznetsov, V.I. Maksimov, T.A. Nagornova, N.I. Kurilenko, G. Ya. Mamontov, J. Engin. Ph. Thermoph. **86**, 3 (2013)
4. Sh. Chen, X.-h. Huang, Ch. Yang, G.-f. Liu, C.-j. Ding, B. Han, Appl. Math. Mech. - Engl. Ed. **34** 11 (2013)
5. H. Sajjadi, R. Kefayatib, Therm. Sci. **19**, 1 (2015)
6. G.V. Kuznetsov, P.A. Strizhak J. Eng. Thermophys-Rus. **17**, 3 (2008)
7. G.V. Kuznetsov, T.A. Nagornova, A.E. Ni, JEPT. **88**, 1 (2015)
8. G.V. Kuznetsov and M.A. Sheremet, Int. J. Heat Mass Transf. **52**, 1-2 (2009)
9. P. J. Roache, *Computational Fluid Dynamics* (Hermosa Publishers, Albuquerque, 1976)
10. H.N. Dixit, V. Babu, Int. J. Heat Mass Transf. **49** (2006)
11. N.C. Markatos, K.A. Pericleous, Int. J. Heat Mass Transf. **27** (1984)

IGNITION OF THE SOARING DROPLET SETS OF WASTE-DERIVED COAL-WATER SLURRY WITH PETROCHEMICALS

Timur R. Valiullin[1,*], *Margarita* A. Dmitrienko[1], *Sergey* A. Shevyrev[1], and *Roman* I. Egorov[1]

[1]National Research Tomsk Polytechnic University, 634050 Tomsk, Russia

Abstract. We have analyzed the ignition of droplet sets of waste-derived coal-water slurry with petrochemicals for the case of their soaring inside special combustion chamber. The fuel composition consists of filter cake of bituminous coal type G, waste turbine oil, water and plasticizer. Features of the ignition process were emphasized for groups of three soaring droplets in comparison with single droplet ignition. The ignition delay times were registered for particles that were deformed or segregated due to the interaction of initial fuel droplets with walls of the combustion chamber.

1 Introduction

The large scale usage of the oil-products in transportation, chemical industry, heat and electricity production, etc. leads to production of the huge amount of different types of wastes (like waste engine oils or oil sludge). Its amount is quickly growing year per year and nowadays it achieves millions tons per year. [1–3]. Together with this the coal-processing wastes breaks records and now it is near billion tons per year. The accumulation of these wastes is an evidently sharp and still growing problem for whole humanity. The development of the optimal method of the effective utilization of such wastes is actual task for different countries. The waste-derived coal-water slurries with petrochemicals (CWSP) become now the perspective type of alternative fuels for industrial heat production [3]. It can produce enough much heat energy together with strong decrease of waste production for different branches of world industry.

There are some well-known approaches for investigations of the peculiarities of initialization of stable combustion [4-7] that typically used for investigation of the composite fuels. The most popular is introduction of the single fuel droplets into the combustion chambers or high-temperature air channels by usage of the different holders like quick-response thermocouple junction, the ceramic thread or thin metal wire.

The main integral parameters of ignition and combustion of fuel suspensions or at least droplet sets have to be investigated for further effective usage of the CWSP at the industrial heat-producing units. The main way of the industrial usage of the CWSP is injection of the

[*] Corresponding author: radisovich@yandex.ru

fuel aerosol flow to the combustion chamber through the nozzle. Therefore the influence of the droplet interaction with walls of combustion chamber and with other droplets, its coagulation, segregation etc. becomes an important factor determining the parameters of the ignition and combustion processes. As result, we offer the method allowing the investigation of ignition of fuel droplet sets as well as single droplets in case of its free soaring in the hot oxidant flow (air) inside the special combustion chamber [8]. The presented approach doesn't need any holders for fuel droplets and allows observations of the combustion process as it is inside industrial chambers.

The aim of this work is determination of the main integral parameters of ignition of the droplet sets (three CWSP droplets in each) together with comparison of these data with corresponding to ignition of the single droplet in soaring regime.

2 Experimental procedure

The experimental setup and the used method are similar to described in [8]. Determination of the integral parameters of the single fuel droplets and droplet groups in soaring regime was done using special combustion chamber. It was made of heat-resistant quartz glass specially developed for optical measurements of the combustion processes. The heating and pumping system allows setting and keeping of the temperature and the velocity of the air (oxidant) flow in ranges 600–900 K and 0.5–3 m/s correspondingly. The flow temperatures at different point of combustion chamber were controlled by two chromel-alumel thermocouple (working range 273–1373 K, bias ± 3 K, response time ~3 s). All the experiments were done at random variations of the initial flow temperature less than 5 K and for velocity less than 0.1 m/s.

Unlike to experiments presented at [8] the introduction of three fuel droplets into the chamber (and their synchronized ejection) was done by specially developed device (fig. 1). It consists of three conjugated similar elements (fig. 1a). The sub-elements were connected at an angle to one another together with small lateral shift to simplify the recharge by the fuel. The CWSP droplets were created by the batcher Finnpipette Novus. Its size was additionally checked before the introduction to combustion chamber by video frames analysis at Tema Automotive software package.

The synchronized droplet ejection was done by motorized system conjugated with system of fuel introduction into the combustion chamber. Cutting element - 3 ejects the droplet (fig. 1a) from the nichrome holder - 2 when it achieves the center of the combustion chamber.

The analysis of the video records of the soaring droplet combustion was done using special abilities of the Tema Automotive package allowing continuous tracking of the moving objects and checking of their brightness to determine the ignition delay times for each fuel droplet (τ_d) like it was done in [8].

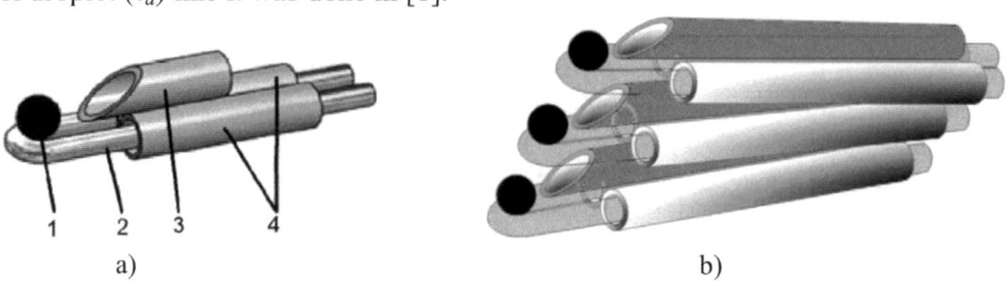

a) b)

Fig. 1. Scheme of the droplet ejection device. The single droplet ejection module (a). There are: 1 – the CWSP droplet; 2 – thin nichrome wire; 3 – cutting element; 4 – metal guiding pipe. The synchronized ejection device for three fuel droplets (b) consists of three modules presented at (a).

The main component of the fuel composition is filter cake of the bituminous coal type "G" which appears during the floatation enrichment of the coals. Nowadays, there are more than 10 coal-enrichment factories at Siberia region produces hundred thousands of filter cakes per year. The filter cake incorporates 40-50% of the solid coal matter and 50-60% of water. The second combustible component of the CWSP fuel is waste turbine oil. The "Neolas" plasticizer was used for stabilization of the fuel composition for long term storage (1-2 weeks) without essential lamination of the slurry. The component ratio of the chosen CWSP is shown in table 1.

Table. 1. The CWSP content (relative mass ratios of the components).

Coal, %	Liquid combustible, %	Water, %	Plasticizer, %
Filter cake of coal type "G"	Waste turbine oil		
50.5	10	39	0.5

3 Results and discussion

The changes of the ignition delay times (τ_d) with oxidant flow temperature are shown at fig. 2. Each of presented curves is approximation of the measured points obtained during the set of experiments. The typical errors are less than 10 % of the measured values.

The differences of the ignition delay times of three droplets inside the certain set are near 1.5 s that could be explained by their random interaction with walls of the combustion chamber. The collisions between droplets were relatively rare due to its low density inside the chamber volume and therefore didn't make essential influence onto the ignition parameters. However, this effect should be taken into account for real combustion chambers which works with the highly dispersed fuel droplet flows.

The droplets inside the chamber fly along different and enough complicated trajectories that leads to random oscillations of the measured ignition delay time due to non-uniform temperature distribution inside. The experimental data analysis makes clear that the temperature dependence of the ignition delay for low density droplet flow is present in between the curves 1 and 4 (the shaded are at fig. 2a).

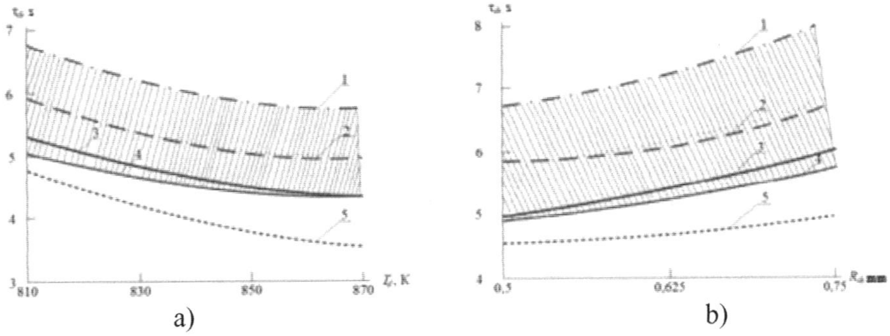

Fig. 2. The dependences (a) of the CWSP ignition delay time with oxidant flow temperature at R_d=0.5 mm and V_g=3.5 m/s. The ignition delay time changes with CWSP droplet size (b) at T_g=813 K. Curves 1, 2, 4 corresponds to ignition of the first, second and third droplets in order of ignition events. The curve 3 corresponds to ignition of the single droplet inside the chamber. The curve 5 corresponds to ignition of typical small secondary droplets segregated from one of the initial set due to collision with chamber walls.

The experiments have demonstrated that droplets collisions with chamber walls can produce small segregated droplets. These fuel particles ignite evidently faster than initial big droplets that could be valuable for real combustion chambers with massive droplet flows decreasing total ignition inertia. These effects could be the subject for future investigation tasks.

The fig. 2b presents the dependences of the ignition delay times for CWSP droplets on droplet sizes. The variations of the droplet radius for 15 % relatively median value ~0.62 mm lead to changes of the τ_d for 0.5–1 s. This median size is close to typically used at industrial combustion chambers. The demonstrated changes of the τd, caused by variations of the droplet size are very similar to thermally induced ones shown at fig. 2a.

Together with all mentioned, the secondary droplets appearing during the destruction of the initial ones, has ignition delay times which almost don't change with initial particle sizes (fig. 2b). Most probably it is possible to make fuel slurries that effectively ignore the random temperature oscillations inside the combustion chamber, using an optimal ratio of the droplet sizes inside the injected fuel flow. Such droplet flows will have more stable ignition and following combustion.

The main result of the investigations is definition of the ranges of the ignition delay times for sets of the droplets which allows the interpolation of obtained data to the case of low-density fuel droplet flows. The mutual arrangement of the shaded ranges at the fig. 2 and the dependencies for ignition parameters of single droplets illustrates weak co-interaction of the droplets inside the set and its influence onto the ignition processes. This conclusion is important for justification of practical applicability of the approach with single CWSP droplet ignition for generalization of the results to the case of processes inside full-scale combustion chamber.

3 Conclusion

1. The ignition delay times inside the set of the fuel droplet have differences ~0.5–1 s with median value ~6 s at observed oxidant flow temperature range. The integral characteristics of ignition of single droplet well correlate with this range. This result illustrates that prediction of the ignition characteristics of droplet set could be effectively done by observation of single droplet ignition.
2. The analysis of influence of the CWSP droplet sizes onto the ignition delay times shows relatively small variation of the ignition delay for certain range around industrially appropriate median size (0.6mm). It allows to make the droplet flows with stable integral parameters of the ignition by optimal mixing of different size droplet fractions.
3. The relatively small ignition delay times detected for small secondary droplets, which always appear from the initial ones inside real chambers, allows estimating the lower threshold of ignition delay at 4–5 s for different sizes of the initial droplets.

Acknowledgments

This work with ignition processes of CWS supported by the Russian Science Foundation, project 15–19–10003. The environmental impact of CWS was investigated within the program of National Research Tomsk Polytechnic University Competitiveness Enhancement in the Group of Top Level World Research and Academic Institutions (project VIU_ENIN_25_2016).

References

1. O.G. Chayka; O.Z. Kovalchuk; Y.A. Chayka, *Monitoring the formation of waste oils. Proceedings Scientifical works* (Lviv Polytechnic, Lviv, 2009) [in Ukrainian]
2. G. Hu, G. Zeng, J. Hazard. Mater. **261** (2013)
3. A.E. Kontorovich, M.I. Epov, L.V. Eder, Russian Geol. Geophys. **55**, 5–6 (2014)
4. W. Gajewski, A. Kijo-Kleczkowska, J. Leszczynski, Fuel **88**, 2 (2009)
5. A. Kijo-Kleczkowska, Fuel **90**, 2 (2011)
6. D.O. Glushkov, D.P. Shabardin, P.A. Strizhak, K.Yu. Vershinina, Fuel Process. Technol. **143** (2016)
7. D.O. Glushkov, P.A. Strizhak, K.Yu. Vershinina, Appl. Therm. Eng. **96** (2016)
8. T.R. Valiullin, P.A. Strizhak, S.A. Shevyrev, Therm. Sci. Online First. 10.2298/TSCI151215221V

MATEC Web of Conferences 72, 01122 (2016) DOI: 10.1051/matecconf/20167201122
HMTTSC-2016

ESTIMATES OF TWO-PHASE FLOW PARAMETERS IN THE CYCLONE CHAMBER

Aleksandr Razva[1,*], *Michail* Vasilevsky[1], *Alice* Pleshco[1], and *Alibek* Asan[1]

[1]National research Tomsk polytechnic university, 634050 Tomsk, Russia

Abstract. It providesanalysis of methods for assessing circumferential speeds distribution of pressure in the cyclone chamber. It is shown that for cylindrical chambers of less than two diameters and the inlet section of at least 0.1 square cross section the maximum peripheral speed and rate distributions circumferential speeds along the current line is not changed. Analysis of the distribution shows that the concentration of small particles that are typical for gas dedusting systems have a significant impact on the distribution of pressure, vacuum and on the axis of the absolute value decreases toward dust of output section unlike the swirl chambers fuel combustion.

1 Introduction

The cyclone chamber with solid particles are widely used in various fields of engineering and technology for the intensification of physical-chemical processes. The critical parameter defining intensity the warmly-mass-exchange of processes and the separation of effects in the swirl chamber is level rotary components of speed of gas and speed of particles The existing methods based on sounding of fields of speeds are [1] labor-consuming and demand introduction of the adjusting coefficients for elimination of influence of a probe on the high-speed field of a flow. Furthermore, they are not suitable for studying swirling two-phase flow. Experience the definition of speeds dust swirling flows cylindrical probe showed unsatisfactory reproducibility of the measurement due to clogging of the probe channels [2]. Therefore, the preferred are more reliable methods that do not require adjustments and are suitable for estimating the parameters of dust swirling flows [3]. The basic geometrical relationships influencing the efficiency and flow resistance of the cyclone chamber, the relationship to the inlet area of the planned area of the cyclone, the ratio of the inlet area to the area gas outlet pipe, the area ratio dust of output holes to the total surface of the chamber.

On the characteristics of flow in cyclones with different geometries can get an idea if we use the data [4, 5], where there are fields of distribution of flow rates in the volume of the cylindrical and conical cyclones. The available numerous pilot studies of distribution of fields of speeds allow to draw a conclusion that in a counter flow cyclone there is an averaging of the moments of speeds and at the relations of the areas of an entrance and an exit less unit distribution on height of radial gradients of the moments of speeds changes

* Corresponding author: razva@mail.ru

slightly. Moreover, the maximum radius of base peripheral speed and the maximum speed of zero radius adjustment for the cylindrical cyclone same [1, 5]. In [1] shows the results of the study chamber length affect the pressure loss of circulation and distribution. Moreover, significant changes in the length of the chamber is not observed, whereas the total level of circumferential speeds within the chamber and the pressure on the wall decreases. This indicates that in the chamber are averaged over the length of the radially circumferential speed distributions due maximum radial and turbulent transfer pulses and angular momentum. The pressure on the wall relative to the pressure input is determined not only vortex design features, but also the relative size of the outlet chamber and a relative length that accounts for the effect of injection indirectly effects in the central area, which extends up to the input splines [4]. Effect of generating turbulence in the forced vortex spreads all over, but the R1/R2<0,2 flow deceleration rotation slightly in the free vortex and circumferential speeds distribution of a free vortex in a potential close to [1].

2 The calculated ratio

The rotational component is determined from the ratio $V^2 = R/\rho \cdot dp/dR$, where ρ – density of the gas, R– the current range, V – circumferential speed of the gas. One of the parameters defining the whirling stream is the intensity of the rotational motion $\Delta p_0 = p(R_2) - p(0) = \rho\int_0^1 (V^2/\bar{r})d\bar{r}$, where $\bar{r}$=R/R2, R2– camera radius [1].

Radial pressure drop in a particular section is determined from the equation of motion in the radial direction.

$$V = V_\Delta\left(\frac{R_2}{r}\right) \text{ при } R_2 > r >; V = V_\Delta\frac{r}{r_m}\left(\frac{R_2}{r_m}\right)^n \text{ at } r < r_m \tag{1}$$

$$\Delta P = \int_0^{R_2}\rho\frac{V^2}{r}dr = \int_0^{r_m}\rho\frac{V^2}{r}dr + \int_{r_m}^{R_1}\rho\frac{V^2}{r}dr + \int_{R_1}^{R_2}\rho\frac{V^2}{r}dr \tag{2}$$

here r_m– the maximum radius of the circumferential speed $V = \upsilon_m$ in particular section, R_1– the radius of the exhaust pipe, V_Δ– circumferential speed at a distance $0{,}03R_2$ from the wall of the outer cylinder of radius R_2. These equations allow to calculate the value of the exponent, the radius rm according to the differential pressure on the wall of the exhaust pipe and the axis of the periphery and axis. At the top of the cyclone pressure on the surface of the radius r_m unknown, it must be determined.

$$\overline{\Delta P}_{0-m} = \frac{\Delta P_{0-m}}{\rho\frac{V_\Delta^2}{2}} = \left(\frac{R_2}{r_m}\right)^{2n}; \qquad \overline{\Delta P}_{1-2} = \frac{\Delta P_{1-2}}{\rho\frac{V_\Delta^2}{2}} = \frac{1}{n}\left[\left(\frac{R_2}{R_1}\right)^{2n} - 1\right], \tag{3}$$

$$\overline{\Delta P}_{0-m} + \overline{\Delta P}_{m-1} = \left(\frac{R_2}{r_m}\right)^{2n} + \frac{1}{n}\left[\left(\frac{R_2}{r_m}\right)^{2n} - \left(\frac{R_2}{R_1}\right)^{2n}\right] = \overline{\Delta P}_{0-m}\left(1+\frac{1}{n}\right) -$$

$$-\frac{1}{n}\left(\overline{\Delta P}_{1-2}\cdot n + 1\right) = \Delta\bar{P}_{0-2} - \Delta\bar{P}_{1-2}$$

from whence

$$\overline{\Delta P}_{0-m} = \frac{n\overline{\Delta P}_{0-2} + 1}{1+n} \tag{4}$$

In these formulas ΔP_{0-m}, ΔP_{1-2}– pressure differentials between the axis and a surface with a radius r_m, and between the radius surface R_1 and the outer wall of the upper part of the cyclone, respectively. Using the first relation (3) is determined by the position of the radius r_m:

$$\frac{R_2}{r_m} = \left(\overline{\Delta P}_{0-m}\right)^{\frac{1}{2n}} \tag{5}$$

In the second equation (3) is numerically define n (table 1). At n>1 relative pressure differences indicated in the table, for $\overline{R}_1$>0.4 impossible.

Table 1. Values *n*.

$\overline{R}_1$	$\overline{\Delta P}_{1-2}$			
	1	3	6	9
0.3	–0.9	0.18	0.7	0.93
0.4	–0.45	0.5	>1	>1
0.6	–0.1	1	>1	>1

3 Experimental setup and measurement

Figure 1 shows the pressure measurement circuit in a cylindrical cyclone 1. Consumption was measured using a pitot tube. For the measurement of small pressure differences used micromanometer. Pressure measured using an axis movable tube outer diameter of 3 mm with drilled holes, which could be isolated. The selection pressure on the periphery of the cyclone was conducted through nozzles communicating with the holes in the cabinet. The diameter of body 100 mm in diameter clean air outlet - 34 mm, pressure in the range of selection of the cyclone cover the surface of the exhaust pipe r_{pr}=20 mm (figure 1), reflector diameter at the bottom of the cylindrical cyclone – 43 mm. F_{in}= 22x50 mm^2, H=200 mm. Distance calculation conducted from dust outlet sections.

Used micropowder M40. The dispenser represented vibrovoronki with different sizes of output sections. The powder is loose material having a particle size of 40 microns, has good flowability.The air flow rate determined by the formula

$$G_в = \rho Q;\ Q = kF_{in}\upsilon_d;\ \upsilon_d = 10\sqrt{\frac{2\Delta p_d g}{\rho}} \approx 40.4\sqrt{\Delta p_d} \tag{6}$$

here ρ – density ρ≈1.2(g/dm^3), Fin – entrance area (dm^3), Δp_d – dynamic pressure (kgs/m^2), $\upsilon_д$ – speed is determined dm/s, $G_в$ – mass flow determined in g / s, k– Nikuradze correction.

Consumption was determined expiration time, in minutes from funnel vibro mass of material in grams $Gm = \frac{M}{60t}\left(\frac{g}{s}\right)$, the concentration of the material is determined by the formula $C = \frac{G_m}{G_v}1000\left(\frac{\text{g material}}{\text{kg air}}\right)$.

DOI: 10.1051/matecconf/20167201122

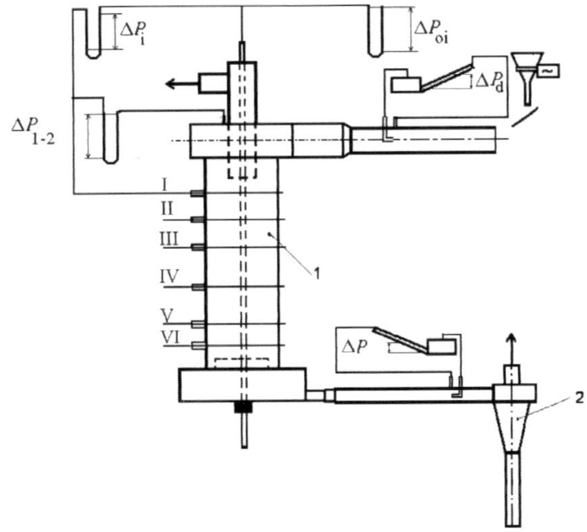

Fig. 1. Diagram of the measured pressure in the cyclone.

In order to reduce pressure fluctuations held retraction some of the air with the dust in the cyclone bypass 2. The influence of drainage on the pressure distribution.Number of exhaust air from the receiver was determined in a similar manner. The relative amount of exhaust air from the receiver determined by the formula $\overline{G}_{отв} = \frac{G_{отв}}{G_в} 100\ \%$.

Figure 2 shows the values of the pressure in the cylindrical cyclone. For clarity tags spaced in height, although they all belong to the same section.

Value Δp

№	Gв = 28 ÷ 29 (г/с)					
	$\overline{G}$ = 5%		$\overline{G}$ = 18%		$\overline{G}$ = 7.5%	
	C=0	C=6.2	C=0	C=6.2	C=0	C=13
	●	■	▲	○	□	+
I 299	170	155	175	167	170	145
II 259	160	145	170	151	160	135
III 219	150	140	170	160	155	130
IV 159	140	125	160	137	145	121
V 104	140	130	160	137	145	124
VI . 74	132	120	150	140	140	126

Value P_0

№	Gв = 28 ÷ 29 (г/с)					
	$\overline{G}$ = 5%		$\overline{G}$ = 18%		$\overline{G}$ = 7.5%	
	C=0	C=6.2	C=0	C=6.2	C=0	C=13
	●	■	▲	○	□	+
I 299	250	240	250	247	250	240
II 259	235	220	245	235	231	230
III 219	230	220	237	227	231	220
IV 159	226	220	230	220	221	220
V 104	220	210	230	217	215	215
VI 74	210	206	226	220	210	210

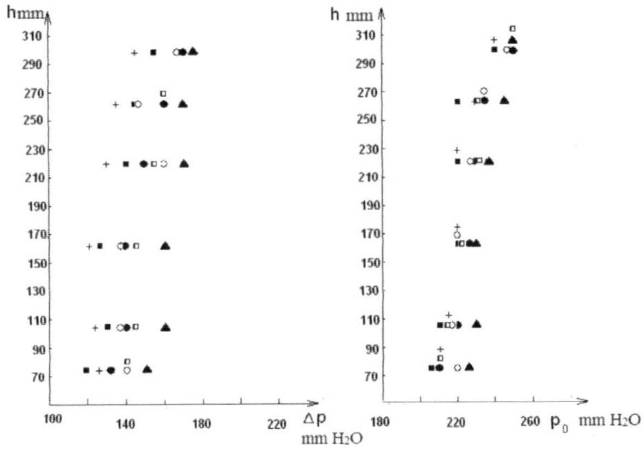

Fig. 2. The pressure distribution in the cylindrical cyclone.

Analysis of the distribution shows that the concentration of small particles that are typical for gas dedusting systems have a significant impact on the distribution of pressure, vacuum and on the axis of the absolute value decreases towards the output section of the dust, unlike vortex chambers, studied in [1]. The gradients of pressure in the axial direction of the dust-laden flow and dusty identical.

Similar distribution in the axial direction of the pressure drop. Considerable importance is the value of the exhaust air in the bypass dust precipitator. With the increase in the relative values - of exhaust dust at the bottom of the thin air on the axis increases, thinning gradients and radial gradients in the axial direction the pressure decreases. Obviously, the value of the relative velocity at the wall is dependent on the concentration of particles.

Consequently, the speed change ratio α is determined on the wall of the particle concentration.

4 Calculated values

Using (2) and (3) the following expressions can be obtained, for example, to section IV

$$\left\{\left(\frac{R_2}{r_m}\right)^{2n^{IV}}+\frac{1}{n^{IV}}\left[\left(\frac{R_2}{r_m}\right)^{2n^{IV}}-1\right]\right\}\frac{V_{\Delta IV}^2}{2}\rho=\Delta P_{0-2}^{IV} \tag{7}$$

$$V_{\Delta IV}=V_{\Delta I}\left(\frac{R_2}{r_m}\right)^{n^I}\left(\frac{R_2}{r_m}\right)^{-n^{IV}} \tag{8}$$

The calculation of the basic values of the radii and circumferential speeds produced by the formulas (1)–(5) at R_2/R_1 = 2.94; F_{in}=1.1·10^{-3} m^2; V_{in} =15.1 m/s; ΔP_{1-2} =400 Pa; ΔP_{0-2} = 1700 Pa; for the upper part of the cyclone (section I) α= 0.88; $V_{\Delta I}$=13.2 m/s. From the second formula of (3) the method of iterations we obtain for the cross section I–I value $n^I\approx$ 0.5; value $\overline{\Delta P_{0-2}}=\dfrac{1700}{1.2\dfrac{(15.1\cdot 0.88)^2}{2}}=16.3$; of the formula (4) we obtain $\overline{\Delta P_{0-m}}=\dfrac{n\overline{\Delta P_{0-2}}+1}{1+n}=\dfrac{0.5\cdot 16.3+1}{1+0.5}=6.1$; of the formula (5) we obtain R_2/r_m=6.1. Throughout this ratio is constant height. From (8) we obtain:

$$V_{\Delta IV}=V_{\Delta I}\left(\frac{R_2}{r_m}\right)^{n^m}\left(\frac{R_2}{r_m}\right)^{-n^{IV}}=13.2\cdot 6.1^{0.5}\left(\frac{R_2}{r_m}\right)^{-n^{IV}}=32.6\left(\frac{R_2}{r_m}\right)^{-n^{IV}}$$,substituting into (7) yields:

$$\left\{(6.1)^{2n^{IV}}+\frac{1}{n^{IV}}\left[6.1^{2n^{IV}}-1\right]\right\}(6.1)^{-2n^{IV}}\cdot 637.7\cdot 1.2=1400;$$

$$\left\{1+\frac{1}{n^{IV}}\left[1-(6.1)^{-n^{IV}}\right]\right\}=2.19.$$

Solving this relation, we obtain $n^{IV}\approx 0.75$.

5 Conclusion

Analysis of the distribution shows that the concentration of small particles that are typical for gas dedusting systems have a significant impact on the distribution of pressure and vacuum on the axis of the absolute value decreases towards the output section of the dust, unlike vortex chambers fuel combustion. Similar distribution in the axial direction of the pressure drop. These relations and the calculations correspond to the experimental data given in [5, 6].

Acknowledgments

The work was held within the research state assignment "Science" 13.1339.2014/K (Code of Federal Target Scientific and Technical Program 2.1410.2014).

References

1. I.I. Smulsky. *Aerodynamics and processes in swirl chambers* (IN "Science", Novosibirsk, 1992)
2. M.V. Vasilevsky, E.Y. Arzamastseva, A.S. Razva, E.G. Zykov, *Determination of parameters of a flow on distribution of pressure in the cyclonic device* (TPU, 2009)
3. M.V. Vasilevsky, V.I. Romandin, E. G. Zykov, V.A. Polyushko, A. S. Razva, Bul. TSU. Math. Mech. **23**, 3 (2013)
4. E.N. Saburov, S.V. Karpov, S. I. Ostashev. *Heat exchange and aerodynamics of the twirled flow in cyclonic devices*, (Publishing house Leningrad University, Leningrad, 1989)
5. M.Vasilevsky, A.S. Razva, A. Pleschko, I. Kadurkin, EPJ Web Conf. **110** (2015)
6. M.V. Vasilevsky, E.G. Zykov, V. S. Loginov *Ustoychivost of gas purification in the cyclonic dust collector,* (News of the Russian Academy of Sciences of the Power engineering specialist, 2005)

MATEC Web of Conferences 72, 01123 (2016) DOI: 10.1051/matecconf/20167201123
HMTTSC-2016

INTEGRAL CHARACTERISTICS OF STABILITY OF COAL-WATER SLURRIES AND COAL-WATER SLURRIES CONTAINING PETROCHEMICALS BY EVALUATION OF SEPARATION INTO LAYERS

Margarita A. Dmitrienko[1*], *Galina* S. Nyashina[1], *Ksenia* Yu. Vershinina[1], and *Sergey* Yu. Lyrschikov[1]

[1]National Research Tomsk Polytechnic University, 634050 Tomsk, Russia

Abstract. For wide use of coal and oil processing wastes in heat power industry we investigated the structural stability of coal-water slurries (CWS) and coal-water slurries containing petrochemicals (CWSP). Three methods of evaluating slurry stability were considered. The method of evaluation of separation (into layers) of the slurry was chosen to determine the stability of fuel compositions based on bituminous and brown coal, filter cakes of bituminous coals, waste motor, turbine, transformer, compressor oils, oil, oiled mixture and mazut. The experiments shown, that the addition of waste oils (with mass fraction of 10 %) to investigated fuel compositions leads to increase of slurry segmental stability is almost by 2 times.

1 Introduction

The technologies of production, transport and combustion of coal-water slurries (CWSs) [1] and coal-water slurries containing petrochemicals (CWSPs) [2–4] are urgent and modern direction of development of thermal power industry. CWS is a composite fuel that has two main components – coal (or wastes of coal processing) and water. The main components of CWSP are water, coal (or wastes of coal processing) and flammable liquid (for example, waste motor oil or oil sludge). The component base for production of CWSs and CWSPs is large and annually intensively increases. The prospects and possibility of using unclaimed coals and coal processing wastes for production of CWSs and CWSPs allow solving the issues of efficient use of resources [5]. In addition, non-energy brown coals, solid residuals of coal and oil processing can be a combustible basis for slurry fuel. These wastes pose a significant environmental risk, and their storage requires considerable expenses. For preparing various compositions of CWSPs [2–4] the wastes of oil production and refining, pyrolysis pitches of various industries, waste motor, transformer, turbine oils and other liquid organic wastes can be used. Liquid fuel components are applied in a composition of

* Corresponding author: mad2@tpu.ru

fuel slurry as for changing the ignition and combustion characteristics, and as a plasticizing and stabilizing additive.

Low rates of slurry stability values, high ash content of fuel components, sufficiently low efficiency of known reagents – plasticizers and stabilizers [6] are one of the basic difficulties for a wide use of CWSs and CWSPs in the energy industry. In terms of the relative coal density is greater than one, and the coal particle diameters are significantly larger than that of colloid, the preservation of stable structure of fuel slurry due to Brownian motion is impossible. Thus, it is difficult to prevent sedimentation of coal particles of CWSs and CWSPs.

Currently quite a lot of experimental studies [7–17] on determining coal slurries stability using different methods for its assessment were carried out. The method of evaluation of separation (into layers) of the slurry [7–10], the method of penetration of the glass rod [11–13] and translucence method [14, 15] are along the best known.

The evaluation of slurry stability by the method of layers separation consists in measuring the volume of the layer separated from the bulk during storage or transport of slurry. Thereby one of the following indicators is calculated:

1) stability index:

$$Y_1 = \frac{v_c}{v_0}, \tag{1}$$

where v_0, v_c, respectively, the volume of fuel filled into vessel (measuring cylinder) and the volume of eventually separated layer;

2) mass fraction [9]:

$$\gamma = \frac{w_2 - w_0}{w_1 - w_0}, \tag{2}$$

where w_1 – initial mass of the slurry together with the measuring cylinder; w_2 – mass of the deposited (dark) layer together with the measuring cylinder; w_0 – mass of the measuring cylinder;

3) the average segmentation rate [10]:

$$r = \frac{\alpha_1 \cdot (c_0 - c_1)}{t \cdot \alpha_0 (100 - c_0)} \cdot 100~\%, \tag{3}$$

where α_0 – initial mass of slurry; α_1 – mass of separated liquid layer; c_0 – initial coal concentration; c_1 – coal concentration in separated liquid layer; t – time of slurry holding.

The method for evaluation of CWS stability, which is based on using glass (or metal) rod (diameter d=4–5 mm, mass m=20 g) [11–13] is consists in the following. The test fuel slurry is poured into a measuring cylinder and kept for 72 hours. Then the glass rod falls spontaneously from the surface of the slurry to the bottom of the cylinder. The separation into layers for slurry with a high stability does not occur and the rod can easily reach the bottom of the cylinder. For suspension with a worse stability the rod penetrates worse the solid layer of sediment. The penetration coefficient is calculated:

$$\gamma = \frac{d}{d_t} \cdot 100~\%, \tag{4}$$

where d – the distance that the rod was moved; d_t – the maximum distance at which the rod can move. This coefficient ranges from 0 to 100 % [11, 12].

The translucence analysis method is based on the change of intensity of the transmitted light over the entire length of the vessel containing fuel slurry [14]. Spectrum of light transmittance can be continuously obtained as along the entire fuel volume, and as in one of its points. In the studies [14–17] the dispersion stability analyzer is used for the translucence analysis method. This analyzer provides a classification and quantitative characterization of the process of separation, which determines the stability of the dispersions and storage time. Such analyzers can be used to optimize processes of CWSs and CWSPs production and quality control, as well as for research.

In the present work the method of evaluation of slurry separation (into layers) [7–10] for determining the stability of CWSs and CWSPs was chosen. The investigated fuel compositions have a low viscosity and using the method of penetration of the glass rod [11–13] is impossible because the rod completely goes down to the bottom of the cylinder for all the investigated slurry compositions. And the main constraint of translucence method [14, 15] is the necessity to purchase specialized expensive equipment (dispersion stability analyzer).

The purpose of the present work is the experimental determination of integral characteristics of stability of CWSs and CWSPs using the evaluation method of slurry separation into layers.

2 Experimental procedure

The determination of slurry stability was carried out for several compositions of CWSs and CWSPs. Waste motor, turbine, transformer, compressor oils, oil ("Gerasimovskoye" oil deposit, Tomsk region, Russia), oiled mixture and mazut were used as a liquid combustible component. Filter cakes (wet residue of bituminous coal enrichment by flotation), brown coal of B2 rank (Borodino deposit of Krasnoyarsk region, Russia), bituminous coal of D rank (Gramoteinskaya mine of Kemerovo Region, Russia). We used filter cakes from coal-preparation plant of Kemerovo region (Russia): filter cake of K rank (from "Severnaya" plant); filter cake of SS rank (from "Chernigovskaya-Koksovaya" plant); filter cake of T rank (from "Kaltanskaya-Energeticheskaya" plant); filter cake of G rank (from plant of the mine named after Kirov); filter cake of D rank (from "Gramoteinskaya" plant).

2.1 The evaluation of stability index

The stability index was determined for the following compositions:

- 100 % filter cake of T rank (moisture content 39.1 %);
- 100 % filter cake of K rank (moisture content 43.5 %);
- 100 % filter cake of SS rank (moisture content 37.9 %);
- 100 % filter cake of G rank (moisture content 53 %);
- 100 % filter cake of D rank (moisture content 56.8 %);
- 90 % filter cake of K rank + 10 % waste turbine oil;
- 90 % filter cake of K rank + 10 % waste transformer oil;
- 90 % filter cake of K rank + 10 % waste motor oil;
- 90 % filter cake of K rank + 10 % oil (ρ=867 kg/m^3);
- 90 % filter cake of K rank + 10 % oiled mixture;
- 90 % filter cake of G rank + 10 % waste compressor oil;
- 90 % filter cake of G rank + 10 % mazut;
- 90 % filter cake of G rank + 10 % waste motor oil;
- 90 % filter cake of G rank + 10 % waste turbine oil.

The stability of the fuel compositions was determined during the monitoring the process of separation of slurries. Tested slurry was poured into the measuring cylinder and was kept for 7–15 days. During the keeping process the transparent layer was formed on the surface of the tested slurry (fig. 1). Volume measurement of the separated transparent layer was carried out and the stability index was determined by the formula (1).

Fig. 1. The appearance of the investigated slurries with separated water layer.

2.2 Evaluation of segmentation rate

The evaluation of segmentation rate was carried out for the following compositions:

- 50 % bituminous coal of D rank + 50 % water;
- 50 % brown coal of B2 rank + 50 % water;
- 40 % bituminous coal of D rank + 50 % water + 10 % waste turbine oil;
- 40 % brown coal of B2 rank + 50 % water + 10 % waste turbine oil.

The procedure of evaluation of segmentation rate consists of several steps:

- preparation of weighed portions of the components in accordance with a given mass fraction in a slurry composition;
- determination of the mass of measuring cylinders for the storage of slurries;
- mixing the components of a slurry by homogenizer;
- the measuring cylinders were filled up with prepared fuel slurries;
- the prepared slurries were kept in the measuring cylinders for three days;
- the separated water layer was removed from the measuring cylinder and then the mass of separated water was measured;
- the separated water was dried and the coal concentration in the separated liquid layer was determined.
- the average segmentation rate was calculated by the formula (3).

3 Result and discussion

Figures 2 and 3 illustrate the experimental dependences of stability index on time of observation for investigated compositions of CWSs and CWSPs.

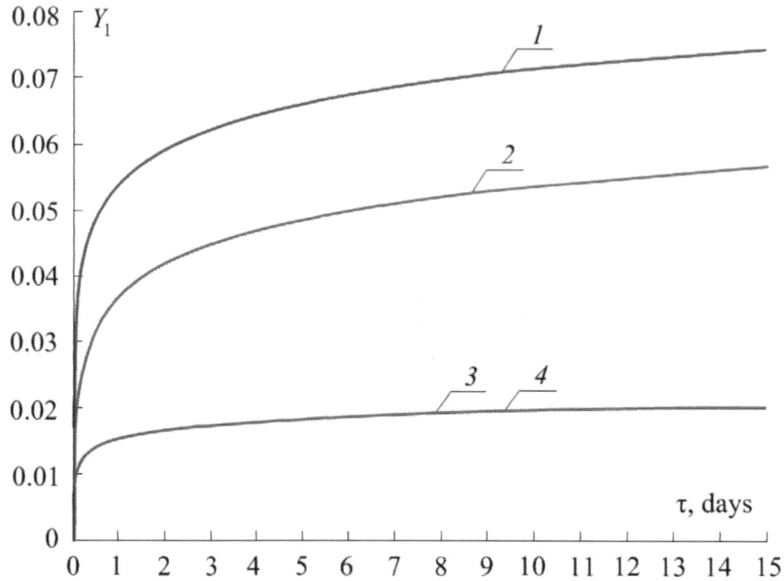

Fig. 2. The change of stability index in time: *1* – 100 % filter cake of T rank; *2* – 100 % filter cake of D rank; *3* – 100 % filter cake of K rank; *4* – 100 % filter cake of SS rank.

As shown in figure 2, the slurries based on filter cakes of K and SS ranks are characterized by the highest structure stability (the lowest values of stability index) in comparison with the rest of the slurries. The intensive separation into layers of these CWS occurred during the first four days with the formation of water layer at the surface. After that the process of separation slowed down. The filter cakes of T and D ranks were less stable. The filter cake of T rank is characterized by the highest value of the stability index. The filter cake of G rank was sufficiently stable and did not separate into layers for 15 days.

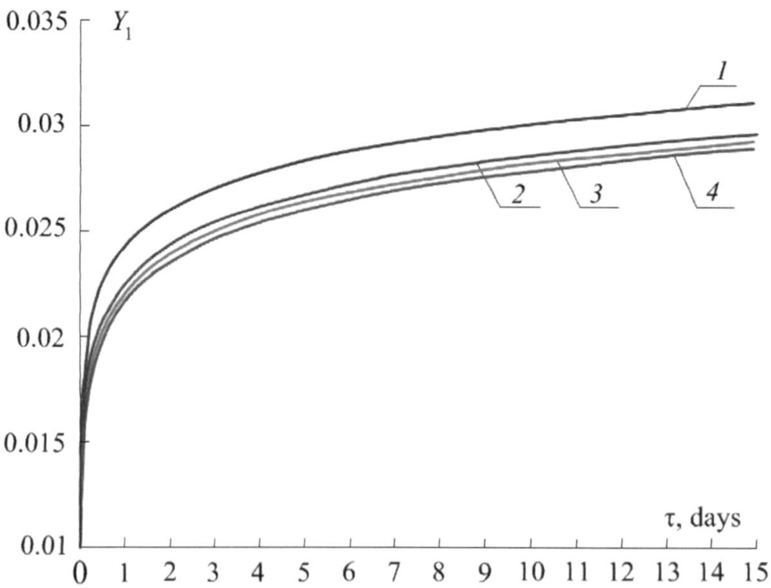

Fig. 3. The change of stability index in time: *1* – 90 % filter cake of G rank, 10 % waste compressor oil; *2* – 90 % filter cake of G rank, 10 % waste motor oil; *3* – 90 % filter cake of G rank, 10 % mazut; *4* – 90 % filter cake of G rank, 10 % waste turbine oil.

The curves shown in figure 3 illustrate that for most compositions the stability index changes in the range from 0.015 to 0.025. It should be noted, that the comparison of the stability indexes shown in fig. 2 and 3 allowed us to conclude, that the slurries based on filter cakes with additives of combustible liquid (fig. 3) are more stable than the CWSs without waste oil or mazut (fig. 2).

The slurries based on filter cake of K rank with additives of waste motor, turbine, transformer oils, oil and oiled mixture were characterized by high stability and did not separate into layers during the time of observation.

The influence of waste oils on the segmentation rate was determined for evaluation of the effectiveness of their using as stabilizing additives for CWSs and CWSPs. The results are presented in table 1.

Table 1. The evaluation of segmentation rate of CWSs and CWSPs.

Compositions of slurries	50 % brown coal of B2 rank, 50 % water	50 % bituminous coal of D rank, 50 % water	40 % bituminous coal of D rank, 50 % water, 10 % waste turbine oil	40 % bituminous coal of D rank, 50 % water, 10 % waste turbine oil
Mass of water, g	8.142	9.127	5.767	7.0814
Mass of coal, g	8.142	9.127	4.612	5.665
Mass of oil, g	–	–	1.153	1.4162
Initial slurry mass α_0, g	16.284	18.254	11.532	14.163
Mass of liquid layer α_1, g	3.26	3.267	1.044	1.456
Initial concentration of coal, C_0, %	50	50	40	40
Ultimate concentration of coal, C_1, %	6.17	17.01	7.372	2.63
Average segmentation rate, r, %/day	5.850	3.936	1.641	2.134

The analysis of table 1 allows making a conclusion that segmentation rates for slurries based on brown or bituminous coal with the additive of waste turbine oil are almost twice smaller than the similar characteristic of slurries without oil additive. This indicates that the use of turbine oil had a stabilizing effect on the fuel composition by increasing its stability.

4 Conclusion

In accordance with the known experimental investigations we identified three main methods of determining stability of composite fuel slurries: the method of evaluation of separation (into layers) of the slurry [7–10], the method of penetration of the glass rod [11–13] and translucence method [14, 15]. In the present study the stability of CWSs and CWSPs compositions were determined by the method of evaluation of separation (into layers). Stability index and segmentation rate were chosen as integral characteristics of fuel stability. We found, that the addition of combustible liquid (for example, waste motor or turbine oil) to CWS provides more stable slurry (with high stability of structure). The limit (maximum) duration of conservation of sedimentation stability of investigated compositions was 15 days. The experiments illustrates that using inexpensive and affordable wastes (oiled mixtures, waste oils of different power plants, turbines, motors, transformers) is advisable for composite fuel slurry production.

Acknowledgments

This work supported by the Russian Science Foundation, project 15-19-10003.

References

1. A. Kijo-Kleczkowska, Fuel **90**, 2 (2011)
2. E.G. Gorlov, Solid Fuel Chem. **38**, 40 (2004)
3. V. Soloiu, J. Lewis, Y. Yoshihara, K. Nishiwaki, Energy **36**, 4353 (2011)
4. P. Feng, L. Hao, C. Huo, Z. Wang, W. Lin, W. Song, Energy **66**, 744 (2014)
5. K.N. Trubetskoi, A.S. Kondrat'ev, V.I. Murko, V.E. Zaidenvarg, G.A. Kassikhin, I.K. Nekhoroshii, Thermal engineering **55**, 5 (2014)
6. S.N. D'yakov, A.V. Papin, A.V. Nevedrov, E.V. Zhbyr", Coke and Chemistry **55**, 10 (2012)
7. Q. He, D. Xie, R. Xu, T. Wang, B. Hu, Fuel **159**, 40 (2015)
8. H. Usui, J. Non-Newtonian Fluid Mech. **60**, 2 (1995)
9. F.Yi, A. Gopan, R. L. Axelbaum, Journal of Fuel Chemistry and Technology **42**, 10 (2014)
10. R. Xu, B. Hu, Q. He, J. Cai, Y. Pan, J. Shen, Fuel **85**, 17 (2006)
11. R. Wang, J. Liu, F. Gao, J. Zhou, K. Cen, Fuel Processing Technology **104**, 57 (2012)
12. X. Zhan, Z. Zhou, W. Kang, F. Wang, Fuel Processing Technology **91**, 10 (2010)
13. M. Zhou, X. Qiu, D. Yang, H. Lou, X. Ouyang, Fuel Processing Technology **88**, 4 (2007)
14. P. Li, D. Yang, H. Lou, X. Qiu, Journal of Fuel Chemistry and Technology **36**, 5 (2008)
15. O. Mengual, G. Meunier, I. Cayre, K. Puech, P. Snabre, Colloids Surf. A **52**, 1 (1999)
16. Z.H. Xia, L.Y. Lu, H. Wang, Spectroscopy and Spectral Analysis **35**, 7 (2015)
17. C.K. Kim, G.J. Lee, M.K. Lee, C.K. Rhee, Arch. Metall. Mater. **60**, 2 (2015)

SOFTWARE PRODUCTS FOR THE NUMERIC EVALUATION OF OSCILLATION OF NO_X IN THE PULVERIZED FUEL COMBUSTOR

E. A. Zueva[1], *N.* V. Vizgavlyust[1,*], *L.* V. Vorobyeva, and *A.* V. Starchenko[2]

[1]Tomsk Polytechnic University, 634050 Tomsk, Russia
[2]Tomsk State University, 634050 Tomsk, Russia

Abstract. In this paper, a numerical study of the formation of nitrogen oxides in the combustion chamber based on the model created by Mitchellom and Terbellom. The distribution of furnace temperature and the concentration of nitrogen oxides, as well as a comparison of numerical results with the data of field experiment.

1. Introduction

One of most crucial problems occurring when organizing burning of fuel is the reduction of toxic substances generation, in particular nitrogen oxide. The requirements to air pollution control became more strict and they require elaborating some new approaches to designing of combustion equipment and chambers, and various power units [1, 2]. At present time for the above purpose numerical simulation is often used which allows not only obtaining numerical evaluation of toxic emissions but developing recommendations on their reduction.

Of late numerous products appeared at the software market which can simulate and evaluate toxic emissions. The software differs regarding its price and accuracy of the final outcome.

The study is aimed to compare the software for numerical evaluation of nitrogen oxide generation when burning the solid fuel. The following software has been chosen for comparison. NO_x and FIRE 3D-NO_x.

NO_x software used for calculation of nitrogen oxide emissions is based on the technique of [3]. This software is one dimensional mathematical model. The results of calibration calculation of the boiler unit can be used as initial data for this software.

* Corresponding author: vizgavljust@tpu.ru

DOI: 10.1051/matecconf/20167201124

2. Description of the model and the object of research

The applied software of FIRE-3D-NO_x is a mathematical model, which task and kinetic scheme are presented in [4, 5].

The studied device was the combustion chamber of steam boiler of BKZ-320-140PT where Irsha-borodinsky coal from Kansk-Achinsk coal field is burnt. Steam boiler of BKZ-320-140-PT is the unit with vertical water pipe with natural circulation of the flat top design. The combustion unit of the boiler is half open with fluid slag removal and consists of two interconnected eight-faced primary furnaces. Each of the primary furnaces is equipped with four straight flow two stage burners. Two discharge burners are located at the controlled zone (fig. 1).

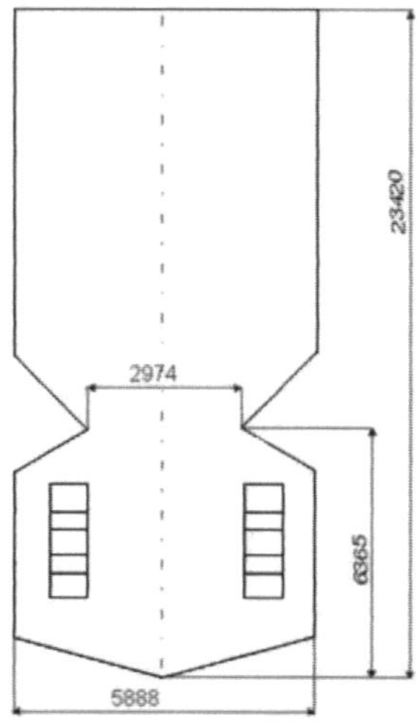

Fig. 1. Geometrical characteristics of the furnace of BKZ-320-140 boiler [6].

This boiler has been selected for the certain purpose. A number of engineering solutions aimed at the intra-furnace emissions inhibition have been implemented and tested in this boiler, which provided achieving the ultimate permissible level for the working boiler with furnace fluid slag removal [6].

3. Results and discussion

When working with the software the following techniques controlling nitrogen oxide emissions were selected: changes in steam capacity of the boiler and staged combustion of the solid fluid.

When changing the load of the boiler (D) within the ranges of 220-320 tons per hour the concentration of nitrogen oxide increases by 120 mg/m^3 (fig. 2) due to the increase of the temperature level in the furnace chamber (the outlet furnace temperature changed from 1100°C to 1217 °C).

When changing the fuel distribution rate by staged combustion from 0.38 to 0.72, the results of numerical testing and actual measurements are of fairly similar values (fig. 3). Increase of the fuel supply to the lower stage of this boiler unit is the best option to suppress the nitrogen oxide in the furnace: NO_x decreases by 20% (100-150 mg/m^3).

The results of numerical simulation (fig. 2 and 3) have values fairly similar to the measurements and the qualitative parameters do not differ greatly.

DOI: 10.1051/matecconf/20167201124

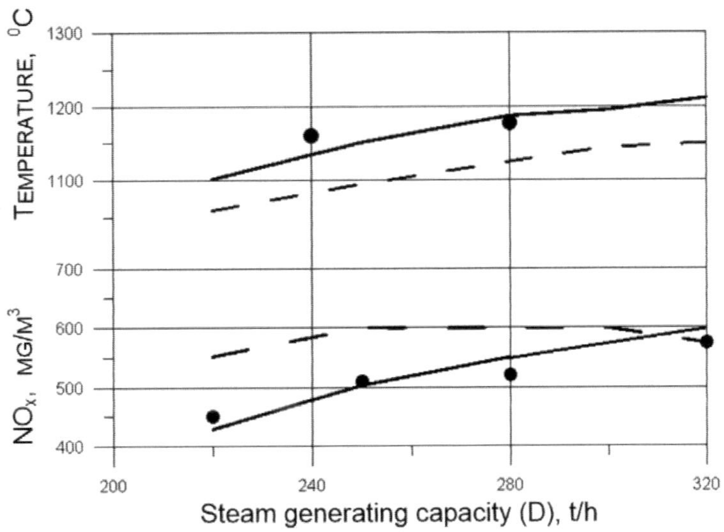

Fig. 2. Changes of the furnace outlet temperature and NO_x concentration depending on the steam capacity: ● – measurements [7], —— – calculation with PPP FIRE 3D–NO_x, — — – calculation as per technique [3].

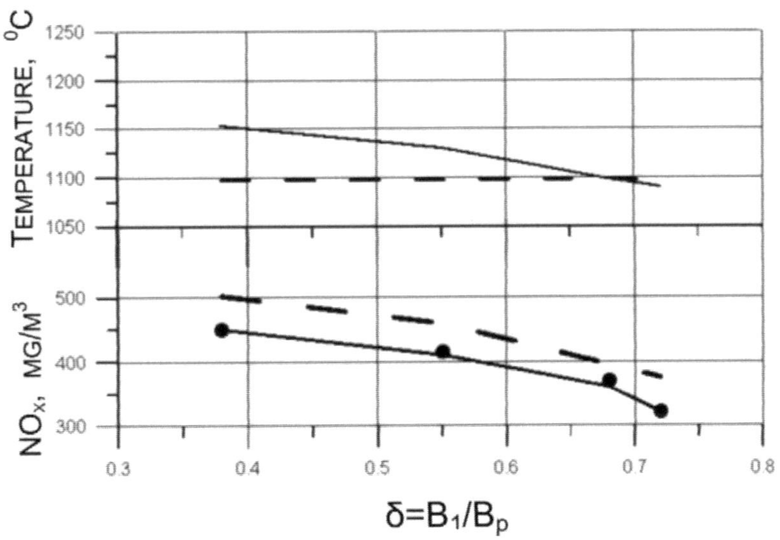

Fig. 3. Changes of the furnace outlet temperature and nitrogen oxide concentration depending on the staging of the fuel supply: ● – measurements [7], —— – calculation with PPP FIRE 3D–NO_x, — — – calculation as per technique [3].

4 Conclusions

The studied ranges of the capacity changes as in any boiler with fluid slag removal limited by temperature level providing the reliable discharge of the fluid slag demonstrates the fairly effective impact of the used means of nitrogen oxide generation inhibition and provides the basis for comparison with the other variable parameters of burning process.

The approach based on the staged combustion provides the temperature in the combustion area on the fairly high level and thus the regular fluid slag removal. Therefore unlike the other techniques it can be used for stabilization of emissions in the combination with changes in the boiler load thus compensating the generation of NO_x when it increases.

Use of FIRE 3D–NO_x software is an effective tool for forecasting the generation of nitrogen oxide in coal-dust furnaces of the boilers compared to NO_x software which has been confirmed by the results of test numerical experiments demonstrating fairly good convergence with the data of actual measurements. FIRE 3D–NO_x software considers the information on the process of discharge and combustion of volatile fuel, aftercombustion of coke residue, to consider non-isothermality, turbulent diffusion and transfer of furnace medium components by the averaged motion, application of the developed model of NO formation when burning coal will allow to optimize process of combustion of dust-like fuel in package boilers of thermal power plants for increase of efficiency of burning out of fuel and reduction of emissions of NO_x.

References

1. *Boiler units. Heat mechanical equipment. General technical requirements* (State Standard R 50831-95)
2. V.R. Kotler, *Nitrogen oxides in the flue gas boilers*, (1987)
3. A.N. Bezgreshnov, Yu.M Lipov, Shleyfer B.M, *Steam boiler calculations* (1991)
4. J.W. Mitchell, J.M. Tarbell, AIChE Journal. **28** (1982)
5. N.V. Vizgavlyust, A.V. Starchenko, A.V. Gil, T.S. Taylasheva, EPJ Web Conf. **82** (2015)
6. O.I. Budilov, A.S. Zavorin, *Experience to improve the environmental performance of the thermal power plant*, (1994)
7. *Study of the burning process on PK-39-2 boiler of 300MW of Yermakovskaya Hydro Power Station when burning ekibaztus coal by the direct injection scheme* (Novosbirsk, 1975)

MODELING OF POLLUTANT DISPERSION IN ATMOSPHERE BASED ON SOFTWARE COMPLEX "SKAT"

Aleksander V. Vorobiev[1,*], *Ekaterina* M. Dutova[1], and *Andrei* S. Latyshev[1]

[1]National Research Tomsk Polytechnic University, 634050 Tomsk, Russia

Abstract. The author of the article analyzes existing methods for calculation of pollutants' dispersion in the atmosphere from a short continuous source under constant meteorological conditions. For this purpose the "SKAT" copyright software system elaborated by author was used. Furthermore, the author makes conclusions about advantages of using „Paskvilla method" for the calculation of gaseous substances' dispersion from sources of this type.

1 Introduction

The ecology of the regions is significantly affected by pollutants from the sources of natural and anthropogenic origin. Pollutant emissions enter various processes of air traffic, are released to the ground, spread in the hydrosphere contaminating the surface and groundwater.

Mathematical models and techniques adequately reflecting the dispersion of these substances in the atmosphere are required to describe and predict pollutant transport in the environment.

2 Description of techniques

In domestic practice, two techniques are currently most widely used, and they are recommended for calculation of air pollution with substances of industrial origin. One of the techniques developed under the guidance of M. Berlyand laid the basis for the regulatory document OND-86 [1]. The second one is the technique for calculating the emission dispersion developed by F. Paskvill (Gaussian model), which is recommended by the World Meteorological Organization to assess the atmospheric pollution by industrial and power plants [2].

2.1 The main concepts of the OND-86 technique

The technique is based on the following assumptions [1]:

* Corresponding author: worob@tpu.ru

- The pollutant dispersion in the atmosphere is affected by meteorological parameters (wind velocity and direction, temperature-stratification of the atmosphere and air temperature);
- The ground-level concentration of pollutants depends on the parameters of the emission source and composition of the air-gas mixture.

The calculation of spatial distribution of the pollutant concentration implies the calculation of the three basic parameters: the maximum ground-level concentration C_m; critical wind velocity u_m; the distance from the emission source to the point of maximum ground-level concentration x_m The highest concentration of the pollutant in the air C_m for a given emission source appears in case of critical wind velocity and only at a certain source distance x_m.

The maximum ground-level concentration of harmful substances due to the release of the air-gas mixture from a single point source with a circular mouth is determined by the formula:

$$C_m = \frac{A \cdot M \cdot F \cdot m \cdot n \cdot \eta}{H^2 \cdot \sqrt[3]{V_1 \cdot \Delta T}},$$

where C_m is measured in mg/m³; A is a dimensionless coefficient that depends on the temperature-stratification of the atmosphere; M is the mass of the pollutant emitted into the atmosphere per unit time, g/s; F is a dimensionless coefficient that takes into account the rate of pollutant sedimentation in the atmospheric air; m, n are dimensionless coefficients that take into account the conditions of the air-gas mixture yield from the mouth of the emission source; η is a dimensionless coefficient that takes into account the effect of the land topography; H is the height of the emission source above the ground level, m; V_1 is the volumetric flow rate of the air-gas mixture, m³/s; $\Delta T = T_g - T_a$ is the differential temperature of the discharged air-gas mixture T_g and the ambient air T_a, ºC.

The OND-86 technique is currently approved at the state level and recommended for use to calculate the emission dispersion in the air while rationing and limiting maximum permissible emissions (MPE). However, this technique is difficult to use for scenario and forecast calculations under specific meteorological conditions for specific emission sources. The OND-86 technique does not fully takes into account the type of the underlying layer.

2.2 The main concepts of the Paskvill's technique

When pollutants are emitted into the atmosphere, a cloud is formed that drifts along with the atmospheric air with the wind. As a result of turbulent exchange with atmospheric air, the cloud expands over a distance and the concentration of pollutants in it changes. This leads to three-dimensional pollutants distribution described by the normal (Gaussian) law in the most general case.

According to this technique, the expression for the ground-level pollutant concentration from a short-time continuously emitting source (hereinafter SPC-source) under similar meteorological conditions is of the form [2]:

$$C(x) = \frac{Q}{\pi \cdot \sigma_y \cdot \sigma_z \cdot u} \cdot \exp\left[-\left(\frac{H_{eff}^2}{2 \cdot \sigma_z^2}\right)\right] \cdot f_F \cdot f_W,$$

where $C(x)$ is the concentration of ground-level concentrations of harmful substances, mg/m³; Q is the source power, g/s; H_{eff} is the effective emission height, m; u is the wind velocity at the effective emission height, m/s; σ_y, σ_z are the dispersion coefficients of the Gaussian distribution (turbulent diffusion) in directions perpendicular to the wind, m; f_f is the empirical depletion factor, which takes into account dry deposition; f_W is the empirical depletion factor, which takes into account washout.

One of the main advantages of the Gaussian model is the possibility to account the impact of the following important factors on the pollutant distribution:

- Atmospheric stability;
- The type (roughness) of the underlying layer;
- Thermal rise of the stream due to overheating of the ejected air-gas mixture;
- Washing ability of precipitations and others.

Software description

We have developed a software package "SKAT" using the licensed Borland Turbo Delphi programming system to implement the algorithms of pollutant dispersion calculation according to the above techniques.

3 Results of the calculations

The software package "SCAT" was used to carry out comparative calculations of gaseous pollutant dispersion by the Gaussian model and the formulas of the OND-86 technique. A short-time continuously emitting source of (SO_2, NO_2) flue gases was chosen as a test object.

The values of the input data taken for calculations correspond to the known [1] test calculation.

Fig. 1 shows the results of the calculation of the maximum values of ground-level concentrations C_m of gaseous pollutants and the distances x_m for the emission source where these values of the ground-level concentration are attained under various weather conditions.

Under favorable weather conditions (fig. 1, a), the modeling results of the distribution of the gaseous pollutant emission concentration along the plume centreline substantially depend on the technique (OND-86 or Paskvill's model) used for the calculation. The position of the point with the maximum ground-level concentration found by the OND-86 formulas virtually does not depend on meteorological parameter that contradicts the common sense and the data [3, 4].

On the contrary, under unfavorable weather conditions (fig. 1, b), the pollutant concentration distribution depends little on the calculation technique used. In particular, this concerns the location of the point with the maximum ground-level pollutant concentration.

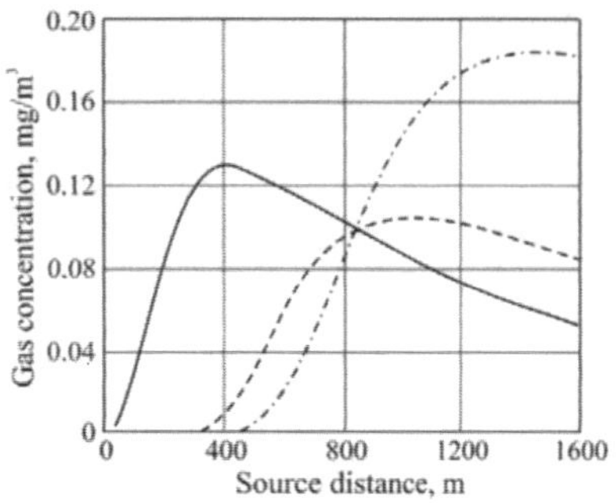

a)

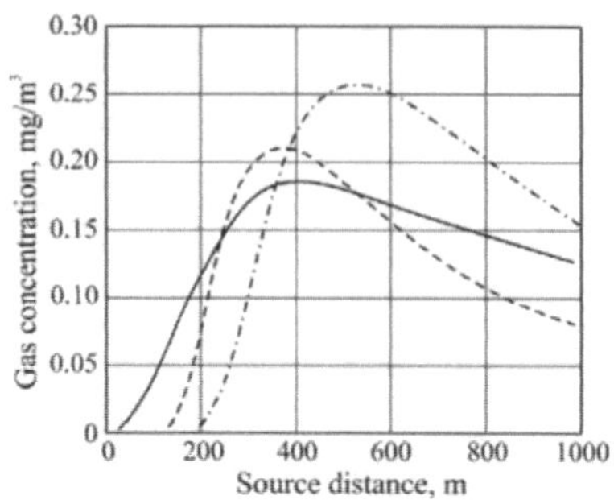

b)

Fig. 1. Distribution of the ground-level concentration of gaseous pollutants along the plume centreline at critical wind velocities corresponding to favorable (a) and unfavorable (b) weather conditions:— — — and — · — · — - calculation by the Paskvill's model (z_0 = 10 cm and 1 cm);—— - calculation by the OND-86 technique.

4 Conclusions

The analysis of the techniques for calculation of gaseous substance dispersion for a short-time continuously emitting source suggests that under stable meteorological factors the Paskvill's model is preferable as it takes into account a wider range of the key parameters which adequately describe the local weather conditions and the type of the underlying surface.

The developed software package "SKAT" can be used for calculating the pollutant dispersion in the atmosphere to solve the monitoring problems of air pollutant emission for currently emitting sources.

References

1. *OND-86. The method of calculating the concentration of harmful substances emitted by industrial enterprises in the atmospheric air* (Gidrometeoizdat, Leningrad, 1987)
2. *Method of development of standards for maximum permissible emissions of radioactive substances in the atmospheric air* (DV, 2010)
3. M. R. Beychock, *Fundamentals of Stack Gas Dispersion* (Author-Published, 2005)
4. S. Arya, *Pal Air Pollution Meteorology and Dispersion* (Oxford University Press, Oxford, 1998)
5. Schnelle, B. Karl, and P. R. Dey, *Atmospheric Dispersion Modeling Compliance Guide* (McGraw-Hill Professional, 1999)
6. D.B. Turner, *Workbook of Atmospheric Dispersion Estimates: An Introduction to Dispersion Modeling* (CRC Press, 1994)
7. K. Wark, C.F. Warner, and W. T. Davis, *Air Pollution: Its Origin and Control* (Pretice-Hall, 1998)
8. A. Riddle, D. Carruthers, A. Sharpe, C. McHugh, J. Stocker, Atm. Env. **38** (2004)
9. R. Barrat, *Atmospheric Dispersion Modelling* (Earthscan Publications, 2001)
10. A. Daly, P. Zannetti. 2007 *Air Pollution Modeling* (The Arab School for Science and Technology, 2007)

DETERMINATION OF NECESSARY AND SUFFICIENT OPERATIONAL CONDITIONS OF THE SLEEVE-TYPE BEARINGS IN AUTONOMOUS POWER PLANTS OF ORMAT TYPE UNDER CONDITIONS OF SIBERIA AND THE FAR EAST

Olga V. Vysokomornaya[1,*], *Vladimir* S. Vysokomorny[1], *Alexander* S. Beck[1], *Andrey* G. Kosintsev[1], and *Vladimir* F. Panin

[1]National Research Tomsk Polytechnic University, 634050 Tomsk, Russia

Abstract. The analysis of integral reliability characteristics for autonomous energy sources in gas pipelines of Russia is carried out. The reasons of the most frequent failures are determined. Recommendations for increase of integral reliability characteristics for autonomous energy sources are formulated.

1 Introduction

No-break supply of manufacturing plant with energy resources is an important factor of stability of production processes and their effective functioning [4]. In most cases, power supply of industrial enterprises is carried out by a centralized power supply networks. However, because of the emergence of new consumers of energy resources in areas geographically remote from centralized power grids, it is necessary to find other cost-effective energy sources [2, 3]. Often the problem of energy supply of remote manufacturing plants is solved through the use of stand-alone power plants. In particular, for the operation of auxiliary equipment of industrial enterprises transporting hydrocarbon raw materials (including remotely controlled items, radio relay stations, etc.) in Eastern Siberia and the Far East of Russia the thermal power stations of small capacity (Mini-TPS) with closed thermodynamic steam cycle (fig. 1) are successfully used.

2 Target of research

Mini-CHP Plant fully autonomous and automate, uses natural gas as fuel, comply with the requirements between periods of planned servicing and repair upon condition running in compliance with demands of factory of origin. Frequently environmental withstand in Eastern Siberia and the Far East of Russia characterized by high differences temperature

* Corresponding author: vysokomornyy@yandex.ru

fall of outward. This application environment may be reasons of appearance of not restricted operating regime in work of mini-CPH-plant connected with temperature disparity of power fluid with factory of origin demands. Above-noted disparity is the cause of failure mini-CPH-plant due to high temperature of leaving aerial condenser organic fluid [1, 6].

Due to the fact that the constructive features of observe autonomous power facilities (fig. 1) was using organic fluid such as oiling matter for turbine shaft bearing, recycled after cooling of the aerial condenser, providing the required parameters of the power fluid is a determining factor for ensuring the smooth operation of the power plant.

When temperature of power fluid for bearing lubrication more, than recommended value of factory [5] take place risk of technical deficiency appearance in work of turbine-generator of mini-CHP plant.

For prevention emergency situation in power plants work it's necessary to determine causes and develop recommendations for decrease failure in work of autonomous mini-CHP plant due to mismatch application environment requirements of the factory.

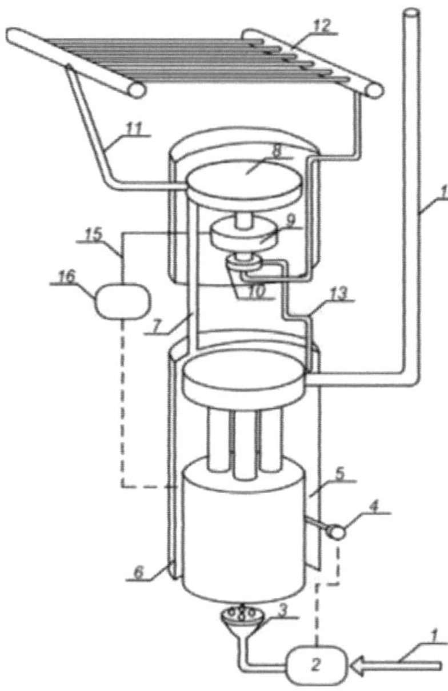

Fig. 1. mini-CHP plant scheme with closed steam cycle: 1 – fossil fuel supply line; 2 – control panel of fuel supply; 3 – fuel burner; 4 – thermostat; 5 – organic liquid; 6 – steam generating unit; 7 – steam line; 8 – turbine; 9 – alternator; 10 – pump; 11 – steam supply tube to condenser; 12 – condenser; 13 – tube of condensate output; 14 – chimney; 15 – AC input cables to the rectifier; 16 – electrical switchboard.

Physical and mathematical formulation of solvable problem of heating transfer present in papers [1, 6]. Vaporizing working agent from turbine 8 pass to aerial condenser 12, where heat transfer occurs from pipe wall to ambient air. Then coolant condensate goes to steam power boiler for heating and vaporation passing and lubricating sleeve-type bearings of turbine-generator. Mathematical model, which describe condensing process of working agent includes energy equation for vapor of working agent, heat-transfer equation for film of condensate and pipe wall, diffusion equation, moving, equation of through flow and constitutive equation of working fluid vapor. Nonstationary differential equation system with appropriate initial and boundary conditions solved by finite difference method with using four-point implicit difference scheme [1, 6].

3 Statistic of the most typical technical deficiency in mini-CPH-plant work

Figure 2 shows the relevant statistics of typical technical accidents in the work of 101 mini-TPS operating in Eastern Siberia and Russian Far East for the period from 2012 to 2015, and earlier accumulated statistics for the period from 2008 to 2012.

The positive dynamics in reducing the number of technical accidents due the "extremely high temperature in the condenser" is associated with the recommendations presented by Vysokomorny V.S. [1] of increase of the reliability and resource efficiency in mini-TPS, operating in cogeneration mode. Because of this recommendations the number of accidents due the "extremely high temperature in the condenser" decreased by 4.9 times. Analysis of the second most important technical accident related to the "mechanical failure of turbogenerator" leads to the conclusion that the probability of its occurance is 35%/ This statistic proves that the need for analysis of occurrence of this accident is an critical task witch requires a detailed study of the appearance causes.

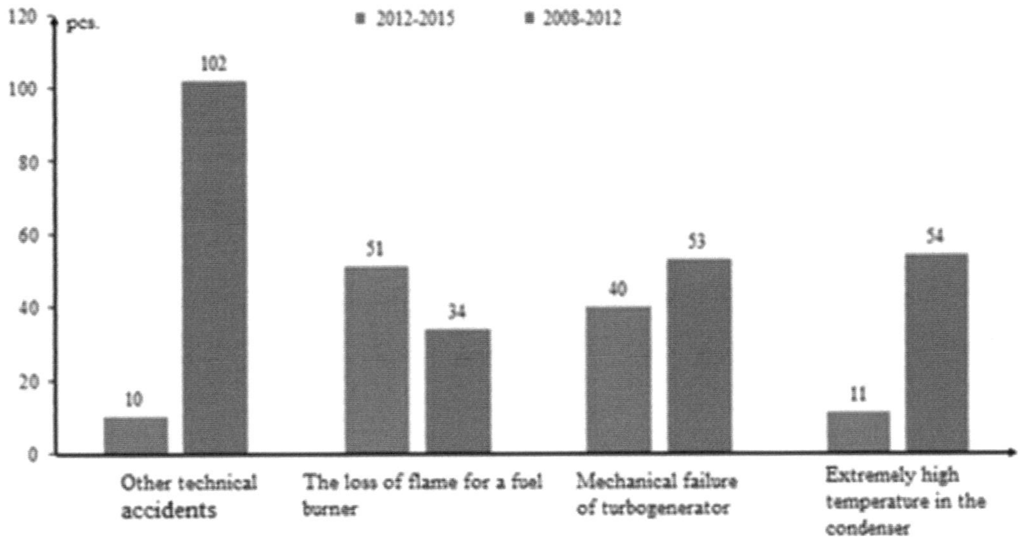

Fig. 2. Statistics of the most typical technical accidents in the work of mini-CHP plant, depending on the year.

4 Results of numerical simulation

Figure 3 shows the numerical results of the integral characteristics of the condensation process in the tube, depending on various external air temperature T_{out} values when the workflow temperature changes entering the tube T_{output}. Numerical simulation of heat and mass transfer processes in the conditions of intensive phase transitions is performed at the following values of thermal characteristics of the interacting substances: C_1=1300 (J/kg·K), ρ_1=5.1 kg/m^3, λ_1=0.1 W/(m·K); C_2=1300 (J/kg·K), ρ_2=1300 kg/m^3, λ_2=0.12 W/(m·K); C_3=460 (J/kg·K), ρ_3=7700 kg/m^3, λ_3=54 W/(m·K); Q_c=311.7·10^3 J/kg,
where C is heat capacity (J/kg·K); ρ is density, kg/m^3; λ is thermal conductivity, W/(m·K); Q_c is the thermal effect of condensing agents, J/kg; codes "1", "2" and "3" correspond to steams, condensate of working substance and tub's sides, respectively.

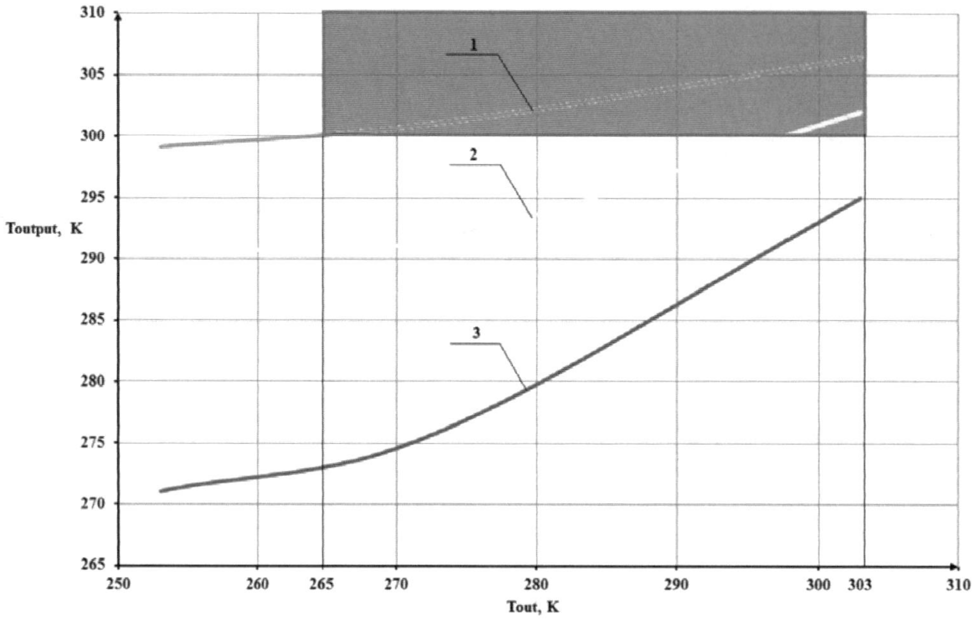

Fig. 3. Dependence of the output workflow T_{output} temperature on the ambient air temperature: 1 – T_{input}=449 K; 2 – T_{input}=426 K; 3 – T_{input}=405 K.

Dependence analysis shows that the change in the output electric power of the considered mini-CHP Plant has a significant impact on the process of condensation in the tube and workflow T_{output} outlet temperature. It directly affects the efficient operation of a separate process and the work of the power plant as a whole. This result is explained by the fact that the speed of two-phase workflow through the condenser unit grows when the power of the power plant increases. It leads to an increase of the initial temperature of the working substance that affects the condensation process in a condenser, and hence the output temperature of the workflow from the condenser. In cases where the workflow appears in high ambient temperatures (265 K and above) there is a significant increase in the temperature of the working substance. Workflow temperature in the range between 300 K and above, is critical for the operation of power plants and may cause unrestricted operational states of mini-CHP plant.

5 Conclusion

Obtained results leads to the following conclusions:

1. It is established that the main parameter that significantly affects the output temperature of the working stream from the condenser T_{output} is the ambient temperature and electric power, as well as the initial temperature of working medium at the condenser intake T_{input} .

2. At high temperatures of outside air the reliability of autonomous energy unit work is reduced due the low efficiency of the method of additional heat transfer process trough the use of the fan for cooling the external surfaces of the condenser tubes. Significantly greater cooling efficiency can be achieved by using water cooling system of surfaces of the tubes of the condenser.

3. A high probability of exceeding the temperature of working medium at the zone of lubrication of slider bearings due to the wide range of ambient temperatures at different

capacity of energy unit. According to the manufacturer the temperature of the working medium at the inlet to a zone of lubrication of the sliding bearings should not exceed 300 K.

The obtained results allow to determine necessary and sufficiency conditions for accident free operation of the sliding bearings of the mini-CHP plant turbogenerator:

1. In case of energy units operation with nominal parameters (3000 W and 4000W) increases the probability of unexpected occurrences associated with the units shutdown in the ranges of the outside temperatures from 283 K and higher and 253 K and higher, respectively.

2. Increasing the area of heat transfer between the working medium and outside air allow to achieve the optimum temperature of the working stream at the outlet of the condenser, which will provide the decrease of the number of technical accidents associated with “mechanical failure of turbogenerator”.

Acknowledgments

This work was financially supported by the Ministry of Education and Science of the Russian Federation (the scholarship of the President of the Russian Federation for young scientists and graduate students SP-1350.2015.1). The work at the field of modeling of heat and mass transfer processes near boundary between two layers was supported by the scientific schools grant NSH-7538-2016.8.

References

1. O.V. Vysokomornaya, V.S. Vysokomorny, P.A. Strizhak, Energy security and energy efficiency **1**, 25 (2014)
2. P.I. Kononenko, V.G. Mikhailuts, A.J. Bezzutsev-Kondakov, Energetik **3**, 43 (2007)
3. A.P. Livinskij, I.J. Red’ko, V.M. Filin, Energetik **4**, 22 (2010)
4. *Energy Strategy of Russia for the period until 2020* (Approved by the RF Government, 2003)
5. *Eps operation & maintenance manual* (powered by CCVT) (ORMAT SYSTEMS LTD, 2006)
6. M.V. Piskunov, I.S. Voytkov, O.V. Vysokomornaya, V.S. Vysokomorny, EPJ Web Conf. **82** (2015)

MATEC Web of Conferences 72, 01127 (2016) DOI: 10.1051/matecconf/20167201127
HMTTSC-2016

USING A HIGH-TEMPERATURE FLUE GASES IN THE TECHNOLOGY OF COMBUSTION NEUTRALIZATION OF WASTEWATER

Olga V. Vysokomornaya[1,*], *Yuliya* S. Zakharevich[1], *Natalya* N. Aykina[1], and *Denis* V. Ushmaev[1]

[1]National Research Tomsk Polytechnic University, 634050 Tomsk, Russia

Abstract. This paper provides an analysis of composition of typical polluting contaminants of industrial wastewater. The study suggest industry branches, wastewater of which contains the greatest amount of organic flammable substanses (petroleum и oil). The present study was conducted to analyze an opportunity to replace fuel torch by the high-temperature flow of flue gases in the realization thermal treatment method (combustion neutralization).

1 Introduction

In today's world, water pollution problem of water objects by industrial and domestic wastewater is very actual [1, 2]. Uncontrolled discharges of the polluted water into natural reservoir have a significant and often irreparable damages to ecosystem of territory, which is close to industry. Chemical composition of polluted substances, which are discharges in reservoirs, vary and depend on specific of technology process in industry, which is made a discharge. During the choice of methods of treatment industry wastewater to consentration of harmful substances, which are not exceed maximum values, are guided by preliminary analysis of wastewater composition.

The aim of this paper is to analyse the composition of flammable substances of industrial wastewater, which are treated by combustion neutralization, and also assessment of using in the technology of high-temperature treatment of flue gases water.

2 Analysis of wastewater composition

The most toxiferous and dangerous compounds for organic reservoirs and ground, which could be in composition of industry wastewater are such organic substances, as: phenol, pitches, oil and etc. This substances are stable for biological decomposition [4], in case wastewater treatment is very important event for saving the ecologic balance.

The Analysis of composition of industrial wastewater shows us that the biggest concentration of toxic organic compounds is charachteristic to discharged of pulp and

* Corresponding author: vysokomornaja@tpu.ru

paper, coal, oil, coke-chemical enterprises, enterprises of mechanical engineering [1–6]. The table below shows us the main types of organic substanses in industry wastewater.

Table 1. A typical concentration of pollutants in industrial wastewaters.

Pulp and Paper Mill (sulphate production) [3] Suspended solids (100÷450 mg/l) BOD_5 (130÷480 mg (O_2) /l) Phenols C_6H_5OH (~0.06 mg/l) Petroleum products (50÷60 mg/l)	**Petroleum Processing Plant** [4] Suspended solids (100÷300 mg/l) Solids (800÷1000 mg/l) BOD_5 (90÷290 mg (O_2) /l) Petroleum products (7000÷15000 mg/l)
Machine Building Plant [5, 6] Solids (100÷200 mg/l) Phenols C_6H_5OH (0.01÷1,0 mg/l) Petroleum products (20÷100 mg/l) Sulphates $CaSO_4$, $MgSO_4$ (150÷250 mg/l)	**Enterprises of Coal Industry** [5, 6] Suspended solids (15000÷25000 mg/l) Solids (5000÷1000 mg/l) Sulphates $CaSO_4$, $MgSO_4$ (100÷150 mg/l)

The majority of the organic pollution presented in the tab. are flammable substances [7]. One of the ways to treat indusrty wastewater of flammable substances is combustion neutralization. [5, 6]. Industrial diagrams for realization of this method usually provides spraying wastewater for dispersible in the area of a flame into specialized combustion chamber (fig. 1). At the same time flammable substances of flow are burned and removed in the form of gas through the flue, unflammable contaminants dropped out in the form of solid residue.

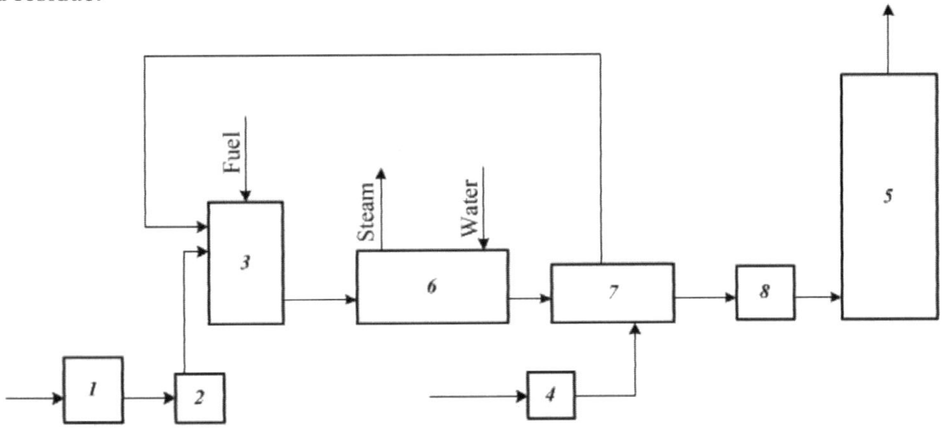

Fig. 1. Scheme of a plant for wastewater combustion neutralization [8]: *1* – wastewater collector; *2* – pump; *3* – combustion chamber; *4* – blower; *5* – chimney; *6* – waste heat recovery boiler; *7* – air heater; *8* – exhauster.

Usually for realization of the combustion neutralization method used special furnace, in which is used fuel torch. At the same time for burning a contaminants fuel is spent and large amount of combustion gases is formed. As one of measures of increase the energy efficiency of thermal treatment could be utilization in diagrams of combustion neutralization of flue gases other industrial diagrams (for example, energy or technology coppers in case of selections of gases with tempetature T>600 K). An interaction of a sprayed liquid flow (technological water with flammable contaminants) the high-temperature gas environment (T≈500–1000 K) there will be an intensive evaporation of water which then can be taken away by a separator and come back to a production cycle. Burning out of combustible impurity will promote maintenance of temperature level of process for a treating process continious. Such method of economy of fuel and utilization of secondary energy resources (combustion gases) can be implemented at the entities operating boiler rooms and power plants of own needs, and also technological coppers of low and average power. However for projection and practical use of this technology of energy saving requires carrying out theoretical and experimental researches for determination of the main features and regularities of evaporation of water and burning out of impurity of technological wastewater at dispersion in the environment of high-temperature combustion gases. At that time there are published results of researches of major mechanisms and regularities of the movement of evaporation of drop streams of liquids at contact with hot gases and the heating surfaces [9, 10]. For development of technology of high-temperature wastewater treatment with using of hot combustion gases similar results of researches are necessary for drop streams of liquid with combustible contaminants.

Acknowledgments

This work was supported by the Russian Foundation for Basic Research (No. 14-08-00057). The work at the field of modeling of heat and mass transfer processes near boundary between two layers was supported by scholarship of the President of the Russian Federation for young scientists and graduate students SP-1350.2015.1.

References

1. B.P. Walsh, S.N. Murray, D.T.J. O'Sullivan, Water Resources and Industry **10**, 15 (2015)
2. M.M. Redina, A.P. Khaustov, *Rationing and the reduction of environmental pollution* (Yurait, Moscow, 2014) [in Russian]
3. M.A. Gusakova, K.G. Bogolitsyn, Water: Chemistry and Ecology **6**, 78 (2014) [in Russian]
4. A.I. Vladimirov, V.V. Remizov, *Ecology Oil and Gas Industry* (Oil and Gas, Moscow, 2003) [in Russian]
5. L.I. Pugach, *Energy and Ecology* (NSTU Publ., Novosibirsk, 2003) [in Russian]
6. V.A. Zaitsev, *Industrial Ecology* (BINOM, Moscow, 2012) [in Russian]
7. O.V. Vysokomornaya, E.Yu. Kurilenko, A.A. Shcherbinina, MATEC Web Conf. **23** 01041 (2015)
8. O.V. Vysokomornaya, J.E. Balakhnina, M.V. Shikhman, MATEC Web Conf. **23** 01065 (2015)
9. O.V. Vysokomornaya, G.V. Kuznetsov, P.A. Strishak, Journ. Eng. Phys. Thermophys. **89**, 1 (2016)
10. K.A. Batischeva, E.G. Orlova, D.V. Feoktistov, EPJ Web Conf. **82** 01055 (2015)

MATEC Web of Conferences **72**, 01128 (2016) DOI: 10.1051/matecconf/20167201128
HMTTSC-2016

IGNITION PARAMETERS OF DIESEL LAYERS UNDER CONDITIONS OF LOCAL HEATING

Arkadiy V. Zakharevich[1,*], *Mikhail* S. Zygin[1], *Pavel* V. Panchenko[1], and *Yekaterina* A. Vlasova[1]

[1]National Research Tomsk Polytechnic University, 634050 Tomsk, Russia

Abstract. Describes the methodology and results of experimental studies of ignition of diesel fuel by a single heated up to high temperatures steel particle. The layer thickness of liquid fuel was changed from 1 mm to 8 mm. It was established dependence of numerical values of the ignition time delay from the thickness layer of diesel fuel. The revealed ignition mode and the limiting layer thickness for diesel fuel when the ignition is missing.

1 Introduction

It is well-known [1-5] that high temperature metal particles can be applied as ignition sources of solid [1], liquid [2], and dispersed [3] fuels and combustible materials of plant origin [4,5]. The theory about local ignition of solid [6.7] and liquid [8,9] fuels under conditions of local heating has been already developed. At the same time possible is another process the ignition of flammable fluids, when a drop of the distillate fuel is fit in the gas stream or in close proximity to heated to high temperature surfaces of the elements of power plants for various purposes. Methods for developed determination of ignition delay time the fluids in the fall of a fuel drop on a heated surface [10,11] for such of the ignition mode. Now also accumulated extensive information about heat and mass transfer during evaporation of pure liquids from the surfaces of the heating [12, 13]. But these models [6-13] did not explain how ignition characteristics depend on layer thickness. However, the impact of subsurface layer heterogeneity on integral characteristics of ignition was presented in studies [14,15], where structurally inhomogeneous condensed matters were considered.

The aim of this experimental study is how layer thickness of typical liquid fuel impacts on its ignition characteristics by local source of energy.

2 Experimental technique

The object of our research is winter diesel fuel. Disc- shaped monolithic steel particles were applied as sources of energy and had fixed sizes: diameter (d_p=6·10^{-3} m) and height (h_p=3·10^{-3} m). Particle geometry was accepted to assure minimum dissipation of parameters, which characterize experimental conditions. Experiments was carried out in

[*] Corresponding author: bet@tpu.ru

normal climatic conditions: temperature 290.15-293.15K, atmospheric moisture capacity 50-70%, atmospheric pressure 700-800 mm.Hg. Experimental stand in fig.1 was used for carried out researches. The conditions of liquid fuel spreading on a solid surface, in the capacity of which was used glass of Petri dish (diameter d_v=10·10^{-2} m and h_v=1.5·10^{-2} m.), was simulated. The heated steel particle was in the solid condition and was not deformed during it fall in diesel fuel. Video recording of studying process carried out on high-speed camera Phantom v411 with software "TemaAutomotive". Ignition delay time was determined from the moment of contact particle - diesel layer and appearance of fire by processing video frames. Experiments series (6-8) performed for each liquid fuel layer (h) when the initial temperature of the steel particle (T_p) is fixed. Random error of determining ignition time delay were not more than ± 6,7%.

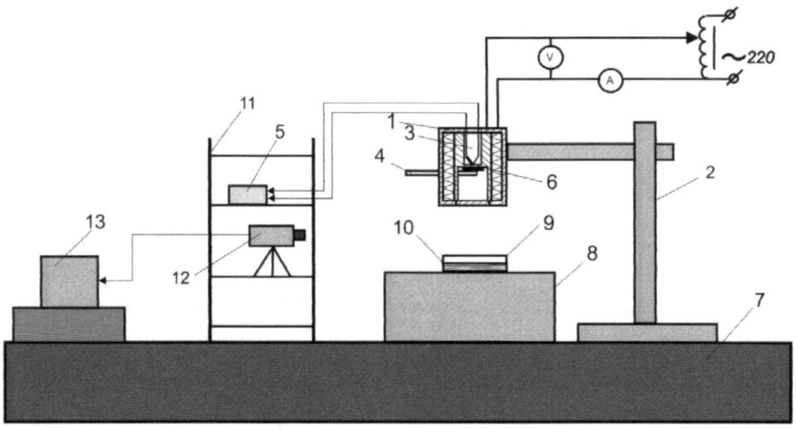

Fig. 1. Scheme of experimental setup: 1 – heating device; 2 – support; 3 – chromel-alumel thermocouple; 4 – ceramic rod; 5 – measuring block with control function electron; 6 – metal particle; 7 - working surface of the experimental setup; 8 – fire-resistant area; 9 – glass dish Petri; 10 – diesel fuel; 11 – shelves metal; 12 – high-speed camera; 13 – touchscreen monitor.

3 The discussion of experiments' results

Fig. 2 shows dependences of ignition time delay of diesel fuel versus thickness layer. Two trends in fig. 2 correspond to different values of steel particle temperature. Minimum and maximum values of layer thickness in fig. 2 correspond to limiting conditions of ignition reaction initiation. Extremely high value of layer thickness (h=7mm) was quite evident. After fuel evaporation diesel vapours, moved to the oxidant (air) area and heated while flowing steel particle. Then, moving to the low dimension channel, formed above the upper bound of particle, vapours became cool (fig. 3, 4). Energy, which was transfer to vapours from particle at h>7mm, was too low to initiate ignition process in consequence of further cool.

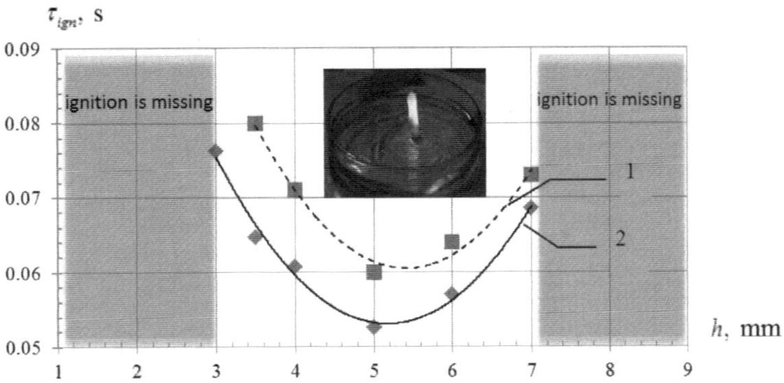

Fig. 2. Experimental dependences of diesel fuel ignition time delay from the thickness of the film: 1 – steel particle with a temperature of 1353 K, 2 – steel particle with a temperature of 1373 K.

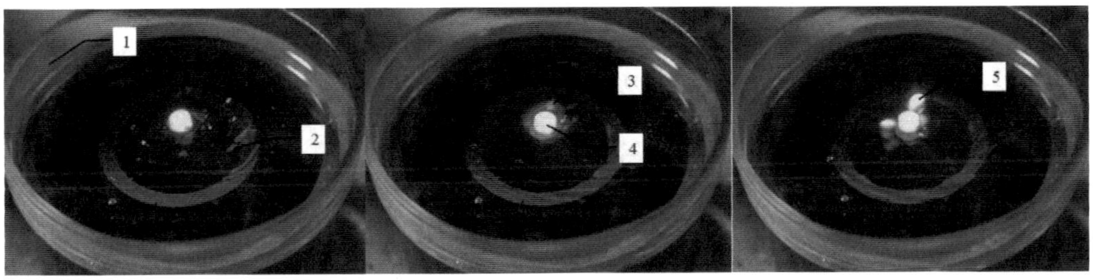

Fig. 3. Video frames of the experiment with ignition of diesel fuel by a steel hot particle: 1 – glass Petri dish; 2 – layer diesel fuel; 3 – pair of diesel fuel; 4 – metal particle; 5 – flame.

Fuel vapours rise up, when the height of vertical channels in the layer is from 3 to 7 mm, and leave out of direct contact with the source of heat energy. As a result fuel vapours shifted toward to particle axis of symmetry and heated by heat exchange with the upper horizontal surface of the particle (fig. 4a). In this case the upper surface was source of active heating of diesel vapours. Experimental video analysis shows that ignition reaction proceeds on symmetry axis of system diesel layer – particle – channel filled with fuel vapours.

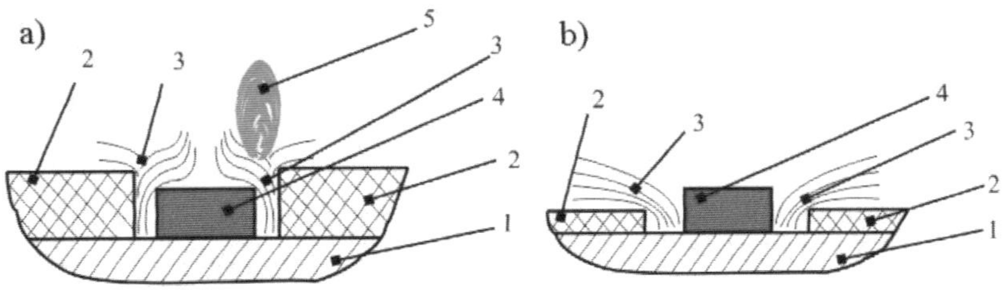

Fig. 4. Options for the location of steel particles in the layer of diesel fuel: a) ignition, b) ignition is missing. 1- the bottom of the Petri dish; 2 – layer diesel fuel; 3 – pair of diesel fuel; 4 – metal particle; 5 – flame.

Probably, left (fig. 2) limited value of *h* was caused by other reasons. When particle height was bigger than fuel layer, fuel vapours spreaded near the top surface of the heat source and to its opposite side. Consequently, fuel concentration was not sufficient for ignition process activation in the area of the highest air temperatures near the particle axis of symmetry.

Moreover, if particle height was smaller than *h*, fuel layer compresses vapours in the gap at the side surface of the particle due to mass forces. Fuel was heated more intensive. When *h* was smaller than h_p, evaporation rate (fig. 4b) of fuel was lower due to an increase in the gap thickness. As the results, dry spot was formed around the particle. Fuel evaporated at a low rate at the spot boundaries. The analysis set out in the laws of the experiments shows that the ignition of flammable liquids films by local sources of heat is only possible at certain ratios of thickness h and h_p. Range relations (h/h_p) $1 \leq h/h_p \leq 2.3$ comply with the conditions of diesel fuel ignition.

4 Conclusions

According to the experimental results, the limiting modes and thickness of diesel layers ignition under conditions of local heating were established. Physical model of heat and mass transfer when igniting the diesel fuel layers by high temperature particles was formulated.

References

1. A.V. Zakharevich, G.V. Kuznetsov, V.I. Maksimov, V.T. Kuznetsov, Combustion, Explosion, and Shock Waves. **44**, 543 (2008)
2. G.V. Kuznetsov, P.A. Strizhak, J. Eng.Thermophys. **17,** 224 (2008)
3. A.V. Zakharevich, A.R. Bogomolov, J. Phys. Chem. B **9**, 907 (2015)
4. A.V. Zakharevich, N.V. Baranovskiy, J. Phys. Chem. B **9**, 911 (2015)
5. A.V. Zakharevich, J. Eng. Phys. Thermophys. **87,** 98 (2014)
6. G.V. Kuznetsov, G.Ya. Mamontov, G.V. Taratushkina, Combustion, Explosion and Shock Waves **40**, 70 (2004)
7. G.V. Kuznetsov, G.Ya. Mamontov, G.V. Taratushkina, Khim. Fiz. **23**, 67 (2004)
8. O.V. Vysokomornaya, G.V. Kuznetsov, P.A. Strizhak, Russ. J. Phys. Chem. B **5**, 668 (2011)
9. D.O. Glushkov, J.-C. Legros, P.A. Strizhak, A.V. Zakharevich, Fuel **175**, 105 (2016)
10. P. Dagaut and M. Cathonnet, Prog. Energy Combust. Sci. **32**, 48 (2016)
11. G.V. Kuznetsov, A.V. Zakharevich, N.S. Belkov, Chem. Pet. Eng. **50**, 424 (2014)
12. G.V. Kuznetsov, D.V. Feoktistov, E.G. Orlova, Thermophysics and Aeromechanics **23**, 17 (2016)
13. Y.V. Lyulin, D.V. Feoktistov, I.A. Afanas'ev, E.S. Chachilo, O.A. Kabov, G.V. Kuznetsov, Tech. Phys. Lett. **41**, 665 (2015)
14. V.L. Strakhov, A.N. Garashchenko, G.V. Kuznetsov and V.P. Rudzinskii, Combustion, Explosion, and Shock Waves **37**, 178 (2001)
15. D.O. Glushkov, G.V. Kuznetsov, P.A. Strizhak, Russ. J. Phys. Chem. B **5**, 1000 (2011)

MATEC Web of Conferences 72, 01129 (2016) DOI: 10.1051/matecconf/20167201129
HMTTSC-2016

RESEARCH OF HEAT AND MASS TRANSFER PROCESSES IN THE NODES OF FREE-FLOW MICRO-HPP WITH THE USE OF 3D TECHNOLOGY

Stanislav Yankovsky[1,*], *German* Kulupanov[1], and *Vladimir* Andriyashchenko[1]

[1]Tomsk Polytechnic University, 634050, Lenin Avenue 30, Tomsk, Russia

Abstract. In the article the analysis of existing micro-hydro turbines. Created 3D experimental model. Designed and developed an experimental stand for testing micro-turbine. Investigated options for generators on permanent magnets, selected the optimum scheme and developed a prototype three phase generator with permanent magnets.

1 Introduction

Production of electrical energy is the basis of life for all industries. The largest share of electricity generation falls on the use of fossil fuels. Trend of continuous growth in electricity consumption and heavy pollution forces international community to muse upon not only traditional energy sources, but also to look closer at renewable energy sources such as wind, solar, biomass and small-scale energy using water.

Small-scale energy – is the most cost-effective solution to energy problems today. According to specialists, technical hydropower potential of small Russian rivers is 170 billion kilowatt-hours per year, which allows building small hydropower plants with total installed capacity of over 40 GW [1]. There are more than 200 hydraulic structures in Europe [2]. The use of mini-hydropower plants in these countries provides an additional amount of electricity without major capital investment.

It follows from the foregoing, that creation of free-flow micro-HPPs is an urgent task. It is necessary to develop and design a prototype of operating autonomous free-flow low power hydro turbine, using modern 3D-printing technology, widely implemented in the modern world.

2 Research and development

Research of sci-tech literature, which describes the existing installations, both in Russia and Europe was conducted to create a turbine model. Analysis of existing prototypes on the following parameters was conducted: the number of revolutions at low flow rates, size, ease of design, reliability and behavior of the water flow.

* Corresponding author: jankovsky@tpu.ru

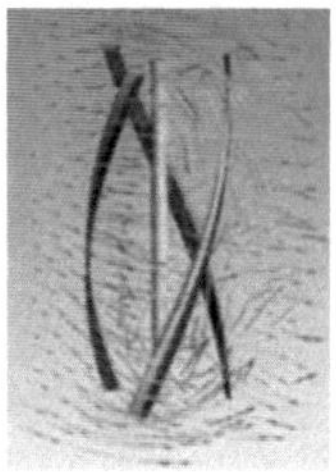

Fig. 1. lines of flux in helical blades turbine [3]. **Fig. 2.** 3D-model of the turbine.

It can concluded analyzing behavior of 3D-model of the turbine (figure 1) with twirled (helicoid) blades, that this model has advantages over other types:

- High speed of uniform rotation at a relatively slow liquid flow (low pressure liquid);
- High efficiency;
- Absence of torque fluctuations;
- Absence of visible signs of cavitation in water at a high speed;
- Self-start in slow waters or wind.

Results of the analysis showed that the model of three-blade helicoid turbine invented by A.M. Gorlov best suites the declared parameters. [5]

Standard profile NACA 4418 is taken as an profile of the blades [5].

Models of blades and two bases for the turbine were designed and visualized with the use of Solid Edge program (figure 2).

Prototype was obtained with the use of 3D-printing technology. Experimental hydraulic test setup was constructed for tests and study of heat and mass transfer processes in the mobile nodes of micro-turbine and subsequent upgrades. This setup allows comparing data of mathematical modeling with natural tests.

Hydraulic test setup allows simulating processes of water flow in rivers by the use of hydraulic pumps with frequency regulation. Created conditions are as realistic as possible (there are portions of fluid flow similar to the laminar, which are used for tests) (figure 4). This stand is applicable not only for scientific purposes, but also for training undergraduates in order to improve their skills. [6-8]

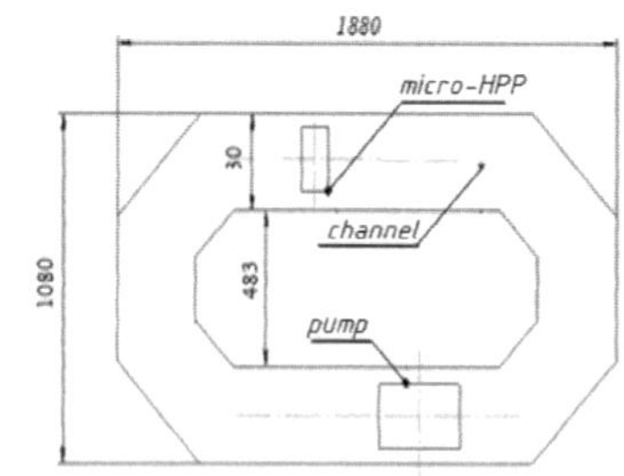

Fig. 3. Hydraulic test setup.

Compact generator is required to convert kinetic energy of the turbine into electrical. It allows to generate electricity capable of charging mobile devices and computers. During the study of generators design, it was found out that output is reduced to zero 3 times per revolution in single-phase generator, while induction generator has a complex and bulky design concept. Three-phase generator with permanent magnets has an optimum design, is easy to operate and manufacture and has small size with good output power. Based on the foregoing, creation of the prototype was based on the three-phase generator design. Modeling and design of the generator was performed in the Solid Edge program (figure 5). Elements of experimental model were created using the 3D-printer. Currently, components are being assembled.

Created model of hydro turbine is in stage of testing. Generator is in assembly stage.

Fig. 4. 3D-model of the generator, a) component parts; b) exploded view; c) assembly of the model.

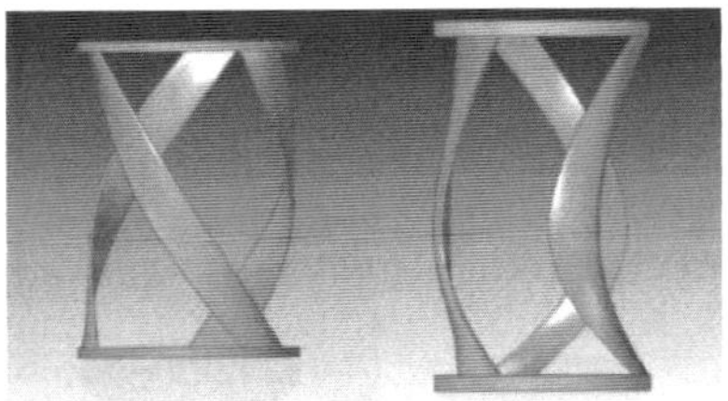

Fig. 5. Possible blades modifications.

3 Conclusion

Unique hydraulic test setup for the tests of micro-turbines in the water flow was created;

Installation of hydro turbine was designed and manufactured;

Three-phase generator was designed.

Main research areas: necessity to improve the blades (figure 6), improve performance of micro hydro turbine and synchronize with the developed generator. Explore the most optimal modes of micro HPP operation. Explore processes of heat and mass transfer in mobile nodes of hydro unit; compare simulation results with real experiments.

Acknowledgments

Work is performed within the program of increasing competitiveness of TPU.

References

1. V.I. Dobrohotov, V.M. Kargiev, A.B. Belov, D.S. Strebkov, A.K. Sokolsky, M.B. Sachs, V.V. Afyan. Renew. En. (2005)
2. S. Yankovsky, *Basics Thermal Energy Technology* (Publishing house TPU, Tomsk, 2010)
3. Taco Anton Niet, *Modelling Renewable Energy at Race Rocks* (University of Victoria, 1988)
4. V.V. Ermolaev, N.N. Gudkov, A.N. Babiev, O.N. Petrova O.N, S.A. Galaev, A.I. Kirillov, E.M. Smirnov, Therm. Eng. **4** (2007)
5. A. M. Gorlov, A. N. Gorban, V. M. Silantyev, J. En. Res. Tech. (2001)
6. K. B. Larionov, S.V. Lavrinenko, V.E. Gubin , G.A. Kitaev, Procedia – Soc. and Beh. Sc. **206** (2015)
7. S. V. Lavrinenko et. al., Procedia – Soc. and Beh. Sc. **206** (2015)
8. S. A. Yankovsky et. al., MATEC Web Conf. **37** (2015)

THE INFLUENCE OF GAS DIFFUSION ON THE DYNAMICS OF A SPHERICAL LAYER OF VISCOUS INCOMPRESSIBLE LIQUID AND HEAT AND MASS TRANSFER IN IT

Ekaterina Rezanova[1,2], and *Alla* Zakurdaeva[1,2,*]

[1]Altai State University, 656049 Barnaul, Russia
[2]Kutateladze Institute of Thermophysics SB RAS, 630090 Novosibirsk, Russia

Abstract. The formation of spherical microballoons in the case of short-term weightlessness is investigated numerically. The algorithm of the numerical solution of the problem is described and the results of numerical studies of the formation of liquid glass microballoons, saturated with carbon dioxide, are presented. The results of calculations for the problem in the full statement (mathematical model includes the influence of inertial, thermal and diffusive factors) and simplified statement, when the process of gas diffusion is not taken into account, are compared.

1 Introduction

The results of numerical investigation of the dynamics of a spherical liquid layer containing a gas bubble within itself are presented in this paper. Some quantity of the gas is dissolved in the liquid, and it is assumed that liquid with dissolved in it gas is viscous and incompressible [1-3]. The study of liquid layers, called microballoons, is connected with the investigation of such materials as sensitizers of emulsion explosives and spheroplast used in various constructions as a reinforcing additive and filler [4, 5].

The mathematical model that describes the processes within the liquid layer includes the Navier-Stokes equations and equations of the heat transfer and gas diffusion [1-3]. Inside the gas bubble the ideal gas law is carried out. The condition of short-term weightlessness, under which the statement of the problem is studying, allows us to consider a spherically symmetric process. Physical quantities depend on time and radial coordinate.

* Corresponding author: alla2300@bk.ru

2 Statement of the problem

Let $R_1(t) < r < R_2(t)$ be a domain, filled with viscous incompressible liquid, where $r = R_1(t)$ and $r = R_2(t)$ are the internal and external free boundaries of the spherical layer. The following equations are obtained as a consequence of the initial mathematical model [2] (see also [6]):

$$\frac{dV}{dt} = \frac{1}{2}V^2 \frac{(R_2^2 + R_1^2)(R_2 + R_1)}{R_2^3 R_1^3} + \frac{1}{\mathrm{Re}}\left[P_g S - P_{ex} S - 2\overline{Si}\sigma(T)\frac{R_2 + R_1}{R_2 R_1}\right]\frac{R_2 R_1}{R_2 - R_1} - \tag{1}$$

$$-\frac{4}{\mathrm{Re}}\nu(T)V\frac{R_2^2 + R_2 R_1 + R_1^2}{R_2^2 R_1^2}, \quad t > 0, \quad V(0) = V_0;$$

$$\frac{dR_1}{dt} = \frac{V}{R_1^2}, \quad t > 0, \quad R_1(0) = R_{10}; \tag{2}$$

$$\frac{d\rho_g}{dt} = -\frac{3V}{R_1^3}\rho_g + \frac{3}{Pe_d R_1}D(T)\frac{\partial C}{\partial r}, \quad t > 0, \quad \rho_g(0) = \rho_{g0}; \tag{3}$$

$$\frac{\partial T}{\partial t} + \frac{V}{r^2}\frac{\partial T}{\partial r} = \frac{1}{Pe}r^{-2}\frac{\partial}{\partial r}(r^2\chi(T)\frac{\partial T}{\partial r}), \quad t > 0, \quad T(0,r) = T_0(r); \tag{4}$$

$$\frac{\partial C}{\partial t} + \frac{V}{r^2}\frac{\partial C}{\partial r} = \frac{1}{Pe_d}r^{-2}\frac{\partial}{\partial r}(r^2 D(T)\frac{\partial C}{\partial r}) \quad t > 0, \quad C(0,r) = C_0(r). \tag{5}$$

Here V is a rate of change of the microballoon's volume, $V = r^2\upsilon$, υ is the radial component of the velocity of the liquid, ρ_g is the density of gas in the bubble, T is the temperature, C is the concentration of gas in the liquid, P_g, P_{ex} are the pressure in the buble and external pressure. It is assumed, that coefficients of kinematic viscosity $\nu(T)$, surface tension $\sigma(T)$, thermal diffusivity $\chi(T)$ and diffusion $D(T)$ depend on the temparature. The equations are presented in dimensionless form with following parameters, emerged during the transition: the Reynolds number $\mathrm{Re} = (\upsilon_* r_*)/\nu_*$, the Peclet number $Pe = (\upsilon_* r_*)/\chi_*$, the diffusive Peclet number $Pe_d = (\upsilon_* r_*)/D_*$ and dimensionless complexes $S = (r_* P_*)/(\rho_* \nu_* \upsilon_*)$, $Si = \sigma_*/(r_* P_*)$, $\overline{Si} = Si \cdot S$. Characteristic values of physical quantities are designated by *. Herewith $r_* = t_* \upsilon_*$, $\chi_* = \kappa_*/(\rho c_1)$, where ρ is the density of the liquid (characteristic density), $\kappa(T)$ is the thermal conductivity coefficient, c_1 is the heat capacity of the liquid.

Let us explain that equations (1), (2) arise out of the Navier-Stokes equations, dynamic and kinematic conditions at the free boundaries, the equation (3) represents the change in mass of the gas in the bubble, the equations (4) and (5) define the heat transfer and gas diffusion processes.

Besides kinematic condition (2) on the internal free boundary $r = R_1(t)$ the following boundary conditions are fulfilled [1-3] (see also [7] about the formulation of boundary conditions at the interface): the condition of temperature continuity $T = T_g$, the energy balance equation $\frac{1}{3}\alpha_1 P_g \frac{dR_1^3}{dt} + \alpha_2 \frac{dT}{dt} = R_1^2 \frac{\partial T}{\partial r} + \alpha_3\{\frac{d}{dt}[R_1^2 T^2 \frac{d}{dT}(\frac{\sigma}{T})] + \sigma\frac{dR_1^2}{dt}\}$, the Henry law $C = HA(T_g)P_g^n$. On the external free boundary $r = R_2(t)$ the kinematic condition $\frac{dR_2}{dt} = \frac{V}{R_2^2}$, $t > 0$, $R_2(0) = R_{20}$, the heat exchange with the atmosphere of the first type (Dirichlet boundary condition) $T = T_{ex}$ and the Henry law $C = HA(T_{ex})P_{ex}^n$ take place. Here $\alpha_1 = P_* r_* \upsilon_* /(\kappa_* T_*)$, $\alpha_2 = c_V m \upsilon_* /(4\pi r_*^2 \kappa_*)$, $\alpha_3 = \upsilon_* \sigma_* /(\kappa_* T_*)$, $H = A_* P_*^n / \rho$ are dimensionless parameters, the Biot number $Bi = \beta r_* / \kappa_*$, T_g, T_{ex} are the temperature of the gas in the bubble and external temperature, m is the mass of gas in the bubble, c_V is the heat capacity of the gas at constant volume, A is the coefficient in Henry law, n is the index in Henry law, β is the heat transfer coefficient.

Inside the gas bubble the ideal gas law is determined: $P_g = \widetilde{R}\rho_g T_g$, where $\widetilde{R} = R\rho T_* / P_*$ is a dimensionless complex, R is the universal gas constant [2].

3 Algorithm of numerical solution

The problem is solved with the help of numerical algorithm, which is described in detail in [6, 8] and includes the solution of the Cauchy problem for a system of ordinary differential equations (1)-(3) by means of the fourth-order Runge-Kutta method [9] (determination of the rate of change of the microballoon's volume V, the density of gas in the bubble ρ_g and the internal radius R_1), the transition from a region with moving boundaries to the fixed area, the construction of finite-difference schemes for the equations (4) and (5), the calculation of the temperature distribution T within the liquid layer by Thomas algorithm with a parameter [10] and the computation of the gas concentration C in the liquid layer by an ordinary Thomas algorithm [11].

4 Results of investigation

Numerical studies on the formation of spherical microballoons of liquid glass containing carbon dioxide gas are conducted. At the initial time the internal radius of the liquid layer $R_{10} = 0.02$ cm, the external radius $R_{20} = 0.05$ cm, the temperature is set as a constant $T_0(x) = 1171.1$ K, the concentration of gas in the liquid $C_0(x)$ is distributed by parabolic law. The temperature of the external atmosphere is increasing from $T_{ex1} = 1171.1$ K to $T_{ex2} = 1673$ K under the law:

$$\begin{cases} T_{ex} = T_{ex1} + \frac{(T_{ex2} - T_{ex1})}{t_2 - t_1}(t - t_1), t_1 \le t \le t_2; \\ T_{ex} = T_{ex2}, t > t_2. \end{cases}$$

Here $t_1 = 0$ s, $t_2 = 0.3$ s.

The values of the physical parameters and characteristic quantities are given in [10].

The calculation results for the system “liquid glass – carbon dioxide gas” in the case when the density of gas in the bubble at the initial time $\rho_{g0} = 0.92 \cdot 10^{-3}$ g/cm^3 and the external pressure $P_{ex} = 0.3$ atm are presented onwards. When the gas diffusion is taken into account (i.e. when the problem is considered in the full statement) the liquid layer expands more intensively, and the steady state is reached much later than in the case of the problem consideration in the simplified statement (the so-called thermal approximation) (fig. 1a). Furthermore, in the case of thermal approximation at the time t_0=*0.45* s the temperature inside the entire layer is already a constant $T=1673$ K (fig. 1b), while when the problem is solving in the full statement the constant distribution of the temperature is not achieved even when $t_0=1.2$ s (fig. 2b). fig. 2a shows that the distribution of gas concentration within the liquid layer also continues to change after the heating has stopped.

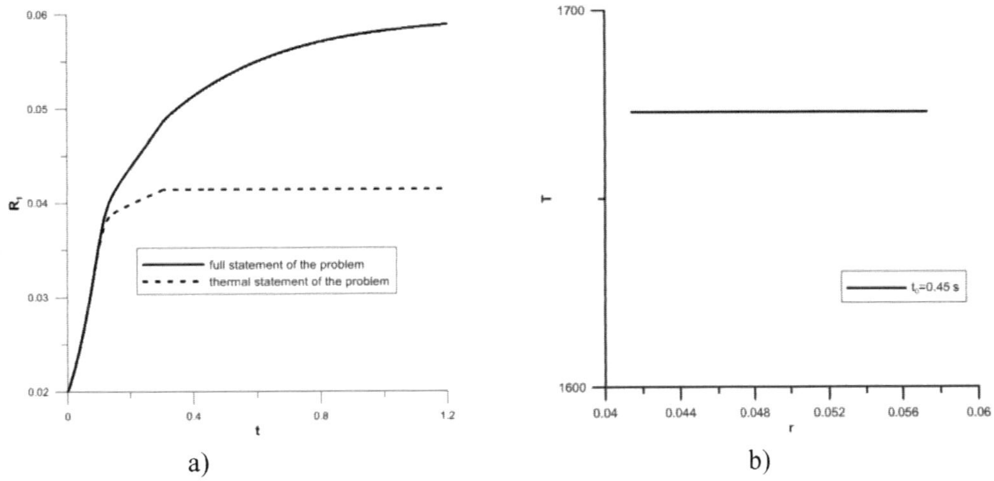

Fig. 1. a) Comparison of the internal radius R_1 variation within time for the problem in the full statement (solid line) and the problem in the thermal approximation, without the consideration of gas diffusion (dashed line), b) Temperature distribution T inside the liquid layer at the moment $t_0=0.45$ s for the thermal statement of the problem.

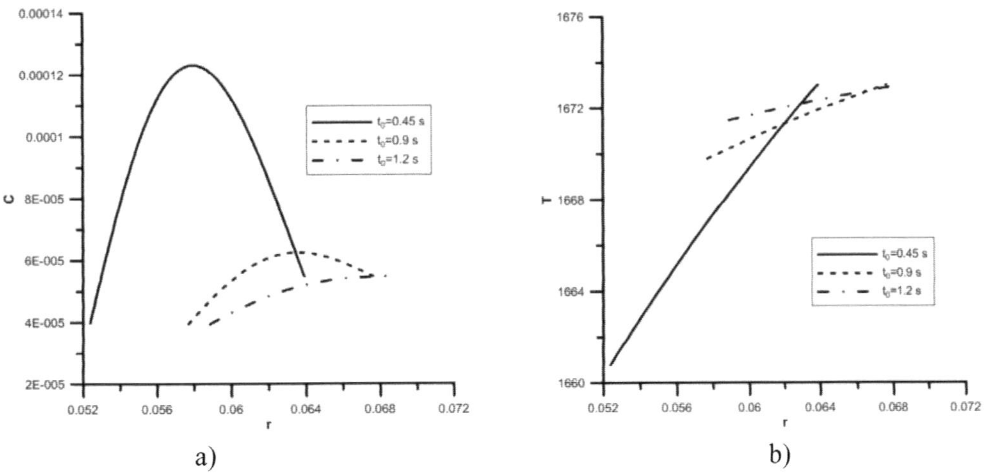

Fig. 2. The distribution of gas diffusion C (a) and the distribution of temperature T (b) inside the liquid layer at different moments of time for the problem in the full statement.

5 Conclusions

The formation of spherical shells in the case of short-term weightlessness is investigated numerically. The mathematical model of the problem and the numerical algorithm of its solving are presented. Numerical studies for the system "liquid glass - carbon dioxide gas" are conducted, and the comparison of results obtained for the problem in full and simplified statements is carried out. The calculation results show that the diffusion of gas has a significant influence on the dynamics of a spherical layer.

Acknowledgments

We gratefully acknowledge the support from the Ministry of Education and Science of Russia (Agreement No 14.613.21.0011, project identifier RFMEFI61314X0011).

References

1. O.N. Goncharova, Dinamika sploshnoy sredi **82** (1987)
2. O.N. Goncharova, V.V. Pukhnachev, Modelirovanie v mehanike **5** (1990)
3. O.N. Goncharova, Dinamika sploshnoy sredi **106** (1993)
4. A.G. Anshits, N N. Anshits, A.A. Deribas, S.M. Karakhanov, N.S. Kasatkina, A.V. Plastinin, A.Yu. Reshetnyak, V.V. Silvestrov, Fizika goreniya i vzryva **41**, 5 (2005)
5. E.V. Karpov, Prikladnaya mehanika i tehnicheskaya fizika **50**, 1 (2009)
6. E.V. Rezanova, Izvestia AltGU **77**, 1 (2013)
7. O.N. Goncharova, Izvestia AltGU **73**, 1 (2012)
8. A.V. Zakurdaeva, MATEC Web Conf. (submitted for publication)
9. N.N. Kalitkin, E.A. Alshina, *Numerical analysis* (Academia, Moscow, 2013)
10. A.V. Zakurdaeva, E.V. Rezanova, Omskiy nauchniy vestnik **3**, 143 (2015)
11. A.V. Zakurdaeva, E.V. Rezanova, Prikladnaya matematika i fundamentalnaya informatika **2** (2015)

MATEC Web of Conferences **72**, 01131 (2016) DOI: 10.1051/matecconf/20167201131
HMTTSC-2016

RESEARCH OF HEAT RATES EFFECT ON THE PROCESS OF FUEL-BED GASIFICATION OF "BALAKHTINSKOE", "OSINNIKOVSKOE", "KRASNOGORSKOE" AND "BORODINSKOE" COAL DEPOSITS

Andrey Zenkov[1], *Stanislav* Yankovsky[1,*], *Anastasia* Matveeva[1], *Sergey* Lavrinenko[1], and *Alexander* Gromov[2]

[1]Tomsk Polytechnic University, 634050 Tomsk, Russia
[2]Technical University Nuremberg Faculty of Chemical Engineering Particle Technology, 90489 Nuremberg, Germany

Abstract. Experimental research of fuel-bed gasification at different heating rates was conducted. Release of four gases (CO, NO, H_2O, CO_2) was determined. Optimal heating rate mode for this method of gasification was established.

1 Introduction

It is known [1], that almost all the coal mined in Russia is used in production of energy by direct combustion in thermal power plants; only 15% is used in metallurgy and 5% in chemical industry [1].

In order to implement the directive №2008/1/EC three directions of efficient use of energy resources are identified in Russia:

- gasification to produce synthesis gas, hydrogen and other reducing gases for chemical industry;
- gasification of low-rank coals, followed by burning of the treated gas at thermal power plants;
- production of natural gas substitute used in industrial, energy and domestic sectors of the economy [1].

Research of gas intensification processes from coal pyrolysis in a dense layer was performed by simultaneous thermal analysis method using Netzsch STA 449 F3 Jupiter and QMS 403 D Aeolos [1-2].

* Corresponding author: jankovsky@tpu.ru

2 Experimental setup

In the preliminary phase of study of initial fuels gasification their elemental analysis was determined using a scanning electron microscope (SEM) JEOL JCM-6000 [1-4]. Preparation of the samples was carried out in accordance with GOST 10742-71, followed by sieving in accordance with GOST 3306-88. Particle size of the test powder was less than 80 microns. Performed chemical analysis of initial coal samples allowed obtaining elemental composition of raw material (Table 1).

Table 1. Elemental composition of raw material.

Element	Composition of element, mass.%			
	Anthracite	Bituminous coal	Brown coal	
	A	T	2B	3B
C	82.14	71.05	65.99	82.65
N	11.26	-	12.55	13.70
O	4.04	19.26	14.25	0.06
Al	0.51	2.02	0.84	0.33
Si	1.36	3.04	2.72	0.29
S	0.08	0.42	0.16	0.17
Ca	0.61	2.63	1.84	2.30
K	-	0.59	0.40	0.04
Fe	-	0.56	1.25	0.46
Na	-	0.03	-	-
Mg	-	0.13	-	-
Ti	-	0.09	-	-
Zr	-	0.18	-	-

Relying on the obtained research results, output of gases that should be recorded in mass-spectrometer during experiments on modeling of fuel-bed gasification was predicted [3-4].

Prepared samples with weight of twenty milligrams were analyzed at different heating rates of 278, 283, 293, 303 K/min. Argon consumption for blowdown was 10 ml/min, air consumption – 50 ml/min. Maximum heating temperature was 1573 K.

3 Results

A series of experiments was performed and dependences of changes of produced synthesis gas recorded mass from time were obtained. Relative area is the ratio of certain integral summation area to weight of coal sample minus the ash residue. The results of the synthesis gas studies are shown in figure 1.

Relative integrand area of gases release A, heating rate B [1].

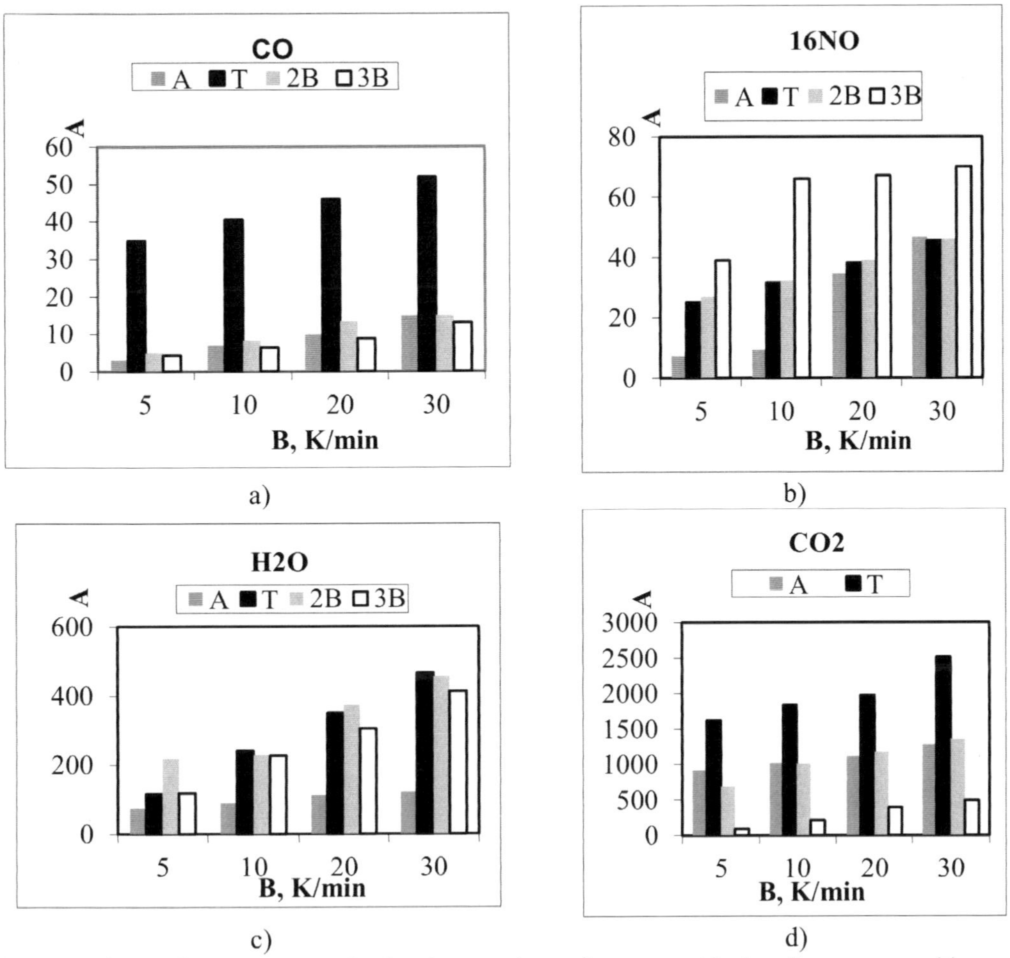

Fig. 1. Release of gases at respective heating rate (a - carbon monoxide, b - nitrogen monoxide, c - water vapor, d - carbon dioxide).

Analysis of the results presented in figure 1 shows that combustible gases CO, NO, H_2O, CO_2 produced at heating rate of 30 K/min have the highest intensification of emission for all grades of coal. Gases emission is lower at other heating rates. figure 1 shows that with an increase of heating rate intensification of gas increases.

These phenomena are explained by the fact that the process of thermal degradation is determined by the simultaneous occurrence of a whole range of consequential-parallel reactions with different activation energies and kinetic parameters.

It should be noted that the influence of heating rate appears as formation of volatile substances.

4 Conclusion

Studies on modeling of raw coal samples fuel-bed gasification allowed determining composition of gases (CO, 16NO, H2O, CO2). It was established that optimal heating rate mode for this method of gasification is 30 K/min. Thus, there is a maximum emission of gas suitable for the future technological applications.

Acknowledgments

The research was realized with financial support of Minobrnauki of Russia in framework of FTP "Research and development in prior direction of scientific-technological complex of Russia in 2014-2020 years", unique R&D identifier RFMEFI58114X0001.

References

1. S. A. Yankovsky et al. MATEC Web Conf. **37** (2015)
2. K. B. Larionov et al. MATEC Web Conf. **37** (2015)
3. G. D. Shiling, B. Bonn, U. Kraus, *Coal Gasification* (Nedra, Moscow, 1986)
4. J. Goulstain *Scanning electron microscopy and X-ray analysis* (Mir, Moscow, 1984)

MATHEMATICAL MODELING OF HEAT FRICTION CONTACT MASTER BELT WITH THE GUN MOUNT BARREL DURING THE PROCESS OF HIGH-SPEED MOTION

Jaroslav Zezulinsky[1,*], *Anatolyi* Guskov[1], and *Konstantin* Milevsky[1]

[1]Novosibirsk State Technical University, Novosibirsk, Russia

Abstract. The friction in the gun mount barrel at sliding speeds of artillery projectile 500 - 700 m/s is not sufficiently studied. The main problem is to increase the efficiency of the master belt with a significant increase of the interaction parameters of the barrel with the projectile. To determine the effect of heating on the change of physical and mechanical properties of the surface layer and friction coefficient on the surface of the master belt were made mathematical modeling of heat transfer.

The friction in the gun mount barrel at sliding speeds of artillery projectile 500 - 700 m/s is not sufficiently studied. A distinctive feature of the system "gun mount barrel - artillery projectile" is an intense plastic deformation in the process of formation of the master belt, which provides a seal during high-speed motion of an artillery projectile on the core surface of the gun mount barrel [1, 2, 3]. The main problem is to increase the efficiency of the master belt with a significant increase of the interaction parameters of the barrel with the projectile. It is known that during the motion of the projectile through the gun mount barrel its contact temperature on the surface of the master belt reaches a considerable value, which leads to melting of the surface layers of the master belt material [4, 5, 6]. To determine the effect of heating on the change of physical and mechanical properties of the surface layer and friction coefficient on the surface of the master belt three areas that interact with the surface of the rifling of the gun mount barrel have been selected. They are R1 –field rifling, R2 – fighting line and R3 - the bottom of the rifling (fig. 1 a, b, c), respectively.

* Corresponding author: stroncyi@yandex.ru

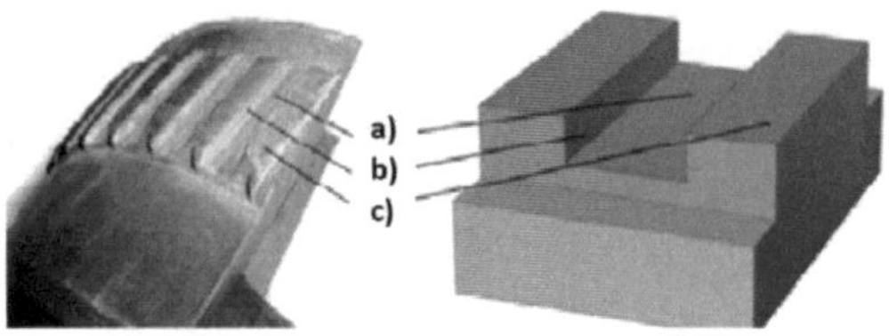

Fig. 1. Areas for the solution of the problem of determining the temperature at the surface contact between the master belt and the barrel.

The surface temperature T at any point overlapping off the heat source is [7, 8, 9]:

$$T = 2Q\,(t/\pi k\rho c)^{\frac{1}{2}}$$

where t - is the propagation time of the heat source (master belt) through the point of the barrel, which is equal to t = B/V; k, ρ, c – thermal conductivity, density and specific heat of the material half-space.

The source of heat is frictional contact of "master belt – gun mount barrel". Power dissipation in the contact area is Q = fpV where p is the specific pressure.

The temperature at the contact surface is described by the expression:

$$T_{max} = 2fP\sqrt{\frac{B/V}{\pi\rho kc}}.$$

To determine the dependence of the speed of projectile on the barrel, the results obtained by the authors were used. The coefficient F of the friction pair "steel – copper" for the initial time is taken as equal to 0.3. The specific pressure is constant and equals the material fluidity strength of the master belt. According to the theory the "flashpoint" for metals is determined by the melting point. For the material M1 of the model master belt of the projectile, it is equal to 1053 °C, which corresponds to the melting point of 1083°C minus the ambient temperature equal 30°C.

To solve the problem of heat conduction in the master belt material, the solution scope was selected. The Cartesian coordinate system was used. The curvature of the surface was not taken into account, as the diameter of the master belt is much greater than the width of the bottom, field and height of rifling. The origin was located at the intersection of the axis of symmetry of the projection and the matching surface of the master belt with the casing of an artillery projectile.

The heat conduction equation has been solved:

$$\rho C_p \frac{\partial T}{\partial t} = \lambda\left(\frac{\partial^2 T}{\partial x^2} + \frac{\partial^2 T}{\partial y^2}\right),$$

The initial (t = 0) and boundary conditions are:

$$\frac{\partial^2 T}{\partial x^2} = 0, \left| \begin{array}{l} X = 0,\ 0 \le Y \le Y_2 \\ X = X_2,\ 0 \le Y \le Y_1 \end{array} \right. \qquad \begin{array}{ll} T = T_{R_1}(t) & Y = Y_1\ X_1 \le X \le X_2 \\ T = T_{R_2}(t) & X = X_1,\ Y_1 \le Y \le Y_2 \end{array}$$

$$\frac{\partial T}{\partial x} = 0,\ Y = 0,\ 0 \le X \le X_2 \qquad T = T_{R_3}(t) \quad Y = Y_2,\ 0 \le X \le X_1$$

From fig. 2 it is seen that in all areas of contact according to temperature changes depending on the time of projectile motion are identical, but the most intensive temperature increase occurred in the initial period of motion of the projectile through the barrel (up to t = 0,023s) in the area R1.

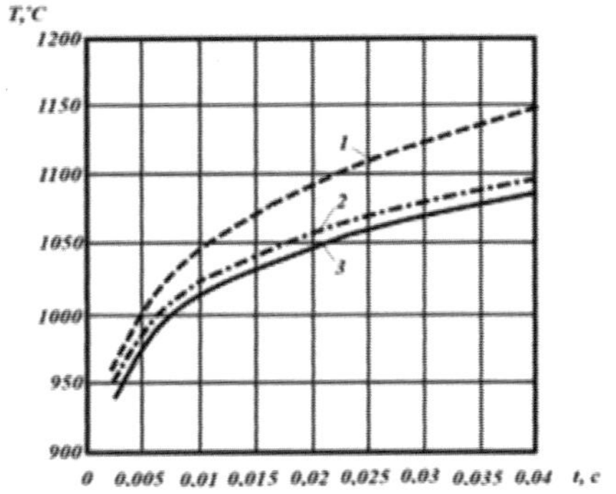

Fig. 2. Temperature variation on the surface areas of master belt during motion of the projectile through gun mount barrel: 1) R1; 2) R2 3) R3.

Thus, it is possible to draw the following conclusions:

• Under the action of frictional contact "Barrel – master belt – artillery projectile" is an intensive heating of near-surface deformed layer of the copper master belt of artillery projectile.

• As a result of intensive heating of the surface material of the master belt is partially recrystallized and becomes molten.

• Change of physical and mechanical properties of the master belt material during recrystallization leads to the increased elongation and reduced tensile strength.

• As a result of changes in physical and mechanical properties and the intense heating the transfer of thin layers of copper on the surface of the barrel occurs.

References

1. A. Gus'kov, Vestnik akademii voennyh nauk (2012)
2. V. Balakin, *Trenie i iznos pri vysokih skorostjah skol'zhenija* (1980)
3. R. Montgomery, Wear **38**, 2 (1976)
4. Jettls, S. *Vlijanie teplovyh jeffektov na trenie pri vysokih skorostjah skol'zhenija* (1986)
5. A. Gus'kov, K.. Milevskij, E. Slesareva, Izv. Vys. Uch. Zav. Fiz. **56**, 6–3 (2013)
6. A. Gus'kov, N. Gorelova, K. Milevskij, A. Sil'chenko, Trudy XIV vseros. Nauch.-tehn. Konf. Nauka Prom. Ob. (2013)
7. H. Blok, Proc. Inst. Mech. Eng. **2** (1937)
8. A. Gus'kov, K. Milevskij, N. Chajka, Ja. Zezjulinski, Mezh. nauch. konf. "SibOptika-2015" (2015)
9. A. Gus'kov, Ja. Zezjulinskij, K. Milevskij, E. Nezhivyh, Conf. Nauka. Promysh. Ob. (2015)

NUMERICAL ANALYSIS OF INFLUENCE OF ENGINEERING CONSTRUCTION IN THE AREA OF PLACING FUEL STORAGE TANKS ON THEIR HEAT LOSS

V. Yu. Polovnikov [1,*], *F.* T. Makhsutbek[1], *Zh.* F. Ozhikenova[1], and *I.* K. Zharova[2]

[1]National Research Tomsk Polytechnic University, 634050 Tomsk, Russia
[2]Scientific Research Institute of Applied Mathematics and Mechanics, 634050 Tomsk, Russia

Abstract. This paper describes the numerical modeling of heat transfer in the area placing of the tank for storage fuel of thermal power plant and boiler with the influence of engineering construction. We have established that the presence of engineering structures in the area of placing of the tank for storage fuel of thermal power plant and boiler have little effect on the change of heat loss.

1 Introduction

The research of thermal modes of different objects (a residential building [1], a hot water pipes [2], a fuel storage tank et al.) is an important problem.

Currently, the method of calculation and design of a fuel storage tank does not take full account of all processes in a tank, that occur during the preparation of a fuel oil for combustion.

One of the most promising approaches to research of different equipment is the use of a computer programs. They take into account various effects and processes in the system of heat transfer.

The objective is the mathematical modeling of the thermal loss of fuel storage tank and numerical analysis of thermal conditions of the tank.

2 Problem statement

We consider a typical vertical cylindrical steel underground tank. Figure 1 shows a schematic representation of the field solution of the problem. For the domain under consideration (fig. 1) we solve a 2D linear and stationary problem of heat transfer in conditions of thawing of the soil.

Formulating the problem, we used the following assumptions:

1. The heat transfer processes in the internal and the external environment are disregarded.

* Corresponding author: polov@tpu.ru

2. Thermophysical characteristics of materials used in the analysis are constant and known values.

3. There is an ideal thermal contact conditions at the boundaries.

4. The heat in the decision domain (fig. 1) is transferred only by conduction.

The listed assumptions, on the one hand, do not impose constrains of principle on the physical model of the system (fig. 1), but, on the other hand, allow one to simplify in a certain manner the algorithm and method for solving the posed problem.

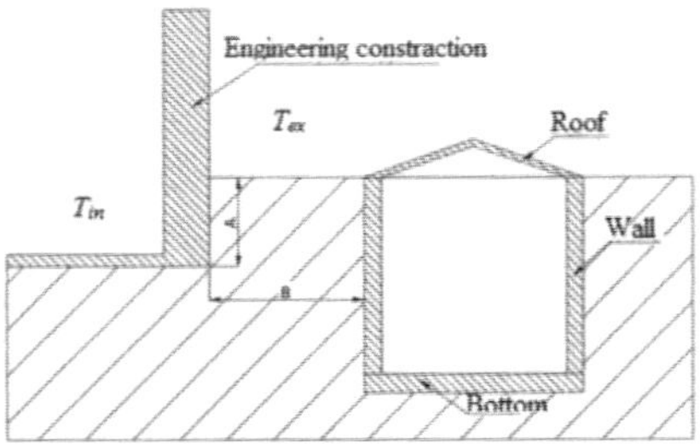

Fig. 1. Schematic representation of the field solution of the problem.

3 Mathematical model

The processes of heat transfer to the object under consideration (fig. 1) will be described by the following equation:

$$\nabla^2 T_{ins} = 0; \tag{1}$$

$$\nabla^2 T_g = 0; \tag{2}$$

$$\nabla^2 T_f = 0; \tag{3}$$

$$T_{ins,1} = T_{oil} = \text{const}; \tag{4}$$

$$\lambda_{ins}\,\text{grad}(T_{ins,2}) = \lambda_g\,\text{grad}(T_{g,2});\; T_{ins,2} = T_{g,2}; \tag{5}$$

$$\lambda_g\,\text{grad}(T_{g,3}) = \lambda_f\,\text{grad}(T_{f,3});\; T_{g,3} = T_{f,3}; \tag{6}$$

$$-\lambda_g\,\text{grad}(T_{g,4}) = \alpha(T_{g,4} - T_{ex}); \tag{7}$$

$$-\lambda_f\,\text{grad}(T_{f,5}) = \alpha(T_{f,5} - T_{ex}); \tag{8}$$

$$-\lambda_f\,\text{grad}(T_{f,6}) = \alpha(T_{f,6} - T_{in})\,; \tag{9}$$

$$-\lambda_f\,\text{grad}(T_{f,7}) = \alpha(T_{f,7} - T_{in}); \tag{10}$$

$$\text{grad}\,(T_g)=0,\; x\rightarrow\pm\infty,\; y\rightarrow-\infty; \tag{11}$$

$$\text{grad}\,(T_f)=0,\; x\rightarrow\pm\infty,\; y\rightarrow+\infty. \tag{12}$$

4 Method of solution and Initial data

The system of equations (1)–(8) was solved by the finite element method [3, 4], using the Galerkin approximation [5]. The investigations were carried out on a nonuniform finite-element mesh. The number of elements was chosen from conditions of convergence of solution, the mesh was made denser by the Delaunay method [5].

In spite of the fact that in statement of the problem it was assumed to use an infinite-size domain (expressions (11) and (12)), in the numerical analysis of heat loss we used a finite size of calculation domain. The sizes of the calculation domain were chosen on the basis of a series of preliminary numerical experiments in such a manner that the relative change of the temperature gradients at the domain boundary does not exceed 0.5%.

The analysis was carried out for a vertical steel tank with the thermal insulation of mineral wool (50 mm thick). The temperature of an oil was $T_{oil} = 353°K$. The ambient temperature T_{ex} taken equal to the average air temperature for the heating period in the city of Tomsk. The air temperature inside the engineering structure was equal to $T_{in} = 293.15$ K. The coefficients of heat transfer in all variants of the numerical analysis were $\alpha = 15$ W/(m^2·K). The change of temperature in depth of soil is 3.6 degrees per 100 meters. The distance from the external tank insulation to the basement was B = 20 m, and the deepening of the foundation was A = 2 m (fig. 1).

Table 1 contains values [2, 4] of thermophysical characteristics, which were used in the numerical investigations of thermal conditions of the system under consideration (fig. 1).

Table 1. Thermal properties of materials.

Material	Density, [kg/m^3]	Thermal conductivity, [W/(m·K)]	Heat capacity, [J/(kg·K)]
Mineral wool	150	0.046	840
Reinforced concrete	2200	1.54	887
Ground	1835	1.42	1155

5 Results of numerical simulation

The main results of numerical modeling of thermal conditions of the system under consideration (fig. 1) are listed in Table 2, 3 and in figures 2, 3.

Table 2 lists the results of numerical experiments of the heat loss of a vertical cylindrical steel underground tank (Q_1 is the heat loss without engineering construction and Q_2 is the heat loss with considering engineering construction).

Validity and reliability of the obtained results follow from tests of the methods for convergence and stability of solutions on multiple meshes, fulfillment of the energy balance conditions at boundaries of the calculation domain. The relative calculation error in all versions of the numerical analysis did not exceed 0.5%, which is acceptable for investigations of thermal conditions of the system under consideration (fig. 1).

Figures 2 and 3 show the typical temperature fields in the area of storage tanks for fuel thermal power plants and boiler plants.

Numerical analysis results indicate that the presence of thawing of soil in the container placement area for the storage of fuel thermal power plants and boiler houses does not lead to significant changes in the thermal regime of the object.

Figures 2 and 3 show the typical temperature fields in the area of storage tanks for fuel thermal power plants and boiler plants, taking into account the thawing of sandy and clay soil.

Table 2. The results of heat loss underground tank without taking a fixed ground temperature.

Month	T_{ex}, [°C]	Q_1, [W]	Q_2, [W]	$\lvert(Q_1 - Q_2)/Q_1\rvert 100\%$
September	+9.2	6519.76	6542.28	0.34
October	+0.9	7666.57	7704.85	0.50
November	–10.4	9228.60	9310.08	0.88
December	–17.5	10233.07	10349.46	1.12
January	–19.2	10449.10	10587.58	1.31
February	–16.7	10121.49	10214.96	0.92
March	–10.1	9182.77	9245.62	0.68
April	–0.1	7782.53	7822.48	0.51
May	+8.6	6603.88	6648.76	0.68
June	+15.3	5665.41	5687.85	0.39

Table 3. The results of heat loss underground tank based on a fixed soil temperature.

Month	T_{ex}, [°C]	Q_1, [W]	Q_2, [W]	$\lvert(Q_1 - Q_2)/Q_1\rvert 100\%$
September	+9.2	6278.26	6323.88	0.72
October	+0.9	7507.07	7574.23	0.89
November	–10.4	9103.06	9185.61	0.90
December	–17.5	10106.08	10214.12	1.06
January	–19.2	10346.24	10519.14	1.64
February	–16.7	9992.97	10102.27	1.08
March	–10.1	9060.87	9157.06	1.05
April	–0.1	7648.47	7714.04	0.85
May	+8.6	6419.66	6469.05	0.76
June	+15.3	5473.19	5504.81	0.57

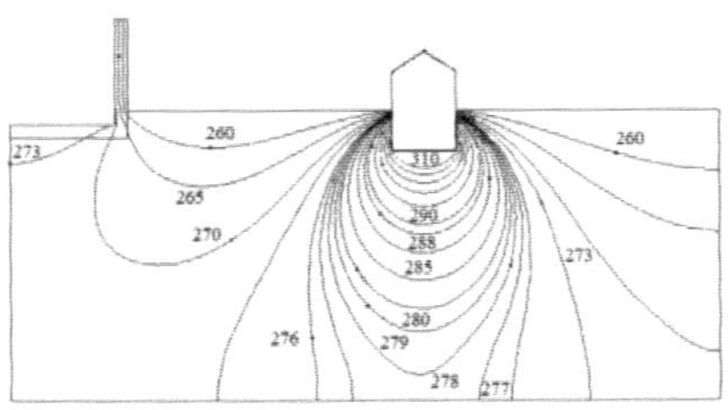

Fig. 2. Typical temperature field in the zone of the storage tanks for fuel thermal power plants and boiler at ambient temperature –40 °C, taking into account the impact of engineering structures.

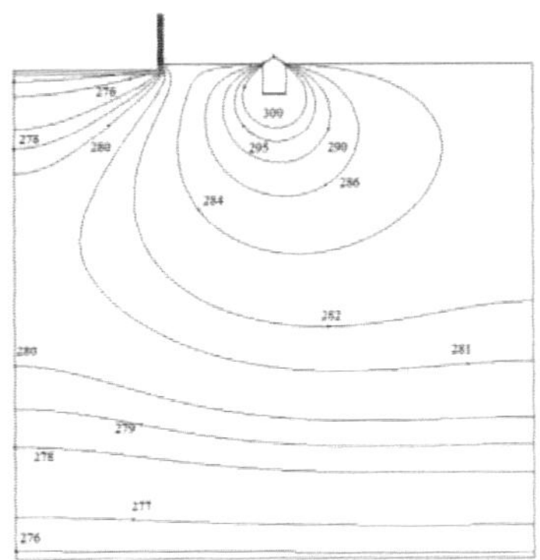

Fig. 3. Typical temperature field in the zone of the storage tanks for fuel thermal power plants and boiler at ambient temperature – 40 °C, taking into account the impact of engineering structures, consider a fixed temperature of the soil.

6 Conclusion

Numerical analysis of heat loss of storage tanks for fuel thermal power plants and boiler plants (Table 2, 3 and Fig. 2, 3) indicate that the presence of engineering structures in the area of their distribution have little effect on the change of heat loss, so you can ignore this indication. The maximum deviation in the calculations (table 2 and 3) is not more than 2 %. Therefore the influence of engineering structures in the area of storage tanks for fuels, power plants and boilers can be reasonably neglected.

The result of numerical analysis of heat loss of storage tanks for fuels, power plants and boilers it is possible to say that the introduction of a fixed temperature on the lower boundary of the solution has little effect on the changes of heat loss (table 3, figure 3), but significantly affect the temperature distribution in the area of the tank.

Notations

T – temperature, K; λ – coefficient of thermal conductivity, W/(m·K); α – the heat transfer coefficient, W/(m^2·K).

Indexes: ins – a layer of heat insulation; g – ground; oil – oil; f – foundation engineering structure; in – internal; ex – external; 1 – the inner surface of the tank insulation; 2 – the boundary between "shell of tank – ground"; 3 – the boundary between "soil – foundation engineering structure"; 4 – the boundary between "soil – environment"; 5 – the boundary between "soil – engineering structure"; 6 – the boundary between "foundation – the air inside the engineering structure" 7 – the boundary between "basement floor – the air inside the engineering structure"

Acknowledgments

The work was supported by the Russian President's grant (Scientific School project - 7538.2016.8).

References

1. S. Korniyenko, Procedia Engineering **117**, 1 (2015)
2. V.Yu. Polovnikov, E.V. Gubina, J. Eng. Phys. Thermophys. **87** (2014)
3. A.L. Garcia, *Numerical methods for physics* (Prentice Hall, New York, 2000)
4. V.Yu. Polovnikov, E.S. Glazyrin, EPJ Web Conf. **82** (2015)
5. J. N. Reddy *An Introduction to Nonlinear Finite Element Analysis* (Oxford University Press, New York, 2004)

MATEC Web of Conferences **72**, 01134 (2016) DOI: 10.1051/matecconf/20167201134
HMTTSC-2016

DETERMINATION OF THE MINIMAL AMOUNT OF WATER FOR EFFECTIVE SUPPRESSION OF THE THERMAL DECOMPOSITION OF FOREST COMBUSTIBLE MATERIALS

Alena O. Zhdanova[1,*], *Galina* S. Nyashina[1], and *Maria* M. Osmolovskaya[1]

[1]National Research Tomsk Polytechnic University, 634050 Tomsk, Russia

Abstract. Forest fires are big problem for whole the world community. The development of new effective methods is needed to increase the efficiency of the firefighting. We have investigated experimentally the suppression of thermal decomposition of different typical forest combustibles using water aerosol. Droplet sizes were 0.02-0.2mm; the concentration - $3.8 \cdot 10^{-5}$ m^3 of water/m^3, the flow rate - 0.00035 l/s, flow velocity - 2 m/s. Registration of the aerosol propagation and interaction with combustibles was done by high-speed video camera using Shadow Photography and Particle Tracking Velocimetry methods. The effective water volumes for fire suppression were determined together with corresponding suppression times. The obtained results could be used for improvement of the fire-fighting technologies.

1 Introduction

Forest fires are sharp problem for whole the world community [1-3]. Determination of the minimal effective water volume for optimal fire-fighting is the key problem in this area. Hundreds or even thousands tons of water are usually needed for suppression of forest fire. However the usage of greater and greater water amounts does not correspond to acceleration of the forest fire suppression. There are some recommendations about water atomization before injection into the fire site to cover widest area. Though there is not an effective realization of the proposed ideas still.

The aim of this work is experimental determination of the minimal water amount that is enough for effective suppression of the thermal decomposition of forest combustible materials (FCM) during pre-defined time interval.

* Corresponding author: zhdanovaao@tpu.ru

DOI: 10.1051/matecconf/20167201134

2 The methods and the experimental setup

The scheme of the experimental setup is shown at fig. 1. The experiments were done in two stages. At the first one we have registered the details of parching of the model fire sites for each type of investigated FCM with determination of characteristic times t_b. At the second one, we have measured the fire suppression times t_e for model fire sites.

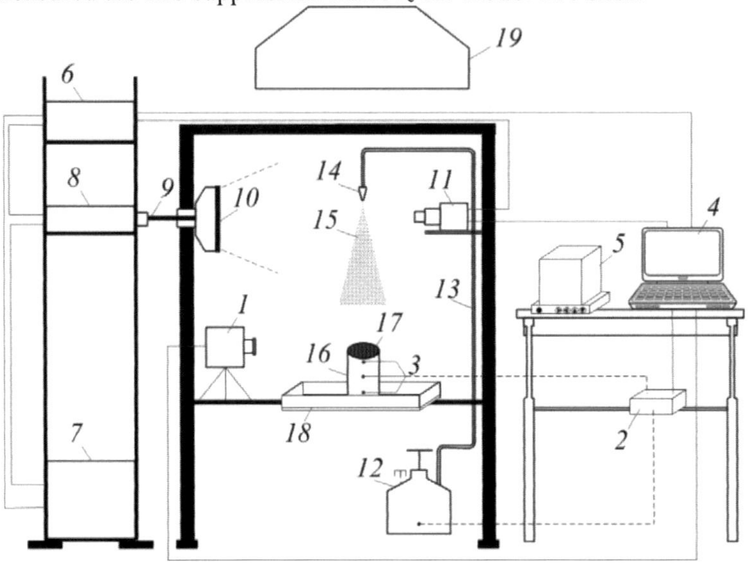

Fig. 1. Scheme of experimental setup: *1* – high-speed camera; *2* – high-speed analog input module; *3* – thermocouples; *4* – workstation (PC); *5* – laboratory microbalance; *6* – signal synchronizer; *7* – laser generator; *8* – solid double-pulse laser; *9* – optical fiber; *10* – diffuse screen; *11* – cross-correlation camera; *12* – container with water; *13* – water supply channel; *14* – spray nozzle; *15* – water mist; *16* – cylinder with forest fuel; *17* – forest fuel; *18* – metal tray; *19* – exhaust ventilation.

Four types of typical FCM were used during experiments: the birch leaves, spruce needles, aspen branches and the mix of them (with mass ratios 25 %, 15 % and 60%).

The first stage includes following steps:

- the cylinder *17* with portion of the FCM was installed onto fire-resisting tray *19*;
- The needle-shape thermocouples *3* (K-type, with working range 223–1473 K, bias ± 3 K andtime lag less than 1 s) were placed at three points on the symmetry axe of the cylinder 17 ;
- The time count was started from the moment of the ignition of the model fire site (with precision 0.01 s);
- The time gap between ignition and the moment of temperature decrease lower than the thermal decomposition threshold (T_f≈370 K) for all three thermocouples was estemated as total parching time t_b for chosen type of the FCM [4–6]. The stable exceed of the layer temperature above mentioned threshold was assumed as stable combustion of the FCM.

The second stage of the experiment (namely extinguishing of the fire site) consists of water aerosol feeding to the FCM portion surface. The water atomization was done by system consisting of nozzle *14*, tank for pressurized water *12* and supporting fittings *13*. The water feeding was started in 10-15 s after ignition (this time is enough for establishment of the thermal decomposition at whole the FCM portion). The water aerosol feeding was supported up to the moment of the total extinguishing of the model fire site.

The water droplet sizes were controled by the optical diagnostic system [7] using “Shadow Photography” (SP) method. The droplet velocities were measured by the same system using “Particle Tracking Velocimetry” (PTV) method [8].

We assumed that the thermal decomposition of the FCM is started if the temperature measured by one of the thermocouples (which are installed at different heights of the FCM portion) was exceeding 370 K. The end of the thermal decomposition was assumed to the moment when the temperature of all the thermocouples become lower than 370 K that is lower than active pyrolysis threshold T_{ef}=370 K. The fig. 2 shows typical views of the FCM portions during the experiment: initial state, active combustion, at the fire extinction and after the fire extinguishing by the water aerosol.

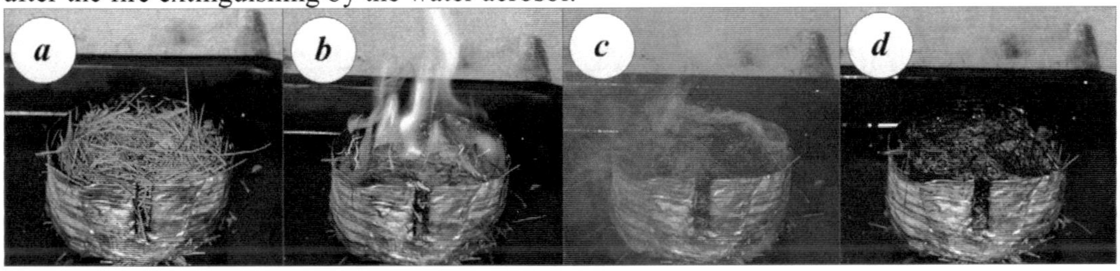

Fig. 2. Typical states of the spruce needles portion during the simulation fo the forest fire extinguishing by aerosol flow (the layer thickness is h_f=40 mm, diameter d_f=100 mm) . The states are: *a* – initial; *b* –during the active combustion; *c* –during the extinguishing; *d* – after the total fire extinction.

The water residue at the tray *19* (under the FCM portion) was collected after each experiment for analysis of its volume V_s. The water amount used for fire extinction was calculated by formula (1):

$$V_w=\mu_w \cdot t_e - V_s. \quad (1)$$

where μ_w·is water flow rate and t_e is fire extinction time. There were 15-20 experiments done for each type and certain portion of the FCM. The results (t_b or t_e,values measured for identical samples) which differ from the average value for more than 5% were excluded from the consideration.

Systematic errors for main parameters were: droplet size (R_d) – $7\cdot10^{-6}$ m, droplet velocity (U_d) – 3.4%, times t_e and t_b – 0.5 s, volume V_e – $5\cdot10^{-4}$ l. The random errors of the temperature (T_f) measurements were less than 30 K.

3 Results and discussion

The table 1 shows the dependencies of the thermal decomposition suppression time on FCM portion diameter. The times of the total parching of the FCM sample without fire extinction efforts are presented too.

Table 1. The dependencies of the FCM parching time (t_b) and thermal decomposition suppression time (t_e) on the FCM portion diameter (for fixed portion height h_f≈40 mm).

d_f, m	birch leaves		spruce needles		aspen branches		mixed FCM	
	t_b	t_e	t_b	t_e	t_b	t_e	t_b	t_e
0.02	103.3	33.5	159.7	6.2	382.4	14.2	134.7	17.0
0.04	147.7	72.0	192.2	10.9	407.3	24.3	194.4	36.9
0.06	189.6	109.1	228.9	19.3	463.2	41.5	258.3	57.5
0.08	229.0	144.8	267.0	34.3	550.1	71.1	326.4	78.6
0.1	266.1	179.0	315.2	60.7	668.1	121.6	398.7	100.4
0.12	300.8	211.80	364.7	107.6	817.2	208.2	475.2	122.6
0.14	333.0	243.2	418.4	190.7	997.3	356.2	555.9	145.5

It is shown that the FCM total parching times t_b are much longer than fire extinction times t_e. It means that water aerosol can be effective tool for forest fire suppression.

The remarkable parameter of the fire extinction efficiency is the water consumption for the chosen method. Table 2 demonstrate the dependence of the water amount used for the model fire site extinguishing on the FCM portion volume V_f. These dependencies were obtained by generalization of the experimental data when FCM portion height (h_f) was in range from 40 mm to 100 mm and the portion diameter d_f from 20 mm to 80 mm, that is for ranges when the irrigation density ξ_f has relatively low influenceonto the experimental results. The thermal effect of the burning layer has the definitive influence at this case.

Table 2. The dependence of water consumption (V_w, l) on the FCM portion volume (V_f, m^3).

V_f, m^3	birch leaves	spruce needles	aspen branches	mixed FCM
0.0002	0.004	0.003	0.005	0.01
0.0004	0.007	0.005	0.011	0.015
0.0006	0.014	0.013	0.02	0.019
0.0008	0.023	0.021	0.034	0.025
0.001	0.036	0.035	0.05	0.03

It was shown that the growth of the FCM layer thickness h_f from 40 mm to 100 mm leads to growth of the V_e for not more than 50–60 %. The growth of the model fire site diameter d_f from 20 mm to 60 mm leads to essential growth (for 3-4 times) of the V_w.
The analysis of table 2 allows to estimate the water volumes needed for effective suppression of the FCM decomposition for surfaces which are much higher than were used in our experiments. It was shown that 1 m^2 of the spruce needles needs in 0.7 l of water aerosol, the birch leaves needs – 0.85 l/m^2 and mixed FCM needs 0.9 l/ m^2. Together with this the traditional techniques needs from 2–3 l/ m^2 to 8–10 l/ m^2.

Acknowledgments

The investigation was supported by Russian Science Foundation (project 14–39–00003).

References

1. X.K. Xiao, B.H. Cong, X.S. Wang, K.Q. Kuang, Richard K.K. Yuen, G.X. Liao, J. Fire Sci. **29**, 4 (2011)
2. S. McAllister, Fire Saf. J. **61** (2013)
3. S. McAllister, I. Grenfell, A. Hadlow, W.M. Jolly, M. Finney, J. Cohen, Fire Saf. J. **51** (2012)
4. A.M. Grishin, *Mathematical modeling of forest fire and new methods of fighting them* (Publishing House of Tomsk State University, Tomsk, 1997)
5. A.M. Grishin, S.P. Sinitsyn, I.V. Akimova, Combust. Explosion Phys. J. **27** (1991)
6. C.H. Lautenberger, C.A. Fernando-Pello, Combust Flame **156** (2009)
7. R.S. Volkov, G.V. Kuznetsov, P.A. Strizhak, Int. J. Therm. Sci. **88** (2015)
8. Y. Hagiwara, S. Sakamoto, M. Tanaka, K. Yoshimura Exp. Therm. Fluid Sci. **26**, 2–4 (2002)

VERIFICATION OF MODEL OF CALCULATION OF INTRA-CHAMBER PARAMETERS IN HYBRID SOLID-PROPELLANT ROCKET ENGINES

Ilya S. Zhukov[1], *Boris* V. Borisov[2,*], *Sergey* S. Bondarchuk[1,3], and *Alexander* S. Zhukov[1]

[1]Tomsk State University, 634050 Tomsk, Russia
[2]Tomsk Polytechnic University, 634050 Tomsk, Russia
[3]Tomsk State Pedagogical University, 634057 Tomsk, Russia

Abstract. On the basis of obtained analytical estimate of characteristics of hybrid solid-propellant rocket engine verification of earlier developed physical and mathematical model of processes in a hybrid solid-propellant rocket engine for quasi-steady-state flow regime was performed. Comparative analysis of calculated and analytical data indicated satisfactory comparability of simulation results.

In order to verify mathematical model for calculation of intra-chamber parameters in a hybrid solid-propellant rocket engine [1, 2] adequacy assessment of calculation methods was performed using analytical estimation of engine's characteristics for quasi-steady-state flow regime, constant flow section area S and perimeter P of cylindrical grain channel with the radius r. Other sufficiently accurate solutions of this problem were not obtained.
For abovementioned assumptions the change of oxidizer flow along the grain channel of a hybrid engine can be described by the following equation:

$$\frac{d}{dx}(\varrho uS) = -\mathrm{P}\alpha_{ox}\varrho_f a(\varrho u)^{\nu}, \qquad (1)$$

where ϱ, u – oxidizer density and flow rate;
a, ν – burning rate law constants; ϱ_f – fuel density; $\alpha_{ox}=m_{ox}/m_f$ – ratio of "consumed" oxidizer mass flow m_{ox} to the mass of fuel m_f in the process of combustion.
Solution of equation (1) regarding oxidizer flow at any channel point (section) is as follows

$$(\varrho u)\big|_x = \left[(\varrho u)_0^{1-\nu} - \varphi x\right]^{1/(1-\nu)}, \quad \varphi = \frac{2}{r}\alpha_{ox}\varrho_f(1-\nu)a \qquad (2)$$

where $(\varrho u)_0 = \left(m_{ox}^{in}\right)/S$ – mass flow rate of oxidizer injection (m_{ox}^{in} – total oxidizer consumption).
As it is implied from equation (2) oxidizer mass consumer for combustion m_{ox} can be determined from the following relation

* Corresponding author: bvborisov@tpu.ru

$$m_{ox}=\begin{cases} m_{ox}^{in}-S\cdot\left[(\varrho u)_0^{1-\nu}-\varphi L\right]^{1/(1-\nu)}, & есл u(\varrho u)_0^{1-\nu}>\varphi L,\\ m_{ox}^{in}, & есл u(\varrho u)_0^{1-\nu}\le\varphi L.\end{cases}$$

Mass of fuel m_f consumed in the process of combustion is equal to $m_f = m_{ox}/\alpha_{ox}$ and the total mass of combustion products and "unused" oxidizer is equal to $m_\Sigma = m_{ox}^{in} + m_f$.
Thermodynamic characteristics of the mixture (gas constant R and isobaric heat capacity C_p) are determined additively using mass individual characteristics of oxidizer gas $\left(R_{ox}, C_{p_{ox}}\right)$ and combustion products $\left(R_f, C_{p_f}\right)$

$$R = c_{ox}R_{ox} + c_f R_f,\ C_p = c_{ox}C_{p_{ox}} + c_f C_{p_f},$$

where mass fractions of "unused" oxidizer c_{ox} and combustion products c_f are determined by the following relations

$$c_{ox} = \frac{m_{ox}^{in} - m_{ox}}{m_\Sigma},\ c_f = \frac{m_f + m_{ox}}{m_\Sigma} = 1 - c_{ox}.$$

The temperature of mixture of combustion products and "unused" oxidizer can be determined from the law of conservation of energy:

$$\frac{\gamma}{\gamma-1}RTm_\Sigma = \left(m_f + m_{ox}\right)\left(\frac{\gamma}{\gamma-1}RT\right)_f - m_{ox}^{in}\left(\frac{\gamma}{\gamma-1}R\right)_{ox}\left(\tilde{T} - T_{ox}\right). \tag{3}$$

The first member in the right part of equation (3) describes energy input due to fuel combustion products with the temperature T_f entering the flow. The last member of the right part of this equation represents energy losses due to heating of oxidizer gas from its initial temperature T_{ox} to some average value $\tilde{T}$. At a first approximation it can be taken as $\tilde{T} = \left(T_f + T_{ox}\right)/2$.

Then from the balance of mass of gas supplied to combustion chamber m_{ox}^{in} and combustion products m_Σ taking into account the fact that they are equal to flow rate G through the nozzle block with throat area S_* in case of critical flow, pressure p in combustion chamber can be determined

$$p = \frac{m_\Sigma}{S_* G_*(\gamma)}\sqrt{RT},\quad G = S_* G_*(\gamma)\cdot\frac{p}{\sqrt{RT}} = m_\Sigma,\quad G_*(\gamma) = \sqrt{\gamma\left(\frac{2}{\gamma+1}\right)^{\frac{\gamma+1}{\gamma-1}}}.$$

Comparison of numerical calculation results and analytical estimation data obtained using the abovementioned relationships was performed for various levels of oxidizer supply to combustion chamber of hybrid solid-propellant rocket engine with the following parameters.

Initial data for calculation

Engine geometry parameters	Oxidizer parameters
Channel radius r = 1 cm;	Gas temperature T_{ox} = 300 K;
Throat area S_* = 0.79 cm^2;	Adiabatic index γ_{ox} = 1.4;
Fuel grain length L = 50 cm.	Gas constant R_{ox} = 296 J/kg/K.

Fuel parameters

Burning rate law a = 0.07 mm/s, ν = 0.65;
Fuel density ϱ_f = 1000 kg/m3;
Gas constant of combustion products R_f = 320 J/kg/K;
Adiabatic index of combustion products γ_f = 1.22;
Fuel-oxidizer ratio α_{ox} = 2.7;
Burning temperature T_f = 3000 K.

Fig.1 illustrates pressure distribution along the relative channel length for m_{ox}^{in}=0.2 kg/s. Relatively smooth pressure profile (power-law type) typical for initial conditions of engine operation is observed. It should be noted that for initial data as well as for other values of oxidizer flow rate the profile is transformed smoothly due to non-uniform fuel burnout and as the front part of fuel grain burns completely a pressure peak is formed and moves together with fuel burnout boundary.

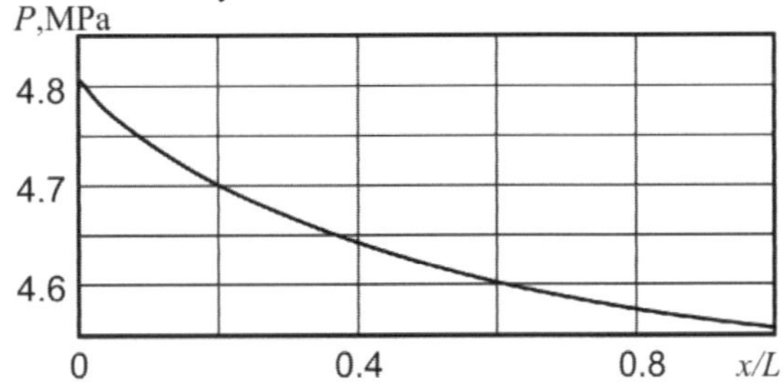

Fig. 1. Pressure distribution along the channel length.

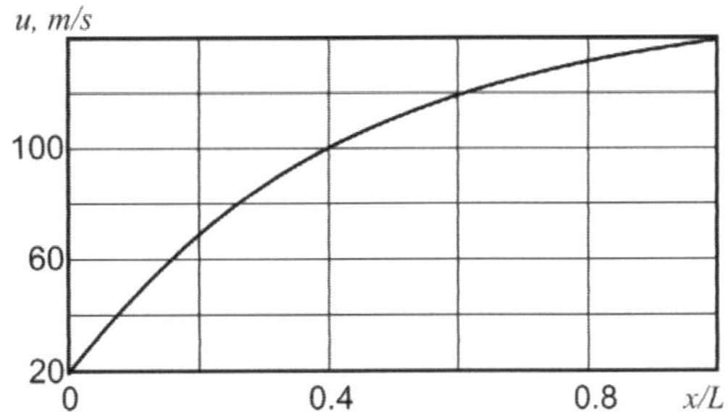

Fig. 2. Gas flow rate distribution along the channel length.

Table 1. – Results of comparison of numerical and analytical data.

m_{ox}^{in}, kg/s	p, MPa		$c_{ox}\|_{x=L}$		$T\|_{x=L}$, K		G, kg/s	
	num	theor	num	theor	num	theor	num	theor
0.10	2.33	2.40	0.002	0.002	2385	2523	0.137	0.137
0.20	4.59	4.71	0.029	0.026	2361	2480	0.271	0.271
0.35	7.84	7.97	0.070	0.068	2326	2404	0.467	0.468
0.80	16.87	16.91	0.163	0.159	2233	2235	1.034	1.035
1.00	20.73	20.63	0.189	0.187	2206	2180	1.281	1.281
2.00	38.88	37.95	0.283	0.281	2095	1992	2.481	2.482
Δ, %	1.70		3.01		3.36		0.06	

Δ – mean difference between parameter values.

The results of comparison of numerical and analytical data ("num" – numerical calculation results; "theor" – analytical estimation data) for various levels of oxidizer supply to combustion chamber of hybrid solid-propellant rocket engine are given in Table 1.

 DOI: 10.1051/matecconf/20167201135

Mean difference between values of parameters determining the process (Table 1) does not exceed 3.5% for selected configuration.

Thus, an analytical estimate of engine's characteristics for verification of developed physical and mathematical model of processes in a hybrid solid-propellant rocket engine for quasi-steady-state flow regime, constant flow section area and constant perimeter of working surface of cylindrical grain channel is given in this paper. Performed parametric comparative analysis of calculated and analytical data indicated more than satisfactory comparability of results of numerical calculation and analytical estimation – mean difference between values of parameters determining the process does not exceed 3.5% for selected configuration.

Acknowledgments

This research was supported by the Ministry of Education and Science of the Russian Federation under the Federal Target Program "Research and development on priority directions of scientific-technological complex of Russia for 2014-2020", the agreement No. 14.578.21.0034, unique identifier PNI RFMEFI57814X0034.

References

1. S.S. Bondarchuk, A.B. Vorozhtsov, A.S. Zhukov, B.V. Borisov., Russian Phys. **57**, 12 (2015)
2. S.S.Bondarchuk, A.B. Vorozhtsov, A.S. Zhukov, I.A. Zhukov, S.A. Vorozhtsov, V.V. Promakhov The Proceedings of the 46th International Annual Conference of the Fraunhofer ICT: Energetic Materials. Performance, Safety and System Applications. Karlsruhe, Germany (2015)

MATEC Web of Conferences **72**, 01136 (2016) DOI: 10.1051/matecconf/20167201136
HMTTSC-2016

DYNAMIC REGIME OF IGNITION OF SOLID PROPELLANT

Nikolay Zolotorev[1,*], *Valeriy* Kuznetsov[1], and *Eugene* Maslov [2]

[1]National Research Tomsk State University, 634050 Tomsk, Russia
[2]National Research Tomsk Polytechnic University, 634050 Tomsk, Russia

Abstract. This article presents a dynamic regime of exposure of the radiant flux on the sample of gun-cotton. Obtained time the ignition of gun-cotton in the heating conditions of increasing heat flux in the range from 0.2 W/cm^2 to 22 W/cm^2. A comparison of the delay times of the ignition when heated variable and constant heat flux.

1 Introduction

Studying the process of ignition of solid fuels is usually carried out under simple conditions of added heat: constant temperature or heat flux on the surface in so-called "static regime"[1–5]. In real conditions, the ignition is carried out at variable, time-dependent values of heat flux. The formation of the warm layer in this case is carried out in complicated dynamic conditions and significantly depends on the regime of change in heat flux during the period of heating of the studied sample [1].

The analysis of dynamic regimes in thermal ignition theory held by A. G. Merzhanov with employees [1], that showed that the characteristics of the ignition process depend on the type of the external heat flux. With increasing flow the influence of kinetic parameters on the ignition timing is smaller than in the case of constant or decreasing heat flux with time. Experimental study of this phenomenon is quite difficult in technical execution. In the available literature the amount of information experimental data are limited [5].

In the theory of solid-phase ignition occurs at the moment when to the velocity of heat generation in the zone of chemical reaction compares heat sink. With history of determines heat the amount of heat that must be expended to carry out fuel ignition. Characteristics of dynamic regimes of ignition are of practical importance in the use of energy sources with the specified content

Gun-cotton was chosen as model of a solid fuel since it is most studied substance with known kinetic and heat physical characteristics [2]. Delayed times of ignition of gun-cotton in the heating conditions of increasing heat flux in the range from 0.2 W/cm^2 to 22 W/cm^2 are obtained.

* Corresponding author: nikzolotorev@mail.ru

DOI: 10.1051/matecconf/20167201136

2 The methodology of the experiments

The experiments were carried out using an optical furnace Uran-1 (fig. 1). The emitter consists of elliptical mirror and hid xenon lamp type DKsR-10000 rated thermal input of 10 kW. [6].

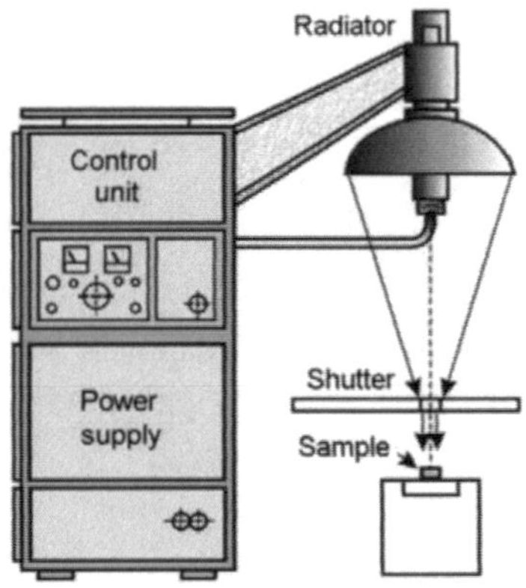

Fig. 1. Installation diagram "Uran-1".

Influence of variable heat flux on the delay time of ignition of gun-cotton in air at atmospheric pressure is studied. The diameter and height of samples are 10 mm, the density is equal to 1.45 g/cm^2. The ignition is determined by the appearance of the flame at the sample surface and recorded by a photoelectric method. Dynamic regime heat flux was provided by moving the test sample during its heating heat flux along the optical axis of the elliptic reflector, the installation of "Uran-1". For this purpose electromechanical device was used that moves the sample at a speed of 4.75 cm/s. The rate of change of the heat flux was determined by the speed of movement of the sample. The measurement error of the heat flux does not exceed 10 %. With the speed of movement of the sample along the optical axis there is a possibility to go from fixed distances to time and to determine the dynamics of the heat flux in this experiment.

3 Experimental results

In the regime of increasing the heat flux, we performed a series of experiments that differed by the rate of change of heat flux density in the process of ignition of gun-cotton. A comparison of the characteristics of ignition of gun-cotton obtained with increasing heat flux with the experimental data under constant flow [2] is shown in fig. 2.

From graphs in fig. 2 it is seen that in the studied range q delay time of ignition of gun-cotton with increasing heat flux with an average value of q is smaller than at a constant heat flux q_0. With increasing density of the heat flux, difference according to the delay time of ignition decreases and asymptotically tends to zero when $q > 8$ W/cm^2. The dependence of the amount of thermal energy Q obtained by means of gun-cotton during the period from induction of the heat flux during heating of constant q_0 (1) and increasing q (2) heat flux is show in fig. 3.

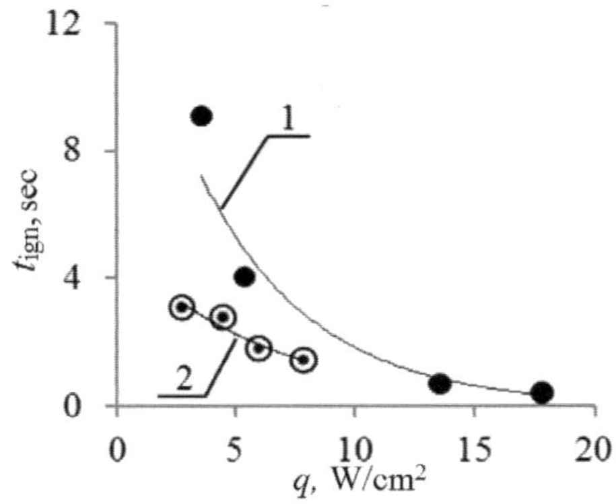

Fig. 2. The delay time of ignition during heating of constant q_0 (1) and increasing q (2) heat flux.

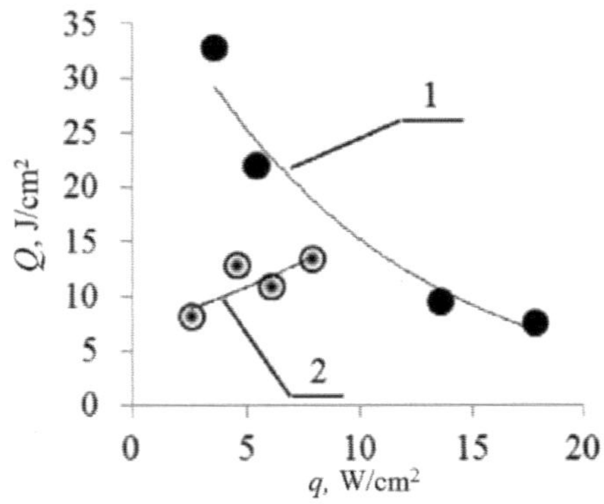

Fig. 3. The dependence of Q on the density of the heat flux at heating of constant q_0 (1) and increasing q (2) heat flux.

The found effect is described with the peculiarities of forming the heated layer of the condensed phase under conditions of static and dynamic regimes of heating of gun-cotton.

4 Conclusion

The results of experimental study of ignition of gun-cotton during the heating with increasing radiation flux in the range (0.2÷22) W/cm^2 are presented. A comparison of the delay times of the ignition at heating the samples by means of variable and constant heat flux was performed. The data obtained has shown that in the investigated range of heat flux, the flammability of gun-cotton in dynamic heating conditions is higher than in static. Physical interpretation of the obtained results requires additional research.

Acknowledgments

This work was supported by Ministry of Education and Science under the Federal Program "Research and development on priority directions of scientific-technological complex of Russia for 2014-2020", the agreement No. 14.578.21.0034, unique identifier PNI RFMEFI57814X0034.

References

1. A. Merzhanov, A. Averson, Combust. Flame **16**, 1 (1971)
2. V. Vilyunov, V. Kuznetsov, A. Skorik, *Ignition of pyroxylin by high-Intensity light flux. in Gorenie i vzryv* [Combustion and Explosion] (Nauka, Moscow, 1977) [in Russian]
3. L. Gusachenko, V. Zarko, A. Rychkov, COMBUST EXPLO SHOCK+ **48**, 1 (2012)
4. V. Arkhipov, S. Bondarchuk, A. Korotkikh, V. Kuznetsov, A. Gromov, S. Volkov, L. Revyagin, COMBUST EXPLO SHOCK+ **48**, 5 (2012)
5. R. Enaleev, V. Mateosov, K. Sinaev, B. Dinovetskiy, R. Gainutdinov, Cheboksary, Chuvash State University Publ. (1973)
6. G. Lopatina, V. Sasorov, B. Spitcyn, D. Fedoseev, *Opticheskie pechi* [Optical furnace] (Metallurgiya Publ., Moscow, 1969) [in Russian]
7. V. Arkhipov, V. Zarko, A. Korotkikh, V. Kuznetsov, A. Razdobreev, Method of determining the characteristics of ignition of samples high-energy materials in radiant heat flux. Patent RF № 2569641 (2015)

MATEC Web of Conferences **72**, 01137 (2016) DOI: 10.1051/matecconf/20167201137
HMTTSC-2016

ANALYSIS OF FRACTIONATED FLY ASH AND SLAG AT KEMEROVO STATE DISTRICT POWER PLANT

Andrey S. Sysolyatin[1], *Ivan* A. Zvingul[1], and *Elena* Yu. Temnikova[1,*]

[1]Kuzbass State Technical University named after T.F. Gorbachev, 650000 Kemerovo, Russia

Abstract. Experimental results of the separation of fly ash into microspheres, magnetic and non-magnetic fractions and analysis of chemical composition of slag and magnetic and non-magnetic fractions of fly ash from Kemerovo SDPS (Russia, Kuzbass) are presented.

1 Introduction

Now the problem of using the wastes of coal generation companies requires immediate solution, because the active ash disposal areas are characterized by limit accumulation of ash and slag waste (ASW), increasing anthropogenic impact on the environment and increasing charges for environmental pollution. The technology for ASW processing into the conditioned ash products is suggested in [1]. The advantage of this technology is ASW maturing, when the contents of sulfur compounds, alkali oxides, humic acids, and heavy metals reduce, but this technology ignores worsening environmental conditions.

The European experience in solving the problem of waste utilizxation shows that the majority of plants use the technologies of dry ash and slag removal, they have separate warehouses for storage of ash and slag, laboratories for hourly monitoring of the produced ash, but not only coal, and systems of ash product making with no ash dumps [2]. In 2010, in all EU countries, the annual amount of ASW was estimated at 100 million tons. They are mainly used in production of building materials, civil engineering, road construction, coal mines, and pit recultivation. This also includes slag production in the amount of about 1 million tons, used mainly for blasting and road construction [3]. In addition, dry ash from Estonia is successfully sold within six years at the Russian market, and this confirms the demand for ash [2]. However, fractionated fly ash of dry sampling is in the most demand.

2 Problem statement

The purpose of this work is to study the chemical and mineralogical composition of fly ash and slag at Kemerovo state district power plant (SDPP) for further industrial fractionation to the valuable components.

The work tasks are separation of fly ash into microspheres, magnetic and non-magnetic fractions and analysis of chemical composition of slag and magnetic and non-magnetic fractions of fly ash.

* Corresponding author: teu.pmahp@rambler.ru

3 Result and discussion

For this purpose, the samples of dry fly ash and slag, obtained at combustion of coal of grade D, were taken at Kemerovo SDPP. Their pictures are shown in fig. 1.

The microspheres were removed from fly ash by wet gravitation deposition of the heavy fraction and microsphere collection from the water surface with their following drying. The magnetic and non-magnetic fractions were separated by magnetic separators in two ways: the dry and wet ones. Wet separation was carried out in water suspension of fly ash, where the magnetic fraction was concentrated in the magnetic field, with simultaneous washing of the non-magnetic fraction and drying of the obtained components. Dry separation involves the collection of the magnetic fraction by the magnetic field and its separation from the magnet [4]. As a result, the following composition of separated fractions of fly ash, presented in Table 1, was obtained. The typical types of these fractions are shown in fig. 2. Slag was not separated because it is composed of the plate-like particles of various sizes (fig. 1).

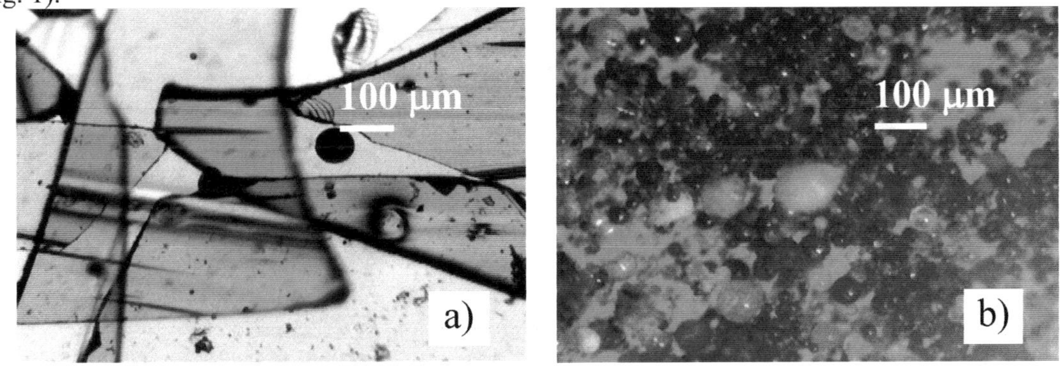

Fig. 1. Photos: a) slag; b) fly ash.

Table 1. Composition of fly ash.

Separation method	Composition of fly ash, % mass.			
	Microspheres	Magnetic fraction	Non-magnetic fraction	Losses
Dry [4]	1.65	4.23	92.73	1.39
Wet	2.46	3.82	89.18	4.54

Table 2. The content of elements in the fractions of fly ash and slag.

Element	Non-magnetic fraction		Magnetic fraction		Slag
	wet	dry	wet	dry	
C	0	0	0	0	17.47
O	47.39	46.67	42.12	42.78	40.48
Na	1.18	1.31	1.01	1.08	1.38
Mg	1.29	1.22	2.3	1.94	1.25
Al	11.13	11.65	7.87	8.62	10.51
Si	28.09	27.18	19.08	20.82	24.9
K	2.11	2.08	1.16	1.38	1.2
Ca	4.15	3.84	4.9	4.33	1.27
Ti	0.59	0.50	0.29	0.35	0.18
Mn	0.06	0.05	0.05	0.07	0
Fe	4.01	4.20	21.22	17.51	1.35

According to Table 1, at wet separation the losses in the material balance are greater than at dry separation, and the share of magnetic fraction at wet separation is less than 11%

The content of elements in the non-magnetic and magnetic fractions of fly ash (for two separation methods) and slag (Table 2) was determined. The results show that despite a smaller share of magnetic fraction (Table. 1) at wet separation, the content of iron in the magnetic fraction is 21% higher than at dry separation. That is, the more concentrated product is obtained at wet separation. This can be explained by the fact that at dry separation a negative effect of static stress is observed. According to Table 2, slag contains 17.47% of carbon and has low iron content of 1.35%.

Fig. 2. Photos: a) microspheres; b) magnetic; c) non-magnetic fractions.

4 Conclusion

In the future trends, these results will contribute to determination of a rational flowchart of the process of separation of the magnetic fraction (iron-bearing concentrate), light fraction of fly ash (hollow microspheres) and non-magnetic fraction. The developed technology of fly ash and slag utilization will allow demonstration of the ways for improving its consumer properties and, consequently, prices. It is known that ASWs are of the industrial interest due to the chemical composition and physical-chemical properties. Ashes and and slag contain the components with unique valuable technological properties that allow their efficient use in many modern technologies. For example, according to the data of Siberian generating company, the ASW output at Kemerovo SDPP in 2012 amounted to 226 thousand tons, in 2013 it was 134 thousand tons, and in 2014, it was 136 tons. The practice shows that ASWs consist of 20% of slag and 80% of dry fly ash. In 2014, the output of dry fly ash was slightly higher than 100 thousand tons, and from this amount, it is possible to obtain about 90 thousand tons of non-magnetic fraction, sent to construction, 4 thousand tons of the magnetic fraction for the production of ferrosilicon, iron, and steel, 1.5 tons of microspheres, used as a filler for various products, in production of spheroplastics, road marking thermoplastics, cementing and drilling fluids, lightweight construction ceramics, heat-insulating roasting-free materials and heat-resistant concretes.

ASW recycling will contribute to solving the environmental problems, creation of new jobs, reduction of the environmental fines and obtaining economic benefit.

Acknowledgments

The studies were financially supported by the Russian Foundation for Basic Research and Administration of Kemerovo region in the framework of scientific project No. 16-48-420871, "r_a".

References

1. A.N. Nabokov, G.N. Freiberg, T.P. Shcheblykina, Energetik **10** (2014)
2. A.I. Kalachev, TEK. Strategy of Development **39**, 4 (2015)
3. A.A. Salamov, Energetik **10** (2014)
4. A.R. Bogomolov, E.Yu Temnikova, S.A. Shevyrev, A.V. Poltavets and A.S. Sysolyatin, MATEC Web Conf. **37** (2015)

MATEC Web of Conferences **72**, 01138 (2016) DOI: 10.1051/matecconf/20167201138
HMTTSC-2016

REGISTRATION OF TIMES OF PRESERVING THE LOWERED TEMPERATURES OF COMBUSTION PRODUCTS IN THE TRACE OF THE AEROSOL FLOW

Ivan S. Vojtkov[1], *Roman* S. Volkov[1, *], and *Galina* A. Chernova[1]

[1]National Research Tomsk Polytechnic University, 634050 Tomsk, Russia

Abstract. This study examines the traces of water droplets moving through high-temperature combustion products (initial temperatures are 430–950 K). The temperature of a gas-vapor mixture in the area of droplets' traces is measured using low-inertia thermocouples (thermal lag is less than 0.1 s). The experiments revealed that the temperature in the trace of water droplets is significantly reduced (from 15 K to 140 K) even when using an aerosol flow with small droplets. The experiments showed that the times of keeping low temperatures in the trace of water droplets may reach 30 s.

1 Introduction

Active development of different industries is inevitably accompanied with the growth of the number of fires and fires in factories, which often cause the most disastrous consequences. These industries primarily include oil and gas production, refineries and energy complexes. Therefore, the issues of fire safety in such enterprises are very important: rules and requirements for personnel are tightened, modern automatic fire extinguishing system based on the microprocessor are implemented, various methods for the elimination of fires and delivery of quenching compounds in the fire zone are offered.

Application of sprayed water for fire localization and liquidation is one of the most common methods of firefighting nowadays. Fire-extinguishing system based on water mist are developed, tested and put into operation. However, there is no an accurate information about the experimental ranges of the temperature of the combustion products when they are impacted by a stream of droplets. So, there is an interest in experimental research of temperature changing of combustion products process in the steam-dripping flow wake.

The purpose of this work - experimental investigation of changing of the combustion products temperature in the trace of a droplet aerosol flow when it moves into the fire.

[*] Corresponding author: romanvolkov@tpu.ru

2 The methods and the experimental setup

In experimental studies, a setup (similar to [1, 2]) equipped with high-speed video recording tools (cameras Phantom V411 and Phantom Miro M310, frame rate is up to $6 \cdot 10^5$ fps, maximum recording resolution is 1280×1280 pix), and a cross-correlation software and hardware complex (double-pulse solid-state Nd:YAG laser Quantel EverGreen 70) was used. The scheme of the experiment is shown at fig. 1.

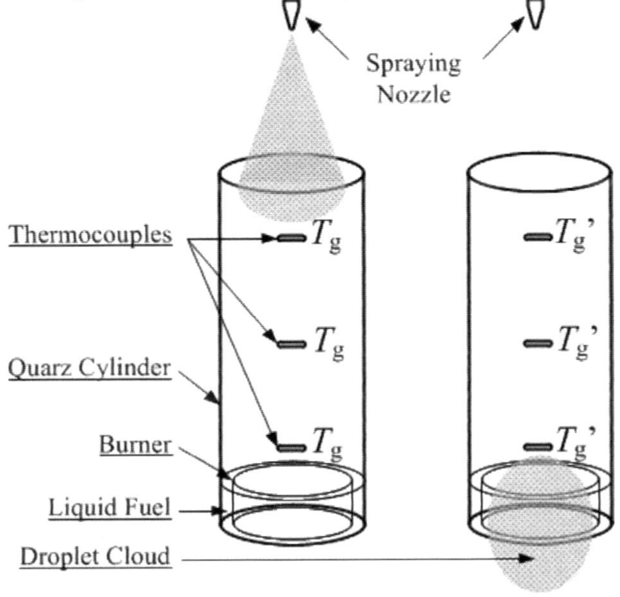

Fig. 1. Scheme of the experiment.

Combustion products were generated as follows. Burner (fig. 1) was filled with fuel in an amount of 250 ml. The ignition of fuel was initiated using a piezoburning system. Glass cylinder was installed to burner. Thus, the upward flow of the kerosene combustion products was formed inside the cylinder.

Thermocouples of type K (for T_g registration) (see fig. 1) were placed at the three points over the height of cylindrical channel (at distances of 0.12, 0.4, and 0.7 m from burner base) on its axis of symmetry.

To generate the droplet cloud, a system was used consisting of tank with water (liquid pressure is 100–300 kPa depending on the experimental conditions), water supply channel, and a set of spray nozzles (fig. 1). We used three types of spray nozzles for generation of different (by range of particle sizes) drip flows: R_d=0.08–0.35 mm – large droplets; R_d=0.06–0.18 mm – average droplets; and R_d=0.04–0.12 mm – fine spray. The initial velocities (U_d) of the drip flow varied in the range of 1–3 m/s in the experiments. The volume concentration of droplets in the sprayed flow (γ_d) varied in the range of $3.8 \cdot 10^{-5}$–$10.3 \cdot 10^{-5}$ (m^3 of water)/(m^3 of gas). For this purpose, several (up to three) spray nozzles were used simultaneously to provide the same range of change in droplets sizes.

A type-L thermocouple (temperature measurement range is 273–473 K, systematic error is ± 1.5 K, thermal lag is less 10 s) monitored the initial water temperature (T_w) which was maintained at 300 K by a set of built-in heating elements in tank *8*. The maximum random error in measuring the temperature T_w was 3 K, T_g – 15 K.

To determine the initial droplet flow velocities (U_d), PIV [3] and PTV [4] methods were used. The PIV method was used for processing the videograms of the drip flow at the input of cylindrical channel *12*. The PTV method was used for the videograms captured at the output of the cylindrical channel. The SP method [5] was used to estimate the initial sizes of water droplets. The droplet flow was monitored at two points: at the input and output of cylindrical channel (see fig. 1). For this, the experimental setup was equipped with diffuse screens, which were connected to Nd:YAG laser via optical fibers. Procedures for measuring velocities U_d and sizes R_d were performed simultaneously.

The error in determining U_d using this approach did not exceed 2%. Systematic error in determining droplet sizes by the SP method was defined using the scale factor (*S*) and did not exceed 3.5% in the experiments.

The following parameters were determined after each experiment: ΔT_g – temperature reduction of combustion products in the wake of the drip flow with respect to its initial values, K; τ – time of keeping low temperatures (relative to initial ones) of combustion products in the wake of the moving liquid (a time interval from the beginning of temperature reduction of combustion products until the complete temperature recovery to the initial value). The error in determining the times τ did not exceed 1 s.

3 Results of research

Temperature recovery times (τ) in the trace of droplet cloud obtained during the processing of the result of experimentation are presented in the Table 1. Established that primary droplet size exert determining influence on the intensity of the temperature drop in the trace of droplet cloud. Most time value of temperature recovery to datum value was observed at relative large liquid spray (R_d=0.08-0.35 mm). For the other two types of spray can be noted minor differences times τ (in the range of 8-10%), which is likely caused by a close range drop size change in the stream of data types of the liquid spray: R_d=0.06-0.18 mm - middle and R_d=0.04-0.12 mm - small, respectively.

Table 1. Time values (τ) of temperature recovery to datum value (to water injection) in the trace of droplet cloud.

Droplet size, mm	T_g=450 K	T_g=550 K	T_g=650 K	T_g=750 K
0.08< R_d <0.35	τ≈16.5 s	τ≈19.5 s	τ≈22 s	τ≈24 s
0.06< R_d <0.18	τ≈16.5 s	τ≈18 s	τ≈19 s	τ≈20 s
0.04< R_d <0.12	τ≈17.5 s	τ≈19 s	τ≈20 s	τ≈21 s

Fig. 2 presents time values of saved low temperature (τ) in the trace of droplet cloud depending on temperature of combustion products which was result of data processing. The fig. 2 shows that peak value time of recovery (τ≈24 s) characteristic of the droplet size 0.25 mm and peak value of the combustion products temperature 850 K.

This result is explained as follows. Droplets with size more than 0.17 mm (R_d>0.17 mm) are evaporated by 20-40% [6] passing through combustion products 1 meter thereby contributing to lowering of combusting products temperature by two mechanism: lowering flames and combustion products temperature by evaporating (endothermic phase transformation with intense absorption of energy); supplantation oxidants and combustion products from combustion zone by the formed water vapor.

Fig. 2 clearly shows that for all the above types of spray value τ are comparable and differ by no more than 1.5-2 s at temperature gas environment 400-500 K. It is obvious that at these temperatures of combustion products is realized the first mechanism to reduce the flame temperature. At temperatures above 550 K, a significant difference in values τ, which is especially noticeable when the T_g values of 850 K (fig. 2).

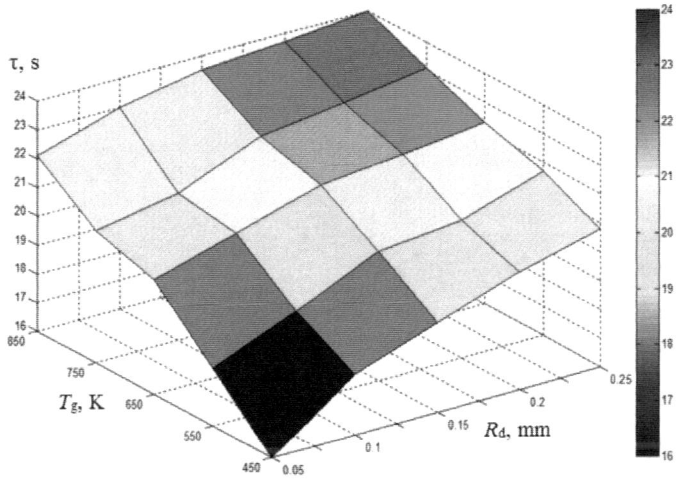

Fig. 2. 3D-diagram time of saving low temperature in the race of droplet cloud depending on droplet size in the aerosol flow at different combustion products temperature and concentration drops in aerosol ($\gamma \approx 7.1 \cdot 10^{-5}$ (m^3 of water)/(m^3 of gas); $U_g \approx 0.8$ m/s; $T_w \approx 300$ K).

This result is due to the fact that droplet sizes R_d<0.16 mm can be completely evaporated by passing a distance of 1 m in the high-temperature combustion products [1, 2]. Thus, there is a decrease in temperature of the combustion products at the expense of the flame energy costs for the phase transition (evaporation).

3 Conclusion

The experiments identified ranges of temperature change, as well as the value of time saving low temperature of the combustion products in the trace of a drop flow. Define values of the time τ for different values of temperature T_g, and droplet sizes R_d and temperature of combustion products can be used for making a choose of sprayers operation condition in water mist fire suppression systems. Using these results we can predict conditions that ensure the effective reduction of combustion products temperature.

Acknowledgments

The investigation was supported by Russian Science Foundation (project 14–39–00003).

References

1. R.S. Volkov, G.V. Kuznetsov, P.A. Strizhak, Int. J. Heat Mass Transfer **96** (2016)
2. R.S. Volkov, G.V. Kuznetsov, J.C. Legros, P.A. Strizhak, Int. J. Heat Mass Transfer **95** (2016)
3. J.V. Simo Tala, S. Russeil, D. Bougeard, J.-L. Harion, Exp. Therm. Fluid Sci. **50** (2013)
4. Y. Hagiwara, S. Sakamoto, M. Tanaka, K. Yoshimura, Exp. Therm. Fluid Sci. **26**, 2 (2002)
5. S. Dehaeck, H. Van Parys, A. Hubin, J.P.A.J. Van Beeck, Exp. Fluids **47**, 2 (2009)
6. O.V. Vysokomornaya, G.V. Kuznetsov, P.A. Strizhak, J. Eng. Phys. Thermophys. **89**, 1 (2016)

AUTHOR INDEX

Abramovskih Alexey....212
Afanaseva Viktoria....570
Andreeva Ksenia N....39
Andriyashchenko Vladimir....643
Andyk Vladimir S....28
Antonov Dmitry V....1
Arkhipov Vladimir....6, 69
Arukov R. N....13
Asan Alibek....610
Astafev Alexander....18, 555
Atroshenko Yuliana K....23, 28
Aykina Natalya N....636
Babihina M.N....59, 276
Barakhnin Vladimir B....39
Barakhovskaya Ella....33
Baranovskiy Nikolay V....39, 121
Bartashevich Maria V....372
Basalayev Sergey....163
Baturin D. A....45
Beck Alexander S....631
Belyaev L. A....49
Belyaev Leonid....582
Bespala Evgeny....54
Bezmaternykh A. O....59
Bogomolov Alexander R....365, 596
Bogomolov Gennady N....64
Bondarchuk Ivan. S....69
Bondarchuk Sergey S....79, 667
Borisov Boris V....79, 667
Boykov Dmitry V....347
Boykova Tamara....83
Brendakov Roman....87, 512
Brendakov Vladimir....87, 512
Breus Sergei....297
Bubenchikov Aleksey....289, 578
Bubenchikov Michael....90, 289
Bugaeva K. A....170
Bukhmirov Vyacheslav....236, 452
Bulba Elena....95
Bychkova Alena A....23, 28
Chemakin Maxim A....551
Cherepanov Roman O....99
Chernetskaya Nelya....103
Chernetskiy Mikhail....103, 507
Chernova Galina A....678
Cheverda Vyacheslav....110
Chuguev Danil N....114
Chupashev A.A....327
Deeva Vera....247
Demikhova Alena N....121
Dmitrienko Margarita A....605, 616
Donskoy Igor G....127
Dutova Ekaterina M....627
Dvorovenko Igor V....551
Dyatchina Anna....95
Egorov Roman I....605
Elistratov Dmitriy S....335
Elistratov Sergey....342, 515
Ershov Vadim....130
Faraponov V.V....327
Feoktistov D.V....403
Feoktistov Dmitriy.V....312
Filimonov Sergey A....347
Fisenko Roman....297
Gabitov Ramil....236
Gal’chenko Nina K....154
Galashov Nikolay....142
Gatapova E. Ya....146, 226
Gavrilov Andrey....503
Gerasimov Alexander V....99
Gil A. V....45, 150
Glazunov Anatolij A....392
Glazyrin Victor P....64
Glotov Oleg....241
Gluzdov Dmitry S....146
Goldaev Sergey V....158
Goldayev Sergey....163
Gorbasenko Stepan....487
Grishchernko Maria V....293
Gromov Alexander....408, 651
Gubin V. E....59
Gubin Vladimir....184
Guk Olga V....216
Gumerov Vladislav M....438
Guskov Anatolyi....655
Gutareva N. Y....170, 545
Guzei Dmitriy....178
Guzey Dmitriy....468
Gvozdyakov D.V....276
Gvozdyakov Dmitry....184
Haymin Vladimir....212
Iashutina Olga....193
Isachenko Ekaterina....188
Ivanova Evgeniya....193
Kabanova Marina F....198
Kabov O. A....231
Kabov Oleg....188
Karelin Vadim....490
Karpenok Viktor I....365
Kats Ilija M....206
Kats Mark D....1, 206
Kazakov Alexander....18, 555
Khaustov Sergei A....216
Khudobina Yulia....222, 289
Khyamyalyainen Veniamin A....365
Kirichenko D. P....231

AUTHOR INDEX

Kirichenko E. O. 226
Kiselev Alexander 142
Kobenko Yuriy V. 198
Koksharev O. M. 150
Kolibaba Olga 236
Kondakov A. A. 49
Korneva O. S. 137
Korotkikh Alexander 241, 519
Kosintsev Andrey G. 631
Kovtun Igor 433
Kozyrev Ilya 247
Krainov A.V. 256
Krainov Alexey 251
Kramshonkov E.N. 256
Krasnoshlykov Alexander 266
Kravchenko Evgeny 83, 570
Kreta Aleksei 271
Kryukov Sergey V. 551
Kudiiarov V.N. 59, 276
Kulesh Roman N. 280, 293
Kulupanov German 643
Kurilenko Elena Yu. 318
Kurilenko Nikolai I. 307
Kuzmin Arian V. 198
Kuznetsov Denis 285, 353
Kuznetsov Valeriy 671
Larionov Kirill 408
Latyshev Andrei S. 627
Lavrenyuk Tatiana 487
Lavrinenko Sergey 212, 428, 651
Lebed D. V. 150
Lebedev V.M. 293, 448
Legros Jean C. 438
Libin Eduard 90, 289
Lihach Snejana A. 293
Lisenkov Artem 216
Loginov V.S. 422
Lyrschikov Sergey Yu. 616
Lyubimova Lyudmila 297
Lyulin Yuriy 271, 533
Makhsutbek F.T. 444, 658
Maksimov Vyacheslav I. 307
Malyhina Ekaterina R. 312
Malyugin Roman V. 392
Mamontov Gennady Ya 114, 318
Marchuk Igor 33
Martynova A. 330
Masanik A.S. 323, 538
Maslov E.A. 327
Maslov Eugene 671
Matskevich V.V. 327
Matveev A. S. 49
Matveev Alexander 408
Matveeva Anastasia 651
Medvedev G. 330
Medvedev Valeriy V. 154
Meleshkin Anton V. 335, 338
Mezentsev Ivan V. 335, 338
Mezentseva Nadezhda N. 338
Mihienkova Evgeniya I. 347
Milevsky Konstantin 655
Minakov Andrey 178, 468
Minkov Leonid 251
Minkova Natalya 163
Mironova Nina V. 342
Mishcharina Elena 90
Misyura Sergei 359, 362
Moiseev Mikhail 285, 353
Moiseeva Ksenia 251
Morozov Maxim N. 397
Morozov Vladimir 359, 362
Morozov Vladimir S. 342
Murko Vasiliy I. 365
Nagornova Tatiana A. 307
Nakoryakov Vladimir 359, 372
Naumkin Aleksandr 472
Nee Alexander 376, 599
Nemova Tatiana 95, 382
Neobyavlyayuschiy Pavel A. 347
Nigay Aleksandr G. 154
Nizkodubov Gavriil A. 481
Novoselov Ivan 54
Nurpeiis Atlant 382
Nurpeiis A.E. 403
Nyashina Galina S. 616, 663
Openyshev P.V. 389
Orlik Evgeniy 188
Orlov Alexey 392, 528
Orlov Anton 455
Orlov Maxim Yu. 64
Orlova E. G. 137, 312
Orlova Ksenia Y. 280, 293
Osipov Vitaliy 408
Osmolovskaya Maria M. 303, 663
Ostapenko Vladimir V. 372
Ovchinnikov Valeriy 515
Ovchinnikov Vyacheslav 412
Ozerova Irina P. 23, 303
Ozhikenova V.Yu. 444
Ozhikenova Zh. F. 658
Panchenko Pavel V. 639
Panin Vladimir F. 631
Parpiev A.T. 422
Paschenko Sergey 490
Patrakov Yuriy 6
Pavelyev Egor 428

AUTHOR INDEX

Peters Andrey N. 596
Petrashchuk Svitlana 433
Piskunov Maxim V. 438
Pleshco Alice 472, 610
Polovnikov V.Yu. 444, 658
Popov V.V 448
Promahov Vladimir V. 73
Prorokova Maria 452
Proshkin Alexander 455
Prosuntsov Pavel V. 461
Pryazhnikov Maxim 178, 468
Ratkov Roman S. 342
Razva Aleksandr 472, 610
Rezanova Ekaterina 477, 646
Riabova Elena S. 198
Romashova Olga 582
Ron'shin Fedor 110
Rostovtseva Veronika M. 481
Rudenko Michael 487
Ruleva Eugenia 412
Ryzhikh Yuliya N. 73
Saigash A. S. 137
Salomatov Vasiliy 490
Salomatov Vladimir 490
Savina Karina 212
Sechin Aleksandr 6
Senchurova Yuliya A. 365
Shatskiy Evgeniy E. 499
Shebelev Alexander 503
Shebeleva Anna 468, 507
Sheremet M.A. 389
Shevelev S. A. 49
Shevyrev Sergey A. 605
Shorohov P.V. 256
Shvab Alexander 87, 512
Shvab Svetlana 184
Shvaljov Y. B. 59
Sivkov Dmitriy S. 154
Slesareva Ekaterina 515
Slyusarskiy Konstantin 519
Solodkin Andrey S. 523
Sorokin Ivan 241, 519
Sovach Victor 528
Spesivtsev Serafim 533
Starchenko A. V. 150, 623
Stuparenko Alexandr A. 523
Subbotin A.N. 13, 538
Syrodoy S. V. 170, 545
Sysolyatin Andrey S. 551, 674
Tabakaev Roman 18, 297
Taburchinov Roman I. 545
Tailakov Oleg V. 365
Tanishev Anton 184
Tarasanov A.S. 538
Taraskin Nikita Yu. 461
Tarasov Egor 222
Tarasova Ekaterina 130
Taratushkina Galina V. 318
Taylasheva T.S. 448
Temnikova Elena Yu. 596, 674
Theeb Maathe. A. 561
Timoshenko Evgenya 570
Tkachenko Egor M. 574
Tretyakova Kseniya B. 158
Tsibulskiy Svyatoslav 142
Tsimbalyuk Alexandr F. 392
Tsydypov Sevan 578
Tsyrenova Valentina 578
Tubolev Alexander 582
Ulchiekov M. A. 586, 590
Usanina Anna 6
Ushakov Anton 528
Ushakov Ivan 54
Ushakov Konstantin Yu. 596
Ushakov Roman B. 397
Ushmaev Denis V. 636
Valieva Liliya 376, 599
Valiullin Timur R. 605
Vasilevskiy A.V. 323
Vasilevsky Michail 472, 610
Vasiltsov Mark 487
Velicanov V. D. 137
Vershinina Ksenia Yu. 616
Vershinina Kseniya 83, 193
Vizgavlyust N. V. 623
Vlasova Yekaterina A. 639
Vojtkov Ivan S. 678
Volkov Roman S. 303, 678
Vorobiev Aleksander V. 627
Vorobyeva L. V. 623
Voznyak Andriy 433
Vympin E.A. 403
Vysokomornaya Olga V. 631, 636
Vysokomorny Vladimir S. 631
Yakimov Anatoliy 412
Yankovsky Stanislav 643, 651
Yuzhakov Aleksey V. 198
Zaitsev A. S. 49
Zaitsev D. V. 231, 574
Zakharevich Arkadiy V. 639
Zakharevich Yuliya S. 636
Zakurdaeva Alla 646
Zavorin Alexander 18, 555
Zenkov Andrey 651
Zezulinsky Jaroslav 655
Zharova I. K. 658

AUTHOR INDEX

Zhdanova Alena O. 1, 663
Zhidkova Larisa.S. 312
Zhigarev Vladimir 178, 468
Zhuikov Andrey 103
Zhukov Alexander S. 69, 667
Zhukov Ilya S. 79, 667
Zhukov Vladimir 285, 353
Zhukova Tatyana V. 73
Zolotorev Nikolay 671
Zueva E. A. 623
Zvingul Ivan A. 674
Zyatikov Pavel 247
Zygin Mikhail S. 639

EDP Sciences – Web of Conferences
17, Avenue du Hoggar
Parc d'Activité de Courtabouf
BP 112
F-91944 Les Ulis Cedex A
France

ISBN 978-1-5108-2870-4

9 781510 828704

CURRAN ASSOCIATES INC.
proceedings.com